ABOUT THIS BOOK

AMSCO's original edition of *Algebra I,* concentrating on down-to-earth exposition, detailed guidance, and abundant exercises, has steadfastly served generations of students. This updated edition, while maintaining AMSCO hallmarks, focuses on implementing the Curriculum Standards of the National Council of Teachers of Mathematics. In addition to teaching students how to *do* mathematics, a major aim of the revised book is to help students develop an appreciation for the power and worth of mathematics in the real world.

AMSCO's *Algebra I: An Integrated Approach*

- presents problem solving as the foundation for the course and introduces algebraic techniques as tools of problem solving
- connects algebra to other mathematical disciplines, integrating it with geometry, trigonometry, and probability
- introduces the concept of algebraic modeling of meaningful real-world situations
- utilizes the technology of the scientific calculator

This program

- supports students in building skill at independent application

  Try It Problems, with complete solutions at the end of a chapter, are offered as an intermediary stage between detailed Model Problems and a variety of do-it-yourself Exercises.

- introduces the scientific calculator

  Calculator Connections teach about the calculator; Calculator Challenges encourage exploration and discovery.

- encourages self-assessment

  Chapter Review Exercises, Spiral Review Exercises, and Self-Tests serve as cumulative checkpoints.

- prepares college-intending students to cope with formal testing

  College Test Preparations, exercises in the style of college entrance examinations, are tied to the topics developed in each chapter.

This contemporary text is offered in the tradition of excellence so highly valued by AMSCO and is dedicated to the memory of Isidore Dressler, a pioneer in the belief that the study of mathematics can be made accessible to all students.

ALGEBRA I
AN INTEGRATED APPROACH

ISIDORE DRESSLER
Former Chairman, Mathematics
Bayside High School
New York City

MITCHELL H. BERNSTEIN
Department of Mathematics
High School for Girls
Philadelphia

MARILYN OCCHIOGROSSO
Former Assistant Principal, Mathematics
Erasmus Hall High School
New York City

When ordering this book, please specify *either* **R 600 S** *or*
ALGEBRA I: AN INTEGRATED APPROACH, SOFTBOUND

Dedicated to serving

AMSCO

our nation's youth

AMSCO SCHOOL PUBLICATIONS, INC.
315 Hudson Street / New York, N.Y. 10013

Contributors to the Project

Joan E. Gucken, Ed.D.
Department Head, Mathematics
High School for Girls
Philadelphia

Lance Modell
Director of Mathematics
West Hempstead Public Schools
West Hempstead, N.Y.

Emily Meadows, Ed.D.
Mathematics Education Consultant
Daniels, W.V.

Alan Barson, Ed.D.
School District of Philadelphia
Philadelphia

Noel Gurin
Former Assistant Principal, Mathematics
Springfield Gardens High School
New York City

ISBN 0-87720-288-5

PRINTED IN THE UNITED STATES OF AMERICA

1 2 3 4 5 6 7 8 9 10   00 99 98 97 96 95 94 93

CONTENTS

Chapter 1 Problem Solving

1-1 Guess and Check. 2
1-2 Draw a Diagram 5
1-3 Make a List or a Table 11
1-4 Find a Pattern 15
1-5 Work Backward. 19
1-6 Choose a Strategy 23
     Chapter Summary 28 • Chapter Review Exercises 29
     Problems for Pleasure 30 • Calculator Challenge 31
     College Test Preparation 32 • Spiral Review Exercises 33
     Solutions to TRY IT! Problems 35

Chapter 2 The Language of Algebra

2-1 Algebraic Expressions. 43
2-2 Terms, Factors, and Exponents 48
2-3 Sets 54
2-4 The Number Line 60
2-5 Using an Equation in an Algebraic Model 66
2-6 Working With an Inequality 74
     Chapter Summary 83 • Chapter Review Exercises 84
     Problems for Pleasure 86 • Calculator Challenge 87
     College Test Preparation 88 • Spiral Review Exercises 90
     Solutions to TRY IT! Problems 92

Chapter 3 Real Numbers: Properties and Operations

3-1 Closure and Substitution 95
3-2 Addition of Real Numbers 98
3-3 Multiplication of Real Numbers. 105
3-4 Subtraction and Division. 110
3-5 The Distributive Property; Summary of Properties 116
3-6 Using the Properties of the Real Number System 119
3-7 Defining an Operation 124
3-8 Another Arithmetic: Matrices 128
     Chapter Summary 136 • Chapter Review Exercises 137
     Problems for Pleasure 139 • Calculator Challenge 140
     College Test Preparation 140 • Spiral Review Exercises 142
     Solutions to TRY IT! Problems 144
     SELF-TEST Chapters 1–3 146

Chapter 4   Equations and Inequalities

4-1	Properties of Equality	149
4-2	Solving an Equation by Applying Inverses	153
4-3	Using an Equation to Solve a Problem	156
4-4	Combining Like Terms in an Equation	160
4-5	Working With a Formula	166
4-6	Solving an Inequality	173
4-7	Absolute Value in an Open Sentence	182

Chapter Summary *187* • Chapter Review Exercises *188*

Problems for Pleasure *191* • Calculator Challenge *191*

College Test Preparation *192* • Spiral Review Exercises *193*

Solutions to TRY IT! Problems *195*

Chapter 5   More Applications of Equations and Inequalities

5-1	Consecutive-Integer Problems	199
5-2	Money-Value Problems	203
5-3	Motion Problems	207
5-4	Lever and Pulley Problems	214
5-5	Angle Problems	217
5-6	Triangle Problems	222
5-7	Perimeter and Area Problems	227
5-8	Surface Area and Volume Problems	236
5-9	A Roundup of Problems	245

Chapter Summary *247* • Chapter Review Exercises *248*

Problems for Pleasure *251* • Calculator Challenge *252*

College Test Preparation *253* • Spiral Review Exercises *254*

Solutions to TRY IT! Problems *256*

Chapter 6   Polynomials

6-1	Adding and Subtracting Polynomials	259
6-2	Properties of Exponents	266
6-3	Scientific Notation	271
6-4	Multiplying Polynomials	276
6-5	Dividing Polynomials	285
6-6	Factoring	291
6-7	More About Factoring	297
6-8	Solving a Quadratic Equation by Factoring	304

Chapter Summary *310* • Chapter Review Exercises *311*

Problems for Pleasure *313* • Calculator Challenge *313*

College Test Preparation *314* • Spiral Review Exercises *315*

Solutions to TRY IT! Problems *317*

SELF-TEST Chapters 1–6 *320*

Chapter 7 Rational Expressions and Open Sentences

7-1 Simplifying a Rational Expression 323
7-2 Multiplying and Dividing Rational Expressions 328
7-3 Adding and Subtracting Rational Expressions 334
7-4 Solving Open Sentences That Contain Fractional Coefficients . . .342
7-5 Applying Open Sentences That Contain Fractional Coefficients . .346
7-6 Solving Equations That Contain Rational Expressions 358
7-7 Applying Equations That Contain Rational Expressions 364

Chapter Summary 374 • Chapter Review Exercises 375
Problems for Pleasure 377 • Calculator Challenge 377
College Test Preparation 378 • Spiral Review Exercises 380
Solutions to TRY IT! Problems 381

Chapter 8 Applying Ratio and Proportion

8-1 Ratio 387
8-2 Proportion 394
8-3 Variation 400
8-4 Similarity 414
8-5 Trigonometric Ratios 420
8-6 Applying the Trigonometric Ratios 427
8-7 Probability 439
8-8 Compound Events 448

Chapter Summary 458 • Chapter Review Exercises 459
Problems for Pleasure 463 • Calculator Challenge 464
College Test Preparation 464 • Spiral Review Exercises 467
Solutions to TRY IT! Problems 468

Chapter 9 Relations, Functions, and Graphs

9-1 Graphing in a Plane 473
9-2 Relations and Their Graphs .479
9-3 Functions and Their Graphs 483
9-4 Lines and Linear Functions .488
9-5 The Slope of a Line 492
9-6 The Slope-Intercept Form of a Linear Equation 499
9-7 Graphing a Linear Inequality in the Plane 507

Chapter Summary 511 • Chapter Review Exercises 512
Problems for Pleasure 515 • Calculator Challenge 515
College Test Preparation 516 • Spiral Review Exercises 518
Solutions to TRY IT! Problems 520
SELF-TEST Chapters 1–9 523

Chapter 10 Systems of Equations and Inequalities

10-1 Solving a System of Linear Equations Graphically 527
10-2 Solving a System of Linear Equations Algebraically 530
10-3 Solving Verbal Problems by Using Two Variables 537
10-4 Solving a System of Linear Inequalities Graphically 554
10-5 Solving a System of Three Linear Equations. 557
10-6 Solving a System of Linear Equations by Using Matrices 561

Chapter Summary 565 • Chapter Review Exercises 566

Problems for Pleasure 567 • Calculator Challenge 568

College Test Preparation 568 • Spiral Review Exercises 570

Solutions to TRY IT! Problems 571

Chapter 11 Irrational Numbers

11-1 Finding the Root of a Number 573
11-2 Simplifying and Combining Radicals 578
11-3 Multiplying and Dividing Radicals 583
11-4 Solving a Radical Equation 591
11-5 Graphing a Quadratic Function. 596
11-6 More Methods for Solving a Quadratic Equation 603
11-7 The Pythagorean Relation 614

Chapter Summary 621 • Chapter Review Exercises 622

Problems for Pleasure 624 • Calculator Challenge 624

College Test Preparation 625 • Spiral Review Exercises 627

Solutions to TRY IT! Problems 628

SELF-TEST Chapters 1–11 633

Glossary 637
Answers to Odd-Numbered Exercises 641
Index 701

ALGEBRA I

AN INTEGRATED APPROACH

CHAPTER 1

PROBLEM SOLVING

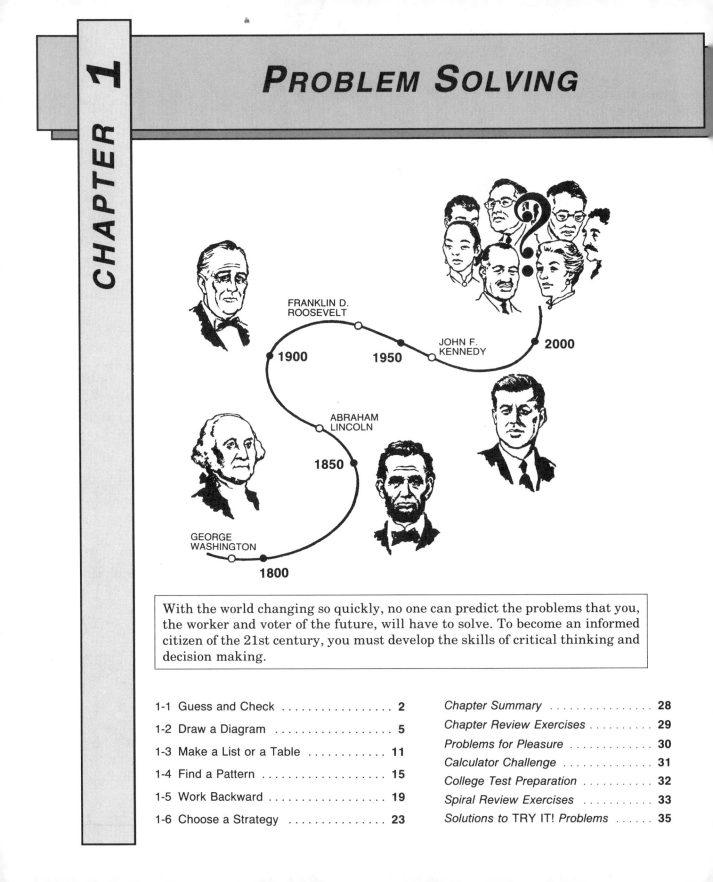

FRANKLIN D. ROOSEVELT

1900

1950

JOHN F. KENNEDY

2000

ABRAHAM LINCOLN

1850

GEORGE WASHINGTON

1800

With the world changing so quickly, no one can predict the problems that you, the worker and voter of the future, will have to solve. To become an informed citizen of the 21st century, you must develop the skills of critical thinking and decision making.

1-1 Guess and Check 2

1-2 Draw a Diagram 5

1-3 Make a List or a Table 11

1-4 Find a Pattern 15

1-5 Work Backward 19

1-6 Choose a Strategy 23

Chapter Summary 28

Chapter Review Exercises 29

Problems for Pleasure 30

Calculator Challenge 31

College Test Preparation 32

Spiral Review Exercises 33

Solutions to TRY IT! Problems 35

What Is Problem Solving?

Everyone is a problem solver. Every time you make a decision, you are solving a problem. In mathematics, you study how to use *strategies*, which are careful plans to help you arrive at solutions.

NASA is planning a manned space station.   You plan the activities of your day.

Whatever the nature of the problem to be solved, planning is important for all.

Problem-solving ability does not come all at once. A good problem solver is one who does not give up easily. Sometimes, several strategies must be tried to find one that works in a particular problem. Here is a general framework for problem solving.

GENERAL FRAMEWORK

1. **ANALYZE THE PROBLEM.** Answer three questions:
   1. *What must you find?* A number? An expression? An age? A distance? A name?
   2. *What do you know?* What information is given in the problem?
   3. *Which strategy will help you find the solution?* Guessing and checking? Drawing a diagram? Making a list or a table? Finding a pattern? Working backward?
2. **SOLVE THE PROBLEM.** Try your strategy. If it does not work, try another.
3. **CHECK YOUR RESULT.** The result of your computation is merely a candidate for a solution. The check determines whether this candidate is in fact the answer.
   1. Be sure your result is *reasonable*.
   2. Check your arithmetic.
   3. See if your solution works in the original problem.
4. **LEARN FROM THE PROBLEM.** Review your solution. Note anything that is different from similar problems you have done. Be sure you can explain your solution.

1-1 GUESS AND CHECK

Guessing and then *checking* the result is particularly effective when the number of possible solutions to a problem is limited, as, for example, in a multiple-choice problem, or in a problem where only a small range of numbers is involved.

Even if your first guesses are far off the mark, each check should enable you to make a better guess on the next try. In many problems, you may be able to first arrive at an *estimate* of what the answer should be. Then, instead of trying solutions at random, you can start with an "educated guess."

MODEL PROBLEMS

1. Find two numbers whose sum is 12 and whose product is 20.

ANALYZE THE PROBLEM.

What must you find?	The two numbers.
What do you know?	(1) The sum (addition) of two numbers is 12.
	(2) Their product (multiplication) is 20.
What strategy should you use?	Since the number of possible solutions is small, you can guess and check.

SOLVE THE PROBLEM. Guess two numbers whose sum is 12.

Guess: 1 and 11. Check: Is their product 20? No, their product is 11.

Guess: 2 and 10. Check: Yes, their product is 20.

CHECK YOUR RESULT. In this strategy, the check is part of *Solve the Problem.*

1, 11	2, 10	All of these number pairs have a sum of 12.
3, 9	4, 8	But only one of these pairs
5, 7	6, 6	also has a product of 20.

ANSWER: The numbers are 2 and 10.

LEARN FROM THE PROBLEM. In this problem, there are only a few combinations of numbers whose sum is 12. This limited number of possibilities permits you to guess. There must also be a way to check your guess. The second condition of the problem (that the product of the two numbers is 20) acts as a check.

2. The span of the Golden Gate Bridge, the longest over San Francisco Bay, is 3 times as long as the span of the nearby San Francisco–Oakland Bridge. The difference between the lengths of these bridges is 2,800 feet. What is the length of the Golden Gate Bridge?

(1) 3,900 feet   (2) 4,200 feet   (3) 5,100 feet   (4) 5,280 feet

ANALYZE THE PROBLEM.

What must you find?   The length of the Golden Gate Bridge.

What do you know?   (1) The length of the Golden Gate Bridge is 3 times that of the San Francisco–Oakland Bridge.
         (2) The difference between the lengths is 2,800 feet.

What strategy?   Multiple-choice allows you to guess and check.

SOLVE THE PROBLEM. Try 3,900.   Check: The Golden Gate Bridge is 3 times as long as the other bridge, which makes the other bridge 3,900 ÷ 3 or 1,300 feet long. But the difference of these lengths is 3,900 − 1,300 or 2,600, which contradicts what you know.

     Try 4,200.   Check: The shorter bridge must measure 4,200 ÷ 3 or 1,400 feet. The difference (4,200 − 1,400) is 2,800, which fits what you know. This guess is correct.

CHECK YOUR RESULT. The check is part of *Solve the Problem.*

     ANSWER: (2) 4,200 feet.

LEARN FROM THE PROBLEM. Guessing and checking can often be used in multiple-choice problems. In order to use this strategy, there must be a way to check each choice.

Compare the length of the Golden Gate Bridge to 1 mile.

Now it's your turn to . . . **TRY IT!**   *(See page **35** for solutions.)*

1. The number of horror movies in this week's offering at Zany Video Shop is 3 times the number of cartoon tapes. If there are 60 horror movies and cartoon tapes altogether, how many cartoon tapes are there?

2. Although Bill Thomas, a worker at Enzio's, could not work full time, Enzio agreed to give Bill a regular weekly paycheck. For each day Bill worked, he received $60 and for each day he did not work, $20 was deducted from his paycheck. At the end of 10 workdays, Bill was paid $40. How many days did he miss?

EXERCISES

1. Juanita has two more nickels than dimes. If she has $1.60 in all, how many dimes are there?

2. If I add my age to my older brother's age, I get 48. The product of our ages is 567. How old is my brother?

3. Rudy is at the market buying dog food and peas. Each can of dog food weighs 7 ounces. Each can of peas weighs 8 ounces. The total weight of the bag of groceries is 75 ounces. How many cans of dog food does it contain?   (1) 1   (2) 2   (3) 4   (4) 5

4. I told each of my friends a secret. They each told 3 other people. Altogether, 29 of us knew the secret. How many friends did I tell?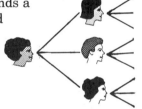

5. Zack had twice as much money as Sonya. After he paid Sonya $14 that he owed her, Zack then had $1 less than she did. How much money did Sonya have before Zack paid his debt?

6. At a hardware store, long nails cost 13 cents each and short nails cost 6 cents each. Ike spent 75 cents buying some of each. How many short nails did Ike buy?

7. Sarah, Li, and Drake each have two U.S. coins. If they put all of their coins together, which of the following amounts can they *not* have?
    (1) 26¢   (2) 38¢   (3) 53¢   (4) 76¢

8. On the 4 major math tests given this term, Awilda had grades of 78, 84, 69, and 83. She would like to raise her test average to 80. What mark must Awilda get on her next test?

9. In a 3-drawer dresser, Matteo has 30% of his T-shirts in the top drawer, 30% in the middle drawer, and the rest in the bottom drawer. What is the smallest number of T-shirts he could have in the bottom drawer?   (1) 1   (2) 6   (3) 8   (4) 12

1-2 DRAW A DIAGRAM

Drawing a diagram is a strategy that is especially useful for problems about shapes or distances. You can show relationships by using a diagram to organize information about the problem. Sometimes, the diagram will even provide the solution directly.

In addition to those involving shapes and distances, there are also other relationships for which you will see how special types of diagrams can lead to a solution.

MODEL PROBLEMS

1. A triangle is placed inside a circle. The triangle may or may not touch the circle. Into how many separate sections can the triangle divide the interior of the circle?

ANALYZE THE PROBLEM.

What must you find?	The number of separate sections inside the circle.
What do you know?	A triangle is inside a circle. It may or may not touch the circle.
What strategy?	Since the problem involves shapes, draw a diagram.

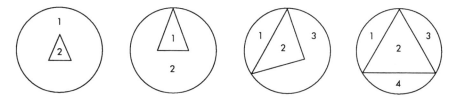

SOLVE THE PROBLEM. The diagrams show that there is more than one possibility. The size and position of the triangle determine the number of points at which it touches the circle.

Touching at 0 or 1 point gives 2 sections.

Touching at 2 points gives 3 sections.

Touching at 3 points gives 4 sections.

CHECK YOUR RESULT. Be sure that you didn't miss any possibilities. The possibilities are that the triangle will touch the circle in 0, 1, 2, or 3 points.

ANSWER: There are 2, 3, or 4 sections.

LEARN FROM THE PROBLEM. Some problems have several possible solutions.

2. The houses on my block are all 24 feet wide and 14 feet apart. At each end of the block, there is a space of 20 feet between the last house and the street. If the entire block is 292 feet long, how many houses are on my side of the block?

ANALYZE THE PROBLEM.

What must you find?   The number of houses on the block.

What do you know?   (1) The houses are 24 feet wide and 14 feet apart.
    (2) There is a space of 20 feet at each end of the block.
    (3) The block is 292 feet long.

What strategy?   Since this problem is about distances, draw a diagram.

SOLVE THE PROBLEM.   Draw some houses. Show the spacing given in the problem.

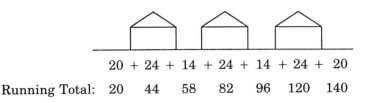

20 + 24 + 14 + 24 + 14 + 24 + 20

Running Total: 20   44   58   82   96   120   140

Continue adding houses until the total length of the block is 292 feet. Then count houses.

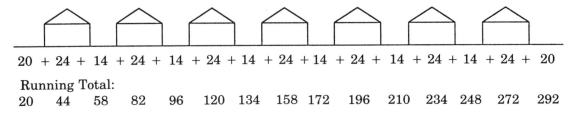

20 + 24 + 14 + 24 + 14 + 24 + 14 + 24 + 14 + 24 + 14 + 24 + 14 + 24 + 20

Running Total:
20   44   58   82   96   120 134 158 172 196 210 234 248 272 292

CHECK YOUR RESULT. Check your addition. The widths of 7 houses and all the spaces add up to 292 feet.

    ANSWER: There are 7 houses.

LEARN FROM THE PROBLEM. A drawing often helps to simplify problems that involve measurements.

Special types of diagrams are used throughout mathematics and science to illustrate situations and help solve problems. A ***Venn diagram*** is useful for showing overlapping sets. A ***tree diagram*** can help to organize a series of successive choices.

3. Nineteen of Miss Fong's students had a hamburger for lunch and 17 had a slice of pizza. Thirteen students had both. If only 26 of Miss Fong's students were in class today, how many students had only a hamburger? How many students had neither?

ANALYZE THE PROBLEM.

What must you find?

   (1) The number of students who ate only a hamburger.
   (2) The number of students who had neither a hamburger nor a slice of pizza.

What do you know?

   (1) 26 students were present.
   (2) 19 students ate hamburger.
   (3) 17 students ate pizza.
   (4) 13 students ate both hamburger and pizza.

What strategy should you use?

Since this problem involves sets of students (those who ate hamburger and those who ate pizza), draw a Venn Diagram.

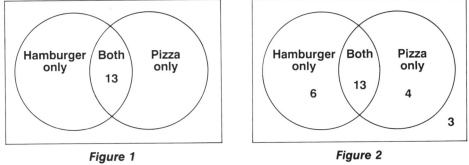

        Figure 1   **Figure 2**

SOLVE THE PROBLEM. Draw a circle to represent the students who ate hamburger. Draw a second, overlapping, circle to represent the students who ate pizza. Draw a rectangle around the circles to represent the entire class.

    Fill in the number of students who ate both hamburger and pizza (Figure 1). Then fill in the number of additional students necessary to complete each set (Figure 2). The number of students who ate only a hamburger is 19 − 13, or 6. The number of students who ate only pizza is 17 − 13, or 4. The total of these students is 6 + 13 + 4, or 23 students. Since there were 26 students in attendance, there were 26 − 23, or 3 students who ate neither hamburger nor pizza.

CHECK YOUR RESULT. The number of students who ate only hamburger is 6. The number of students who ate pizza and hamburger is 13. The number of students eating only pizza is 4. The number of students who ate neither is 3. 6 + 13 + 4 + 3 = 26

    ANSWER: 6 students had only hamburger. 3 students had neither.

LEARN FROM THE PROBLEM. Venn diagrams are useful when items belong to two or more sets. They provide a visual "membership list."

4. In how many ways can the letters, *a*, *b*, and *c* be arranged so that all 3 letters are used exactly once in each arrangement?

ANALYZE THE PROBLEM.

What must you find?   The number of possible arrangements.

What do you know?   (1) There are 3 different letters.
(2) Each letter is to be used exactly once in each arrangement.

What strategy?   Draw a tree diagram to show the possibilities.

SOLVE THE PROBLEM.   Think of the problem in parts:

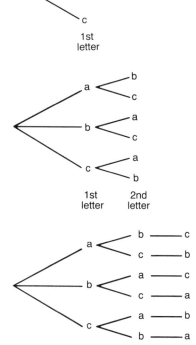

(1) Begin by showing that the first position has 3 choices.

(2) For each letter in the first position, show the choices for the second position.

(3) Complete the tree with the choices in the third position.
    Read the tree by tracing along each branch from beginning to end.

There are 6 possible arrangements:

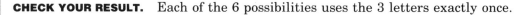

a b c	*b a c*	*c a b*
a c b	*b c a*	*c b a*

CHECK YOUR RESULT.   Each of the 6 possibilities uses the 3 letters exactly once.

ANSWER: There are 6 possible arrangements.

LEARN FROM THE PROBLEM.   A tree diagram shows the branchings that are possible in a series of choices. You can list the different arrangements and then count them.

1. Each side of a square has the same length as each side of a triangle. The two figures are positioned so that one side of the triangle is also a side of the square, and the other sides of the triangle are outside the square. The **perimeter** of (distance around) the resulting figure is 35 inches. What is the perimeter of the triangle?

2. In a 26-mile race, a runner may drink one bottle of water at each checkpoint. The first checkpoint is at the end of the first mile. After that, checkpoints are 3 miles apart. What is the maximum number of bottles of water that a runner may drink during the race?

3. In a quiz, questions 1–3 in one column are to be matched with choices *A*, *B*, *C*, or *D* in the other column. If each answer choice can only be used once, how many question-answer combinations are possible?

(See pages 36–37 for solutions.)

EXERCISES

1. Two triangles have the same shape, but the sides of one are twice as long as the sides of the other. How many of the smaller triangles will fit inside the larger triangle?

2. The towns of Dexter, Midville, and Sinistra lie in a straight line. The distance from Sinistra to Dexter is 23 kilometers. The distance from Midville to Dexter is 14 kilometers. The distance from Midville to Sinistra is 37 kilometers. Which town lies between the others?

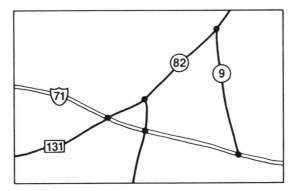

3. Twenty students enrolled in a foreign language program are studying either French or Spanish or both. If 15 students study French and 7 students study both French and Spanish, how many students study Spanish only?

4. Figures that have the same size and shape are called **congruent**. Edna put two congruent squares side by side and then measured the perimeter of the rectangle that was formed. It was 48 inches. How long was a side of one of the squares?

5. Dominic's Delivery Service has routes that connect four cities. If there is a direct route between each pair of cities, how many different routes are there altogether?

6. Paul's Pizza Parlor prepares pepperoni pizza, plain pizza, and mushroom pizza. Of 53 pizzas prepared, 31 had pepperoni topping and 33 had mushroom topping. If 19 of these pizzas had both toppings, how many pizzas were plain?

7. The tree diagram shows the different ways Dr. Johnson can commute to work. From home, he can walk (W), drive (D), or ride his motorbike (M) to the depot, then take a train (T) or a bus (B) to Midtown. Finally, he can walk by either of two routes (X or Y) to his clinic. Write all of the ways Dr. Johnson can get to work. (For example, the first one shown is W T X.)

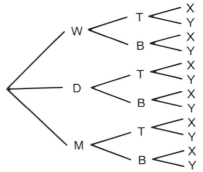

8. In a multiple-choice test, there are 4 possible answers for the first question, 3 choices for the second question, and 2 choices for the third question. If the test consists of only three questions, how many different answer combinations are possible?

9. At an 800-foot-long road construction site, safety cones were placed on the street at 50-foot intervals. There are cones at both ends of the site. How many cones were needed?

10. Into how many separate sections can three straight lines drawn across a sheet of paper divide the paper?

11. A new five-story building in my town has a first floor that is 3 feet above ground level and 12 feet high. The ceiling height is 9 feet on the other floors. The roof is flat, but is surrounded by a wall that adds another 4 feet to the building's height. The steel beams used for the frame add 1 foot between floors and on the roof. What is the height of the top of the building from the ground?

12. Nine congruent squares are put together to form a large square. The sides of the small squares are still visible. Including the large square, how many squares can be found in all?

13. A restaurant offers 2 appetizers, 3 main dishes, and 3 desserts. Assuming that diners order one from each category, how many different meals does the restaurant offer?

1-3 MAKE A LIST OR A TABLE

Making a list is a strategy that is often used in problems that ask "how many?" or require all possibilities to be shown. This strategy is used to arrange or organize existing information so that conclusions can be drawn from it. Note that a table is a type of list.

MODEL PROBLEMS

1. Using only whole numbers, find all of the two-number combinations whose product is 24 and whose sum is less than 13.

ANALYZE THE PROBLEM.

What must you find?	All combinations that fit the conditions.
What do you know?	(1) The product of two numbers is 24.
	(2) The sum of the numbers is less than 13.
What strategy?	You want to find all the possibilities. Make a list.

SOLVE THE PROBLEM. To be sure that you list every possibility, you should make the list in a systematic (organized) way. For instance, begin with 1 as the first number, and list the numbers in increasing order. When the same numbers start to appear in reverse order, your list is complete. There is no need to list both 4×6 and 6×4.

Number pair whose product is 24	Sum of the number pair
1, 24	25
2, 12	14
3, 8	11
4, 6	10

CHECK YOUR RESULT. Be sure all possibilities are included. Check the arithmetic. The only pairs whose sum is less than 13 are 3, 8 and 4, 6.

ANSWER: 3, 8 and 4, 6.

LEARN FROM THE PROBLEM. When listing all possibilities, use a systematic approach to be sure you don't miss anything.

2. At a paperback book sale, Yim bought 3 books at $1.95 each, George bought 6 books at $.98 each, and Henry bought 2 books at $2.95 each. Who spent the least?

ANALYZE THE PROBLEM.

What must you find?	Who spent the least.
What do you know?	(1) How many books each person bought.
	(2) The price of each book.
What strategy?	To organize the information and find the totals, make a table.

	Yim	George	Henry
Price of book	$1.95	$ .98	$2.95
How many books	3	6	2
Money spent	$5.85	$5.88	$5.90

CHECK YOUR RESULT.   Verify the multiplications.

ANSWER: Yim spent the least.

LEARN FROM THE PROBLEM.   When you have a lot of information to keep track of, a table can help you organize it.

3. Every page of the manuscript an author typed on her word processor had a page number printed at the top. The spell-checker counted the number of *digits* (0, 1, 2, 3, 4, 5, 6, 7, 8, and 9 are digits) in the manuscript and displayed a message that there were 660 digits. Since all of the digits were page numbers, starting with page 1, how many pages were in the manuscript?

ANALYZE THE PROBLEM.

What must you find?   The number of pages in the manuscript.

What do you know?   (1) Starting with page 1, there are 660 digits.
(2) Pages 1 through 9 have 1-digit numbers.
(3) Pages 10 through 99 (90 pages) have 2-digit numbers.

What strategy?   To organize the information, make a table.

SOLVE THE PROBLEM.   Select appropriate headings for your table, and fill in the given information.

To find out how many 3-digit numbers there are, subtract 189 from 660 and get 471 digits. Divide 471 by 3 to get the number of pages, 157, and fill in the table.

	No. of Pages	No. of Digits	Total No. of Digits So Far
1-digit numbers	9	9	9
2-digit numbers	90	180	189
3-digit numbers	157	471	660
Total	256		

Verify the computations in the table.

There are	9 pages with 1-digit numbers:	$1 \times 9 =$	9
There are	90 pages with 2-digit numbers:	$2 \times 90 =$	180
There are	157 pages with 3-digit numbers:	$3 \times 157 =$	471

Totals: 256 pages   660 digits

ANSWER: There are 256 pages in the manuscript.

LEARN FROM THE PROBLEM. Once information is organized in a table, it is easy to see what data (information) is needed to complete the problem.

Now it's your turn to . . . **TRY IT!**

1. How many English words (including plurals) can be made using all of the letters in STOP? Use each letter just once in each arrangement.

2. Three friends—Cassius, Clay, and Muhammed—are different ages. Cassius is a bachelor. Clay earns less than the youngest of the three. The oldest of the three earns most, but has the cost of supporting his wife and son. Who is youngest? Who is oldest?

(See pages 37–38 for solutions.)

EXERCISES

In 1 and 2, copy the table. Use your calculator for computation and fill in the missing information.

1. Von saves 30¢ from each dollar he earns babysitting.

Amount earned	1.00	1.50	3.00	?	7.75	?
Amount saved	.30	?	?	1.35	?	2.93

2. The students at Lake City High School had a friendly competition to see which grade would raise the most money in ticket sales for the school play. A total of 546 tickets were sold at $2.25 per ticket.

Grade Level	Number of Tickets Sold	Value of Tickets Sold
9th	175	?
10th	136	?
11th	?	?
12th	116	?
Total	546	?

3. The following table shows the number of years of formal education completed by voting-age residents of Halliwell County. Determine the number of people in each category.

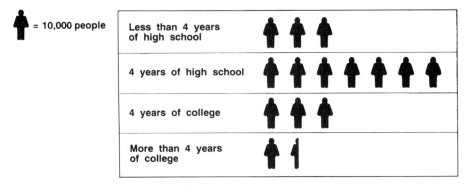

4. Find all of the odd numbers between 10 and 100 that are divisible by 7.

5. Frank, Sam, and André are married to Gert, Rhonda, and Anna, but not necessarily in that order. Sam isn't married to Anna, and Rhonda isn't married to André. Sam's wife is Gert's best friend. Who is Sam's wife?

6. For 5 days, Arnold Jones was paid $63 per day as a temporary employee at the Crown Company. Then he was taken on as a permanent employee and was paid $71 per day for the next 7 days. Then everyone in the company got a raise of $5 per day. How much was Mr. Jones paid for his first 20 days of work?

7. Val P., Lana Q., and Zoe R. are friends. They work as a meteorologist, a therapist, and an administrator. Ms. R. doesn't know much about clouds, Ms. P. doesn't know about muscles, and Ms. Q. doesn't know about muscles or clouds. What work does each do?

8. Arthur, Bertha, and Conchetta tend full-grown animals at Cooper County Zooland. Their favorite animals are monkeys, elephants, and giraffes, but not necessarily in that order. Bertha likes animals that are more than four feet tall. Arthur likes animals that weigh less than 300 pounds. Conchetta likes heavy animals with short necks. Who likes giraffes?

9. Find all of the whole-number pairs whose product is 360.

10. How many different combinations of U.S. coins can you have to get a total of 39 cents?

11. At a fair, I won a prize of $37. I was paid with several tokens, each worth $3 or $7. In how many different ways could I have been paid?

12. Three students each spent a quarter on colored sticker dots to use in flagging research information for their history reports. The school store had large yellow sticker dots at 3 for a nickel, medium red dots at 4 for a nickel, and small blue dots at a penny each. Each student chose a different variety of dots, but each had exactly 20 dots. What were their selections?

1-4 FIND A PATTERN

Finding a pattern is a strategy that you can use to make a general rule from a sequence of specific numbers. Once the rule has been stated, other numbers that fit the pattern can then be found.

The three steps are:

    1. Find a pattern.

    2. State a rule that describes the pattern.

    3. Use the rule to find other numbers.

MODEL PROBLEMS

1. Fill in the missing numbers: 2, 5, 10, 17, —, —, 50

ANALYZE THE PROBLEM.

What must you find?	The missing numbers.
What do you know?	2, 5, 10, 17, —, —, 50 form a pattern.
What strategy should you use?	This problem is stated as a sequence of numbers. In order to fill in the missing numbers, you should find a pattern.

SOLVE THE PROBLEM. Subtract successive numbers to see if a pattern emerges.

Numbers: 2   5   10   17   —   —   50

Differences: 3   5   7

The pattern is that the differences are successive odd numbers. Continue this pattern to fill in the missing numbers.

Numbers: 2   5   10   17   26   37   50

Differences: 3   5   7   9   11   13

CHECK YOUR RESULT. Verify the additions:
$17 + 9 = 26$, $26 + 11 = 37$,   and $37 + 13 = 50$. This last sum, 50, verifies that the pattern found fits the numbers given.

    ANSWER: The missing numbers are 26 and 37.

LEARN FROM THE PROBLEM. Given a sequence of numbers, you can sometimes find a pattern by subtracting successive numbers.

2. Take any odd or even number and multiply it by itself. Now subtract the original number from the result. The answer will be
(1) always odd   (2) always even   (3) sometimes odd, sometimes even

ANALYZE THE PROBLEM.

What must you find?	If the answer is odd or even, or sometimes odd and sometimes even.
What do you know?	You must multiply a number by itself and then subtract the original number from the result.
What strategy?	To find a general rule, try to find a pattern.

SOLVE THE PROBLEM.   Try an odd number, 3. Following the instructions, you see that $3 \times 3 = 9$ and $9 - 3 = 6$. Now try an even number, 4. You get a result of 12. Try another odd number and another even number. All of your results will be even. Your best assumption is that this procedure gives you an answer that will always be even.

CHECK YOUR RESULT.   Later, you will learn how to verify algebraically that you will always get an even number. Meanwhile, after trying several examples, this conclusion seems most reasonable.

   ANSWER: (2) always even

LEARN FROM THE PROBLEM.   For some problems, you will not be certain that your answer is correct. All that you can do is make your best assumption from the evidence.

3. When working on a problem, I added the even numbers, $2 + 4 + 6 + 8 + 10 + 12$, and so forth. By the time I added the number 20, I noticed that sometimes the result was divisible by 4 and sometimes it wasn't. If I continue adding the even numbers through 30, will the result be divisible by 4?

ANALYZE THE PROBLEM.

What must you find?	If the sum of even numbers from 2 through 30 is divisible by 4.
What do you know?	The sum of even numbers is sometimes divisible by 4.
What strategy?	Since this problem is about a number sequence, find a pattern.

SOLVE THE PROBLEM.   Look for a pattern by examining the sums in order.

Numbers:	2	+ 4	+ 6	+ 8	+ 10	+ 12	+ 14	+ 16	+ 18	+ 20
Sums:	2	6	12	20	30	42	56	72	90	110
Sum divisible by 4?	N	N	Y	Y	N	N	Y	Y	N	N

One pattern is that N (No) and Y (Yes) occur in alternating pairs.

***Solution* 1:** Continue the pattern of alternating pairs of No and Yes.

Numbers:	22	24	26	28	30
Sum divisible by 4?	Y	Y	N	N	Y

***Solution* 2:** Look at the pairs of numbers associated with sums that are divisible by 4, namely, 6, 8 and 14, 16. Notice that the number associated with the second Yes in each pair (8 in the first pair, and 16 in the second pair) is divisible by 8. To get a Yes result, either the last number added or the one following it must be divisible by 8.

The number after 30 is 32, which is divisible by 8. Thus, you get a Yes result, and the sum of the even numbers from 2 up to and including 30 is divisible by 4.

CHECK YOUR RESULT. Verify that you made no mistakes in applying the pattern. Note that two different solutions give the same result.

ANSWER: Yes, the result will be divisible by 4.

LEARN FROM THE PROBLEM. The first solution applies only to the specific numbers for which you continue the pattern. The second solution is more general and can be applied to any stopping point.

Now it's your turn to . . . **TRY IT!**

1. Numbers that follow each other in order are *consecutive*. For example, 3, 4, and 5 are consecutive numbers. Adding any three consecutive numbers, the result will be
   (1) always odd   (2) always even   (3) sometimes odd, sometimes even

2. Fill in the missing numbers: 11, 15, 22, 32, __, __, 80

(See pages 38–39 for solutions.)

EXERCISES

1. Mr. Boyd started saving for a gift for his wife 12 weeks before her birthday. The table shows the amount he saved in the first 4 weeks. If Mr. Boyd continued his pattern of saving over the next 8 weeks:
   a. Identify the pattern Mr. Boyd is following.
   b. How much did he save during week 9?
   c. How much did he save altogether?

Week	Amount Saved Each Week	Total Amount Saved
1	$1	$1
2	$3	$4
3	$5	$9
4	$7	$16

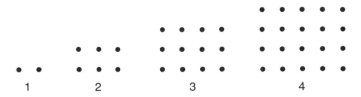

2. **a.** Identify the pattern of the dots in the numbered figures.
   b. Describe the pattern that would appear in the 15th figure.
   c. How many dots would be in the 15th figure?

3. It is now 2 P.M. **a.** What time will it be in 103 hours? **b.** What time was it 46 hours ago?

4. My local video store charges $3.00 for a first overnight video rental. The cost of each rental after that decreases by 5 cents. Thus, the second rental is $2.95, the third rental is $2.90, and so forth, to a minimum of $.75 per rental. How much would the 45th rental cost?

5. Take half of any even number and then add 1. Your answer will be
   (1) always odd   (2) always even   (3) sometimes odd, sometimes even

6. Fill in the missing numbers: 2, 3, 5, 8, ___, ___, 23

7. Choose a two-digit number. Switch the digits and then add your original number to the new number. Your result will always be a multiple of what number?

8. I mowed my neighbor's lawn for the first time this week and earned $3.50. We agreed that every time I mow the lawn I will get 15 cents more than the last time. How much money will I earn the 13th time I mow the lawn?

9. Keith started adding the even numbers from 2 up. By the time he added the number 20, he noticed that sometimes the result was divisible by 5 and sometimes it wasn't. If he continues adding the even numbers through 38, will the result be divisible by 5?

10. Fill in the missing number: 143, 142, 139, 130, ___, 22

11. An inchworm at the bottom of a wall 17 feet high and 70 feet long is climbing the wall as shown. If his climb continues to follow this pattern, how far along the length of the wall will he have traveled when he reaches the top?

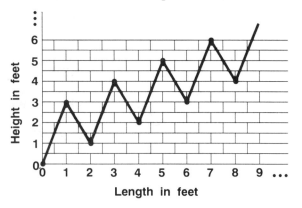

1-5 WORK BACKWARD

Working backward is a strategy for solving problems in which the result is given and you must determine something that has gone before. To do this, you do the opposite, or *inverse*, of what was done to get the result. For example, if you added 3, the inverse is to subtract 3. Begin with the last step and work backward to the beginning.

You have used this strategy to check arithmetic problems. When you did a subtraction problem, you checked your solution by changing the problem into addition. You began with the result of the subtraction and worked backward. For example:

$$\text{\textit{Problem}:} \quad 17 - 9 = 8 \qquad \text{\textit{Check}:} \quad 8 + 9 = 17$$

This strategy will also be useful in solving many of the algebra problems that will be presented in this course.

MODEL PROBLEMS

1. I am thinking of a number. If I add 7 and then multiply by 2, I get 26. What is the number?

ANALYZE THE PROBLEM.

What must you find?	The number.
What do you know?	Start with the number. Add 7. Multiply by 2. You get 26.
What strategy should you use?	Since you already know the result of several computations and you want to find the original number used in the computation, work backward.

SOLVE THE PROBLEM.   Begin with the result, 26.

Since the last operation in the problem was multiplying by 2, do the inverse: divide by 2. You get $26 \div 2 = 13$.

Now do the inverse of adding 7, which is subtracting 7. You get $13 - 7 = 6$.

CHECK YOUR RESULT.   Start with 6, and work in the order given in the problem.

$$\text{Add 7:} \quad 6 + 7 = 13$$

$$\text{Multiply by 2:} \quad 13 \times 2 = 26$$

ANSWER: The number is 6.

LEARN FROM THE PROBLEM.   When you work backward, you begin at the end and reverse, or "undo," everything that was done to get the result.

2. Every family on my block was asked to contribute $3 for Halloween decorations. Seven families didn't contribute, but $81 was collected from the others. How many families live on my block?

ANALYZE THE PROBLEM.

What must you find?	The number of families on the block.
What do you know?	(1) All the families except 7 contributed $3.
	(2) The total collected was $81.
What strategy should you use?	Since you know the total amount collected and you must find a number used to get the total, work backward.

SOLVE THE PROBLEM.   The number of donating families × $3 = $81.
Working backward: $81 ÷ $3 = 27, which means that the number of contributing families is 27. Since 7 families didn't contribute: 27 + 7 = 34, the total number.

CHECK YOUR RESULT.   Since there are 34 families: 34 − 7 = 27 and 27 × $3 = $81

ANSWER: There are 34 families on the block.

LEARN FROM THE PROBLEM.   You can work backward to find numbers used in computations for which the result is known. Sometimes, when you are given an answer to a problem that you could not solve, you can work backward to see how that answer was obtained.

3. *Cryptarithms* or *alphanumerics* are problems in which letters are used to stand for digits. Every time a letter is used in a problem, it stands for the same digit. Different letters stand for different digits. For instance, if *A* stands for 3 and *B* stands for 5, then *BBABA* represents 55353.

In this cryptarithm, what digit does each letter stand for?

$$\begin{array}{r} A\ A \\ +\ A\ B \\ \hline C\ B\ A \end{array}$$

ANALYZE THE PROBLEM.

What must you find?	Digits that make the problem true.
What do you know?	The addition problem uses three different digits.
What strategy should you use?	By inspection, you can determine some of the digits. Then you can work backward to find others.

SOLVE THE PROBLEM.   You know that if you add two numbers with two digits and you get an answer with three digits, you must have carried a number. Since the most you can get when you add two digits that are the same is 18 (9 + 9), the number you carried must be 1. Therefore, *C* = 1.

In the ones column, you see that $A + B = A$. The only way you can start and end with the same number in addition is by adding 0. Thus, you know that $B = 0$.

$$\begin{array}{r} A\,A \\ +\ A\,B \\ \hline 1\,B\,A \end{array}$$

In the tens column, you see that $A + A = 10$. Thus, $A = 5$.

$$\begin{array}{r} A\,A \\ +\ A\,0 \\ \hline 1\,0\,A \end{array}$$

CHECK YOUR RESULT.

$$\begin{array}{r} 5\,5 \\ +\ 5\,0 \\ \hline 1\,0\,5 \end{array}$$

ANSWER: $A = 5$ $B = 0$ $C = 1$

LEARN FROM THE PROBLEM. Working backward starts from something you know. In this type of problem, you know an addition pattern, rather than specific numbers. You have to find your own place to start, and you must rely on your general knowledge of addition to guide you to the next step.

Now it's your turn to . . . **TRY IT!**   *(See pages 39–40 for solutions.)*

1. Phil spent $23 on Christmas gifts for his two sisters. He gave Carmella a record and he gave Janice three books. Phil forgot the exact prices he paid, but remembered that the books were on a table with the sign "All books on this table are $3, $4, or $5." How much did Phil pay for Carmella's record?

2. What digit does each letter stand for?

$$\begin{array}{r} A\,C \\ +\ C\,A \\ \hline D\,A\,D \end{array}$$

CALCULATOR CONNECTION

The calculator is an important tool as you learn more mathematics. In addition to providing speed and accuracy in basic computation, calculators will be used later in the text to perform new types of calculations.

Another reason for using a calculator is to aid in problem solving. A calculator is especially helpful in (1) checking a guess or an estimate, (2) finding a pattern, since routine computations can be done quickly, and (3) working backward, by performing calculations in reverse order. Using a calculator may help you solve some of the following exercises.

1. If you double a number and then add 7, you get 29. What is the number?

2. A woman picked a basket of apples. On her way home, she gave half of the apples in the basket to each person she met. She met three people and arrived home with 3 apples. How many apples did she start with?

3. In Mountaintop High School, a grade of 90 or more is an A. I already have 79, 94, and 98 in math. What grade do I need on my next math test to average exactly 90?

4. The income for Peachie Pie Products has grown by $1,000 per month for each of their three kinds of pies. How many months ago was the income from apple pie twice as much as the income from cherry pie? The graph shows the income for this month.

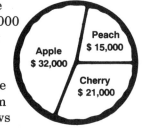

5. Todd left home with tomatoes from his garden for his friends. He met Ramon first and gave him 3 tomatoes. Sylvia took half of what was left, leaving 7 tomatoes for Rudy. How many tomatoes did Todd have when he started?

6. In the cafeteria, some ketchup dripped onto Di's math worksheet. Judging from the numbers that can still be seen, which of the answers provided *cannot* be an answer to the problem?

    a.   6 5▨
        + 7▨

    (1) 725   (2) 73,942   (3) 6,281   (4) 7,200

    b. 4 0▨9
        − 4▨

    (1) 3,725   (2) 3,611   (3) 3,592   (4) 3,650

    c. 6▨
        × 9▨

    (1) 54,270   (2) 59,892   (3) 66,445   (4) 71,201

7. Alphonse Navarro is a Wall Street stock trader. He always uses half of his total assets to fund each new venture. The rest is put away in case his investment doesn't work out. Alphonse had 3 spectacular successes in which he doubled his money each time. In his most recent venture, however, he lost the money he invested. He has only the $54,000 that he put away. How much money did he have when he made his first investment?

In 8–10, what digit does each letter stand for?

8.   E F
     + G G
     ―――――
     G H E

9.   A B B
     + C B D
     ―――――
     C D D B

10.   A B C D
     + A B C D
     ―――――
     E E A B C

1-6 CHOOSE A STRATEGY

The five strategies that you have learned are:

1. Guess and check.
2. Draw a diagram.
3. Make a table or a list.
4. Find a pattern.
5. Work backward.

> THE SUREST WAY
> TO MAKE A PROBLEM GO AWAY
> IS TO . . . SOLVE IT!

Working within the general framework, you must now choose a strategy to solve each problem. For some problems, any of several strategies may be appropriate, or you may combine strategies.

Although the five strategies set forth in this chapter are suitable for a great variety of problems, no list of strategies is all-inclusive. Even in this chapter, along with the strategies given, you have used other strategies that were not identified by name. These include the *process of elimination*, *breaking a problem into smaller problems*, and *relating the problem to a simpler problem*.

In combination with many strategies, it is helpful to also use *estimation*. For example, when you guess and check, estimating can help you make a good guess. Estimating is often useful in judging whether an answer is reasonable.

MODEL PROBLEMS

1. For which of the following numbers can you start with the number, add 4, then multiply by 3, and finally subtract 2 to get 25?   (1) 4   (2) 5   (3) 6   (4) 7

ANALYZE THE PROBLEM.

What must you find?　　The starting number.

What do you know?　　To a number: add 4, multiply by 3, subtract 2, and get 25.

What strategy
should you use?　　In a multiple-choice problem, you can guess and check.

or

Since the problem gives a result and asks you to find a number
used in the computation, you can work backward.

SOLVE THE PROBLEM.

By guessing and checking

Try 4; you get 22. Wrong.

Try 5; you get 25. Correct.

By working backward

Start with 25 and reverse the operations.

Add 2:	Divide by 3:	Subtract 4:
$25 + 2 = 27$	$27 \div 3 = 9$	$9 - 4 = 5$

CHECK YOUR RESULT. Work in the order given in the problem. Start with 5.

$$5 + 4 = 9 \qquad 9 \times 3 = 27 \qquad 27 - 2 = 25$$

ANSWER: (2) 5

LEARN FROM THE PROBLEM. This problem can be solved by either of two strategies.

2. In running errands, Joan went north 8 blocks, then south 6 blocks, then north 3 blocks, then south 9 blocks, then finally north 3 blocks. Where did she finish with respect to her original starting point?

ANALYZE THE PROBLEM.

What must you find?	Where Joan finished.
What do you know?	She went 8 blocks north, 6 blocks south, 3 blocks north, 9 blocks south, then 3 blocks north.
What strategy should you use?	Since distance is involved, you can draw a diagram.

or

You can arrange the information by making a table.

SOLVE THE PROBLEM.

By drawing a diagram

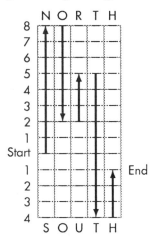

By making a table

Joan went	Joan is now
8 blocks north	8 blocks north
6 blocks south	2 blocks north
3 blocks north	5 blocks north
9 blocks south	4 blocks south
3 blocks north	1 block south

CHECK YOUR RESULT. Be sure you didn't skip or repeat anything.

ANSWER: Joan finished 1 block south of where she started.

LEARN FROM THE PROBLEM. Some problems can be solved by any of several strategies. When more than one strategy can be applied, use an alternate solution as a check.

Now it's your turn to ... **TRY IT!**

1. Five people in a room each shake hands with all of the others. How many handshakes are there altogether?

2. This problem may seem difficult or even impossible. It isn't. Stick with it.

   Two acquaintances met on a street corner. They exchanged greetings and each asked about the other's family. The woman, who knew that the man had three children, asked how old the kids were now. The man replied, "I won't tell you, but I'll give you a hint. The product of their ages is 36."

   The woman said that she still didn't know their ages.

   "I'll give you another hint," the man said. "The sum of their ages is equal to your house number."

   The woman again responded that she didn't know the children's ages.

   So the man said, "I'll give you one more hint. My youngest one is blond."

   To which the woman replied, "Now I know."

   How old are the children and what is the woman's house number?

   (See pages 40–41 for solutions.)

EXERCISES

In 1–16, solve each problem and tell which strategy you used to solve it.

1. I am thinking of two numbers whose sum is 10 and whose product is 24. What are the numbers?

2. If you add three odd numbers, your answer is
   (1) always even   (2) always odd   (3) sometimes even, sometimes odd

3. Alice and Faye each work 5 days every week, and Doreen works 4 days every week. Together, the three women must fill two jobs 7 days a week. Alice works on days that begin with *M*, *T* and *S*. Faye has Monday and Tuesday off. What are Doreen's days off?

4. You have two congruent squares side by side so that the right edge of one is the left edge of the other. If you draw an *X* in each square from corner to corner, how many triangles will be formed altogether?

5. Pick a whole number from 3 through 14. Multiply it by itself and then subtract the original number. Your result can always be divided by
   (1) 5   (2) 4   (3) 3   (4) 2

6. What digit does each letter stand for?

$$\begin{array}{r} P\,Q\,R \\ +\ S\,T\,R \\ \hline S\,T\,P\,T \end{array}$$

7. "If I take a two-digit number with two different digits and switch the digits around, then subtract the smaller number from the larger, I always get a multiple of 9." Is this statement true?

8. Yvonne spent half of her money at Norma's Yarn Shop, and then $10 more for gasoline. This left her with only $3. How much money did she start with?

9. Madison and Jefferson have adjoining square gardens that are the same size. If they cooperate and put one fence around the two gardens with no fence between, they need exactly 72 feet of fencing altogether. How long is one side of Madison's square garden?

10. At the hardware store, Mai bought some screws for 9 cents each and some washers for 7 cents each. She spent 83 cents. How many washers did she buy?

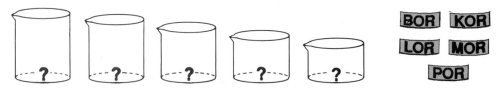

11. Five containers were arranged in size order on a laboratory shelf. When Mel was cleaning, the labels fell off. Help Mel replace the labels on the proper containers, using these clues:
    (1) The LOR container is larger than the BOR.
    (2) The MOR is not the smallest.
    (3) The LOR is not the largest.
    (4) Only one container is smaller than the KOR.
    (5) The POR is larger than the MOR.
    (6) There are at least two containers larger than the MOR.

12. Mrs. Kalil's class was collecting money for disaster relief and for the local hospital. Of the 21 students who donated, 15 gave money for disaster relief and 13 gave money for the hospital. Some of the students donated to both. How many donated only for disaster relief?

In September, 1985, a massive earthquake, registering a staggering 8.1 on the Richter scale, shook Mexico City. Ten thousand people died, and the estimates of damage ran into the billions of dollars.

The inscriptions on the jackets show that the disaster relief workers in the photograph are associated with the Cruz Roja, or Red Cross. Red Cross and allied societies help victims of natural disasters in about 150 countries around the world.

13. When Mr. Reston checked in at Newark airport for a flight to Denver, he was told that his flight had been overbooked, and the airline offered him a free flight if he would change to a later plane. Mr. Reston had arranged to meet a business acquaintance at the Denver airport. If the flight time was $4\frac{1}{2}$ hours, and Denver time was 2 hours earlier than Newark time, what was the latest he could leave (on Newark time) and still make his 7:00 P.M. appointment in Denver?

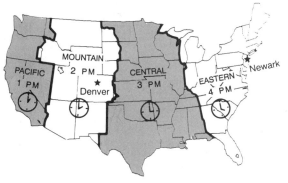

14. Twenty cupcakes have vanilla icing and 18 have chocolate icing. If 7 of these cupcakes have both chocolate and vanilla icing, how many cupcakes are there altogether?

15. Yoko started adding the even numbers, from 2 up. By the time she added the number 20, she saw that sometimes the result was divisible by 3 and sometimes it wasn't. If she continued adding the even numbers through 30, would the result be divisible by 3?

16. Taking part in a local recycling drive, the three neighbors Ava, Boris, and Cindy spent a Saturday morning collecting cans and bottles. When they counted their collections, they discovered that even though Ava had collected 3 times as many cans as Boris, and Boris had collected 4 times as many bottles as Cindy, each had collected exactly the same number of items. Also, the three as a group had collected exactly as many cans as bottles. How many cans and bottles did Ava collect?
   (1) 9 cans, 6 bottles   (2) 12 cans, 8 bottles
   (3) 15 cans, 14 bottles   (4) 18 cans, 16 bottles

Specially-designed recycling equipment handles cans, bottles, plastics, and paper separately. To maintain a clean environment, more and more communities have recycling programs, and it is each person's responsibility to help solve the problem of waste disposal.

Studies have suggested new ways of using recycled materials—waste plastics can be made into park benches; crushed glass is an ingredient of a paving material for city streets.

CHAPTER SUMMARY

A general approach can be applied to solving all types of problems. The methods used to solve particular types of problems are called *strategies.*

General Framework

1. **ANALYZE THE PROBLEM.** Ask yourself:

   a. What am I asked?   b. What is given?   c. Which strategy will work best?

2. **SOLVE THE PROBLEM.** Put your strategy into action. If your first approach doesn't work, try a new one. Stay with a problem.

3. **CHECK YOUR RESULT.** Is your answer reasonable? Go back to the statement of the problem to see if your answer makes sense.

4. **LEARN FROM THE PROBLEM.** Other problems may be like the one you just solved. For a new problem, try a strategy that has worked with a similar problem.

Specific Problem-Solving Strategies

1. *Guess and Check* when a problem has only a few possible solutions.

2. *Draw a Diagram* for problems involving shapes or distances.
   In overlapping sets, a *Venn diagram* shows where the members are located.
   A *tree diagram* shows all the results of a series of choices.

3. *Make a Table or List* to help solve problems that ask "how many?"

4. *Find a Pattern* to arrive at a general rule from a sequence of numbers.

5. *Work Backward* to solve problems in which the end result is known but you must find a fact that led to it.

You can use different strategies to solve the same problem. You may use a combination of strategies. You may even invent other strategies. The best strategy is the one that enables you to solve the problem.

In this course, you will learn a powerful new strategy. You will see that, if problems are expressed in the language of mathematics, it is easier to find a solution. Using algebra as a tool, you will be able to solve problems of greater and greater complexity.

VOCABULARY CHECKUP

SECTION

Intro. *strategy*

1-2 *Venn diagram / tree diagram / perimeter / congruent*

1-3 *digit*   **1-4** *consecutive numbers*   **1-5** *inverse*

In 1 and 2, solve by guessing and checking. (Section 1-1)

1. Nicola has been present in school 5 more than twice as many days as she has been absent. If school has been in session for 38 days, how many days has Nicola missed?

2. Heidi has 50 U.S. coins with a total value of $1.00. If she has the same number of nickels as dimes, how many dimes does she have?
   (1) 1   (2) 2   (3) 3   (4) 4

In 3 and 4, solve by drawing a diagram. (Section 1-2)

3. Mr. Sagot wants to plant flower seedlings at 7-inch intervals on both sides of a path in front of his house. If the length of the path is 7 feet and he begins at one end of the path, how many seedlings will he need?

4. Of the 132 instrumental music students in Washington High School, 85 are in the band, 97 are in the orchestra, and 12 are in neither band nor orchestra. How many students are in both band and orchestra?

In 5 and 6, solve by making a list or table. (Section 1-3)

5. Michelle went to a restaurant where a hamburger costs $3.35, a cheeseburger costs $3.70, a soft drink costs $.95, and french fries cost $1.10. If Michelle bought either two or three different items and gave the cashier a $10 bill, which of the following amounts of change could she receive?
   (1) $3.40   (2) $4.45   (3) $5.35   (4) $6.55

6. The numbers 1, 2, 4, and 8 are exact divisors of 8. The numbers 1, 2, 5, and 10 are exact divisors of 10. How many exact divisors does 144 have?

In 7 and 8, solve by finding a pattern. (Section 1-4)

7. Find the next two numbers in the sequence: $1\frac{1}{2}$, 4, $7\frac{1}{2}$, 12, $17\frac{1}{2}$, 24, ___, ___

8. Mina planned regular increases in the time she spent practicing the piano. She started by practicing for 15 minutes on a Monday, and increased the time by 5 minutes each weekday thereafter. If she continues this pattern, in what week will the 5 days' practice total 500 minutes?

In 9 and 10, solve by working backward. (Section 1-5)

9. Mrs. Melendez took her daughter shopping. She spent half her money for clothes, one-third of the remaining money for groceries, and half of what was then left in the drug store. She then gave one-third of what remained to her daughter, and had $11.32 left over. How much money did she start with?

10. In a multiplication of a 1-digit number by a 2-digit number, the product was 335. What numbers were multiplied?

In 11–14, choose a strategy. (Section 1-6)

11. A sales manager is conducting his annual sales meeting and is undecided about the seating arrangements. He has fewer than 100 salespeople and can seat them either 3, 5, or 7 to a table. If he seats them 3 to a table, then 1 person remains to be seated; if he seats them 5 to a table, 2 remain to be seated; if he seats them 7 to a table, 3 remain to be seated. How many salespeople are expected to attend the meeting?

12. In a league of 6 soccer teams, how many league games will be played in a season if each team plays 2 games with every other team?

13. If I have 2 more dimes than nickels, 3 more quarters than dimes, and no pennies, which of the following amounts could I have?   (1) $2.05   (2) $2.15   (3) $2.25   (4) $2.35

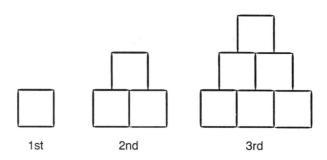

1st   2nd   3rd

14. Toothpicks are placed to form squares, and the squares in turn form pyramids. Given the pattern shown:
   a. Draw or make a toothpick figure of the 4th pyramid.
   b. Find how many squares would be in
      (1) the 5th pyramid   (2) the 10th pyramid
   c. Find how many toothpicks would be in
      (1) the 5th pyramid   (2) the 10th pyramid

PROBLEMS FOR PLEASURE

In 1 and 2, what digit does each letter stand for?

1. F O R T Y
   +   T E N
   +   T E N
   ─────────
    S I X T Y

2.   S E N D
   + M O R E
   ─────────
    M O N E Y

3. The foreign language program at my school has 25 students studying French, 23 studying Spanish, and 28 studying Italian. Of these, 4 students are studying all three languages, 5 are studying only Spanish and Italian, 8 are studying only French, and 11 are studying only Italian. How many students are in the foreign language program?

4. Each of the 16 students in Mrs. Faust's class decorated a campaign poster for school elections. Ten used red markers, 8 used green markers, and 9 used yellow markers. Only 4 students used all three colors. How many students used two colors?

5. A ***network*** is a figure containing paths and intersections. One application for a network is in mapping routes. In the Air America network, 4 cities on the east coast are connected by nonstop flights, as are 3 cities on the west coast. All of these cities offer nonstop flights to and from Denver, the Air America hub city. Cross-country flights originating at any of the coast cities make a single stop, at Denver. Use a network diagram to determine the number of nonstop routes flown by Air America. (Count each direction; Boston to Miami is different from Miami to Boston.)

Network theory was originated by the 18th-century Swiss mathematician Leonhard Euler, in his exploration of the Seven-Bridges problem. Konigsberg, a town on the banks of a river, included two islands, with seven bridges joining the parts of the town.

The townspeople asked Euler whether he could devise a walk that would cross each bridge exactly once. Euler concluded that it could not be done. However, his study of the problem laid the foundation for network theory.

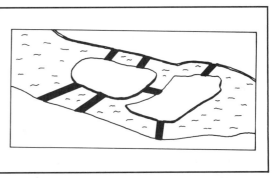

CALCULATOR CHALLENGE

1. Do the first three problems in each set using a calculator. See if you can find a pattern that will give the answer to the fourth problem without using a calculator. Check your answer with a calculator.

a.	**b.**	**c.**	**d.**
$101 \times 85 =$	$111 \times 111 =$	$6 \div 99 =$	$12 \times 9 + 3 =$
$101 \times 43 =$	$111 \times 212 =$	$7 \div 99 =$	$123 \times 9 + 4 =$
$101 \times 57 =$	$111 \times 313 =$	$9 \div 99 =$	$1,234 \times 9 + 5 =$
$101 \times 89 =$	$111 \times 414 =$	$12 \div 99 =$	$12,345 \times 9 + 6 =$

2. Bridge stickers cost $15 each month. Bridge tolls are usually 90 cents, but tolls for cars with bridge stickers are 25 cents. How many trips must you make each month to save money by buying a bridge sticker?

3. The phone company charges $13.45 per month for basic local service. This provides 20 free local calls each month, with a charge of 17 cents for each additional call. Extended phone service provides unlimited local calls each month for $20.00. How many calls must you make in a month for the extended service to save you money?

4. Luke used his calculator to do some homework problems. On his way to school the next morning, the rain spattered his notebook, washing away part of his work. Find the three missing digits.

Questions in this section were selected from recent *SATs* published by the College Entrance Examination Board, and are reprinted by permission of Educational Testing Service, the copyright owner of the sample questions. Permission to reprint this material does not constitute endorsement by Educational Testing Service or the College Board of this publication as a whole or of any other testing information it may contain.

In 1–13, choose the correct answer.

1. The cube above has a number on each of its six faces. If the sum of the numbers on each pair of opposite faces is 10, what is the sum of the numbers on the faces *not* shown?
   (A) 8 (B) 10 (C) 12 (D) 14 (E) 16

2. A man has 15 bags of grain in his barn. Given that he can carry at most 2 bags of grain at a time, what is the least number of trips he must make from the barn to the truck in order for him to carry all of the bags of grain to his truck?
   (A) 7 (B) 7.5 (C) 8 (D) 8.5 (E) 9

X	Y
1	4
2	5
3	6

3. If x is a number from column X and y is a number from column Y in the table above, how many different values are possible for $x + y$?
   (A) Nine   (B) Eight   (C) Seven
   (D) Six   (E) Five

4. A month with 5 Wednesdays could start on a
   (A) Sunday   (B) Monday (C) Thursday
   (D) Friday   (E) Saturday

5. A small bus has 3 empty seats, 6 seated passengers, and 2 standing passengers. If 3 passengers get off, 4 get on, and everyone on the bus is seated, how many empty seats would there be?
   (A) None   (B) One   (C) Two
   (D) Three   (E) Four

6. John had exactly $7 before Bill paid him a $26 debt. After the debt was paid, Bill had $\frac{1}{3}$ the amount that John then had. How much did Bill have before the debt was paid?
   (A) $33   (B) $35   (C) $36
   (D) $37   (E) $47

7. Several people are standing in a straight line. Starting at one end of the line, Bill is counted as the 5th person and, starting at the other end, he is counted as the 12th person. How many people are in the line?
   (A) 15 (B) 16 (C) 17 (D) 18 (E) 19

8. The correct addition problem shows the sum of two 2-digit numbers. If X and Y represent different nonzero digits, then Y equals
$$\begin{array}{r} 7X \\ +X1 \\ \hline 1Y7 \end{array}$$
   (A) 1 (B) 3 (C) 6 (D) 7 (E) 8

$$A = \{3, 6, 9\}$$
$$B = \{5, 7, 9\}$$
$$C = \{7, 8, 9\}$$

9. If three *different* numbers are selected, one from each of the sets shown above, what is the greatest sum that these three numbers could have?
   (A) 22 (B) 23 (C) 24 (D) 25 (E) 27

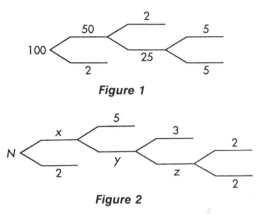

Figure 1

Figure 2

10. Figure 1 is an example of a "factor diagram" of 100. What is the value of N if Figure 2 is a "factor diagram" of N?
(A) 70 (B) 120 (C) 150 (D) 240 (E) 300

11. Colored banners are strung on a cord to advertise a carnival. If the colors form a repeating pattern starting with white, red, yellow, blue, green, purple; white, red, yellow, blue, green, purple, and so on, what is the color of the 76th banner?
(A) Red   (B) Yellow   (C) Blue
(D) Green   (E) Purple

12. Twenty-five squares, each painted one of the solid colors red, green, yellow, or blue, are lined up side by side in a single row so that no two adjacent squares are the same color. If there is at least one square of each color, what is the maximum possible number of blue squares?
(A) 9 (B) 10 (C) 11 (D) 12 (E) 13

Number of Donuts	Total Price
1	$0.40
Box of 6	$1.89
Box of 12	$3.59

13. What would be the *least* amount of money needed to purchase exactly 21 donuts?
(A) $5.88   (B) $6.68   (C) $7.19
(D) $7.38   (E) $8.40

SPIRAL REVIEW EXERCISES

The Spiral Review Exercises at the end of each chapter will help you remember material that you learned earlier. Some questions, particularly in the first few chapters, are on topics you studied in previous courses.

1. Round 964,825 to the nearest ten thousand.

2. Evaluate:

$$\frac{1}{2} \times \frac{1}{2} \times \frac{1}{2}$$

3. Subtract:

9 feet 4 inches
− 3 feet 7 inches

4. Evaluate:

$$\frac{10}{5} - \frac{5}{10}$$

5. Cindi got 74 on her last algebra test. What grade must she get on the next test to bring her average for the two tests up to 80?

6. Find the sum of 2.006, 1.135, 6, 0.5, and 3.5.

7. The Sharks won $\frac{2}{3}$ of the basketball games they played. If they lost 6 games, how many games did they play?

A   B C D

8. If segment *AB* is twice the length of segment *BC* and segment *BC* is twice the length of segment *CD*, then what fractional part of segment *AD* is segment *BD*?

9. The price of soap that normally costs 3 bars for $1 is reduced to 8 bars for $2. How much do you save if you buy 24 bars of soap?

10. Alex arrived home from school at 3:20 P.M. He spent 45 minutes on his math homework and 40 minutes on his social studies homework. He then spent 35 minutes playing ball with his friends. If dinner begins promptly at 6:00 P.M., how much time is left before dinner?

11. How much greater than 17 is the sum of $3\frac{2}{3}$, $5\frac{1}{6}$, and $8\frac{1}{4}$?

12. A telephone call costs $1.25 for the first 3 minutes and $0.25 for each minute thereafter. At these rates, what is the cost of a 12-minute phone call?

13. The raisins in a bag weigh $\frac{9}{16}$ of a pound. If the raisins are divided into 6 equal shares, how much will 4 shares weigh?

14. This graph of the federal government dollar shows where the money comes from.

   a. What fractional part of federal government revenue is made up of the combined income from individual income taxes and social insurance and retirement programs?

   b. The money that the government gets by borrowing is how many times as much as the money it collects from excise taxes?

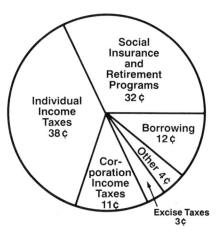

15. Bell *A* rings every 4 minutes and Bell *B* rings every 3 minutes. If the bells ring simultaneously at 2:00 P.M., what is the total number of times the bells will ring simultaneously between 2:00 P.M. and 3:00 P.M., inclusive?

16. In Lower Krilovia, 40 sheep can be traded for 80 goats and 40 goats can be traded for 80 pigs. How many sheep can be traded for 40 pigs?

17. There are fewer than 33 eggs in a basket. If they are counted out 6 at a time, there are 4 left over. If they are counted out 5 at a time, there are 3 left. How many eggs are in the basket?

1-1 GUESS AND CHECK

TRY IT! *Problem 1 on page 4*

ANALYZE THE PROBLEM.

What must you find?
   The number of cartoon tapes.

What do you know?
   The sum of the two numbers is 60. The number of horror movies is 3 times the number of cartoon tapes.

What strategy should you use?
   The number of possible solutions is small. Guess and check.

SOLVE THE PROBLEM.

   Guess: 10 cartoon tapes. Check: The number of horror movies is $3 \times 10 = 30$. The sum of these numbers, $10 + 30$, is only 40. This is too low. Guess higher.

   Guess: 12 cartoon tapes. Check: The number of horror movies is $3 \times 12 = 36$. The sum here is 48. Still too low; guess higher.

   Guess: 16 cartoon tapes. Check: The number of horror movies is $3 \times 16 = 48$. The sum here is 64. Too high; guess lower.

   Guess: 15 cartoon tapes. Check: The number of horror movies is $3 \times 15 = 45$. The sum, $15 + 45 = 60$, agrees with what you know.

CHECK YOUR RESULT.   The check is part of the strategy in *Solve the Problem*.

   ANSWER: There are 15 cartoon tapes.

LEARN FROM THE PROBLEM.   Use the result from each guess to guide you toward your next guess.

TRY IT! *Problem 2 on page 4*

ANALYZE THE PROBLEM.

What must you find?
   The number of days Bill missed.

What do you know?
   Bill gets $60 per day worked. He loses $20 per day missed. His pay for 10 workdays was $40.

What strategy should you use?
   The number of solutions is limited. Guess and check.

SOLVE THE PROBLEM.   It is clear from the amount he was paid that Bill missed some days and worked some days. Guess that he missed 5 days. Check: The salary for working 5 days is $300 and the loss for missing 5 days is $100. The pay would be $200. Since his actual pay was only $40, he missed more than 5 days.

   Guess: 6 Check: The salary for working 4 days is $240 and the loss for missing 6 days is $120. The pay would be $120. He missed more than 6 days.

   Guess: 7 Check: The salary for working 3 days is $180 and the loss for missing 7 days is $140. The pay would be $40. This agrees with what you know.

CHECK YOUR RESULT.   The check is included in *Solve the Problem* as part of the strategy.

   ANSWER: Bill missed 7 days.

LEARN FROM THE PROBLEM. When making guesses, use the result from each guess to make the next guess a better guess. In this problem, guessing 5 missed days gave a result of $200 pay. Since this was too much pay, it led to a guess of more days missed, which would result in less pay.

1-2 DRAW A DIAGRAM

TRY IT! *Problem 1 on page 9*

ANALYZE THE PROBLEM.

What must you find?
  The perimeter of the triangle.

What do you know?
  (1) All sides of the square and triangle are the same length.
  (2) The perimeter of the resulting figure is 35 inches.

What strategy should you use?
  Since the problem involves shapes, draw a diagram.

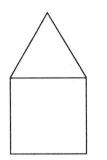

SOLVE THE PROBLEM. The diagram shows that the total perimeter is formed by 5 sides. Since all the sides are the same length, divide the 35 inches into 5 equal parts to find that each side is 7 inches long. Since the triangle contains three sides, its perimeter is 7 × 3, or 21 inches.

CHECK YOUR RESULT. The total perimeter consists of 5 sides, each 7 inches long. The result, 5 × 7 or 35 inches, matches the given information.

> **ANSWER:** The perimeter of the triangle is 21 inches.

LEARN FROM THE PROBLEM. A diagram can help you get more information in problems about shapes.

TRY IT! *Problem 2 on page 9*

ANALYZE THE PROBLEM.

What must you find?
  The maximum number of bottles that a runner may drink.

What do you know?
  The race is 26 miles long. Checkpoints are 3 miles apart. The first checkpoint is after the first mile. The maximum number of bottles is the same as the number of checkpoints.

What strategy should you use?
  Since this problem involves distance, draw a diagram.

SOLVE THE PROBLEM. Draw and label the diagram. Then count the checkpoints.

Start	1	3	3	3	3	3	3	3	3	1	End
Miles:	1	4	7	10	13	16	19	22	25	26	

  There are 9 checkpoints.

CHECK YOUR RESULT. Be sure the distances add up to 26 miles.

> **ANSWER:** A runner may drink a maximum of 9 bottles of water.

LEARN FROM THE PROBLEM. A drawing can provide the solution to some problems.

TRY IT! *Problem 3 on page 9*

ANALYZE THE PROBLEM.

What must you find?
  The number of possible question-answer combinations.

What do you know?
  (1) The matching test has 3 questions.
  (2) There are 4 answers.
  (3) Answers may only be used once.

What strategy should you use?
  Since this problem requires finding all possible combinations of questions and answers, draw a tree diagram.

SOLVE THE PROBLEM. There are 4 possible choices for the first question. Then 3 choices remain for the second question, and 2 for the third question.

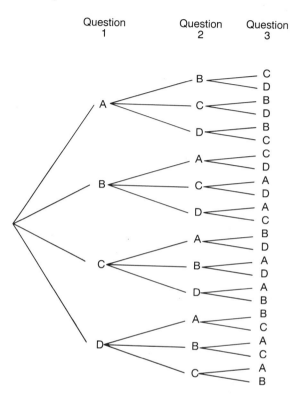

Question 1   Question 2   Question 3

CHECK YOUR RESULT. For each of the three questions, a different answer was chosen. All possible combinations of first, second and third answers were tried.

ANSWER:
There are 24 possible combinations.

LEARN FROM THE PROBLEM. The number of combinations may be more than you might expect from such a small number of questions and answers. A tree diagram provides a way to display all of the possibilities. However, since a diagram with a great many branches is awkward to draw, the usefulness of a tree diagram is limited to situations in which the number of choices is small.

1-3 MAKE A LIST OR A TABLE

TRY IT! *Problem 1 on page 13*

ANALYZE THE PROBLEM.

What must you find?
The number of 4-letter English words you can make from the letters in STOP.

What do you know?
(1) You must use all the letters.
(2) You may use plurals.
(3) You may not repeat letters.

What strategy should you use?
To show all possibilities, make a list.

SOLVE THE PROBLEM. A list should be made systematically to be sure that all possibilities are included. You might use each letter of STOP, in order, as the 1st letter. Then any of 3 letters could be in 2nd place, and the remaining 2 letters could alternate in 3rd place.

S: sopt, sotp, spot, spto, stop, stpo
T: tops, tosp, tpos, tpso, tsop, tspo
O: opst, opts, ospt, ostp, otps, otsp
P: post, pots, psot, psto, ptos, ptso

S-words: spot, stop
T-words: tops
O-words: opts
P-words: post, pots

CHECK YOUR RESULT. Check to be sure you didn't miss any words.

ANSWER: There are 6 words.

LEARN FROM THE PROBLEM. When listing all possibilities, use a system to be sure you don't miss anything.

TRY IT! *Problem 2 on page 13*

ANALYZE THE PROBLEM.

What must you find?
Who is youngest? Who is oldest?

What do you know?
Cassius is a bachelor.
Clay earns less than the youngest does.
The oldest earns most, but must support his wife and son.

What strategy should you use?
To organize the information, make a table.

SOLVE THE PROBLEM. Set up a table first. Use information from the problem and the process of elimination to cross off combinations that are not true, writing ×. When there is only one choice left in a column or row, write "Yes."

You will have to go through the problem several times to get all of the information you need. The numbers (1)–(9) in the following table show one order in which the entries might be made.

	Cassius	**Clay**	**Muhammed**
Youngest	Yes (8)	× (1)	× (7)
Middle	× (9)	Yes (5)	× (6)
Oldest	× (2)	× (3)	Yes (4)

CHECK YOUR RESULT. Be sure that your answer doesn't contradict any information from the problem. It is often wise to try this type of problem again to be sure you get the same result.

> ***ANSWER:*** Cassius is the youngest.
> Muhammed is the oldest.

LEARN FROM THE PROBLEM. In this type of problem, the information is not put directly into the table. Rather, your conclusions from the information are put into the table.

1-4 FIND A PATTERN

TRY IT! *Problem 1 on page 17*

ANALYZE THE PROBLEM.

What must you find?
If the answer is odd, even, or sometimes odd and sometimes even.

What do you know?
Three consecutive numbers are to be added.

What strategy should you use?
This problem uses a sequence of numbers to form a general rule. In order to make such a rule, you must first find a pattern.

SOLVE THE PROBLEM. Since no numbers are given, you must choose numbers and add them to see what you get. You must try enough examples to see if the result is always odd, always even, or if it changes.

$$3 + 4 + 5 = 12 \text{ even}$$
$$4 + 5 + 6 = 15 \text{ odd}$$

Here, two tries are enough to see that the result is sometimes odd, sometimes even.

CHECK YOUR RESULT. Check the arithmetic:

$$3 + 4 + 5 = 12 \qquad 4 + 5 + 6 = 15$$

> ***ANSWER:***
> (3) sometimes odd, sometimes even

LEARN FROM THE PROBLEM. When no numbers are given and you must choose your own numbers to find a pattern, remember two things:

1. Choose numbers that are easy to work with, usually small numbers.

2. Choose different types of numbers. For example, in this problem two odd and one even number were chosen, then one odd and two even numbers.

ANALYZE THE PROBLEM.

What must you find?
The missing numbers.

What do you know?
11, 15, 22, 32, ___, ___, 80 form a sequence.

What strategy should you use?
When you have numbers in a sequence, try to find a pattern.

SOLVE THE PROBLEM. Subtract successive numbers to find the pattern.

Numbers: 11　15　22　32　___　___　80
Differences:　4　7　10

The pattern is that each difference is 3 more than the one before. Continue the pattern of adding 3 to each of the differences. Then, fill in the missing numbers.

Numbers: 11　15　22　32　45　61　80
Differences:　4　7　10　13　16　19

CHECK YOUR RESULT. Verify the additions.

Thus, 32 + 13 = 45,
　　　45 + 16 = 61, and
　　　61 + 19 = 80.

The pattern fits what you know.

> **ANSWER:** The missing numbers are 45 and 61.

LEARN FROM THE PROBLEM. Review the three parts of this strategy: find the pattern, state the rule, use the rule to find other numbers.

1-5 WORK BACKWARD

TRY IT! *Problem 1 on page 21*

ANALYZE THE PROBLEM.

What must you find?
The cost of Carmella's record.

What do you know?
(1) Carmella got a record.
(2) Janice got 3 books.
(3) A book costs $3, $4, or $5.
(4) The gifts cost $23 in all.

What strategy should you use?
Since you know the total cost and are trying to find the individual costs, work backward.

SOLVE THE PROBLEM. By adding the costs of Janice's books and Carmella's record, you get a total cost of $23. Working backward from the $23 total, subtract the cost of the 3 books to get the cost of the record.

In this problem, the books cost $3, $4, or $5 each. Therefore, the least they might have cost is 3 × $3, or $9. The most they might have cost is 3 × $5, or $15. Thus, $23 − $9 = $14, and $23 − $15 = $8. Carmella's record cost an even dollar amount from $8 to $14.

CHECK YOUR RESULT. Use the strategy of making a table. List all possible combinations of book and record prices.

Book	5	5	5	5	5	5	4	4	4	3
Book	5	5	5	4	4	3	4	4	3	3
Book	5	4	3	4	3	3	4	3	3	3
Record	8	9	10	10	11	12	11	12	13	14
Total	23	23	23	23	23	23	23	23	23	23

> **ANSWER:** Carmella's record cost an even dollar amount from $8 through $14.

LEARN FROM THE PROBLEM. Some problems have a range of answers rather than a single answer. Also, a problem solved by one strategy can sometimes be checked using a different strategy.

ANALYZE THE PROBLEM.

What must you find?
A value for each digit that will make the addition true.

What do you know?
The addition uses three different digits.

What strategy should you use?
Work backward.

SOLVE THE PROBLEM. If you add two 2-digit numbers and get a 3-digit answer, the first digit of the answer must be 1. Therefore, $D = 1$.

$$\begin{array}{r} A\,C \\ +\ C\,A \\ \hline D\,A\,D \end{array}$$

In the ones column, it appears that $C + A = 1$. This cannot be true because one of the numbers would have to be 0 and the other would have to be 1. Since $D = 1$, neither C nor A can be 1. Therefore, $C + A = 11$. Write 1; carry 1 to the tens column.

$$\begin{array}{r} A\,C \\ +\ C\,A \\ \hline 1\,A\,1 \end{array}$$

Since $C + A = 11$, in the tens column $1 + A + C = 1 + 11 = 12$. Write 2, and carry 1 to the hundreds column. Therefore, $A = 2$.

$$\begin{array}{r} 1 \\ A\,C \\ +\ C\,A \\ \hline 1\,A\,1 \end{array}$$

Look again at the ones column. $C + A = 11$.

Since $A = 2$, $C + 2 = 11$, and $C = 9$.

$$\begin{array}{r} 1\ 1 \\ 2\,C \\ +\ C\,2 \\ \hline 1\,2\,1 \end{array}$$

CHECK YOUR RESULT.

$$\begin{array}{r} 2\,9 \\ +\ 9\,2 \\ \hline 1\,2\,1 \end{array}$$

ANSWER: $A = 2 \quad C = 9 \quad D = 1$

LEARN FROM THE PROBLEM. This problem required two strategies. The main strategy was working backward. Within the problem, you also used guessing and checking to find out if $C + A = 1$ or if $C + A = 11$. Several strategies may be used within a single problem.

1-6 CHOOSE A STRATEGY

ANALYZE THE PROBLEM.

What must you find?
The total number of handshakes.

What do you know?
Five people all shake hands with the others.

What strategy should you use?
The number of people is small enough that you can make a list or draw a diagram to show all of the handshakes. To get a more general solution, you could find a pattern.

SOLVE THE PROBLEM.

By making a list

1st–2nd	2nd–3rd	3rd–4th	4th–5th
1st–3rd	2nd–4th	3rd–5th	
1st–4th	2nd–5th		
1st–5th			

The list above shows 10 handshakes.

By drawing a diagram

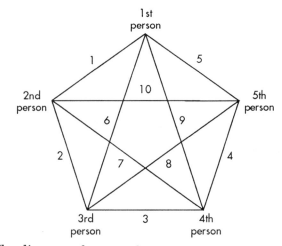

The diagram shows 10 handshakes.

By finding a pattern

Each person shakes hands with everyone but himself. This is 1 less than the total number of people. If there are 5 people, each shakes hands with the 4 others. This suggests that the number of handshakes can be found by multiplying the number of people by the number of handshakes each person makes, which is 1 less than the number of people.

However, you do not want to count a handshake twice. If the 1st person shakes hands with the 2nd person and the 2nd person shakes hands with the 1st person, it is the same handshake. Therefore, the previous result must be divided by 2.

The rule is to multiply the number of people by 1 less than the number of people, then divide the result by 2.

In the problem, there are 5 people. Using the rule: $5 \times 4 = 20$ and $20 \div 2 = 10$

CHECK YOUR RESULT. The result is the same using any of the three strategies.

ANSWER: There are 10 handshakes.

LEARN FROM THE PROBLEM. This problem can be solved by several different strategies. If there had been a large number of people, however, finding a pattern and making a rule would be the easiest to use.

TRY IT! *Problem 2 on page 25*

ANALYZE THE PROBLEM.

What must you find?
The children's ages and the woman's house number.

What do you know?
(1) The product of the ages of the three children is 36. (2) The sum of their ages is equal to the woman's house number. (3) The youngest child is blond.

What strategy should you use?
Make a list of the possibilities.

SOLVE THE PROBLEM.

Ages	Sum
1, 1, 36	38
1, 2, 18	21
1, 3, 12	16
1, 4, 9	14
1, 6, 6	13
2, 2, 9	13
2, 3, 6	11
3, 3, 4	10

The list shows all combinations of ages with 36 as their product. Since there are eight possibilities, the woman would not know which is the correct combination.

The second hint is that the sum of the ages is equal to the woman's house number. (The sums are listed in the column on the right.) Although the woman knows her own house number, she still says that she doesn't know the ages. This could only happen if there is more than one combination of ages whose sum is her house number. Thus, her house number must be 13, and you can eliminate all combinations whose sum is not 13.

The final hint is that the youngest one is blond. The remaining choices are 1, 6, 6 and 2, 2, 9. Of these, the expression "youngest one" can apply only to the combination with a single youngest child, as opposed to the combination with twins as the youngest. So the ages must be 1, 6, and 6.

CHECK YOUR RESULT. The combination 1, 6, 6 definitely satisfies the first hint, and is the only possibility that satisfies the rest of the conditions in the problem.

ANSWER: The house number is 13. The children's ages are 1, 6, and 6.

LEARN FROM THE PROBLEM. Sometimes a problem can be solved by simply listing all the possibilities and then eliminating the ones that do not fit all of the given information. It is important to stick with a problem even if you cannot do it immediately.

CHAPTER 2

THE LANGUAGE OF ALGEBRA

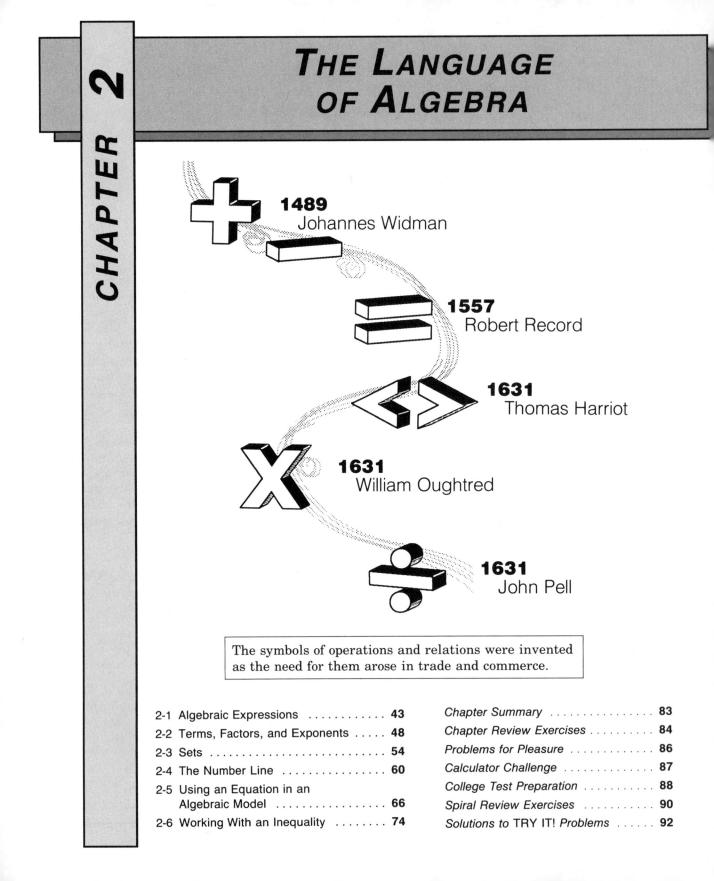

1489
Johannes Widman

1557
Robert Record

1631
Thomas Harriot

1631
William Oughtred

1631
John Pell

The symbols of operations and relations were invented as the need for them arose in trade and commerce.

2-1 Algebraic Expressions **43**

2-2 Terms, Factors, and Exponents **48**

2-3 Sets **54**

2-4 The Number Line **60**

2-5 Using an Equation in an
Algebraic Model **66**

2-6 Working With an Inequality **74**

Chapter Summary **83**

Chapter Review Exercises **84**

Problems for Pleasure **86**

Calculator Challenge **87**

College Test Preparation **88**

Spiral Review Exercises **90**

Solutions to TRY IT! *Problems* **92**

Many of the solutions to the problems seen in Chapter 1 can be performed more efficiently by using algebraic methods. In this course, you will learn underlying properties, procedures, and agreements that are used in all of mathematics. You will also learn some of the mathematics vocabulary that is used to communicate these ideas. In ordinary speech, words often have more than one meaning, but the language of algebra is very specific.

2-1 ALGEBRAIC EXPRESSIONS

Operations

Operation	Symbol	Result
addition	+	sum
subtraction	−	difference
multiplication	×	product
division	÷	quotient

Addition, subtraction, multiplication, and *division* are all instructions that tell what to do with two numbers. In algebra, the single word **operation** refers to all such instructions. The table shows the symbols used for these operations and the names given to the results.

Expressions for Numbers

Number is the word that tells "how many." The symbols you write to represent numbers are called **numerals**, **numerical expressions**, or simply **expressions**.

A number may be represented by many different expressions. For example, $3 + 2$, $6 − 1$, and 5 are all expressions representing the same number. The number named by an expression is called the **value** of the expression. To **simplify** a numerical expression means to find its value.

Conventions

The need for agreements in mathematics becomes apparent when you use operations to simplify expressions. Consider $3 + 4 \times 5$.

If you add first: $\qquad 3 + 4 \times 5 = 7 \times 5 = 35$

If you multiply first: $\quad 3 + 4 \times 5 = 3 + 20 = 23$

In order to get the same result, everyone must agree to do the problem the same way. Such agreements are called **conventions**. The convention here is to do multiplication before addition. Thus, the correct value is 23.

Grouping Symbols

Grouping symbols can be used to change the order of computation. If you rewrite the problem $3 + 4 \times 5$ as $(3 + 4) \times 5$, the grouping symbols show that the 3 and the 4 must be treated as a single quantity and added first, before multiplying by 5. Thus, the value of $(3 + 4) \times 5$ is 35. The most common grouping symbols are parentheses, ( ). Brackets, [ ], are also used as grouping symbols.

A bar is another symbol that can be used to show grouping. In the fraction $\frac{20 - 8}{2 + 1}$, the bar, or fraction line, indicates that the number $(20 - 8)$ is to be divided by $(2 + 1)$. Therefore, $\frac{20 - 8}{2 + 1} = \frac{12}{3} = 4$.

PROCEDURE

To simplify an expression containing more than one operation:

1. Simplify any expression within grouping symbols. If one set of grouping symbols is inside another set, simplify the innermost expression first.

2. Do all multiplications and divisions in order from left to right.

3. Do all additions and subtractions in order from left to right.

This sequence of steps is called the *order of operations*.

MODEL PROBLEMS

1. Simplify: $3 \times 5 + 4 \times 2$

How to Proceed	Solution
(1) Write the expression.	$3 \times 5 + 4 \times 2$
(2) Multiply first.	$= \quad 15 \quad + \quad 8$
(3) Then add.	$= \qquad\qquad 23$ **ANSWER**

2. Simplify: $3 + 8 - 5 + 9$

(1) Write the expression.	$3 + 8 - 5 + 9$
(2) Do the first addition on the left.	$= \quad 11 \quad - 5 + 9$
(3) Do the subtraction.	$= \qquad 6 \quad + 9$
(4) Do the remaining addition.	$= \qquad\qquad 15$ **ANSWER**

NOTE. Doing additions and subtractions in order from left to right means doing them in the order in which they appear, not all additions first and then all subtractions. The same applies to multiplication and division.

3. Simplify: $9 - [8 - (4 + 2)]$

How to Proceed	*Solution*
(1) Write the expression.	$9 - [8 - (4 + 2)]$
(2) Simplify the innermost expression.	$= 9 - [8 - \quad 6 \quad]$
(3) Simplify the expression in brackets.	$= 9 - \quad 2$
(4) Do the subtraction.	$= \quad 7$ ***ANSWER***

Evaluating an Algebraic Expression

In algebra, a letter that is used to stand for a number is called a **variable**. The algebraic expression $3n + 1$ represents an unspecified number. When you replace the variable n with a number, $3n + 1$ represents a specific number. For example:

If $n = 1$: $3n + 1 = 3(1) + 1 = 3 + 1 = 4$ If $n = 3$: $3n + 1 = 3(3) + 1 = 9 + 1 = 10$

When you determine the number that an algebraic expression represents for specific values of its variables, you are *evaluating* the algebraic expression. Note that $3n$, $3(n)$, and $3 \cdot n$ all mean the same thing, the product of 3 and n.

PROCEDURE

To evaluate an algebraic expression:

1. Replace each variable with its specific value.

2. Simplify the resulting expression. Be sure to follow the order of operations.

MODEL PROBLEM

Angles are measured in **degrees**. The greater the opening between the sides, the greater the number of degrees in the angle. A full rotation measures 360 degrees, or $360°$.

If the number of degrees in angle A is represented by $6a - 3bc$, find the measure of angle A when $a = 12$, $b = 4$, and $c = 2$.

How to Proceed	*Solution*
(1) Write the expression.	$6a - 3bc$
(2) Replace the variables by the given values.	$= 6(12) - 3(4)(2)$
(3) Do the multiplication.	$= 72 - 24$
(4) Do the subtraction.	$= 48$

ANSWER: The measure of angle A is $48°$.

Now it's your turn to . . . **TRY IT!**

In 1–3, simplify the expression.

1. $(5 - 2) \times (3 + 4)$ **2.** $12 \div 3 \times (2 + 2) - 1$ **3.** $5 + 2 \times [6 + (3 - 1) \times 4]$

4. Evaluate $2ab + 3b$ when $a = 5$ and $b = 3$.

5. Evaluate $\dfrac{a - c + d}{b - c}$ when $a = 9$, $b = 5$, $c = 3$, and $d = 4$.

(See page 92 for solutions.)

CALCULATOR CONNECTION

The calculator sections of this text are written for scientific calculators, which do more than just the basic arithmetic operations. Scientific calculators have *algebraic logic*, which means that they follow the order of operations.

Check Your Calculator

Test your calculator for algebraic logic. Enter: **3** $\boxed{+}$ **4** $\boxed{\times}$ **5** $\boxed{=}$

A display of $\boxed{ \textbf{23.}}$ shows algebraic logic.

A display of $\boxed{ \textbf{35.}}$ shows no algebraic logic.

Grouping Symbols

For grouping, parentheses are also used on a calculator.

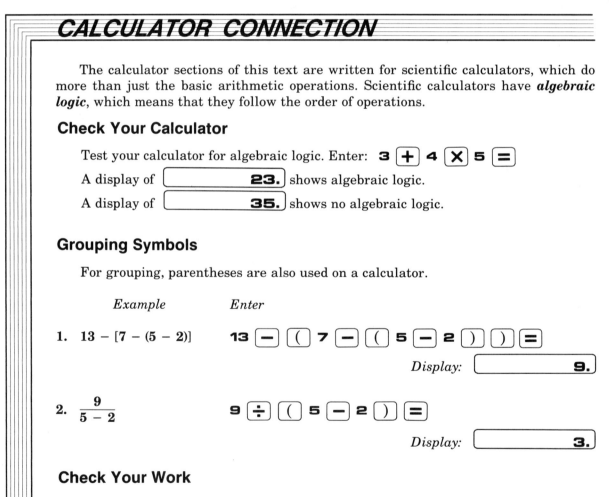

	Example	*Enter*
1.	$13 - [7 - (5 - 2)]$	**13** $\boxed{-}$ $\boxed{(}$ **7** $\boxed{-}$ $\boxed{(}$ **5** $\boxed{-}$ **2** $\boxed{)}$ $\boxed{)}$ $\boxed{=}$

Display: $\boxed{ \textbf{9.}}$

	Example	*Enter*
2.	$\dfrac{9}{5 - 2}$	**9** $\boxed{\div}$ $\boxed{(}$ **5** $\boxed{-}$ **2** $\boxed{)}$ $\boxed{=}$

Display: $\boxed{ \textbf{3.}}$

Check Your Work

To develop computational skill, it is important that you first do the following exercises without a calculator. Then you can check them with a calculator.

In 1–33, simplify the expression.

1. $5 - 1 + 2$　　　　　　**2.** $12 \div 2 \times 3 \div 2$　　　　**3.** $3 + 5 - 1 + 2$

4. $6 + 8 \div 2 + 2$　　　　**5.** $4 \times 3 + 12 \div 3$　　　　**6.** $8 - 2 \times 3 + 1$

7. $(4 - 1) \times 5 - 3$　　　**8.** $4 - 1 \times (5 - 3)$　　　**9.** $3 \times (2 + 6) - 5$

10. $5 - (9 - 7) - 3$　　　**11.** $9 - 7 - (5 - 3)$　　　**12.** $9 - (7 - 5) - 3$

13. $10 + (1 + 4)$　　　　**14.** $13 - (9 + 1)$　　　　**15.** $36 - (10 - 8)$

16. $7 \times (5 + 2)$　　　　**17.** $(6 - 1) \times 10$　　　　**18.** $20 \div (7 + 3)$

19. $48 \div (15 - 3)$　　　**20.** $(24 - 8) \div 2$　　　　**21.** $(17 + 13) \div 10$

22. $26 - 4 \times (7 - 5)$　　**23.** $25 \div (6 - 1) + 3$　　**24.** $3 \times (6 + 4) \times (6 - 4)$

25. $\dfrac{39}{8 + 7 - 2}$　　　　**26.** $\dfrac{7 \times (6 + 14)}{2}$　　　**27.** $\frac{1}{2} \times 8 \times (12 + 14)$

28. $10 \times [4 + (7 - 1) \times 5]$　　　　　　**29.** $75 - [6 \times (8 - 6) + 3]$

30. $36 + 4 \times [1 + (12 - 8) \times 2]$　　　**31.** $5 \times 7 - 3 - (2 + 3) \times 2$

32. $5 \times [(7 - 3) - 2 + 3] \times 2$　　　　**33.** $7 \times [20 - 2 \times (4 + 3)] + 12 \div 4$

In 34–51, find the numerical value if $a = 8$, $b = 6$, $d = 3$, $x = 4$, $y = 5$, and $z = 1$.

34. $5a$　　　**35.** $\frac{1}{2}x$　　　**36.** $3y$　　　**37.** $\frac{2}{3}b$　　　**38.** $3xy$　　　**39.** $2x + 9$

40. $3y - b$　　　**41.** $20 - 4z$　　　**42.** $5x + 2y$　　　**43.** $ab - dx$

44. $a + 5d + 3x$　　**45.** $9y + 6b - d$　　**46.** $ab - d - xy$　　**47.** $\dfrac{x + 2a}{4}$

48. $\dfrac{3a - 2x}{3x - a}$　　**49.** $\dfrac{a + b + x}{y - d + z}$　　**50.** $\dfrac{7y}{5} + \dfrac{b}{2}$　　**51.** $\dfrac{dx - yz + a}{bx - ad + dy}$

In 52–63, find the numerical value if $w = 10$, $x = 8$, $y = 5$, and $z = 2$.

52. $2(x + 5)$　　　　　**53.** $x(y - 2)$　　　　　**54.** $3(2x + z)$

55. $\frac{5}{9}(4y - z)$　　　　**56.** $\frac{x}{2}(y + z)$　　　　**57.** $\frac{1}{2}w(y + z)$

58. $4x + (y + z)$　　　　**59.** $3y - (x - z)$　　　　**60.** $2x + 5(y - 1)$

61. $30 - 4(x - y)$　　　**62.** $(w + x)(y + z)$　　　**63.** $(w - z)x - y$

64. The degree measures of the three angles of a triangle are represented by $5a + b$, $10a - 3b$, and $3a + 2b$, respectively. If $a = 10$ and $b = 12$:

　　a. Find the measure of each angle.

　　b. Find the sum of the angle measures.

2-2 TERMS, FACTORS, AND EXPONENTS

Term

A *term* is a numeral, a variable, or both numerals and variables that are connected by multiplication or division. For example, 5, x, $4y$, a^2b^3, and $\dfrac{2x}{3y}$ are terms.

In an algebraic expression such as $4a + 2b - 5c$, which has more than one term, the terms $4a$, $2b$, and $5c$ are separated by $+$ and $-$ signs.

Factors of a Product

If two or more numbers are multiplied, each of the numbers, as well as the product of any of them, is a *factor* of the product. The number 1 is always understood to be a factor. For example, in the product $3xy$, the factors are 1, 3, x, y, $3x$, $3y$, xy, and $3xy$.

Coefficient

In a product, any factor or group of factors is the *coefficient* of the remaining factor or factors. In the product $4ab$, 4 is the coefficient of ab, $4a$ is the coefficient of b, and $4b$ is the coefficient of a.

A *constant* is an expression that always has the same value. All numbers are constants. When a constant and variables are factors of a product, the constant is called the *numerical coefficient* of the product. For example, in $4ab$, the numerical coefficient is 4.

Since x names the same number as $1 \cdot x$, the coefficient of x is understood to be 1.

Base, Exponent, Power

In general, a *base* is a number that is used as a factor. An *exponent* is a number that tells how many times the base is to be used as a factor. A *power* is a number that can be expressed as a product in which all the factors are the same.

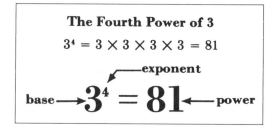

The Fourth Power of 3

$$3^4 = 3 \times 3 \times 3 \times 3 = 81$$

$$\text{base} \longrightarrow 3^{\overset{\text{exponent}}{4}} = 81 \longleftarrow \text{power}$$

When no exponent is written, the exponent is 1. Thus, 4 means 4^1 and a means a^1.

When no grouping symbols are used, an exponent refers only to the base that is directly to the left of it:

$$5d^2 \text{ means } 5dd, \text{ whereas } (5d)^2 \text{ means } (5d)(5d).$$

$$-3^2 \text{ means } -(3)(3) = -9, \text{ whereas } (-3)^2 \text{ means } (-3)(-3) = 9.$$

1. Name the coefficient, base, and exponent.

Term	Coefficient	Base	Exponent
a. $5x^4$	5	x	4
b. y^3	1	y	3
c. $\frac{1}{2}m$	$\frac{1}{2}$	m	1

2. Write each product, using exponents.

	ANSWERS
a. $2 \cdot d \cdot d \cdot d \cdot d \cdot d$	$2d^5$
b. $8 \cdot 8 \cdot 8 \cdot 8$	8^4
c. $3 \cdot x \cdot x \cdot y \cdot y \cdot y$	$3x^2y^3$

Expressions Containing Exponents

To simplify expressions that contain powers, you simplify the powers before following the usual order for the other operations. Be sure to apply the exponent only to its base, and not to any other symbols.

MODEL PROBLEMS

	How to Proceed	*Solution*
1. Simplify $5(6 - 4)^3 - 5$.	(1) Write the expression.	$5(6 - 4)^3 - 5$
	(2) Simplify within parentheses.	$= 5(2)^3 - 5$
	(3) Simplify the power.	$= 5(8) - 5$
	(4) Do the multiplication.	$= 40 - 5$
	(5) Do the subtraction.	$= 35$
2. Evaluate $4x^2 + 9y^2$ if $x = 1$ and $y = 4$.	(1) Write the expression.	$4x^2 + 9y^2$
	(2) Replace the variables.	$= 4(1)^2 + 9(4)^2$
	(3) Simplify the powers.	$= 4(1) + 9(16)$
	(4) Do the multiplications.	$= 4 + 144$
	(5) Do the addition.	$= 148$

Now it's your turn to . . . **TRY IT!**   *(See page 92 for solutions.)*

1. Write the product, using exponents: $53 \cdot g \cdot g \cdot g \cdot g \cdot g \cdot g \cdot g$

2. Simplify each expression: **a.** $5^3 - 4^2$ **b.** $(7 - 1)^2 \div 3^2$

3. Evaluate each expression when $a = 4$ and $b = 3$:   **a.** $(a + b)^2$ **b.** $a^2 + b^2$

A **flowchart** is a diagram to guide you step by step through a procedure. Connecting arrows direct you from one step to the next. Flowchart instructions are enclosed in standard shapes.

 indicates START or STOP.   ◇ requires you to make a decision.

▱ shows input or output.   ▭ gives instructions for action.

ORDER OF OPERATIONS

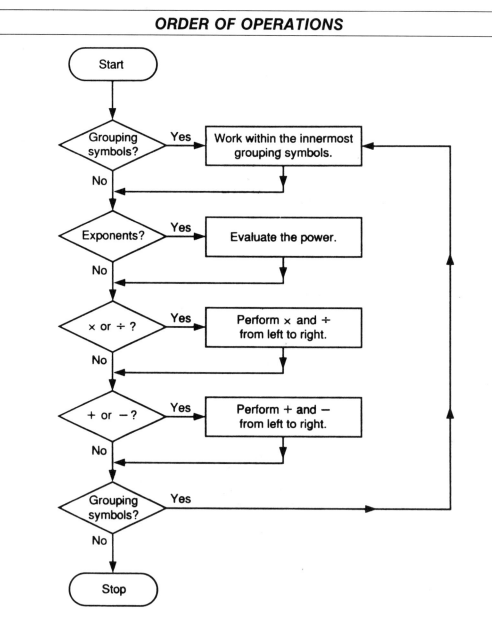

MODEL PROBLEM

Write a flowchart procedure to evaluate $2x^2 - 3(y + 1)$ when $x = 5$ and $y = 7$.

Solution:

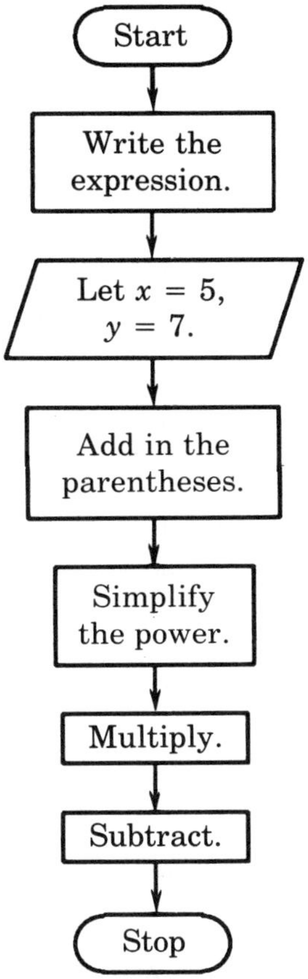

Start

| Write the expression. | $2x^2 - 3(y + 1)$ |

| Let $x = 5$, $y = 7$. | $2 \cdot 5^2 - 3(7 + 1)$ |

| Add in the parentheses. | $2 \cdot 5^2 - 3(8)$ |

| Simplify the power. | $2 \cdot 25 - 3(8)$ |

| Multiply. | $50 - 24$ |

| Subtract. | 26 |

Stop

CALCULATOR CONNECTION

Scientific calculators have an exponent key, usually labeled y^x, that does a calculation such as 3^5 with four keypresses: $3 \boxed{y^x} 5 \boxed{=}$

Because the exponent 2 appears so frequently in computation, most calculators have a separate key, labeled x^2. Thus, you can calculate 8^2 by entering $8 \boxed{x^2}$.

y^x and x^2 may be used in the same computation. To simplify $(6^2 - 2^5)^4$,

enter $\boxed{(} 6 \boxed{x^2} \boxed{-} 2 \boxed{y^x} 5 \boxed{)} \boxed{y^x} 4 \boxed{=}$, displaying $\boxed{ 256.}$.

1. Name the factors of each product. **a.** $5n$ **b.** $7mn$ **c.** $13xy$
2. Name the numerical coefficient of x. **a.** $8x$ **b.** $(5 + 2)x$ **c.** x
3. Name the base and exponent in each term. **a.** t **b.** 10^6 **c.** $(5y)^4$
4. Write each term without using exponents. **a.** $5x^4$ **b.** $4a^4b^2$ **c.** $(3y)^5$
5. Write each product, using exponents.
    a. $10 \cdot 10 \cdot 10 \cdot 10$ **b.** $a \cdot a \cdot a \cdot a \cdot b \cdot b$ **c.** $(6a)(6a)(6a)$
6. A printed page has an area of $3w$ square inches. Name the factors of $3w$.
7. A plane is traveling at $52p$ miles per hour. Name the numerical coefficient of p.
8. A new model sports car is selling for $750y$ dollars. Name the numerical coefficient of y.
9. A missile is falling at the rate of $9k^2$ feet per minute. Name the exponent.
10. An amount of P dollars invested at r percent for a period of n years grows to a value of $P(1 + r)^n$. If P is the coefficient, find the base and exponent of this value.
11. The intensity of an earthquake measured $3n^5$ on the Richter scale. Write this intensity without using exponents.

A *seismograph* records the ground motion of an earthquake. The strength of the earthquake is rated on the Richter scale, a system developed in 1935 by the American seismologist Charles Richter. The numbers on the Richter scale differ by a factor of 10; that is, an earthquake measuring 7 is 10 times as strong as an earthquake measuring 6.

In 12–41, simplify the numerical expression.

12. 9^2	**13.** 12^2	**14.** 10^3	**15.** 2^6	**16.** 1^3	**17.** 0^4
18. 2×3^2	**19.** 4×5^2	**20.** $2^3 \times 1^2$	**21.** $10^2 \times 3^3$	**22.** $1^4 \times 9^2$	**23.** $5^2 + 12^2$
24. $16^2 + 9^2$		**25.** $13^2 - 5^2$		**26.** $20^2 - 12$	
27. $6 + 4(5)^2$		**28.** $3(2)^2 + 6$		**29.** $120 - 6(2)^4$	

30. $(4 + 6)^2$

31. $(5 + 8)^2$

32. $(20 - 15)^2$

33. $(7 - 2 \times 3)^3$

34. $2(4 + 6)^2 - 10$

35. $200 - 3(5 - 1)^3$

36. $12(5^2 - 4^2)$

37. $(7^2 - 6^2)(1^2 + 2^2)$

38. $(3^2 - 2^3)^4$

39. $\dfrac{8^2 - 6^2}{2}$

40. $\dfrac{12^2 + 9^2 + 7^2}{4}$

41. $\dfrac{15^2 + 25^2}{2(4 + 1)}$

In 42–44, write a flowchart to evaluate the expression when $a = 5$, $b = 3$, and $c = 2$.

42. $6(a - b)$ **43.** $4(b + c)^2$ **44.** $4a + b^2(10 - c)$

In 45–78, evaluate the expression. Use $a = 8$, $b = 6$, $d = 3$, $x = 4$, $y = 5$, and $z = 1$.

45. a^2 **46.** b^3 **47.** d^4 **48.** $4d^3$ **49.** $6z^5$ **50.** $\dfrac{b^2}{9}$

51. $\frac{3}{4}x^3$ **52.** xy^2 **53.** $2a^2b^3$ **54.** $\frac{1}{5}x^2y^3z^3$ **55.** $a^2 + b^2$ **56.** $b^2 - y^2$

57. $a^2 + b^2 - d^2$ **58.** $x^2 + x$ **59.** $b^2 + 2b$ **60.** $y^2 - 4y$

61. $2b^2 + b$ **62.** $2y^2 - y$ **63.** $9a - a^2$ **64.** $5z - 3z^2$

65. $x^2 + 4y^2$ **66.** $x^2 + 3x + 5$ **67.** $y^2 + 2y - 7$ **68.** $2a^2 - 4a + 6$

69. $2b^2 - 5b - 10$ **70.** $15 + 5z - z^2$ **71.** $36 + 5y - 2y^2$ **72.** $\dfrac{d^2 + x^2 - y^2}{2dx}$

73. $\dfrac{b^3 - d^3}{(b - d)^3}$ **74.** $\dfrac{(4y)^2 - (3x)^2}{4y^2 - 3x^2}$ **75.** $4y^2 - 3(2y + 1)(3y - 12)$

76. $4(2a^2 + b^2) - 6(4a - 3b)$ **77.** $\dfrac{x^2 + 2xy + y^2 - z^2}{(x + y + z)(x + y - z)}$ **78.** $\dfrac{x^4 - 2x^2 + 1}{(x - 1)^2(x + 1)^2}$

A story in the Bible tells of an ancient people who were building a mighty tower to reach into heaven. Because the plan showed pride and conceit, the builders were made to speak different languages, so that they could no longer communicate, and the Tower of Babel was never completed.

The story illustrates the importance of communication. Today, while over 700 million people speak Mandarin Chinese, over 300 million speak English, and well over 200 million speak Spanish, the language of algebra is a universal language used around the world to set down mathematical ideas.

2-3 SETS

Set Notation

A group of numbers or other items that share common characteristics form a **set**. The members, or **elements**, of a set may be shown within braces, { }.

One method of representing a set is to write a list of all its elements, as in $A = \{2, 4, 6, 8\}$, which is read "A is the set whose elements are 2, 4, 6, 8." To show that 4 is an element of A, write $4 \in A$. Since 5 is not an element of A, write $5 \notin A$.

When listing a set, an element is named only once. To write the word KEEP, four symbols are used. However, the set of these symbols has only three elements: {K, E, P}.

The elements of a set may be listed in any order: {2, 4} and {4, 2} represent the same set.

In a set with many elements, three dots may be used to indicate that a pattern continues, as in {2, 4, 6, . . . , 998}. Another method of representing this set is by a description: {even numbers between 1 and 999}

Still another way of representing a set is by **set-builder notation**. "The set of all elements n such that n is an even number between 1 and 999" is represented by:

$$\{n \mid n \text{ is an even number between 1 and 999}\}$$

A set whose elements can be counted is called a **finite set**. In an **infinite set**, the list of elements continues without end.

A set that contains no elements, written { }, is also represented by the symbol $\emptyset$, which means the **empty set**, or **null set**.

Sets that name the same elements are **equal sets**.

{odd numbers between 0 and 8} and {1, 3, 5, 7} are equal sets.

Sets such as {1, 2, 3} and {a, b, c}, that have the same number of elements, are **equivalent sets**.

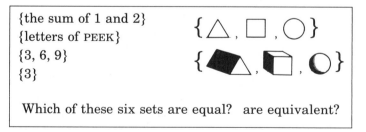

{the sum of 1 and 2}
{letters of PEEK}
{3, 6, 9}
{3}

Which of these six sets are equal? are equivalent?

If every element of one set is also a member of a second set, then the first set is a **subset** of the second: {1, 5} is a subset of {1, 3, 5}.

1. List the elements of each set.   **ANSWERS**

   a. $\{w \mid w$ is an odd number between 10 and 20$\}$ $\{11, 13, 15, 17, 19\}$

   b. {letters in the word BOOKKEEPER} $\{\text{B, O, K, E, P, R}\}$

   c. {whole numbers between 2 and 3} $\{\ \}$ or $\varnothing$

2. Give a description of each set.

   a. $\{4, 3, 2\}$ {natural numbers between 5 and 1}

   b. $\{7, 9, 11, 13\}$ {odd whole numbers between 6 and 14}

Number Sets to Remember

Natural Numbers

The members of $\{1, 2, 3, \ldots\}$ are the **natural numbers**, or **counting numbers**. Each natural number is the **successor** of the number that comes before, and the **predecessor** of the number that follows; that is, 2 is the successor of 1, and the predecessor of 3.

Whole Numbers

The set of **whole numbers**, $\{0, 1, 2, \ldots\}$, includes 0 and the set of natural numbers.

Integers

When negative numbers are included with the whole numbers, the resulting set is the set of **integers**:

$$\{\ldots, -3, -2, -1, 0, 1, 2, 3, \ldots\} \quad or \quad \{0, 1, -1, 2, -2, 3, -3, \ldots\}$$

Rational Numbers

A **rational number** is a number that can be expressed as a quotient of integers. Unlike the preceding number sets, the set of rational numbers cannot be represented by a listing. The reason, as you will see, is that between any two rational numbers, there are infinitely many other rational numbers.

When writing a number in the quotient form $\frac{a}{b}$, the denominator b cannot be 0. To understand why this is so, consider the following:

$$\text{If } \frac{6}{2} = 3, \text{ then } 6 = 3 \times 2. \quad \text{Similarly, if } \frac{6}{0} = ?, \text{ then } 6 = ? \times 0.$$

Since multiplying by 0 always gives a product of 0, there is no number that could replace the question mark to make the statement true. An expression such as $\frac{6}{0}$ is **undefined**.

Some examples of rational numbers are -4, $\frac{2}{5}$, $.3$, and $1\frac{1}{3}$. Observe that although every rational number *can* be written in the form $\frac{a}{b}$, where b is not 0, not every rational number need be a fraction. Similarly, you will later see fractions that are not rational numbers.

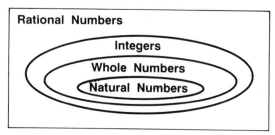

Rational Numbers
Integers
Whole Numbers
Natural Numbers

The natural numbers, whole numbers, and integers are all subsets of the rational numbers. The diagram shows which of these three sets are subsets of each other.

MODEL PROBLEM

Match each number with one set to which it belongs. (Use each set once.)

		ANSWERS
a. $\frac{1}{3}$	(1) Natural numbers	**a.** (4)
b. -2	(2) Integers	**b.** (2)
c. 0	(3) Whole numbers	**c.** (3)
d. 5	(4) Rational numbers	**d.** (1)

Rational Numbers as Decimals

To express a rational number in decimal form, divide the denominator into the numerator. If the division has no remainder, the result is a *terminating decimal*. Examples are:

$$\frac{1}{2} = .5 \qquad -\frac{3}{4} = -.75 \qquad \frac{1}{16} = .0625$$

Other rational numbers are expressed as *repeating decimals*:

$$\frac{1}{3} = .333\ldots \qquad \frac{2}{11} = .181818\ldots \qquad \frac{1}{6} = .1666\ldots$$

A repeating decimal can also be written by placing a bar over the repeating digits:

$$\frac{1}{3} = .\overline{3} \qquad \frac{2}{11} = .\overline{18} \qquad \frac{1}{6} = .1\overline{6}$$

Thus: **Every rational number can be expressed as either a terminating decimal or as a repeating decimal.**

It is also true that every terminating or repeating decimal can be written as the quotient of two integers. You already can do this with terminating decimals, as in $.37 = \frac{37}{100}$. Later, you will learn to do so for repeating decimals.

Irrational Numbers

A decimal that neither terminates nor repeats cannot be expressed in the rational-number form $\frac{a}{b}$, where a and b are integers. Such a decimal is an ***irrational number***. An example is .121121112 . . . , where each group of digits is different from those before. Later, you will work with other irrational numbers that are not decimals.

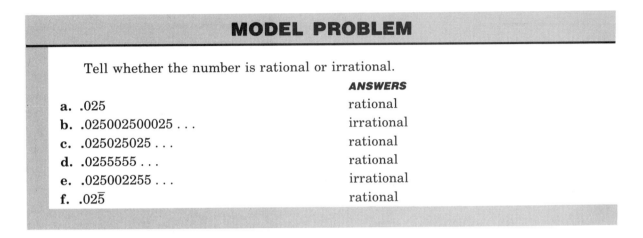

MODEL PROBLEM

Tell whether the number is rational or irrational.

		ANSWERS
a.	.025	rational
b.	.025002500025 . . .	irrational
c.	.025025025 . . .	rational
d.	.0255555 . . .	rational
e.	.025002255 . . .	irrational
f.	.02$\overline{5}$	rational

Real Numbers

The set of all rational numbers and all irrational numbers, taken together, is called the set of ***real numbers***. Throughout this course, the set of real numbers is the most inclusive set.

The real numbers are sometimes called ***signed numbers***. Just as a negative number is indicated by a minus sign, a positive number can be indicated by a plus sign. Thus, 6 can be written as +6 or $^+$6. In this text, positive numbers will be written without plus signs.

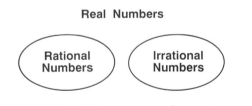

Real Numbers

Rational Numbers

Irrational Numbers

On her wedding day ten centuries ago, a 12-year-old Hindu girl watched for the exact moment at which the astrologers said she could safely be married. By a fateful accident, a pearl fell into the water clock, the clock malfunctioned, and the moment passed. Lilavati remained unmarried all her life.

To comfort his daughter, Bhaskara, a great mathematician, promised that her name would live on. He kept his promise; the Algebra book that he wrote, and named The Lilavati, is known to this day.

In 1–4, list the members of the set, or write $\varnothing$ to represent an empty set.

1. $\{p \mid p$ is an odd natural number$\}$ **2.** $\{$natural numbers less than 1$\}$

3. $\{g \mid g$ is an odd number less than 11$\}$ **4.** $\{r \mid r$ is a vowel in HEADACHE$\}$

In 5–8, state whether the set is a finite set, an infinite set, or the empty set.

5. $\{$real numbers$\}$ **6.** $\{p \mid p$ is an even number between 10 and 11$\}$

7. $\{$students in your school$\}$ **8.** $\{5, 10, 15, 20, \ldots, 1{,}000{,}000\}$

In 9–12, give a description of the set.

9. $\{15, 17, 19\}$ **10.** $\{2, 4, 6, 8, \ldots\}$ **11.** $\{r, a, c\}$ **12.** $\{0, 1, 2, 3, 4, 5\}$

In 13–17, each set in Column II is either equal or equivalent to a set in Column I. Match the sets of the two columns, and specify whether they are equal or equivalent.

Column I   *Column II*

13. $\{0, 1, 2, 3, 4\}$ (a) $\{$zero$\}$

14. $\{\triangle, \square, \bigcirc\}$ (b) $\{$natural numbers less than 5$\}$

15. $\varnothing$ (c) $\{$whole numbers less than 5$\}$

16. $\{0\}$ (d) $\{a, b, c\}$

17. $\{1, 2, 3, 4\}$ (e) $\{$whole numbers between 1 and 2$\}$

In 18–23, tell whether the sentence is true or false.

18. 4 is an element of $\{14, 24, 34, 44\}$.   **19.** $12 \in \{10, 12, 14, 16, 18, 20\}$

20. $\triangle \in \{\bigcirc, \square, \square, \square\}$ **21.** bat $\notin \{$base, glove, bat, ball$\}$

22. Claire $\notin \{$Sue, Sally, Bess, Helen$\}$ **23.** $0 \in \{n \mid n$ is a whole number$\}$

In 24–26, name the number described.

24. the first counting number   **25.** a whole number that is not a natural number

26. the successor of the natural number: **a.** 120   **b.** 999   **c.** 64,499

In 27–33, write the rational number as a terminating decimal or as a repeating decimal.

27. $\frac{5}{8}$ **28.** $\frac{5}{3}$ **29.** $\frac{9}{4}$ **30.** $\frac{7}{9}$ **31.** $5\frac{1}{2}$ **32.** $\frac{7}{6}$ **33.** $\frac{7}{12}$

In 34–39, express the decimal as a fraction or a mixed number.

34. .5   **35.** .81   **36.** .009   **37.** 3.2   **38.** 5.23   **39.** $-.233$

In 40–47, state whether the sentence is *always, sometimes,* or *never* true.

40. A natural number is a whole number.   **41.** A rational number is a real number.

42. A whole number is a rational number.

43. An integer is an irrational number.

44. There is a first counting number.

45. There is a last natural number.

46. A rational number is an integer.

47. A real number is a rational number.

48. Draw a Venn diagram to place the following sets in their proper relationship: whole numbers, irrational numbers, integers, real numbers, rational numbers, natural numbers.

A number whose digits read the same forward and backward, such as 181, is a *palindrome*. In 49–51, find the numbers described. Omit palindromes such as 444, which repeat one digit.

49. Find two palindromic numbers between 1,000 and 3,000 whose difference is one of the numbers and whose sum is also a palindrome.

50. Find two palindromic numbers between 300 and 500 whose sum and difference are both divisible by 11.

51. For how many 3-digit palindromic numbers is the sum of the digits equal to 20?

In 52–54, find the numbers that satisfy the conditions.

52. Find four natural numbers, each less than 5 and not necessarily all different, such that their sum is equal to their product.

53. a. Write the next row in the triangular arrangement of natural numbers shown at the right.
b. Write a rule that describes how to obtain each successive row of this arrangement.

$$1$$
$$1 \quad 2 \quad 1$$
$$1 \quad 3 \quad 3 \quad 1$$
$$1 \quad 4 \quad 6 \quad 4 \quad 1$$

This number arrangement, with interesting applications, is called Pascal's Triangle, after a 17th-century French scientist. Though Pascal created the pattern when he originated probability theory, history shows it as far back as 13th-century China.

54. a. Write the next natural number in the infinite sequence:

1, 1, 2, 3, 5, 8, 13, 21, 34, _?_, . . .

b. Write a rule that describes how to obtain each successive number in this sequence.

The above sequence of numbers was formulated in the 13th century by the Italian mathematician Fibonacci. Centuries later, scientists found that many things in nature, such as patterns of rabbit births or leaf growth, relate to terms of the *Fibonacci sequence*.

The center of a daisy exhibits a growth pattern that shows two opposite sets of rotating spirals. Mysteriously, the daisy's spiral pattern corresponds to the two adjacent terms 21 and 34 of the Fibonacci sequence.

2-4 THE NUMBER LINE

The Real Number Line

A *number line* pairs a set of numbers with a set of points on a straight line. On the number line below, the integers correspond to evenly-spaced points, and the arrowheads indicate that the line continues without end.

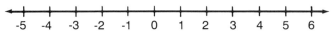

The number that corresponds to a point is the **coordinate** of that point. The point is the **graph** of the number.

If you divide the intervals between integers into halves, thirds, quarters, etc., you can label additional points, as shown on this part of the number line:

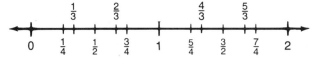

The coordinates of points on the number line may be irrational as well as rational numbers. Since each point can be paired with a real number, and each real number can be paired with a point, the points on the line are in **one-to-one correspondence** with the real numbers.

MODEL PROBLEMS

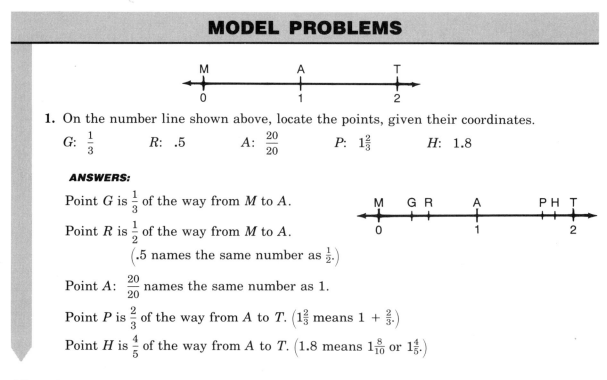

1. On the number line shown above, locate the points, given their coordinates.

G: $\frac{1}{3}$ R: .5   A: $\frac{20}{20}$ P: $1\frac{2}{3}$ H: 1.8

ANSWERS:

Point G is $\frac{1}{3}$ of the way from M to A.

Point R is $\frac{1}{2}$ of the way from M to A.

     (.5 names the same number as $\frac{1}{2}$.)

Point A: $\frac{20}{20}$ names the same number as 1.

Point P is $\frac{2}{3}$ of the way from A to T. ($1\frac{2}{3}$ means $1 + \frac{2}{3}$.)

Point H is $\frac{4}{5}$ of the way from A to T. (1.8 means $1\frac{8}{10}$ or $1\frac{4}{5}$.)

2. a. List the whole numbers between 3 and 6.

   b. List four numbers between 3 and 6.

   c. List four numbers between .5 and .6. (Think of .5 and .6 as .50 and .60.)

     ANSWERS: **a.** 4, 5

             b. $3.5, \frac{13}{3}, 5, 5\frac{1}{2}$
There are infinitely many real numbers between 3 and 6.

             c. Any of the numbers .51, .52, .53, . . . , .59.
Also, infinitely many other numbers, such as .515, .5263.

3. Find a fraction between $\frac{1}{3}$ and $\frac{1}{2}$.

    Solution: Write the fractions with a common denominator: $\frac{1}{3} = \frac{2}{6}$ and $\frac{1}{2} = \frac{3}{6}$

      Since a fraction between $\frac{2}{6}$ and $\frac{3}{6}$ cannot yet be determined,

    use a larger common denominator, any multiple of 6, such as 12: $\frac{1}{3} = \frac{4}{12}$ and $\frac{1}{2} = \frac{6}{12}$

      Since 5 is between 4 and 6, $\frac{5}{12}$ is between $\frac{4}{12}$ and $\frac{6}{12}$.

    ANSWER: $\frac{5}{12}$ is a fraction between $\frac{1}{3}$ and $\frac{1}{2}$.

    NOTE. To find other numbers between $\frac{1}{3}$ and $\frac{1}{2}$, continue to increase the denominator.

    For example, between $\frac{1}{3} = \frac{8}{24}$ and $\frac{1}{2} = \frac{12}{24}$, there are the fractions $\frac{9}{24}, \frac{10}{24},$ and $\frac{11}{24}$.

Now it's your turn to . . . **TRY IT!**   *(See page 92 for solutions.)*

1. Use the number line shown.

   a. Name the coordinate of point *M*.

   b. Name the point that is the graph of $\frac{1}{2}$.

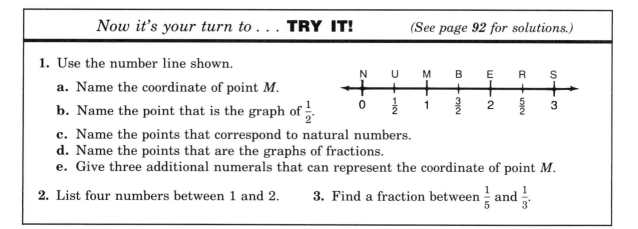

   c. Name the points that correspond to natural numbers.

   d. Name the points that are the graphs of fractions.

   e. Give three additional numerals that can represent the coordinate of point *M*.

2. List four numbers between 1 and 2.   **3.** Find a fraction between $\frac{1}{5}$ and $\frac{1}{3}$.

Opposites

In daily life, we speak of opposite situations, such as traveling east or west, and gaining or losing weight. Such situations can be represented by signed numbers.

If 10 miles east is indicated by 10, then 10 miles west is -10.
Negative 10 is the opposite of positive 10.

If losing 5 pounds is represented by -5, then gaining 5 pounds is 5.
Positive 5 is the opposite of -5.

On the real number line, a pair of numbers that are the same distance from 0 but on opposite sides of 0 are called *opposites*.

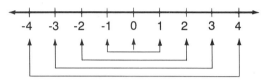

Think of the way a mirror reflects an image. The graph of an opposite is the *reflection* of the graph of the number, with 0 as the point of reflection. For example, -1 is the opposite of 1 and 2 is the opposite of -2. To show that the opposite of negative 2 is positive 2, write $-(-2) = 2$. The number 0 is its own opposite.

The opposite of a number may be positive, negative, or 0, depending upon the value of the given number.

The image of India's Taj Mahal is shown in a reflecting pool. There are many real-life occurrences of reflection. A mirror reflects light. An echo is a reflection of sound waves. Reflection of radio waves has been used in radar. In mathematics, reflection is a basic concept of a branch of geometry called transformation geometry.

Adding Opposites

Compare a sum of numbers with the sum of their opposites.

Numbers	Opposites	Numbers	Opposites	Numbers	Opposites
2	−2	5	−5	4	−4
3	−3	−7	7	6	−6
5	−5	−2	2	−3	3
				7	−7

In each case, the sum of the opposites of the numbers is the opposite of the sum of the original numbers. In general, if a and b are real numbers: $(-a) + (-b) = -(a + b)$

Absolute Value

Two numbers that are opposites are the same distance from 0 on a number line.

That distance represents the **absolute value** of each number of the pair. For example, the absolute value of both 10 and −10 is 10. In symbols, $|10| = |-10| = 10$.

The absolute value of 0 is 0, written $|0| = 0$.

The absolute value of every nonzero number is positive.

MODEL PROBLEMS

1. Write the simplest symbol that represents the opposite of the number.

Number	**a.** 15	**b.** −10	**c.** (4 + 8)	**d.** −[−(9 − 3)]
Opposite:	**a.** −15	**b.** 10	**c.** −12	**d.** −6

2. Find the value of the expression $|12| + |-3|$.

   Solution: $|12| + |-3| = 12 + 3 = 15$

Now it's your turn to . . . **TRY IT!**　　(See page **92** for solutions.)

1. Write the simplest symbol that represents the opposite of the number.
   a. −3.2　　**b.** 10 − 3　　**c.** −[−(12 + 4)]

2. State whether the sentence is true or false.
   a. $|-12| = -12$　　**b.** $|10| - |-10| = 0$

In 1 and 2, name the number that corresponds to each of the lettered points on the number line. Assume the markings are uniformly spaced.

In 3–5, draw a number line, and locate the points whose coordinates are given.

3. 4, −2, 0, −5, 3

4. .2, −1, 0, 3, −4.1

5. 3, $-\frac{2}{3}$, $\frac{5}{6}$, $-2\frac{1}{3}$, 0

In 6–9, list three numbers between the given numbers.

6. 5 and 6

7. $\frac{5}{100}$ and $\frac{6}{100}$

8. 3.21 and 3.22

9. $\frac{1}{3}$ and .6

In 10–15, use the number line below to name the point that is described.

10. the graph of the number: **a.** $\frac{8}{4}$ **b.** $1\frac{3}{4}$ **c.** $2\frac{1}{2}$ **d.** .5   **e.** 1.25

11. one-half the distance from A to G **12.** one-third the distance from B to K

13. two-fifths the distance from C to M

14. between E and K and twice as far from K as it is from E

15. to the right of C and twice as far from C as it is from D

16. How does the numeral for a number reveal whether the number is to the right or to the left of 0 on a number line?

17. If two different numbers are the same distance from 0 on a number line, what must be true?

In 18–21, select the number of the pair that is farther from 0 on a number line.

18. −6, −8   **19.** .01, .1   **20.** $\frac{1}{2}, \frac{1}{5}$ **21.** $-\frac{1}{2}, -\frac{1}{5}$

22. If two numbers are different distances from 0, how do their absolute values compare?

In 23–28, tell which number of the pair is to the right of the other on a number line.

23. 5, −2   **24.** −7, −3   **25.** 0, −4   **26.** $-5\frac{1}{2}$, −6   **27.** −.1, −.01   **28.** $-\frac{1}{3}$, −.6

29. If one number is to the right of another on a number line, how do their values compare?

In 30–32, describe the opposite of the situation.

30. a rise in price   **31.** south of the equator   **32.** below average

33. If sea level is represented by 0, represent 500 feet below sea level.

34. Explain what is meant by the statement:
    a. The altitude of the Dead Sea is $-1,300$ feet.
    b. The net change in the value of a stock is $+2$ dollars.

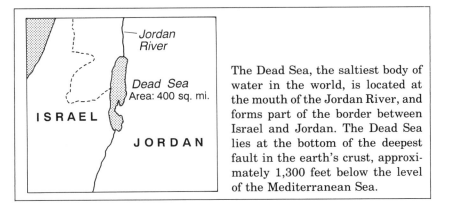

The Dead Sea, the saltiest body of water in the world, is located at the mouth of the Jordan River, and forms part of the border between Israel and Jordan. The Dead Sea lies at the bottom of the deepest fault in the earth's crust, approximately 1,300 feet below the level of the Mediterranean Sea.

In 35–40, write the simplest symbol that represents the opposite of the number.

35. 8   **36.** -6.5 **37.** 8×0 **38.** $-(-7)$ **39.** $-\left(-\frac{3}{4}\right)$ **40.** $-[-(6 + 8)]$

In 41–46, tell whether the statement is true or false. If the statement is false, give a *counterexample* (an example that makes it false).

41. If a is a real number, then $-a$ is always a negative number.

42. If a is a negative number, then $-a$ is always a positive number.

43. The opposite of a number is always a different number.

44. On a number line, the opposite of a positive number is to the left of the number.

45. On a number line, the opposite of any number is always to the left of the number.

46. The opposite of the opposite of a number is that number itself.

In 47–51:   **a.** Give the absolute value of the number.
                     b. Give another number that has the same absolute value.

47. 3   **48.** -5 **49.** $1\frac{1}{2}$ **50.** $-3\frac{3}{4}$ **51.** 2.7

In 52–57, state whether the sentence is true or false.

52. $|20| = 20$ **53.** $|5| - |-5| = 0$ **54.** $|9| + |-9| = 0$

55. $|-13| = 13$ **56.** $|-15| = -15$ **57.** $|-9| = |9|$

In 58–63, find the value of the expression.

58. $|9| + |3|$ **59.** $|-6| + |4|$ **60.** $|4.5| - |-4.5|$

61. $|6| \times |-4|$ **62.** $|(8 - 4)| + |-3|$ **63.** $-(|-9| - |7|)$

2-5 USING AN EQUATION IN AN ALGEBRAIC MODEL

Often, the best way to understand a new idea is to relate it to something you already know; your past experience provides a *model* for the new situation. In many fields, people use mathematical models to describe complex conditions.

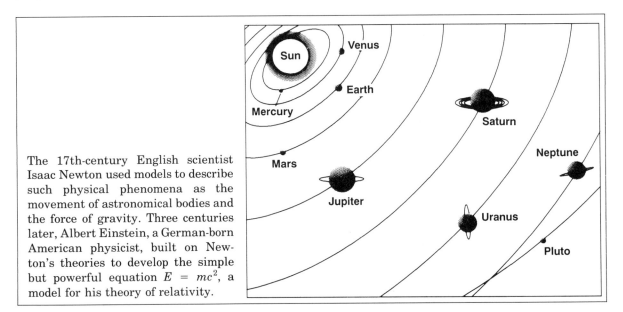

The 17th-century English scientist Isaac Newton used models to describe such physical phenomena as the movement of astronomical bodies and the force of gravity. Three centuries later, Albert Einstein, a German-born American physicist, built on Newton's theories to develop the simple but powerful equation $E = mc^2$, a model for his theory of relativity.

The important problem-solving strategy that you will now be learning to apply is to set up and solve an algebraic model for a verbal problem. A first step is to translate English phrases into mathematical language.

Translating Words to Symbols

A single algebraic phrase may represent a variety of English phrases, as shown in the table.

$a + b$	$a - b$	$a \cdot b$	$\dfrac{a}{b}$
the sum of a and b	the difference of a and b	the product of a and b	the quotient of a and b
b more than a	b less than a	a times b	a divided by b
a increased by b	a decreased by b	Twice a is written $2a$.	Half of a is written $\dfrac{a}{2}$ or $\dfrac{1}{2}a$.

Note. Alternate forms for $a \cdot b$ are $a(b)$ or ab.

In verbal phrases involving two or more operations, a comma can act as a grouping symbol. For example, in "the product of x and y, decreased by 2," the comma after y tells you that the phrase means $(xy) - 2$ and not $x(y - 2)$.

If there is no comma, the convention is that the first operation mentioned determines the grouping. For example, the phrase "the difference of fifteen and seven increased by two" could mean $(15 - 7) + 2$ or $15 - (7 + 2)$. Since "difference" appears before "increased by," $15 - (7 + 2)$ is the correct form.

MODEL PROBLEMS

1. Translate the verbal phrase, using mathematical symbols.

   ANSWERS

   a. four increased by the product of two and one $\qquad$ $4 + 2 \times 1$

   b. w more than 3 $\qquad$ $3 + w$

   c. 2 less than r $\qquad$ $r - 2$

   d. the sum of t and u, divided by 6 $\qquad$ $\dfrac{t + u}{6}$

   e. 100 decreased by twice the quantity $(x + 5)$ $\qquad$ $100 - 2(x + 5)$

   When a verbal expression involves a real-life application, a useful strategy is to relate the expression to a similar numerical expression.

2. Represent algebraically the value of n hats, each worth d dollars.

How to Proceed	*Solution*
(1) Write a similar expression using numbers.	the value of 5 hats, each worth 10 dollars
(2) Represent the desired quantity using numbers.	(5×10) dollars
(3) Represent the expression using variables.	$(n \times d)$ dollars or nd dollars

   ANSWER: nd dollars

Now it's your turn to . . . **TRY IT!** $\qquad$ *(See page 92 for solutions.)*

1. Translate the verbal phrase, using mathematical symbols.
   a. ten decreased by the product of eight and four
   b. the difference of s and 4, divided by 7

2. Represent the number of cents in d dimes.

There are many ways in which models are designed to give useful information about the real thing. There are physical models, like model airplanes that can be tested in a wind tunnel, and there are mathematical models, which may be used to make predictions about population growth or changes in the atmosphere.

Some very complex models can be designed only with the help of a computer. A model of the "greenhouse effect," the warming of the earth's atmosphere caused by accumulated gases, may require over 100,000 computer instructions. Since one scientist's model differs from that of another, there is a great deal of controversy about the greenhouse effect on our planet.

A view of the western hemisphere from the Apollo 8 spacecraft.

Open Sentences

An *equation* is a mathematical sentence that uses the symbol $=$ to show that two expressions are equal. An equation such as $2 + 3 = 5$ contains only numerical expressions. An equation such as $n + 3 = 5$ contains a variable, and is called an *open sentence*.

To *solve* an open sentence, or to find its *solution set*, is to find a value for the variable that makes the sentence true. The value comes from the *replacement set*, or *domain*, of the variable.

In this section, you will see how to represent a given situation by an algebraic model that uses an equation with one variable. Algebraic methods for solving an equation are discussed later, in Chapter 4.

PROCEDURE

To write a model for a verbal problem:

1. Choose a variable to represent the unknown number.

2. Use the variable to express the verbal sentence as an equation.

MODEL PROBLEMS

1. Write the following verbal sentence as an equation:

Five times a number decreased by 7 equals 13.

Solution: Let x represent the number.

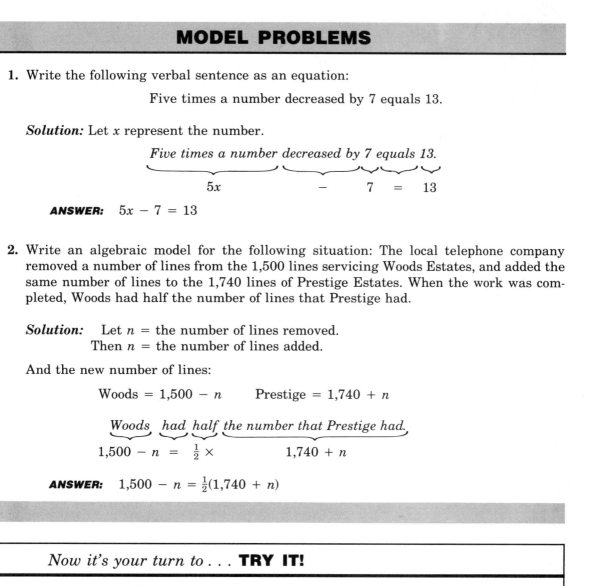

Five times a number decreased by 7 equals 13.

$5x$ $-$ 7 $=$ 13

ANSWER: $5x - 7 = 13$

2. Write an algebraic model for the following situation: The local telephone company removed a number of lines from the 1,500 lines servicing Woods Estates, and added the same number of lines to the 1,740 lines of Prestige Estates. When the work was completed, Woods had half the number of lines that Prestige had.

Solution: Let n = the number of lines removed.
            Then n = the number of lines added.

And the new number of lines:

Woods $= 1,500 - n$ Prestige $= 1,740 + n$

Woods had half the number that Prestige had.

$1,500 - n \; = \; \frac{1}{2} \times$ $1,740 + n$

ANSWER: $1,500 - n = \frac{1}{2}(1,740 + n)$

Now it's your turn to . . . TRY IT!

1. Write each verbal sentence as an equation. Let p represent the number.

a. Eight less than 7 times a number equals 6.

b. Five more than the quotient of a number and 9 equals 3.

c. Four less than a number, divided by 9, equals 3.

2. Write an algebraic model for the situation:
The width of a rectangle is 3 cm less than its length. The perimeter of the rectangle is 26 cm.

(See page 93 for solutions.)

1. Use mathematical symbols to write each phrase as a numerical expression.
   a. seven increased by the product of three and nine
   b. the product of five and the difference of eight and six
   c. twenty decreased by the quotient of thirty and five
   d. eight less than the product of three and seven

2. Use mathematical symbols to translate each verbal phrase into algebraic language.
   a. the product of $2c$ and $3d$
   b. twice the difference of p and q
   c. x subtracted from 5
   d. one-half of the sum of L and W

3. If n represents "a number," write each verbal phrase in algebraic language.
   a. 2 less than a number
   b. three-fourths of a number
   c. 4 times a number, increased by 3
   d. 8 more than the product of 3 and a number

   In 4–23, write an algebraic expression, using the variable(s) mentioned.

4. Warren traveled m miles. If Wanda traveled 100 miles farther than Warren, represent the number of miles Wanda traveled.

5. Mary is y years old. If Troy is 5 years older than Mary, represent his age in years.

6. Mr. Poore invested $1,000 in stocks. If he lost d dollars when he sold them, represent the amount he received for them.

7. Pecans weighing 10 kilograms were taken from a shipment weighing k kilograms. Represent the weight of the remaining pecans.

8. A crate of vegetables weighed 90 pounds. Represent the weight after x pounds of vegetables were removed.

9. A jockey, who weighs c pounds, is d pounds overweight. Represent the number of pounds she should weigh.

The common units of weight in the metric system are the gram (about the weight of a dollar bill), the kilogram (1,000 grams), and the milligram ($\frac{1}{1,000}$ of a gram).

Many scales, used commercially and at home, show weight in both customary units and metric units.

1 kilogram ≈ 2.2 pounds
1 pound ≈ .45 kilogram

10. Ahmad is x years old. Represent his age: **a.** r years from now   **b.** t years ago

11. Elvin bought an article for c dollars and sold it at a profit of \$25. Represent the amount for which he sold it.

12. A total of f feet of fencing was installed around the perimeters of two triangular garden plots. If p feet were used for one plot, represent the amount left for the other.

13. Represent the number of cents in: **a.** n nickels   **b.** d dimes and h half-dollars

14. Represent the number of:
    a. inches in f feet   **b.** feet in i inches   **c.** days in w weeks
    d. weeks in d days   **e.** grams in k kilograms   **f.** kilograms in g grams

15. If one book costs d dollars, represent the cost of n of these books.

16. Represent the number of posters Di bought for c dollars if each poster cost m dollars.

17. For each week of a 2-year union contract, a shipping clerk at Coolidge Company will earn m dollars. Represent the clerk's wage for: **a.** h weeks   **b.** 2 years

18. Nelson bought k pounds of bananas at f cents per pound and t pounds of grapes at m cents per pound. Represent the: **a.** total cost of his purchases   **b.** change from a \$5 bill

19. Mrs. Ramirez earns d dollars a month and spends s dollars a month. Represent the number of dollars she will save in 3 years.

20. The base of a rectangle measures 90 cm. Represent the:
    a. height, if it measures x cm less than the base
    b. perimeter, in terms of x
    c. area (base × height), in terms of x

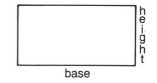

base

21. If t compact discs cost q dollars, represent the cost of 1 compact disc.

22. If n shirts cost c dollars, represent the cost of d shirts.

23. If Omar has f dollars and Raoul has g dollars and each of them earns an additional w dollars, represent the total amount of money that they have.

In 24 and 25, let n represent any natural number.

24. Represent the successor of n.   25. Represent the predecessor of n.

26. Write a verbal phrase for each expression if n represents "a number."
    a. $n - 3$ **b.** $2n + 1$ **c.** $\dfrac{n}{4}$ **d.** $\dfrac{n + 2}{2}$

In 27–30, write the verbal sentence as an equation. Use n to represent the number.

27. The product of 7 and a number equals 70.   28. Four less than a number equals 32.

29. Twice a number, increased by 7, equals 27.   30. The sum of 3 times a number and 7 is 22.

In 31–37, select the equation that represents, in terms of the given variable, the relationship expressed in the sentence.

31. Three times Harold's height is 108 inches. Let h = Harold's height.
(1) $h + 3 = 108$ (2) $3h = 108$ (3) $h - 3 = 108$ (4) $\frac{1}{3}h = 108$

32. A number increased by 7 equals 28. Let n = the number.
(1) $7n = 28$ (2) $n + 7 = 28$ (3) $n - 7 = 28$ (4) $\frac{1}{7}n = 28$

33. A number decreased by 5 equals 15. Let x = the number.
(1) $x + 5 = 15$ (2) $5x = 15$ (3) $\frac{1}{5}x = 15$ (4) $x - 5 = 15$

34. If 7 is subtracted from a number, the result is 8. Let x = the number.
(1) $7 - x = 8$ (2) $x - 7 = 8$ (3) $8 - x = 7$ (4) $x - 8 = 7$

35. Jim spent $6 for tokens that cost $2 each. Let t = the number of tokens that Jim bought.
(1) $t + 2 = 6$ (2) $t - 2 = 6$ (3) $2t = 6$ (4) $\frac{1}{2}t = 6$

36. Tom, who is 15 years old, is one-third as old as his father. Let f = the father's age.
(1) $3f = 15$ (2) $f + 3 = 15$ (3) $\frac{1}{3}f = 15$ (4) $f - 3 = 15$

37. A merchant bought 15 suits and now has 75 suits. Let s = the original number of suits.

(1) $s + 15 = 75$ (2) $s - 15 = 75$ (3) $15s = 75$ (4) $\dfrac{s}{15} = 75$

38. Match the items in Column A with those in Column B.

Column A	Column B
a. $3 - 2n = 1$	(1) Twice the difference of n and 3 is equal to 1.
b. $3(n - 2) = 1$	(2) When twice n is subtracted from 3, the result is 1.
c. $\dfrac{n - 3}{2} = 1$	(3) The product of 2 and n, decreased by 3, equals 1. (4) Three times the difference of n and 2 equals 1.
d. $2(n - 3) = 1$	(5) Half the difference of n and 3 is 1.
e. $2n - 3 = 1$	

In 39–42, write a verbal sentence that gives a meaning of the equation, if the variable represents "a number."

39. $x + 7 = 12$ **40.** $s - 5 = 15$ **41.** $\dfrac{x}{4} = 8$ **42.** $2c + 4 = 12$

In 43–50, name the set of numbers that would be an appropriate domain for the situation. For example, for the divisions on a ruler, you would use positive rational numbers.

43. a shoe size   **44.** the size of a population

45. the length of a line segment   **46.** the weight of an object

47. a number of hours   **48.** a distance traveled

49. a number of telephone calls   **50.** a numerical value

In 51–57, write an algebraic model for the situation.

51. When 9 is subtracted from 5 times a number, the result is 31.

52. In one day, the Sun and Spa Pool Corporation sold 12 more than 3 times the number of pools that it sold on the same date last year. If twice last year's number were increased by 24, the sales figures would be the same.

53. Huck worked part-time as a lifeguard at the Frog Park Pool. He worked twice as many hours the second week on the job as he had the first week. The total for the two weeks was 45 hours.

54. Scott took some tennis balls from a bucket that contained 20 balls. The number of balls remaining in the bucket was 4 more than the number taken out.

55. Eight pounds more than half the weight of Elizabeth's backpack is the same as 4 pounds less than its weight.

56. The length of a rectangle is 5 inches greater than its width. The perimeter of the rectangle is 42 inches.

57. Shanta added the crystals she had brought home from her field trip to her collection of 100 crystals. The total number of crystals in her collection was now the same as 3 times the number she had added.

The physical properties of a quartz crystal make it uniquely useful as microchip material for watches and computers. A chip that is no wider than the eye of a needle can hold a complex circuit.

2-6 WORKING WITH AN INEQUALITY

Symbols of Inequality

To change the false statement $3 + 6 = 5$ to a true statement, replace the symbol $=$ with the symbol $\neq$, read "is not equal to." Thus, $3 + 6 \neq 5$ is a true statement. A statement that one number is not equal to another number is called an ***inequality***. An open sentence may be an inequality as well as an equation.

For any two real numbers, one must be greater than the other, equal to the other, or less than the other. To compare unequal numbers, use the following symbols.

Symbol	Read	Example
>	is greater than	$4 > 2$
<	is less than	$1 < 3$
≯	is not greater than	$2 \ngtr 7$
≮	is not less than	$5 \nless 4$
≥	is greater than or equal to is at least	$6 \geq 3$
≤	is less than or equal to is at most	$5 \leq 5$

Notice that in an inequality that is a true sentence involving the symbol $>$ or $<$, such as $4 > 2$ or $1 < 3$, the symbol points to the *smaller* number.

An inequality symbol such as $\geq$ allows the choice of $>$ *or* $=$. The number sentence $6 \geq 3$ is true because $6 > 3$. Tell why the sentence $5 \leq 5$ is true.

Ordering Numbers on a Number Line

Any number on a number line is greater than every number to its left and less than every number to its right. For example, 4 is greater than 2, written $4 > 2$, and 4 is to the right of 2. Thinking in reverse, 2 is less than 4, written $2 < 4$, and 2 is to the left of 4.

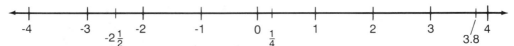

You can also describe the position of a number as *between* two other numbers. For example, you can say that 1 is between -2 and 3. This means that -2 is less than 1, written $-2 < 1$, *and* that 1 is less than 3, written $1 < 3$. Combining these two statements, write $-2 < 1 < 3$.

In reverse, $3 > 1 > -2$ means 3 is greater than 1 *and* 1 is greater than -2. Of course, you can locate 1 between other pairs of numbers. Name some.

MODEL PROBLEMS

1. State whether the sentence is true or false.

   a. $6 + 7 \neq 15$ True   **b.** $0 \leq 8 - 8$ True

   c. $-8 + 6 > -10$ True   **d.** $-3\frac{1}{2} \not> -8$ False

2. Order the given set of numbers by first using the symbol $>$ and then $<$.

Numbers	Using $>$	Using $<$
$7, -2$	$7 > -2$	$-2 < 7$
$-5, -3$	$-3 > -5$	$-5 < -3$
$9, -2, 1$	$9 > 1 > -2$	$-2 < 1 < 9$

   Observe that a positive number is always greater than a negative number.

3. Which is the smaller of the numbers $-\frac{2}{3}$ and $-\frac{5}{3}$?

   Solution: A fraction is negative if either its numerator or its denominator (but not both) is negative. Thus, $-\frac{2}{3} = \frac{-2}{3}$ and $-\frac{5}{3} = \frac{-5}{3}$. Compare fractions that have the same positive denominator by examining the numerators.

   Since $-5 < -2$, then $\frac{-5}{3} < \frac{-2}{3}$.

4. Which is the greater of the numbers $\frac{3}{4}$ and $\frac{5}{8}$?

   Solution: Compare fractions that have different denominators by first rewriting them with a common denominator.

   Since $\frac{3}{4} = \frac{6}{8}$ and $\frac{6}{8} > \frac{5}{8}$, then $\frac{3}{4} > \frac{5}{8}$.

   Fractions can also be compared if they are rewritten as decimals.

   $$\frac{3}{4} = .75 \qquad \frac{5}{8} = .625 \qquad \text{and } .75 > .625 \qquad \text{thus } \frac{3}{4} > \frac{5}{8}$$

   Finally, numbers can be ordered according to their placement on a number line. To locate $\frac{3}{4}$ and $\frac{5}{8}$ on a number line, divide the interval between 0 and 1 into eighths.

   Since $\frac{3}{4}$ is to the right of $\frac{5}{8}$ on the number line, $\frac{3}{4} > \frac{5}{8}$.

In the last section, you learned to represent a given situation by writing an equation. An inequality can also be used to represent a situation.

In 5 and 6, write an algebraic model for the situation.

5. To buy a blouse and a skirt, Rivka plans to spend less than $50. She thinks the blouse will cost about half as much as the skirt.

Solution: Identify the variable, and write an open sentence.

Let x = the approximate cost of the skirt.
Then $\frac{1}{2}x$ = the approximate cost of the blouse.

$$\overbrace{\text{The total cost}}^{x + \frac{1}{2}x} \overbrace{\text{is less than}}^{<} \overbrace{\$50}^{50}.$$

ANSWER: $x + \frac{1}{2}x < 50$

6. For tonight's homework, Alonso has assignments in three subjects: Math, French, and Biology. He plans to devote equal amounts of time to French and Biology, and twice as much time to Math. He has at least $2\frac{1}{2}$ hours to work.

Solution:

Let x = the time for French.
Then x = the time for Biology.
And $2x$ = the time for Math.

$$\overbrace{\text{The total time}}^{x + x + 2x} \overbrace{\text{is at least}}^{\geq} \overbrace{2\frac{1}{2} \text{ hours}}^{2\frac{1}{2}}.$$

ANSWER: $x + x + 2x \geq 2\frac{1}{2}$

Now it's your turn to . . . TRY IT!

Write an algebraic model for the situation.

1. In her collection of baseball cards, Idalia has the same number of National League cards as American League. The total number of cards in the collection is more than 250.

2. The average weight of the bulls that Rancho Rio is raising for the rodeo is now 750 pounds. The average weight of a ranch worker is 160 pounds. On a freight elevator that has a maximum capacity of 20,000 pounds, there is a worker with each bull.

(See page 93 for solutions.)

Graphing a Set of Numbers

When you graph a set of numbers on the real number line, distinguish between a set of separate points and a set of continuous points, as in:

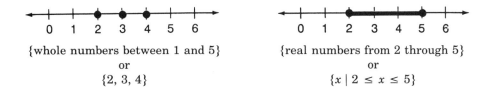

{whole numbers between 1 and 5}	{real numbers from 2 through 5}
or	or
{2, 3, 4}	$\{x \mid 2 \leq x \leq 5\}$

The graph of {2, 3, 4} shows exactly the three points that correspond to the elements of the set. The graph of {real numbers from 2 through 5} shows the continuous set of points that begins at the point with coordinate 2 and ends at the point with coordinate 5.

Note how the notation describing this set, $\{x \mid 2 \leq x \leq 5\}$, relates to the graph: the real numbers between 2 and 5, including 2 and 5. On the graph, a darkened line ▬▬ represents the continuous set.

Suppose a set of continuous points excludes one or both of the endpoints:

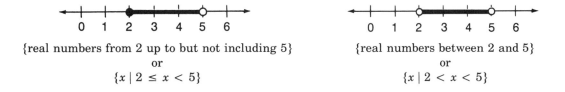

{real numbers from 2 up to but not including 5}	{real numbers between 2 and 5}
or	or
$\{x \mid 2 \leq x < 5\}$	$\{x \mid 2 < x < 5\}$

Note how to distinguish between a point that is included in the set as opposed to a point that is excluded from the set. In the description of the set, $\{x \mid 2 \leq x < 5\}$, the symbol $\leq$ includes the point 2, while the symbol $<$ excludes the point 5. On the graph, a darkened circle ● includes a point and an open circle ○ excludes a point.

Consider also those graphs that continue infinitely to the right or to the left on a number line, as in:

| $\{x \mid x \geq 1\}$ | $\{x \mid x < 1\}$ |

The decisions you need to make when graphing a set of numbers on a number line are summarized in the accompanying flowchart.

GRAPHING A SET OF NUMBERS

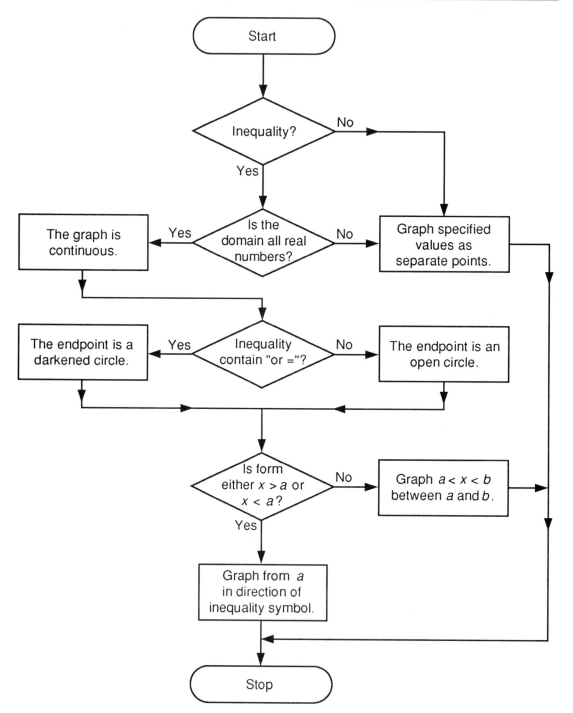

1. Graph the given set of numbers on the real number line.

   a. {natural numbers between 1 and 4}

   b. $\{x \mid x \geq 20\}$

   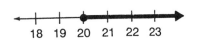

   It is important to show an appropriate portion of the scale for your graph.

2. Using {0, 1, 2, 3} as the replacement set, find and graph the solution set for the open sentence $2x + 1 = 5$.

   Solution: Replace x by each element of the replacement set, and determine if the resulting statement is true or false.

   If $x = 0$, then $2 \times 0 + 1 = 5$ is false. If $x = 1$, then $2 \times 1 + 1 = 5$ is false.

   If $x = 2$, then $2 \times 2 + 1 = 5$ is *true*. If $x = 3$, then $2 \times 3 + 1 = 5$ is false.

   Since 2 is the only element that makes the sentence true, the solution set is {2}.

3. Using the set of real numbers as the replacement set, graph the solution set of the given open sentence. Explain the graph.

   a. $m \leq -7$

   Explanation: The darkened circle at -7 shows that -7 is included. The darkened line and arrow to the left represent the infinite set of real numbers less than -7.

   b. $-3 \leq t < 1$

   Explanation: The darkened circle at -3 shows -3 is included. The open circle at 1 shows 1 is not included. The darkened line represents the infinite set of real numbers between -3 and 1.

Now it's your turn to . . . **TRY IT!**   (*See page 93 for solutions.*)

1. Find and graph the solution set of $n + 1 > 2$ when the domain is {0, 1, 2, 3, 4}.
2. If the domain is {real numbers}, graph the solution set of $x \geq 2$. Explain the graph.
3. Graph the solution set of $1 < x \leq 4$ when $x \in$ {real numbers}. Explain the graph.

In 1–10, tell whether the inequality is true or false.

1. $8 + 5 \neq 6 + 4$ **2.** $9 + 2 \not> 2 + 9$ **3.** $6 \times 0 \not< 4 \times 0$

4. The sum of 8 and 12 is not equal to the product of 24 and 4.

5. $-3 < 0 < 1$ **6.** $7 > 1 > -3$ **7.** $-4 > -3 > -2$

8. $4.6 - 2.1 > 1.5 + .9$ **9.** $\frac{1}{2} + \frac{1}{8} \leq .8$ **10.** $3 - .25 > 2\frac{1}{2}$

In 11–15, order the numbers using the symbol: **a.** $<$ **b.** $>$

11. $-4, 8$ **12.** $-1\frac{1}{2}, 0$ **13.** $3, -2, -4$ **14.** $-3\frac{1}{2}, 6, 2\frac{1}{2}$ **15.** $-1.5, 3\frac{1}{2}, -2\frac{1}{2}$

In 16–18, write the inequality in symbols.

16. The sum of 9 and 4 is less than the product of 10 and 5.

17. 9 is less than 11, and 11 is less than 20.

18. 17 is greater than 1, and 1 is greater than -4.

In 19–22, express the inequality in words.

19. $9 + 8 > 16$ **20.** $12 - 2 \leq 4 \times 7$ **21.** $-2 < -1 < 0$ **22.** $2 < 3 + 4 < 10$

23. If x is a positive number and y is a negative number ($x > 0$ and $y < 0$), tell whether each statement is true or false.
  a. $x = y$ **b.** $x > y$ **c.** $y > x$ **d.** $x < 0$ **e.** $y < 0 < x$

24. Explain why it is incorrect to write $3 < 5 > 1$.

In 25–32, tell which of the given numbers is the greater.

25. $\frac{5}{2}, \frac{7}{2}$ **26.** $\frac{5}{2}, \frac{7}{4}$ **27.** $\frac{5}{8}, \frac{5}{12}$ **28.** $1.4, 1\frac{3}{5}$

29. $-3.4, -3\frac{1}{3}$ **30.** $2, 2.25$ **31.** $-5.7, -5.9$ **32.** $.7, .\overline{7}$

In 33–36, arrange the given set of real numbers in order from smallest to largest.

33. $\{.3, .31, .333\ldots, .313113111\ldots\}$ **34.** $\{.\overline{25}, .20, .\overline{2}, .202002000\ldots\}$

35. $\{\frac{2}{7}, .27, .\overline{27}, .272272227\ldots\}$ **36.** $\{-\frac{3}{5}, -.\overline{61}, -.\overline{6}, -.60616263\ldots\}$

In 37–42, name the points on the line that are the graph of the given set of numbers.

A B C D E F G H I J K L M N P Q
0 1 2 3 4 5 6 7 8 9 10 11 12 13 14 15

37. {whole numbers between 1 and 10} **38.** {even whole numbers between 1 and 10}

39. {whole numbers between 7 and 8}   **40.** {multiples of 3 between 7 and 14}

41. $\{x \mid 10 < x < 15$ where $x \in$ whole numbers$\}$

42. $\{y \mid 3 \le y \le 13$ where $y \in$ odd integers$\}$

In 43–49, draw the graph of the given set of numbers.

43. $\{-2, 0, 2\frac{2}{3}\}$ **44.** $\{1\frac{3}{4}, 2\frac{1}{2}, 3.5\}$ **45.** {whole numbers between 3 and 10}

46. a. {whole numbers less than 7}   **47. a.** {real numbers greater than 4}

  b. {real numbers less than 7}   **b.** {real numbers less than 4}

48. {real numbers less than or equal to -3}   **49.** {real numbers between 4 and 6}

In 50–57, find and graph the solution set when the replacement set is $\{0, 1, 2, 3, 4, 5\}$.

50. $2n = 4$ **51.** $n + 3 = 6$ **52.** $2x - 1 = 7$ **53.** $5x = 0$

54. $y > 4$ **55.** $t \le 1$ **56.** $t - 3 > 1$ **57.** $3t + 1 < 8$

In 58–65, if the replacement set is $\{-4, -3, -2, -1, 0, 1, 2, 3, 4\}$, graph the solution set of the open sentence.

58. $x > 0$ **59.** $x \le 0$ **60.** $y > -1$ **61.** $t < 2$

62. $-m \ge 1$ **63.** $-1 < d < 3$ **64.** $-1 \le x < 2$ **65.** $-3 \le t \le 3$

In 66–73, the domain is the set of real numbers.

  a. Graph the solution set of the sentence.

  b. Explain the graph.

66. $x > 6$ **67.** $s \le 0$ **68.** $m \ge -1$ **69.** $x \ne 2$

70. $z \ge 2\frac{1}{2}$ **71.** $-3 < x < 2$ **72.** $-1 < y \le 3$ **73.** $-3 \le m \le 3$

In 74–78, graph the solution set on the real number line.

74. There are always at least 3 dogs in Kevin's Kennel's main section, which holds as many as 10 dogs. How many dogs might be in the kennel's main section today?

75. Mr. Moses sells at least 4 cars daily. His goal is to sell 8 cars each day in order to maximize his commissions. How many cars might Mr. Moses sell today?

76. The temperature in Pike's Valley during the winter is at most $36°F$ and at least $-12°F$. What is the possible range of temperatures in Pike's Valley on February 8?

77. Baby Jane lost 6 rattles this week. How many rattles could she have lost on Monday?

78. An *acute angle* measures more than 0 degrees but less than 90 degrees. What measures are possible for acute angle A?

In 79–86, write an algebraic model for the situation.

79. Two triangles are congruent. The total area of the two figures is less than 250 square cm.

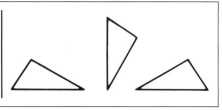

Congruent triangles have 6 parts equal, 3 pairs of sides and 3 pairs of angles. However, a match of just the 3 pairs of sides is enough to guarantee congruence. Whatever their position, triangles with matching side lengths will have the same shape and size.

80. Melba and Risa have newspaper routes after school. This week, Melba collected twice as much money in tips as Risa. Their combined tips amounted to more than $80.

81. The Hank Corporation and the Newton Corporation are hiring members of the graduating class of Drake University for their accounting divisions. The Hank Corporation will add 7 more employees than will the Newton Corporation. The total number of graduates needed by these corporations is at least 41.

82. Hans and Ali both work out on a bench press at the Century III Health Club. Hans presses 35 more pounds than Ali. Together, they press at least 315 pounds.

83. Each month, Mr. and Mrs. Dakota save a sum of money for their children's college education. Their savings for February exceeded the January amount by $50. Their March savings were 3 times their January savings. The total for this winter period was at least $450.

84. Colby, Margarita, and Howard work for the Able Insurance Company. Colby earned twice as much as Margarita in commissions this month, and Howard earned $600 more than Margarita. The total of their monthly commissions was greater than $10,200.

85. For the off-season weekend before Memorial Day, the Vanguard Hotel and Country Club will offer special room rates. At these rates, they will reserve half as many poolside rooms as beachfront rooms, and 30 fewer shoreside rooms than poolside rooms. The total number of rooms offered at these special rates is at most 300.

86. Professor Digby, a geologist, is measuring layers of rock. In the layers shown, the volume of layer I is equal to the volume of layer IV, the volume of layer II is equal to three-fourths of the volume of layer IV, and the volume of layer III is equal to twice the volume of layer II. The volume of the four layers is at most 570 cubic meters.

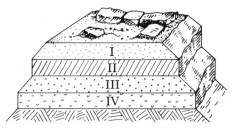

CHAPTER SUMMARY

Algebra has its own language and conventions. A letter that is used to represent a number is called a *variable*. To *evaluate* an algebraic expression, each variable in the expression is replaced by a number; then the arithmetic operations are performed, according to the *order of operations*.

A *term* in an algebraic expression is the product of its *factors*. Factors that are repeated can be written compactly by using *exponents*.

A *graph* on a *number line* provides a visual representation of a set of numbers. Some important number sets, in addition to the familiar counting numbers and whole numbers, are the *integers*, the *rational numbers*, and the *irrational numbers*. The rational and irrational numbers together comprise the *real numbers*. The real numbers are in *one-to-one correspondence* with the points on the number line.

The number assigned to a point on the number line is its *coordinate*. The positive and negative coordinates of two points that are the same distance from 0 are *opposites*. The numbers in a pair of opposites have the same *absolute value*, which can be defined as their distance from 0.

A mathematical or real-life situation can be represented by an *algebraic model*, which uses a variable in an *open sentence* as the symbol of an unknown quantity. An open sentence may be either an *equation* or an *inequality*. Later chapters will deal with algebraic methods of solving open sentences.

A *flowchart* helps you organize your thinking and guides you through a procedure. In science and industry, flowcharts can be useful in organizing complicated computer programs that contain a number of miniprograms.

VOCABULARY CHECKUP

SECTION

2-1 *simplify / order of operations / evaluating / variable*

2-2 *term / factor / coefficient / constant / base / exponent / power*

2-3 *set / element / finite, infinite sets / subset / natural, whole numbers / integers / rational, irrational, real numbers*

2-4 *number line / one-to-one correspondence / coordinate / graph / opposites / reflection / absolute value / counterexample*

2-5 *open sentence / equation / algebraic model / solution set / replacement set / domain*

2-6 *inequality*

Symbols $\in, \notin$ $\neq$ $>, \not>$ $<, \not<$ $\geq, \leq$ $|a|$

CHAPTER REVIEW EXERCISES

In 1–4, simplify the expression. (Section 2-1)

1. $4 + 6 \times 3$ **2.** $15 \div 3 \times 4 - 1$ **3.** $15 - 9 + 2$ **4.** $4 \times (10 - 4) - 4$

In 5–8, find the value of the expression. Use $a = 6$, $b = 9$, and $c = 4$. (Section 2-1)

5. $ab - c$ **6.** $5c - 3a$ **7.** $ab - ac + b$ **8.** $(a + b)(b - c)$

In 9–12, simplify the expression. (Section 2-2)

9. 3^3 **10.** $4^3 - 2^6$ **11.** $5^3 - 2(4)^2$ **12.** $5^2 \times 3^2$

In 13–16, evaluate the expression if $a = 6$, $b = 3$, and $c = 4$. (Section 2-2)

13. $a^2 - b^3$ **14.** $ab^2 + bc^2$ **15.** $\dfrac{a^2 - c^2}{a - c}$ **16.** $\dfrac{2b^2c - a^2}{bc}$

In 17–20, list the members of the set. (Section 2-3)

17. {letters of the word BANANA}
18. {whole numbers less than 5}
19. {odd natural numbers greater than 1}
20. {$a | a$ is a number that is neither positive nor negative}

In 21–24, give a description of the set. (Section 2-3)

21. {1, 3, 5, 7, . . . } **22.** {2, 4, 6, 8, 10} **23.** {r, a, e} **24.** {7, 9, 11, 13, 15}

In 25–28, indicate to which of the given set(s) the number belongs. (Section 2-3)

Write: N if the number is a natural number
W if the number is a whole number
I if the number is an integer
R if the number is a rational number

25. -5 **26.** $\dfrac{4}{5}$ **27.** 0 **28.** 17

In 29–33, write the rational number as a decimal. (Section 2-3)

29. $\dfrac{5}{8}$ **30.** $\dfrac{7}{4}$ **31.** $5\frac{1}{6}$ **32.** $\dfrac{13}{5}$ **33.** $\dfrac{5}{12}$

In 34–37, express the decimal as a quotient of two integers or as a mixed number. (Section 2-3)

34. .21 **35.** .017 **36.** 1.7 **37.** 8.73

In 38–40, use the given number line. (Section 2-4)

$$\begin{array}{ccccc} A & B & C & D \\ \end{array}$$

```
    A   B   C   D
 ───┼───┼───┼───┼───
    0   1   2   3
```

38. Name the coordinate of point C.

39. Which points have coordinates that are natural numbers?

40. Name the point that is the graph of the number 3.

In 41–44, write the simplest symbol that represents the opposite of the number. (Section 2-4)

41. -8 **42.** $4 - 1$ **43.** $-(4 + 6)$ **44.** $-(-(-4))$

In 45–47, find the value of the expression. (Section 2-4)

45. $|6| - |-5|$ **46.** $5 - |-3|$ **47.** $|5| \times |-2| - 1$

In 48–52, rewrite the verbal phrase using symbols. (Section 2-5)

48. five more than the difference of twelve and three

49. eight less than the product of seven and six

50. a number that is 4 less than t **51.** the product of x and y **52.** 12 increased by t

In 53–58, write an algebraic expression. (Section 2-5)

53. An angle measures p degrees. Represent the measure of an angle that is 7 degrees smaller.

54. Hans bought a painting for d dollars. If he sold it for a $40 profit, represent the amount of money he received when he sold the painting.

55. Represent the number of cents in d dimes.

56. Lunch costs d dollars each week. Represent the cost of lunch for w weeks.

57. If b books cost d dollars, represent the cost of 1 book.

58. If 3 apples cost c cents, represent the cost of 5 apples.

In 59–62, write the sentence as an equation. Let p represent the number. (Section 2-5)

59. Five times the difference of a number and 7 equals 3.

60. Five times a number, decreased by 7, equals 3.

61. Sixteen decreased by twice a number equals 9.

62. The product of 8 and a number is the same as the difference between 18 and the number.

In 63–66, indicate whether the inequality is true or false. (Section 2-6)

63. $\dfrac{3}{5} > \dfrac{1}{2}$ **64.** $6 + 1 < 5 + 3$ **65.** $6 \geq 6$ **66.** $.7 + 1 > .8$

In 67–70, use the symbol > to order the numbers. (Section 2-6)

67. $-4, -7$ **68.** $.9, .89, -.91$ **69.** $-\frac{1}{3}, -\frac{1}{2}, -\frac{1}{4}$ **70.** $-.6\overline{7}, -.\overline{67}$

In 71–74, state which of the numbers is greater. (Section 2-6)

71. $.9, .09$ **72.** $-.03, -.2$ **73.** $\frac{3}{7}, \frac{3}{8}$ **74.** $-1, -\frac{3}{4}$

In 75–77, write an algebraic model for the situation. (Section 2-6)

75. Mr. Tomkins sewed a blouse and shirt for his daughter. To make the skirt, he used twice as much fabric as he did for the blouse. In all, he used more than 4 yards of fabric.

76. Paul, Roy, and Andrea sold tickets to their school play. Paul sold 3 times as many tickets as Roy, and Andrea sold 10 fewer tickets than Paul. Together, they sold at least 60 tickets.

77. Farmer Sahadi raises sheep and chickens. At this time, he has equal numbers of lambs and chicks. In all, the number of legs among these baby animals is at most 150.

In 78–80, name the points on the line that are the graphs of the given set of numbers. (Section 2-6)

78. {integers between -2 and 4}   **79.** {odd whole numbers less than 7}

80. $\{x \mid -1 < x \le 3 \text{ and } x \text{ is an integer}\}$

In 81–85, draw the graph of the given set of real numbers. (Section 2-6)

81. {numbers between -3 and -6}   **82.** {odd whole numbers between 14 and 21}

83. $\{x \mid x \ge -8\}$ **84.** $\{y \mid 7 < y \le 13 \text{ and } y \in \text{whole numbers}\}$ **85.** $\{z \mid 4 \le z \le 5\}$

In 86–89, graph the solution set of the open sentence if the domain of the variable is $\{-3, -2, -1, 0, 1, 2, 3, 4\}$. (Section 2-6)

86. $y > -2$ **87.** $-a \le 3$ **88.** $-w > -1$ **89.** $-2 < -x < 1$

PROBLEMS FOR PLEASURE

In 1–3, evaluate the expression if $a = 2$ and $b = 5$.

1. $a^b - b^a - a^a$ **2.** $(ab)^a - a(a + b)^a - a$ **3.** $[(b - a)^a - b - a]^b$

4. Arthur and Paul were planning an 8,000-mile automobile trip across the country. Bad news: a mechanic told them that their tires would only last 4,000 miles. They had a new spare tire in the trunk, but since they had little money, they wanted to buy as few new tires as possible. What is the smallest number of tires they must buy to complete their trip? Explain.

5. If you subtract one irrational number from another irrational number, will your result always be irrational, or will it always be rational, or will it be sometimes irrational and sometimes rational? Give examples to support your answer.

6. In the Venn diagram, each of the letters a through m stands for an integer from 1 through 8. Each circle contains a set of numbers, and the sum of the numbers in each set is 16. Each number from 1 through 8 is used either once or twice. Numbers in the same set cannot be repeated. Use the clues to find the numbers.

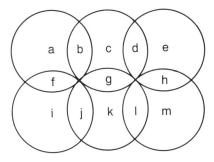

1. $a + b = 8$ 2. $d + e = 8$
3. $k > i > m$ 4. $g > l > j > d$
5. $b > a$ 6. $g < m$
7. $b < c$

CALCULATOR CHALLENGE

1. ESTIMATION GAME: Requires 3 to 5 players. One player reads the problem to the other players. The other players estimate the answer and write it down. The reader then checks the correct answer, using a calculator. All answers will be whole numbers.

Scoring: 1 point for finishing first. 3 points for the answer closest to the correct answer. (If in doubt, the reader can subtract the correct answer from the written answers to check which is closest.) The highest score after 10 questions wins.

GAME 1	GAME 2	GAME 3
1. $42{,}596 \div 23$	1. $542{,}190 \div 62$	1. $193{,}374 \div 54$
2. 534×83	2. 758×87	2. 345×45
3. $84 \times .5$	3. $55 \times .2$	3. $65 \times .6$
4. $2{,}091 \div 17$	4. $684 \div 19$	4. $896 \div 28$
5. $560 \times .3$	5. $285 \times .2$	5. $875 \times .4$
6. $768 \div .5$	6. $286 \div .4$	6. $957 \div .2$
7. $60 \times .8$	7. $45 \times .6$	7. $25 \times .4$
8. $60 \div .8$	8. $45 \div .6$	8. $25 \div .4$
9. 69×71	9. 74×66	9. 33×27
10. $1{,}360 \div 8$	10. $1{,}404 \div 9$	10. $2{,}422 \div 7$

2. The square of every whole number larger than 2 can be written as the difference of squares of two other whole numbers. For instance, $16 = 25 - 9$, which can also be written $4^2 = 5^2 - 3^2$. Find the two whole-number squares whose difference is:
a. 7^2 **b.** 18^2

3. Using only the given numbers and any operations (you may repeat both numbers and operations), write an expression that produces the required result. Try to find the expression that uses the fewest operations.

	Numbers	Result
a.	3, 4	78
b.	2, 3, 4, 5	352

4. **a.** Find all the multiples of 11 less than 400 that will give a palindromic number when multiplied by 11.

   b. Consider palindromic numbers such as 202 or 5,005, where the first and last digits are the same and the inner digits are 0's. If you multiply these numbers by 11, will the result always be a palindromic number?

COLLEGE TEST PREPARATION

In 1–16, select the correct answer.

1. If $a = 1$, $b = 2$, and $c = 3$, then
$a^2b^3c - c^2 - a^3 =$
(A) 12   (B) 14   (C) 16   (D) 38   (E) 40

2. A theater sells admission tickets at the rate of $4.00 per adult and $2.00 per child. Which of the following is a formula for the total receipts, T dollars, if a adult tickets and c children's tickets were sold?
(A) $T = a + c$ (B) $T = 2a + 2c$
(C) $T = 4a + 4c$ (D) $T = 2a + 4c$
(E) $T = 4a + 2c$

3. If r is 10 less than a number, then 10 more than the number is
(A) $10 - r$ (B) $r - 10$ (C) $r + 10$
(D) $r - 20$ (E) $r + 20$

4. If $a = 3$, $b = 2$, and $c = 4$, then a factor of $a^b + b^c$ is
(A) 2   (B) 3   (C) 5   (D) 7   (E) 11

5. Add 5 to the number n. Multiply the result by 5. Subtract 5, then divide this result by 5. What is the result?
(A) n (B) $n + 1$ (C) $n + 2$
(D) $n + 3$ (E) $n + 4$

6. If $a = 3$ and $b = 2$, then the expression
$$\frac{5a}{b} - \frac{3b}{a} + \frac{1}{ab} \text{ equals}$$
(A) $\frac{13}{2}$ (B) $\frac{32}{3}$ (C) $\frac{15}{2}$ (D) $\frac{17}{3}$ (E) $\frac{5}{3}$

7. If $r = 2$, $s = 3$, and $t = 5$, which expression has a value that is a factor of $2 \cdot 3^2 \cdot 5 \cdot 7$?
(A) $2rst$ (B) $3rs^2t$ (C) $7rs^2t^2$
(D) $\dfrac{rs^2t}{r + 7}$ (E) $\dfrac{r^2st}{r + s}$

8. Which of the following expressions has the greatest value when $x = 3$ and $y = 1$?
(A) $\dfrac{(x + y)^2}{(x - y)^2}$ (B) $\left(\dfrac{x + y}{x - y}\right)^2$ (C) $x + y$
(D) $x^2 - y^2$ (E) $x^2 + y^2$

9. Ron was y years old n years ago. How old will he be in x years?
(A) $y + n + x$ (B) $y + n - x$
(C) $y - n + x$ (D) $y - n - x$
(E) $n - y + x$

10. If $w = 1$, $x = 2$, $y = 3$, and $z = 4$, which expression has a value different from the others?
(A) $w + x + y + z$ (B) $w + y^x$
(C) $yz - wx$ (D) $yw^x + w$
(E) $z^x - xy$

11. The sum of two numbers is twice the difference of the numbers. The equation that states this is
(A) $a + b = 2a - b$
(B) $2a + b = a - b$
(C) $2(a + b) = a - b$
(D) $a + b = 2(a - b)$
(E) $a + b = a - 2b$

12. The members of a scout troop brought 5 boxes containing 10 trash bags each to their picnic. They used b bags and the number remaining was 6 more than the number used. Which equation applies?
(A) $10(5 - b) = b + 6$
(B) $5(10 - b) = b + 6$
(C) $b = 5(10) + 6$
(D) $b + 6 = 5(10)$
(E) $50 - b = b + 6$

13. The fraction $\frac{5}{4}$ is between
(A) $1\frac{1}{2}$ and 2 (B) $\frac{2}{3}$ and 1 (C) $\frac{1}{2}$ and $\frac{9}{8}$
(D) $\frac{4}{5}$ and $\frac{9}{5}$ (E) $\frac{7}{10}$ and $\frac{11}{10}$

14. If $W = \{$whole numbers $< 10\}$, what is the sum of the least even number in W and the greatest odd number in W?
(A) 12 (B) 11 (C) 10 (D) 9 (E) 8

15. Which set of numbers is listed in *ascending order* (lowest to highest)?
(A) $-\frac{1}{3}, -\frac{1}{2}, \frac{2}{3}, \frac{3}{4}$ (B) $-\frac{1}{2}, -\frac{1}{3}, \frac{2}{3}, \frac{3}{4}$
(C) $-\frac{1}{3}, -\frac{1}{2}, \frac{3}{4}, \frac{2}{3}$ (D) $-\frac{1}{2}, -\frac{1}{3}, \frac{3}{4}, \frac{2}{3}$
(E) $-\frac{1}{2}, \frac{1}{3}, \frac{3}{4}, \frac{2}{3}$

16. Which of the following will give the least value if $x = 3$ and $y = 2$?
(A) $\frac{x}{y^2}$ (B) $\frac{y}{x}$ (C) $\frac{y^3}{x^3}$ (D) $\frac{x^2}{y^3}$ (E) $\frac{y^2}{x^2}$

Questions 17–26 each consist of two quantities, one in Column A and one in Column B. You are to compare the two quantities and choose:

A   if the quantity in Column A is greater;
B   if the quantity in Column B is greater;
C   if the two quantities are equal;
D   if the relationship cannot be determined from the information given.

1. In certain questions, information concerning one or both of the quantities to be compared is centered above the two columns.
2. In a given question, a symbol that appears in both columns represents the same thing in Column A as it does in Column B.
3. x, n, and k, etc. stand for real numbers.

Column A	Column B
17. x^3 if $x = \frac{3}{4}$	$y \cdot y \cdot y$ if $y = 0.75$

$x = 10$

18. $\dfrac{10x + 10}{10}$	$10 + 10$

$x = \frac{1}{2}$

19. $x\left(x + \frac{1}{2}\right)$	$x^2 + \frac{1}{2}x$

20. $\left(2 + \frac{1}{10}\right)^2$	$(2)^2 + \left(\frac{1}{10}\right)^2$

21. $-.024$	$-.076$

22. $.002 + .3$	$.003 + .2$

23. $\left(\frac{1}{2}\right)^3$	$\left(\frac{1}{3}\right)^2$

$x \neq 0$

24. x	x^2

$x = 3$

25. $\dfrac{5 - x}{\lvert 5 + x \rvert}$	$\dfrac{4 - x}{\lvert 6 \rvert - x}$

$x = 2$

26. $x^3 + 3x^2 + 2x$	$2x^2 + 6x + 4$

Air Pollutant Emissions

1. Refer to the table of air pollutants.
   a. How many metric tons of sulfur oxides were emitted?
   b. By how many metric tons did the carbon monoxide emissions exceed the emissions from volatile organic compounds?
   c. How many metric tons were emitted from lead?

Pollutant	Total Emissions (in millions of metric tons)
Carbon monoxide	60.9
Sulfur oxides	21.2
Volatile organic compounds	19.5
Particulates	6.8
Nitrogen oxides	19.3
Lead	.0086

2. How many fifths of an inch are there in 25 inches?

3. If the numerals in the tenths and hundredths places are interchanged in the number 248.395, what will the resulting number be?

4. Find two numbers such that their sum is 30 and one number is 4 times the other.

5. Find the next two numbers in the sequence: 3, 4, 7, 12, 19, __, __

6. Express the result of $(8 \div 2) \div 5$ as a decimal.

7. If 36 chairs are arranged in 6 parallel rows of 6 chairs each, what is the least number of chairs that must be moved in order to get 4 parallel rows of 9 chairs each?

8. Find two numbers that have a sum of 12 and a product of 32.

9. If $\frac{1}{2}$ of a number is 9, what is $\frac{2}{3}$ of the number?

10. If A, B, and C are 3 points on a line and segment AB is twice the length of segment BC, then which of the following is not true?

    A •————————————• B —————• C

    (1) BC is $\frac{1}{2} AB$.   (2) BC is $\frac{1}{3} AC$.   (3) AC is 3 times BC.   (4) AC is twice BC.

11. Shredded bran comes in an 8-ounce box and a 12-ounce box. If the small size sells for $1.60, what is the most that the large box could sell for and still be a better buy than the small size?

12. The number 6, 8 $\underline{?}$ 4 is missing a digit. If the 4-digit number is divisible by 11, what is the missing digit?

13. A food store received a shipment of 12 dozen cartons of oat bran and sold $\frac{3}{4}$ of the shipment within the first five hours and a dozen cartons more in the next hour. How many dozen cartons of oat bran were then left to be sold?

14. If a peach costs half as much as an apple and an apple costs twice as much as a plum, then a peach and a plum together cost as much as
    (1) $\frac{1}{2}$ apple   (2) 1 apple   (3) $1\frac{1}{2}$ apples   (4) 2 apples

In 15–17, a muffin costs more than 50 cents, and orange juice costs less than a dollar (but is not free).

15. What is the least amount that a muffin and orange juice could cost?

16. If a muffin and orange juice together cost $1.00, what is the most that orange juice could cost?

17. What is the average of the minimum price of a muffin and the maximum price of orange juice?

18. If a number is doubled, what is the percent of increase?

Calories Consumed per Hour of Activity

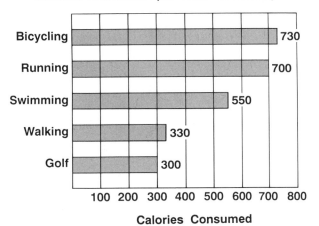

Calories Consumed

19. The graph shows the number of calories consumed in various activities by an average 150-pound person. A heavier person burns more calories, and a lighter person burns fewer calories. If there is a 10% difference in calorie consumption for each 15-pound difference in weight, how many calories per hour will a:
a. 180-pound person consume by running?
b. 135-pound person consume by swimming?
c. 165-pound person consume by walking?

20. Find all the numbers from 1 to 60 that are divisible by both 3 and 8.

21. If $\frac{2}{3}$ of a number is 12, what is $\frac{1}{2}$ of the number?   **22.** What is 20% of 20% of 20?

23. In a group of 140 students, 60 are enrolled in a math class and 128 are enrolled in a science class. If every student is enrolled in at least one of the classes, how many students are enrolled in both?

24. If A, B, and C represent different digits in the addition problem shown, what fraction is equivalent to 0.ABC?

$$\begin{array}{r} 4\,A\,B\,C \\ +\ 5\,A\,B\,C \\ \hline 1\,0\,5\,0\,0 \end{array}$$

2-1 ALGEBRAIC EXPRESSIONS

TRY IT! *Problems 1–5 on page 46*

1. $(5 - 2) \times (3 + 4)$
$= \quad 3 \quad \times \quad 7$
$= \qquad 21$

2. $12 \div 3 \times (2 + 2) - 1$
$= 12 \div 3 \times \quad 4 \quad - 1$
$= \quad 4 \quad \times \quad 4 \quad - 1$
$= \qquad 16 \qquad - 1$
$= \qquad\qquad 15$

3. $5 + 2 \times [6 + (3 - 1) \times 4]$
$= 5 + 2 \times [6 + \quad 2 \quad \times 4]$
$= 5 + 2 \times [6 + \qquad 8 \quad]$
$= 5 + 2 \times \quad 14$
$= 5 + \quad 28$
$= \quad 33$

4. $2ab + 3b$
$= 2 \times 5 \times 3 + 3 \times 3$
$= 30 + 9$
$= 39$

5. $\dfrac{a - c + d}{b - c}$

$= \dfrac{9 - 3 + 4}{5 - 3}$

$= \dfrac{10}{2} = 5$

2-2 TERMS, FACTORS, AND EXPONENTS

TRY IT! *Problems 1–3 on page 49*

1. $53g^7$

2. a. $5^3 - 4^2$
$= 125 - 16$
$= 109$

b. $(7 - 1)^2 \div 3^2$
$= 6^2 \div 3^2$
$= 36 \div 9$
$= 4$

3. a. $(a + b)^2$
$= (4 + 3)^2$
$= (7)^2$
$= 49$

b. $a^2 + b^2$
$= (4)^2 + (3)^2$
$= 16 + 9$
$= 25$

2-4 THE NUMBER LINE

TRY IT! *Problems 1–3 on page 61*

1. a. 1 **b.** U **c.** M, E, S
d. U, B, R **e.** $\frac{2}{3}, \frac{3}{3}, \frac{4}{4}$

2. $1.1, 1\frac{1}{2}, \frac{17}{10}, 1.65$
(Infinitely many answers are possible.)

3. $\frac{4}{15}$ is between $\frac{1}{5} = \frac{3}{15}$ and $\frac{1}{3} = \frac{5}{15}$.

TRY IT! *Problems 1 and 2 on page 63*

1. a. 3.2 **b.** -7 **c.** -16

2. a. false **b.** true

2-5 USING AN EQUATION IN AN ALGEBRAIC MODEL

TRY IT! *Problems 1 and 2 on page 67*

1. a. $10 - 8 \times 4$

b. $\dfrac{s - 4}{7}$

2. There are 10 cents in 1 dime;
(10×5) cents in 5 dimes.
Then there are $(10 \times d)$ cents,
or $10d$ cents, in d dimes.

ANSWER: $10d$ cents

TRY IT! *Problems 1 and 2 on page 69*

1. a. $7p - 8 = 6$ **b.** $\dfrac{p}{9} + 5 = 3$

   c. $\dfrac{p - 4}{9} = 3$

2.   Let n = the length of the rectangle.
         Then $n - 3$ = the width.

$$\underbrace{\text{The perimeter (distance around) is 26.}}$$
$$\underbrace{n + n - 3 + n + n - 3}\ \ \underbrace{= 26}$$

ANSWER: $n + n - 3 + n + n - 3 = 26$

2-6   WORKING WITH AN INEQUALITY

TRY IT! *Problems 1 and 2 on page 76*

1.   Let x = number of National League.
      Then x = number of American League.

$$\underbrace{\text{The total number}}\ \underbrace{\text{is more than}}\ \underbrace{250}.$$
$$\qquad x + x \qquad\qquad > \qquad\quad 250$$

ANSWER: $x + x > 250$

2.   Let x = number of workers on elevator.
      Then x = number of bulls on elevator.
   And $160x$ = weight of workers on elevator.
   And $750x$ = weight of bulls on elevator.

$$\underbrace{\text{The total weight}}\ \underbrace{\text{is at most}}\ \underbrace{\text{20,000 pounds.}}$$
$$160x + 750x \qquad \leq \qquad 20,000$$

ANSWER: $160x + 750x \leq 20,000$

TRY IT! *Problems 1–3 on page 79*

1. $\{2, 3, 4\}$

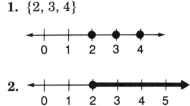

2.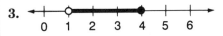

The darkened circle at 2 shows that 2 is
included. The darkened line and arrow to
the right represent the inifinite set of numbers greater than 2.

3.

The open circle at 1 shows that 1 is not
included. The darkened circle at 4 shows
that 4 is included. The darkened line represents the infinite set of real numbers
between 1 and 4.

CHAPTER 3

REAL NUMBERS: PROPERTIES AND OPERATIONS

Technological aids for performing operations with real numbers are becoming more and more sophisticated. But some people can still compute more quickly on an abacus than you can on a calculator.

3-1 Closure and Substitution 95
3-2 Addition of Real Numbers 98
3-3 Multiplication of Real Numbers 105
3-4 Subtraction and Division 110
3-5 The Distributive Property;
    Summary of Properties 116
3-6 Using the Properties of the
    Real Number System 119
3-7 Defining an Operation 124
3-8 Another Arithmetic: Matrices 128

Chapter Summary 136
Chapter Review Exercises 137
Problems for Pleasure 139
Calculator Challenge 140
College Test Preparation 140
Spiral Review Exercises 142
Solutions to TRY IT! Problems 144

SELF-TEST   Chapters 1–3 146

Your experiences in arithmetic have led you to notice certain patterns of behavior among numbers. For example, the order in which you add two numbers is not important but the order in which you subtract is important. As you continue to work with operations on sets of numbers, you will come to see the logical structure of the real number system.

This chapter concentrates on the behavior patterns, called *properties*, of different sets of numbers under the four operations of arithmetic, and makes some generalizations. These general statements are important in the structure of algebra.

3-1 CLOSURE AND SUBSTITUTION

The Closure Property

When you add any two natural numbers, the result is always another natural number. Further, that result is also *unique*; there is one—and only one—natural number that is the sum of two natural numbers; for example, $6 + 2 = 8$, and only 8. When a set of numbers behaves in this way under an operation, it has a *closure property*. The set of natural numbers is *closed* under the operations of addition and multiplication.

In general, for all natural numbers a and b:

There is a unique natural number c such that $a + b = c$.

There is a unique natural number d such that $a \times b = d$.

A set can be closed under some operations and not others. The set of natural numbers is not closed under subtraction or division; neither $2 - 8$ nor $2 \div 8$ represents a natural number.

Think of examples to show that the set of odd natural numbers is closed under the operations of multiplication and squaring, but not under the operations of addition, subtraction, or division.

The Substitution Property

A numerical value can be written in more than one way, and is not limited to particular numerals. For example, $2(3 + 5) = 16$ and $2(8) = 16$.

This example illustrates the *substitution property:*

For all numbers a and b, if $a = b$, then b may be substituted for a, and a may be substituted for b, in any expression.

MODEL PROBLEMS

1. Tell if {natural numbers that are multiples of 4} is closed under:
   a. addition **b.** subtraction **c.** multiplication **d.** division
   If it is not, give a counterexample.

   Solution: {natural numbers that are multiples of 4} is the same as {4, 8, 12, . . . }.

Operation	Closed?	Reason
a. addition	yes	When the operation is performed on any two elements of the set, the result is a unique number that is also in the set.
b. subtraction	no	The result is not always in the set. (Counterexample: $4 - 8$)
c. multiplication	yes	Same as part **a.**
d. division	no	The result is not always in the set. (Counterexample: $12 \div 12$)

   Notice, from the counterexample $12 \div 12$ in part **d,** that it is important to examine different possibilities when trying to determine if a given set has a particular property. The property must be true for *all* elements a and b of the set, even for $a = b$.

2. Tell if {0, 1} is closed under: **a.** addition **b.** multiplication

   Answer:
   a. Not closed under addition since $(1 + 1)$ does not name an element of the set.
   b. Closed under multiplication since the product of any two elements is in the set.

3. Write 3 equivalent expressions that may be substituted for the given expression.

Expression	Equivalent Expressions		
a. 10	$12 - 2$	$30 \div 3$	5×2
b. $5z$	$(6 - 1)z$	$(10 \div 2)z$	$(5 \times 1)z$

From ancient times, people have devised manual aids for performing operations on numbers. Arabs, Greeks, and Chinese may have independently invented forms of the abacus.

In the western hemisphere, early Peruvians used an arrangement of cords called a *quipu,* for historical records as well as for calculating. Knots in the cords represented units, tens, and hundreds, and different colors represented different activities. In remote regions of the Andes, quipus are still used for counting herds.

In 1–8, state whether the numerical phrase names a natural number.

1. $8 + 4$ **2.** $4 + 8$ **3.** $8 - 4$ **4.** $4 - 8$

5. 8×4 **6.** 4×8 **7.** $8 \div 4$ **8.** $4 \div 8$

9. If x and y are natural numbers, does the phrase always represent a natural number?
   a. $x + y$ **b.** $x - y$ **c.** xy **d.** $x \div y$

In 10–19, state whether the set is closed under the indicated operation.

10. $\{2, 4, 6, 8, \ldots\}$; addition   **11.** $\{1, 3, 5, \ldots\}$; multiplication

12. $\{2, 4, 8\}$; division   **13.** $\{0, 1, 2\}$; multiplication

14. $\{0, 1\}$; multiplication   **15.** $\{0, 2, 4\}$; subtraction   **16.** $\{1\}$; multiplication

17. $\{4\}$; subtraction   **18.** $\{0\}$; addition   **19.** $\{1\}$; division

In 20–28, tell if the set is closed under:
   a. addition   **b.** subtraction   **c.** multiplication   **d.** division   Explain your answers.

20. {even numbers}   **21.** {odd numbers}   **22.** {numbers that are multiples of 3}

23. $\{1, 3, 5\}$ **24.** $\{2, 4, 6\}$ **25.** $\{0, 1\}$ **26.** $\{0\}$ **27.** $\{1\}$ **28.** $\{10\}$

In 29–33, write three equivalents that may be substituted for the given expression.

29. 14 **30.** $\frac{1}{2}(20 - 4)$ **31.** $18x$ **32.** $\frac{1}{3}y$ **33.** $(12 - 3)t$

In 34–40, use the given variable(s) to substitute an algebraic expression for the situation described.

34. Norman drove at a rate of 55 miles per hour for h hours. Represent the distance he drove. (Remember: distance = rate $\times$ time)

35. Dr. Blank flew a distance of 40 miles in t hours on her first solo flight. Represent the rate at which she flew.

36. Imelda is now 60 years old. Represent her age x years from now.

37. Represent the successor of the natural number n.

38. At a 20%-off sale, Connie bought a sweater whose original price was d dollars. Represent the sale price.

39. Luis can clean his entire apartment in c hours. Represent the part of the apartment he can clean in 1 hour.

40. The measure of the width of a rectangle is w inches and the measure of the length is ℓ inches. Represent the perimeter of the rectangle.

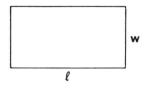

3-2 ADDITION OF REAL NUMBERS

Adding on a Number Line

Adding numbers can be thought of as a series of directed moves, or ***translations,*** on a number line. A translation to the right will represent addition of a positive number, and a translation to the left will represent addition of a negative number.

PROCEDURE

To add two numbers on a number line:

1. Start at 0, and graph the first number.
   a. If the number is positive, move to the right the indicated number of units.
   b. If the number is negative, move to the left the indicated number of units.
   c. If the number is 0, do not move.

2. Continue from the graph of the first number, and graph the second number. Follow the procedure in a–c above.

3. The coordinate of the point reached is the sum of the two numbers.

MODEL PROBLEMS

Using a number line, add the given pairs of numbers. Explain the graphs.

a. Add 3 and 2.

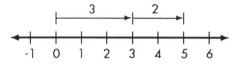

Explanation: Starting from 0, the graph shows a translation of 3 units to the right followed by a translation of 2 units to the right, arriving at 5.

ANSWER: $3 + 2 = 5$

b. Add -3 and -2.

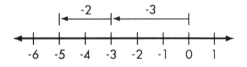

Explanation: Starting from 0, the graph shows a translation of 3 units to the left followed by a translation of 2 units to the left, arriving at -5.

ANSWER: $(-3) + (-2) = -5$

c. Add 3 and −2.

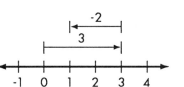

Explanation: Starting from 0, the graph shows a translation of 3 units to the right followed by a translation of 2 units to the left, arriving at 1.

ANSWER: $3 + (-2) = 1$

d. Add −3 and 2.

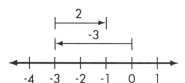

Explanation: Starting from 0, the graph shows a translation of 3 units to the left followed by a translation of 2 units to the right, arriving at −1.

ANSWER: $(-3) + 2 = -1$

Rules of Addition

The results of number-line addition suggest general rules that hold true for addition of any two positive or negative real numbers a and b.

Case	Sum in Symbols	Example
a and b are both positive.	$a + b = \lvert a \rvert + \lvert b \rvert$	$5 + 7 = 12$
The sum of two positive numbers is a positive number.		
a and b are both negative.	$a + b = -(\lvert a \rvert + \lvert b \rvert)$	$(-5) + (-7) = -12$
To find the sum of two negative numbers: *1. Add the absolute values of the given numbers.* *2. Place a negative sign before the result.*		
a and b have different signs, say a is positive and b is negative.	**If** $\lvert a \rvert > \lvert b \rvert$, $a + b = \lvert a \rvert - \lvert b \rvert$ **If** $\lvert b \rvert > \lvert a \rvert$, $a + b = -(\lvert b \rvert - \lvert a \rvert)$	$7 + (-5) = 2$ $5 + (-7) = -2$
To find the sum of a positive number and a negative number: *1. Subtract the smaller absolute value from the larger.* *2. Use the same sign as that of the number with the larger absolute value.*		

In combination with the rules, you can use your experience with a number line to form a mental image of the result of adding two numbers. In particular, when adding a positive number and a negative number, the number represented by the longer arrow will determine whether the result is on the positive or the negative side of 0.

The flowchart summarizes the decisions to make when adding real numbers.

ADDING TWO REAL NUMBERS

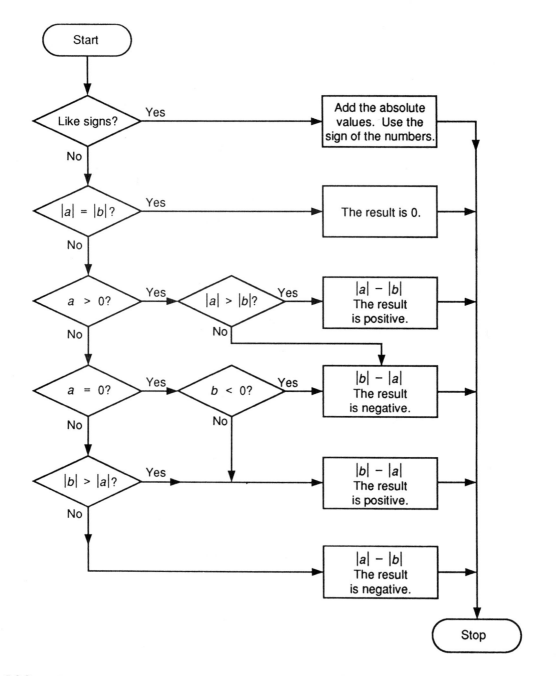

Properties of Addition

The table summarizes addition properties of the real numbers.

Name of Property	Representation for All Real Numbers a, b, c	Example
Closure Property	$a + b$ **is a unique real number.**	$-4 + 6 = 2$
The sum of two real numbers is a unique real number.		
Commutative Property	$a + b = b + a$	$3 + (-1) = -1 + 3$
Real numbers may be added in any order.		
Associative Property	$a + (b + c) = (a + b) + c$	$-2 + (3 + 6) = (-2 + 3) + 6$
Real numbers may be regrouped for addition.		
Addition Property of 0	$a + 0 = a$	$7 + 0 = 7$
*The number 0 is the **additive identity** because the sum of 0 and any real number is the number itself.*		
Addition Property of Opposites	$a + (-a) = 0$	$5 + (-5) = 0$
*The sum of a real number and its **additive inverse**, or **opposite**, is 0.*		
Property of the Opposite of a Sum	$-(a + b) = (-a) + (-b)$	$-(1 + 4) = (-1) + (-4)$
The opposite of a sum of two real numbers is equal to the sum of the opposites.		

Applying the Properties of Addition

In problems containing more than one addition, you have learned to follow the order of operations and do the additions in order from left to right. For example:

$$6 + (-2) + 7 + (-4) = 4 + 7 + (-4)$$
$$= \quad 11 \quad + (-4)$$
$$= \qquad 7$$

The commutative and associative properties of addition allow you to reorder and regroup. You can first add all the positive numbers, next add all the negative numbers, and then add the two results.

$$6 + (-2) + 7 + (-4) = 6 + 7 + (-2) + (-4)$$
$$= \quad 13 \quad + \quad (-6)$$
$$= \qquad 7$$

When you reorder and regroup the terms of an expression, you should know the property that justifies each step.

MODEL PROBLEMS

1. The chain of equations shows that $[(-3) + x] + 3 = x$. Give a reason for each step.

Solution:

Step		*Reason*
$[(-3) + x] + 3$		
(1) $= [x + (-3)] + 3$		Commutative property of addition.
(2) $= x + [(-3) + 3]$		Associative property of addition.
(3) $= x + 0$		Addition property of opposites.
(4) $= x$		Addition property of 0.

2. The sum $82 + 68$ can be performed mentally as: $(80 + 2) + (70 + (-2)) = 150$
Write the steps, with their reasons, that justify this calculation.

Solution:

Step		*Reason*
$82 + 68$		
(1) $= (80 + 2) + (70 + (-2))$		Substitution property.
(2) $= (80 + 2 + 70) + (-2)$		Associative property of addition.
(3) $= (80 + 70 + 2) + (-2)$		Commutative property of addition.
(4) $= (80 + 70) + (2 + (-2))$		Associative property of addition.
(5) $= (80 + 70) + 0$		Addition property of opposites.
(6) $= 80 + (70 + 0)$		Associative property of addition.
(7) $= 80 + 70$		Addition property of 0.
(8) $= 150$		Substitution property.

Now it's your turn to . . . **TRY IT!**   (*See page 144 for solutions.*)

1. Add on a number line, and explain your graph: $5 + (-4)$

2. Add: $-7 + 5 + (-4) + 1$

3. Give the reason for each step of the following proof.

(1) $[a + (-5)] + (-a) = (-a) + [a + (-5)]$
(2) $= [(-a) + a] + (-5)$
(3) $= 0 + (-5)$
(4) $= -5$

In 1–11, use a number line to find the sum of the numbers. In 1–4, explain the graph.

1. $3 + 4$ **2.** $-2 + (-4)$ **3.** $7 + (-4)$ **4.** $-6 + 5$

5. $4 + (-4)$ **6.** $-7 + 7$ **7.** $0 + (-6)$ **8.** $(3 + 4) + 2$

9. $[8 + (-4)] + (-6)$ **10.** $[-7 + (-3)] + 6$ **11.** $(-5 + 2) + 3$

In 12–51, add.

12. $\begin{array}{r} 15 \\ 9 \\ \hline \end{array}$ **13.** $\begin{array}{r} -17 \\ -\ 8 \\ \hline \end{array}$ **14.** $\begin{array}{r} -15 \\ -15 \\ \hline \end{array}$ **15.** $\begin{array}{r} 6\frac{2}{3} \\ 1\frac{1}{3} \\ \hline \end{array}$ **16.** $\begin{array}{r} -5\frac{1}{2} \\ -3\frac{1}{2} \\ \hline \end{array}$ **17.** $\begin{array}{r} 9\frac{1}{2} \\ 8\frac{3}{4} \\ \hline \end{array}$

18. $\begin{array}{r} -6\frac{5}{6} \\ -1\frac{2}{3} \\ \hline \end{array}$ **19.** $\begin{array}{r} -5.6 \\ -2.2 \\ \hline \end{array}$ **20.** $\begin{array}{r} 6.8 \\ 3.2 \\ \hline \end{array}$ **21.** $\begin{array}{r} 70 \\ -20 \\ \hline \end{array}$ **22.** $\begin{array}{r} -55 \\ 20 \\ \hline \end{array}$ **23.** $\begin{array}{r} -15 \\ 42 \\ \hline \end{array}$

24. $\begin{array}{r} 10 \\ -10 \\ \hline \end{array}$ **25.** $\begin{array}{r} 9\frac{1}{2} \\ -3 \\ \hline \end{array}$ **26.** $\begin{array}{r} 7 \\ -8\frac{3}{4} \\ \hline \end{array}$ **27.** $\begin{array}{r} 7.9 \\ -5.6 \\ \hline \end{array}$ **28.** $\begin{array}{r} -8.7 \\ 3.7 \\ \hline \end{array}$ **29.** $\begin{array}{r} 7.1 \\ -9.4 \\ \hline \end{array}$

30. $13 + (-32)$ **31.** $-41 + (-9)$ **32.** $8 + (-14)$ **33.** $(-12) + 37$

34. $40 + (-17)$ **35.** $0 + (-28)$ **36.** $15 + (-15)$ **37.** $-19 + 7$

38. $|-34| + |20|$ **39.** $-|7| + (-10)$ **40.** $|15| + (-|-15|)$ **41.** $|4 - 7| + |7 - 4|$

42. $18 + (-15) + 9$ **43.** $-19 + 8 + (-15)$ **44.** $12 + (-18) + (-4) + 7$

45. $-19 + 8 + (-5) + 16$ **46.** $-1.5 + 3.1 + 6.8 + (-3.4)$

47. $9.6 + (-7.7) + (-5.6) + 2.2$ **48.** $5\frac{1}{4} + (-8) + 6\frac{3}{4} + \left(-1\frac{1}{2}\right)$

49. $-4\frac{1}{3} + 7 + 8\frac{1}{3} + (-11)$ **50.** $|7| + |-8| + |0|$ **51.** $|-13| + |7| + (-\,|-20|)$

In 52–60, write the number that represents the sum of the quantities.

52. a rise of 4 meters and a rise of 6 meters

53. a distance of 50 miles north and a distance of 30 miles north

54. a loss of 6 yards and a loss of 2 yards

55. a rise of 7 feet and a fall of 5 feet

56. a loss of $10 and a profit of $8

57. a deposit of $8 and a withdrawal of $8

58. a loss of $20 and a profit of $20

59. a gain of 8 kg, a gain of 4 kg, and a loss of 7 kg

60. a rise of 4°, a drop of 3°, and a drop of 5°

A compass is an instrument used to indicate direction. A simple compass has a magnetic needle that always points to the north pole.

In 61–63, use positive and negative numbers to solve the problem.

61. In the first quarter, the Jets gained 10 yards, then gained 6 yards, then lost 4 yards, and then gained 3 yards. What was their total gain or loss for the quarter?

62. The temperature dropped 18° between 9:00 P.M. and 5:00 A.M. From 5:00 A.M. to 11 A.M., the temperature rose 12°. Find the temperature change from 9:00 P.M. to 11:00 A.M.

63. From my home in Delmar City, I can reach shore points by traveling either 15 miles east or 13 miles west. Find the distance between the two shore points.

In 64–69, state the additive inverse of the given expression.

64. 10   **65.** -8 **66.** -2.5 **67.** C **68.** $-d$ **69.** $|5|$

In 70–73, replace the question mark to make the resulting sentence true.

70. $4 + \,? = 0$ **71.** $-y + \,? = 0$ **72.** $-6 + \,? = 2$ **73.** $|-7| + \,? = 0$

In 74–77, replace the variable to make the resulting sentence true.

74. $x + (-12) = 0$ **75.** $5 + c = 1$ **76.** $x + 4 = -2$ **77.** $d + (-5) = -3$

In 78–82, name the addition property that makes each sentence true.

78. $-3 + 8 = 8 + (-3)$ **79.** $50 + (-50) = 0$ **80.** $-8 + 0 = -8$

81. $-[8 + 9] = -8 + (-9)$ **82.** $-(|5| + |-3|) = (-5) + (-3)$

In 83 and 84, state the reason for each step.

83. (1) $9 + (-5) = (4 + 5) + (-5)$
    (2) $= 4 + [5 + (-5)]$
    (3) $= 4 + 0$
    (4) $= 4$

84. (1) $[(-2) + a] + 2 = [a + (-2)] + 2$
    (2) $= a + [(-2) + 2]$
    (3) $= a + 0$
    (4) $= a$

In 85 and 86, use a chain of equations such as in Exercise 83 to show that the statement is true. Give the reason for each of your steps.

85. $15 + (-8) = 7$ **86.** $12 + (-3) = 9$

In 87–89, state whether the sentence is true or false.

87. $|x| + |-x| = 0 \ (x \neq 0)$ **88.** $(-c) + (-d) = -(c + d)$ **89.** $-(-b) = b$

In 90–94, is the statement true or false? If it is false, give a counterexample.

90. The sum of a real number and its additive inverse is equal to the additive identity.

91. The opposite of the difference of two real numbers is equal to the difference of the opposites.

92. If an operation is commutative on the members of a set, then the set is closed under that operation.

93. If a set is closed under multiplication, then it must be closed under addition.

94. If a set contains the additive identity 0, then every element has an additive inverse.

3-3 MULTIPLICATION OF REAL NUMBERS

Modeling the Rules for Multiplication

To illustrate multiplication of real numbers, consider the following model:

Let a weight gain be represented by a positive number and a weight loss by a negative number, a number of weeks in the future by a positive number and a number of weeks in the past by a negative number.

There are 4 possible cases to consider:

1. *Multiplying a Positive Number by a Positive Number*

   If Jo gains 1 pound each week, in 4 weeks she will weigh 4 pounds more, represented by: $1 \times 4 = 4$

   Observation: The product of two positive numbers is a positive number.

2. *Multiplying a Negative Number by a Positive Number*

   If Jo loses 1 pound each week, in 4 weeks she will weigh 4 pounds less, represented by: $-1 \times 4 = -4$

   Observation: The product of a negative and a positive number is negative.

3. *Multiplying a Positive Number by a Negative Number*

   If Jo has gained 1 pound each week, 4 weeks ago she weighed 4 pounds less, represented by: $1 \times (-4) = -4$

   Observation: The product of a positive and a negative number is negative.

4. *Multiplying a Negative Number by a Negative Number*

   If Jo has lost 1 pound each week, 4 weeks ago she weighed 4 pounds more, represented by: $(-1) \times (-4) = 4$

   Observation: The product of two negative numbers is positive.

These examples illustrate the reasonableness of the following rules for multiplication:

1. **The product of two positive numbers, or of two negative numbers, is positive.**

2. **The product of a positive number and a negative number is negative.**

Properties of Multiplication

Many properties that you have seen for addition also apply to multiplication.

Name of Property	Representation for All Real Numbers a, b, c	Example
Closure Property	ab **is a unique real number.**	$-3 \cdot 2 = -6$
The product of two real numbers is a unique real number.		
Commutative Property	$ab = ba$	$5 \cdot (-2) = -2 \cdot 5$
Real numbers may be multiplied in any order.		
Associative Property	$a(bc) = (ab)c$	$2 \cdot (3 \cdot 4) = (2 \cdot 3) \cdot 4$
Real numbers may be regrouped for multiplication.		
Multiplication Property of 1	$a \cdot 1 = a$	$3 \cdot 1 = 3$
*The number 1 is the **multiplicative identity** because the product of 1 and any real number is the number itself.*		
Multiplication Property of Reciprocals	$a \cdot \dfrac{1}{a} = 1$	$3 \cdot \dfrac{1}{3} = 1$
*The product of a real number and its **multiplicative inverse**, or **reciprocal**, is 1.* *		
Multiplication Property of 0	$a \cdot 0 = 0$	$-4 \cdot 0 = 0$
The product of 0 and any real number is 0.		

*Since $\dfrac{1}{0}$ is undefined, 0 has no reciprocal.

Applying the Properties of Multiplication

Since the commutative and associative laws of multiplication hold for real numbers, numbers may be arranged and multiplied in any order you choose. If you are multiplying more than two numbers, first multiply any two of them, then multiply this product by one of the remaining factors. Continue this procedure until all factors have been used.

Example: You may want to find the product of -2, 3, and -4 in different ways.

By multiplying in order: $(-2)(3)(-4) = (-6)(-4) = 24$

By regrouping the negatives: $(-2)(3)(-4) = (-2)(-4)(3) = 8(3) = 24$

Let us examine the following pattern of products:

$$(3)(-1)(3) = -9 \quad \text{(1 negative factor)}$$
$$(-3)(-1)(3) = 9 \quad \text{(2 negative factors)}$$
$$(-3)(-1)(-3) = -9 \quad \text{(3 negative factors)}$$
$$(-1)(-3)(-1)(-3) = 9 \quad \text{(4 negative factors)}$$

This pattern establishes a relationship between the number of negative factors and the nature of the product:

1. **When a product contains an *odd* number of negative factors, the product is *negative*.**

2. **When a product contains an *even* number of negative factors, the product is *positive*.**

MODEL PROBLEMS

1. Find the product of the numbers.

   a. $\begin{array}{r} -13 \\ -5 \\ \hline 65 \end{array}$ **b.** $\begin{array}{r} -15 \\ 6 \\ \hline -90 \end{array}$ **c.** $\begin{array}{r} 3.4 \\ -3 \\ \hline -10.2 \end{array}$ **d.** $\begin{array}{r} 7\frac{1}{8} \\ 3 \\ \hline 21\frac{3}{8} \end{array}$

2. Evaluate the expression, and explain the sign of the product.

Expression	*Evaluation*	*Explanation*
a. $(-2)^3$	$(-2)(-2)(-2) = -8$	An odd number of negative factors results in a negative product.
b. $(-3)^4$	$(-3)(-3)(-3)(-3) = 81$	An even number of negative factors results in a positive product.

3. The chain of equations shows that $[(-3) \cdot x] \cdot \left(-\frac{1}{3}\right) = x$.

   Give a reason for each step.

Step		*Reason*
	$[(-3) \cdot x] \cdot \left(-\frac{1}{3}\right)$	
(1)	$= [x \cdot (-3)] \cdot \left(-\frac{1}{3}\right)$	Commutative property of multiplication.
(2)	$= x \cdot \left[(-3) \cdot \left(-\frac{1}{3}\right)\right]$	Associative property of multiplication.
(3)	$= x \cdot 1$	Multiplication property of reciprocals.
(4)	$= x$	Multiplication property of 1.

In 1–4, multiply.

1. $(-9)(4)$ **2.** $-6 \cdot \frac{2}{3}$ **3.** $(-4)(-1)(-3)$ **4.** $(-2)^4$

5. Give the reason for each step of the following proof.

(1) $[a \cdot (-5)] \cdot \dfrac{1}{a} = [(-5) \cdot a] \cdot \dfrac{1}{a}$

(2) $\qquad\qquad = (-5) \cdot \left[a \cdot \dfrac{1}{a}\right]$

(3) $\qquad\qquad = (-5) \cdot 1$

(4) $\qquad\qquad = -5$

EXERCISES

In 1–12, find the product of the numbers.

1. $\begin{array}{r} 23 \\ 5 \\ \hline \end{array}$ **2.** $\begin{array}{r} 36 \\ -2 \\ \hline \end{array}$ **3.** $\begin{array}{r} -15 \\ -8 \\ \hline \end{array}$ **4.** $\begin{array}{r} -24 \\ 8 \\ \hline \end{array}$ **5.** $\begin{array}{r} 0 \\ -5 \\ \hline \end{array}$ **6.** $\begin{array}{r} -75 \\ -3 \\ \hline \end{array}$

7. $\begin{array}{r} 1.5 \\ -2.4 \\ \hline \end{array}$ **8.** $\begin{array}{r} -.25 \\ 80 \\ \hline \end{array}$ **9.** $\begin{array}{r} 8 \\ \frac{1}{2} \\ \hline \end{array}$ **10.** $\begin{array}{r} -15 \\ \frac{3}{5} \\ \hline \end{array}$ **11.** $\begin{array}{r} -\frac{1}{2} \\ -\frac{1}{3} \\ \hline \end{array}$ **12.** $\begin{array}{r} 16 \\ -2\frac{1}{4} \\ \hline \end{array}$

In 13–30, multiply.

13. (8) by (6)

14. (-12) by (-5)

15. (11) by (-7)

16. (-10) by (9)

17. (0) by (-3)

18. (15) by $(-.6)$

19. $(8)\left(\frac{1}{4}\right)$

20. $\left(-\frac{3}{5}\right)(-20)$

21. $(2)\left(\frac{1}{2}\right)$

22. $\left(\frac{3}{8}\right)\left(-\frac{32}{27}\right)$

23. $\left(-4\frac{1}{2}\right)\left(\frac{2}{3}\right)$

24. $|-15| \cdot \left(-3\frac{1}{5}\right)$

25. (4)(3)(2)

26. $(-1)(-7)(-8)$

27. $(-3)(-5)(4)(-1)$

28. $(-7)(2)(0)$

29. $|10| \cdot |-3| \cdot (-4)$

30. $(8)(-9)(0)(-10)$

In 31–36, simplify.

31. $5 + (-2)(-3)$

32. $2(-4) + 3(-2)$

33. $|-10| + 6(-1)$

34. $(-3)(8) + (-3)(-2)$

35. $-6 + (-8)\left(\frac{3}{4}\right) + 20$

36. $(-7)\left(-\frac{1}{7}\right) + [4 + (-4)]$

37. Replace the question mark to make the resulting sentence true.

    a. $5 \times ? = 1$ **b.** $\frac{3}{4} \times ? = 1$ **c.** $\frac{5}{2} \times ? = 1$ **d.** $-7 \times ? = 1$

38. Replace the variable to make the resulting sentence true.

    a. $3x = 1$ **b.** $-2y = 1$ **c.** $\frac{4}{5}w = 1$ **d.** $-\frac{1}{8}z = 1$

In 39–50, find the value of the expression.

39. $(4)^2$ **40.** $(-3)^2$ **41.** $(5)^3$ **42.** $(-4)^3$ **43.** $(-5)^3$ **44.** $(-1)^4$

45. $\left(\frac{1}{2}\right)^2$ **46.** $\left(-\frac{1}{2}\right)^2$ **47.** $\left(\frac{2}{3}\right)^3$ **48.** $\left(-\frac{3}{5}\right)^3$ **49.** $\left(-\frac{1}{4}\right)^3$ **50.** $\left(-\frac{1}{5}\right)^4$

In 51–54, name the multiplication property illustrated.

51. $(-6) \times (-5) = (-5) \times (-6)$ **52.** $[(-3) \cdot 4] \cdot 7 = (-3) \cdot [4 \cdot 7]$

53. $(5)(1) = 5$ **54.** $0 \cdot (-31) = 0$

In 55 and 56, write 4 cases to illustrate the given model for multiplication. Refer to the 4 cases described on page 105. Describe the cases both verbally and numerically.

55. Let money saved be represented by a positive number, and money spent be represented by a negative number. Let future time, in months, be represented by a positive number, and past time, in months, be represented by a negative number.

56. Let a rise in temperature be represented by a positive number, and a fall in temperature by a negative number. Let a number of hours in the future be represented by a positive number, and a number of hours in the past be represented by a negative number.

Over the years, the technique of using a model has helped people deal with problems. Sometimes, models must be adjusted as new information becomes available. For example, centuries ago the earth was thought to be flat, but today a sphere is recognized to be an approximate model of our planet.

3-4 SUBTRACTION AND DIVISION

The Meaning of Subtraction

To subtract 3 from 7, you find a number that, when added to 3, will give 7. You know that $7 - 3 = 4$ because $3 + 4 = 7$. Because of this relation to addition, subtraction is called the *inverse operation* of addition.

To subtract -2 from 3 means to find a number that, when added to -2, will give 3. The number-line graph shows the answer to $-2 + ? = 3$. From 2 units to the left of 0, you must move 5 units to the right to arrive at 3. The addition statement $-2 + 5 = 3$ is equivalent to the subtraction statement $3 - (-2) = 5$.

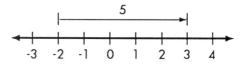

Adding the Opposite

Now compare the addition and subtraction examples at the right. Recall the names you learned for the parts of a subtraction problem. Observe that adding the opposite of a number gives the same result as subtracting that number. Therefore, subtraction is defined as adding the opposite.

Add	*Subtract*	
9	9	minuend
-6	6	subtrahend
3	3	difference

For all real numbers a and b: $a - b = a + (-b)$

The set of real numbers is closed under subtraction.

MODEL PROBLEMS

In 1 and 2, perform the indicated subtraction by adding an opposite.

1. $30 - 45 = 30 + (-45) = -15$ **2.** $-19 - (-27) = -19 + 27 = 8$

3. A pilot flying at an altitude of 32,000 feet rose 4,000 feet to pass over a storm, and then reduced the plane's altitude by 6,000 feet after clearing the storm area. At what altitude did the plane continue its flight?

Solution: $32,000 + 4,000 - 6,000 = 30,000$

ANSWER: The plane continued its flight at 30,000 feet.

Uses of the Symbol $-$

In $7 - (-5)$, the symbol $-$ is used in two ways. The $-$ between 7 and (-5) indicates subtraction. The $-$ that is part of the numeral (-5) indicates the opposite of 5.

Since subtraction is equivalent to adding the opposite, you can shorten notation.

$7 + (-5)$ can be written $7 - 5$ and $7 - (-5)$ can be written $7 + 5$.

The Meaning of Division

Division is the inverse operation of multiplication. To divide 6 by 2 means to find a number that, when multiplied by 2, gives 6. Since $3 \cdot 2 = 6$, then $\frac{6}{2} = 3$. The number 6 is the **dividend,** 2 is the **divisor,** and 3 is the **quotient.** Remember that although 0 cannot be used as a divisor $\left(\frac{3}{0} \text{ is undefined}\right)$, 0 may be a dividend. The quotient of 0 and another number, such as $\frac{0}{3}$, is 0.

Since division can be expressed in terms of multiplication, similar rules apply:

1. **The quotient of two positive numbers, or of two negative numbers, is positive.**

2. **The quotient of a positive number and a negative number is negative.**

Multiplying by the Reciprocal

Compare the following division and multiplication examples:

Divide	*Multiply*
$\dfrac{12}{3} = 4$	$12 \cdot \dfrac{1}{3} = 4$

Observe that dividing by a number gives the same result as multiplying by the reciprocal of that number. Therefore, division is defined as multiplying by the reciprocal. For real numbers a and b:

$$\frac{a}{b} = a \cdot \frac{1}{b} \quad (b \neq 0)$$

Except for division by 0, the set of real numbers is closed under division.

NOTE. When a negative quotient is written as a fraction, the negative sign may appear in the numerator, the denominator, or before the fraction. For example, $\frac{-2}{3}$, $\frac{2}{-3}$, and $-\frac{2}{3}$ all represent the same value.

MODEL PROBLEMS

In 1–4, perform the division.

1. $\dfrac{90}{-10} = -9$ **2.** $\dfrac{-27}{-3} = 9$ **3.** $\dfrac{0}{-3} = 0$ **4.** $18 \div \left(-\frac{1}{2}\right) = 18\,(-2) = -36$

Now it's your turn to . . . **TRY IT!** *(See page 144 for solutions.)*

In 1–3, perform the operations.

1. $5 - 12 + 2$ **2.** $\dfrac{-28}{7}$ **3.** $12 \div \left(-\dfrac{1}{4}\right) - 8 \div (-2)$

CALCULATOR CONNECTION

You can use calculators to perform operations with negative numbers. Almost all calculators can subtract a larger number from a smaller one and display a negative answer. To evaluate **5(3 − 9)**, enter **3** $\boxed{-}$ **9** $\boxed{\times}$, obtaining $\boxed{\qquad\qquad -6.}$. Now press **5** $\boxed{=}$, displaying $\boxed{\qquad\quad -30.}$.

Entering a Negative Number

You can enter a negative number directly by using the subtraction key. For instance, pressing $\boxed{-}$ **3** $\boxed{=}$, displays $\boxed{\qquad\quad -3.}$. The limitation of this procedure is that the negative number must be the first number you enter. Thus, to multiply −**3** by **5**, press $\boxed{-}$ **3** $\boxed{\times}$ **5** $\boxed{=}$, displaying $\boxed{\qquad\quad -15.}$.

Try to multiply **5** by −**3** by pressing **5** $\boxed{\times}$ $\boxed{-}$ **3** $\boxed{=}$. Instead of −**15**, you will get the incorrect value **2**. The reason is that if several operation keys are pressed consecutively, your calculator will use only the last operation pressed before a number is entered. Thus, **5** $\boxed{\times}$ $\boxed{-}$ **3** $\boxed{=}$ is interpreted by your calculator as **5** $\boxed{-}$ **3** $\boxed{=}$.

The $\boxed{+/-}$ Key

The $\boxed{+/-}$ key enables calculators to perform operations with negative numbers. Enter −**3** by pressing **3** $\boxed{+/-}$, which will change the positive **3** to −**3**. Thus, to multiply **5** by −**3**, press **5** $\boxed{\times}$ **3** $\boxed{+/-}$ $\boxed{=}$, displaying $\boxed{\qquad\quad -15.}$.

Use the $\boxed{+/-}$ key to perform other calculations with negative numbers.

To calculate **5** − (−**6**), enter **5** $\boxed{-}$ **6** $\boxed{+/-}$ $\boxed{=}$, displaying $\boxed{\qquad\quad 11.}$.

To multiply −**3** by −**3**, enter **3** $\boxed{+/-}$ $\boxed{\times}$ **3** $\boxed{+/-}$ $\boxed{=}$, displaying $\boxed{\qquad\quad 9.}$.

To divide −**20** by −**4**, enter **20** $\boxed{+/-}$ $\boxed{\div}$ **4** $\boxed{+/-}$ $\boxed{=}$, displaying $\boxed{\qquad\quad 5.}$.

In 1–4, perform the subtraction by adding an opposite.

1. $5 - (-3)$ **2.** $-1 - 2$ **3.** $-3 - (-4)$ **4.** $-3 - 3$

In 5–16, subtract the lower number from the upper number.

5. $\begin{array}{r} 18 \\ 29 \\ \hline \end{array}$ **6.** $\begin{array}{r} 36 \\ -15 \\ \hline \end{array}$ **7.** $\begin{array}{r} -39 \\ 15 \\ \hline \end{array}$ **8.** $\begin{array}{r} -26 \\ -18 \\ \hline \end{array}$ **9.** $\begin{array}{r} -6 \\ 6 \\ \hline \end{array}$ **10.** $\begin{array}{r} -8 \\ -8 \\ \hline \end{array}$

11. $\begin{array}{r} 8.3 \\ -6.2 \\ \hline \end{array}$ **12.** $\begin{array}{r} -6.9 \\ 3.7 \\ \hline \end{array}$ **13.** $\begin{array}{r} -3.6 \\ -5.2 \\ \hline \end{array}$ **14.** $\begin{array}{r} -3\frac{1}{4} \\ -7\frac{3}{4} \\ \hline \end{array}$ **15.** $\begin{array}{r} 7\frac{3}{4} \\ -2\frac{1}{4} \\ \hline \end{array}$ **16.** $\begin{array}{r} -6\frac{5}{6} \\ 3\frac{1}{3} \\ \hline \end{array}$

In 17–24, perform the subtraction.

17. $19 - 30$ **18.** $-12 - (-25)$ **19.** $22 - (-8)$ **20.** $6.4 - 8.1$

21. $-3.7 - (-5.2)$ **22.** $-9.2 - 8.3$ **23.** $5\frac{1}{3} - 3\frac{1}{3}$ **24.** $6\frac{1}{2} - \left(-2\frac{1}{4}\right)$

25. How much greater than -15 is 12?   **26.** How much greater than -4 is -1?

27. How much less than 6 is -3?   **28.** What number is 6 less than -6?

29. Subtract 8 from the sum of -6 and -12.   **30.** Subtract -3 from the sum of 5 and -8.

31. Subtract the product of 7 and -4 from 32.   **32.** Subtract 6 from the product of 0 and 6.

In 33–42, find the value of the given expression.

33. $7 + 9 - (-4)$ **34.** $-12 - 9 + (-20)$

35. $8.9 - 5.2 + 6.7$ **36.** $-5.1 - (-8.4) - (-1.7)$

37. $6\frac{1}{4} + 9\frac{1}{2} - 7\frac{3}{4}$ **38.** $-8\frac{5}{6} - 2\frac{1}{3} - \left(-5\frac{2}{3}\right)$

39. $32 - 49 - 21 + 10$ **40.** $-15 + 8 - 5 + 12$

41. $9(-4) - 2$ **42.** $6 - 8 - 3(-10)$

43. Find the difference when the temperature changes from:
    a. $5°$ to $8°$ **b.** $-10°$ to $18°$ **c.** $-6°$ to $-18°$ **d.** $12°$ to $-4°$

44. Ramp Road rises at an angle of $10°$ from the horizontal. Ravine Road rises at an angle of $37°$. How many degrees steeper is Ravine Road than Ramp Road?

45. An elevator started on the ground floor and rose 30 floors. Then it came down 12 floors. At which floor was it at that time?

46. A plane started in Chicago and flew 600 miles south in one hour. During the next hour, it flew 575 miles farther south. At the end of two hours, where was the plane with reference to Chicago?

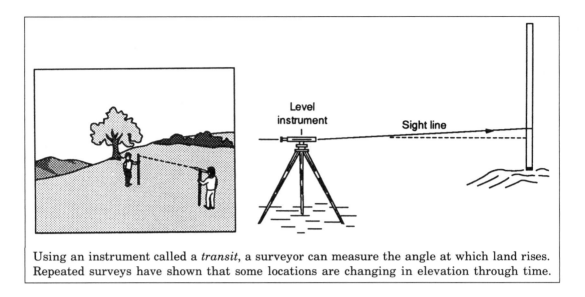

Using an instrument called a *transit*, a surveyor can measure the angle at which land rises. Repeated surveys have shown that some locations are changing in elevation through time.

47. Find the change in elevation from a place that is 15 feet below sea level to a place that is 95 feet above sea level.

48. A football team gained 7 yards on the first play, lost 2 yards on the second, and lost 8 yards on the third. What was the net result of the three plays?

49. Scott deposited $250 in a bank. During the next month, he made a deposit of $60 and a withdrawal of $80. How much money did Scott have in the bank at the end of that month?

50. During a four-day period, the dollar value of a share of stock rose $1\frac{1}{2}$ on the first day, dropped $\frac{5}{8}$ on the second day, rose $\frac{1}{8}$ on the third day, and dropped $1\frac{3}{4}$ on the fourth day. What was the net change in the value during this period?

51. True or false?   **a.** $5 - (-3) = -3 - 5$ **b.** $-7 - (-4) = -4 - (-7)$

52. **a.** Does $x - y = y - x$ for all replacements of x and y?
b. Does $x - y = y - x$ for any replacements of x and y? If yes, give an example.
c. What is the relation between $x - y$ and $y - x$ for all replacements of x and y?
d. Is the operation of subtraction commutative?

53. True or false?   **a.** $(15 - 9) - 6 = 15 - (9 - 6)$
b. $(-10 - 4) - 8 = -10 - (4 - 8)$
c. The operation of subtraction is associative.

In 54–59, name the reciprocal of the given number.

54. 6   **55.** -5 **56.** -1 **57.** $-\frac{1}{10}$ **58.** $\frac{3}{4}$ **59.** $-x \ (x \neq 0)$

In 60–85, find the quotient.

60. $\dfrac{18}{6}$ **61.** $\dfrac{52}{-4}$ **62.** $\dfrac{100}{-25}$ **63.** $\dfrac{55}{-11}$ **64.** $\dfrac{-84}{-12}$ **65.** $\dfrac{-144}{9}$

66. $\dfrac{75}{15}$ **67.** $\dfrac{-100}{-20}$ **68.** $\dfrac{-108}{9}$ **69.** $\dfrac{-65}{5}$ **70.** $\dfrac{4}{-8}$ **71.** $\dfrac{-15}{-12}$

72. $\dfrac{18}{-4}$ **73.** $\dfrac{100}{-8}$ **74.** $\dfrac{-36}{-8}$ **75.** $\dfrac{0}{-4}$ **76.** $\dfrac{8.4}{-4}$ **77.** $\dfrac{-3.6}{1.2}$

78. $48 \div (-6)$ **79.** $-75 \div (-15)$ **80.** $-50 \div 10$ **81.** $12 \div \left(-\frac{1}{3}\right)$

82. $-\frac{3}{4} \div 6$ **83.** $-\frac{3}{4} \div \left(-\frac{2}{3}\right)$ **84.** $9.6 \div (-3)$ **85.** $-1.8 \div 9$

In 86–97, simplify.

86. $40 - 18 + 2$ **87.** $-8 - 4 - (12)(\frac{1}{2})$ **88.** $25 - 3 - (-30)$

89. $5(15) - 5(12) + 5(2)$ **90.** $-6 \div \left(-\frac{1}{3}\right) - 6\left(-\frac{1}{3}\right)$ **91.** $1.4 - (2.8)(\frac{1}{4})$

92. $-6.2 - (-8.4) - (-5.6)$ **93.** $8\frac{11}{12} - 5\frac{2}{3} - 2\frac{1}{4}$ **94.** $6\frac{5}{6} + 3\frac{1}{2} - 4\frac{1}{3}$

95. $\dfrac{18}{-3} - \left(\dfrac{-42}{7}\right)$ **96.** $\dfrac{-55}{-11} - 6 \div \left(\dfrac{-3}{4}\right)$ **97.** $100 - \left(\dfrac{100}{-5}\right) + 5 \div \left(-\dfrac{5}{8}\right)$

98. Find the value of x for which the fraction $\dfrac{1}{x - 2}$ is not defined.

In 99–102: **a.** Give the multiplicative inverse of the expression.
b. State the value of x for which the multiplicative inverse is not defined.

99. $x - 5$ **100.** $x + 3$ **101.** $2x - 1$ **102.** $3x + 1$

103. True or false? **a.** $10 \div (-5) = -5 \div 10$ **b.** $-16 \div (-2) = -2 \div (-16)$

104. a. Does $x \div y = y \div x$ for all replacements of x and y?
b. Does $x \div y = y \div x$ for any replacements of x and y? If yes, give an example.
c. What is the relation between $x \div y$ and $y \div x$ when $x \neq 0$ and $y \neq 0$?
d. Is the operation of division commutative?

105. True or false? **a.** $(16 \div 4) \div 2 = 16 \div (4 \div 2)$
b. $[-36 \div 6] \div (-2) = -36 \div [6 \div (-2)]$
c. The operation of division is associative.

106. On her last 4 math tests, Carla scored 85, 92, 96, and 89.

a. Explain how you would find the average of Carla's 4 scores.

b. Consider the following method for finding the average of these 4 scores.
(1) Observing that the 4 scores cluster around 90, assume an average of 90.
(2) Using signed numbers, determine the difference between each score and 90.
(3) Averaging these differences, adjust the assumed average of 90.

c. Do the arithmetic for the method of finding the average in part **a**.

d. Comment on the advantages and disadvantages of the methods in parts **a** and **b**.

3-5 THE DISTRIBUTIVE PROPERTY; SUMMARY OF PROPERTIES

The Distributive Property

Observe that $2(20 + 3) = 2(23) = 46$ and $2 \times 20 + 2 \times 3 = 40 + 6 = 46$. Since both computations give the same result, it follows that $2(20 + 3) = 2 \times 20 + 2 \times 3$.

This example illustrates the ***distributive property of multiplication over addition***. In general, for all real numbers a, b, and c:

$$a(b + c) = ab + ac$$

The distributive property can be applied to multiplication over subtraction, as well as addition, and can be applied to any number of terms.

The distributive property is used in different ways.

1. To perform numerical calculations, as in:

$$20\left(\tfrac{1}{4} + \tfrac{1}{5}\right) = 20 \times \tfrac{1}{4} + 20 \times \tfrac{1}{5} = 5 + 4 = 9$$

2. To change the form of an algebraic expression.
   a. When a given expression is written as a product, it can be transformed into an equivalent expression that is a sum, as in:

   $$6(3x + 5) = 6 \cdot 3x + 6 \cdot 5 = 18x + 30$$

   b. When a given expression is written as a sum, it can be transformed into an equivalent expression that is a product, as in:

   $$xa + xb + xc = x(a + b + c)$$

MODEL PROBLEMS

1. Find the value of the numerical expression by using the distributive property.

   a. $6 \times 23 = 6 \times 20 + 6 \times 3 = 120 + 18 = 138$

   b. $9 \times 3\tfrac{1}{3} = 9\left(3 + \tfrac{1}{3}\right) = 9 \times 3 + 9 \times \tfrac{1}{3} = 27 + 3 = 30$

   c. $6.5 \times 8 = (6 + .5) \times 8 = 6 \times 8 + .5 \times 8 = 48 + 4 = 52$

2. Transform the expression to an equivalent expression without parentheses.

   a. $5(a + b) = 5a + 5b$

   b. $r(m + n - t) = rm + rn - rt$

3. Express the sum (or difference) as a product.

a. $3c + 3d = 3(c + d)$ **b.** $9y - 4y = (9 - 4)y = 5y$

c. $5x + 20 = 5 \cdot x + 5 \cdot 4 = 5(x + 4)$

4. Write the distributive property in a form that can be used to simplify $3x + 5x$.

Solution:

Write the sum form on the left. $a(b + c) = ab + ac$
Then use the commutative property. $ab + ac = a(b + c)$
Let $a = x$, $b = 3$, and $c = 5$. $ba + ca = (b + c)a$
     $3x + 5x = (3 + 5)x = 8x$

Now it's your turn to . . . **TRY IT!**   *(See page 144 for solutions.)*

1. Compute, using the distributive property: **a.** $56 \times 3 + 56 \times 7$ **b.** $60 \left(\frac{1}{2} - \frac{1}{6}\right)$

2. Write an equivalent expression without parentheses: **a.** $3x(y + 4)$ **b.** $-5(a - 2b)$

3. Express as a product: **a.** $5w + 3w$ **b.** $12x - 2y$

Summary of Properties of the Real Number System

In the following statements, a, b, and c represent any real numbers.

Property	*Symbolic Representation*
1. Closure under addition.	1. $a + b$ is a unique real number.
2. Addition is commutative.	2. $a + b = b + a$
3. Addition is associative.	3. $(a + b) + c = a + (b + c)$
4. 0 is the additive identity.	4. $a + 0 = a$
5. Every a has an additive inverse.	5. $a + (-a) = 0$
6. Closure under multiplication.	6. ab is a unique real number.
7. Multiplication is commutative.	7. $ab = ba$
8. Multiplication is associative.	8. $(ab)c = a(bc)$
9. 1 is the multiplicative identity.	9. $a \times 1 = a$
10. $a \neq 0$ has a multiplicative inverse.	10. $a \times \frac{1}{a} = 1$
11. Multiplication distributes over addition.	11. $a(b + c) = ab + ac$

In addition to the above properties, recall that any form of a number may be substituted for its equal. (This substitution property is not given a symbolic representation.)

In 1–6, state whether the sentence is a correct application of the distributive property. If you believe that it is not, state a reason.

1. $6(5 + 8) = 6 \times 5 + 6 \times 8$ **2.** $2(y + 6) = 2y + 6$ **3.** $(b + 2)a = ba + 2a$

4. $4a(b + c) = 4ab + 4ac$ **5.** $14x - 4x = (14 - 4)x$ **6.** $\frac{1}{2}hb + \frac{1}{2}hc = \frac{1}{2}h(b + c)$

In 7–9, complete the sentence so that it is an application of the distributive property.

7. $-4(p + q) = $ ____   **8.** ____ $= 2x - 2y$ **9.** ____ $= (15 - 7)m$

In 10–15, use the distributive property to find the value.

10. $15 \times 36 + 15 \times 64$ **11.** $128 \times 615 - 28 \times 615$ **12.** $1\frac{3}{4} \times 576 + 8\frac{1}{4} \times 576$

13. $937 \times .8 + 937 \times .2$ **14.** $36\left(\frac{1}{3} + \frac{1}{4}\right)$ **15.** $50 \times 8\frac{3}{5}$

In 16–20, use the distributive property to write the expression without parentheses.

16. $4(m + n)$ **17.** $-7(2x + 1)$ **18.** $a(x - 4)$ **19.** $\left(\frac{1}{3}m - \frac{3}{5}n\right)(-30)$ **20.** $-3a(7 - b)$

In 21–25, use the distributive property to express the sum as a product.

21. $2p + 2q$ **22.** $8p - 2m$ **23.** $at + ar$ **24.** $12y - 5y$ **25.** $7x + 21$

In 26 and 27, name the property that justifies each step in the set of related equations.

26. (1) $s(m + n) = s(n + m)$ **27.** (1) $5x + (3x + 4) = (5x + 3x) + 4$
    (2) $\qquad\qquad = sn + sm$ (2) $\qquad\qquad\qquad = (5 + 3)x + 4$
    (3) $\qquad\qquad = ns + ms$ (3) $\qquad\qquad\qquad = 8x + 4$

28. Eleven properties of real numbers are listed in this section. State which, if any, of these eleven properties do *not* hold for the set of integers.

29. Consider the set of positive integers, the set of negative integers, the set of odd integers, the set of even integers, the set of rational numbers, and the set of real numbers. Which of these sets are closed under:
(a) addition **(b)** subtraction **(c)** multiplication **(d)** division (excluding division by zero)?

30. The given chain of equations can be used to show that $x(yz) = (xz)y$ when x, y, and z are members of the set of real numbers. State the reason for each step.
(1) $x(yz) = x(zy)$ (2) $x(zy) = (xz)y$ (3) $x(yz) = (xz)y$

In 31–36, all variables represent members of the set of real numbers. Use the properties of real numbers to write a step-by-step proof that the sentence is true. State reasons.

31. $(ab)c = a(cb)$ **32.** $(a + b) + c = c + (b + a)$ **33.** $a(b + c) = ab + ca$

34. $\dfrac{1}{n}(mn) = m \quad (n \neq 0)$ **35.** $m + n + (-m) = n$ **36.** $\dfrac{1}{n}(m + n) = 1 + m \cdot \dfrac{1}{n}$

3-6 USING THE PROPERTIES OF THE REAL NUMBER SYSTEM

Combining Like Terms

To add or subtract terms, the terms must have the same variables as factors, with corresponding variables having the same exponent.

like terms: $3a$ and $5a$; x^2 and $7x^2$ *unlike terms*: $3a$ and $3b$; $5a$ and $7a^2$

When you use the distributive property to express the sum of like terms as a single term, you **combine like terms**. For example, $9x + 2x = (9 + 2)x = 11x$.

The sum or difference of two unlike terms cannot be expressed as a single term. For example, $2x + 3y$ cannot be simplified.

In some problems of simplification, the distributive property is used in two ways:

Simplify: $3x + 7(2x + 3)$

First multiply: $3x + 7(2x + 3) = 3x + 7(2x) + 7(3) = 3x + 14x + 21$

Then combine like terms: $3x + 14x + 21 = (3 + 14)x + 21 = 17x + 21$

Sometimes the multiplier is not shown. To apply the distributive property, you use the multiplication property of 1.

$3x + (2x + 3)$ is equivalent to $3x + 1(2x + 3)$.
$9y - (4 + 2y)$ is equivalent to $9y - 1(4 + 2y)$.

MODEL PROBLEMS

In 1–6, use the distributive property to simplify the expression.

1. $9t + 4t - t$
$= (9 + 4 - 1)t$
$= 12t$

2. $7a + 6b + 5a - 2b$
$= 7a + 5a + 6b - 2b$
$= (7 + 5)a + (6 - 2)b$
$= 12a + 4b$

3. $3(x + 5) - 10$
$= 3x + 15 - 10$
$= 3x + 5$

4. $2c + (7c - 4)$
$= 2c + 1(7c - 4)$
$= 2c + 7c - 4$
$= 9c - 4$

5. $-2(3 - 2x) - (6 - 5x)$
$= -2(3 - 2x) - 1(6 - 5x)$
$= -6 + 4x - 6 + 5x$
$= 9x - 12$

6. $6x - [3x - 2(x - 5)]$
$= 6x - [3x - 2x + 10]$
$= 6x - [x + 10]$
$= 6x - 1[x + 10]$
$= 6x - x - 10$
$= 5x - 10$

> ### Now it's your turn to . . . **TRY IT!**
>
> In 1–3, use the distributive property to simplify the expression
>
> **1.** $9a + 5 - 7a - 8$ **2.** $4(3d - 5) - 9$ **3.** $4w - 3(4 - w)$
>
> *(See page **144** for solutions.)*

Evaluating Algebraic Expressions

Recall the order of operations you learned in Chapter 2:

1. Work within grouping symbols.
2. Simplify powers.
3. Do all multiplications and divisions.
4. Do all additions and subtractions.

At that time, you used only natural numbers, but this order is still applicable for the entire set of real numbers. Now that you have worked with both positive and negative numbers, you must be prepared to use signed numbers in your calculations. Also, since you have worked with the properties of the real number system, you can use these properties in evaluating algebraic expressions.

MODEL PROBLEMS

1. Find the value of $10 - x^2(3 - x)$ when $x = 5$.

How to Proceed	*Solution*
(1) Write the expression.	$10 - x^2(3 - x)$
(2) Replace the variable by its given value.	$= 10 - 5^2(3 - 5)$
(3) Work within the parentheses.	$= 10 - 5^2(-2)$
(4) Simplify the power.	$= 10 - 25(-2)$
(5) Do the multiplication.	$= 10 - (-50)$
(6) Do the subtraction.	$= 60$ **ANSWER**

2. Show that $4x + 5x = 9x$ is a true sentence when $x = 10$.

How to Proceed	*Solution*
(1) Write the sentence.	$4x + 5x = 9x$
(2) Replace the variable by its value.	$4 \times 10 + 5 \times 10 \stackrel{?}{=} 9 \times 10$
(3) Do the multiplications.	$40 + 50 \stackrel{?}{=} 90$
(4) Do the addition. The result is a true sentence.	$90 = 90 ✓$

In 1–4, find the value of each expression if $a = -5$ and $b = -2$.

1. $2ab$ **2.** $-3ab^2$ **3.** $b - a^2$ **4.** $100 - a^2(2 - b)$

(See page 144 for solutions.)

CALCULATOR CONNECTION

Some computations involve a single number that must be entered repeatedly. If the number has several digits, requiring multiple keypresses, you can save time by using the calculator's memory.

There are three steps in using the memory: entering (using a key such as STO, Min, or M+), recalling (MR), and clearing. A calculator with the STO key automatically clears the memory when new numbers are entered; some calculators use M- to remove existing values.

Example 1. Evaluate $x^2 - x(5 - x)$ when $x = -5.2$.

Solution: Store **−5.2** in memory.

Example 2. Evaluate $4x + 6y$ when $x = -5.2$ and $y = 3$.

Solution: Use the **−5.2** that is already in memory.

$$4 \; \boxed{\times} \; \boxed{MR} \; \boxed{+} \; 6 \; \boxed{\times} \; 3 \; \boxed{=} \qquad \boxed{-2.8}$$

Exercises

Evaluate when $x = -5.2$ and $y = 3$.

1. $x + y$ **2.** $2x - y$ **3.** $y - 2x$ **4.** $2xy$ **5.** xy^2

Answers: **1.** -2.2 **2.** -13.4 **3.** 13.4 **4.** -31.2 **5.** -46.8

In 1–12, simplify the expression by combining like terms.

1. $7x + 3x$

2. $9t - 5t$

3. $10c + c$

4. $\frac{5}{2}d + \frac{3}{2}d$

5. $3.4y + 1.3y$

6. $6.2r - r$

7. $9ab + 2ab$

8. $8z^2 + z^2$

9. $8m + 5m + m$

10. $8.2b + 3.8b - 12b$

11. $\frac{3}{4}xy - 2xy + \frac{1}{2}xy$

12. $9d^2 - 6d^2 - d^2$

13. Express the perimeter and simplify the result by combining like terms:

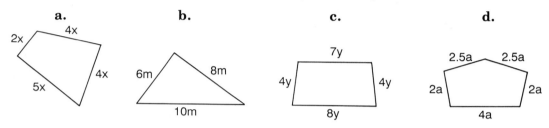

a.   **b.**   **c.**   **d.**

In 14–16, show that the sentence is true for the given values of the variables.

14. $3x + 5x = 8x$ when x equals: **a.** 7   **b.** -3 **c.** $\frac{1}{2}$

15. $5y^2 - 2y^2 = 3y^2$ when y equals: **a.** 12   **b.** -2 **c.** .2

16. $6c + 9d + c + 2d = 7c + 11d$ when: **a.** $c = 10, d = -7$ **b.** $c = 1.2, d = .1$

In 17–24, simplify the expression by combining like terms.

17. $8m + 5 + 7m + 6$

18. $a + 5 + 2a - 4$

19. $1.5 + 7.2b + 5.1 + 8.6b$

20. $5a + 3m + a + m$

21. $5a + b + 3c + \frac{4}{3}a + b + \frac{3}{4}c$

22. $5y^2 + 25 + 2y^2 - 10$

23. $7(m + 5) + 4(2m + 1)$

24. $5(2c + 9) + 3(4c - 7)$

25. Express the perimeter and simplify the result by combining like terms:

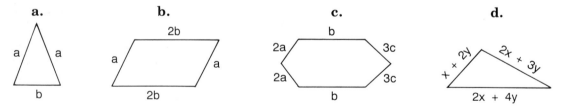

a.   **b.**   **c.**   **d.**

In 26–40, simplify the expression.

26. $5(d + 3) - 10$

27. $7 + 2(7x - 5)$

28. $-2(x - 1) + 6$

29. $5 - 4(3e - 5)$

30. $8 + (4e - 2)$

31. $(6b + 4) - 2b$

32. $9 - (5t + 6)$

33. $4 - (2 - 8s)$

34. $-(6x - 7) + 4$

35. $7x + 3(2x - 1) - 8$ **36.** $7c - 4d - 2(4c - 3d)$ **37.** $(a + 3b) - (a - 3b)$

38. $4(2x + 5) - 3(2 - 7x)$ **39.** $7[5x + 2(x - 3) + 4]$ **40.** $-4[8y - 7 - 3(2y - 1)]$

41. A carpenter has a piece of lumber that is $(x + 2)$ yards in length and another piece that is $(2x - 1)$ feet in length. Represent, in simplest form, the total number of inches of lumber.

42. Tom has $(3x - 4)$ nickels and $(2x + 2)$ dimes. Represent, in simplest form, the total number of cents he has.

43. On their vacation, the Chans stayed with relatives for 10 days and traveled for the remaining $(2x - 1)$ weeks. Represent in simplest form the total number of their vacation days.

In 44–89, find the numerical value of the expression if $a = -8$, $b = 6$, $d = -3$, $x = -4$, $y = 5$, and $z = -1$.

44. $6a$ **45.** $-5b$ **46.** ab **47.** $2xy$ **48.** $-4bz$ **49.** $\frac{1}{3}d$

50. $-\frac{2}{3}b$ **51.** $\frac{3}{8}a$ **52.** $\frac{1}{2}xy$ **53.** $-\frac{3}{4}ab$ **54.** a^2 **55.** d^3

56. $-y^2$ **57.** $-d^2$ **58.** $-z^3$ **59.** $2x^2$ **60.** $-3y^2$ **61.** $-3b^2$

62. $4d^2$ **63.** $-2z^3$ **64.** xy^2 **65.** $2d^2y^2$ **66.** $\frac{1}{2}db^2$ **67.** $-2d^3z^2$

68. $a + b$ **69.** $2x + z$ **70.** $3y - b$ **71.** $a - 2d$ **72.** $7b - 5x$ **73.** $x^2 + x$

74. $2d^2 - d$ **75.** $2a + 5d + 3x$ **76.** $8y + 5b - 6d$ **77.** $x^2 + 3x + 5$

78. $z^2 + 2z - 7$ **79.** $a^2 - 5a - 6$ **80.** $2x^2 - 3x + 5$ **81.** $15 + 5z - z^2$

82. $3(2x - 1) + 6$ **83.** $10 - 3(x - 4)$ **84.** $(x + 2)(x - 1)$ **85.** $(a - b)(a + b)$

86. $(x + d)(x - 4z)$ **87.** $x(x^2 - 2x)$ **88.** $12 + d(3d - 5)$ **89.** $20 - xy(a + 3)$

In 90–94, find the value of the expression if $a = -12$, $b = 6$, and $c = -1$.

90. $\dfrac{2a}{b}$ **91.** $\dfrac{ac}{-3b}$ **92.** $\dfrac{3a^2c^3}{b^3}$ **93.** $\dfrac{a - b^2}{-2c^2}$ **94.** $\dfrac{b^2 - a^2}{b^2 + a^2}$

95. There were b commemorative stamps issued in one year. The next year, there were c fewer stamps issued. Find the number of commemoratives issued in the latter year if $b = 40$ and $c = 9$.

96. In the Grand City orchestra, there are m violinists. The number of violinists is to be increased to 5 more than twice the present number. Find the new number of violinists if $m = 9$.

UNITED STATES IN SPACE

Commemorative stamps call attention to a person or an event, subjects usually proposed by groups of citizens. The United States often issues more than 30 such stamps in a year.

97. Gina spent $(y - 4)$ dollars on groceries for each of 6 weeks. How much did she spend for groceries for the 6 weeks if $y = 52$?

98. In two years, the price of a computer chip increased to $(x^2 - 3x + 5)$ dollars. Find the new price of a chip if $x = 5$.

3-7 DEFINING AN OPERATION

You are now familiar with four operations for the set of real numbers. Are other operations possible? Yes, and you will learn about them as you progress in your studies.

Not only are other operations possible, but you too can invent an operation. All you need do is define it; that is, you must specify, in words or in symbols, exactly how the operation is to be performed for all real numbers. By following your definition, anyone can then carry out this operation on any particular real numbers. Note that a new operation requires a new symbol, and a new operation can include familiar operations.

MODEL PROBLEMS

1. For all real numbers a and b, let $a \odot b = a + 2b$. Find the value of $3 \odot 5$.

  Solution: Carry out the operation $\odot$ on the real numbers 3 and 5 exactly as it is defined for all real numbers a and b. That is, substitute 3 for a and 5 for b in the definition, and perform the familiar operations on which the new operation is defined.

$$a \odot b = a + 2b$$
$$3 \odot 5 = 3 + 2(5)$$
$$= 3 + 10 = 13 \quad \textbf{\textit{ANSWER}}$$

2. For all real numbers a, b, c, d, let $\begin{array}{c|c} a & b \\ \hline c & d \end{array} = ad - bc$.

  a. Evaluate: $\begin{array}{c|c} 1 & 5 \\ \hline -2 & -4 \end{array}$
  b. If $\begin{array}{c|c} 7 & 6 \\ \hline x & x \end{array} = 8$, find x.

  c. $\begin{array}{c|c} 4 & 5 \\ \hline y & x \end{array} + \begin{array}{c|c} -2 & y \\ \hline 3 & x \end{array}$ is equivalent to
  (1) $\begin{array}{c|c} 8 & 2 \\ \hline x & y \end{array}$
  (2) $\begin{array}{c|c} 2 & 8 \\ \hline y & x \end{array}$
  (3) $\begin{array}{c|c} 6 & y \\ \hline 2 & x \end{array}$

  Solution:
  a. Substitute the given values into the definition.

$$\begin{array}{c|c} a & b \\ \hline c & d \end{array} = ad - bc$$

$$\begin{array}{c|c} 1 & 5 \\ \hline -2 & -4 \end{array} = 1(-4) - 5(-2)$$

$$= -4 - (-10)$$

$$= 6 \quad \textbf{\textit{ANSWER}}$$

b. Carry out the new operation and simplify the resulting open sentence by combining like terms.

$$\frac{7 \uparrow 6}{x \downarrow x} = 8$$

$$7x - 6x = 8$$

$$x = 8 \quad \textbf{\textit{ANSWER}}$$

c. Evaluate the expression according to the definition, and simplify the result by combining like terms. Examine the choices to determine which of those expressions leads to the same result.

$$\frac{4 \uparrow 5}{y \downarrow x} \; + \; \frac{-2 \uparrow y}{3 \downarrow x}$$

$$= (4x - 5y) + (-2x - 3y)$$

$$= 4x - 5y - 2x - 3y$$

$$= 2x - 8y$$

\textit{ANSWER:}   Choice (2) because $\dfrac{2 \uparrow 8}{y \downarrow x} = 2x - 8y.$

Now it's your turn to . . . **TRY IT!**

1. If $a \star b = a + b - ab$, evaluate: **a.** $5 \star 2$ **b.** $(-3) \star (-1)$ **c.** $4 \star .1$

2. For all real numbers a, b, c, and d, let $d\begin{array}{c} a \\ \square \\ c \end{array}b = ac - bd$.

Then $4\begin{array}{c} -3 \\ \square \\ -2 \end{array}1 + 2\begin{array}{c} 3 \\ \square \\ 4 \end{array}-1$ is equivalent to

(1) $4\begin{array}{c} 4 \\ \square \\ 4 \end{array}4$ (2) $.5\begin{array}{c} .5 \\ \square \\ 8 \end{array}24$ (3) $.5\begin{array}{c} .5 \\ \square \\ -8 \end{array}24$ (4) $.5\begin{array}{c} .5 \\ \square \\ 24 \end{array}-8$

(See page 145 for solutions.)

In 1–4, the operation is defined for all real numbers a and b. Apply the definition to the real values $a = 3$ and $b = -2$.

1. $a * b = a^2 - b$ **2.** $a \diamond b = \dfrac{a + b}{a - b}$ **3.** $a \odot b = a^2b - b$ **4.** $a \triangle b = \dfrac{a^2}{b^2}$

In 5–8, the operation is defined for all real numbers a and b.

5. a *max* b means the greater of the two values a or b.
   Evaluate: **a.** (-3) *max* 7 **b.** (-8) *max* (-10) **c.** $\left(-\frac{1}{2}\right)$ *max* $\left(-\frac{1}{7}\right)$ **d.** $.4$ *max* $\frac{1}{5}$

6. a *min* b means the smaller of the two values a or b.
   Evaluate: **a.** $.1$ *min* $.01$ **b.** (-3) *min* (-7) **c.** $(-.9)$ *min* $(-.99)$ **d.** $.\overline{3}$ *min* $\frac{1}{3}$

7. $[a]$ means the greatest integer that is not greater than a.
   Evaluate: **a.** $[1.2]$ **b.** $[10.333 \ldots]$ **c.** $[-4.2]$ **d.** $[-7.\overline{12}]$

8. $\llcorner a \lrcorner$ means the least integer that is greater than or equal to a.
   Evaluate: **a.** $\llcorner 4.3 \lrcorner$ **b.** $\llcorner 7 \lrcorner$ **c.** $\llcorner -5.2 \lrcorner$ **d.** $\llcorner .6 \lrcorner$

In 9–13, the following operations are defined for all sets A and B.

   $A \cap B$ means {elements common to both A and B}, called the **intersection** of the sets.

   $A \cup B$ means {elements that appear in either A or B}, called the **union** of the sets.

For the given sets, list: **a.** the intersection $\cap$ **b.** the union $\cup$

9. $A = \{2, 5, 7, 10\}$ $B = \{5, 10, 12\}$

10. $A = \{\triangle, \bigcirc, \diamondsuit, \square, \square\!\!\!\!/\ \}$ $B = \{\triangle, \bullet, \blacklozenge, \square, \blacksquare, \star\}$

11. $A = \{$even integers$\}$ $B = \{$odd integers$\}$

12. $A = \varnothing$ $B = \{$natural numbers$\}$

13. $A = \{$integral multiples of 3$\}$ $B = \{$integral multiples of 5$\}$

In 14 and 15, use the Venn diagram to list the specified combination of the given sets.

14. List: **a.** $M \cap N$
             b. $M \cup N$

15. List: **a.** $A \cap B$
             b. $A \cup C$
             c. $(A \cap B) \cap C$
             d. $(A \cup B) \cap C$
             e. $(A \cap B) \cup C$

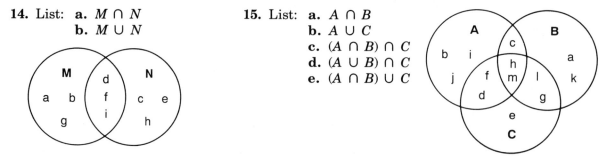

In 16 and 17, for all real numbers a, let $\boxed{a}$ represent the area of a square whose side measures a feet.

16. Evaluate: **a.** $\boxed{10}$ **b.** $\boxed{2.5}$ **c.** $\boxed{\frac{1}{9}}$

17. $\boxed{3} + \boxed{4}$ is equivalent to (1) $\boxed{7}$ (2) $\boxed{49}$ (3) $\boxed{25}$ (4) $\boxed{5}$

In 18 and 19, let $\boxed{\begin{matrix} a \\ b \\ c \end{matrix}} = a \div b \div c$ for all real numbers a, b, c where b and $c \neq 0$.

18. Evaluate: **a.** $\boxed{\begin{matrix} 9 \\ 3 \\ 3 \end{matrix}}$ **b.** $\boxed{\begin{matrix} 12 \\ 4 \\ 6 \end{matrix}}$ **c.** $\boxed{\begin{matrix} 5 \\ 10 \\ 2 \end{matrix}}$

19. $\boxed{\begin{matrix} .8 \\ 4 \\ 1 \end{matrix}}$ is equivalent to (1) $\boxed{\begin{matrix} 8 \\ 4 \\ .1 \end{matrix}}$ (2) $\boxed{\begin{matrix} 10 \\ 5 \\ .1 \end{matrix}}$ (3) $\boxed{\begin{matrix} 10 \\ 5 \\ 10 \end{matrix}}$ (4) $\boxed{\begin{matrix} 10 \\ 10 \\ .2 \end{matrix}}$

In 20–22, let $\boxed{\begin{matrix} a & b \\ c & d \end{matrix}} = bc - ad$ for all real numbers a, b, c, and d.

20. Evaluate: $\boxed{\begin{matrix} -2 & -4 \\ 3 & -2 \end{matrix}}$ **21.** If $\boxed{\begin{matrix} 7 & 8 \\ x & x \end{matrix}} = 12$, find x.

22. $\boxed{\begin{matrix} 3 & 7 \\ x & y \end{matrix}} + \boxed{\begin{matrix} 1 & 2 \\ y & x \end{matrix}}$ is equivalent to

(1) $\boxed{\begin{matrix} -6 & -1 \\ x & y \end{matrix}}$ (2) $\boxed{\begin{matrix} 6 & 1 \\ y & x \end{matrix}}$ (3) $\boxed{\begin{matrix} 1 & 6 \\ x & y \end{matrix}}$ (4) $\boxed{\begin{matrix} -1 & -6 \\ x & y \end{matrix}}$

3-8 ANOTHER ARITHMETIC: MATRICES

As an inventor, you would be most successful if your inventions proved useful to others. So it is with an arithmetic. An arithmetic that can be used to work out the problems we face in our world is a successful invention. The familiar basic arithmetic of the real number system is so useful that every person needs to know it just to get along. There are other arithmetics that are useful in solving our problems.

Defining a Matrix

A *matrix* (plural: matrices), organizes numbers into horizontal rows and vertical columns and encloses these numbers with brackets, [ ]. A matrix may have any number of rows and columns.

$$\begin{bmatrix} 3 & 4 \\ 1 & -2 \\ -5 & 0 \end{bmatrix}$$

3 rows, 2 columns

$$\begin{bmatrix} -2 \\ 7 \\ .5 \end{bmatrix}$$

3 rows, 1 column

$$\begin{bmatrix} \frac{1}{2} & 7 & -2 \end{bmatrix}$$

1 row, 3 columns

MODEL PROBLEM

A matrix can be used to organize data. The given matrix tells about the stock of red exercise wear in Ed's Emporium.

	small	medium	large
red shirts	10	7	8
red pants	0	16	22

Determine what each of the following tells about this stock.

a. The number in row 1, column 2.
b. The sum of the elements of row 2.
c. The sum of the elements of column 3.
d. The sum of all the elements of the matrix.

ANSWERS:
a. There are 7 red shirts in the medium size.
b. In all, there are 38 pairs of red pants.
c. In the large size, there are 30 pieces of red exercise wear.
d. Ed has a total stock of 63 pieces of red exercise wear.

Adding Matrices

To add two matrices, add the *corresponding* elements. That is, add the elements that occupy the same position in each of the two matrices, and position their sum in that same position in the resulting matrix.

$$\begin{bmatrix} 1 & -5 \\ -2 & 3 \\ -3 & -2 \end{bmatrix} + \begin{bmatrix} -4 & 7 \\ 0 & 6 \\ 3 & -6 \end{bmatrix} = \begin{bmatrix} 1 + (-4) & -5 + 7 \\ -2 + 0 & 3 + 6 \\ -3 + 3 & -2 + (-6) \end{bmatrix}$$

$$= \begin{bmatrix} -3 & 2 \\ -2 & 9 \\ 0 & -8 \end{bmatrix}$$

Note that you can add two matrices only if they have the same *dimensions*. That is, the two matrices must have the same number of rows and the same number of columns.

The preceding problem about the stock of red exercise wear suggests a practical application of matrix addition. Suppose Ed stocks two colors of exercise wear, red and grey. There would then be two matrices, one for each color, and the sum of the matrices would give information about all the exercise wear that Ed stocks. Since Ed sells many other items as well, you realize that a store's stock is complicated and that an efficient way to organize and control the information is important.

Multiplying Matrices

If you had 2 nickels, 3 dimes, and 2 quarters, how would you compute the total amount of money?

You would first multiply: 2 nickels = 2(5) = 10 cents

3 dimes = 3(10) = 30 cents

2 quarters = 2(25) = 50 cents

Then you would add: 10 + 30 + 50 = 90 cents

Matrix multiplication is performed in much the same way, using both multiplication and addition. To represent the preceding example, one matrix shows the quantities of coins, and the other matrix shows the value of each coin.

$$[2 \quad 3 \quad 2] \begin{bmatrix} 5 \\ 10 \\ 25 \end{bmatrix}$$

$$= 2(5) + 3(10) + 2(25)$$

$$= 10 + 30 + 50 = 90$$

Observe that, in one matrix, the elements are arranged in a row, and in the other matrix, in a column. Matrix multiplication is also called *row-by-column* multiplication.

Matrix multiplication can be extended to as many rows and columns as needed, as is shown in the model problems and exercises that follow. When the matrices have more than one row or column, the product will have the same number of rows as the first matrix and the same number of columns as the second.

Since the preceding coin problem involves a matrix with 1 row multiplying a matrix with 1 column, the resulting product matrix is [90], containing only a single element.

MODEL PROBLEMS

1. Mona and her business partner Pam make decorative boxes from two types of wood veneers, walnut and oak. The costs of the wood veneers differ; walnut is $2 per square foot and oak is $1.50 per square foot. For the boxes they made for the Christmas season, Mona used 40 square feet of walnut and 60 square feet of oak, and Pam used 30 square feet of walnut and 70 square feet of oak.

Find the cost of the wood used by each partner, and find the total cost.

Solution:

Set up matrix A to display the data about how much wood was used, and matrix B to show the costs. To determine the total costs, multiply the two matrices.

$$A = \begin{array}{c} \text{Mona} \\ \text{Pam} \end{array} \begin{bmatrix} \overset{\text{walnut}}{40} & \overset{\text{oak}}{60} \\ 30 & 70 \end{bmatrix} \qquad B = \begin{array}{c} \text{walnut} \\ \text{oak} \end{array} \begin{bmatrix} \overset{\text{cost}}{2} \\ 1.50 \end{bmatrix}$$

To find AB, multiply each row of A by each column of B. In this problem, there is only one column in matrix B.

row 1 of A by column 1 of B: $40(2) + 60(1.50) = 80 + 90 = 170$

row 2 of A by column 1 of B: $30(2) + 70(1.50) = 60 + 105 = 165$

$$AB = \begin{array}{c} \text{Mona} \\ \text{Pam} \end{array} \begin{bmatrix} \overset{\text{cost}}{170} \\ 165 \end{bmatrix}$$

The product matrix AB shows that Mona spent $170 and Pam spent $165, a total cost of $335. ***ANSWER***

Observation: In an application problem, the column headings of the first matrix are the same as the row headings of the second matrix. In the product, these headings drop out. The product then has the row headings of the first matrix and the column headings of the second matrix.

2. Given: $A = \begin{bmatrix} 2 & -1 \\ -3 & 4 \end{bmatrix}$ and $B = \begin{bmatrix} 5 & 6 \\ 0 & -7 \end{bmatrix}$ Find: AB

Solution: Multiply each row of matrix A by each column of matrix B.

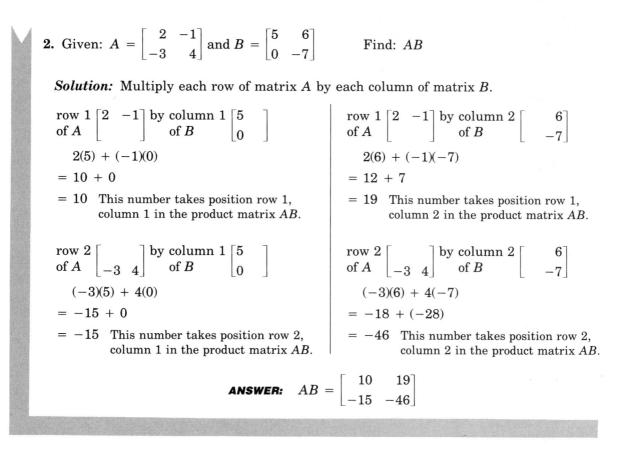

row 1 $\begin{bmatrix} 2 & -1 \end{bmatrix}$ by column 1 $\begin{bmatrix} 5 \\ 0 \end{bmatrix}$
of A of B

$2(5) + (-1)(0)$

$= 10 + 0$

$= 10$ This number takes position row 1, column 1 in the product matrix AB.

row 1 $\begin{bmatrix} 2 & -1 \end{bmatrix}$ by column 2 $\begin{bmatrix} 6 \\ -7 \end{bmatrix}$
of A of B

$2(6) + (-1)(-7)$

$= 12 + 7$

$= 19$ This number takes position row 1, column 2 in the product matrix AB.

row 2 $\begin{bmatrix} -3 & 4 \end{bmatrix}$ by column 1 $\begin{bmatrix} 5 \\ 0 \end{bmatrix}$
of A of B

$(-3)(5) + 4(0)$

$= -15 + 0$

$= -15$ This number takes position row 2, column 1 in the product matrix AB.

row 2 $\begin{bmatrix} -3 & 4 \end{bmatrix}$ by column 2 $\begin{bmatrix} 6 \\ -7 \end{bmatrix}$
of A of B

$(-3)(6) + 4(-7)$

$= -18 + (-28)$

$= -46$ This number takes position row 2, column 2 in the product matrix AB.

ANSWER: $AB = \begin{bmatrix} 10 & 19 \\ -15 & -46 \end{bmatrix}$

NOTE.   You can multiply two matrices only if the rows of the first matrix have the same number of elements as the columns of the second matrix.

Now it's your turn to . . . **TRY IT!**

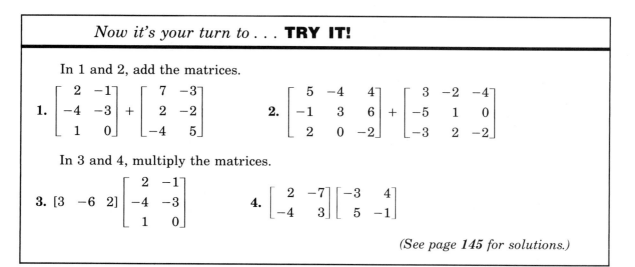

In 1 and 2, add the matrices.

1. $\begin{bmatrix} 2 & -1 \\ -4 & -3 \\ 1 & 0 \end{bmatrix} + \begin{bmatrix} 7 & -3 \\ 2 & -2 \\ -4 & 5 \end{bmatrix}$

2. $\begin{bmatrix} 5 & -4 & 4 \\ -1 & 3 & 6 \\ 2 & 0 & -2 \end{bmatrix} + \begin{bmatrix} 3 & -2 & -4 \\ -5 & 1 & 0 \\ -3 & 2 & -2 \end{bmatrix}$

In 3 and 4, multiply the matrices.

3. $\begin{bmatrix} 3 & -6 & 2 \end{bmatrix} \begin{bmatrix} 2 & -1 \\ -4 & -3 \\ 1 & 0 \end{bmatrix}$

4. $\begin{bmatrix} 2 & -7 \\ -4 & 3 \end{bmatrix} \begin{bmatrix} -3 & 4 \\ 5 & -1 \end{bmatrix}$

(See page 145 for solutions.)

1. At Hoover High School, students may choose one of three languages. A total of 85 freshmen, 112 sophomores, 96 juniors, and 78 seniors are studying Spanish. Of the students studying French, 78 are freshmen, 96 are sophomores, 72 are juniors, and 63 are seniors. Russian is being offered for the first time this year and 34 freshmen are enrolled.

   From the matrix, state what the number tells about this high school.

   a. The number in row 1, column 3.
   b. The sum of the elements in row 2.
   c. The sum of the elements in column 1.
   d. The sum of all the elements of the matrix.

$$\begin{array}{l} \quad\quad\quad\quad \text{Fr.} \quad \text{Soph.} \quad \text{Jr.} \quad \text{Sr.} \\ \begin{array}{l} \text{Spanish} \\ \text{French} \\ \text{Russian} \end{array} \left[\begin{array}{cccc} 85 & 112 & 96 & 78 \\ 78 & 96 & 72 & 63 \\ 34 & 0 & 0 & 0 \end{array}\right] \end{array}$$

2. The bar graph displays the data for the sale of sandwiches over a 3-month period at the cafeteria of Ryder High School.

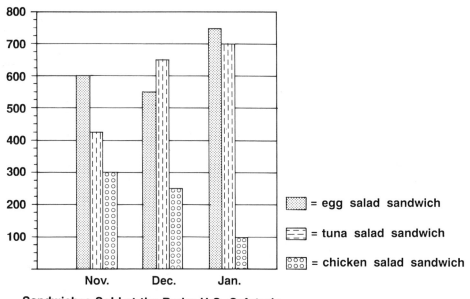

= egg salad sandwich

= tuna salad sandwich

= chicken salad sandwich

Sandwiches Sold at the Ryder H.S. Cafeteria

   a. Using the types of sandwiches as headings for the rows and the names of the months as headings for the columns, write a matrix to display the data shown in the bar graph.

   b. For each part below, tell whether you find the bar graph or the matrix a more efficient way of displaying the data, and give your reason.
   (1) In which month was the greatest number of tuna salad sandwiches sold?
   (2) Over the 3-month period, what was the total number of tuna salad sandwiches sold?
   (3) Over the 3-month period, what was the total number of sandwiches sold?
   (4) In January, how many fewer chicken salad than egg salad sandwiches were sold?

   c. From the data, what might the cafeteria director conclude about offering these types of sandwiches?

3. Last weekend, Dan's Diner included the following total numbers of rolls of coins among its bank deposits in the night depository: on Saturday, 15 rolls of pennies, 20 rolls of nickels, 18 rolls of dimes, and 12 rolls of quarters, and on Sunday, 13 rolls of pennies, 22 rolls of nickels, 16 rolls of dimes, and 20 rolls of quarters.

   a. Write a matrix to display this data. Use *Saturday* and *Sunday* as row headings, and *pennies*, *nickels*, *dimes*, and *quarters* as column headings.

   b. Find the specified value and tell what it represents.
      (1) The number in row 2, column 4.   (3) The sum of the elements in column 3.
      (2) The sum of the elements in row 1.   (4) The sum of all the elements of the matrix.

   c. Explain how you would find the total value of these coin deposits.

4. Matrices with equal corresponding elements are called **equal matrices**.

   a. Which matrix equals $\begin{bmatrix} 6-3 & 5 \div 10 \\ 5+2 & 4 \times 1 \end{bmatrix}$?   (1) $\begin{bmatrix} 3 & 5 \\ 7 & 4 \end{bmatrix}$ (2) $\begin{bmatrix} 3 & .5 \\ 7 & 4 \end{bmatrix}$ (3) $\begin{bmatrix} 3 & .5 \\ 7 & 1 \end{bmatrix}$

   b. If $\begin{bmatrix} 1 & x \\ -2 & 4 \end{bmatrix} = \begin{bmatrix} 1 & 5 \\ -2 & 4 \end{bmatrix}$, find x.

   In 5–7, add the matrices.

5. $\begin{bmatrix} -4 \\ 1 \end{bmatrix} + \begin{bmatrix} 3 \\ -2 \end{bmatrix}$ 6. $\begin{bmatrix} 2 & -3 \\ -1 & 5 \end{bmatrix} + \begin{bmatrix} -2 & 5 \\ -5 & 4 \end{bmatrix}$ 7. $\begin{bmatrix} 5 & -4 & 4 \\ -1 & 3 & 6 \\ 2 & 0 & -2 \end{bmatrix} + \begin{bmatrix} 3 & -2 & -4 \\ -5 & 1 & 0 \\ -3 & 2 & -2 \end{bmatrix}$

8. For a holiday weekend, Happy Rest Motel will be making its facilities available to the Kings Hotel so that the Kings can accommodate a convention group.

   a. At its highway site, the Happy Rest has 15 single, 26 double, and 12 king-size rooms. At its lake site, the Happy Rest has 18 single, 32 double, and 16 king-size rooms.
      Write matrix A to display the data about the rooms at Happy Rest. Use *highway* and *lake* as row headings, and *single*, *double*, and *king* as column headings.

   b. In its main building, the Kings Hotel has 65 single, 140 double, and 80 king-size rooms. In its annex, the Kings has 41 single, 85 double, and 50 king-size rooms.
      Write matrix B to display the data about the rooms at the Kings. Use *main* and *annex* as row headings, and column headings as in part **a**.

   c. Find $A + B$, the sum of the two matrices.

   d. Use $A + B$ to determine: (1) The number of each type of room available.
                                  (2) The total number of rooms available.

9. The true statement at the right describes an operation involving the number 4.
      Write a rule to describe this operation with a matrix.

   $$4\begin{bmatrix} 2 & -3 \\ -2 & 1 \end{bmatrix} = \begin{bmatrix} 4(2) & 4(-3) \\ 4(-2) & 4(1) \end{bmatrix}$$

   In 10–12, write the given expression as a single matrix in simplest form.

10. $-2\begin{bmatrix} -4 & 7 \\ 3 & 5 \\ 0 & -1 \end{bmatrix}$ 11. $.5[1 \quad 2 \quad 0 \quad -3]$ 12. $3\begin{bmatrix} 4 & -2 \\ -2 & 1 \end{bmatrix} + \begin{bmatrix} 7 & 9 \\ 3 & -5 \end{bmatrix}$

In 13–17, find the product of the matrices.

13. $[4 \quad -2] \begin{bmatrix} 2 & -3 \\ -1 & 5 \end{bmatrix}$

14. $\begin{bmatrix} 7 & 4 \\ -3 & 2 \end{bmatrix} \begin{bmatrix} 0 & -2 \\ -1 & -5 \end{bmatrix}$

15. $\begin{bmatrix} 3 & -4 & 0 \\ -2 & 2 & 5 \\ -1 & 0 & -6 \end{bmatrix} \begin{bmatrix} -2 \\ 1 \\ -3 \end{bmatrix}$

16. $\begin{bmatrix} 4 & 2 & -3 \\ -6 & -2 & 5 \\ -1 & -2 & 0 \end{bmatrix} \begin{bmatrix} -3 & -1 \\ -3 & 2 \\ -3 & 0 \end{bmatrix}$

17. $-2 \begin{bmatrix} 2 & 6 \\ -3 & -4 \\ 1 & 5 \end{bmatrix} \begin{bmatrix} 2 & -3 & 5 \\ -3 & 1 & 0 \end{bmatrix}$

18. a. Multiply: **(1)** $\begin{bmatrix} -2 & 5 \\ 3 & -1 \end{bmatrix} \begin{bmatrix} 1 & 0 \\ 0 & 1 \end{bmatrix}$ **(2)** $\begin{bmatrix} 1 & 0 \\ 0 & 1 \end{bmatrix} \begin{bmatrix} -2 & 5 \\ 3 & -1 \end{bmatrix}$

b. How do the results of part **a** relate to the given matrices?

c. What multiplication property of the real number system is suggested?

d. Given: $A = \begin{bmatrix} 4 & -2 \\ -3 & 1 \end{bmatrix}$ What matrix should multiply matrix A so that the product is matrix A? Verify your answer by doing the multiplication.

19. a. Explain why it is not possible to find the product $\begin{bmatrix} 1 & 0 & 0 \\ 0 & 1 & 0 \\ 0 & 0 & 1 \end{bmatrix} \begin{bmatrix} 1 & 2 & 3 \\ 3 & 2 & 1 \end{bmatrix}$.

b. Describe a matrix for which a multiplicative identity can exist.

c. Write the matrix that would be the multiplicative identity for a matrix with 4 rows and 4 columns.

20. If $A = \begin{bmatrix} 1 & 2 \\ -1 & 0 \end{bmatrix}$ and $B = \begin{bmatrix} 3 & -2 \\ 0 & 5 \end{bmatrix}$, find the product: **a.** AB **b.** BA

c. What multiplication property of the real number system fails for matrix multiplication?

In 21 and 22, write the given expression as a single matrix in simplest form.

21. $\left(3 \begin{bmatrix} 5 & -1 \\ -2 & 0 \end{bmatrix} \right) \begin{bmatrix} 3 \\ -4 \end{bmatrix}$

22. $3 \left(\begin{bmatrix} 5 & -1 \\ -2 & 0 \end{bmatrix} \begin{bmatrix} 3 \\ -4 \end{bmatrix} \right)$

23. Name a multiplication property of the real number system that is demonstrated by the results of Exercises 21 and 22.

In 24 and 25, use a calculator.

24. Miguel and Carl will work together to build a wooden shed that requires both a light wood and a heavy wood in its construction.

a. For his part of the work, Miguel needs 20 sq. ft. of the light wood and 14 sq. ft. of the heavy. Carl needs 16 sq. ft. of the light and 12 sq. ft. of the heavy.

Write matrix A to display the information about the amount of wood needed. Use *Miguel* and *Carl* as row headings and *light* and *heavy* as column headings.

b. If the boys purchase the wood at the McKay Company, the cost will be $2.50 per sq. ft. for the light and $4 per sq. ft. for the heavy. If they purchase the wood at the Dale Company, the cost will be $2 per sq. ft. for the light and $4.50 per sq. ft. for the heavy.

Write matrix *B* to display the information about the cost. Use *light* and *heavy* as row headings, and *McKay* and *Dale* as column headings.

c. Find *AB*, the product of the two matrices.

d. Use *AB* to determine:
   (1) Miguel's projected cost at each company.
   (2) Carl's projected cost at each company.
   (3) The company at which the boys should purchase their wood to obtain the better buy. How much money will they save?

25. There are 200 subscribers to a theater group.

a. Of the 200 subscribers, 83 request orchestra seats, 55 mezzanine seats, and 62 balcony seats.

Write matrix *A* to display the data about seat requests. Use 1 row, headed *requests*, and use *orchestra*, *mezzanine*, and *balcony* as column headings.

b. The managers of the group collect price data at different theaters.
Astro: $35 orchestra, $25 mezzanine, $15 balcony
Lido: $40 orchestra, $30 mezzanine, $10 balcony
Vox: $37 orchestra, $32 mezzanine, $20 balcony

Write matrix *B* to display the data about ticket costs. Use *orchestra*, *mezzanine*, and *balcony* as row headings, and use *Astro*, *Lido*, and *Vox* as column headings.

c. Find *AB*, the product of the two matrices.

d. Use *AB* to determine:
   (1) The theater that would mean the highest total cost.
   (2) The theater that would mean the lowest total cost.
   (3) The difference in total cost if the group attends the Astro rather than the Vox.

歌 舞 伎 Kabuki (the three syllables mean music, dancing, and skill) is a form of Japanese theater that has been kept essentially unchanged for centuries. Traditionally, both male and female roles are performed by men. The exaggerated style of presentation accentuates the action, color, and drama.

CHAPTER SUMMARY

The properties of operations indicate what you can and cannot do when you are working with real numbers. For example, the commutative property says that it is permissible to switch the positions of two real numbers in an addition problem; this is not true in a subtraction problem. Many of these properties were already familiar to you, although you may not have seen them stated formally.

Here are some important rules:

- Subtracting a number is equivalent to adding the opposite of that number.
- Dividing by a number is equivalent to multiplying by the reciprocal.
- The product of two negative numbers is positive.
- The sum of two negative numbers is negative.
- The product of a negative number and a positive number is always negative. The sum could be zero, or it could be positive or negative, depending on which of the two numbers has greater absolute value.

You can use a number line to help you add and subtract real numbers. There is an inverse relationship between addition and subtraction, and between multiplication and division. You will use these relationships when you solve equations.

The properties of operations apply to operations with algebraic as well as numerical expressions. In operating with algebraic expressions, *like terms* and *unlike terms* are particularly important because like terms can be combined through addition or subtraction but unlike terms cannot.

You have seen how the arithmetic of real numbers is extended by defining new operations, and by using arrangements called *matrices*.

VOCABULARY CHECKUP

SECTION

3-1 *closure / substitution property*

3-2 *translation /*
commutative property of addition /
associative property of addition /
additive identity /
additive inverse / opposite

3-3 *commutative property of multiplication /*
associative property of multiplication /
multiplicative identity /
multiplicative inverse / reciprocal

SECTION

3-4 *inverse operation / minuend /*
subtrahend / difference /
dividend / divisor / quotient

3-5 *distributive property*

3-6 *like terms / unlike terms*

3-7 *intersection of sets /*
union of sets

3-8 *matrix*

In 1–4, state whether the set is closed under the indicated operation. (Section 3-1)

1. $\{0, 2, 4, 6, \ldots\}$; subtraction

2. $\{0, 2, 4, 6, \ldots\}$; addition

3. $\{5, 6, 7\}$; multiplication

4. $\{5, 6, 7, \ldots\}$; multiplication

In 5–21, perform the indicated operation. (Sections 3-2 through 3-4)

5. $-3 - 3$

6. $-8 + 3$

7. $-4(-2)$

8. $-7 - (-4)$

9. $-12 \div 6$

10. $-9 + 4 - 7$

11. $-5 + (-2)$

12. $6(-5)$

13. $-7 - 2 - (-5)$

14. $48 \div (-16)$

15. $2(-3)(-4)$

16. $-3 - (-5)$

17. $\frac{7}{4} \div (-3)$

18. $-2(4)$

19. $-2(-3)(-2)(3)$

20. $-2 - 3 - 1 + (-2)$

21. $3 - 2 - (-5)$

In 22–27, write the name of the property illustrated. (Sections 3-2 and 3-3)

22. $3 + (7 + 4) = (3 + 7) + 4$

23. $5 + (3 + 4) = 5 + (4 + 3)$

24. $-4(0) = 0$

25. $-7 + 0 = -7$

26. $-9 + 9 = 0$

27. $(2)(-5) = (-5)(2)$

In 28–33, use the distributive property to remove parentheses. (Section 3-5)

28. $4(a - 2)$

29. $(2a + 5)3$

30. $2a(5 + w)$

31. $(4 + 3x)3b$

32. $5(4w - 3h)$

33. $h(m - n)$

In 34–39, express the sum as a product. (Section 3-5)

34. $3a + 3w$

35. $5d + ad$

36. $3x - 12$

37. $9g - 12$

38. $15h - 10y$

39. $ax + 12a$

In 40–49, write an example of the property. (Section 3-5)

40. distributive property

41. commutative property of multiplication

42. multiplicative inverse property

43. multiplicative identity property

44. commutative property of addition

45. associative property of multiplication

46. additive inverse property

47. zero property of multiplication

48. associative property of addition

49. additive identity property

In 50–58, simplify each expression by combining like terms. (Section 3-6)

50. $4a + 7a$

51. $3b - 5b$

52. $8c + c$

53. $3a + 2w - w + 4a$

54. $5r - 12 - 3r + 6$

55. $7p - 9 - 5 - 2p - 4 - p$

56. $4ab - 5ac - ab + 9ac$

57. $4(h + 3) + 6(2h - 1)$

58. $5(x - y) + 4(3y - 2x)$

In 59–67, find the value of the expression if $a = -5$, $b = 4$, $p = -3$, and $r = -2$. (Section 3-6)

59. $-4a$ **60.** $-2p^2$ **61.** $-5r^3$ **62.** $a - b - p$ **63.** $3ab$

64. $\dfrac{b - a}{p}$ **65.** p^2r^3 **66.** $(pr)^2$ **67.** $2a - 3b - pr$

In 68–70, the operation is defined for all real numbers a and b. (Section 3-7)

68. a *max* b means the greater of the two values a and b.
Evaluate: **a.** 5 *max* 3 **b.** (-9) *max* (-11) **c.** $\frac{1}{3}$ *max* $.3$

69. $[a]$ means the greatest integer that is not greater than a.
Evaluate: **a.** $[-3.3]$ **b.** $[5.7]$ **c.** $[.5]$

70. $a \triangle b$ means $(a - b)^2$.
Evaluate: **a.** $4 \triangle 7$ **b.** $(-7) \triangle (-3)$ **c.** $(-5) \triangle 9$

In 71 and 72, find: **a.** $A \cap B$ **b.** $A \cup B$ (Section 3-7)

71. $A = \{1, 3, 5, \ldots, 13\}$ $B = \{$odd numbers between 10 and 20$\}$
72. $A = \{$the vowels in CANNIBAL$\}$ $B = \{$the vowels in CONTINENT$\}$

In 73 and 74, use the Venn diagram to list the element(s) of the indicated set. (Section 3-7)

73. $A \cap (B \cup C)$
74. $A \cup (B \cap C)$

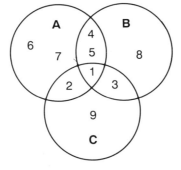

In 75 and 76, use the matrix to answer the question. (Section 3-8)

$$\begin{array}{c} \quad\ \text{boys}\ \ \text{girls} \\ \begin{array}{r} \text{1st grade} \\ \text{2nd grade} \\ \text{3rd grade} \end{array} \left[\begin{array}{cc} 24 & 33 \\ 32 & 30 \\ 31 & 29 \end{array}\right] \end{array}$$

75. What does the number in row 3, column 2 tell you?

76. How many boys are there in grades 1 through 3?

In 77 and 78, find the value of a. (Section 3-8)

77. $\begin{bmatrix} a & b \\ c & d \end{bmatrix} = \begin{bmatrix} 4 & 3 \\ 5 & -7 \end{bmatrix}$

78. $3[a \quad 5] = [-6 \quad 15]$

In 79 and 80, perform the indicated operation. (Section 3-8)

79. $\begin{bmatrix} 4 & 3 \\ -2 & 0 \\ 1 & 5 \end{bmatrix} + \begin{bmatrix} 0 & -7 \\ 1 & 8 \\ 6 & -4 \end{bmatrix}$

80. $\begin{bmatrix} 3 & -8 \\ 1 & -1 \end{bmatrix}\begin{bmatrix} 2 & 0 \\ 3 & -5 \end{bmatrix}$

1. a. If the table defines a commutative operation, copy and complete the table.
b. Find the value of $P \# R \# I \# M \# E$.
c. What is the identity element for operation #?
d. What is the inverse of R under #?
e. Are there any elements that do not have an inverse under #? Explain.

#	P	R	I	M	E
P	P	P		R	P
R		R			R
I	R	M	I	I	I
M		E		M	M
E	P	R	I	M	E

2. You can make an addition table that uses letters instead of numbers, as in the tables shown. According to Table 1, $a \oplus b = c$ and $b \oplus a = c$. Similarly, $(a \oplus b) \oplus c = c$ and $a \oplus (b \oplus c) = c$. If you try all possible additions with two and three letters, you will find that additions from this table are both commutative and associative.

⊕	a	b	c
a	b	c	a
b	c	a	b
c	a	b	c

⊕	a	b	c
a	b	a	c
b	c	a	b
c	a	b	c

Table 1　　　　**Table 2**

Table 2 is arranged so that $a \oplus b = a$ and $b \oplus a = c$. Similarly, $(a \oplus b) \oplus c = c$ and $a \oplus (b \oplus c) = a$. Thus, in Table 2, addition is neither commutative nor associative.

Can you fill in a third table in a similar manner, using a, b, and c each 3 times so that the resulting additions will always be commutative, but not always associative? Specifically, $(a \oplus b) \oplus c$ should not give the same result as $a \oplus (b \oplus c)$.

3. Pat is opening a delicatessen and plans to sell cold cuts weighed by the ounce. She cannot afford an electronic scale, so she plans to use a balance scale, on which she wants to be able to weigh any number of ounces from 1 through 40 in one weighing.

When Pat is buying weights for her scale, she finds that every weight costs $10, regardless of the number of ounces in the weight. If she bought one of every weight, she would spend $400. To be economical, she wants to buy the smallest possible number of weights. The salesperson tells her that she could manage with a 27-ounce weight and only 3 other weights. What are they?

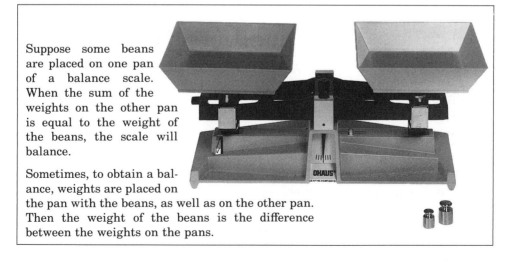

Suppose some beans are placed on one pan of a balance scale. When the sum of the weights on the other pan is equal to the weight of the beans, the scale will balance.

Sometimes, to obtain a balance, weights are placed on the pan with the beans, as well as on the other pan. Then the weight of the beans is the difference between the weights on the pans.

1. Find the two integers that have the following sums and products:
   a. sum: -4 **b.** sum: -33 **c.** sum: 75   **d.** sum: 104
   product: -117 product: 162   product: $-1,786$ product: 1,743

2. Together, you and your friends have $7.35 to spend at a local burger place. You also have, from that place, two coupons for 25¢ off any item over $1 and two coupons for 15¢ off any item less than $1.

   Selecting from the listed items, you plan to use all your coupons and spend all your money, and there is no food tax. What combination of foods could you buy?

Item	Cost
hamburger	2.20
hot dog	1.35
french fries	.90
large soda	1.40
small soda	.80
ice cream	.95
cupcake	.55

In 3 and 4, find the highest scoring route from START to END. The sign in front of each number indicates the operation that must be performed with the number. Perform each operation as you encounter it. You may move up or down or left or right, but not diagonally. You may not enter the same square twice. You need not enter every square.

3.

START	$+15$	$\times 3$	-51
-71	$+18$	$+26$	$\times 2$
$+42$	-67	-3	-98
$\times 4$	$+34$	-45	END

4.

START	$+37$	-12	$\times 3$
$+19$	$+18$	$+26$	-19
$+42$	-71	-4	-98
-11	$\times 2$	-39	END

In 1–10, choose the correct answer.

1. The product of a negative odd number and a positive even number
   (A) is negative and odd
   (B) is positive and even
   (C) is negative and even
   (D) is positive and odd
   (E) cannot be determined

2. If $x + 7 = 3$, then $3x + 21$ equals
   (A) 9   (B) 11   (C) 17   (D) 19   (E) 33

3. If $x = -2$, which of the following gives the largest value?
   (A) $2x^3$ (B) $(2x)^3$ (C) $(-2x)^3$
   (D) $-2x^3$ (E) $(2x)^2$

4. If x is a number between -1 and 2, and y is a number between 4 and 10, then the expression $\frac{xy}{2}$ represents a number between
   (A) -2 and 10   (B) -4 and 20
   (C) -4 and 10   (D) -5 and 20
   (E) -5 and 10

5. Which of the following will give the least value if $x = -1$ and $y = -2$?
   (A) $|x + y|$ (B) $|x - y|$
   (C) $|x| + |y|$ (D) $|x| - |y|$
   (E) $x - y$

6. If $c{\overset{\overset{a}{\bullet}}{\underset{\underset{b}{\bullet}}{\times}}}d = \dfrac{a}{b} - \dfrac{c}{d}$, then $3{\overset{\overset{5}{\bullet}}{\underset{\underset{6}{\bullet}}{\times}}}4 =$

(A) $\frac{1}{12}$ (B) $\frac{1}{6}$ (C) $\frac{1}{5}$ (D) $\frac{1}{3}$
(E) none of these

7. The reciprocal of n is multiplied by the reciprocal of q. By what must this product be multiplied to give 1 as the result?

(A) $\dfrac{1}{nq}$ (B) $\dfrac{1}{n+q}$ (C) $\dfrac{n+q}{nq}$

(D) $n + q$ (E) nq

8. Which of the following sets is not closed under both addition and multiplication?
(A) $\{0, 2, 4, 6, \ldots\}$ (B) $\{2, 4, 6, 8, \ldots\}$
(C) $\{3, 6, 9, 12, \ldots\}$ (D) $\{1, 3, 5, 7, \ldots\}$
(E) $\{0\}$

9. If $\underset{b\quad c}{\overset{a}{\triangle}}$ is defined to be $ab + bc - ac$, for

what value of y will $\underset{2\quad y}{\overset{3}{\triangle}}$ equal zero?

(A) -6 (B) -5 (C) 4   (D) 5   (E) 6

10. If $\textcircled{k} = (k-1)^2$ and $\textcircled{\textcircled{k}} = (k-2)^2$,

then $\textcircled{\textcircled{\textcircled{7}}}$ equals

(A) 33   (B) 9   (C) 4
(D) 24   (E) none of these

Questions 11–22 each consist of two quantities, one in Column A and one in Column B. You are to compare the two quantities and choose:

A if the quantity in Column A is greater;
B if the quantity in Column B is greater;
C if the two quantities are equal;
D if the relationship cannot be determined from the information given.

1. In certain questions, information concerning one or both of the quantities to be compared is centered above the two columns.

2. In a given question, a symbol that appears in both columns represents the same thing in Column A as it does in Column B.

3. x, n, and k, etc., stand for real numbers.

Column A	Column B
11. The multiplicative identity	The additive identity
12. The product of the reciprocals of 2 and 3	The reciprocal of the product of 2 and 3
13. $-1 - 2 - 3$	$(-1)(-2)(-3)$
14. $3^2(n^2) - (n^2)$	$2^3(n^2)$

Let $\lceil x \rceil$ = the least integer greater than x.

15. $\lceil -4 \rceil$	$\lceil -4.5 \rceil$

$$y = 0$$

16. $-5axy^3$	$axy^3 - 5$

$$a > 0, b < 0$$

17. ab	$-|a| \cdot |b|$

$$x = -3$$

18. $\dfrac{x-2}{|x-2|}$	$\dfrac{x-2}{|x|-2}$

$$x - y = 5$$
$$x - z = 3$$

19. y	z

$$a = 9, b = -3$$

20. $|a + b|$	$||b| - |a||$

$$y < 0$$

21. y^2	y^3

$$x > 0$$

22. x^2	x^3

1. Match the symbol with the description.

   a. $<$
   b. $\geq$
   c. $\neq$
   d. $\varnothing$
   e. $|a|$
   f. a^3

   (1) is equal to
   (2) is greater than
   (3) is less than
   (4) is greater than or equal to
   (5) is less than or equal to
   (6) 3 times a
   (7) null set
   (8) absolute value of a
   (9) a squared
   (10) a cubed
   (11) is not equal to

2. How many natural numbers are between 3 and 4?

3. Round 3,426,815 to the nearest hundred thousand.

4. In the expression $y^2 + 12x$, the numerical coefficient of x is:
   (1) y^2 (2) 1   (3) 12   (4) $y^2 + 12$

5. Find the perimeter of a rectangular kitchen that measures 9 feet by 12 feet.

6. Evaluate $\frac{1}{8}x^2$ when $x = 12$.

7. Using n to represent a number, write as an equation:
   The product of eight and a number, decreased by 3, equals 29.

8. If you add any four consecutive whole numbers, your result will be:
   (1) always odd   (2) always even   (3) sometimes odd, sometimes even

9. Arrange $-.43, -.4, -.428, -.401$ in ascending order, using $<$.

10. If $\frac{x}{3}, \frac{x}{4}$, and $\frac{x}{5}$ are integers, then x could be:
    (1) 30   (2) 40   (3) 50   (4) 60

11. Vertex A below is called *even* since it is the endpoint of an even number of segments. Vertex B is called *odd* since it is the endpoint of an odd number of segments. The vertices in the third figure are:
    (1) 4 even, 3 odd   (2) 5 even, 2 odd   (3) 2 even, 5 odd   (4) 4 odd, 3 even

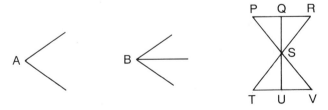

In 12 and 13, use the following graph:

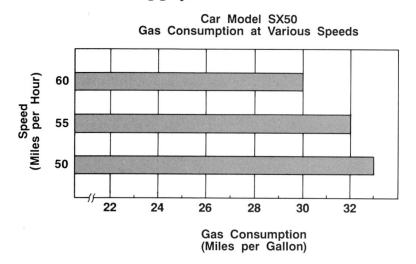

Car Model SX50
Gas Consumption at Various Speeds

12. How many miles per gallon will model *SX*50 get if its speed is 55 m.p.h.?

13. On a 500-mile trip, how many more gallons of fuel will model *SX*50 use traveling at 60 m.p.h. than at 50 m.p.h.? (Round answer to the nearest tenth of a gallon.)

14. If *f* fans pay *d* dollars each for admission to a soccer game, what will be the total number of dollars paid by these people?

15. The sum of the ages of Tom and his mother is 48. If his mother's age is 3 times Tom's age, how old is Tom?

16. For how many different integer values of *d* is $\frac{d}{8}$ both greater than $\frac{1}{3}$ and less than $\frac{1}{2}$?

17. For centuries, the average person has seen no more than one palindromic year (1881 was a palindromic year) in his or her lifetime. In your lifetime, you are likely to encounter two palindromic years. What are they?

18. The number of dimes in a half-dollar can be represented by:
    (1) $\frac{1}{2} \div 10$ (2) $\frac{1}{2} \div \frac{1}{10}$ (3) $.5 \div 10$ (4) $50 \div .1$

19. Evaluate:
    a. $3 + 4 \times 7$
    b. $(3 + 4) \times 7$
    c. Which expression, **a** or **b**, could be used to answer each of the following?
       (1) In the past 7 days, an Olympic trainee spent 3 hours a day lifting weights, and 4 hours a day on aerobic exercises. What was the total number of hours spent on both activities?
       (2) From January 14th, how many days is it until Valentine's Day, February 14th?

3-2 ADDITION OF REAL NUMBERS

TRY IT! *Problems 1–3 on page 102*

1.

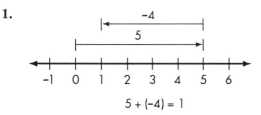

$$5 + (-4) = 1$$

Explanation: Starting from 0, the graph shows a translation of 5 units to the right, followed by a translation of 4 units to the left, arriving at 1.

$$
\begin{aligned}
2. \quad & -7 + 5 + (-4) + 1 \\
= \quad & -2 + (-4) + 1 \\
= \quad & -6 + 1 \\
= \quad & -5
\end{aligned}
$$

3. (1) Commutative property of addition.
   (2) Associative property of addition.
   (3) Addition property of opposites.
   (4) Addition property of 0.

3-3 MULTIPLICATION OF REAL NUMBERS

TRY IT! *Problems 1–5 on page 108*

1. $(-9)(4) = -36$

2. $-6 \cdot \frac{2}{3} = -4$

3. $(-4)(-1)(-3) = -12$

4. $(-2)^4 = (-2)(-2)(-2)(-2) = 16$

5. (1) Commutative property of multiplication.
   (2) Associative property of multiplication.
   (3) Multiplication property of reciprocals.
   (4) Multiplication property of 1.

3-4 SUBTRACTION AND DIVISION

TRY IT! *Problems 1–3 on page 112*

1. $5 - 12 + 2 = -5$ 2. $\dfrac{-28}{7} = -4$

3. $12 \div \left(-\frac{1}{4}\right) - 8 \div (-2)$

$$= 12(-4) - (-4) = -48 + 4 = -44$$

3-5 THE DISTRIBUTIVE PROPERTY; SUMMARY OF PROPERTIES

TRY IT! *Problems 1–3 on page 117*

1. **a.** $56 \times 3 + 56 \times 7 = 56(3 + 7)$
$$= 56(10) = 560$$

   b. $60\left(\frac{1}{2} - \frac{1}{6}\right) = 60 \times \frac{1}{2} - 60 \times \frac{1}{6}$
$$= 30 - 10 = 20$$

2. **a.** $3x(y + 4) = 3x \cdot y + 3x \cdot 4$
$$= 3xy + 12x$$

   b. $-5(a - 2b) = (-5)(a) + (-5)(-2b)$
$$= -5a + 10b$$

3. **a.** $5w + 3w = (5 + 3)w = 8w$

   b. $12x - 2y = 2 \cdot 6x - 2 \cdot y = 2(6x - y)$

3-6 USING THE PROPERTIES OF THE REAL NUMBER SYSTEM

TRY IT! *Problems 1–3 on page 120*

1. $9a + 5 - 7a - 8 = 9a - 7a + 5 - 8$
$$= (9 - 7)a - 3$$
$$= 2a - 3$$

2. $\quad 4(3d - 5) - 9$ 3. $\quad 4w - 3(4 - w)$
   $= 12d - 20 - 9$ $= 4w - 12 + 3w$
   $= 12d - 29$ $= 7w - 12$

TRY IT! *Problems 1–4 on page 121*

1. $\quad 2ab$ 2. $\quad -3ab^2$
   $= 2(-5)(-2)$ $= -3(-5)(-2)^2$
   $= 20$ $= -3(-5)(4)$
                       $= 60$

3. $b - a^2$
  $= (-2) - (-5)^2$
  $= (-2) - (25)$
  $= -27$

4. $100 - a^2(2 - b)$
  $= 100 - (-5)^2[2 - (-2)]$
  $= 100 - 25(4)$
  $= 100 - 100 = 0$

3-7   DEFINING AN OPERATION

TRY IT! *Problems 1 and 2 on page 125*

1. $a \star b = a + b - ab$

  a. $5 \star 2 = 5 + 2 - 5 \cdot 2 = 7 - 10 = -3$
  b. $(-3) \star (-1) = (-3) + (-1) - (-3)(-1)$
  $= -4 - 3 = -7$
  c. $4 \star .1 = 4 + .1 - 4(.1)$
  $= 4.1 - .4 = 3.7$

2.

$$d \begin{array}{c} a \\ \boxed{} \\ c \end{array} b = ac - bd$$

$$4 \begin{array}{c} -3 \\ \boxed{} \\ -2 \end{array} 1 + 2 \begin{array}{c} 3 \\ \boxed{} \\ 4 \end{array} -1$$

$= [(-3)(-2) - 1(4)] + [3(4) - (-1)(2)]$
$= [6 - 4] + [12 - (-2)] = 2 + 14 = 16$

  ANSWER: (4)
  $.5(24) - .5(-8) = 12 + 4 = 16$

3-8   ANOTHER ARITHMETIC: MATRICES

TRY IT! *Problems 1–4 on page 131*

1. $\begin{bmatrix} 2 & -1 \\ -4 & -3 \\ 1 & 0 \end{bmatrix} + \begin{bmatrix} 7 & -3 \\ 2 & -2 \\ -4 & 5 \end{bmatrix}$

$= \begin{bmatrix} 2 + 7 & -1 + (-3) \\ -4 + 2 & -3 + (-2) \\ 1 + (-4) & 0 + 5 \end{bmatrix} = \begin{bmatrix} 9 & -4 \\ -2 & -5 \\ -3 & 5 \end{bmatrix}$

2. $\begin{bmatrix} 5 & -4 & 4 \\ -1 & 3 & 6 \\ 2 & 0 & -2 \end{bmatrix} + \begin{bmatrix} 3 & -2 & -4 \\ -5 & 1 & 0 \\ -3 & 2 & -2 \end{bmatrix}$

$= \begin{bmatrix} 5 + 3 & -4 + (-2) & 4 + (-4) \\ -1 + (-5) & 3 + 1 & 6 + 0 \\ 2 + (-3) & 0 + 2 & -2 + (-2) \end{bmatrix}$

$= \begin{bmatrix} 8 & -6 & 0 \\ -6 & 4 & 6 \\ -1 & 2 & -4 \end{bmatrix}$

3. $[3 \quad -6 \quad 2] \begin{bmatrix} 2 & -1 \\ -4 & -3 \\ 1 & 0 \end{bmatrix}$

  row 1 × column 1
  $3(2) + (-6)(-4) + 2(1) = 32$

  row 1 × column 2
  $3(-1) + (-6)(-3) + 2(0) = 15$

$= [32 \quad 15]$

4. $\begin{bmatrix} 2 & -7 \\ -4 & 3 \end{bmatrix} \begin{bmatrix} -3 & 4 \\ 5 & -1 \end{bmatrix}$

  row 1 × column 1
  $2(-3) + (-7)(5) = -41$

  row 1 × column 2
  $2(4) + (-7)(-1) = 15$

  row 2 × column 1
  $(-4)(-3) + 3(5) = 27$

  row 2 × column 2
  $(-4)4 + 3(-1) = -19$

$= \begin{bmatrix} -41 & 15 \\ 27 & -19 \end{bmatrix}$

SELF-TEST Chapters 1-3

Part I

1. Evaluate $4x^2y^3$ when $x = 3$ and $y = 2$.

2. List the members of the set {whole numbers less than 3}.

3. Write the rational number $\frac{3}{11}$ as a decimal.

4. Use $>$, $<$, or $=$ to replace ? to make the sentence true.
$|3| - |-5|$? $5 - |3|$

5. Find a number on the number line that is the same distance from 0 as $-1\frac{1}{3}$, but on the opposite side of 0.

6. Find the solution set of $|y| = 5$ if the domain is the set of real numbers.

7. Find the value of $(-2)^3$.

8. Simplify: $x - 3y + 4x + 2y$.

9. From 25, subtract -9.

10. State which of the numbers is greater, $-\frac{2}{3}$ or $-\frac{4}{5}$.

11. Evaluate $\dfrac{a - b + c}{b - c}$ when $a = 10$, $b = 4$, and $c = 2$.

12. Using the domain $\{2, 2\frac{1}{2}, 3, 3\frac{1}{2}, 4, 4\frac{1}{2}\}$, find the solution set of $7 - w = 3\frac{1}{2}$.

In 13-20, choose the correct answer.

13. $(3 \times 4) \times 5 = 3 \times (4 \times 5)$ is an illustration of which property?
(1) associative
(2) commutative
(3) distributive
(4) identity

14. The value of $-18 - 3 + (-10)$ is:
(1) -31
(2) -25
(3) -11
(4) -5

15. Under which operation is the set $\{0, 1\}$ closed?
(1) addition
(2) subtraction
(3) multiplication
(4) division

16. Solve the inequality $m + 3 < 5$, given the domain $\{0, 1, 2, 3, 4, 5, 6, 7, 8, 9\}$.
(1) $\{m \,|\, m < 2\}$
(2) $\{0, 1, 2\}$
(3) $\{0, 1\}$
(4) $\{3, 4, 5, 6, 7, 8, 9\}$

17. Write, using mathematical symbols: When eight is subtracted from twice a number, the result is forty.

(1) $8 - 2n = 40$
(2) $2n - 8 = 40$
(3) $2(n - 8) = 40$
(4) $(2n)(-8) = 40$

18. Divide: $36 \div \left(-\frac{2}{3}\right)$

(1) -54 (2) -24 (3) 24 (4) 54

19. What number represents the result of a loss of \$250 and a profit of \$200?

(1) 450 (2) -450 (3) 50 (4) -50

20. If the domain is the set of real numbers, which is the graph of $-1 \le x < 3$?

(1)

(2)

(3)

(4)

Part II

21. Show by counterexample that if a and b are integers and $a \star b = ab - (a - b)$, the operation $\star$ is not commutative.

22. Use a property of the real numbers to justify each step of the following proof.

$$(1) \quad (ba) \cdot \frac{1}{b} = (ab) \cdot \frac{1}{b}$$

$$(2) \qquad\qquad = a \cdot \left(b \cdot \frac{1}{b}\right)$$

$$(3) \qquad\qquad = a \cdot 1$$

$$(4) \qquad\qquad = a$$

23. Write each sentence in algebraic notation.
  a. The product of 2 and n, increased by 5, is equal to 40.
  b. The sum of n and 12 is equal to the difference of 4 and n.
  c. When the product of n and 6 is divided by 4, the quotient is greater than 8.

24. The given expressions represent the lengths of the sides of a rectangle. If $a = 1$, $b = -2$, and $c = 4$, find the value of:
  a. $2a - bc$ **b.** ac^2
  c. b^2c **d.** $-(ab + bc)$
  e. the perimeter of the rectangle

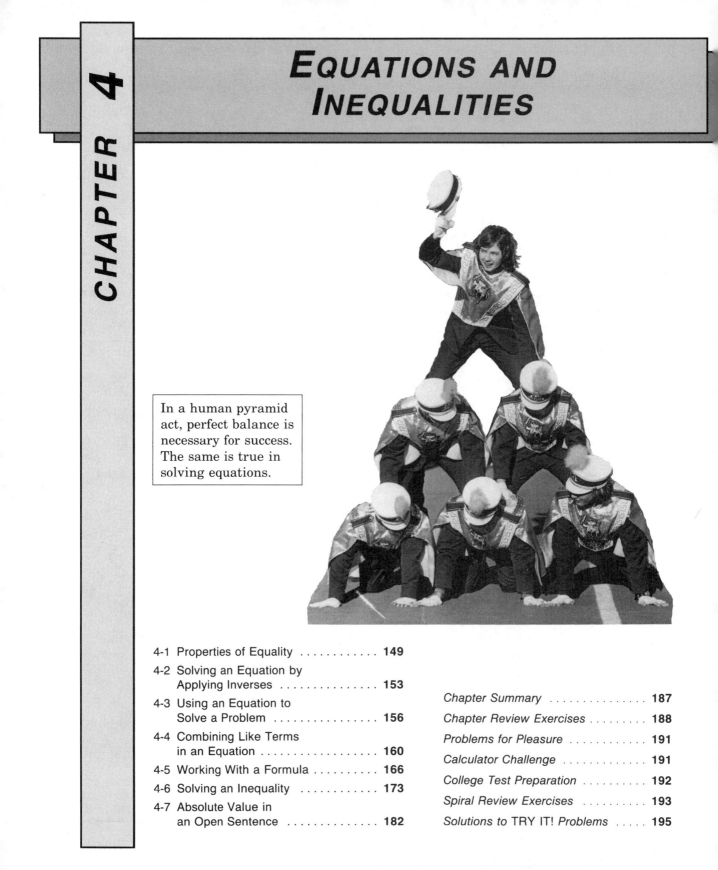

CHAPTER 4

EQUATIONS AND INEQUALITIES

In a human pyramid act, perfect balance is necessary for success. The same is true in solving equations.

4-1 Properties of Equality 149

4-2 Solving an Equation by
Applying Inverses 153

4-3 Using an Equation to
Solve a Problem 156

4-4 Combining Like Terms
in an Equation 160

4-5 Working With a Formula 166

4-6 Solving an Inequality 173

4-7 Absolute Value in
an Open Sentence 182

Chapter Summary 187

Chapter Review Exercises 188

Problems for Pleasure 191

Calculator Challenge 191

College Test Preparation 192

Spiral Review Exercises 193

Solutions to TRY IT! Problems 195

In this chapter, you will learn procedures for solving open sentences of equality and inequality. The steps of the procedures are justified by the properties of equality and inequality, and by the properties of the real numbers.

4-1 PROPERTIES OF EQUALITY

The following properties of equality tell how an equality may be operated upon and still remain an equality.

Name of Property	Representation for All Real Numbers a, b, c	Example
Addition Property of Equality	**If $a = b$, then $a + c = b + c$.**	If $x - 2 = 8$, add 2 to both sides so that $x - 2 + 2 = 8 + 2$ and $x = 10$.
If the same number is added to equal quantities, the equality is retained.		
Subtraction Property of Equality	**If $a = b$, then $a - c = b - c$.**	If $y + 3 = 9$, subtract 3 from both sides so that $y + 3 - 3 = 9 - 3$ and $y = 6$.
If the same number is subtracted from equal quantities, the equality is retained.		
Multiplication Property of Equality	**If $a = b$, then $a \times c = b \times c$.**	If $\dfrac{z}{5} = 10$, multiply both sides by 5 so that $5 \cdot \dfrac{z}{5} = 5 \cdot 10$ and $z = 50$.
If equal quantities are multiplied by the same number, the equality is retained.		
Division Property of Equality	**If $a = b$, then $a \div c = b \div c$.** $(c \neq 0)$	If $2t = 12$, divide both sides by 2 so that $\dfrac{2t}{2} = \dfrac{12}{2}$ and $t = 6$.
If equal quantities are divided by the same number, the equality is retained.		

Recall the Substitution Property (a number may be replaced by its equivalent), and note this property is also applicable to the equality relation. Later in your studies, you will learn other properties of equality.

In contrast to the preceding properties, which involve arithmetic operations, other properties of equality are independent of the operations. The first two of the following properties allow you to view an equality in different ways. The final property presented here connects two numbers because of their equality relationship to a third number.

Name of Property	Representation for All Real Numbers a, b, c	Example
Reflexive Property of Equality	$a = a$	$-2 = -2$
Any number is equal to itself.		
Symmetric Property of Equality	**If $a = b$, then $b = a$.**	If $4 = x$, then $x = 4$.
An equality may be reversed.		
Transitive Property of Equality	**If $a = b$ and $b = c$, then $a = c$.**	If $x = 5$ and $5 = y$, then $x = y$.
If one number is equal to a second number, and the second number equals a third number, then the first number equals the third number.		

MODEL PROBLEMS

In 1–3, name the property of equality that the situation describes.

Situation   *Property of Equality*

1. If a dozen audio tapes cost $10.50, then   Multiplication property.
   three dozen of those tapes cost $31.50.

2. If AB measures 5 cm, and 5 cm is also   Transitive property
   the measure of CD, then $AB = CD$.   or Substitution property.

   A   B   C   D

3. If $x - 7\frac{1}{2} = 12\frac{1}{2}$, then $x = 20$.   Addition property.

4. Write the properties of equality that justify the steps in the chain of equivalent equations that lead to the solution of the given equation.

   Equivalent Equations   *Reasons*
   $6x + 5 = 17$ Given.
   $6x = 12$ Subtraction property.
   $x = 2$ Division property.

In 1–11, name the property of equality that the situation describes.

1. Since an hour contains 60 minutes, then 4 hours contain 240 minutes.

2. If Sandy measures 5′3″, and 5′3″ is Beth's height, then Sandy is as tall as Beth.

3. After his purchases were totaled at $27.40, Lou decided to include a pack of gum that cost 50 cents. Lou should pay $27.90.

4. If a dozen pencils cost $1.80 and a dozen file tabs cost $1.80, then the supplier can exchange a dozen tabs for a dozen pencils.

5. Mwela bought some items totaling $32.90 from a supplier. The purchase included a dozen pencils that cost $1.80, tax included. If Mwela decided not to take the pencils, then he should pay $31.10.

6. If 3 brothers equally share a $15,000 inheritance from their grandmother, then each man receives $5,000.

7. Since a quart contains 32 ounces, and a gallon contains 4 quarts, then a gallon contains 128 ounces.

8. If $RT = SU$, then $RS = TU$.

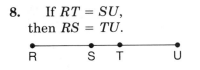

9. If arc LM = arc NO, then arc LN = arc MO.

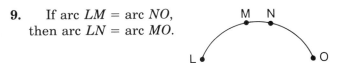

10. If the measures of angles ABD and EBC are equal, then the measures of angles ABE and DBC are equal.

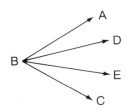

11. If the area enclosed by $ABCH$ is equal to the area enclosed by $FGHE$, then the area of $ABDE$ equals the area of $FGCD$.

In 12–23, name the property of equality that leads to the solution of the equation.

12. If $x - 1 = 7$, then $x = 8$.

13. If $y + 4 = 9$, then $y = 5$.

14. If $7z = 14$, then $z = 2$.

15. If $\dfrac{t}{3} = 18$, then $t = 54$.

16. If $17 = a + 3$, then $14 = a$.

17. If $20 = b - 4$, then $24 = b$.

18. If $15 = \dfrac{c}{5}$, then $75 = c$.

19. If $64 = 16d$, then $4 = d$.

20. If $y + 2.3 = 4.7$, then $y = 2.4$.

21. If $3x = 15.6$, then $x = 5.2$.

22. If $q - 1\frac{1}{2} = 3\frac{1}{2}$, then $q = 5$.

23. If $\frac{1}{4}m = \frac{1}{4}$, then $m = 1$.

In 24–31, write the properties of equality that justify the steps in the chain of equivalent equations that lead to the solution of the given equation.

24. $3x + 2 = 14$
$3x = 12$
$x = 4$

25. $5y - 4 = 11$
$5y = 15$
$y = 3$

26. $7 + 4z = 27$
$4z = 20$
$z = 5$

27. $30 = 9 + 7k$
$21 = 7k$
$3 = k$

28. $-7t + 14 = 0$
$-7t = -14$
$t = 2$

29. $-8f - 5 = 11$
$-8f = 16$
$f = -2$

30. $6 - 5y = 16$
$-5y = 10$
$y = -2$

31. $6 = 18 - 4x$
$-12 = -4x$
$3 = x$

32. Other relations may have some of the properties that the relation *equality* has for the set of real numbers. For example, the relation "sits next to" when applied to {people in a row} satisfies the symmetric property. That is, it is true for any two persons in that row that if person 1 is sitting next to person 2, then person 2 is also sitting next to person 1.

Does the relation "sits next to" when applied to {people at a round table} satisfy the symmetric property? Explain.

An object or a figure has **symmetry** if its image on one side of a central point or dividing line is reflected on the other side.

In 33–36, tell if the relation for the given set satisfies the symmetric property. If your answer is "no," give an example of why the property fails.

33. "is the brother of"   {Al, Ed}

34. "is the brother of"   {Al, Ed, Sue}

35. "is less than"   {real numbers}

36. "has the same area as"   {shapes}

In 37 and 38, tell if the relation for the given set satisfies the reflexive property. If your answer is "no," give an example of why the property fails.

37. "is greater than"   {real numbers}

38. "is the square of"   {0, 1}

In 39–42, tell if the relation for the given set satisfies the transitive property. If your answer is "no," give an example of why the property fails.

39. "is the son of"   {humans}

40. "is a factor of"   {natural numbers}

41. "is parallel to"   {lines}

42. "is perpendicular to"   {lines}

4-2 SOLVING AN EQUATION BY APPLYING INVERSES

Equivalent Equations

Equations that have the same solution set are called ***equivalent equations***. For example, $3x + 4 = 16$ and $3x = 12$ and $x = 4$ are equivalent equations because they have the same solution set, $\{4\}$.

When you solve an equation, you transform it into a chain of progressively simpler equivalent equations. The simplest of these equations then identifies the replacement for the variable that makes the open sentence a true statement. Thus, in the example above, each succeeding equation is simpler than the one before it, and the last equation, $x = 4$, shows the solution. Note that you can read the solution when the variable stands alone on one side of the equation.

Applying an Inverse Operation

Your objective, then, is to simplify an equation so that the variable stands alone on one side. To do this, you use the properties of equality, applying the same operation to both sides to arrive at a simpler equivalent equation.

In deciding which operations to apply, note that you must undo the effects of the operations that exist in the given equation. Applying an ***inverse operation*** will undo the effect of an existing operation. Thus, in the equation $x + 2 = 10$, 2 is added to x. Subtracting 2 from both sides results in the equivalent equation $x = 8$. Recall that the basic operations are inverses in pairs: addition and subtraction, multiplication and division.

MODEL PROBLEMS

1. Solve and check: $x - 5 = 4$

Solution: Add 5 to both sides.

$$x - 5 = 4$$
$$x - 5 + 5 = 4 + 5$$
$$x = 9$$

Check: Substitute 9 for x in the given equation, and evaluate the two sides of the equation separately.

$$x - 5 = 4$$
$$9 - 5 \overset{?}{=} 4$$
$$4 = 4 \checkmark$$

ANSWER: $x = 9$

2. Solve and check: $22 = 4y$

Solution: Divide both sides by 4.

$$22 = 4y$$
$$\frac{22}{4} = \frac{4y}{4}$$
$$5\tfrac{1}{2} = y$$

Check: $22 = 4y$

$$22 \overset{?}{=} 4\left(5\tfrac{1}{2}\right)$$
$$22 = 22 \checkmark$$

ANSWER: $y = 5\tfrac{1}{2}$

3. Solve and check: $\frac{3}{4}x - 4 = 17$

Solution: When there are two operations to be undone, you will find it easier to apply the addition or subtraction inverses first, before those of multiplication or division.

$$\frac{3}{4}x - 4 = 17$$

$$\frac{3}{4}x - 4 + 4 = 17 + 4 \quad \text{Add 4 to both sides.}$$

$$\frac{3}{4}x = 21$$

$$\frac{4}{3}\left(\frac{3}{4}x\right) = \frac{4}{3}(21) \quad \text{Multiply by } \frac{4}{3}.$$

$$x = 28$$

ANSWER: $x = 28$

Check:

$$\frac{3}{4}x - 4 = 17$$

$$\frac{3}{4}(28) - 4 \stackrel{?}{=} 17$$

$$21 - 4 \stackrel{?}{=} 17$$

$$17 = 17 ✓$$

Now it's your turn to . . . **TRY IT!**   (See page **195** for solutions.)

In 1–3, solve and check.

1. $m - 3\frac{2}{5} = 1$ **2.** $12 = -3h$ **3.** $4a + 9 = -19$

EXERCISES

In 1–76, solve the equation and check.

1. $x - 7 = 14$ **2.** $7 + x = 14$ **3.** $7x = 14$ **4.** $14x = 7$

5. $7 = 14 + x$ **6.** $x - 14 = 7$ **7.** $\dfrac{x}{14} = 7$ **8.** $\dfrac{x}{7} = 14$

9. $9 = 18 + y$ **10.** $9 = y - 18$ **11.** $18y = 9$ **12.** $-9y = 18$

13. $\dfrac{y}{18} = 9$ **14.** $\dfrac{y}{9} = 18$ **15.** $18 = \frac{1}{9}y$ **16.** $-18 = y - 9$

17. $-z = 6$ **18.** $z - 6 = 6$ **19.** $6 + z = -6$ **20.** $-6 = z - 6$

21. $-15r = 3$ **22.** $3r = -15$ **23.** $15 + r = -3$ **24.** $r - 15 = -3$

25. $t + \frac{1}{2} = 2$ **26.** $2 = t - \frac{1}{2}$ **27.** $\frac{1}{2}t = 2$ **28.** $2t = \frac{1}{2}$

29. $\frac{1}{2}t = \frac{1}{2}$ **30.** $2 + t = \frac{1}{2}$ **31.** $t - 2 = \frac{1}{2}$ **32.** $\frac{1}{2} = -2t$

33. $\dfrac{m}{5} = -5$ **34.** $0 = r - 8$ **35.** $n + 15 = 15$ **36.** $\frac{1}{10}x = 0$

37. $0 = 2b$ **38.** $c + 0 = -4$ **39.** $-q = -24$ **40.** $\frac{1}{2}p = 8$

41. $d - 5 = 2.3$ **42.** $m + .7 = 2.9$ **43.** $4x = 3.2$ **44.** $9 = .3y$

45. $c - 1\frac{1}{4} = 6\frac{1}{2}$ **46.** $9\frac{1}{4} = d + 3\frac{1}{2}$ **47.** $\frac{1}{8}x = \frac{1}{4}$ **48.** $x + 1.5 = \frac{1}{2}$

49. $\dfrac{1}{3} = \dfrac{m}{4}$ **50.** $\dfrac{c}{9} = -\dfrac{2}{3}$ **51.** $\dfrac{2x}{3} = \dfrac{4}{9}$ **52.** $\dfrac{-3y}{2} = -\dfrac{15}{4}$

53. $3m + 5 = 35$ **54.** $5a + 17 = 47$ **55.** $4x - 1 = 15$ **56.** $16 - z = 12$

57. $55 = 7 - 6a$ **58.** $17 = 8c - 4$ **59.** $15x + 14 = 19$ **60.** $11 = 20 - 3z$

61. $14 = 12b + 8$ **62.** $8 = 18c - 1$ **63.** $11 = 15t + 1$ **64.** $11 = 1 - 16d$

65. $2c + 1 = -31$ **66.** $32 - 7s = 4$ **67.** $2y + 18 = 8$ **68.** $2x + 37 = 9$

69. $5r + 15 = 0$ **70.** $4t + 8 = 8$ **71.** $-5.4 = 2.6 + 2z$ **72.** $1.2h - 5.6 = 1.6$

73. $.1p + .1 = 1$ **74.** $\frac{1}{2}z + 7 = 15$ **75.** $-25 = \frac{7}{3}r - 11$ **76.** $\frac{2}{5}w - 9 = -19$

In 77–85, determine the element(s) in the solution set if:
   a. $x \in \{\text{natural numbers}\}$ **b.** $x \in \{\text{real numbers}\}$

77. $\{x \mid x + 5 = 17\}$ **78.** $\{x \mid 48x = 12\}$ **79.** $\{x \mid x - 3 = -4\}$

80. $\{x \mid 2x + 11 = 15\}$ **81.** $\{x \mid 8x - 1 = -1\}$ **82.** $\{x \mid .2x - 4.9 = -3.1\}$

83. $\{x \mid \frac{2}{3}x - 12 = 60\}$ **84.** $\{x \mid 7.5 - \frac{1}{2}x = 8\}$ **85.** $\left\{x \mid 80 = 20 - \dfrac{3x}{5}\right\}$

In 86–93, solve the equation to evaluate the expression.

86. If $g + 9 = 11$, evaluate $7g$.   **87.** If $t - .5 = 2.5$, evaluate $t + 7$.

88. If $9x = 36$, evaluate $\frac{1}{2}x$.   **89.** If $\dfrac{r}{2} = 12$, evaluate $5r$.

90. If $\frac{2}{3}z = 18$, evaluate $3z + 7$.   **91.** If $9 - t = 6$, evaluate $2t$.

92. If $\frac{1}{2}q + 10 = 20$, evaluate $5 - q$.   **93.** If $2.4 + .1m = 3.8$, evaluate $5m - 7$.

94. Recall that *equal matrices* are matrices whose corresponding elements are equal.

   a. Which of the following matrices is equal to $\begin{bmatrix} 2 & -3 \\ 0 & 4 \end{bmatrix}$?

   I. $\begin{bmatrix} 2 & 4 \\ 0 & -3 \end{bmatrix}$ II. $\begin{bmatrix} -3 & 2 \\ 4 & 0 \end{bmatrix}$ III. $\begin{bmatrix} 2 & -3 \\ 0 & 4 \end{bmatrix}$

   (1) I, II, and III   (2) only I and II   (3) only III   (4) none

   b. If $\begin{bmatrix} 2 & -3 \\ 0 & 4 \end{bmatrix} = \begin{bmatrix} x - 5 & -3 \\ 0 & 4 \end{bmatrix}$, find the value of x.

In 95–98, two equal matrices are given. Find the value of: **a.** x **b.** y

95. $\begin{bmatrix} x + 7 & 2 \\ 15 & 8 \end{bmatrix} = \begin{bmatrix} -12 & 2 \\ 15 & y \end{bmatrix}$ **96.** $\begin{bmatrix} 10 & 2 \\ 3x - 2 & -1 \end{bmatrix} = \begin{bmatrix} -y & 2 \\ 16 & -1 \end{bmatrix}$

97. $\begin{bmatrix} 5x & -2 \\ 9.1 & 7y \\ 2 - 1 & 3 \end{bmatrix} = \begin{bmatrix} -10 & -2 \\ 9.1 & 0 \\ 1 & 3 \end{bmatrix}$ **98.** $\begin{bmatrix} 12 \div 4 & 8 - 0 \\ \frac{1}{2}x + 6 & 8(0) \\ .1(20) & 8 + 0 \end{bmatrix} = \begin{bmatrix} 2 + 1 & 8 \\ 16 & 2y + 2 \\ 2 & 8 \end{bmatrix}$

4-3 USING AN EQUATION TO SOLVE A PROBLEM

Earlier, you learned a general framework and some strategies for problem solving. Then you saw how an algebraic model can be used to represent a real-life situation. Now that you can solve an equation, you can apply the general framework to carry an algebraic model through to a solution, thus adding an *algebraic strategy* to those you already know. Study the following procedure for solving verbal problems.

PROCEDURE

To solve a verbal problem by using an equation containing one variable:

1. Select a variable to represent the basic unknown number of the problem.
2. Represent all the other unknown numbers in terms of that variable.
3. Write an equation that represents the relationships stated in the problem.
4. Solve the equation.
5. Check the value you obtained for the variable by testing it in the original verbal problem to see that it satisfies all stated conditions.

MODEL PROBLEM

Ned traveled $\frac{1}{4}$ of the distance that Ben traveled. If Ned traveled 12 miles, how far did Ben travel?

How to Proceed	*Solution*
(1) Represent Ben's distance by a variable.	Let d = distance Ben traveled.
(2) Using the same variable, represent Ned's distance.	Then $\frac{1}{4}d$ = distance Ned traveled.
(3) Write the statement as an equation.	$\frac{1}{4}d = 12$ (Ned traveled 12 miles.)
(4) Solve the equation.	$d = 48$ Multiply by 4.
(5) Check in the original problem.	If Ben went 48 miles and Ned 12 miles, did Ned go $\frac{1}{4}$ of Ben's distance? Yes.

ANSWER: Ben traveled 48 miles.

Now it's your turn to . . . **TRY IT!**

1. Ming is 5 years older than Tom. If Ming is 31 years old, how old is Tom?
2. Zeb owns 3 times as many sheep as Al. If Zeb owns 84 sheep, how many does Al own?

(See page 195 for solutions.)

EXERCISES

In 1–40, solve the problem using an equation containing one variable.

1. If 7 is subtracted from a number, the result is 46. Find the number.
2. If 18 is added to a number, the result is 32. Find the number.
3. Ten less than a number is 42. Find the number.
4. After Rolf spent $.25, he had $.85 left. How much money did he have originally?
5. Eight years ago, Teresita was 7 years old. How old is she now?
6. Twelve years from now, Paul will be 30 years old. How old is Paul now?
7. After the price of a boat rose $225, it was sold for $2,670. What was the original price?
8. After a car increased its rate of speed by 15 miles per hour, it was traveling 48 miles per hour. What was its original rate of speed?
9. During a charity drive, the boys in a class contributed $3.75 more than the girls. If the boys contributed $8.25, how much did the girls contribute?
10. The height of a parallelogram measures 8 cm less than its length. If the height measures 9.5 cm, find the length of the parallelogram.

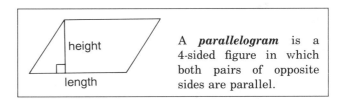

A ***parallelogram*** is a 4-sided figure in which both pairs of opposite sides are parallel.

11. A high school admitted 1,125 sophomores, which was 78 fewer than the number admitted the year before. How many sophomores were admitted the year before?
12. A dealer sold an electric broiler for $39.98. This sum was $12.50 more than the broiler cost him. How much did the broiler cost the dealer?
13. After using a basketball for some time, Carla sold it for $7.25 less than she paid for it. If Carla sold the ball for $3.50, how much did she pay for it originally?
14. Of the food processors produced by Apex Company last month, 4 hundredths were defective. If 16 processors were defective, how many did Apex produce?
15. Bill traveled 5 times as far as Holly. If Bill traveled 150 miles, how far did Holly travel?
16. If Arnon earned $600 in 2 weeks, what was his weekly salary?

17. The twelve signs of the Zodiac are depicted in a circle. The distance around a circle, the **circumference**, contains 360 degrees. If the twelve arcs are equal in measure, find the number of degrees in each arc.

18. Three-fourths of the daily amount of oxygen that an average adult needs to keep alive is 21 pounds. How much oxygen does an adult need daily?

19. Wilma deposited $90 in the bank last month. This was $2\frac{1}{2}$ times as much as Robin deposited. How much did Robin deposit?

20. Sandy cut a piece of lumber into 6 equal pieces. If each piece was $1\frac{1}{2}$ feet long, what was the length of the original piece of lumber?

21. Six months after Mr. Doyle bought a used car, he sold it, taking a loss of $\frac{1}{5}$ of the price he paid. If he lost $550, what had he paid for the car?

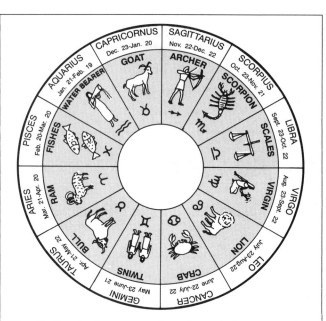

The *Zodiac* is the name given by astrologers to that portion of the sky through which the sun, moon, and planets move from east to west. The Zodiac is divided into 12 equal portions called *houses*, each of which is named for a constellation of stars.

22. George has walked $\frac{1}{4}$ of the distance from his home to school. If he has walked $\frac{1}{2}$ mile, find the distance from his home to school.

23. The Newtown Bolts won $\frac{3}{4}$ of the games played this season. If they won 18 games, how many games did they play this season?

24. Ms. Marcos gave a test to her 5th-period math class. The table shows the distribution of the 28 test scores. How many students scored in the interval 81–90?

Interval	Frequency
61–70	6
71–80	9
81–90	?
91–100	5

25. Last week, Revival Records reissued 41 versions of five songs that had been popular during the '60s. Three of these songs accounted for 22 of the current revivals. How many versions did Revival do of the other two songs?

26. The time it takes to digest a food depends on its fat and fiber content, how well the food is chewed, and the amount consumed. It takes about 2 hours and 5 minutes to digest a serving of beef. If it takes 50 minutes less to digest a serving of chicken, about how long does it take to digest chicken?

27. If 6 times a number is decreased by 4, the result is 68. Find the number.

28. When 12 is subtracted from 3 times a number, the result is 24. Find the number.

29. If 38 is added to $\frac{5}{9}$ of a number, the result is 128. Find the number.

30. In a municipal parking lot, the number of unrestricted parking spaces is 12 more than 7 times the number of spaces reserved for the handicapped. If there are 52 parking spaces altogether, how many are set aside for the handicapped?

31. Last month, Ace Construction completed 115 houses, which was 5 less than twice the number that Build-Right completed. How many houses did Build-Right complete?

32. Of the 48 continental United States, the number of states that share a border with five states is 2 more than four times the number of states that share a border with eight states. If there are 10 states of the five-border type, how many states are of the eight-border type?

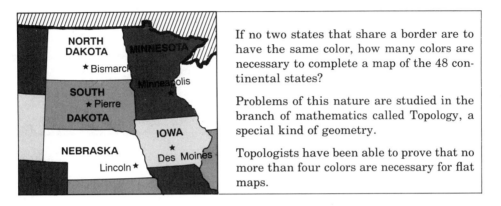

If no two states that share a border are to have the same color, how many colors are necessary to complete a map of the 48 continental states?

Problems of this nature are studied in the branch of mathematics called Topology, a special kind of geometry.

Topologists have been able to prove that no more than four colors are necessary for flat maps.

33. For today's performance, Comic Carnival sold 130 children's tickets. If this number exceeded four times the number of adult tickets by 10, how many adult tickets were sold?

34. Chapter 2 of Susan's Biology textbook has 7 less than twice the number of pages contained in Chapter 1. If Chapter 2 has 39 pages, how many are there in Chapter 1?

35. In December, the Bovins made a number of telephone calls to their family in France. As a result, their December telephone bill exceeded three times their November bill by $15. If the December bill was $120.60, how much was the November bill?

36. Mr. Berges went twice a week to a physical therapist to whom he paid $60 per week. At the last session, Mr. Berges purchased a set of weights for $75. If his total cost was $435, including the weights, for how many weeks did Mr. Berges go to the therapist?

37. On her job at Multi-Media Mail Order, Patrice took two days to complete the labeling of a mailing of envelopes. On the second day, she labeled 30 less than $\frac{3}{4}$ of the number she had labeled on the first day. If Patrice labeled 1,980 envelopes on the second day, how many envelopes were in the mailing?

38. Lloyd needs $90 to buy a camera. He has only $12 in his bank account. If he saves $13 each week, how long will it take him to accumulate the $90 for the camera?

39. At a conference, $\frac{2}{3}$ of the scheduled participants had arrived by 8 A.M. The arrival of 3 more participants brought the number present to 25. What was the total number expected?

40. On a stock exchange, 3 times as many shares of *PDQ* Electronics were traded today as yesterday. If another 300 shares of the stock had been traded, the total number of shares traded for the two days would have been 6,000. How many shares were traded today?

4-4 COMBINING LIKE TERMS IN AN EQUATION

When an equation contains grouping symbols, or more than one variable term, the solution procedure changes as follows.

PROCEDURE

To solve an equation containing grouping symbols or like terms:

1. If the equation contains grouping symbols, use the distributive property to eliminate the grouping symbols, and combine like terms.

2. If the equation has the variable on both sides, use an inverse operation to collect the variable terms on one side, and combine like terms.

MODEL PROBLEMS

1. Solve: $5(3y - 2) = 5$

Solution

$5(3y - 2) = 5$
$15y - 10 = 5$ Distributive property.
$15y = 15$ Add 10.
$y = 1$ Divide by 15.

Check by substituting $y = 1$ into the original equation.

ANSWER: $y = 1$

2. Solve: $7x = 63 - 2x$

Solution

$7x = 63 - 2x$
$7x + 2x = 63 - 2x + 2x$ Collect the variable terms.
$9x = 63$ Combine.
$x = 7$ Divide by 9.

Check by substituting $x = 7$ into the original equation.

ANSWER: $x = 7$

3. The larger of two numbers is 4 times the smaller. If the sum of the two numbers is 55, find the numbers.

Solution

Let x = the smaller number.
Then $4x$ = the larger number.
The sum of the two numbers is 55.
$x + 4x = 55$
$5x = 55$
$x = 11$
$4x = 44$

Check

Is the larger number 4 times the smaller?
$44 \overset{?}{=} 4 \times 11$ Yes.
Is the sum of the two numbers 55?
$11 + 44 \overset{?}{=} 55$ Yes.

ANSWER: The smaller number is 11 and the larger number is 44.

The flowchart summarizes the decisions you make when solving equations.

SOLVING LINEAR EQUATIONS

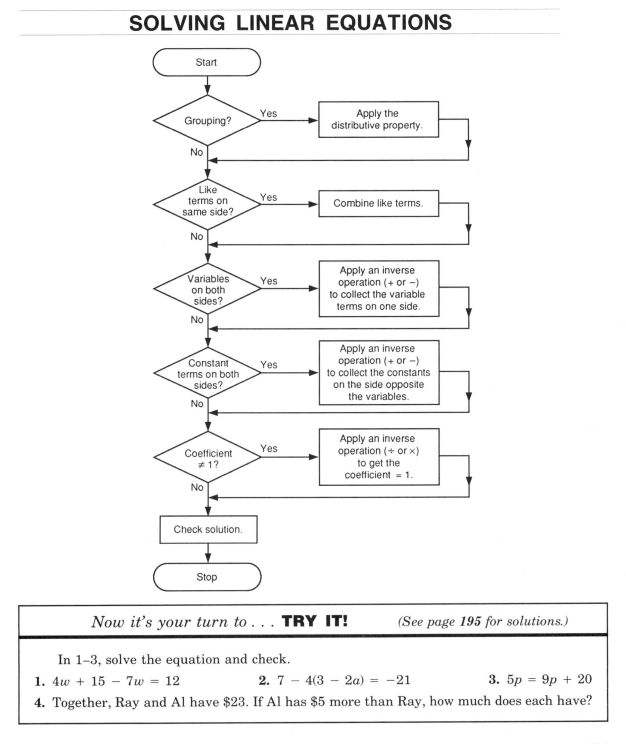

Now it's your turn to . . . **TRY IT!**   *(See page 195 for solutions.)*

In 1–3, solve the equation and check.

1. $4w + 15 - 7w = 12$ **2.** $7 - 4(3 - 2a) = -21$ **3.** $5p = 9p + 20$

4. Together, Ray and Al have $23. If Al has $5 more than Ray, how much does each have?

In 1–100, solve the equation and check.

1. $2a + 2a = 50$

2. $144 = 9b + 3b$

3. $12x - 4x = 108$

4. $5x - 3x = 22$

5. $18 = 7x - x$

6. $3.6d - 2.4d = 24$

7. $7x = 10 + 2x$

8. $9x = 44 - 2x$

9. $12y = 3y + 27$

10. $9x = 3x - 54$

11. $y = 4y + 30$

12. $2d = 36 + 5d$

13. $8x + 3x + 4x = 60$

14. $3e - e + 4e = 90$

15. $39 = 8c + 6c - c$

16. $.8m = .2m + 24$

17. $8y = 90 - 2y$

18. $4a - 55 = 9a$

19. $6y + 2y - 3 = 21$

20. $5y + 7 + y = 37$

21. $26 = 3y + 2y - 9$

22. $4 - 2y = 6y$

23. $3 - y = 8y$

24. $2.3x + 36 = .3x$

25. $8y - 3y + 7 = 87$

26. $6x - x + 12 = 52$

27. $95 = 8c - 3c + 15$

28. $5a - 40 = 3a$

29. $x = 9x - 72$

30. $y = 9y - 56$

31. $\frac{1}{4}x + \frac{1}{2}x = 18$

32. $\frac{2}{3}x - \frac{1}{3}x = 17$

33. $.5m - 30 = 1.1m$

34. $x - (12 - x) = 38$

35. $5y + (2y - 7) = 63$

36. $10y - (5y + 8) = 42$

37. $(15x + 7) - 12 = 4$

38. $7r + 10 = 3r + 50$

39. $8t + 17 = 5t + 35$

40. $4y + 20 = 5y + 9$

41. $8s + 56 = 14s + 26$

42. $(14 - 3c) + 7c = 94$

43. $x + (4x + 32) = 12$

44. $37 + x = 5x + 9$

45. $6b + 11 = 2b + 47$

46. $7x - (4x - 39) = 0$

47. $5(x + 2) = 20$

48. $7x + 8 = 6x + 1$

49. $x + 4 = 9x + 4$

50. $3(y - 9) = 30$

51. $7(a + 3) = 28$

52. $9x - 3 = 2x + 46$

53. $y + 30 = 12y - 14$

54. $4(b - 6) = 44$

55. $8(2c - 1) = 56$

56. $6x - 7 = 4x + 3$

57. $2z + 1 = 10z - 1$

58. $c + 20 = 55 - 4c$

59. $9a - 23 = 5a - 11$

60. $5(6y - 2) = 50$

61. $2(10 - 3d) = 80$

62. $2d + 36 = -3d - 54$

63. $-4d - 37 = 7d + 18$

64. $7y - 5 = 9y + 29$

65. $2m - 1 = 6m + 1$

66. $6(3c - 1) = -42$

67. $3(2b + 1) - 7 = 50$

68. $4x - 3 = 47 - x$

69. $5c - 13 = 43 - 2c$

70. $3b - 8 = 14 - 8b$

71. $\frac{2}{3}t - 11 = 64 - 4\frac{1}{3}t$

72. $5(3c - 2) + 8 = 43$

73. $4(y - 3) + 3y = 16$

74. $18 - 4n = 6 - 16n$

75. $-2y - 39 = 5y - 18$

76. $6(2s + 3) - 2s = 28$

77. $8y - (6y - 3) = 9$

78. $7x - 4 = 5x - x + 35$

79. $10 - x - 3x = 7x - 23$

80. $7r - (6r - 5) = 7$

81. $8y - (5y + 2) = 16$

82. $8a - 15 - 6a = 85 - 3a$

83. $5d + 9 - 4d = 51 - 5d$

84. $8w - (3w + 6) = 19$

85. $10z - (3z - 11) = 17$

86. $5x - 2x + 13 = x + 1$

87. $8c + 1 = 7c - 14 - 2c$

88. $8b - 4(b - 2) = 24$

89. $15a - 2(a + 6) = 14$

90. $9c - 2c + 8 = 4c + 38$

91. $12x - 5 = 8x - x + 50$

92. $15c - 4(3c + 2) = 13$

93. $5m - 2(m - 5) = 17$

94. $6d - 12 - d = 9d + 53 + d$

95. $8x - 4 + 7 = 6x + x + 9$

96. $22s - 3(5s + 4) = 16$

97. $28r - 6(3r - 5) = 40$

98. $3m - 5m - 12 = 7m - 88 - 5$

99. $5 - 3z - 18 = z - 1 + 8z$

100. $4(2r + 1) - 3(2r - 5) = 29$

101. a. Simplify the equation $x - (x + 2) = -2$.

    b. Explain why the simplified equation does not show a solution set.

    c. Check 3 different numbers in the original equation. How many numbers do you think would check in this equation?

    d. Draw a conclusion about an equation that is equivalent to a statement that is always true. Note that an equation that is true for all replacements of the variable is called an **_identical equation_** or, simply, an **_identity_**.

102. a. Simplify the equation $x + 2 + 2x = 3x + 7$.

    b. Explain why the simplified equation does not show a solution set.

    c. Compare the nature of the last statement obtained for this equation to the nature of the last statement obtained for the equation in Exercise 101.

    d. Draw a conclusion about an equation that is equivalent to a false statement.

In 103–106, two equal matrices are given. Find the value of: **a.** x **b.** y

103. $\begin{bmatrix} x + 6 & 3 \\ 10 & -4 \end{bmatrix} = \begin{bmatrix} 2x & 3 \\ 10 & -y \end{bmatrix}$

104. $\begin{bmatrix} 9 + y & 2 \\ 2x - 5 & -1 \end{bmatrix} = \begin{bmatrix} -2y & 6 \div 3 \\ 3 + 4x & -1 \end{bmatrix}$

105. $\begin{bmatrix} 6 + 5x & -2 \\ 9(.1) & \frac{1}{2}y \\ 0 \div 1 & 3 \end{bmatrix} = \begin{bmatrix} 2x + 6 & -2 \\ .9 & 10 \\ 0 & 3 \end{bmatrix}$

106. $\begin{bmatrix} 2 + 3(4) & 0 \\ x - (2 - x) & 14 \\ y + 3(y - 1) & 2 \end{bmatrix} = \begin{bmatrix} 14 & -1 + 1 \\ 2 - 2x & 14 \\ 2y - 3 & 2 \end{bmatrix}$

In 107 and 108, find the value of x for the given perimeter.

107.

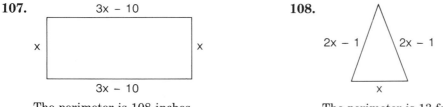

The perimeter is 108 inches.

108.

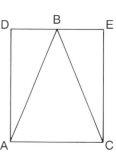

The perimeter is 13 feet.

109. A triangle that has two sides equal in measure is called an **_isosceles triangle_**. The diagram shows isosceles triangle ABC nested in rectangle $ADEC$ so that point B of the triangle divides width DE of the rectangle in half.

    a. Use a variable to represent the width of the rectangle. If the height is 2 inches longer than the width, write an expression for the height.

    b. Use the same variable to write expressions representing the lengths of the three sides of the isosceles triangle, given that each of the equal sides of the triangle is 1 inch longer than the height of the rectangle.

    c. If it would take 70 inches of tape to outline all the segments of the diagram, what is the width of the rectangle?

In 110–134, use an equation containing one variable to solve the problem.

110. The larger of two numbers is twice the smaller. If the sum of the two numbers is 96, find the numbers.

111. One number is 5 times another. If their difference is 96, find the numbers.

112. Sy is 5 times as old as Mike. If the sum of their ages is 18 years, find the age of each boy.

113. Bob and Diane earned a total of $24 shoveling snow. If Diane earned 3 times as much as Bob, how much did each earn?

114. If 6 times a number is increased by 9, the result is the same as the number increased by 34. Find the number.

115. If 4 times a number is decreased by 9, the result is the same as 3 times the number decreased by 1. Find the number.

116. Carl and Richard earned $10.50 delivering packages. If they agreed that Carl should get 1.5 times as much as Richard gets, how much did each boy receive?

117. Caia's height is $\frac{3}{4}$ of Sylvia's height. If the difference between their heights is 15 inches, find the height of each girl.

118. Steve goes to the Movieplex once a month and Joy goes to the Movieplex $\frac{1}{4}$ as often as Steve. If Joy had spent $27 more for tickets last year, the amount they spent for admissions would have been identical. What was the ticket price at the Movieplex?

119. A bed costs $15 more than a rug. Two beds and 4 rugs cost $300. Find the cost of each.

120. The Pentagon Building, with its parking and planted areas, covers a total of 296 acres. The number of acres reserved for parking areas is 9 more than twice the number of acres occupied by the building itself. The number of acres devoted to lawns and terraces is 90 less than ten times the number of acres occupied by the building. Find the number of acres occupied by the building.

The Pentagon Building, across the Potomac River from Washington, D.C., is the headquarters of the U.S. Department of Defense. Built in the form of a five-sided figure, a ***pentagon***, the building's five rings of pentagons are connected by ten spokelike corridors. The perimeter of the building is about a mile.

121. The Greenbacks lead the Bullets by 3 points after the first half of the football game. If the Greenbacks score 3 points and the Bullets score 24 points in the second half, the Bullets' score will be twice that of the Greenbacks. Find the score at halftime.

122. Mrs. Powers travels 12 miles less each day in going to and from her job than Mr. Clay does. The difference between the distance Mr. Clay travels in 6 days and the distance that Mrs. Powers travels in 5 days is 96 miles. How far does each one travel each day?

123. A group of students signed up for a class trip. If 4 more students go on the trip, the cost will be $6 each. If 2 fewer students go on the trip, the cost will be $8 each. If the total sum is the same in either case, how many students have signed up so far?

124. Don and Nat work a 40-hour week for Delco Corp. Don's hourly pay is half of Nat's. Their combined gross incomes last week totaled $720. Find their hourly wages.

125. The price of gasoline increased by 4¢ a gallon during a cold snap. Cindy bought 8 gallons of gasoline before the price increase, and bought 8 gallons again after the increase. If she spent a total of $19.20, find the price of gasoline before the increase.

126. Mrs. Ortiz buys 3 packages of bagels and Mrs. Cavanaugh buys 4 packages of bagels. If Mrs. Ortiz takes 2 extra bagels and Mrs. Cavanaugh removes 5 bagels, they will have the same number of bagels. How many bagels are in each package?

127. The sum of $\frac{1}{2}$ of Lee's earnings and $\frac{1}{3}$ of her earnings is $100. The sum of $\frac{1}{2}$ of Lisa's earnings and $\frac{1}{3}$ of her earnings is $90. Find the difference in their earnings.

128. Mrs. Lyons put down a new kitchen floor consisting of tiles with the designs △ and □. Design □ cost $6 more per tile than design △. The floor required 140 □ tiles and 80 △ tiles. Mrs. Lyons paid $1,940 for the tiles. What was the cost of each △ tile?

129. Team A and Team B are tied this year. Last year, Team A scored 9 fewer points than this year, and Team B scored 6 more points than this year. In order to have tied Team B last year, Team A would have needed 4 times the number of points they actually scored. How many points did each team score last year?

130. Mrs. Best bought some rolls at a bakery and gave half of them to her sister. At home, she made some sandwiches, using one roll for herself and one for each of her 3 children. If there were 2 rolls left when Mr. Best came home, how many rolls had Mrs. Best bought?

131. In his will, Mr. Sampson left 50% of his estate to his wife, 20% to each of his two children, and the remaining $10,000 to charity. How much was in the estate?

132. Juanita spent half of her savings on a present for her mother, and half of what was left on a book for herself. If $6 remained, how much did Juanita have originally?

133. Rich's Electronics offers a CD player at 10% off the regular price. In addition, the Sunday paper has a $5-off coupon for this product. If Neil paid $211 for a player, taking advantage of both the 10% discount and the coupon, what was the regular price?

134. When Steve bought some apples at the supermarket early one morning, the new checker mistook the weight for the cost; that is, if an item weighed 1.30 pounds, he would charge $1.30. Steve saw that the checker had given him the wrong change from a $10 bill, but, by coincidence, the amount of change he received was the same as the actual correct cost of his purchase. If the apples cost 50 cents a pound, how many pounds did Steve buy?

4-5 WORKING WITH A FORMULA

A *formula* is an equation that expresses a relationship between two or more variables.

Translating a Verbal Sentence Into a Formula

MODEL PROBLEMS

1. Write a formula to express each of the following relationships:

   a. The perimeter, P, of a square is equal to 4 times the length of each side, s.

   ANSWER: $P = 4s$

   b. The cost, C, of a number of articles is the product of the number of articles, n, and the price, p, of each article.

   ANSWER: $C = np$

2. Write a formula to express the number of months, m, that there are in y years.

   Solution: Look for a rule that states the relation between the variables m and y.

   Since there are 12 months in one year, the number of months, m, that there are in y years is equal to 12 times the number of years, y.

   ANSWER: $m = 12y$

Evaluating a Formula

The leading variable of a formula is called the *subject* of the formula. For example, D is the subject of the formula $D = R \times T$.

If the values of all except one of the variables in a formula are known, you can compute the value of the remaining variable.

PROCEDURE

To evaluate a variable in a formula:

1. Replace the other variables in the formula by their given values.
2. Perform the indicated operations.

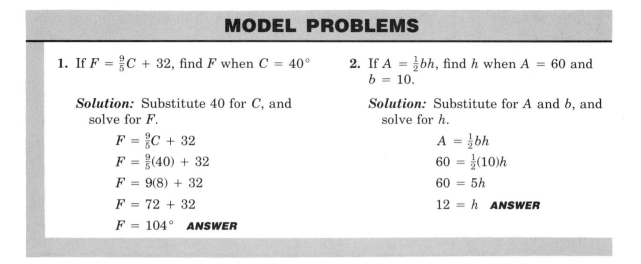

MODEL PROBLEMS

1. If $F = \frac{9}{5}C + 32$, find F when $C = 40°$

Solution: Substitute 40 for C, and solve for F.

$$F = \frac{9}{5}C + 32$$
$$F = \frac{9}{5}(40) + 32$$
$$F = 9(8) + 32$$
$$F = 72 + 32$$
$$F = 104° \quad \textbf{ANSWER}$$

2. If $A = \frac{1}{2}bh$, find h when $A = 60$ and $b = 10$.

Solution: Substitute for A and b, and solve for h.

$$A = \frac{1}{2}bh$$
$$60 = \frac{1}{2}(10)h$$
$$60 = 5h$$
$$12 = h \quad \textbf{ANSWER}$$

Now it's your turn to . . . **TRY IT!**   *(See page 196 for solutions.)*

1. Write a formula to express the relationship: The total cost, C, of 8 hats, each of which costs d dollars, equals 8 times the cost of one hat.

2. If $P = 2b + 2h$, find P when $b = 3$ and $h = 7$.

3. If $p = a + b + c$, find b when $p = 23$, $a = 7$, and $c = 11$.

Rewriting a Formula

Sometimes you can make computation easier by rewriting a formula to make a different variable the subject. More generally, an equation containing more than one variable can be written to express any one variable in terms of the others. The method is the same as in solving a similar equation that contains only one variable.

MODEL PROBLEMS

1. Solve for x: $ax = b$

Solution: Compare with $2x = 7$.

$$2x = 7 \qquad\qquad ax = b$$
$$\frac{2x}{2} = \frac{7}{2} \qquad\qquad \frac{ax}{a} = \frac{b}{a}$$
$$x = \frac{7}{2} \quad \textbf{ANSWER} \quad x = \frac{b}{a}$$

2. Solve for x: $x + a = b$

Solution: Compare with $x + 5 = 9$.

$$x + 5 = 9 \qquad\qquad x + a = b$$
$$x + 5 - 5 = 9 - 5 \qquad x + a - a = b - a$$
$$x = 4 \qquad \textbf{ANSWER} \quad x = b - a$$

3. Solve for x: $2ax = 10a - 3ax$

Solution: Compare with $2x = 10 - 3x$.

		Check
$2x = 10 - 3x$	$2ax = 10a - 3ax$	$2ax = 10a - 3ax$
$2x + 3x = 10 - 3x + 3x$	$2ax + 3ax = 10a - 3ax + 3ax$	$2a(2) \stackrel{?}{=} 10a - 3a(2)$
$5x = 10$	$5ax = 10a$	$4a \stackrel{?}{=} 10a - 6a$
$\dfrac{5x}{5} = \dfrac{10}{5}$	$\dfrac{5ax}{5a} = \dfrac{10a}{5a}$	$4a = 4a \checkmark$
$x = 2$	**ANSWER** $x = 2$	

> ### Now it's your turn to . . . **TRY IT!**   *(See page 196 for solutions.)*
>
> **1.** Solve for x: $x - b = c$ **2.** Solve for y: $ay + b = c$
> **3.** Solve the formula $V = \frac{1}{3}Bh$ for h.

EXERCISES

In 1–12, write a formula to express the relationship.

1. The total length, l, of 10 pieces of lumber, each f feet in length, is 10 times the length of each piece of lumber.

2. The total weight, w, of n cars, each weighing p pounds, is n times the weight of each car.

3. The selling price of an article, s, equals its cost, c, plus profit, p.

4. The average, M, of three numbers, a, b, c, is their sum divided by 3.

5. The Fahrenheit temperature, F, is 32 more than nine-fifths of the Celsius temperature, C.

6. The Celsius temperature, C, is equal to five-ninths of the difference between the Fahrenheit temperature, F, and 32.

7. The average rate of speed, R, is equal to the distance that is traveled, D, divided by the time spent on the trip, T.

8. The average weight, w, of 2 boys is equal to one-half the sum of their weights, a and b.

9. The average weight, w, of p people is the total of all their weights, T, divided by p.

10. The number of cents change, c, to be received from a one-dollar bill is equal to 100 decreased by the price of the article purchased, p, when p is less than \$1.00.

11. To estimate the number of bushels, n, in a bin, multiply the length, l, by the width, w, by the height, h, each expressed in feet, and divide this product by 1.25.

12. A worker's earnings during a week, E, are equal to her weekly salary, S, increased by $1\frac{1}{2}$ times her hourly rate of pay, P, times the number of hours she works overtime, H.

In 13–18, write a formula to express the number of:

13. centimeters, c, in m meters

14. meters, m, in k kilometers

15. inches, i, in y yards

16. feet, f, in i inches

17. ounces, o, in p pounds.

18. days, n, in w weeks and 5 days

19. Write a formula for the number of trees, n, in an orchard containing r rows of t trees each.

20. Write a formula for the total number of seats, n, in the school auditorium, if it has two sections, each with r rows having s seats in each row.

21. Write a formula for the number of students, n, that may be seated in a room in which there are s single seats and t double seats.

22. A group of n persons in an automobile crosses a river on a ferry. Write a formula for the total ferry charge, c, in cents, if the charge is 90 cents for the car and driver and t cents for each additional person.

23. Write a formula for the cost in cents, c, of a telephone conversation lasting 9 minutes if the charge for the first 3 minutes is x cents and the cost for each additional minute is y cents.

24. Write a formula for the distance, d, that a body will fall from rest, if d is one-half the product of the gravitational constant, g, and the square of the time, t.

Once a satellite is rocket-propelled into space, the law of inertia would maintain the satellite in its outward path, but gravity draws it toward earth. A balance of the forces of inertia and gravity keeps the satellite in orbit.

25. Write a formula for the cost in cents, c, of sending a telegram of 18 words if the cost of sending the first 10 words is a cents and each additional word costs b cents.

26. A gasoline dealer is allowed a profit of 8 cents a gallon for each gallon she sells. If she sells more than 25,000 gallons in a year, she is given an additional profit of 5 cents for every gallon over that number. Assuming that she always sells more than 25,000 gallons a year, express as a formula the number of dollars, D, in her yearly income in terms of N, the number of gallons sold.

27. If $E = \dfrac{360}{n}$, find E when n equals: **a.** 8   **b.** 12

28. If $S = \frac{1}{2}gt^2$, find S when $g = 32$ and t equals: **a.** 3   **b.** 1.5   **c.** $1\frac{3}{4}$

29. If F represents a Fahrenheit temperature and C represents the equivalent Celsius temperature, $F = \frac{9}{5}C + 32$.

Find F when C equals:

  a. $20°$ **b.** $35°$

  c. $0°$ **d.** $-5°$

  e. $-20°$

30. If C represents a Celsius temperature and F represents the equivalent Fahrenheit temperature, $C = \frac{5}{9}(F - 32)$.

Find C when F equals:

  a. $50°$ **b.** $86°$

  c. $32°$ **d.** $-4°$

  e. $-40°$

The 16th-century Italian astronomer Galileo is credited with having invented the thermometer. A major improvement in the accuracy of temperature measurement occurred in 1714, when the German-Dutch physicist Gabriel Fahrenheit created the mercury thermometer. About 20 years later, the Swedish astronomer Anders Celsius devised a convenient new scale for temperature readings.

The Fahrenheit scale is customarily used in the United States; the Celsius scale is used by scientists, and in countries that use the metric system of measurement. Other scales used in science are the Kelvin and Rankine scales, both of which are based on absolute zero, the point at which all heat is removed.

31. If $S = \dfrac{n}{2}(a + l)$, find S when:

  a. $n = 20$, $a = 2$, $l = 40$ **b.** $n = 11$, $a = 4$, $l = 44$

  c. $n = 7$, $a = 1.3$, $l = 1.9$ **d.** $n = 9$, $a = -17$, $l = 1$

32. The lifting force L on an airfoil is given by the formula $L = KAV^2$. Find L when $K = .0025$, $A = 350$, and $V = 100$.

33. The horsepower H required for flight is given by the formula $H = \dfrac{DV}{375}$.

Find H when $D = 187.5$ lb. and $V = 200$ mph.

34. In the formula $L = a + (n - 1)d$, find the value of L when $a = 7$, $n = 13$, and $d = 3$.

35. In the formula $K = 2a - 5(n - 1)$, find the value of K when $a = 8$ and $n = 3$.

36. If $S = \dfrac{a}{1 - r}$, find S when $a = 8$ and $r = .5$.

37. If $S = \dfrac{rl - a}{r - 1}$, find S when $r = 3$, $l = 15$, and $a = 5$.

38. If $p = 4s$, find s when $p = 20$.   **39.** If $f = \frac{1}{3}y$, find y when $f = 27$.

40. If $p = a + b + c$, find c when $p = 80$, $a = 20$, and $b = 25$.

41. If $D = RT$, find R when $D = 40$ and $T = \frac{1}{2}$.

42. If $A = \frac{1}{2}bh$, find h when $A = 24$ and $b = 8$.

43. If $F = \frac{9}{5}C + 32$, find C when $F = 95°$.

44. If $S = \dfrac{n}{2}(a + l)$, find:

    a. n when $S = 30$, $a = 4$, $l = 6$
    b. l when $S = 36$, $n = 4$, $a = 5$
    c. a when $S = 42$, $n = 14$, $l = 2$

45. If $A = \frac{1}{2}h(b + c)$, find:

    a. h when $A = 24$, $b = 9$, $c = 3$
    b. b when $A = 50$, $h = 4$, $c = 11$
    c. c when $A = 54$, $h = 12$, $b = 5.5$

46. In physics, kinetic energy is expressed in units of work, or *joules*. The formula $K = \frac{1}{2}mv^2$ gives the kinetic energy K of a mass m moving at a velocity v. Find the number of joules in K when $m = 10$ kilograms and $v = 15$ meters per second.

47. Boyle's law in chemistry states that pressure and volume at a constant temperature are related by the formula $PV = k$. If the units of pressure are given in atmospheres, and the units of volume are liters, then the constant k represents atmosphere-liters. Find V when $k = 3{,}200$ atm-l and $P = 80$ atm.

48. Simple interest is calculated by the formula $I = Prt$, where the interest I is the product of the principal P, rate r, and time t. If the interest is \$300, the rate is 5%, and the principal is \$2,000, find the time.

When evaluating formulas, the units in the result can often be found by "cancelling" given units. For example, the formula $\dfrac{S}{r} = h$ states that a salary S, divided by an hourly rate r, results in h, the number of hours worked.

To find the number of hours worked to earn \$195 at an hourly rate of \$6.50:

$$S \div r = 195 \text{ dollars} \div \frac{6.50 \text{ dollars}}{1 \text{ hour}}$$

$$= 195 \; \cancel{\text{dollars}} \times \frac{1 \text{ hour}}{6.50 \; \cancel{\text{dollars}}}$$

$$= \frac{195}{6.50} \times 1 \text{ hour} = 30 \text{ hours}$$

In 49 and 50, "cancel" units to answer the question.

49. Use the formula $d = rt$ to find the displacement d of a moving object if the rate r is 30 meters per second and the time t is 5 seconds.

50. At a temperature T_1 of 300 degrees in the Kelvin scale, the volume V_1 of gas in a balloon is 10 cubic meters. Use the formula $V_2 = \dfrac{V_1 T_2}{T_1}$ to find the new volume V_2 if the temperature T_2 is 360 degrees.

In 51–78, solve for x or y and check.

51. $5x = b$ **52.** $sx = 8$ **53.** $ry = s$ **54.** $3y = t$

55. $cy = 5$ **56.** $hy = m$ **57.** $x + 5 = r$ **58.** $x + a = 7$

59. $y + c = d$ **60.** $4 + x = k$ **61.** $d + y = 9$ **62.** $3x - q = p$

63. $x - 2 = r$

64. $y - a = 7$

65. $x - c = d$

66. $4x - 5c = 3c$

67. $bx = 9b$

68. $cx + c = 5c - 3cx$

69. $bx - 5 = c$

70. $a = by + 6$

71. $ry + s = t$

72. $abx - d = 5d$

73. $rsx - rs = 0$

74. $m^2x - 3m^2 = 12m^2$

75. $9x - 24a = 6a + 4x$

76. $5y + 2b = y + 6b$

77. $8ax - 7a = 19a - 5ax$

78. $5by - 3b = 2by + 6b$

In 79–99, transform the given formula by solving for the indicated quantity.

79. $P = 4s$ for s

80. $A = bh$ for h

81. $C = \pi D$ for π

82. $V = lwh$ for h

83. $C = 2\pi r$ for r

84. $i = prt$ for p

85. $CN = 360$ for N

86. $A = \frac{1}{2}bh$ for h

87. $A = \frac{1}{3}BH$ for H

88. $K = \dfrac{AP}{2}$ for A

89. $S = c + g$ for g

90. $l = c - s$ for c

91. $S = \frac{1}{2}gt^2$ for g

92. $S = \pi \dfrac{R^2A}{90}$ for A

93. $P = 2l + 2w$ for l

94. $F = \frac{9}{5}C + 32$ for C

95. $2S = n(a + l)$ for a

96. $2S = n(a + l)$ for l

97. $A = \dfrac{h}{2}(b + c)$ for b

98. $T = m(g - b)$ for g

99. $E = I(R + r)$ for R

100. If $A = BH$, express H in terms of A and B.

101. If $P = 2a + b$, express b in terms of P and a.

102. If $A = \frac{1}{2}rp$, express r in terms of A and p.

103. If $P = 2a + b + c$, express a in terms of the other quantities.

In 104–109: **a.** Transform the given formula by solving for the quantity to be evaluated.
b. Substitute the given values in the result obtained in part **a** to find the value of this quantity.

104. If $LWH = 144$, find W when $L = 3$ and $H = 6$.

105. If $A = \frac{1}{2}bh$, find h when $A = 15$ and $b = 5$.

106. If $F = \frac{9}{5}C + 32$, find C when $F = 95$.

107. If $P = 2L + 2W$, find L when $P = 64$ and $W = 13$.

108. If $S = \dfrac{n}{2}(a + l)$, find l when $S = 36$, $n = 4$, and $a = 5$.

109. If $A = P(1 + rt)$, find r when $A = 200$, $P = 100$, and $t = 10$.

110. If $A = bh$, express A in terms of h if $b = 4h$.

111. If $A = \frac{1}{2}bh$, express A in terms of b if $h = 4b$.

112. In the formula $A = \dfrac{h}{2}(b + c)$, express A in terms of b if $h = 3b$ and $c = 5b$.

4-6 SOLVING AN INEQUALITY

Properties of Inequality

Like the relation *equality*, the relation *inequality* has certain properties.

The order of an inequality does not change when the same number is added to or subtracted from both sides. Similarly, when multiplying or dividing both sides of an inequality by a positive number, the order of the inequality is unchanged. However, when multiplying or dividing by a negative number, the order of the inequality is affected.

	$20 > -10$			$20 > -10$	
multiply by 2:	$40 > -20$	(true)	divide by 2:	$10 > -5$	(true)
multiply by -2:	$-40 > 20$	(false)	divide by -2:	$-10 > 5$	(false)
reverse order:	$-40 < 20$	(true)	reverse order:	$-10 < 5$	(true)

The following tables summarize the properties of inequality.

Name of Property	Representation for All Real Numbers a, b, c	Example
Addition Property of Inequality	If $a < b$, then $a + c < b + c$. If $a > b$, then $a + c > b + c$.	If $x - 2 < 6$, then $x < 8$. If $x - 2 > 6$, then $x > 8$.
	If the same number is added to unequal quantities, the order of the inequality is retained.	
Subtraction Property of Inequality	If $a < b$, then $a - c < b - c$. If $a > b$, then $a - c > b - c$.	If $y + 3 < 9$, then $y < 6$. If $y + 3 > 9$, then $y > 6$.
	If the same number is subtracted from unequal quantities, the order of the inequality is retained.	
Multiplication and Division Properties of Inequality	If $a < b$ and c is positive, then $ac < bc$; also $\dfrac{a}{c} < \dfrac{b}{c}$.	If $\dfrac{z}{2} < 10$, then $z < 20$. If $2z < 10$, then $z < 5$.
	If $a > b$ and c is positive, then $ac > bc$; also $\dfrac{a}{c} > \dfrac{b}{c}$.	If $\dfrac{z}{2} > 10$, then $z > 20$. If $2z > 10$, then $z > 5$.
	If unequal quantities are multiplied or divided by the same positive number, the order of the inequality is retained.	
	If $a < b$ and c is negative, then $ac > bc$; also $\dfrac{a}{c} > \dfrac{b}{c}$.	If $\dfrac{z}{-2} < 10$, then $z > -20$. If $-2z < 10$, then $z > -5$.
	If $a > b$ and c is negative, then $ac < bc$; also $\dfrac{a}{c} < \dfrac{b}{c}$.	If $\dfrac{z}{-2} > 10$, then $z < -20$. If $-2z > 10$, then $z < -5$.
	If unequal quantities are multiplied or divided by the same negative number, the order of the inequality is changed.	

Although the reflexive and symmetric properties are not true for inequality, the transitive property is true.

Name of Property	Representation for All Real Numbers a, b, c	Example
Transitive Property of Inequality	**If $a > b$ and $b > c$, then $a > c$.**	If $x > 5$ and $5 > y$, then $x > y$.
	If one number is greater than a second number, and the second number is greater than a third number, then the first number is greater than the third number.	
	If $a < b$, and $b < c$, then $a < c$.	If $x < 5$ and $5 < y$, then $x < y$.
	If one number is less than a second number, and the second number is less than a third number, then the first number is less than the third number.	

Finally, the Substitution Property holds: In an inequality, a quantity may be substituted for its equal.

Applying the Properties to Solve an Inequality

Solving an inequality is much like solving an equation: apply inverse operations and combine like terms to write a chain of progressively simpler equivalent inequalities. Remember, however, to reverse the order of an inequality when multiplying or dividing by a negative number.

MODEL PROBLEMS

1. Find and graph the solution set of the inequality: $\quad -3x + 2 > 14$

Solution: $-3x + 2 > 14$

$\qquad\qquad -3x > 12 \quad$ Subtract 2 from both sides.

$\qquad\qquad\quad x < -4 \quad$ Divide both sides by -3, reversing the inequality.

Check: Choose a test value that is less than -4, say -5.

$$-3x + 2 > 14$$
$$-3(-5) + 2 \overset{?}{>} 14$$
$$15 + 2 \overset{?}{>} 14$$
$$17 > 14 ✓$$

ANSWER: $\{x \mid x < -4\}$

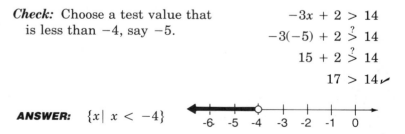

2. The entertainment committee for the school dance has to make a decision about which band to hire. The Far-Outs will play for $300 plus 30% of the gate, and the Way-Ins will play for a flat fee of $540. After an audition, the committee prefers the Far-Outs, but still wants to go with the more economical choice.

   The committee expects about 200 students to attend. What is the most they can charge per ticket and still make the Far-Outs an economical choice? How much would be left for other expenses?

Solution: Let t = the cost per ticket.

   Then $200t$ = the gate receipts.

   And $300 + .30(200t)$ = the Far-Outs' fee.

   The Far-Outs' fee must be at most the Way-Ins' fee.

$$300 + .30(200t) \le 540$$
$$300 + 60t \le 540 \quad \text{Multiply.}$$
$$60t \le 240 \quad \text{Subtract 300.}$$
$$t \le 4 \quad \text{Divide by 60.}$$

Check: At $4 per ticket, the gate would be 200(4) or $800.
The Far-Outs' fee would be $300 + .30(800)$ or $540, the same as the Way-Ins'.
There would then be $800 - 540$ or $260 for the other expenses.

ANSWER: To make the Far-Outs an economical choice, the cost per ticket could be at most $4 and then there would be $260 for the other expenses.

Compound Inequality

   In English grammar, a sentence that expresses two thoughts that are connected by the word *and* or *or* is called a ***compound sentence***. For example, the day before a test, you might think, "I will study hard and I will do well on the test."

   In algebraic language, you have seen compound sentences expressed and illustrated in different ways. Here are ways of expressing an ***and sentence.***

Word form
x is greater than 2 and less than 6

Symbolic form
$x > 2$ and $x < 6$, written $2 < x < 6$

Set notation
$\{x \mid 2 < x < 6\}$

Graph

Venn diagram

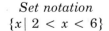

$A \cap B$
{reals between 2 and 6}

Here are ways of expressing an *or sentence*.

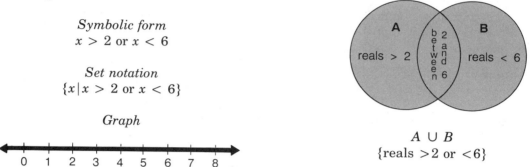

Word form
x is greater than 2 or less than 6

Symbolic form
x > 2 or x < 6

Set notation
{x | x > 2 or x < 6}

Graph

Venn diagram

$A \cup B$
{reals >2 or <6}

Now that you can solve a simple inequality, you are ready to move on to problems that can be modeled algebraically by a compound inequality.

MODEL PROBLEM

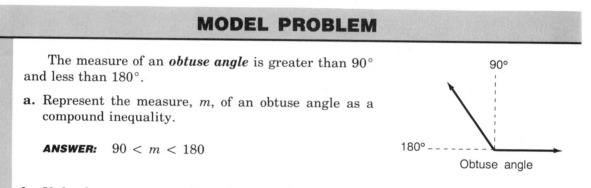

The measure of an *obtuse angle* is greater than 90° and less than 180°.

a. Represent the measure, *m*, of an obtuse angle as a compound inequality.

ANSWER: 90 < m < 180

b. If the degree measure of an obtuse angle is represented by 3x − 30, find the possible values of *x*.

Solution: The measure 3x − 30 of the angle is both greater than 90° and less than 180°. This can be written 90 < 3x − 30 < 180, and can be solved as two simple inequalities.

$$90 < 3x - 30 \qquad\qquad 3x - 30 < 180$$
$$120 < 3x \qquad\qquad\qquad 3x < 210$$
$$40 < x \qquad\qquad\qquad\quad x < 70$$

Check: Test a value of *x* that is between 40 and 70, say 50, to see that the value of 3x − 30 turns out to be between 90 and 180.

ANSWER: 40 < x < 70

<table>
<tr><td>

Now it's your turn to . . . **TRY IT!**

---

1. Find and graph the solution set of the inequality $-x + 1 \leq 4$.

2. Mr. Collins expects to pay about $2\frac{1}{2}$ times as much for a suit as for a pair of shoes, and he has at most $250 to spend for both. To the nearest ten dollars, what is the maximum amount (including tax) he should consider for the shoes?

3. Given: The compound sentence "x is less than 4 or x is greater than 7."
Represent the compound inequality:
   a. symbolically   **b.** graphically   **c.** in set notation   **d.** as a Venn diagram

*(See page **196** for solutions.)*
</td></tr>
</table>

EXERCISES

In 1–7, name the property of inequality that the situation describes.

1. Since an hour is less time than a day, then 3 hours is less time than 3 days.

2. With each girl wearing 2″ heels, Mona is taller than Fran. In bare feet, Mona is taller than Fran.

3. Rayback short-style sneakers cost less than the high sneakers. If the price of each kind of sneaker is increased by $3 per pair, the short sneakers cost less than the high ones.

4. If $AB > CD$, then $AC > BD$.

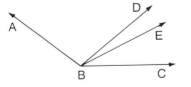

5. If arc $QR <$ arc ST, then arc $QS <$ arc RT.

6. If measure $\angle ABD >$ measure $\angle EBC$, then measure $\angle ABE >$ measure $\angle DBC$.

7. If area of $MNRQ <$ area of $TSQP$, then area of $MNOP <$ area of $TSRO$.

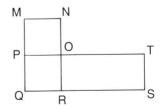

In 8–11, replace ? by $>$ or $<$ to make a true sentence.

8. If $x < 5$ and $5 < y$, then x ? y.

9. If $m > -7$ and $-7 > a$, then m ? a.

10. If $x < 10$ and $z > 10$, then x ? z.

11. If $a > b$ and $c < b$, then a ? c.

In 12–19, name the property of inequality that leads to the solution of the inequality.

12. If $x - 1 < -7$, then $x < -6$.

13. If $y + 4 > -9$, then $y > -13$.

14. If $-7z > 14$, then $z < -2$.

15. If $\dfrac{t}{3} < 18$, then $t < 54$.

16. If $17 \le a + 3$, then $14 \le a$.

17. If $-20 \ge b - 4$, then $-16 \ge b$.

18. If $15 > \dfrac{c}{-5}$, then $-75 < c$.

19. If $64 > 16d$, then $4 > d$.

In 20–25, write the properties of inequality that justify the steps that lead to the solution of the given inequality.

20. $3x + 2 < 17$
$3x < 15$
$x < 5$

21. $5y - 4 > 11$
$5y > 15$
$y > 3$

22. $7 - 4z < 47$
$-4z < 40$
$z > -10$

23. $30 \ge 9 - 7k$
$21 \ge -7k$
$-3 \le k$

24. $-7t + 14 < 0$
$-7t < -14$
$t > 2$

25. $-4w + 6 > 6$
$-4w > 0$
$w < 0$

In 26–28, find the solution set if the domain is $\{-4, -3, -2, \ldots, 4\}$.

26. a. $w > -2$
  b. $-w > -2$

27. a. $y \le 0$
  b. $-y \le 0$

28. a. $-4 < x < 4$
  b. $-4 < -x < 4$

In 29–82, find and graph the solution set of the inequality.

29. $y - \frac{1}{2} > 2$

30. $x - 1.5 < 3.5$

31. $5\frac{3}{4} > w - 1\frac{1}{4}$

32. $y - 4 \ge 4$

33. $y + 3 \le 8$

34. $25 \le d + 22$

35. $3t > 6$

36. $4s > -8$

37. $2x \le 12$

38. $15 \le 3y$

39. $-24 > 6r$

40. $-10 \le 4h$

41. $-3x > 21$

42. $-6y < 24$

43. $27 > -9y$

44. $\frac{1}{3}x > 2$

45. $\frac{1}{2}y < -3$

46. $-\frac{2}{3}z \ge 6$

47. $\dfrac{x}{2} > 1$

48. $\dfrac{y}{3} \le -1$

49. $\dfrac{1}{2} \le \dfrac{z}{4}$

50. $1.5x > 6$

51. $-.4y \le 4$

52. $-10 \ge 2.5z$

53. $2x - 1 > 5$

54. $4c - 3 > 17$

55. $3y - 6 \ge 12$

56. $5x - 1 > -31$

57. $2x - 3 < 12$

58. $-5 \le 3y - 2$

59. $3x + 4 > 10$

60. $2y + 7 < 17$

61. $5y + 3 \ge 13$

62. $6c + 1 > -11$

63. $4d + 3 \le 17$

64. $8h + 5 \ge -23$

65. $5x + 3x - 4 > 4$

66. $8y - 3y + 1 \le 29$

67. $6x + 2 - 8x < 14$

68. $3x + 1 > 2x + 7$

69. $7y - 4 < 6 + 2y$

70. $4 - 3x \ge 16 + x$

71. $2x - 1 > 4 - \frac{1}{2}x$

72. $2c + 5 \ge 14 + \frac{7}{3}c$

73. $4(x - 1) > 16$

74. $8x < 5(2x + 4)$ **75.** $\dfrac{x}{3} - 1 \le \dfrac{x}{2} + 3$ **76.** $12\left(\dfrac{1}{4} + \dfrac{x}{3}\right) > 15$

77. $8m - 2(2m + 3) \ge 0$ **78.** $12r - (8r - 20) > 12$

79. $3y - 6 \le 3(7 + 2y)$ **80.** $5x \le 10 + 2(3x - 4)$

81. $-3(4x - 8) > 2(3 + 2x)$ **82.** $4 - 5(y - 2) \le -2(-9 + 2y)$

83. a. Simplify the inequality $x + 5 > x$.
   b. Explain why the simplified inequality does not show a solution set.
   c. Tell what you think the solution set is.
   d. Draw a conclusion about an inequality that leads to a statement that is always true. Note that an inequality that is true for all replacements of the variable is called an *absolute inequality*.

84. a. Simplify the inequality $x + 2 + 2x > 3x + 7$.
   b. Explain why the simplified inequality does not show a solution set.
   c. Compare the nature of the simplified inequality with the nature of the simplified inequality in Exercise 83.
   d. Draw a conclusion about an inequality that leads to a statement that is never true.

In 85–98, solve algebraically.

85. Twice a number, increased by 6, is less than 48. What numbers satisfy this condition?

86. Five times a number, decreased by 24, is greater than 3 times the number. What numbers satisfy this condition?

87. If Debbie loses 7 pounds, she will still weigh over 136. What is Debbie's weight now?

88. A winter coat can sell for at least $200 if it is sold at a reduction of $\frac{1}{6}$ of its present price. What is the present price of the coat?

89. A worker does not want to spend more than $360 for wallpaper, paint, and supplies. If the paint and supplies cost half as much as the paper, how much can the paper cost?

90. The larger of two integers is 6 times the smaller. The sum of the two integers is less than 49. Find the largest possible values for the integers.

91. ABC Car Rental charges $39 per day, plus 35 cents per mile driven. If Gladys rents a car for 1 day, what distance can she drive and keep her total rental charge under $60?

92. Perry, Jon, Mike, and Seth are going on their spring college break together. Mike is bringing $200 more than Jon, Seth is bringing $300 more than Mike, and Perry is bringing twice as much as Jon. What is the greatest amount of money each should bring if they do not want the total amount they spend to exceed $4,700?

93. It costs $2.50 to rent a Spiderman Video. If Brian wanted to buy the video, it would cost him $29.95. Find the greatest number of times Brian can rent this video without exceeding the purchase price.

94. The Mr. and Mrs. Club of the local Y has decided to hold the annual dinner in the Y's Community Room, for which the rental fee is $500. Additional charges per couple include $40 for dinner and $10 for souvenirs and miscellaneous expenses. How many couples must attend in order to keep the cost per couple to no more than $70?

95. Gem Studios wants $200, plus $5 per couple, to videotape the event in Exercise 94. Photo Plus wants a flat fee of $500. Since the studios are known to be equally satisfactory, the dinner committee will use the one that costs less.

  a. For what number of couples will Gem Studios cost less?

  b. For what number of couples will Photo Plus cost less?

  c. Using the information from Exercises 94 and 95, find how many couples must attend the dinner to keep the cost per couple to no more than:

   (1) $70   **(2)** $60   **(3)** $50

96. Joseph is preparing to plant some vegetables in his family's garden. His father gave Joseph a roll of wire containing 180 feet, and told him to mark off a rectangular plot that is twice as long as it is wide. What are the dimensions of the largest rectangle Joseph can make if he uses as much of the wire as possible?

97. The slot in the mail chute of Yael's apartment building will accept an envelope with a maximum width of 8 inches. The length of the envelope that Yael has prepared for mailing is $1\frac{1}{2}$ times its width, and she used 35 inches of tape to reinforce its four borders. Will Yael's envelope fit in the slot? Explain.

98. A rectangle and an isosceles triangle have bases of equal measure. The height of the rectangle measures 2 feet less than twice its base, and each of the two equal sides of the triangle measures 4 feet more than its base. If the base of each figure is an integral number of feet, and the perimeter of the rectangle is less than the perimeter of the triangle, what are all the possible values for the measure of each base?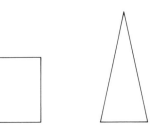

In 99 and 100, write a compound sentence of inequality.

99. Mr. Simmons straightened up his apartment while his dinner was cooking in his microwave oven. If he started cleaning after 6:15 and sat down to eat before 7 P.M., represent the number of minutes, m, he spent cleaning.

A microwave oven cooks by heat generated when the waves penetrate the food. Electromagnetic waves fall into categories according to their lengths.

Microwaves, between 1 millimeter and 1 meter in length, are long when compared to light waves, but short in comparison with radio waves.

100. Donna began reading page 1 of the novel she had borrowed from the library. She intended to read through the end of Chapter 1, which contained 32 pages, but fell asleep before she finished. Represent the number of pages, p, she read.

101. When Mabel left home to go marketing, she had more than $10, but not more than $30. When she returned, she had $5.92 left. Using m to represent the cost of her purchases, write an inequality that shows the amount Mabel spent.

102. Ari is using 5 equal strips of reflecting tape, each at least 6 inches long, to make a star on the back of his jacket. He started with a 5-foot roll of tape, but has already used more than 10 inches of it on his jogging shoes. If he uses the remaining tape to make the star, write an inequality to represent the length s of each strip.

103. Alex received a $10,000 student loan for college tuition. Now that he has graduated and has a job, he is repaying $65 a month.
  a. Write an expression to show the amount Alex still owes after m months.
  b. If he works for at least 2 years, what will be the remaining balance?
  c. What is the least (whole) number of months Alex must work to get the amount he owes under $5,000?

104. Woodville Bowling Lanes charges $1.80 a game and $1.00 to rent a pair of bowling shoes.
  a. Using g to represent the number of games you play, write an expression for the total amount you pay if you rent shoes.
  b. Write and solve an inequality to find the greatest number of games you can play if you have no more than $11.80.

105. Burt's Service Station charges $48 an hour for labor in addition to the cost of parts. Mr. Jason needed $232 worth of parts for his car.
  a. Using h for the number of hours the mechanic worked, write an expression for the amount of Mr. Jason's bill.
  b. If Mr. Jason's bill is between $400 and $520, write and solve an inequality to find the number of hours the mechanic worked.

106. In a race on an 18-mile course, Sophia is running at the rate of .2 mile a minute.
  a. Let m be the number of minutes since the race began. Write an expression in terms of m for the distance Sophia is from the finish line.
  b. If it is at least 40 minutes since the race began, how far is she from the finish line?
  c. For how many minutes will she be between 2 and 5 miles from the finish line?

4-7 ABSOLUTE VALUE IN AN OPEN SENTENCE

Absolute Value in an Equation

Since $|8| = 8$ and $|-8| = 8$, both 8 and -8 can replace x in the equation $|x| = 8$ and make the resulting sentence true. The equation $|x| = 8$ is equivalent to saying that $x = 8$ or $x = -8$. Since these are the only solutions, the solution set is $\{8, -8\}$.

In general when a is positive or 0:

$$|x| = a \text{ is equivalent to } x = a \text{ or } x = -a.$$

PROCEDURE

To solve an equation containing the absolute value of a variable.

1. Isolate the absolute-value expression on one side of the equation.
2. Write two separate equations that do not contain the absolute-value symbol.
3. Solve the resulting equations.

MODEL PROBLEMS

Find the solution set for each equation.

1. $|5x| + 3 = 23$

$|5x| + 3 - 3 = 23 - 3$ Isolate the absolute value.

$\qquad |5x| = 20$

$5x = 20 \quad or \quad 5x = -20$ Write two equations.

$\quad x = 4 \qquad\qquad x = -4$

2. $|2t - 3| = 13$

$2t - 3 = 13 \quad or \quad 2t - 3 = -13$

$\qquad 2t = 16 \qquad\qquad\quad 2t = -10$

$\qquad\quad t = 8 \qquad\qquad\qquad t = -5$

Check: Substitute each value into the original equation.

ANSWER: $\{4, -4\}$ or $\{\pm 4\}$ **ANSWER:** $\{8, -5\}$

Absolute Value in an Inequality

For absolute-value inequalities, it is helpful to recall the definition of $|x|$ as the distance of x from 0. Thus, the solution set of $|x| < 2$ is the set of all numbers whose distance from 0 is less than 2 units. Consider the following graphs.

$|x| < 2$ $|x| > 2$

Just as an equation containing absolute value leads to two equations, an inequality containing absolute value leads to two inequalities.

In general, when a is positive:

$|x| < a$ **is equivalent to** $-a < x$ *and* $x < a$, **also written** $-a < x < a$.

$|x| > a$ **is equivalent to** $x < -a$ *or* $x > a$.

The method of solution for an inequality is similar to that for an equation.

MODEL PROBLEMS

Find and graph the solution set for each inequality.

Solution: After isolating the absolute value, write and solve two separate inequalities.

1.
$$|3x| + 5 < 14$$
$$|3x| + 5 - 5 < 14 - 5$$
$$|3x| < 9$$
$$-9 < 3x \quad and \quad 3x < 9$$
$$\frac{-9}{3} < \frac{3x}{3} \qquad \frac{3x}{3} < \frac{9}{3}$$
$$-3 < x \quad and \quad x < 3$$

Check: Choose one test value between -3 and 3, say 0, to substitute into the given sentence.

ANSWER: $\{x \mid -3 < x < 3\}$

2.
$$|2x - 1| > 5$$

$$2x - 1 < -5 \qquad or \qquad 2x - 1 > 5$$
$$2x - 1 + 1 < -5 + 1 \qquad 2x - 1 + 1 > 5 + 1$$
$$2x < -4 \qquad\qquad 2x > 6$$
$$\frac{2x}{2} < \frac{-4}{2} \qquad\qquad \frac{2x}{2} > \frac{6}{2}$$
$$x < -2 \qquad or \qquad x > 3$$

Check: Choose two test values, one value < -2, say -3, and the other value > 3, say 4, to substitute into the given sentence.

ANSWER: $\{x \mid x < -2 \text{ or } x > 3\}$

Now it's your turn to . . . **TRY IT!**

In 1–4, find, check, and graph the solution set.

1. $\left|\dfrac{y}{3}\right| = 7$ **2.** $-5|w - 7| = -30$ **3.** $|x - 1| > 4$ **4.** $|3t| - 8 < 4$

(See page 197 for solutions.)

In 1–23, find and check the solution set of the equation.

1. $|x| = 14$ **2.** $|c| = 0$ **3.** $|3r| = 36$ **4.** $2|m| = 32$

5. $|4b| + 3 = 7$ **6.** $|w + 8| = 17$ **7.** $|r - 6| = 5$ **8.** $|r| - 6 = 5$

9. $2|y + 1| = 8$ **10.** $4|4 - x| = 16$ **11.** $-3|d + 9| = -3$

12. $|3w - 7| = 5$ **13.** $|7 + 5y| = 32$ **14.** $|5 - 2c| - 8 = 7$

15. $\left|\dfrac{x}{2}\right| = 10$ **16.** $4\left|\dfrac{y}{5}\right| = 20$ **17.** $\dfrac{1}{2}\left|\dfrac{t}{3}\right| = 8$

18. $\left|\dfrac{a}{4}\right| - 7 = -4$ **19.** $\left|\dfrac{5w}{2}\right| + 8 = 13$ **20.** $-\left|\dfrac{2c}{3}\right| = -6$

21. $20 - |y + 4| = 10$ **22.** $8 - |6 - x| = 1$ **23.** $16 - \frac{1}{2}|z - 3| = 4$

24. a. Explain why the solution set for the equation $|x| = -4$ is empty.

     b. Find the solution set for each equation. Check.

         (1) $|x| = -x$ **(2)** $|x| = x$ **(3)** $|2x + 1| = x$ **(4)** $|6x - 10| = x$

In 25–28, two equal matrices are given. Find: **a.** x **b.** y

25. $\begin{bmatrix} |x| & -2 \\ 5 & |7| \end{bmatrix} = \begin{bmatrix} 3 & -2 \\ 5 & y \end{bmatrix}$ **26.** $\begin{bmatrix} 6 - 4 & |3x| \\ 7 & |-3| \end{bmatrix} = \begin{bmatrix} |-2| & 9 \\ |y + 1| & 3 \end{bmatrix}$

27. $\begin{bmatrix} |3 - y| & 3(0) \\ -2 & \frac{1}{2}(7) \end{bmatrix} = \begin{bmatrix} 7 - 1 & 0 \\ -|x| & \frac{7}{2} \end{bmatrix}$ **28.** $\begin{bmatrix} |0| & 2 + 3(4) \\ |2| & 6 - |x| \end{bmatrix} = \begin{bmatrix} |y| & 14 \\ |-2| & -1 \end{bmatrix}$

In 29 and 30, choose the inequality whose solution set is represented by the graph.

29.

     (1) $|x| > 2$ **(2)** $|x| < 2$ **(3)** $|x| \geq 2$ **(4)** $|x| \leq 2$

30.

     (1) $|x| < 1$ **(2)** $|x| \leq 1$ **(3)** $|x| > 1$ **(4)** $|x| \geq 1$

In 31–54, find, check, and graph the solution set of the inequality.

31. $|x| > 4$ **32.** $|y| < 5$ **33.** $|3z| \geq 12$ **34.** $|r| - 3 \leq 3$

35. $\frac{1}{2}|t| > 4$ **36.** $4|p| + 1 < 13$ **37.** $|-5m| \leq 10$ **38.** $|3 - x| > 6$

39. $|3x - 2| > 7$ **40.** $\left|\dfrac{t - 1}{2}\right| < 3$ **41.** $\left|\dfrac{3y}{4} - 1\right| \leq 2$ **42.** $\left|\dfrac{4 - 2w}{5}\right| > 4$

43. $-|b| > -4$ **44.** $13 \le |4x + 1|$ **45.** $3 - 2|r| < -3$

46. $13 < |6y + 7|$ **47.** $15 > |3 - 6b|$ **48.** $-4 < 1 - |r - 1|$

49. $1 > 12 - |3j - 2|$ **50.** $-5 \le 1 - |2q - 2|$ **51.** $-5 > 2 - |3 - 2h|$

52. $|x| < 0$ **53.** $|x| \ge 0$ **54.** $-|x| \le 0$

In 55–57, to solve the inequality, consider positive, negative, and zero values of the variable.

55. $|2x + 3| > x$ **56.** $|2 + 3y| < y$ **57.** $|5z - 12| \ge z$

In 58–60, solve algebraically.

58. If 1 is added to a number, the absolute value of the sum will be 3. Find the original number(s).

59. Find all numbers for which 1 more than 3 times the absolute value of the number is greater than 7.

60. The distance between two points on the number line is the absolute value of the difference of their coordinates. Thus, if a and b are coordinates of points that are 5 units apart, then $|a - b| = 5$, and if x is the coordinate of a point that is less than 8 units from 15, then $|x - 15| < 8$. Find x if:
a. x is 9 units from 0. **b.** x is 3 units from -8.
c. x is more than 5 units from -7. **d.** x is less than 12 units from 4.

61. A turnpike map indicates that Midway Plaza at milepost 13J of the turnpike is the only food stop on the highway.
a. Using the variable m to represent the milepost past which Mr. Cross is driving, write an absolute-value expression to represent his distance from Midway Plaza.
b. Between what two mile markers must he be to be within 7 miles of Midway Plaza?

62. The Smallville town council hired a construction company to build a new town hall. The estimate for the project was $2,402,000, and the actual cost might vary from this figure.
a. Using the variable c to represent the actual cost of construction, write an absolute-value expression that represents the difference between the estimated cost and the actual cost.
b. If the difference between the estimate and the actual cost is guaranteed not to exceed $42,000, what is the range of possible costs for the new town hall?

The one-mile-square village of Cold Spring, New York, was reportedly given its name by George Washington after he enjoyed a refreshing drink of its spring water. The village, on the Hudson River not far from the U.S. Military Academy at West Point, is on the National Register of Historic Places. The Phillipstown town hall shown in the photograph was built in 1865 and is in the Cold Spring historic district.

63. At the County Fair, you want to win the $250 prize for correctly guessing the number of jellybeans in a jar. If nobody guesses correctly, the prize will be given to the person closest to the correct number, provided that the guess is within 15 of the exact number.

    a. If the number of jellybeans is actually 2,384, and the variable n represents your guess, write an absolute-value expression that represents the difference between your guess and the exact number.

    b. Write and solve an absolute-value inequality to find the range of guesses for which the prize will be paid.

64. Dr. Quack's Quickloss Diet specifies 1,450 calories per day plus or minus 200 calories.

    a. Write an absolute-value expression that represents the difference between the actual number of calories per day consumed, represented by the variable c, and the specified number of calories per day.

    b. Was Jack Sprat's food intake within the acceptable range if he consumed:
      (1) 1,257 calories in one day?   **(2)** 1,657 calories in one day?

65. A door must be 34.5 inches wide and 68 inches high with a tolerance of less than .3 inches on both dimensions to fit properly in its frame. (*Tolerance* is an acceptable variation from the specified value.)

    a. Write an absolute-value inequality that gives the acceptable range of values for:
      (1) the width w **(2)** the height h

    b. Could a door that is 34.7 inches wide and 67.8 inches high be used? If not, why?

    c. Could a door that is 34.2 inches wide and 68 inches high be used? If not, why?

66. A replacement glass for a window must be 23.6 inches wide and 35.6 inches high. The acceptable variation (tolerance) for both width and height is .2 inch.

    a. Write an absolute-value inequality that gives the acceptable range of values for:
      (1) the width w **(2)** the height h

    b. Could a pane of glass that measures 35.3 inches by 23.7 inches be used? If not, why?

    c. Could a pane of glass that measures 35.6 inches by 23.6 inches be used? If not, why?

67. On Church Road, all of the odd-numbered houses are on one side of the street, the even-numbered houses are on the other side, and no numbers are skipped. Sandy Davis lives at 6417 Church Road.

    a. Let w represent Art Benjamin's house number.
    If Art is Sandy's next-door neighbor, write an absolute-value equation that represents the possible values for Art's house number.

    b. Let x represent Mike Demchenko's house number.
    If Mike lives not more than 5 houses away from Sandy on the same side of Church Road, write an absolute-value inequality that represents the range of possible values for Mike's house number, and state the domain of x.

    c. Let y represent Antonio Bucci's house number.
    If Antonio is Mike's next-door neighbor, write an absolute-value inequality that represents the range of possible values for Antonio's house number, and state the domain of y.

CHAPTER SUMMARY

Algebra is a powerful tool for solving problems using an ***open sentence of equality or inequality***.

The ***properties*** of equality and inequality provide you with guidelines as to what you may or may not do in solving an open sentence. The notion of "undoing" things that were done to the variable is central to solving: adding 7 undoes subtracting 7; multiplying by 4 undoes dividing by 4. Keep in mind the inverse relationships between addition and subtraction and between multiplication and division.

When solving an equation or inequality, your goal at each step is to transform the problem into a simpler ***equivalent*** statement. Remember that the equality and inequality properties involve *both* sides of the equation or inequality. That is, if you need to add 5, be sure to add 5 to both sides; if you need to divide by 6, be sure to divide both sides by 6. Remember, also, that when you multiply or divide an inequality by a negative number, you must reverse the order of the inequality.

A ***formula***, a type of equation with particular real-life applications, is simply a shorthand method of expressing how two or more quantities are related. For example, the distance formula (distance = rate $\times$ time, or more concisely, $d = rt$) tells you how the distance you travel is related to your rate of speed and the amount of time you travel.

An open sentence that involves the absolute value of a variable expression is equivalent to a combination of two sentences that do not involve absolute value.

VOCABULARY CHECKUP

SECTION

4-1 *addition, subtraction, multiplication, and division properties of equality / reflexive, symmetric, and transitive properties of equality*

4-2 *equivalent equations / inverse operations*

4-3 *algebraic strategy*

4-4 *identity*

4-5 *formula / subject*

4-6 *addition, subtraction, multiplication, and division properties of inequality / transitive property of inequality / compound inequality*

In 1–3, name the property of equality that the situation describes. (Section 4-1)

1. If 3 bars of soap cost $1.20, then 1 bar of soap costs 40 cents.

2. Since Jack and Jill are the same age, and Jill is the same age as Rudolf, then Jack is the same age as Rudolf.

3. The restaurant bill totaled $27.90 before dessert. Then Joe ordered a dessert for $1.45. The final bill came to $29.35.

In 4–7, name the property of equality that leads to the solution of the equation. (Section 4-1)

4. If $x + 6 = 19$, then $x = 13$.

5. If $8 = 2p$, then $4 = p$.

6. If $w - 7 = 11$, then $w = 18$.

7. If $\frac{y}{3} = 5$, then $y = 15$.

In 8 and 9, name the properties of equality that justify the steps that lead to the solution of the given equation. (Section 4-1)

8. $3x - 4 = 11$
$3x = 15$
$x = 5$

9. $-6t + 9 = 21$
$-6t = 12$
$t = -2$

In 10 and 11, tell if the relation for the set {Sarah, Dan} satisfies: (Section 4–1)
    a. the symmetric property
    b. the reflexive property

10. "is the mother of"

11. "is the same age as"

In 12–17, solve the equation and check. (Section 4-2)

12. $w - 13 = -22$

13. $-4m = -36$

14. $\frac{n}{3} = 8$

15. $\frac{2}{5}w = \frac{6}{7}$

16. $13 = 7 - 4h$

17. $\frac{2}{3}f - 9 = -1$

In 18 and 19, solve the equation to evaluate the expression. (Section 4-2)

18. If $3w = 18$, evaluate $5w$.

19. If $r - 7 = -3$, evaluate $2r - 1$.

In 20 and 21, two equal matrices are given. Find the value of: **a.** x **b.** y (Section 4-2)

20. $\begin{bmatrix} x - 3 & -7 \\ 4 & 9 \end{bmatrix} = \begin{bmatrix} 5 & -7 \\ 4 & 3y \end{bmatrix}$

21. $\begin{bmatrix} 19 & y - 9 \\ -5 & 18 \end{bmatrix} = \begin{bmatrix} 18 + 1 & -7 \\ -5 & -3x \end{bmatrix}$

In 22–26, use an equation containing one variable to solve the problem. (Section 4-3)

22. After 7 pounds of provisions were used, Elsa's camping gear weighed 103 pounds. How much did it weigh before?

23. In a 100-km bicycle race, Ilya has already traveled 37 km. How far must he still travel to finish the race?

24. Friedrich was paid $121 for one painting job. This brought his earnings for the week to $537. How much did he earn from his other jobs?

25. The larger of two numbers is 21 more than twice the smaller. Their sum is 84. Find the larger number.

26. Erika and Antonio made 31 posters altogether. If Erika made 2 less than twice as many as Antonio, how many posters did Antonio make?

In 27–34, solve the equation and check. (Section 4-4)

27. $4a - 7 + 5a = 47$

28. $4(3x + 8) = 6 - 22$

29. $3 - (2t - 7) = -10$

30. $4 - 2(3 - 4y) = 2$

31. $4h = 30 - 2h$

32. $6b - 31 = 4b - 37$

33. $3 - x - 2x = 14 - 5x + 1$

34. $2(3 + t) = 5 - 6t - 12$

In 35 and 36, two equal matrices are given. Find the value of: **a.** x **b.** y (Section 4-4)

35. $\begin{bmatrix} x + 4 & 13 \\ 4 & y - 9 \end{bmatrix} = \begin{bmatrix} -5x & 13 \\ 4 & 3y \end{bmatrix}$

36. $\begin{bmatrix} 14 & 2y - 9 \\ -5 & 5x - 8 \end{bmatrix} = \begin{bmatrix} 14 & 7y + 11 \\ -5 & -8 - 3x \end{bmatrix}$

In 37–39, use an equation containing one variable to solve the problem. (Section 4-4)

37. If 3 times a number is decreased by 19, the result is the same as when 5 is added to the number. Find the number.

38. In an algebra textbook, Chapter 2 contains 4 times as many pages as Chapter 1, and Chapter 3 contains 9 fewer than Chapter 2. If the combined number of pages for the three chapters is 108, how many pages does Chapter 1 contain?

39. Noises louder than 85 decibels can be harmful to the eardrums. Twice the average noise level at a recent rock concert, increased by 40 decibels, is the same as 5 times the average noise level decreased by 200 decibels. What was the average noise level at the rock concert?

In 40–42, write a formula to express the relationship. (Section 4-5)

40. The total weight, w, of n identical cartons is n times the weight, c, of 1 carton.

41. If two angles are complementary, then the number of degrees, d, in the first angle equals 90 minus the measure, m, of the second angle.

42. Express the number of minutes, m, in h hours.

In 43–45, find the value of the indicated variable. (Section 4-5)

43. If $E = IR$, find I when $E = 240$ and $R = 8$.

44. If $C = np + st$, find s when $C = 90$, $n = 5$, $p = 12$, and $t = 6$.

45. If $D = RT$, find R when $D = 81$ and $T = 3$.

In 46–49, solve for x or y and check. (Section 4-5)

46. $3x = 2a$

47. $2ay = 16a$

48. $x + b = c$

49. $9a - 3y = a + y$

In 50 and 51, transform the given formula by solving for the indicated variable. (Section 4-5)

50. $P = 2\ell + 2w$, for w.

51. $V = \ell wh$, for h.

In 52 and 53, name the property of inequality that the situation describes. (Section 4-6)

52. Since 5 cans of chicken soup cost less than 3 cans of cream of mushroom soup, 15 cans of chicken soup will cost less than 9 cans of cream of mushroom soup.

53. Since Ron is older than Selma, 8 years from now Ron will still be older than Selma.

In 54 and 55, replace the question mark with the symbol $>$ or the symbol $<$ so that the resulting sentence will be true. All variables represent nonzero numbers. (Section 4-6)

54. If $y < x$ and $x < 9$, then y ? 9.   **55.** If $x > 24$ and $y < 24$, then x ? y.

In 56 and 57, name the property of inequality that leads to the solution of the inequality. (Section 4-6)

56. If $x - 9 < 23$, then $x < 32$.   **57.** If $8 > -4x$, then $-2 < x$.

In 58 and 59, name the properties of inequality that justify the steps that lead to the solution of the inequality. (Section 4-6)

58. $6x + 4 < 16$
$$6x < 12$$
$$x < 2$$

59. $-6t + 9 \geq 21$
$$-6t \geq 12$$
$$t \leq -2$$

In 60–63, find and graph the solution set of the inequality. (Section 4-6)

60. $4h > -12$ **61.** $w + 8.4 \leq 11$

62. $a - 3 \leq 3a + 11$ **63.** $4 > -12n - 7$

In 64–66, solve algebraically. (Section 4-6)

64. Even when the dealer and the manufacturer give $2,500 in rebates and discounts, the new car still costs more than $12,400. What was the sticker price?

65. When 3 times a number is decreased by 13, the result is less than or equal to 14. What is the number?

66. If the Book Barn takes in an average of $800 per day that it can use for salaries, what is the maximum number of salespeople that can be hired for 8-hour shifts at $9.00 per hour?

In 67–69, solve the equation and check. (Section 4-7)

67. $|3x| = 27$ **68.** $|2a - 5| = 9$ **69.** $4|x - 3| = 32$

In 70 and 71, two equal matrices are given. Find the value(s) of: **a.** x **b.** y (Section 4-7)

70. $\begin{bmatrix} |2x| & 4 \\ 5 & |y| \end{bmatrix} = \begin{bmatrix} 16 & 4 \\ 5 & 3 \end{bmatrix}$ **71.** $\begin{bmatrix} 24 & |y + 1| \\ 25 & 8 \end{bmatrix} = \begin{bmatrix} 24 & 11 \\ 25 & |x - 3| \end{bmatrix}$

In 72–74, find, check, and graph the solution set of the inequality. (Section 4-7)

72. $|t| > 5$ **73.** $|2r - 1| < 9$ **74.** $2|s| - 10 \geq -4$

PROBLEMS FOR PLEASURE

1. Two numbers have a difference that is twice their sum, although their sum is 4 more than their difference. Find the numbers.

2. If dividing a number by 3 gives a larger result than multiplying the number by 3, what must be true of the number?

3. If $x - y = y - x$, then which of the following is true?
   (There may be more than one correct answer.)
   (A) $2xy + y = 2y^2 + x$
   (B) $x^2 - y^2 = x - y$
   (C) $y^2 - x^2 = y - x$
   (D) $y^2 - x^2 = x - y$
   (E) $x + y = 4x - 2y$

CALCULATOR CHALLENGE

1. Have teams try to solve the same five problems, one team using equations, the other guessing and checking using calculators. Score one point for finishing first and three points for each correct answer.
   a. Two positive integers differ by 5. Their sum is 237. Find the larger number.
   b. A man is 56 years older than his granddaughter. If the man is 5 times as old as his granddaughter, how old is he?
   c. One number is 5 more than 3 times another number. Their sum is 145. Find the numbers.
   d. The product of 5 and a number is 13 less than the sum of 5 and the number. Find the number.
   e. Multiplying a number by 5 gives the same value as adding 12 to the number and doubling the result. Find the number.

2. Guessing and checking can be used to solve problems in which the numbers involved are whole numbers. When the solutions may include rational numbers, working backward is a more effective strategy. Use your calculator to solve these problems by working backward. Recall that this means doing the inverse operations in reverse order.

   Example: A number is multiplied by 8, and 19 is subtracted from the result, leaving 42. Find the number.
   Solution: Begin with 42 and add 19. You get 61. Now divide by 8. The number is 7.625.

   a. A number is multiplied by 5, and 27 is subtracted from the result. This leaves 83. Find the number.
   b. If 27 is added to twice a number, the result is -64. Find the number.

   Since the verbal expressions above can also be written as equations, it follows that some equations can be solved by working backward and using a calculator. Solve the following equations using your calculator.
   c. $3x - 16 = -73$
   d. $23a - 446 = 336$
   e. $32x + 129 = -775$

3. The difference between the boiling and freezing points of water is $212 - 32$ or 180 degrees in the Fahrenheit scale and $100 - 0$ or 100 degrees in the Celsius scale. Thus, a Celsius degree is $\frac{180}{100}$ or $\frac{9}{5}$ of a Fahrenheit degree. A formula for converting from C to F temperatures accounts for both the 32-degree difference in freezing points and the relative sizes of the degrees: $F = \frac{9}{5}C + 32$

  a. Determine the calculator keys you would press to change a given C temperature to F.

    (1) Find the F temperature corresponding to these C temperatures: 0, 5, 10, 15, 20

    (2) For each 5-degree rise in C temperature, what is the corresponding rise in F temperature?

  b. Solve the formula $F = \frac{9}{5}C + 32$ to obtain C in terms of F.

  c. Determine the keystrokes needed to change a given F reading to C.

    (1) Find the C temperatures that correspond to these F temperatures: 212, 203, 194, 185, 176

    (2) For each 9-degree drop in F temperature, what is the corresponding drop in C temperature?

COLLEGE TEST PREPARATION

In 1–16, choose the correct answer.

1. If $33 \times 4 \times p = 12$, then $p =$
   (A) $\frac{1}{11}$ (B) $\frac{1}{10}$ (C) 9 (D) 11 (E) 13

2. If $3x - 5 = 10$, then $2x + 3$ is equal to
   (A) 10 (B) 12 (C) 13
   (D) 15 (E) 18

3. If $5(p + 3) = 40$ and $q = p - 3$, then q is
   (A) 1 (B) 2 (C) 4 (D) 6 (E) 7

4. If 20 is $\frac{2}{25}$ of t, then $t =$
   (A) 10 (B) 25 (C) 50
   (D) 200 (E) 250

5. If $3n - 5 = 10$, then $7n =$
   (A) 30 (B) 35 (C) 40
   (D) 49 (E) 56

6. If $3x + 5 = 7$ and $3x + 10 = 2 + y$, then $y =$
   (A) 5 (B) 9 (C) 10 (D) 12 (E) 14

7. How much is twice a number increased by 5, if half the same number decreased by 5 is 1?
   (A) 29 (B) 21 (C) 18 (D) 12 (E) 6

8. The formula $C = 120 + 15f$ gives the cost, in cents, of developing and printing f films. If Jill paid \$4.50 to develop and print some films, how many films did she have?
   (A) 25 (B) 24 (C) 22 (D) 20 (E) 16

9. If $0.1x = 2$ and $2y = 0.1$, then xy is
   (A) 1 (B) 2 (C) 10 (D) 20 (E) 100

10. If x is $\frac{1}{3}$ of y and y is $\frac{3}{5}$ of z and $5x + 3 = 4$, then $z + 5$ is
    (A) 3 (B) 4 (C) 5 (D) 6 (E) 7

11. If $1 \cdot 2 \cdot 3 \cdot x = 1 + 2 + 3 + x$, then how much is $(1 + 2 + 3 + 4)\,x$?
    (A) 6 (B) 10 (C) 12 (D) 24 (E) 20

12. If $3x + 1 = 5$, then $1 \cdot 2 \cdot 3 \cdot x$ equals
    (A) 3 (B) 4 (C) 6 (D) 8 (E) 10

13. The product of 4 and a number is 4 more than 12. What is half of the number?
    (A) 2 (B) 3 (C) 6 (D) 8 (E) 9

14. If $3(7) = 7y$ and $y + 3 = z - 3$, then z equals
    (A) 0 (B) 3 (C) 4 (D) 6 (E) 9

15. If $4 + x = 1 - 3x$, then $4x + 4$ equals
(A) 1 (B) 2 (C) 3 (D) 4 (E) 5

16. If $\frac{2}{3}x = 5$ and $3y = 7$, then xy equals
(A) 70 (B) 35 (C) $17\frac{1}{2}$ (D) $10\frac{2}{3}$ (E) $\frac{14}{35}$

Questions 17–25 each consist of two quantities, one in Column A and one in Column B. You are to compare the two quantities and choose:

A if the quantity in Column A is greater;
B if the quantity in Column B is greater;
C if the two quantities are equal;
D if the relationship cannot be determined from the information given.

1. In certain questions, information concerning one or both of the quantities to be compared is centered above the two columns.

2. In a given question, a symbol that appears in both columns represents the same thing in Column A as it does in Column B.

3. x, n, k, etc. stand for real numbers.

Column A	**Column B**
17.	$\frac{1}{2}(8 - x) = \frac{1}{4}(x + 10)$
$|2x|$	3

18.	$3(4 - x) = 6(x + 1)$
$|12 - 3x|$	$|4 + 6x|$

19.
$$3(2x - 4) + 1 > 7$$
$$7y - 5 > 9$$

x	y

20.
$$4c - 7 > 9$$
$$1 - 2(d - 3) \geq 3$$

$2d$	c

21.
$$|y| < \frac{7}{2}$$
$$|z| < \frac{5}{3}$$

$y + z$	$\frac{31}{6}$

22.
$$|5a| = 4.5$$
$$|b + 3| = 0$$

$a + b$	$a - b$

23.
$$|t - 7| = |3t + 1|$$
$$t > 0$$

$2t$	3

24.
$$\left|\frac{d}{6} + 5\right| = 1$$

d	-30

25.
$$|5x + 2| = |14 - x|$$

$|x + 1|$	3

SPIRAL REVIEW EXERCISES

1. Divide:
$$\frac{15}{-30}$$

2. Add:
$$-18 + 23 + (-7)$$

3. Evaluate:
$$\begin{bmatrix} 3 & 2 \\ -4 & 1 \end{bmatrix} + \begin{bmatrix} 5 & -3 \\ 0 & 1 \end{bmatrix}$$

4. Simplify:
$$|-3| + (-8) - |5|$$

5. Find the additive inverse of $-\frac{1}{2}$.

6. State whether $0.\overline{18}$ is rational or irrational.

7. If s pears cost m cents, represent the cost of r pears.

8. Which expression is not equal to $-(x + y)$?
(1) $-x + (-y)$ (2) $-x - y$ (3) $-x + y$ (4) $-y - x$

9. If $A = \{$letters in BARN$\}$ and $B = \{$letters in BURNING$\}$, find: **a.** $A \cap B$ **b.** $A \cup B$

10. If the four integers represented by x, 3, $x + 3$, and $4x$ are all different and in increasing order, find a value of x among the first five positive integers.

11. If $\boxed{n} = 180 - \dfrac{360}{n}$ for all positive integers n, find the value of $\boxed{8}$.

12. The perimeters of a pool and deck to be built for each house in a development will be constructed in the sequence shown at the right. If the pattern continues from house to house, which would be the 102nd section completed?

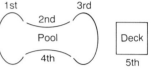

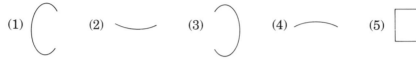

13. Graph the solution set of $-y < -5$ if the domain of y is $\{3, 4, 5, 6, 7, 8\}$.

14. Evaluate $x^2 + \dfrac{y}{x}$ when $x = -4$ and $y = 4$.

15. If $x = 1 - \frac{9}{10}$, $y = 1 - 0.99$, and $z = \frac{1}{9}$, which is the order of x, y, and z from least to greatest? (1) x, y, z (2) x, z, y (3) y, z, x (4) y, x, z

16. Write and solve an equation:
Larry has $2 less than three times the amount of money that Kevin has. If the sum of money both boys have is $10, how much money does each boy have?

17. If $7 + x + y = 21$, and if x and y are 1-digit positive integers, the least possible value of x is (1) 0 (2) 1 (3) 3 (4) 5

18. On a day that includes a 2-hour morning sales meeting, a sales manager must see that the following tasks are accomplished. The tasks are not listed in order. Draw a network to organize the work.

  1. Manager previews advertising campaign before meeting.
  2. Manager consults with ad agency after meeting.
  3. Lunch.
  4. Secretary types minutes of meeting.
  5. Accounting department gives manager a statement of delinquent accounts.
  6. Manager has meeting with sales staff to discuss advertising and sales.
  7. Manager compiles sales data for presentation at meeting.
  8. Assistant inputs manager's sales data into computer.
  9. Manager dictates letters to secretary, *re* delinquent accounts.
  10. Mailroom sends manager's letters to delinquent accounts.
  11. Secretary types letters to delinquent accounts.
  12. Assistant places printouts of computerized sales data in salespersons' portfolios before the meeting.
  13. Manager signs letters to delinquent accounts.

4-2 SOLVING AN EQUATION BY APPLYING INVERSES

TRY IT! *Problems 1–3 on page 154.*

1.

$$m - 3\tfrac{2}{5} = 1$$
$$m - 3\tfrac{2}{5} + 3\tfrac{2}{5} = 1 + 3\tfrac{2}{5}$$
$$m = 4\tfrac{2}{5}$$

Check

$$m - 3\tfrac{2}{5} = 1$$
$$4\tfrac{2}{5} - 3\tfrac{2}{5} \overset{?}{=} 1$$
$$1 = 1 ✓$$

ANSWER: $m = 4\tfrac{2}{5}$

2.

$$12 = -3h$$
$$\frac{12}{-3} = \frac{-3h}{-3}$$
$$-4 = h$$

Check

$$12 = -3h$$
$$12 \overset{?}{=} -3(-4)$$
$$12 = 12 ✓$$

ANSWER: $h = -4$

3.

$$4a + 9 = -19$$
$$4a + 9 - 9 = -19 - 9$$
$$4a = -28$$
$$a = -7$$

Check

$$4a + 9 = -19$$
$$4(-7) + 9 \overset{?}{=} -19$$
$$-28 + 9 \overset{?}{=} -19$$
$$-19 = -19 ✓$$

ANSWER: $a = -7$

4-3 USING AN EQUATION TO SOLVE A PROBLEM

TRY IT! *Problems 1 and 2 on page 157.*

1.

Let x = Tom's age.
Then $x + 5$ = Ming's age.
$$x + 5 = 31$$
$$x = 26$$

Check:
If Tom is 26 and Ming is 31, is Ming 5 years older than Tom? Yes.

ANSWER: Tom is 26 years old.

2.

Let x = number of Al's sheep.
Then $3x$ = number of Zeb's sheep.
$$3x = 84$$
$$x = 28$$

Check:
If Al owns 28 sheep and Zeb owns 84, does Zeb own 3 times as many as Al? Yes.

ANSWER: Al owns 28 sheep.

4-4 COMBINING LIKE TERMS IN AN EQUATION

TRY IT! *Problems 1–4 on page 161.*

1.
$$4w + 15 - 7w = 12$$
$$-3w + 15 = 12$$
$$-3w = -3$$
$$w = 1$$

Check
$$4w + 15 - 7w = 12$$
$$4(1) + 15 - 7(1) \overset{?}{=} 12$$
$$4 + 15 - 7 \overset{?}{=} 12$$
$$12 = 12 ✓$$

ANSWER: $w = 1$

2.
$$7 - 4(3 - 2a) = -21$$
$$7 - 12 + 8a = -21$$
$$-5 + 8a = -21$$
$$8a = -16$$
$$a = -2$$

Check
$$7 - 4(3 - 2a) = -21$$
$$7 - 4[3 - 2(-2)] \overset{?}{=} -21$$
$$7 - 4[3 - (-4)] \overset{?}{=} -21$$
$$7 - 4[3 + 4] \overset{?}{=} -21$$
$$7 - 4(7) \overset{?}{=} -21$$
$$7 - 28 \overset{?}{=} -21$$
$$-21 = -21 ✓$$

ANSWER: $a = -2$

3.

$$5p = 9p + 20$$
$$5p - 9p = 9p + 20 - 9p$$
$$-4p = 20$$
$$p = -5$$

Check

$$5p = 9p + 20$$
$$5(-5) \stackrel{?}{=} 9(-5) + 20$$
$$-25 \stackrel{?}{=} -45 + 20$$
$$-25 = -25 ✓$$

ANSWER: $p = -5$

4.

Let x = the amount Ray has.
Then $x + 5$ = the amount Al has.
Ray and Al have \$23 altogether.

$$x + x + 5 = 23$$
$$2x + 5 = 23$$
$$2x = 18$$
$$x = 9$$
$$x + 5 = 14$$

Check:
 Together, do Ray and Al have \$23?
\$9 + \$14 = \$23. Yes.
 Does Al have \$5 more than Ray?
\$9 + \$5 = \$14. Yes.

ANSWER: Ray has \$9 and Al has \$14.

4-5 WORKING WITH A FORMULA

TRY IT! *Problems 1–3 on page 167.*

1. $C = 8d$

2. $P = 2b + 2h$
$$P = 2(3) + 2(7)$$
$$= 6 + 14 = 20 \quad \textbf{ANSWER}$$

3.

$$p = a + b + c$$
$$23 = 7 + b + 11$$
$$23 = 18 + b$$
$$5 = b$$

Check

$$p = a + b + c$$
$$23 \stackrel{?}{=} 7 + 5 + 11$$
$$23 = 23 ✓$$

ANSWER: $b = 5$

TRY IT! *Problems 1–3 on page 168.*

1.

$$x - b = c$$
$$x - b + b = c + b$$
$$x = c + b$$

Check

$$x - b = c$$
$$c + b - b \stackrel{?}{=} c$$
$$c = c ✓$$

ANSWER: $x = c + b$

2.

$$ay + b = c$$
$$ay + b - b = c - b$$
$$ay = c - b$$
$$\frac{ay}{a} = \frac{c - b}{a}$$
$$y = \frac{c - b}{a}$$

Check

$$ay + b = c$$
$$a\left(\frac{c - b}{a}\right) + b \stackrel{?}{=} c$$
$$c - b + b \stackrel{?}{=} c$$
$$c = c ✓$$

ANSWER: $y = \dfrac{c - b}{a}$

3.

$$V = \tfrac{1}{3}Bh$$
$$3 \cdot V = 3 \cdot \tfrac{1}{3}Bh$$
$$3V = Bh$$
$$\frac{3V}{B} = \frac{Bh}{B}$$
$$\frac{3V}{B} = h$$

Check

$$V = \tfrac{1}{3}Bh$$
$$V \stackrel{?}{=} \tfrac{1}{3}B\left(\frac{3V}{B}\right)$$
$$V = V ✓$$

ANSWER: $h = \dfrac{3V}{B}$

4-6 SOLVING AN INEQUALITY

TRY IT! *Problems 1–3 on page 177.*

1.

$$-x + 1 \le 4$$
$$-x \le 3 \quad \text{Subtract 1.}$$
$$x \ge -3 \quad \text{Divide by } -1,$$
$$\text{reversing order.}$$

Check

If $x = -3$, does $-x + 1 = 4$?

$$-(-3) + 1 \stackrel{?}{=} 4$$
$$3 + 1 \stackrel{?}{=} 4$$
$$4 = 4 ✓$$

Choose a test value > -3, say 0.

$$-x + 1 < 4$$
$$0 + 1 \stackrel{?}{<} 4$$
$$1 < 4 ✓$$

ANSWER: $\{x \mid x \ge -3\}$

2. Let x = the cost of the shoes.
Then $\frac{5}{2}x$ = the cost of the suit.

For both items, spend at most $250.

$$x + \frac{5}{2}x \le 250$$
$$\frac{7}{2}x \le 250 \quad \text{Combine.}$$
$$x \le 71\tfrac{3}{7} \quad \text{Divide by } \tfrac{7}{2}.$$

To the nearest ten dollars, x can be 70.

ANSWER: About $70 for the shoes.

3. a. $x < 4$ or $x > 7$

b.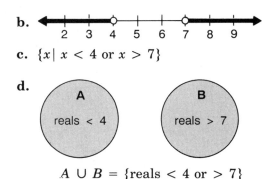

c. $\{x \mid x < 4 \text{ or } x > 7\}$

d.

A	B
reals < 4	reals > 7

$$A \cup B = \{\text{reals} < 4 \text{ or} > 7\}$$

4-7 ABSOLUTE VALUE IN AN OPEN SENTENCE

TRY IT! *Problems 1–4 on page 183.*

1.
$$\frac{y}{3} = 7 \qquad or \qquad \frac{y}{3} = -7$$
$$y = 21 \qquad\qquad y = -21$$

Check: Substitute each value into the given.

$$\left|\frac{21}{3}\right| \stackrel{?}{=} 7 \qquad\qquad \left|\frac{-21}{3}\right| \stackrel{?}{=} 7$$

$$|7| \stackrel{?}{=} 7 \qquad\qquad |-7| \stackrel{?}{=} 7$$
$$7 = 7 ✔ \qquad\qquad 7 = 7 ✔$$

ANSWER: $a = 21$ or -21

2.
$$-5|w - 7| = -30$$
$$\left(-\tfrac{1}{5}\right)(-5)|w - 7| = -30\left(-\tfrac{1}{5}\right)$$
$$|w - 7| = 6$$
$$w - 7 = 6 \qquad or \qquad w - 7 = -6$$
$$w = 13 \qquad\qquad\qquad w = 1$$

Check: Substitute each value into the given.

$$-5|13 - 7| \stackrel{?}{=} -30 \qquad\quad -5|1 - 7| \stackrel{?}{=} -30$$
$$-5|6| \stackrel{?}{=} -30 \qquad\qquad -5|-6| \stackrel{?}{=} -30$$
$$-5(6) \stackrel{?}{=} -30 \qquad\qquad -5(6) \stackrel{?}{=} -30$$
$$-30 = -30 ✔ \qquad\qquad -30 = -30 ✔$$

ANSWER: $w = 13$ or 1

3.
$$|x - 1| > 4$$
$$\begin{array}{ccc} x - 1 < -4 & & x - 1 > 4 \\ x - 1 + 1 < -4 + 1 \ \ or \ \ & & x - 1 + 1 > 4 + 1 \\ x < -3 & & x > 5 \end{array}$$

Check: Choose two test values, say -4 and 6.

ANSWER: $\{x \mid x < -3 \text{ or } x > 5\}$

4.
$$|3t| - 8 < 4$$
$$|3t| - 8 + 8 < 4 + 8$$
$$|3t| < 12$$
$$\begin{array}{ccc} -12 < 3t & and & 3t < 12 \\ \dfrac{-12}{3} < \dfrac{3t}{3} & & \dfrac{3t}{3} < \dfrac{12}{3} \\ -4 < t & & t < 4 \end{array}$$

Check: Choose a test value between -4 and 4, say 0.

ANSWER: $\{t \mid -4 < t < 4\}$

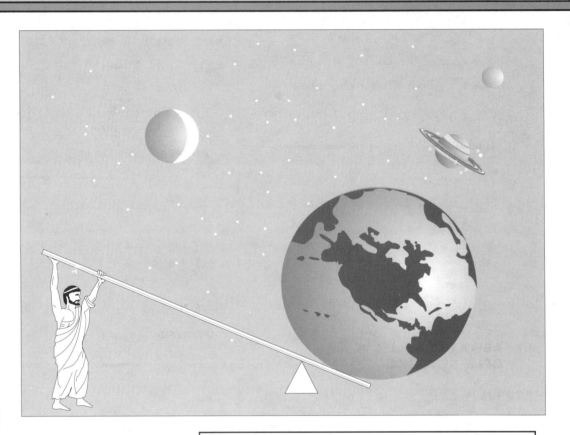

MORE APPLICATIONS OF EQUATIONS AND INEQUALITIES

CHAPTER 5

Over 22 centuries ago, the Greek mathematician Archimedes knew the power of the lever principle. He said, "Give me a place to stand, and I will move the world."

5-1 Consecutive-Integer Problems **199**
5-2 Money-Value Problems **203**
5-3 Motion Problems **207**
5-4 Lever and Pulley Problems **214**
5-5 Angle Problems **217**
5-6 Triangle Problems **222**
5-7 Perimeter and Area Problems **227**
5-8 Surface Area and Volume **236**
5-9 A Roundup of Problems **245**

Chapter Summary **247**
Chapter Review Exercises **248**
Problems for Pleasure **251**
Calculator Challenge **252**
College Test Preparation **253**
Spiral Review Exercises **254**
Solutions to TRY IT! Problems **256**

In Chapter 4, as you learned algebraic procedures for solving equations and inequalities, you were able to apply the procedures to a wide variety of problems. Now you will learn to solve still other kinds. The sections in this chapter group problems according to the type of situation, and include the background information that you need for each specialized type.

Remember to use the general framework for problem solving: analyze the problem, solve the problem (now you are able to apply an algebraic strategy), check your result, and learn from the problem. Recall that it is often convenient to combine strategies; tables and diagrams are commonly used in combination with the algebraic strategy.

5-1 CONSECUTIVE-INTEGER PROBLEMS

Preparing to Solve Consecutive-Integer Problems

You will need to be sure of the vocabulary summarized in the following table.

Vocabulary	Description	Examples	Representation
even integer	An integer that is exactly divisible by 2.	$6, -10, 0$	
odd integer	An integer that is not exactly divisible by 2.	$7, -5$	
consecutive integers	Each successor is 1 more than its predecessor.	$5, 6, 7, 8$ $-5, -4, -3$	$x, x + 1, x + 2$ where $x \in \{\text{integers}\}$
consecutive even integers	Each even successor is 2 more than its even predecessor.	$2, 4, 6, 8$ $-10, -8, -6$	$x, x + 2, x + 4$ where $x \in \{\text{even integers}\}$
consecutive odd integers	Each odd successor is 2 more than its odd predecessor.	$5, 7, 9, 11$ $-3, -1, 1, 3$	$x, x + 2, x + 4$ where $x \in \{\text{odd integers}\}$

EXERCISES

1. Write 4 consecutive integers beginning with the given integer.
   a. 15   **b.** 0   **c.** -10 **d.** y **e.** $2y$ **f.** $3y - 2$

2. If x is even, write 4 consecutive even integers beginning with the given integer.
   a. 0   **b.** -12 **c.** x **d.** $2x$ **e.** $2x + 2$ **f.** $2x - 2$

3. If z is odd, write 4 consecutive odd integers beginning with the given integer.
   a. 13   **b.** -15 **c.** -3 **d.** z **e.** $2z + 1$ **f.** $2z - 1$

4. For each value of x in **(1)–(5)**, evaluate the expressions in **a–c.**
  a. $x + 1, x + 2, x + 3$
  b. $x + 2, x + 4, x + 6$
  c. $x - 3, x - 2, x - 1$

  (1) $x = 3$
  (2) $x = 2$
  (3) $x = -5$
  (4) $x = -4$
  (5) $x = 0$

5. Tell if each expression in **(1)–(6)** represents an odd integer or an even integer:
  a. when n is even
  b. when n is odd

  (1) $n + 1$ **(5)** $n + 3$
  (2) $n - 1$ **(6)** $n + 4$
  (3) $2n$
  (4) $3n$

6. Tell if the expression $x + y$ represents an odd integer or an even integer, given:
  a. x and y are both odd   **b.** x is odd and y is even   **c.** x and y are both even

7. Replace ? by *odd* or *even* to make a true statement.
  a. The sum of an even number of consecutive odd integers is an ? integer.
  b. The sum of an odd number of consecutive odd integers is an ? integer.
  c. The sum of any number of consecutive even integers is an ? integer.

Solving Consecutive-Integer Problems

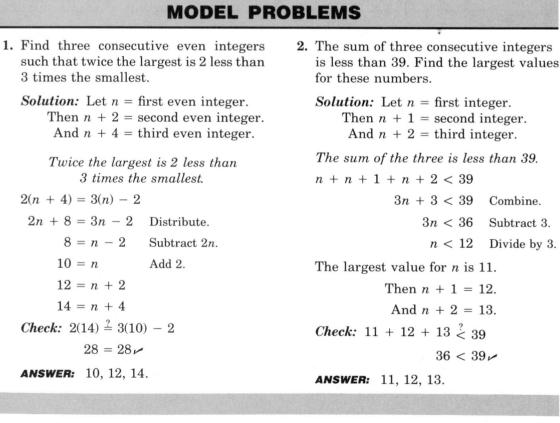

MODEL PROBLEMS

1. Find three consecutive even integers such that twice the largest is 2 less than 3 times the smallest.

Solution: Let n = first even integer.
  Then $n + 2$ = second even integer.
  And $n + 4$ = third even integer.

Twice the largest is 2 less than 3 times the smallest.

$2(n + 4) = 3(n) - 2$

$2n + 8 = 3n - 2$ Distribute.

$8 = n - 2$ Subtract $2n$.

$10 = n$ Add 2.

$12 = n + 2$

$14 = n + 4$

Check: $2(14) \stackrel{?}{=} 3(10) - 2$

$28 = 28$ ✔

ANSWER: 10, 12, 14.

2. The sum of three consecutive integers is less than 39. Find the largest values for these numbers.

Solution: Let n = first integer.
  Then $n + 1$ = second integer.
  And $n + 2$ = third integer.

The sum of the three is less than 39.

$n + n + 1 + n + 2 < 39$

$3n + 3 < 39$ Combine.

$3n < 36$ Subtract 3.

$n < 12$ Divide by 3.

The largest value for n is 11.

Then $n + 1 = 12$.

And $n + 2 = 13$.

Check: $11 + 12 + 13 \stackrel{?}{<} 39$

$36 < 39$ ✔

ANSWER: 11, 12, 13.

EXERCISES

In 1–5, solve algebraically to find:

1. the integers whose sum is as described.
   a. 3 consecutive integers whose sum is: **(1)** 99   **(2)** −12
   b. the smallest 2 consecutive integers whose sum is greater than: **(1)** 31   **(2)** −5
   c. 3 consecutive even integers whose sum is: **(1)** 48   **(2)** −60
   d. the largest 4 consecutive even integers whose sum is less than: **(1)** 60   **(2)** −8
   e. the largest 3 consecutive odd integers whose sum is at most: **(1)** 27   **(2)** −6
   f. 4 consecutive integers such that the sum of the second and fourth is 132.
   g. the largest 3 consecutive integers whose sum is negative.
   h. the smallest 3 consecutive even integers whose sum is more than 49.

2. 3 consecutive integers such that twice the smallest is 12 more than the largest.

3. the smallest 3 consecutive even integers for which twice the sum of the second and third is at least 33 more than the first.

4. 2 consecutive integers such that 4 times the larger exceeds 3 times the smaller by 23.

5. the largest 3 consecutive integers for which the first increased by twice the second exceeds the third by less than 25.

A counting game played by moving beans into consecutive hollows may be the world's oldest game. Under hundreds of different names, and in many versions, it is played by everyone from children to heads of state, and now even by computers. At the Massachusetts Institute of Technology, researchers use the game to study how machines make decisions.

6. Is it possible to find 3 consecutive even integers whose sum is 40? Explain.

7. Is it possible to find 3 consecutive odd integers whose sum is 59? Explain.

8. How many sets of 3 consecutive integers are there for which the sum of the three integers does not equal 3 times the middle integer? Explain.

9. The top 3 hitters on the Hawks baseball team have a total of 246 hits so far this season. If the individual totals for the three players are consecutive even integers, how many hits does each of the 3 players have?

10. An old warehouse is to be divided into 5 sections whose areas are consecutive odd integers. If one of the sections has an area of 175 square feet, find **(a)** the maximum and **(b)** the minimum possible total area of the 5 sections.

11. On a cold January morning in Fairbanks, Alaska, the temperatures were taken 3 times between 6 A.M. and 7 A.M. If the temperature readings were 3 consecutive even integers between −40 and −30 degrees, what temperatures could represent the middle reading?

12. Both rectangles shown in the diagram have a length and a width that are consecutive even integers. The length and the width of the larger rectangle are both two inches more than the length and the width of the smaller rectangle. If the sum of the perimeters of the two rectangles is at most 64 inches, find the greatest possible perimeter of the larger rectangle.
(Hint: Express all dimensions in terms of w, the width of the smaller rectangle.)

13. A man had $100 to distribute among his 3 sons. He gave the sons amounts that were the largest possible consecutive even integers. The remainder of the $100 went to charity.
    a. What amounts were given to the sons?
    b. How much went to charity?

14. Three of the house numbers on Sheridan Street are consecutive odd integers. Their sum contains three consecutive digits such that the middle digit is equal to the sum of the other two. Find the three house numbers on Sheridan Street.

15. The lengths of the sides of $\triangle ABC$ are consecutive even integers, and its perimeter is 24 cm. If the three sides of $\triangle RST$ are equal in length, and if each side of $\triangle RST$ is at least half the length of the longest side of $\triangle ABC$, what is the minimum perimeter of $\triangle RST$?

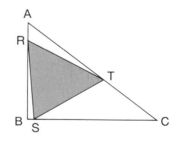

5-2 MONEY-VALUE PROBLEMS

Preparing to Solve Money-Value Problems

When working with items of different money values, such as coins of different denominations (pennies, nickels, dimes, and quarters), it is helpful to represent the values of the items in the same unit of money (the common money unit for coins is cents).

The basic relation of money value is expressed by the formula.

$$\begin{pmatrix} \text{Value of} \\ \text{one item} \end{pmatrix} \times \begin{pmatrix} \text{Number of} \\ \text{those items} \end{pmatrix} = \begin{pmatrix} \text{Total money value} \\ \text{of those items} \end{pmatrix}$$

The formula may be applied several times in a single computation. Thus, the total money value of 6 nickels and 4 dimes is $5(6) + 10(4) = 30 + 40 = 70$ cents.

EXERCISES

In 1–5, represent the total money value, in cents, of the given number of items.

1. pennies
  a. 16
  b. p
  c. $2p$

2. nickels
  a. 8
  b. x
  c. $3x$

3. dimes
  a. 3
  b. y
  c. $y + 2$

4. quarters
  a. 6
  b. q
  c. $2q - 3$

5. dollars
  a. 4
  b. 15
  c. D

6. Represent the value of each of the following in cents:

  a. $13.00
  b. $8.75
  c. 8 pennies and 6 nickels
  d. 8 nickels and 7 dimes
  e. 13 nickels and 7 quarters
  f. 3 dollars and 5 half-dollars
  g. x pennies and $2x$ nickels
  h. n nickels and $(2n - 1)$ dimes
  i. q quarters and $(n + 5)$ dimes
  j. x dollars and $(3x - 2)$ dimes
  k. x nickels and $(15 - x)$ dimes
  l. y dimes and $(20 - y)$ quarters
  m. x pennies, $3x$ dimes, and $(x + 3)$ quarters
  n. y nickels, $2y$ dimes, and $(y - 3)$ dollars

Solving Money-Value Problems

To solve a money-value problem, you can organize the facts in a table or a pair of matrices.

MODEL PROBLEM

Kaya does community service by working in a lunch program for senior citizens. He checked the rice supply and found that the number of pounds of brown rice was twice that of yellow rice, and the number of pounds of white rice was 3 more than that of yellow rice. The prices per pound of rice were 50¢ for brown, 45¢ for yellow, and 40¢ for white. If the total value of the current supply is $38.20, how many pounds of each kind of rice are on hand?

Solution: Let p = the number of pounds of yellow rice.

Then $2p$ = the number of pounds of brown rice.

And $p + 3$ = the number of pounds of white rice.

In a table, fill in the types of rice, the value of each type, and the amount of each type. Then fill in the total value for each type.

Kind	Cent value per lb.	Number of lbs.	Total value
Yellow	45	p	$45p$
Brown	50	$2p$	$50(2p)$
White	40	$p + 3$	$40(p + 3)$

The sum of the entries in the last column of the table represents the total money value, in cents, of the rice supply.

The total value of the rice supply is $38.20.

$$45p + 50(2p) + 40(p + 3) = 3{,}820$$
$$45p + 100p + 40p + 120 = 3{,}820$$
$$185p + 120 = 3{,}820$$
$$185p = 3{,}700$$
$$p = 20$$
$$2p = 40$$
$$p + 3 = 23$$

To use matrices, write one matrix to show the value of each type of rice, and a second matrix to show the number of pounds.

$$\begin{array}{cc} \begin{array}{ccc} Y & B & W \end{array} & \\ \text{Value} \begin{bmatrix} 45 & 50 & 40 \end{bmatrix} & \begin{array}{c} Y \\ B \\ W \end{array} \begin{bmatrix} p \\ 2p \\ p + 3 \end{bmatrix} \end{array}$$

No. of lbs.

The product of the two matrices represents the total money value, in cents, of the rice supply. Setting this product equal to $38.20, or 3,820 cents, results in the same equation obtained from the table.

Check:

Value of yellow = $45(20)$ =   900
Value of brown = $50(40)$ = 2,000
Value of white = $40(23)$ =   920
                    Total value = 3,820
                          or $38.20 ✓

ANSWER: The rice supply has 20 lbs. of yellow, 40 lbs. of brown, and 23 lbs. of white.

As improvements in health care are extending the life span of Americans, a growing number of government agencies and private groups are offering programs that provide recreation and education for older citizens. In the photograph, a volunteer is serving lunch to a client at a Japanese Senior Center in Berkeley, CA.

Now it's your turn to . . . **TRY IT!**   *(See page 256 for solution.)*

Of $1.35 in nickels and dimes, there are 15 coins in all. How many of each are there?

EXERCISES

In 1–17, solve algebraically.

1. Bill has 4 times as many quarters as dimes. In all he has $2.20. How many coins of each type does he have?

2. Roberto changed a $5 bill for dimes and quarters. He received 8 more dimes than quarters. How many coins of each type did he receive?

3. A purse contains at most $1.40 in nickels and quarters. If, in all, there are 15 coins, what is the least number of nickels there could be?

4. James bought 80 postage stamps for which he paid $23.00. Some were 40-cent stamps and some were 25-cent stamps. How many of each kind did he buy?

5. The cashier in a movie box office sold 200 more adult admission tickets at $4 each than children's admission tickets at $2 each. What is the minimum number of each type of ticket that the cashier had to sell for the total receipts to be at least $1,700?

6. Perry's Pizza Parlor sells pepperoni pizza for $1.15 per piece and plain pizza for $.90. Paul purchases 37 pieces of pizza for Pam's party, paying $39.05. How many pieces of pepperoni pizza did Paul purchase?

7. Marie has more than $5.00 in quarters and dimes. The number of quarters is 1 more than twice the number of dimes. Find the smallest number of each coin that Marie could have.

8. Mildred bought 10-cent stamps, 20-cent stamps, and 40-cent stamps for $21.00. The number of 10-cent stamps exceeded the number of 40-cent stamps by 50. The number of 20-cent stamps was 10 less than twice the number of 40-cent stamps. How many stamps of each kind did she buy?

9. A class contributed $3.50 in nickels and dimes to the Red Cross. In all there were 45 coins. How many were there of each kind of coin?

10. A vendor sells bags of popcorn for $.95. She sells bags of cheese twists for $1.25. If she sold 7 more bags of popcorn than bags of cheese twists and took in less than $80.00, what is the most of each she could have sold?

11. A postal clerk sold 70 stamps for $17.00. Some were 35-cent and some were 20-cent stamps. How many of each kind did he sell?

12. Mr. Perkins cashed a $185 check in his bank. He received $1 bills, $5 bills, and $10 bills. In this order, the numbers of the three types of bills he received were three consecutive integers. How many bills of each type did he receive?

13. Marjorie went to the market with $50 to stock up on dog food during a sale of canned goods. All-Beef was selling for 70 cents a can and Balanced-Meal cost 60 cents a can. Marjorie bought 8 more cans of All-Beef than Balanced-Meal. What is the maximum number of cans of All-Beef that she could have bought?

14. Harry's Heavenly Hamburgers sell for $1.50 each. Sam's small sodas sell for 70 cents each. If Sam sold 7 more sodas than Harry sold hamburgers and their combined total sales came to $44.50, how many hamburgers were sold?

15. An amusement park sells three types of tickets: individual rides cost 40 cents, the Special Combination costs $3.00, and the Jumbo Combination costs $4.40. If twice as many Jumbo Combinations have been sold as Special Combinations, and a total of 336 tickets have been sold for $982.40, how many of each type of ticket have been sold?

Water parks, first introduced about 1980, are a popular and fast-growing addition to the amusement park industry, attracting tens of millions of people a year and offering white-water adventure to thrill-seekers who may live far from natural water locations.

In wave pools, surfers can ride 6-foot waves. On water slides 60 feet high that provide momentum of over 40 miles per hour, riders become airborne, shooting forward like cannon balls. Scientific know-how, using physics, computers, and electronic sensors, is applied by designers to make water rides more and more hair-raising while at the same time reducing the risk of injury.

16. Dave Deal sells his famous Big Deal hamburger for $1.70. If you want to add tomato, lettuce, cheese, and Dave's Deal pickles, you can buy a Super Deal for only $1.85. Dave sold a total of 150 hamburgers for $268.95. How many of these were Super Deals?

17. A travel agency sold a 7-day cruise package for $2,350 per person (double occupancy) and a 4-day cruise package for $1,650 per person (double occupancy). Twelve couples bought cruises for a total price of $42,400. How many couples bought the 7-day cruise package?

18. Is it possible to have $4.50 in dimes and quarters and have twice as many quarters as dimes? Explain.

19. Is is possible to spend $10 for 100 stamps consisting of 20¢ and 7¢ stamps? Explain.

5-3 MOTION PROBLEMS

Preparing to Solve Motion Problems

The basic relationship of distance, rate, and time is given in the formula:

$$\text{Distance} = \left(\begin{array}{c}\textbf{Rate per unit}\\\textbf{of time}\end{array}\right) \times \left(\begin{array}{c}\textbf{Number of units}\\\textbf{of time}\end{array}\right) \ \ or \ \ \boldsymbol{D = RT}$$

Since a car does not usually travel at the same uniform rate throughout a trip, a stated rate of speed will represent an average. Rates are generally in miles per hour (mph) or in kilometers per hour (km/h). Note how units "cancel."

A car traveling at x miles per hour for 6 hours will go a distance of:

$$\frac{x \text{ miles}}{1 \text{ hour}} \times 6 \text{ hours} = 6x \text{ miles}$$

The following diagrams show some of the situations that arise in motion problems. The stars indicate the starting points.

Starting at same time and place, going in opposite directions. Distance apart is increasing.

Starting at same time, meeting at a point between. Distance apart is decreasing.

Starting at same place, later starter goes faster to catch up. Distances are the same.

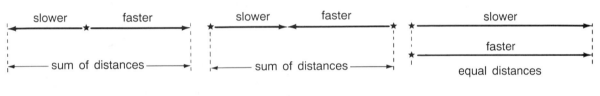

EXERCISES

1. If a car is traveling 80 km per hour, represent how far it will go in:
   a. 5 hr.   **b.** 3 hr. 30 min.   **c.** x hr.   **d.** $(2x + 1)$ hr.   **e.** $(10 - x)$ hr.

In 2–5, draw a diagram to model the problem. Using x to represent a number of hours, show the distance traveled by each person or vehicle.

2. Two cars started from the same place at the same time and traveled in opposite directions at rates of 60 km/h and 80 km/h.

3. Rosa and Nicki started on bicycles at the same time from two different places on a straight road and traveled toward each other. Rosa traveled 8 mph and Nicki traveled 10 mph.

4. Mr. Sands left his home by car, traveling at the rate of 75 km/h. One hour later, his son Barry left home and started after him on the same road, traveling at the rate of 80 km/h.

5. At 3 P.M., Jeff and Katie, both traveling at 28 mph, started their motorboats toward each other from opposite sides of a lake.

6. At 10 A.M., Jim left his home, walking at 3 mph toward the lake 3 miles away. At 10:30 A.M., Jim's sister Alice left their home to try to overtake Jim. If Alice can walk at $4\frac{1}{2}$ mph, can she catch Jim before he reaches the lake? Explain.

7. Nadine and Cyril arranged to meet in Wall, South Dakota. From 200 miles away, Nadine drove east toward Wall at 40 mph. Cyril left at the same time and drove west toward Wall at 60 mph from 250 miles away. Who got there first? Explain.

8. A Greek slave named Aesop, who lived about 600 B.C., told animal stories to teach lessons about human behavior. In Aesop's fable about the race of the hare (rabbit) and the tortoise, suppose the hare had to run a mile and the tortoise had to run only $\frac{1}{10}$ of a mile. Who would win if the hare ran at 20 mph (no napping) and the tortoise ran at $\frac{1}{5}$ mph?

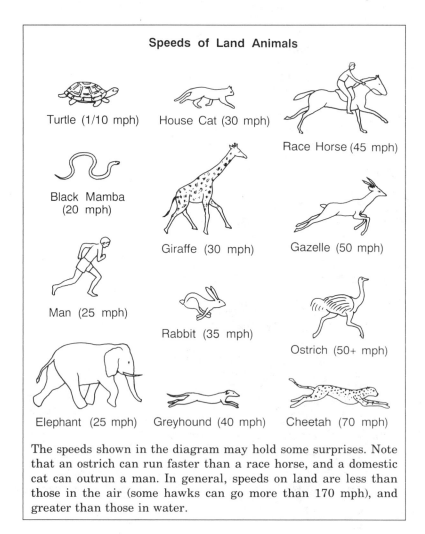

Speeds of Land Animals

Turtle (1/10 mph) House Cat (30 mph) Race Horse (45 mph)

Black Mamba (20 mph) Giraffe (30 mph) Gazelle (50 mph)

Man (25 mph) Rabbit (35 mph) Ostrich (50+ mph)

Elephant (25 mph) Greyhound (40 mph) Cheetah (70 mph)

The speeds shown in the diagram may hold some surprises. Note that an ostrich can run faster than a race horse, and a domestic cat can outrun a man. In general, speeds on land are less than those in the air (some hawks can go more than 170 mph), and greater than those in water.

9. If Lynette drove north from Culver City at noon traveling 55 mph and, at 1 P.M., Gerri took a northbound train traveling 85 mph from Culver City, who would be farther from Culver City at 3 P.M.? How far apart would Lynette and Gerri be?

10. Fred drove to his mother's house for $2\frac{1}{2}$ hours at an average speed of 40 mph. If he returned home at 50 mph, how long did it take him?

11. On a rainy day, Mr. Coles is considering whether he should walk the 1.1 miles from Madison Ave. to West End Ave., or take the M18 crosstown bus. If the bus averages 33 mph, and he walks at 3 mph, how much longer will it take him to walk?

Solving Motion Problems

When solving motion problems, it is helpful to draw a diagram. It is also helpful to organize the facts in the problem by using a table.

MODEL PROBLEMS

1. A passenger train and a freight train start at the same time from stations that are 405 miles apart and travel toward each other. The rate of the passenger train is twice the rate of the freight train. In 3 hours, the trains pass each other. Find the rate of each.

Solution: Let r = the rate of the freight train.

Then $2r$ = the rate of the passenger train.

Draw a diagram to model the problem.

In a table, fill in the rate and time for each train. Then represent each distance.

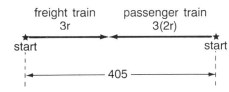

	Rate (mph)	Time (h)	Distance (m)
Freight train	r	3	$3r$
Passenger train	$2r$	3	$3(2r)$

The total distance was 405 miles.

$$3r + 3(2r) = 405$$
$$3r + 6r = 405$$
$$9r = 405$$
$$r = 45$$

Check: $3(45) = 135$
$$3(90) = \underline{270}$$
$$\text{Total} = 405 \checkmark$$

ANSWER: Rate of freight train was 45 mph; rate of passenger train was 90 mph.

2. Two cars left Rapid Falls at the same time, one traveling west at 40 mph and the other east at 50 mph. After how many hours will the cars be at least 270 miles apart?

Solution: Let h = the number of hours each car traveled.

Draw a diagram to model the problem.

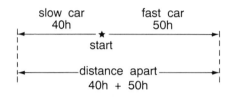

In a table, fill in the rate and time for each car. Then represent each distance.

	Rate (mph)	Time (h)	Distance (m)
Slow car	40	h	$40h$
Fast car	50	h	$50h$

The distance apart will be at least 270 miles.

$$40h + 50h \geq 270 \qquad \textbf{\textit{Check:}}\ 40(3) = 120$$
$$90h \geq 270 \qquad\qquad\qquad 50(3) = \underline{150}$$
$$h \geq 3 \qquad\qquad\qquad\quad \text{Total} = 270✔$$

ANSWER: In 3 hours or more, the two cars will be at least 270 miles apart.

Now it's your turn to . . . **TRY IT!**

1. Martin left his house by car, traveling on a certain road at the rate of 50 km/h. Two hours later, his brother William left the same house and started after him on the same road, traveling at the rate of 70 km/h. In how many hours did William overtake Martin?

2. How far can Mrs. Chen drive out into the country at the average rate of 40 mph and return over the same road at the average rate of 30 mph if she travels a total of 7 hours?

(See page 256 for solutions.)

In 1–26, solve algebraically.

1. A destroyer traveling at 40 knots (nautical miles per hour) and a battleship traveling at 30 knots left the same naval base at the same time and sailed in opposite directions. In how many hours were the ships 350 miles apart?

Starboard bow view of the U.S.S. John F. Kennedy

The aircraft carrier, known as a flat-top, is a floating airfield. Below its flight deck, for takeoff and landing of aircraft, are a hangar deck, maintenance shops, storage space for supplies and equipment, and living quarters for the air and ship crews. Among the largest ships afloat, they can carry almost 100 aircraft and over 5,000 personnel.

2. Two planes left an airport at the same time, one flying east at 180 mph and the other flying west at 320 mph. For how many hours of traveling time were they no more than 1,000 miles apart?

3. At 8 A.M., two cars started from the same place, one traveling north at 35 mph and the other traveling south at 40 mph. At what time were they 300 miles apart?

4. Two planes flying in opposite directions, one traveling at 650 mph and the other at 550 mph, passed each other in flight. When within no more than 60 miles of each other, they could make radio contact. For how many minutes was radio contact possible?

5. Two trains are at stations that are 800 miles apart. If the trains start traveling toward each other at the same time, one at 67 mph and the other at 53 mph, in how many hours will they pass each other?

6. Two trains are 680 km apart. At 10 A.M., they start traveling toward each other at rates of 75 and 95 km/h. At what time will they pass each other?

7. Saratoga is 180 miles north of New York City. A truck traveled from New York toward Saratoga at 44 mph. Another truck traveled from Saratoga toward New York on the same highway at 36 mph. How many miles did each truck travel before they met?

8. Two planes started at the same time from the same airport and flew in opposite directions, one flying 90 km/h faster than the other. In 5 hours, they were 4,200 km apart. Find each rate.

9. A northbound and a southbound train left Tootleville station at the same time. The southbound train was traveling 20 mph faster than the northbound train. After an hour, they were not more than 100 miles apart. What is the maximum possible speed for the northbound train?

10. At 7 A.M., two cars started from the same place, one traveling east and the other traveling west. At 10:30 A.M., they were 287 miles apart. If the rate of the fast car exceeded the rate of the slow car by 6 mph, find the rate of each.

11. A salesman made a trip of 375 miles by bus and train. He traveled 3 hours by bus and 4 hours by train. If the train averaged 15 mph more than the bus, find the rate of each.

12. Two planes left at the same time from two airports that are 6,600 kilometers apart and flew toward each other. The rate of the faster plane was twice the rate of the slower plane. If they passed each other no more than 4 hours after they left, find the minimum possible rate for the slower plane.

13. A motorist made a trip of 275 miles in 8 hours. Before noon she averaged 40 mph, and after noon she averaged 25 mph. At what time did she begin her trip and when did she end it?

14. A motorized wheelchair, fully charged, can travel for several miles at 6 mph on level ground. Nora can average 10 mph on level ground in her manual wheelchair, but she must stop to rest for 1 minute after each quarter mile of exertion. If Nora wanted to travel 1 mile, would it be faster to use a manual or a motorized wheelchair? Explain.

Persons with disabilities are now able to participate in an ever-widening range of activities and employment.

Improved conditions are possible because of heightened public awareness and advances in technology, which have increased access to public facilities.

This photograph shows a student at the Human Resources Center in Albertson, New York.

15. Ricky and Jordan were both driving to a regional tennis match in Milltown, and happened to meet when they stopped for breakfast along the way. Ricky had left home at 7 A.M., driving at 55 mph along Route 1. Jordan, who lived at least 75 miles closer to Milltown, left home an hour later, and drove along the same highway at 45 mph. What is the earliest time they could have met?

16. A destroyer traveling at 60 knots and a battleship traveling at 50 knots leave the same base at the same time and sail in the same direction. In how many hours will they be 100 miles apart?

17. A freight train left a station and traveled at 30 mph. Two hours later, an express train left the same station and traveled in the same direction at 50 mph. In how many hours did the express train overtake the freight train?

18. Susan left her home at 7 A.M., driving her car at 30 mph. At 9 A.M., her sister Marion drove after her along the same highway, traveling at 45 mph. In how many hours did Marion pass Susan?

19. Marie drove on a country road from Centerville to Hurleyville, a distance of 118 miles. Her rate for the first 2 hours was 14 mph more than her rate on the remaining part of the trip, which required no more than an hour. Find the minimum rate possible on the first part of Marie's trip.

20. Mr. Stone started on a trip, planning to average 30 mph. How fast must his son Carl travel in order to overtake him in 3 hours if Carl started 30 minutes after his father?

21. A cargo plane left JFK airport at noon and flew toward Seattle at 300 mph. At 2 P.M., a jet plane left JFK and flew the same route as the cargo plane at 500 mph. How many miles did the jet plane fly before it overtook the cargo plane?

22. A round trip in a helicopter lasted 4 hours and 30 minutes. If the helicopter flew away from the airport at 100 mph and returned at the rate of 50 mph, what was its greatest distance from the airport?

23. A car and a train both leave Motorville at the same time traveling in opposite directions. The speed of the train is 10 mph greater than the speed of the car. After 5 hours, they are between 500 and 600 miles apart. Find the range of possible speeds of the train.

24. A flyer on reconnaissance duty spent $4\frac{1}{2}$ hours on a mission. He flew out from his base with the wind at the rate of 500 mph and returned to his base over the same route, flying against the wind at the rate of 400 mph. How many miles did he fly out before he turned back?

25. A pilot plans to make a flight lasting 2 hours and 30 minutes. How far can she fly from her base at 600 km/h and return over the same route at 400 km/h?

26. Al Unser, Sr. and Bobby Rahal were contenders in the Indy 500 automobile race. In the first 2 hours, Bobby went 9 miles an hour faster than Al, and the total distance covered by both was 666 miles. Find the rate at which Bobby was racing.

5-4 LEVER AND PULLEY PROBLEMS

Lever Problems

A *lever* is a bar that can rotate about a fixed point called the *fulcrum*.

A weight w_1 is placed on one arm of a lever at a distance d_1 from the fulcrum, and a second weight w_2 is placed on the other arm at a distance d_2 from the fulcrum. When the lever is in balance, the following relationship, called the *law of the lever*, is true:

$$w_1 \times d_1 = w_2 \times d_2$$

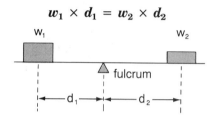

To adjust for different weights, some seesaws are designed so that the board can be placed in different positions over the fulcrum. On other seesaws, springs provide balance.

MODEL PROBLEM

A 14-foot plank is used as a seesaw by Holly, who weighs 120 lbs., and Tanya, who weighs 90 lbs. If the girls balance each other, how far from the fulcrum is each sitting?

Solution: Let x = Holly's distance from the fulcrum.

Then $14 - x$ = Tanya's distance from the fulcrum.

$$w_1 \cdot d_1 = w_2 \cdot d_2$$
$$120x = 90(14 - x)$$
$$120x = 1{,}260 - 90x$$
$$210x = 1{,}260$$
$$x = 6$$
$$14 - x = 8$$

Check:

$$120(6) \stackrel{?}{=} 90(8)$$
$$720 = 720 ✓$$

ANSWER: Holly, 6 feet from the fulcrum; Tanya, 8 feet from the fulcrum.

Now it's your turn to . . . **TRY IT!**   *(See page 257 for solution.)*

Sylvia, who weighs 60 pounds, sits 3 feet from the fulcrum of a seesaw. Max just balances her when he is sitting 4 feet from the fulcrum. How much does Max weigh?

Pulley Problems

A *pulley* is a wheel used to do work, such as lifting weights, by means of a rope or belt passing over its rim.

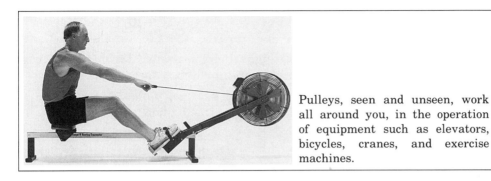

Pulleys, seen and unseen, work all around you, in the operation of equipment such as elevators, bicycles, cranes, and exercise machines.

Pulling power can be increased, when a belt connects two pulleys of different sizes, by raising the turning speed of a pulley. Similar to the law of the lever, the relationship for pulleys involves the speeds, s_1 and s_2 (in revolutions per minute, or rpm), and the lengths of the diameters, d_1 and d_2.

$$s_1 \times d_1 = s_2 \times d_2$$

EXERCISES

In 1–16, solve algebraically.

1. Sue, who weighs 80 pounds, sits 4 feet from the fulcrum of a seesaw and balances Lillian, who is sitting 5 feet from the fulcrum. Find Lillian's weight.

2. Robert weighs 180 lb. and Gene weighs 120 lb. How far from the fulcrum of a seesaw must Robert sit to balance Gene, who is sitting 6 feet from the fulcrum?

3. A pulley with a diameter of 14 inches is running at 140 rpm while belted to a 28-inch pulley. Find the speed of the larger pulley.

4. A pulley with a diameter of 8 inches, that is turning at 270 rpm, is belted to another pulley turning at 360 rpm. Find the diameter of the second pulley.

5. A pulley with a diameter of 8 inches is running at 1,452 rpm while belted to a 12-inch pulley. Find the speed of the larger pulley.

6. The figure shows how a bar 8 ft. long is used to lift a rock that weighs 450 lb. The fulcrum is placed 2 ft. from the rock. What weight must be applied at the upper end of the bar to lift the rock?

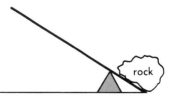

7. Larry can exert a force of 180 lb. How heavy a rock can he lift if he uses a bar that is 5 ft. long and places the fulcrum so that it is 6 in. from the rock?

8. An automobile fan belt works on the pulley principle. The pulley turning at 360 rpm has a diameter of 6 inches. Find the diameter of the second pulley if it is turning at 270 rpm.

9. Martha and Ann use a plank 10 ft. long as a seesaw. Ann weighs 40 lb. and Martha weighs 60 lb. If the girls are to balance each other, how far from the fulcrum must each sit?

10. Fred, who weighs 100 lb., and Jack, who weighs 140 lb., use a 12-ft. plank as a seesaw. Where should the fulcrum be placed if the boys are to balance each other?

11. A mechanic in an elevator shaft is repairing a 5-foot-diameter pulley that runs at 540 rpm. If the pulley that it is belted to runs at 675 rpm, find the diameter of the second pulley.

12. Bill wished to carry two bundles, one weighing 60 lb. and the other weighing 40 lb. He put one of them at each end of a bar 5 ft. long and placed the bar on one shoulder. If he balanced the weights, where did he place the fulcrum (his shoulder)?

13. A block-and-tackle arrangement of two pulleys is used to raise a heavy load. The diameters of the pulleys are 5 feet and 3 feet, respectively. If the larger pulley runs at 18 rpm, find the speed of the smaller pulley.

14. The blades of a fan are operating by means of two pulleys. One is a 9-inch pulley turning at 1,200 rpm. Find the speed of the other pulley if it has a 6-inch diameter.

15. Ronald, who can exert a force of 200 lb., wishes to raise an object that weighs 600 lb. using a 6-ft. bar as a lever and a block as a fulcrum. Where should he place the fulcrum?

16. Together, Mindy and Norma weigh 140 pounds. They balance on a seesaw when Mindy is 8 feet from the fulcrum and Norma is 6 feet from the fulcrum. Find the weight of each.

Without knowing the law of the lever in mathematical terms, people have since ancient times used the principle in practice, to balance heavy loads. The photograph shows a modern-day woman in Guizhou Province, China.

17. A man lifts a concrete block with a 6-foot bar. If he places the fulcrum 6 inches from the block, and has to exert a force of 200 pounds, how heavy is the block?

18. A 200-pound weight is raised from the ground to the roof of a building 50 feet high, using a system of two pulleys. The pulley on the roof has a 27-inch diameter while the one on the ground has a 36-inch diameter. If the speed of the smaller pulley is 85 rpm, find the speed of larger pulley.

Preparing to Solve Angle Problems

Angles are named according to their degree measure, m.

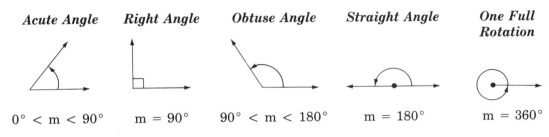

Acute Angle	*Right Angle*	*Obtuse Angle*	*Straight Angle*	*One Full Rotation*
$0° < m < 90°$	$m = 90°$	$90° < m < 180°$	$m = 180°$	$m = 360°$

The point where the sides of an angle meet is the *vertex*. With the symbol $\angle$ representing *angle*, the angle at the right can be named $\angle ABC$ or simply $\angle B$, using the vertex letter. $m\angle B$ is read *the measure of angle B*.

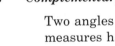

$$m\angle ABC = 63°$$
$$\text{or } m\angle B = 63°$$

Here are descriptions of some special angle pairs:

Vertical Angles

When two lines intersect, the opposite angles formed are equal in measure.

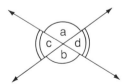

$$m\angle a = m\angle b \text{ and } m\angle c = m\angle d$$

Supplementary Angles

Two angles whose measures have a sum of 180°.

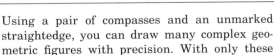

$$m\angle 1 + m\angle 2 = 180°$$

Complementary Angles

Two angles whose measures have a sum of 90°.

$$m\angle x + m\angle y = 90°$$

Using a pair of compasses and an unmarked straightedge, you can draw many complex geometric figures with precision. With only these instruments, it is easy to bisect an angle, but trisecting an angle is impossible.

1. Name the kind of angle illustrated by the stickperson's flag signals.

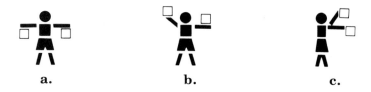

        a.   **b.**   **c.**

2. Name the kind of angle that is formed between the hands of a clock when they show:
   a. 3 o'clock   **b.** 1 o'clock   **c.** 5 o'clock   **d.** 6 o'clock

3. Order the measures of the following types of angles, naming the smallest angle first:
   right angle, acute angle, straight angle, obtuse angle

4. Find the number of degrees there are in:
   a. $\frac{1}{3}$ of a right angle   **b.** $\frac{3}{5}$ of a straight angle   **c.** $\frac{1}{4}$ of a full rotation

5. Find the range of values in the measure of:
   a. $\frac{1}{3}$ of an acute angle   **b.** $\frac{1}{5}$ of an obtuse angle

6. Find the number of degrees through which the earth rotates in:
   a. 24 hours   **b.** 12 hours   **c.** 1 hour   **d.** 4 minutes

7. Find the number of degrees in the angle formed by the hands of a clock at:
   a. 1 P.M.   **b.** 3 P.M.   **c.** 4 P.M.
   d. 6 P.M.   **e.** 12:30 P.M.   **f.** 5:30 P.M.

8. Find the measure of the angle whose supplement measures:
   a. 50°   **b.** 120°   **c.** $y°$

9. Find the measure of the angle whose complement measures:
   a. 30°   **b.** $z°$ **c.** $(90 - z)°$

10. Two lines intersect to form angles 1, 2, 3, and 4. If $m\angle 1 = 30°$, find the measures of the other angles.

11. At the end of a run, a turntable turns an electric trolley car around, to head it back for the return run. Through how many degrees must the car be turned?

Solving Angle Problems

MODEL PROBLEMS

1. Find the number of degrees in an angle that is four times as large as its complement.

Solution: Let x = the number of degrees in the complement of the angle.
Then $4x$ = the number of degrees in the angle.

The sum of the measures of the two complementary angles is 90°.

$$x + 4x = 90$$
$$5x = 90 \qquad \qquad \textit{Check:}$$
$$x = 18 \qquad \quad 72° = 4 \times 18°$$
$$4x = 72 \qquad 72° + 18° = 90° ✔$$

ANSWER: The angle measures 72°.

2. The lines AB and CD intersect at a point E. If $m\angle BED = (7x - 9)°$ and $m\angle AEC = (5x + 3)°$, find the number of degrees in $m\angle BED$.

Solution: Draw a diagram to see the relationship of the angles.

Vertical angles are equal in measure.

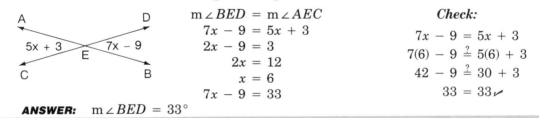

$$m\angle BED = m\angle AEC \qquad \qquad \textit{Check:}$$
$$7x - 9 = 5x + 3 \qquad \qquad 7x - 9 = 5x + 3$$
$$2x - 9 = 3 \qquad \qquad 7(6) - 9 \overset{?}{=} 5(6) + 3$$
$$2x = 12 \qquad \qquad 42 - 9 \overset{?}{=} 30 + 3$$
$$x = 6 \qquad \qquad \qquad 33 = 33 ✔$$
$$7x - 9 = 33$$

ANSWER: $m\angle BED = 33°$

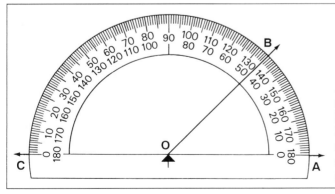

A *protractor* is an instrument used to measure angles. Since the upper scale reads from left to right and the lower scale from right to left, you can measure an angle starting from either of its sides. Each upper-lower pair of numbers are measures of supplementary angles.

In 1 and 2, lines MN and RS intersect at T.

1. If $m \angle RTM = 5x°$ and $m \angle NTS = (3x + 10)°$, find the number of degrees in $\angle RTM$.

2. If $m \angle MTS = (4x - 60)°$ and $m \angle NTR = 2x°$, find the number of degrees in $\angle MTS$.

3. When Brett cut an orange across the middle, he noticed that the membranes separated the fruit into 12 equal sections. If all of the angles at the center were the same size, what was the measure of each central angle?

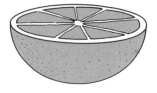

In 4–22, solve algebraically.

4. Two angles are complementary. If the measure of one angle is twice that of the other, find the number of degrees in each angle.

5. The measure of an angle is twice that of its supplement. Find the number of degrees in each angle.

6. If the measure of the complement of an angle is 8 times that of the angle, find the number of degrees in the complement.

7. If the measure of the supplement of an angle is 5 times that of the angle, find the number of degrees in the angle.

8. The measure of the supplement of an angle is one-half of the measure of the angle. Find the number of degrees in the angle.

9. If the measure of the complement of an angle is one-fifth that of the angle, find the number of degrees in the complement.

10. Find the number of degrees in an angle whose measure exceeds twice that of its complement by 36°.

11. How many degrees are in an angle whose measure exceeds that of its supplement by 10°?

12. An angle measures 30° less than 5 times its complement. Find the number of degrees in each angle.

13. The number of degrees in two complementary angles can be represented by two consecutive even numbers. Find the number of degrees in each angle.

14. Find the number of degrees in an angle that measures 20° less than 4 times its supplement.

15. An angle is 36° less than twice its supplement. Find the number of degrees in each angle.

Look at a parking meter to see supplementary angles formed in the time zone.

An opened pair of scissors shows vertical angles.

A partially open door in the corner of a room forms complementary angles with the adjoining walls.

Can you think of other examples of angles in action?

16. Two angles are complementary. The larger exceeds three times the smaller by 10°. Find the number of degrees in each angle.

17. Find two complementary angles such that $\frac{2}{3}$ of the smaller is equal to $\frac{1}{2}$ of the result obtained when the larger is diminished by 20°.

18. Find two supplementary angles such that $\frac{1}{4}$ of the smaller is equal to $\frac{1}{3}$ of the result obtained when the larger is diminished by 110°.

19. In a mountaineering technique called a Tyrolean traverse, a rope is stretched tightly between two points and the climber slides across the rope. In the diagram, the climber's weight has pulled the rope down to form an angle at point C, and $m\angle A = m\angle B$. If the measure of $\angle C$ is 10 times that of $\angle A$, and the sum of the measures of the three angles is 180°, find $m\angle A$.

20. Imaginary lines are drawn from the earth's center to four points on the equator, in Kenya, Borneo, the Gilbert Islands, and Ecuador, respectively, forming four angles. The angle between Borneo and the Gilbert Islands is the smallest, containing 15° less than the angle between Borneo and Kenya. The angle between Kenya and Ecuador is twice the measure of the smallest angle, and the angle between Ecuador and the Gilbert Islands measures 15° less than twice the smallest angle. Find the measures of the four angles.

21. An offshore oil well is located at point A and the oil can be piped from A to either Terminal C or Terminal D, which are located on shoreline BE. If $m\angle DAE$ is 2 degrees less than $m\angle BAC$ and $m\angle CAD$ is 5 degrees more than twice the measure of $\angle BAC$, and if $m\angle BAE = 35°$, find $m\angle BAC$.

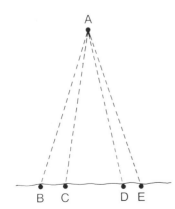

22. Winochee Waterfall drops from a height in three successive leaps. To see the top of the falls, a tourist looks up at an angle of 72° from the horizontal. The three angles through which he views the three sections of the waterfall have measures that are consecutive even integers. Find the measures.

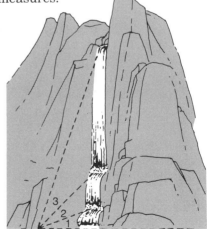

5-6 TRIANGLE PROBLEMS

Preparing to Solve Triangle Problems

Triangles are classified according to the number of sides that are equal in measure. The number of equal angles depends on the number of equal sides.

Scalene Triangle

No sides equal.
No angles equal.

Isosceles Triangle

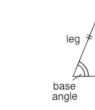

Two sides equal.
Two angles equal.

Equilateral Triangle

Three sides equal.
Three angles equal.

Triangles are also classified by the nature of the angles they contain.

Acute Triangle

Three acute angles.

Right Triangle

One right angle and
two acute angles.

Obtuse Triangle

One obtuse angle and
two acute angles.

All triangles, regardless of size or shape, have the following properties:

1. The sum of the measures of the angles of a triangle is 180°.

You can see that this is so by tearing off two angles of any paper triangle and placing them adjacent to the third angle as is shown in the diagram.

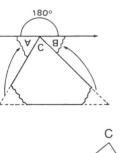

2. The sum of the measures of two sides of a triangle is greater than the measure of the third side.

This property of a triangle stems from the fact that the shortest distance between two points is a straight line. Thus, in the diagram, any way other than the line segment between A and B must be a longer way to get from A to B.

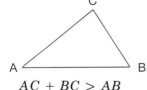

$$AC + BC > AB$$

In 1–4, find the measure of the third angle of the triangle if the first two angles contain:

1. $60°$, $40°$ **2.** $100°$, $20°$ **3.** $30°$, $60°$ **4.** $45°$ each

5. Find the number of degrees in each angle of an equilateral triangle.

6. Explaining your conclusion, determine if it is possible for a triangle to have:
    a. two right angles   **b.** two obtuse angles   **c.** one right and one obtuse angle

7. Consider the two acute angles of a right triangle.
    a. What is the sum of their measures?
    b. What kind of angle pair are they?

8. If two angles of a triangle measure $50°$ and $80°$, what kind of triangle is it?

9. If two angles of a triangle measure $38°$ and $52°$, what kind of triangle is it?

10. Find the measure of the vertex angle of an isosceles triangle if each base angle measures:
    a. $80°$ **b.** $45°$ **c.** $22\frac{1}{2}°$

11. Find the measure of each base angle of an isosceles triangle if the vertex angle measures:
    a. $40°$ **b.** $90°$ **c.** $100°$

12. Explaining your conclusion, determine if a base angle of an isosceles triangle can be:
    a. a right angle   **b.** an obtuse angle

13. A diagonal of a **quadrilateral** (4 straight sides) divides the figure into two triangles. Draw a conclusion about the sum of the measures of the four angles of a quadrilateral.

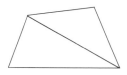

14. If the lengths of two sides of a triangle are 10 and 14, the length of the third side may be
    (1) 22   (2) 2   (3) 24   (4) 4

15. The lengths of two sides of an isosceles triangle are 8 and 10. The length of the third side could be   (1) 8 only   (2) 10 only   (3) either 8 or 10   (4) 6

16. The lengths of two sides of an isosceles triangle are 3 and 7. Find the length of the third side.

17. Which set of numbers could represent the lengths of the sides of a triangle?
    (1) $\{1, 2, 3\}$ (2) $\{2, 4, 6\}$ (3) $\{3, 5, 7\}$ (4) $\{5, 10, 20\}$

18. In which diagram does $AB + BC - AC = 0$?

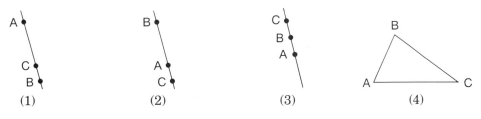

Solving Triangle Problems

MODEL PROBLEMS

1. In isosceles triangle ABC, the measure of vertex angle C exceeds the measure of each base angle by $30°$. Find the number of degrees in each angle of the triangle.

Solution:

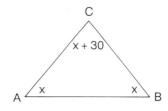

Let x = the number of degrees in $\angle A$.

Then x = the number of degrees in $\angle B$.

And $x + 30$ = the number of degrees in $\angle C$.

The sum of the measures of the angles of a triangle is $180°$.

$$x + x + x + 30 = 180$$
$$3x + 30 = 180$$
$$3x = 150$$
$$x = 50$$
$$x + 30 = 80$$

Check

$$50° + 50° + 80° = 180° ✓$$

ANSWER: $m\angle A = 50°$, $m\angle B = 50°$, $m\angle C = 80°$

2. A carpenter built a triangular wooden brace. If a second side of the triangle exceeded twice the length of the first side by 2 feet, and the third side was 1 foot more than the second, what was the least integral number of feet of wood he could have used?

Solution: Let x = the length of the first side.
Then $2x + 2$ = the length of the second side.
And $2x + 3$ = the length of the third side.

The sum of the measures of two sides is greater than the third side.

$$x + 2x + 2 > 2x + 3$$
$$3x + 2 > 2x + 3$$
$$x + 2 > 3$$
$$x > 1$$

Thus, the smallest possible integral value of x is 2. If $x = 2$, then $2x + 2 = 6$ and $2x + 3 = 7$.

Check: Is it true that the sum of any two of these possible sides is greater than the third? Yes, $2 + 6 > 7$ and $2 + 7 > 6$ and $6 + 7 > 2$.

Therefore, the smallest possible integral perimeter is $2 + 6 + 7$ or 15 feet.

ANSWER: The carpenter used at least 15 feet of wood.

1. In $\triangle RST$, m $\angle S$ is 15° less than m $\angle R$, and m $\angle T$ is 25° more than 3 times m $\angle R$. Find the number of degrees in each angle.

2. A triangular pen is enclosed by 50 feet of fence. The shortest side measures 12 feet less than the longest side. The length of the third side is 8 feet more than the length of the shortest side. How long are the sides of the pen?

(See page 257 for solutions.)

EXERCISES

In 1–10, solve algebraically to find the measure of each angle of the triangle.

1. In $\triangle RST$, the measure of $\angle R$ is one-half the measure of $\angle S$, and the measure of $\angle T$ is 3 times the measure of $\angle S$.

2. In $\triangle ABC$, the measure of $\angle B$ is twice the measure of $\angle A$, and the measure of $\angle C$ is 3 times the measure of $\angle B$.

3. In a triangle, the measure of the second angle is 4 times the measure of the first angle. The measure of the third angle is equal to the sum of the measures of the first two angles.

4. In $\triangle LMN$, the measure of $\angle M$ is 6 times the measure of $\angle L$. The measure of $\angle N$ is equal to the difference between the measures of $\angle M$ and $\angle L$.

5. Each base angle of an isosceles triangle measures 7 times the measure of the vertex angle.

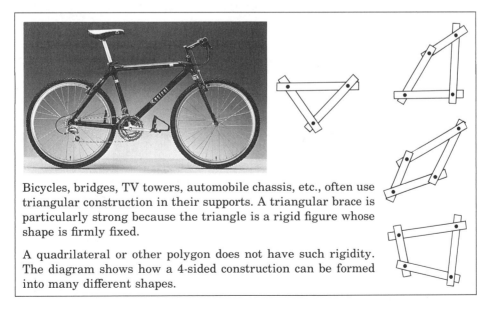

Bicycles, bridges, TV towers, automobile chassis, etc., often use triangular construction in their supports. A triangular brace is particularly strong because the triangle is a rigid figure whose shape is firmly fixed.

A quadrilateral or other polygon does not have such rigidity. The diagram shows how a 4-sided construction can be formed into many different shapes.

6. The measure of each of the equal angles of an isosceles triangle is one-half of the measure of the vertex angle.

7. The measure of the vertex angle of an isosceles triangle is one-fourth that of a base angle.

8. The measure of each of the equal angles of an isosceles triangle is 9° less than 4 times the measure of the vertex angle.

9. The measure of each of the equal angles of an isosceles triangle exceeds twice the measure of the vertex angle by 15°.

10. The measures of the angles of a triangle are three consecutive even integers.

11. The base of an ornamental candelabra is in the shape of an isosceles triangle, measuring 15 inches across the bottom. Find the least possible integral length of one of the equal sides of the triangle.

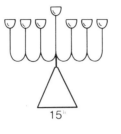

12. A stock clerk is arranging a display in the shape of a triangle. Two sides of the display measure 3 feet and 4 feet, respectively. If the third side measures a whole number of feet, find:
   a. its shortest possible length
   b. its longest possible length

13. Mrs. Sanchez has a vegetable garden in the shape of a quadrilateral, with a fence placed diagonally across. If the lengths of the fence and two sides are represented by $2x + 1$, x, and $2x - 8$, as shown:

   a. Find the least integral value of x.

   b. If the lengths of the remaining two sides are 8 feet and 14 feet, find the least integral value of the perimeter of either triangular section.

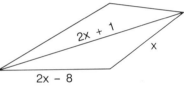

14. The two legs of a triangular archway are of equal length. If the angle that each leg makes with the ground is 15° less than the angle at the top of the arch, find the number of degrees in each of the angles of the triangle.

15. Two legs of a tripod form an isosceles triangle with the ground as the base. The base angles are each 4 times the measure of the vertex angle. Find the angle formed by each leg of the tripod with the ground.

16. A diagram of a ground ball hit to the shortstop in a baseball game can be drawn as a triangle in which the vertices are home plate, first base, and the shortstop. The angle at home plate is 10° more than the angle at first base. The angle at first base is 10° more than the angle at the shortstop position. Through what angle must the shortstop turn to pick up the ground ball and throw it to first base?

17. A cruise ship travels a triangular route that takes it to 2 islands and then back to its mainland port. The measure of the angle at the first island is 12 degrees more than 3 times the measure of the angle at the mainland. The measure of the angle at the second island is twice the measure of the angle at the mainland. What are the measures of the 3 angles of this triangle?

18. In $\triangle DEF$, side EF measures 1 inch less than twice side DE, and side DF measures 1 inch more than twice side DE. If each measure is a whole number, and the longest side measures no more than 7 inches, find the length of each side.

5-7 PERIMETER AND AREA PROBLEMS

Perimeter

The **perimeter** of a flat, two-dimensional shape is the distance around the figure.

The perimeter of this **polygon**, a closed two-dimensional shape that has straight sides, is the sum of the measures of its 7 sides.

While the perimeter of any polygon can be found simply by adding the lengths of its sides, the perimeters of certain figures can be conveniently expressed by formula, as shown in the exercises.

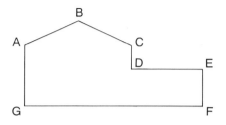

Perimeter problems that are largely numerical can be solved by strategies you learned earlier, such as breaking a problem into smaller problems. Other perimeter problems are best approached by an algebraic strategy.

MODEL PROBLEM

The length of a rectangle measures 2 feet less than 5 times its width. If the perimeter is 32 feet, find the dimensions of the rectangle.

Solution: Write and solve an equation that models the information of the problem. Draw a diagram to aid in the solution.

Let w = the width of the rectangle.

Then $5w - 2$ = the length of the rectangle.

The perimeter of the rectangle is 32.

$$2(\text{length}) + 2(\text{width}) = 32$$
$$2(5w - 2) + 2w = 32$$

$10w - 4 + 2w = 32$	Distributive property.
$12w - 4 = 32$	Collect like terms on left side.
$12w = 36$	Add 4 to both sides.
$w = 3$	Divide both sides by 12.
$5w - 2 = 5(3) - 2 = 13$	

Check:

$$2(13) + 2(3) \overset{?}{=} 32$$
$$26 + 6 \overset{?}{=} 32$$
$$32 = 32 ✓$$

ANSWER: The width is 3 feet and the length is 13 feet.

Now it's your turn to . . . **TRY IT!**   *(See page 257 for solution.)*

The perimeter of a triangle is 46 cm. If the length of the second side is twice the length of the first side and the length of the third side is one cm more than the length of the second side, find the lengths of the sides of the triangle.

EXERCISES

1. Use the given formula to find the perimeter of the triangle.

   a. The formula for the perimeter of a triangle is $P = a + b + c$. Find P when:

   (1) $a = 12$ ft. $b = 8$ ft. $c = 6$ ft.
   (2) $a = 4.5$ m $b = 1.7$ m $c = 3.8$ m
   (3) $a = 7\frac{1}{2}$ ft. $b = 5\frac{3}{4}$ ft. $c = 6\frac{1}{2}$ ft.
   (4) $a = 36$ in. $b = 5$ ft. $c = 4$ ft.

   b. The formula for the perimeter of an equilateral triangle is $P = 3s$.
   Find P when s equals:
   (1) 6 cm
   (2) 4.8 in.
   (3) $9\frac{1}{3}$ ft.

   c. The formula for the perimeter of an isosceles triangle is $P = 2a + b$.
   Find P when:
   (1) $a = 6'$, $b = 4'$
   (2) $a = 3\frac{1}{2}''$, $b = 5''$
   (3) $a = 12.6''$, $b = 7.3''$

2. The length and width of a rectangle are also called the **base** and the **height**. The height, or **altitude**, is **perpendicular** to the base (forms right angles). Use the given formula to find the perimeter of the quadrilateral.

   a. A formula for the perimeter of a rectangle is $P = 2b + 2h$.
   Find P when:
   (1) $b = 8.2''$, $h = 9.3''$
   (2) $b = 5\frac{1}{2}''$, $h = 5\frac{1}{2}''$
   (3) $b = 5\frac{1}{3}''$, $h = 6\frac{1}{2}''$

   b. The formula for the perimeter of a square is $P = 4s$.
   Find P when s equals:
   (1) 7 in.
   (2) 3.5 ft.
   (3) $8\frac{3}{4}$ in.

3. The letter in the diagram is composed of congruent squares. If a side of a square measures $7''$, find the perimeter of the letter.

   a.   **b.**   **c.**

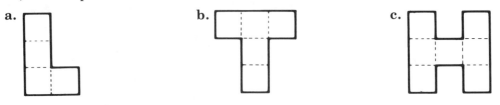

4. The **circumference** is the distance around a circle. The **diameter** is a segment through the center that has both endpoints on the circle. The **radius** extends from the center to the circle. Use the given formula to find the circumference of the circle.

   a. A formula for the circumference of a circle is $C = \pi d$. Find C when $\pi = \frac{22}{7}$ and d equals:
      (1) 14 ft.　　　　**(2)** 21 in.　　　　**(3)** 3 m
   b. A formula for the circumference of a circle is $C = 2\pi r$. Find C when $\pi = 3.14$ and r equals:
      (1) 40 cm　　　　**(2)** 13 ft.　　　　**(3)** 5.6 in.

5. Tell what happens to the given measure under the transformation described.
   a. perimeter of an equilateral triangle if the lengths of the sides are doubled
   b. circumference of a circle if the radius is tripled

In 6–16, solve algebraically.

6. The length of a rectangle is 3 times its width. The perimeter of the rectangle is 24 meters. Find the dimensions of the rectangle.

7. The length of a rectangle is 5 centimeters more than its width. The perimeter is 66 centimeters. Find the dimensions of the rectangle.

8. The perimeter of a rectangular garden is 110 meters. Find its dimensions if the length is 5 meters less than twice the width.

9. In an isosceles triangle whose perimeter is 144 centimeters, the length of each of the equal sides is 4 times the length of the third side. Find the length of each side of the triangle.

10. The perimeter of a triangle is 40 inches. The length of the second side exceeds twice the length of the first side by 1 inch, and the length of the third side is 2 inches less than the length of the second side. Find the length of each side of the triangle.

11. The length of a rectangle is twice the width. If the length is increased by 4 inches and the width is decreased by 1 inch, a new rectangle is formed whose perimeter is 198 inches. Find the dimensions of the original rectangle.

12. If the length of one side of a square is increased by 4 centimeters and the length of an adjacent side is multiplied by 4, the perimeter of the resulting rectangle is 3 times the perimeter of the square. Find the length of a side of the original square.

13. The length of a rectangle exceeds its width by 4 feet. If the width is doubled and the length is diminished by 2 feet, a new rectangle is formed whose perimeter is 8 feet more than the perimeter of the original rectangle. Find the dimensions of the original rectangle.

14. A side of a square measures 10 meters more than the side of an equilateral triangle. The perimeter of the square is 3 times the perimeter of the equilateral triangle. Find the length of a side of the triangle.

15. A rectangle and an isosceles triangle have bases that are equal in measure. The height of the rectangle measures 5 feet less than 12 times its base. Each equal side of the isosceles triangle measures 4 feet more than 8 times its base. If the perimeters of the two figures are equal, find the measure of the base of each figure.

16. Each side of a hexagon (a polygon that has 6 sides) measures 4 inches less than the side of a square. The perimeter of the hexagon is equal to the perimeter of the square. Find the length of a side of the hexagon and the length of a side of the square.

17. Mrs. Cuervo used some empty cartons from the supermarket to mail clothes to her daughter at college. She measured the cartons to find one that met the post office requirement that the length and girth (perimeter of an end) combined should be no more than 108 inches.

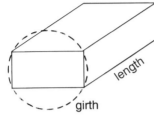

    a. Which dimensions satisfy the requirements?
       (1) 16 in. by 22 in. by 40 in.
       (2) 25 in. by 22 in. by 30 in.
       (3) 15 in. by 20 in. by 36 in.
       (4) 18 in. by 18 in. by 38 in.

    b. When it was packed and ready to mail, the carton weighed 16 lb. 10 oz. If it was sent fourth-class to Zone 7, use the accompanying table to find the postal rate.

Weight, up to but not exceeding— (pounds)	Local	1 & 2	3	4	Zones 5	6	7	8
2	$1.63	$1.69	$1.81	$1.97	$2.24	$2.35	$2.35	$2.35
3	1.68	1.78	1.95	2.20	2.59	2.98	3.42	4.25
4	1.74	1.86	2.10	2.42	2.94	3.46	4.05	5.25
5	1.79	1.95	2.24	2.65	3.29	3.94	4.67	6.25
6	1.85	2.04	2.39	2.87	3.64	4.43	5.30	7.34
7	1.91	2.12	2.53	3.10	4.00	4.91	5.92	8.30
8	1.96	2.21	2.68	3.32	4.35	5.39	6.55	9.26
9	2.02	2.30	2.82	3.55	4.70	5.87	7.17	10.22
10	2.07	2.38	2.97	3.78	5.05	6.35	7.79	11.18
11	2.13	2.47	3.11	4.00	5.40	6.83	8.42	12.14
12	2.19	2.56	3.25	4.22	5.75	7.30	9.03	13.09
13	2.24	2.64	3.40	4.44	6.10	7.78	9.65	14.03
14	2.28	2.69	3.48	4.56	6.27	8.02	9.96	14.50
15	2.32	2.75	3.55	4.67	6.44	8.24	10.24	14.94
16	2.35	2.79	3.63	4.78	6.60	8.45	10.52	15.35
17	2.39	2.84	3.70	4.88	6.75	8.66	10.77	15.74
18	2.42	2.89	3.76	4.98	6.90	8.85	11.02	16.11
19	2.46	2.93	3.83	5.07	7.03	9.03	11.25	16.45
20	2.49	2.98	3.89	5.16	7.16	9.20	11.47	16.79

18. Li Po bought 24 fence posts to use in enclosing a rectangular pen. The length of the pen will be 28 feet and the posts are to be placed at 4-foot intervals. Find the width of the pen.

19. Two people are watching a projection TV screen from opposite sides of a room. If they are both 3 times as far from the screen as from each other and the imaginary triangle formed by the TV and the 2 viewers has a perimeter of 35 feet, how far are the viewers from the TV screen?

20. A fenced rectangular garden is divided into two identical triangular gardens by adding fencing from one corner of the garden to the opposite corner. The diagonal fence is 10 yards longer than the length of the rectangle, which, in turn, is 10 yards longer than the width. If the total length of all the fencing is 190 yards, what are the lengths of the three sides of each triangle?

21. The diagram shows a 22-foot-square maze that was planted at Hedgerow Gardens. If each path is 2 feet wide, and each plant covers a 1-foot-square section of the borders separating the paths, how many plants were needed to form the maze?

Mazes are studied in a branch of mathematics called *topology*. All the paths in a true maze connect to the outside without crossing any boundary. There is a general formula for getting out of any maze, but the formula is as complicated as a maze.

To reach the center of some mazes, such as the garden maze in Williamsburg, Virginia, a simple method can be used: keep one hand on a wall of the maze, and continue to the center without taking the hand from the wall.

22. The tower for a radio transmitter is secured by guy wires anchored in the ground. On one side of the tower, 2 wires form two sides of a triangle, with a section of the tower as the third side. One wire is attached to the tower at a height of 40 feet. The second wire is attached at a height of 75 feet and is 20 feet longer than the other wire. If the perimeter of the triangle is 215 feet, how long are the wires?

23. A cruise ship travels a triangular route that takes it to 2 islands and then back to its mainland port. The trip from the mainland to the first island is twice the distance between the islands, and the trip back to the mainland is 80 nautical miles more than the first leg of the trip. If the entire cruise covers 1,000 nautical miles, how long is each leg of the cruise?

24. In a baseball game, the distance the ball traveled from the batter to the shortstop was 20 feet less than the distance it traveled when the shortstop threw it to first base. If the perimeter of the triangle formed by home plate, first base, and the shortstop was 330 feet, how far did the shortstop throw the ball? (The distance from home plate to first base is 90 feet.)

25. A traveler flew from Philadelphia to St. Louis on a connecting flight through Chicago. The return flight was direct from St. Louis to Philadelphia. The total mileage traveled was 1,978 miles. If the return flight was 37 miles more than 3 times the Chicago-St. Louis flight, and 119 miles more than the Philadelphia-Chicago flight, how many miles was the flight from Chicago to St. Louis?

26. In kitchen design, the triangle formed by the refrigerator, stove, and sink should be as small as possible. In a model kitchen, the perimeter of this triangle is only 8 feet, with the distance from sink to refrigerator being 18 inches more than the distance from sink to stove. The distance from refrigerator to stove is longest, being only 16 inches less than the sum of the other two sides of the triangle. Find the lengths of all 3 sides of this triangle in inches.

Area

The **area** of a flat, two-dimensional shape is the number of **unit squares** that are needed to fill the figure.

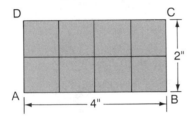

The area of this rectangle is 4 inches × 2 inches or 8 square inches.

EXERCISES

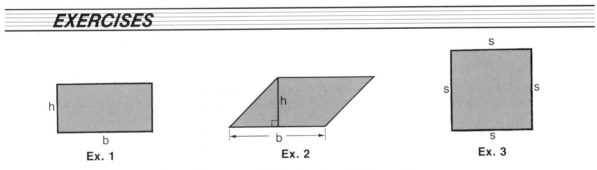

Ex. 1   Ex. 2   Ex. 3

1. The formula for the area of a rectangle is $A = bh$. Find A when:
   a. $b = 10$ m, $h = 8$ m   **b.** $b = 7.5$ yd., $h = 3.4$ yd.
   c. $b = 8\frac{1}{2}$ ft., $h = 6$ ft.   **d.** $b = 4$ ft., $h = 10$ in.

2. A **parallelogram** is a quadrilateral, with both pairs of opposite sides parallel. The formula for the area of a parallelogram is $A = bh$. Find A when:
   a. $b = 8$ cm, $h = 12$ cm   **b.** $b = 3.5$ ft., $h = 6.4$ ft.
   c. $b = 3$ ft., $h = 10$ in.   **d.** $b = 2\frac{1}{3}$ yd., $h = 6$ ft.

3. The formula for the area of a square is $A = s^2$. Find A when s equals:
   a. 9 yd.   **b.** 25 cm   **c.** 32 ft.   **d.** $2\frac{1}{2}$ m

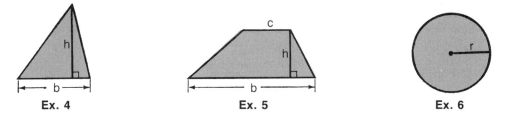

Ex. 4   **Ex. 5**   **Ex. 6**

4. The formula for the area of a triangle is $A = \frac{1}{2}bh$. Find A when:
  a. $b = 10'$, $h = 6'$
          b. $b = 3$ yd., $h = 5$ yd.
  c. $b = 8.2''$, $h = 14''$
          d. $b = 1'$, $h = 5\frac{1}{2}''$

5. A *trapezoid* is a quadrilateral with one pair of opposite sides parallel. The parallel sides are called *bases*. The nonparallel sides are called *legs*. The formula for the area of a trapezoid is $A = \frac{1}{2}h(b + c)$. Find A when:
  a. $h = 10''$, $b = 8''$, $c = 6''$
      b. $h = 9''$, $b = 14''$, $c = 8''$
  c. $h = 5''$, $b = 3\frac{3}{4}''$, $c = \frac{3}{4}''$
      d. $h = 9''$, $b = 7.6''$, $c = 4.4''$

6. The formula for the area of a circle is $A = \pi r^2$. Find A when $\pi = \frac{22}{7}$ and r equals:
  a. 7 ft.
      b. 5.6 cm
      c. $3\frac{1}{2}$ yd.
      d. 10 in.

7. Find the area of the colored region.

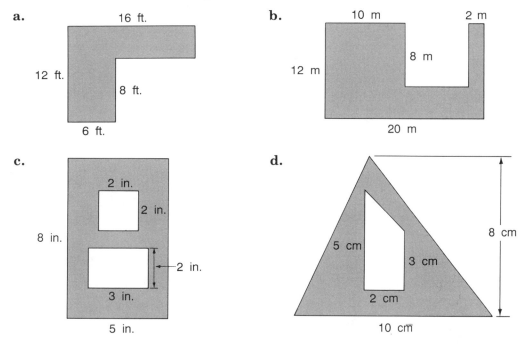

8. Tell what happens to the area of the given figure under the transformation described.
  a. area of a square if the lengths of its sides are doubled
  b. area of a rectangle if the length is tripled

9. When it is fully opened, Mrs. Trent's folding aluminum table measures 24″ by 72″. What is the area of the table top?

10. The triangular sail on Pat Dorn's sailboat has a base of 9 feet and a height of 14 feet. Find the area of the sail.

11. The two main pattern pieces for a barbecue apron consist of a square top and a trapezoidal bottom. The square measures 12 inches on a side. The trapezoid has bases of 14 inches and 20 inches, and a height of 16 inches. Find the combined area of the two pieces.

12. A teacher pastes a 3″ by 5″ index card onto a sheet of paper that measures 8″ by 10″. What area of the paper is not covered by the card?

13. In Joshua's coin collection, the radius of one coin is 1.4 cm. Find the area of the face of the coin.

14. The face of a parking meter is a semicircle whose diameter is 12 inches. Find the area of the semicircular face to the nearest tenth.

15. The windows of the Livingston Library are in the shape of a tall rectangle topped by a semicircle. If the rectangular portion is 3 feet wide and 5 feet high, find, to the nearest tenth, the area of the entire window.

16. Cindi and Hank want to carpet their new video shop, which measures 18 feet by 30 feet. Carpeting will not be laid under the 6 display cases, each of which occupies 4 feet by 12 feet of floor space. How many square yards of carpeting will be needed?

17. A kitchen measures 9 feet by 12 feet. How many square tiles are needed to cover the floor if the tiles measure:
    a. 1 foot on a side?   **b.** 9 inches on a side?

18. A rectangular plywood board covers a box that measures 14 inches long by 3 inches wide by 4 inches deep. If the board extends 1 inch uniformly over each edge of the box, as shown, what is the area of the board?

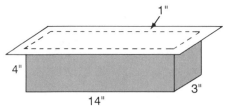

19. A large square is filled with small squares measuring 1 cm by 1 cm. Is it possible for the large square to have an area of 40 square cm? Explain.

20. A lantern used by a railroad maintenance crew has circular lamps set into a frame that has square bases and rectangular sides. The dimensions are shown in the diagram.

a. Find the area of a base.

b. Find the area of a lighting surface.

c. Find the total area of the portions of the two bases that are not part of the lighting surfaces.

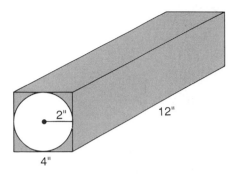

21. At the Ardsley Art Museum, a painting 30 cm high is in a frame that has a uniform width of 4 cm. The area of the frame is 448 square centimeters.

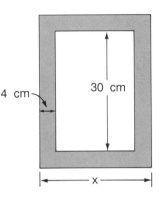

a. If the overall width of the framed painting is represented by x, write and solve an equation to find the value of x.

b. Find the area of the painting without the frame.

22. Mr. Meyers is building a storage shed for garden supplies in back of his house. The floor is to be 4 feet wide, and the height of the shed is to be 6 feet. The floor, walls, and ceiling are rectangular. If the length of the floor is represented by x, and the area of the four walls, including a door, is 160 sq. ft., write and solve an equation to find the length of the floor.

23. Jackie is using a 36-square-foot rectangular area for her garden, and wants to completely enclose it with a fence.

a. How much fencing will be needed if the width of the garden is:
(1) 2 feet   (2) 3 feet   (3) 4 feet   (4) 6 feet

b. If fence posts may be placed either 2 feet or 3 feet apart, what is the smallest number of posts required for the dimensions that use the least amount of fencing?

24. Using the information in Exercise 23, Jackie's husband decides to place the garden against the garage, so that fencing is needed on only 3 sides.

a. How much fencing will be needed if the side of the garden along the garage measures:
(1) 18 feet   (2) 12 feet   (3) 9 feet   (4) 6 feet

b. What is the smallest number of posts required for the dimensions that use the least amount of fencing?

5-8 SURFACE AREA AND VOLUME PROBLEMS

Solid Figures

A three-dimensional geometric figure is called a *solid*. A solid, all of whose sides are flat surfaces, or *faces*, is a *polyhedron*.

A *prism* is a polyhedron with bases that are congruent (have the same size and shape). A *right prism* has a height perpendicular to the bases. A prism may be named according to the shape of its bases.

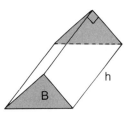
Triangular Right Prism

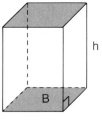
Rectangular Right Prism

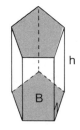
Pentagonal Right Prism

A *pyramid* has triangular faces on a base that may be a square or other polygon.

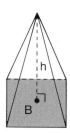

Other solids are related to a circle.

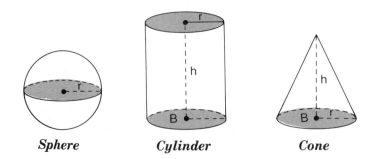
Sphere   *Cylinder*   *Cone*

1. Describe the faces of the solid figure.

Example:

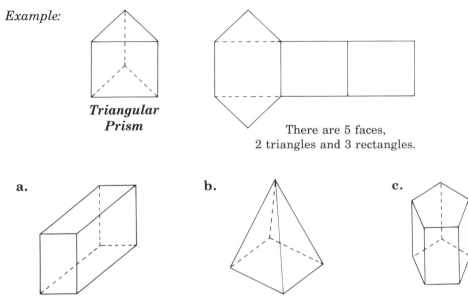

Triangular Prism

There are 5 faces,
2 triangles and 3 rectangles.

a.   **b.**   **c.**

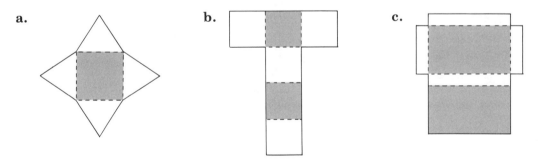

2. Name the polyhedron that would be formed if you folded the diagram along the dotted lines.

a.   **b.**   **c.**

3. For decorations, the prom committee plans to cover cardboard cubes with colored foil. Draw two different patterns that can be used to cut foil in a shape that will just cover a cube.

4. The Corn Crunchies cereal box is glued along one side, and has flaps on the bottom matching those on top. Draw a rectangle, and show how the box can be formed from the rectangle by using dotted lines to indicate folds, and solid lines to show where the rectangle must be cut to make the flaps.

Corn Crunchies

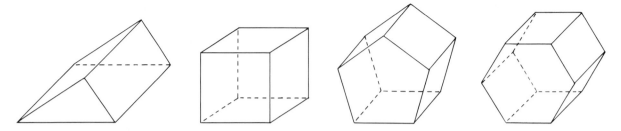

5. The diagrams show polyhedrons whose bases have 3, 4, 5, and 6 sides, respectively. A polyhedron whose base has 3 sides has a total of 5 *faces*. The line segments where the faces meet are the *edges*, and the points where the edges meet are the *vertices*.

	Number of sides in a base				
	3	**4**	**5**	**6**	***n***
Vertices					
Faces					
Edges					

a. Copy the table and fill in the number of vertices in each of the polyhedrons shown. Find a pattern and fill in the number of vertices in a polyhedron whose base has *n* sides.

b. Fill in the row labeled *Faces*. A pattern should show how to fill in the column headed *n*.

c. In the same way, fill in the row labeled *Edges*.

d. Now look at the columns in your table. Find a general formula that relates the numbers of vertices *v*, faces *f*, and edges *e*, in the column headed 3. The same formula should also work in the columns headed 4, 5, and 6. Finally, check that your formula works in the column headed *n*. What is the formula?

Surface Area

The **surface area** of a solid, three-dimensional figure is the sum of the areas of all its faces.

EXERCISES

1. The formula for the surface area of a rectangular solid is
$S = 2\ell w + 2h\ell + 2hw$. Find S when:
a. $\ell = 6'$ $w = 5'$ $h = 3'$
b. $\ell = 7''$ $w = 6''$ $h = 9''$
c. $\ell = 4.5'$ $w = 1.4'$ $h = 2.6'$
d. $\ell = 3\frac{1}{2}''$ $w = 4''$ $h = 4\frac{1}{4}''$

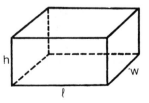

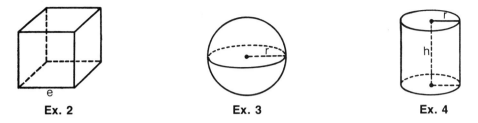

Ex. 2　　　　　　　　Ex. 3　　　　　　　Ex. 4

2. The formula for the surface area of a cube is $S = 6e^2$. Find S when e equals:

   a. 6 m　　　　　　**b.** 7 yd.　　　　　　**c.** $\frac{1}{4}$ ft.　　　　　　**d.** 4.5 in.

3. The formula for the surface area of a sphere is $S = 4\pi r^2$. Find S when:

   a. $\pi = \frac{22}{7}, r = 14$ m　　　**b.** $\pi = \frac{22}{7}, r = 3\frac{1}{2}$ ft.　　　**c.** $\pi = 3.14, r = 10$ cm

   d. $\pi = 3.14, r = 20$ in.　　　**e.** $\pi = \frac{22}{7}, r = 2.8$ cm

4. The formula for the surface area of a cylinder is $S = 2\pi r(r + h)$. Find S when:

   a. $\pi = \frac{22}{7}, r = 4'', h = 3''$　　　　　**b.** $\pi = \frac{22}{7}, r = 2', h = 1\frac{1}{2}'$

   c. $\pi = 3.14, r = 5'', h = 5''$　　　　　**d.** $\pi = 3.14, r = 2.5'', h = 7.5''$

   e. $\pi = \frac{22}{7}, r = 7$ mm, $h = 23$ mm

5. What happens to the surface area of a rectangular solid if each dimension is doubled?

6. Which would cause the greater increase in the surface area of a cylinder, doubling the height or doubling the radius of the base? Explain.

7. If the length of each side of the base of a square pyramid were doubled, and the height of each triangular face were doubled, how would the surface area change?

8. The edge of a cube measures $4''$. A rectangle has dimensions $9''$ by $24''$.
   a. Which is greater, the surface area of the cube or the area of the rectangle?
   b. In order for the surface area of the cube to be the same as the area of the rectangle, what must be the length of an edge of the cube?

9. A label is removed from a cylindrical can and opened to form a rectangle. The can is 15 cm high and its base has a radius of 7 cm. Using $C = 2\pi r$, $A = \pi r^2$, and $\pi = \frac{22}{7}$, find the:
   a. area of the base　　　　　**b.** area of the label　　　　　**c.** total surface area

10. A roll of gift wrap paper holds 25 square feet of paper. Alex is wrapping a package that measures 2 ft. by 1 ft. by 4 in., and is allowing an extra square foot of paper for overlap. If the entire roll of paper costs $1.75, what is the cost of the paper used to wrap the package?

11. In Mr. Child's kindergarten room, nine $16'' \times 16'' \times 16''$ hollow cubes are stacked against a wall, 3 across and 3 high, to form a storage unit. The fronts of the cubes are open, and all the exposed tops and sides are painted. What is the total area of the painted surfaces?

12. The Simpsons want to convert a bedroom closet to a bathroom. The closet is 7 feet long by 5 feet deep by 8 feet high.
    a. Including the door, what is the total interior surface area of the closet?
    b. Except for a 2-foot-by-3-foot window and the 3-foot-by-7-foot door, the walls will be tiled. How much wall space remains to be tiled?

13. The upholstered footrest in the Browns' den is in the shape of a rectangular solid. The top and bottom are 2-foot squares, and the total surface area is 20 square feet.

  a. Find the height of the footrest.

  b. Mr. Brown has chosen fabric that costs $18 a square yard with which to reupholster the top and sides of the footrest. If he buys a whole number of square yards of fabric, what will the fabric cost?

14. The school bookstore, named The Orange and Blue for the school colors, has 8 closed plywood cubes, measuring 2 feet on an edge, with which to create a display unit. The manager wants to stack the cubes in the middle of the room, with at least one face of each cube covered by a face of another cube, and then coat the exposed surfaces with an adhesive fabric.

  a. She has a limited supply of orange fabric on hand. Sketch a stacking arrangement that would have the minimum exposed surface area. Is there more than one possibility?

  b. There is a large amount of blue fabric available. Sketch an arrangement that would have the maximum amount of exposed surface area. Is there more than one possibility?

  c. Determine the number of square feet of fabric needed in part **a** and in part **b**.

Volume

The **volume** of a solid, three-dimensional figure is the number of **unit cubes** that are needed to fill the figure.

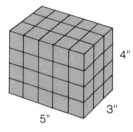

The volume of this rectangular solid is 5 inches × 3 inches × 4 inches or 60 cubic inches.

4"

3"

5"

EXERCISES

1. The formula for the volume of a rectangular solid is $V = \ell w h$.
Find V when:
  a. $\ell = 5'$ $w = 4'$ $h = 7'$
  b. $\ell = 8''$ $w = 7''$ $h = 5''$
  c. $\ell = 8.5''$ $w = 4.2''$ $h = 6.0''$
  d. $\ell = 2\frac{1}{2}''$ $w = 8''$ $h = 5\frac{1}{4}''$

2. The formula for the volume of a cube is $V = e^3$. Find V when e equals:
   a. 2 m   **b.** 3 yd.   **c.** 8 cm   **d.** $\frac{1}{3}$ ft.

3. The formula for the volume of a sphere is $V = \frac{4}{3}\pi r^3$. Find V when:
   a. $\pi = \frac{22}{7}$, $r = 2\frac{1}{3}$ cm   **b.** $\pi = 3.14$, $r = 30$ m   **c.** $\pi = \frac{22}{7}$, $r = 3.5$ mm

4. The formula for the volume of a cylinder is $V = \pi r^2 h$. Find V when $\pi = \frac{22}{7}$ and:
   a. $r = 35''$, $h = 12''$ **b.** $r = 28''$, $h = \frac{3}{4}''$ **c.** $r = 9''$, $h = 5.6''$

5. The altitude of a right prism measures 6 cm. Find the volume of the prism:
   a. if the base is a square of side 9 cm.
   b. if the base is a triangle of base 8 cm and altitude 5 cm.
   c. if the base is a parallelogram of base 10 cm and altitude 4.5 cm.
   d. if the base is a trapezoid of bases 12 cm and 16 cm and altitude 5 cm.

6. How is the volume of a cylinder changed if its altitude is doubled?

7. Formulas you have seen for the volumes of special solids are related to the general formula $V = Bh$, where the capital letter B represents the area of the base of the solid.

Thus, for a rectangular prism: $V = \underbrace{\ell \times w}_{\substack{\text{area of} \\ \text{base}, B}} \times h$

Explain how the general formula $V = Bh$ is related to the volume formula for:
   a. a cube   **b.** a right cylinder

8. Since Paul was saving to buy a car, his family gave him cash for his birthday on August 2nd. The deposit of his birthday money doubled the balance in Paul's savings account, as shown by the graphs.

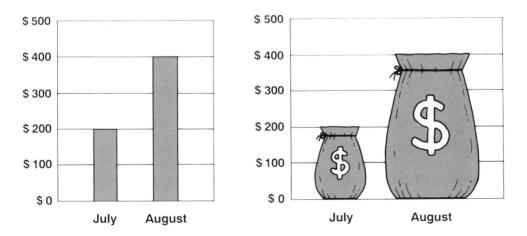

     Compare the bar graph and the pictograph. Why does the pictograph give the impression of a much greater difference between the July and August balances?

Two different shapes that have congruent bases and equal altitudes may have a relationship in their volumes.

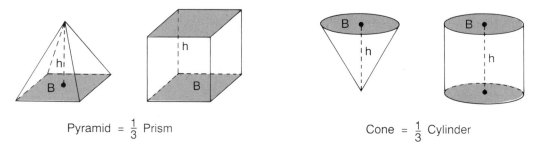

Pyramid = $\frac{1}{3}$ Prism   Cone = $\frac{1}{3}$ Cylinder

9. The altitude of a pyramid measures 7 in. Find the volume of the pyramid if the base is:
   a. a square of side 8 in.
   b. a rectangle of length 1 ft. and height 6 in.
   c. a trapezoid of height 6 in. and bases 4 in. and 12 in.

10. The altitude of a cone measures 6 in. Find the volume of the cone if the base has:
    a. a radius of 4 in.   **b.** a radius of 3 in.   **c.** a diameter of 1 ft.

11. The radius, r, of a sphere is also the radius of the base of a cone. If the height of the cone is equal to the diameter of the sphere, what is the relationship between the volume of the cone and the volume of the sphere?

12. If two cones have exactly the same shape, and one is twice as high as the other, how do their volumes compare? (The volume of the taller cone is *not* twice that of the smaller.)

13. A pyramid with a square base of side 6 in. has the same height as a cone whose base has diameter 6 in. Which has the greater volume?

Of the seven wonders of the ancient world, only one remains virtually intact today: the pyramids at Giza in Egypt.

The Great Pyramid was built under a grand design, with astonishing accuracy. Its measurements represent facts about our universe, such as the circumference of its base is 36,524″, which is exactly equivalent to the 365.24 days of our earth year. This pyramid also acts as a perfect almanac, registering the seasons of the year.

The pyramids were built as tombs for Egyptian rulers, and are guarded by the great stone sphinx, a creature with the head of a man and the body of a lion.

14. The Olde Ice Cream Shoppe sells novelty cones that are completely filled with ice cream from bottom to top. The small cone, which costs $1.25, is 5 in. high and has a 2 in. diameter at the top. The large cone, which costs $2.50, is 7 in. high and has a 3 in. diameter at the top. Which is the better buy?

15. At Schmidt's farm, a silo has a volume of 2,156 cubic feet. If the radius of the base is 7 feet, find the height of the silo.

16. Stan took a cube of ice from the freezer for his soft drink. If the edge of the cube measured 3 cm, what was the volume of the cube?

17. Mirla opened a can of tuna fish for lunch. If the can is 2 in. high, and the radius of the base is $1\frac{3}{4}$ in., what is the volume of the can?

18. In a crafts workshop, Sean made a plastic bird feeder in the shape of a cylinder. If the feeder is 14 inches in diameter and 7 inches high, what volume of birdseed can it hold?

19. Matt is offered his choice of two cylindrical beverage cups at a picnic. The yellow cup is 5 in. high and 4 in. in diameter. The green cup is 8 in. high and 3 in. in diameter. Matt is thirsty and wants the larger amount to drink.
    a. Which cup should he choose?
    b. How much more juice will the larger cup hold?

20. A study suggests that each student in a classroom needs at least 6 cubic meters of airspace. What is the greatest number of students that should be in a room that is 7 meters long by 5 meters wide by 4 meters high?

21. A carton is 16 inches long, 9 inches wide, and 8 inches high.
    a. Find the volume of the carton.
    b. How many boxes with dimensions 2 in. × 3 in. × 8 in. will the carton hold?

22. An ice cream scoop is in the shape of a hemisphere. A gallon of ice cream has a volume of approximately 234 cu. in. Find the number of level scoops that there are in a gallon of ice cream if the scoop has a diameter of: a. 2 in.   b. 3 in.

23. Mr. Phipps is sending a carton of clothing to an area that was hit by a hurricane. The carton measures 18″ by 16″ by 12″. Find the: a. surface area of the carton   b. volume

24. An antique globe of the world is on display at the Hampshire Library. If the radius of the globe is 14 centimeters, find the: a. surface area   b. volume

25. A rectangular solid has a base with dimensions 25 cm and 15 cm. Its height is 20 cm.
    a. If adhesive paper is used to cover the visible faces of the solid, but not the base, how many square centimeters of paper are needed?
    b. If the weight of the solid is 6 grams per cubic centimeter of volume, how much does it weigh?

26. A sandbox with interior dimensions 4 ft. × 5 ft. is 9 inches high and open at the top.
    a. What is the surface area of the inside of the sandbox?
    b. If the sandbox is to be evenly filled with sand to a height of 6 inches, what volume of sand will be needed?

27. A hollow cube with sides measuring 16 inches is mounted on wheels for use as a moving chair. Since the top lifts off, it can also be used as a storage bin. Vicki covered the 4 sides and top of the cube with an adhesive vinyl that costs $12 per square yard.

   a. How much surface area is covered with vinyl?

   b. Victor's Variety Store sells the vinyl by the square yard only (no fractional quantities). What was the cost of Vicki's vinyl?

   c. If the cube is made of $\frac{1}{2}$-inch-thick plywood, what is the volume inside the cube?

28. A hollow cylindrical shell is shown in the diagram. Its volume can be determined by taking the difference of the volumes of the two cylinders that form the shell. The radius of the outer cylinder is 6 cm, the radius of the inner cylinder is 4 cm, and the height is 10 cm.

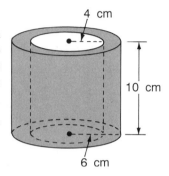

   a. Write a formula that can be used to find the volume of the shell, using R for the outer radius and r for the inner radius.

   b. Find the volume of the shell.

In 29–31, write the equation described, and solve algebraically.

29. Loren's tent has rectangular sides and triangular ends. The base of each triangle measures 10 feet, the height of the tent is 6 feet, and the volume is 420 cubic feet. If x represents the length of the tent, write and solve an equation to find the value of x.

30. Billy wants to fill a cylindrical container by pouring water from a cone-shaped cup. Both the cup and the container have the same height, and the diameter of both the container and the top of the cup is 7 cm.

   a. How many cups of water will be needed?

   b. If the container holds 231 cubic cm of water when filled, write and solve an equation to find the height h.

31. Crates of medical supplies are to be shipped by long-haul trucking. Each crate is 8 feet long and 6 feet wide, and has a volume of 240 cubic feet.

   a. Write and solve an equation to find the height h of a crate.

   b. How many such crates will fit into a van that is 45 ft. long by 8 ft. wide by 10 ft. high?

In 32 and 33, write the equation described, and solve using any problem-solving strategy.

32. Estelle Perkins has a camera that uses a flash cube. The volume of a cube is 27 cm³.

   a. Write an equation that can be used to find the length of an edge e of the flash cube.

   b. Find the value of e.

33. The height of a cylindrical oatmeal box is 3 times the radius of its base. The volume of the box is 81π cubic inches.

   a. If r represents the radius of the base, write an equation that can be used to find r.

   b. Find the value of r.

5-9 A ROUNDUP OF PROBLEMS

1. In a right triangle, the measure of one acute angle is 4 times that of the other. Find the number of degrees in the smaller angle.

2. The seat of a stool is triangular, with one side that is 5 inches longer than either of the other sides. If the perimeter measures less than 40 inches, find the longest possible integral lengths for the sides.

3. A girder for a bridge forms a pair of supplementary angles with the ground. The larger angle measures 68° more than the smaller. What are the measures of the two angles?

4. In Central City, there are three major newspapers. The News Today has been publishing 10 years less than The Herald, and The Timely News has been publishing 30 years longer than The Herald. In 10 years, the combined ages of The Herald and The Timely News will be 3 times the age of The News Today. How long has The Herald been publishing?

5. Mr. Desmond takes a bus and a train each day to and from work. The fare on the train is 30¢ more than the bus fare. Mr. Desmond works Monday to Friday each week and spends $38 weekly for fare. Find the amount of each fare on the bus and on the train.

6. Stacey needs $1.95 for a pack of baseball cards. She has money consisting of nickels, dimes, and quarters. She has 3 times as many dimes as nickels and the number of dimes exceeds the number of quarters by 1. If the money will exactly cover the cost of the baseball cards, how many of each coin does she have?

7. A winning 3-digit lottery number contains 3 consecutive odd integers whose sum is 15. The numbers are arranged in descending order. What is the winning number?

8. On a map, one angle of the Bermuda Triangle measures three degrees more than the smallest angle. The largest angle measures 15 degrees more than the smallest angle. Find the measure of each angle.

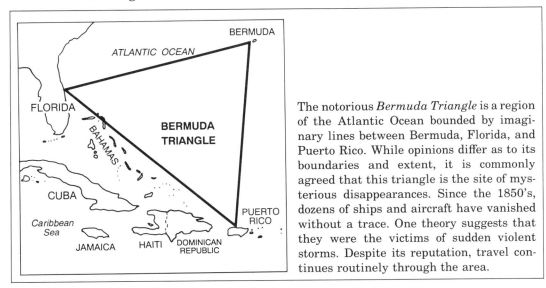

The notorious *Bermuda Triangle* is a region of the Atlantic Ocean bounded by imaginary lines between Bermuda, Florida, and Puerto Rico. While opinions differ as to its boundaries and extent, it is commonly agreed that this triangle is the site of mysterious disappearances. Since the 1850's, dozens of ships and aircraft have vanished without a trace. One theory suggests that they were the victims of sudden violent storms. Despite its reputation, travel continues routinely through the area.

9. Troy does 1 hour of exercise each day by combining bicycling and walking. He bicycles at 12 miles per hour and then walks the same distance at 4 miles per hour. How much of the time does Troy spend walking?

10. Mr. Ross is 4 years older than his wife. They had their first child 3 years after they were married, at which time the sum of their ages was 60. Find the ages of Mr. and Mrs. Ross at the time of their marriage.

11. Fudge sells for 30 cents per ounce and chocolates cost 42 cents per ounce. If 87 ounces of the two varieties were sold for less than $33, what is the largest whole number of ounces of chocolates that could have been sold?

12. The costs of putting on several performances of a musical comedy at Riverview High School total $5,500. The business manager anticipates a total attendance of 1,350 students and 450 adults, with adult tickets selling at twice the price of student tickets. What is the smallest whole-number amount that can be charged for a student ticket in order to make a profit of at least $3,000?

13. On a business trip to Sacramento, Ms. Cooper drove a round-trip distance of 778 miles. It took her 7 hours to get there and 6 hours to return by a different route. If her average rate of travel going to Sacramento was 4 miles per hour less than that on her return trip, find the rate of travel for each part of her trip.

14. Jenny had 3 times as much money as Craig. For doing chores, Jenny got $3 more, and Craig got $6 more. Jenny now has twice as much money as Craig. How much money did each have originally?

15. Michelle has 25 coins, consisting of pennies, nickels, and dimes. The number of nickels is 2 more than the number of dimes. If she has a total of $1.37, how many of each type of coin does she have?

16. The length of a rectangle is 2 centimeters more than its width. If the width were doubled and the length were increased by 4 centimeters, then the perimeter would be increased by 20 centimeters. Find the dimensions of the original rectangle.

17. Jack and his sister Jill are seated on a teeter-totter that is 7 feet long. Jack weighs 120 pounds and Jill weighs 90 pounds. How far from the fulcrum should each of them sit to balance the teeter-totter?

18. Heather and Ryan Robinson drove separately to Austin. Heather left at 9 A.M. and averaged 52 miles per hour. Ryan left an hour later, driving at 54 miles per hour. Traveling by a shorter route, he arrived at their destination at the same time as Heather. Their combined total mileage was 370 miles.
    a. At what time did they arrive in Austin?   **b.** How much shorter was Ryan's route?

19. Three buses running on the Lexington Line have consecutive license numbers. If the sum of the 3 numbers is the lowest 6-digit number that can be obtained from 3 consecutive integers, what are the license numbers?

20. A girl weighing 72 pounds wants to move a rock that weighs 280 pounds. She has a sturdy 6-foot wooden beam to use as a lever and a smaller rock for a fulcrum. To the nearest whole number, how many inches from the rock should the girl place the fulcrum?

CHAPTER SUMMARY

This chapter continues to focus on an ***algebraic strategy*** as a problem-solving technique. In using this strategy, you still need to analyze a problem by asking yourself:

- What do I have to find in the problem?
- What information do I have?
- What is the best approach to finding a solution?

Once you have analyzed the problem, the algebraic strategy requires you to:

- Represent the unknown values in the problem in terms of one variable.
- Write a statement (an equation or inequality) that expresses the relationship between the values.
- Use your algebraic skills to find and check the solution.

As in earlier chapters, different problem-solving strategies can frequently be combined. For example, the algebraic strategy works well with the strategies of making a table or drawing a diagram, which may suggest an approach to finding a solution.

The nature of the problems in this chapter gives you a broader idea of the scope of problem situations. Still other types of situations will be presented as you progress.

VOCABULARY CHECKUP

SECTION

5-1 *consecutive integers*

5-4 *lever / fulcrum / pulley*

5-5 *vertex of an angle / acute, right, obtuse, straight angles / vertical angles / supplementary, complementary angles / protractor*

5-6 *scalene, equilateral, isosceles triangles / vertex angle, base angles / right triangle / quadrilateral*

5-7 *perimeter / polygon / base, height, altitude of a rectangle / perpendicular / circumference, diameter, radius of a circle / area / parallelogram / trapezoid*

5-8 *surface area / volume / solid / prism / pyramid / sphere / cylinder / cone / faces, edges of a polyhedron*

CHAPTER REVIEW EXERCISES

(Section 5-1)

1. Find 3 consecutive integers that have a sum of 84.

2. Find the smallest pair of consecutive even integers for which 3 times the larger is less than 4 times the smaller.

3. Find 3 consecutive odd integers such that 3 times the largest is 2 less than the smallest.

(Section 5-2)

4. Mr. Boyce deposited $170 in his bank. The number of $5 bills was 3 times the number of $10 bills, and the number of $1 bills was 30 more than the number of $5 bills. How many bills of each type did he deposit?

5. The school cafeteria sold 31 more slices of pizza at $1.25 each than hamburgers at $1.50 each. If the total sales were more than $1,000, what is the smallest number of hamburgers that could have been sold?

6. A coin bank contains nickels, dimes, and quarters amounting to $28.60. There are 6 more dimes than nickels and 5 times as many quarters as nickels. How many coins of each kind are there?

(Section 5-3)

7. Cleveland and Milwaukee are 420 miles apart. One car left Cleveland for Milwaukee and averaged 50 mph. At the same time, on the same highway, another car left Milwaukee for Cleveland and averaged 55 mph. How far were the cars from Cleveland when they passed?

8. Two trains started from the same station at the same time and traveled in opposite directions. The rate of the faster train exceeded the rate of the slower train by 15 mph. After traveling for 7 hours, they were still within 525 miles of each other. What was the maximum possible whole-number rate of the slower train?

9. An airplane made a flight of 1,600 miles in 5 hours. During the first 3 hours of the trip, it had good weather. It then ran into bad weather, which decreased its rate by 75 mph for the rest of the flight. Find the rate on each part of the flight.

10. Rosita left home on her bicycle, traveling at the rate of 6 miles per hour. One hour later, her friend Carla set out to overtake her, traveling at the rate of 8 miles per hour. In how many hours will Carla overtake Rosita?

11. Michael spent 6 hours walking out into the country and back. He walked out at the rate of 4 mph and walked back at the rate of 2 mph. How far out into the country did he go?

12. Miss Lee plans to drive her car from her home out into the country, traveling at the rate of 30 mph and returning at the rate of 40 mph. What is the greatest distance that she can drive out and then return to her home if she has at most 7 hours to spend on the trip?

(Section 5-4)

13. A 140-pound acrobat wishes to do a handstand on the end of a 10-foot board at a distance of 3 feet from a raised fulcrum. His daughter will do a handstand at the same time at the other end of the board. In order for the father and daughter to balance, what must be the weight of the daughter?

14. To lift a crate, the edge of a metal bar is placed under its edge. A block placed under the bar as a fulcrum is 6 inches from the end of the bar that is under the crate. If the weight on the crate end of the bar is 350 pounds, and a 50-pound weight at the other end is enough to lift the crate, how long is the bar?

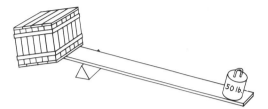

15. A pulley turning at 270 rpm is belted to a pulley with a 9-inch diameter that is turning at 360 rpm. Find the diameter of the first pulley.

(Section 5-5)

16. Give the number of degrees in the angle formed between the hands of a clock at:
    a. 8 o'clock **b.** 11 o'clock **c.** 12 o'clock

17. Give the number of degrees through which the earth rotates in 10 hours.

18. Two angles are complementary. One measures 5 times the other. Find the measure of the larger angle.

19. An angle measures 15° less than twice its supplement. Find the measures of the angles.

20. The measures of a pair of supplementary angles are consecutive multiples of 12. Find the measures of the angles.

21. Lines AB and CD intersect at point E. If $m\angle AEC = 5x - 18$ and $m\angle BED = 2x$, find the number of degrees in $\angle AEC$.

22. The pair of vertical angles formed inside the blades of a pair of scissors measure at most $(x + 100)$ and $(4x + 10)$ degrees respectively. What is the largest angle to which the scissors can open?

(Section 5-6)

23. Find the measure of the third angle of a triangle if 2 angles contain 18° and 87°.

24. In a triangle, one angle measures twice the second, and the third angle measures 16° less than the second. Find the measures of the angles.

25. The angles of a triangle are consecutive multiples of 15. Find the measures of the angles.

26. The vertex angle of an isosceles triangle measures 52°. What is the measure of a base angle?

27. The vertex angle of an isosceles triangle measures 4 times the sum of the measures of the base angles. Find the measure of the vertex angle.

28. Mrs. Finnegan plans to pick up her daughter at school, go to her parents' house for dinner, and then return home. She must travel 17 miles more from the school to her parents' house than she travels to get to the school. The trip home from her parents' house is 2 miles less than the total distance traveled to get there. If all distances are whole numbers and the entire trip is less than 45 miles, what is the maximum distance from Mrs. Finnegan's home to that of her parents?

(Section 5-7)

29. Find the area of the figure shown.

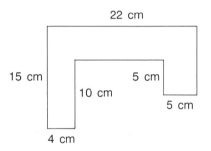

30. If the length of a rectangular garden is 8 m more than the width and the perimeter is 100 m, find the dimensions of the garden.

31. The isosceles triangle formed by 2 stereo speakers and Sara in her favorite seat in her living room puts her 2 feet closer to each of the speakers than the distance of the speakers from each other. If the perimeter of this triangle is 32 feet, how far is Sara from each speaker?

32. The length of a rectangle exceeds 3 times its width by 1 inch. If the length of the rectangle is diminished by 3 inches and the width is doubled, a new rectangle is formed whose perimeter is 46 inches. Find the dimensions of the original rectangle.

33. The length of a side of an equilateral triangle exceeds the length of a side of a square by 8 inches. The perimeter of the equilateral triangle exceeds the perimeter of the square by 20 inches. Find the length of a side of the square and the length of a side of the triangle.

34. Tell what would happen to the area of a rectangle if the length and width were both doubled.

35. What is the largest rectangular area that can be enclosed with 68 feet of fencing, if the length and width are both multiples of two feet?

(Section 5-8)

36. Find the surface area of the figure described.
  a. A rectangular right prism with bases measuring 4 cm by 12 cm, and a height of 10 cm.
  b. A triangular right prism. The height of the prism is 8.5 inches and the triangular bases have base 4 inches and altitude 2.5 inches.
  c. A 3-meter-high cylinder whose base has a radius of 1 meter.

37. The volume of a pyramid is $\frac{1}{3}$ the volume of a prism that has the same base and altitude as the pyramid. Find the volume of a pyramid that has an altitude of 9 feet and a 4-foot-square base.

Geometric forms are often used to add interest to architecture. The main section of this structure in St. Petersburg, Florida, is an upside-down pyramid.

38. A pyramid has a square base whose side measures 3 cm. The height of the pyramid is 4 cm. If a cube has edges of length 6 cm, the volume of the cube is how many times that of the pyramid?

39. The volume of a cone is $\frac{1}{3}$ the volume of a cylinder with the same base and altitude. Find the volume of a cone that has an altitude of 10 in. and a base whose radius is 3 in.

PROBLEMS FOR PLEASURE

1. If you must use at least 1 quarter, 1 dime, 1 nickel, and 1 penny, how can you make change of $1.00 using exactly 50 coins?

2. Use toothpicks to make each of the following figures. Then make the indicated changes so that the result represents a cube.

    a. Add 4 more toothpicks.   **b.** Remove 3 of the 12 toothpicks.

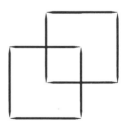 

3. The triangular sail on a sailboat has a perimeter of 47 feet, with the shortest side 5 feet less than the longest side. If the 3 sides are all different lengths and each is a whole number of feet, how long are the sides of the sail?

4. A farmer needs a chain fifteen links long and finds that he has available five pieces of chain of three links each. The blacksmith tells him that it will cost 50 cents to cut a link and a dollar to weld a link that has been cut. If the blacksmith does the job in the cheapest possible way, how much will it cost to have the five pieces joined into one continuous chain?

1. Have teams try to solve the same five problems, one team using equations, the other guessing and checking using calculators. Score one point for finishing first and three points for each correct answer.

   a. Gail buys 23 stamps for $4.95. Some are 25-cent stamps and the rest are 15-cent stamps. How many of each did Gail buy?

   b. Maisha sold 41 tickets to the school play for $84. If adult tickets sell for $3.00 and children's tickets sell for $1.50, how many of each kind did Maisha sell?

   c. Three consecutive odd integers add up to −243. Find the integers.

   d. A large pizza costs $2.00 more than a small pizza. If two large pizzas and one small pizza cost $17.80, what is the cost of a large pizza?

   e. The sum of the largest two of three consecutive integers is 735. Find the smallest integer.

2. Find three consecutive integers whose product is closest to the given number without exceeding it.

   a. 4,597   **b.** 39,392   **c.** 213,508   **d.** 388,047

3. A box is being designed for packaging a new product. The box is to have a volume of 48 cubic inches. The only restriction on its length, width, and height is that each dimension must be a whole number. The packaging material costs 1 cent for each 8 square inches. To keep costs to a minimum, the manufacturer would like to find dimensions that would require the least material.

   a. Prepare a table, as shown, to find the dimensions of the least expensive package.

W	L	H	Area of Front and Back	Area of Sides	Area of Top and Bottom	Total Area	Cost per Box (Nearest Cent)
1	1	48					
1	2	24					
1	3	16					
⋮	⋮	⋮					
3	4	4					

   b. If the manufacturer decided to make one of the dimensions 2 inches so that the package would be easy to hold, what would be the dimensions and cost of the cheapest box?

In 1–8, select the letter of the correct answer.

1. A rectangle has a perimeter of $4x + 20$ and a width of $x + 4$. What is the length of the rectangle?
   (A) $2x + 12$ (B) $2x + 18$
   (C) $x + 6$ (D) $x + 4$
   (E) $2x + 4$

2. Three consecutive even integers are increased by 3, 4, and 5 respectively. If their new sum is 84, the largest integer originally was:
   (A) 24   (B) 26   (C) 28
   (D) 30   (E) 32

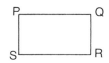

3. The perimeter of rectangle $PQRS$ is 24. If $x = SP$ and $SP < PQ$, then which represents all possible values of x?
   (A) $0 < x < 4$ (B) $0 < x < 6$
   (C) $0 < x < 8$ (D) $0 < x < 12$
   (E) $0 < x < 24$

4. Two cars start along the same route at the same time from the same place, one at 50 miles per hour and the other at 40 miles per hour. The faster car reaches the destination 6 minutes before the slower car. How many miles did the cars travel?
   (A) 15   (B) 20   (C) 30
   (D) 36   (E) 42

5. The supplement of an angle measures 11 times its complement. Then the measure of the angle is:
   (A) $9°$ (B) $99°$ (C) $81°$
   (D) $90°$ (E) none of these

6. Phil has 20 coins, nickels, dimes, and quarters, totaling $2.30. If he has twice as many nickels as quarters, how many quarters does Phil have?
   (A) 12   (B) 6   (C) 2   (D) 8   (E) 4

7. If $\angle X$ and $\angle Y$ are complementary, and $\angle X$ and $\angle Z$ are supplementary, which of the following is true?
   (A) $m \angle X + m \angle Y + m \angle Z = 270°$
   (B) $2(m \angle X) = m \angle Y + m \angle Z$
   (C) $m \angle Z - m \angle X = 90°$
   (D) $m \angle Z - m \angle Y = 90°$
   (E) $m \angle Y + m \angle Z = 180°$

8. The average of the measures of angle X, its complementary angle, and its supplementary angle is 80 degrees. Then the measure of the complement of angle X is
   (A) $30°$ (B) $60°$ (C) $80°$
   (D) $90°$ (E) $120°$

Questions 9–16 each consist of two quantities, one in Column A and one in Column B. You are to compare the two quantities and choose:

A   if the quantity in Column A is greater;
B   if the quantity in Column B is greater;
C   if the two quantities are equal;
D   if the relationship cannot be determined from the information given.

1. In certain questions, information concerning one or both of the quantities to be compared is centered above the two columns.
2. In a given question, a symbol that appears in both columns represents the same thing in Column A as it does in Column B.
3. x, n, and k, etc. stand for real numbers.

Column A	Column B

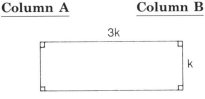

A rectangle has a length that is 3 times its width.

9. The area of the rectangle when its length is doubled | The area of the rectangle when its width is doubled.

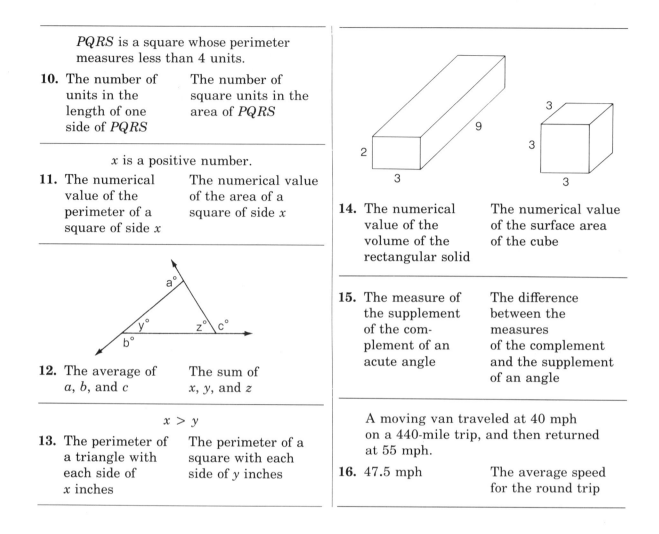

PQRS is a square whose perimeter measures less than 4 units.

10. The number of units in the length of one side of PQRS | The number of square units in the area of PQRS

---

x is a positive number.

11. The numerical value of the perimeter of a square of side x | The numerical value of the area of a square of side x

---

12. The average of a, b, and c | The sum of x, y, and z

$x > y$

13. The perimeter of a triangle with each side of x inches | The perimeter of a square with each side of y inches

---

14. The numerical value of the volume of the rectangular solid | The numerical value of the surface area of the cube

---

15. The measure of the supplement of the complement of an acute angle | The difference between the measures of the complement and the supplement of an angle

---

A moving van traveled at 40 mph on a 440-mile trip, and then returned at 55 mph.

16. 47.5 mph | The average speed for the round trip

SPIRAL REVIEW EXERCISES

1. Represent the additive inverse property using algebraic symbols.
2. Evaluate $(a - b)^2$ when $a = 5$ and $b = -8$.
3. Use the distributive property to express $24y - 3$ as a product.
4. Using the real numbers as the replacement set, graph the solution set for $t < -3$.
5. A packing box contains p jars of peanut butter. Represent the number of jars of peanut butter in s packing boxes.
6. Solve the equation and check:
   a. $9c - 132 = 5c - 48$ **b.** $|4x| - 3 = 9$ **c.** $.06z - .03z = 9$

7. If $\begin{bmatrix} 2 & a \\ 5 & b \end{bmatrix}$ and $\begin{bmatrix} c & -3 \\ 5 & 1 \end{bmatrix}$ are equal matrices, what is the value of c?

8. In a 3-digit palindromic number, the middle digit is 3 times the first digit. If the sum of the digits is 15, find the number.

9. Write the rational number $\frac{5}{12}$ as a decimal.

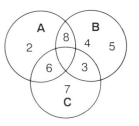

10. Use the Venn diagram to list the element(s) of the indicated set.

  a. $(A \cap B) \cup C$

  b. $A \cap (B \cup C)$

11. If $a < b < c < 12$ and a, b, and c are three different integers, find the greatest possible value of $2a$.

12. If $(x + 3) + (x + 4) + (x + 5) + (x + 6) + (x + 7) = 0$, find the value of x.

13. By inspection, determine which fraction has the greatest value.

  a. $\dfrac{16}{17}, \dfrac{8}{9}, \dfrac{14}{15}, \dfrac{9}{10}, \dfrac{11}{12}$ **b.** $\dfrac{2}{5}, \dfrac{3}{8}, \dfrac{4}{7}, \dfrac{4}{9}, \dfrac{1}{2}$

14. In the sequence 1, 5, 14, 30, y, 91, find the value of y.
  (1) 30   (2) 33   (3) 53   (4) 55

15. If $0 < p < q < 1$ and $pq = r$, then:
  (1) $r < p$ (2) $p < r < q$ (3) $r > q$ (4) $q < r < 1$

16. There are between 20 and 40 books in a box. If books were removed 9 at a time, there would be 3 books left in the box. If the books were removed 12 at a time, there would be 3 books left in the box. How many books are in the box?
  (1) 21   (2) 27   (3) 30   (4) 39

17. A sequence consists of arrangements of squares with sides 1 inch long, as shown. Determine the area and perimeter of the next figure in the sequence.

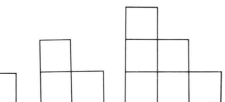

18. If the average of a, b, and c is 9, then what is the average of $3a$, $3b$, and $3c$?

19. Let n be the greatest 3-digit integer such that the product of its digits is 80. What is the units digit of n?

5-1 CONSECUTIVE INTEGER PROBLEMS

TRY IT! *Problems 1 and 2 on page 201*

1. Let n = the first odd integer.
Then $n + 2$ = the second odd integer.
And $n + 4$ = the third odd integer.

The sum of the first and third is 26.

$$n + (n + 4) = 26$$
$$n + n + 4 = 26$$
$$2n + 4 = 26$$
$$2n = 22$$
$$n = 11$$
$$n + 2 = 13$$
$$n + 4 = 15$$

Check: $11 + 15 = 26 \checkmark$

ANSWER: 11, 13, 15

2. Let n = the first integer.
Then $n + 1$ = the second integer.
And $n + 2$ = the third integer.

When the smallest integer is subtracted from twice the largest, the result is more than 24.

$$2(n + 2) - n > 24$$
$$2n + 4 - n > 24$$
$$n + 4 > 24$$
$$n > 20$$

The smallest values are
$n = 21, n + 1 = 22, n + 2 = 23.$

Check: $2(23) - 21 \overset{?}{>} 24$
$$25 > 24 \checkmark$$

ANSWER: 21, 22, 23

5-2 MONEY-VALUE PROBLEMS

TRY IT! *Problem on page 205*

Let d = the number of dimes.
Then $15 - d$ = the number of nickels.

Kind of coin	Number of coins	Value of each coin in cents	Total value in cents
Dime	d	10	$10d$
Nickel	$15 - d$	5	$5(15 - d)$

The total value of all the coins is 135 cents.

$$10d + 5(15 - d) = 135$$
$$10d + 75 - 5d = 135$$
$$5d + 75 = 135$$
$$5d = 60$$
$$d = 12$$
$$15 - d = 3$$

Check

$$12 + 3 = 15$$
Value of 12 dimes = $1.20
Value of 3 nickels = $ .15
Total value = $1.35 \checkmark

ANSWER:
There are 12 dimes and 3 nickels.

5-3 MOTION PROBLEMS

TRY IT! *Problems 1 and 2 on page 210*

1. Let h = hours William traveled.
Then $h + 2$ = hours Martin traveled.

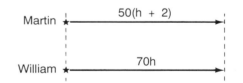

The distance traveled by William is the same as the distance traveled by Martin.

$$70h = 50(h + 2)$$
$$70h = 50h + 100$$
$$20h = 100$$
$$h = 5$$
$$h + 2 = 7$$

Check

Martin traveled 7(50) = 350 km.
William traveled 5(70) = 350 km.
Both brothers traveled the same distance. ✔

ANSWER:
William overtook Martin in 5 hours.

2.   Let h = hours driving out.
Then $7 - h$ = hours driving back.

	Rate (mph)	Time (h)	Distance (mi)
Trip out	40	h	$40h$
Trip back	30	$7 - h$	$30(7 - h)$

*The distance out is the same as
the distance back.*

$$40h = 30(7 - h)$$
$$40h = 210 - 30h$$
$$70h = 210$$
$$h = 3$$
$$7 - h = 4$$
$$40h = 120$$

Check: 3(40) = 120 miles out
  4(30) = 120 miles back
The distances are the same. ✔

ANSWER: She can go out 120 miles.

5-4 LEVER AND PULLEY PROBLEMS

TRY IT! *Problem on page 214*

Let x = Max's weight.
$$w_1 \cdot d_1 = w_2 \cdot d_2$$
$$60(3) = (x)(4)$$
$$180 = 4x$$
$$45 = x$$

Check

$$60(3) \overset{?}{=} 45(4)$$
$$180 = 180 ✔$$

ANSWER: Max weighs 45 pounds.

5-6 TRIANGLE PROBLEMS

TRY IT! *Problems 1 and 2 on page 225*

1.   Let x = m∠R.
Then $x - 15$ = m∠S.
And $3x + 25$ = m∠T.

*The sum of the measures of the angles of
a triangle is 180°.*

$$x + x - 15 + 3x + 25 = 180$$
$$5x + 10 = 180$$
$$5x = 170$$
$$x = 34$$

ANSWER:
m∠R = 34°, m∠S = 19°, m∠T = 127°

2.   Let x = the length of the longest side.
Then $x - 12$ = the length of the shortest side.
And $x - 4$ = the length of the third side.

*A triangular pen is enclosed
by 50 feet of fence.*

$$x + x - 12 + x - 4 = 50$$
$$3x - 16 = 50$$
$$3x = 66$$
$$x = 22$$
$$x - 12 = 10$$
$$x - 4 = 18$$

Check: 22 + 10 + 18 = 50 ✔

ANSWER: The sides are 22 ft., 10 ft., 18 ft.

5-7 PERIMETER AND AREA PROBLEMS

TRY IT! *Problem on page 228*

Let s = the length of the first side.
Then $2s$ = the length of the second side.
And $2s + 1$ = the length of the third side.

The sum of the lengths of all the sides is 46.

$$s + 2s + 2s + 1 = 46$$
$$5s + 1 = 46$$
$$5s = 45$$
$$s = 9$$
$$2s = 18$$
$$2s + 1 = 19$$

Check
9 + 18 + 19 = 46
18 = 2(9)
19 = 18 + 1 ✔

ANSWER: The sides are 9 cm, 18 cm, 19 cm.

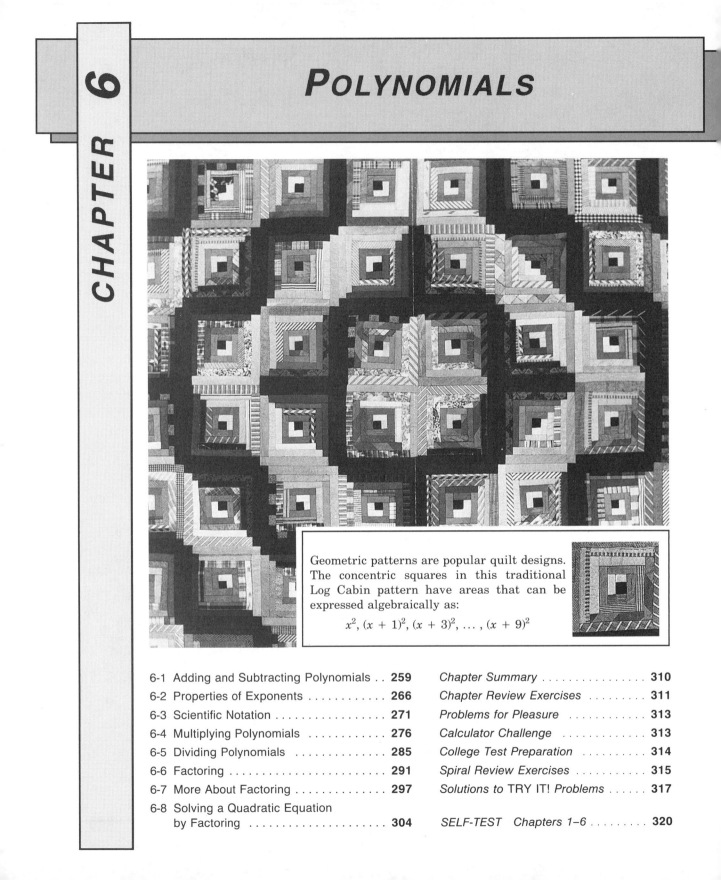

CHAPTER 6

POLYNOMIALS

Geometric patterns are popular quilt designs. The concentric squares in this traditional Log Cabin pattern have areas that can be expressed algebraically as:

$$x^2, (x + 1)^2, (x + 3)^2, \ldots, (x + 9)^2$$

6-1 Adding and Subtracting Polynomials . . **259**

6-2 Properties of Exponents **266**

6-3 Scientific Notation **271**

6-4 Multiplying Polynomials **276**

6-5 Dividing Polynomials **285**

6-6 Factoring **291**

6-7 More About Factoring **297**

6-8 Solving a Quadratic Equation
by Factoring **304**

Chapter Summary **310**

Chapter Review Exercises **311**

Problems for Pleasure **313**

Calculator Challenge **313**

College Test Preparation **314**

Spiral Review Exercises **315**

Solutions to TRY IT! Problems **317**

SELF-TEST   Chapters 1–6 . . . . **320**

A *polynomial* consists of one or more terms. Polynomial operations are extensions of operations with single terms. You will see similarities between algebraic operations with polynomials and the corresponding arithmetic operations with multidigit numbers.

6-1 ADDING AND SUBTRACTING POLYNOMIALS

The Vocabulary of Polynomials

A polynomial can be classified according to the number of terms it has or according to its *degree,* which is the same as the greatest exponent it contains.

Example	Number of Terms	Name	Degree
$2x^3$	1	*monomial*	3
$5y^4 + 7y$	2	*binomial*	4
$4z^2 - 9z + 6$	3	*trinomial*	2
8	1	*constant*	0

A polynomial may be arranged in *descending order* of the exponents, as in $3x^2 - 7x + 4$, or in *ascending order*, as in $4 - 7x + 3x^2$.

MODEL PROBLEM

Answers

a. Arrange the polynomial $5x + 4x^3 - 6 - x^2$ in descending order.

$4x^3 - x^2 + 5x - 6$

b. What is the degree of the polynomial?

degree 3

c. How many terms does the polynomial have?

4 terms

d. Evaluate the polynomial for $x = 2$.

$4x^3 - x^2 + 5x - 6$

$= 4(2^3) - 2^2 + 5(2) - 6$

$= 4(8) - 4 + 10 - 6$

$= 32$

Adding Polynomials

Addition of polynomials builds on the idea of combining like terms.

PROCEDURE

To add polynomials:

1. Arrange them in descending (or ascending) powers of the variable and write them vertically, aligning like terms in columns.
2. Combine like terms.
3. Check by adding again in the opposite direction.

MODEL PROBLEM

Add and check: $4x + 3y - 5z, 3x - 5y - 6z, -2x - y + 3z$

Solution:

$$\begin{array}{r} 4x + 3y - 5z \\ 3x - 5y - 6z \\ -2x - y + 3z \\ \hline 5x - 3y - 8z \end{array} \quad \textbf{ANSWER}$$

Check: Add again in the opposite direction from the way you did the problem. If you added from top to bottom, now add from bottom to top.

If you compare adding polynomials with adding numbers, you can see that it is actually easier to add polynomials.

Compare the arithmetic sum $15 + 2{,}134 + 879$ with the algebraic sum $(2a + 3b) + (9a - 5b) + (b + 7a)$.

	Arithmetic	*Algebra*
Similarity	You add by same place value (ones, tens, hundreds, etc.).	You add like terms.

Arithmetic:
$$\begin{array}{r} 15 \\ 2{,}134 \\ 879 \\ \hline 3{,}028 \end{array}$$

Algebra:
$$\begin{array}{r} 2a + 3b \\ 9a - 5b \\ 7a + b \\ \hline 18a - b \end{array}$$

	Arithmetic	*Algebra*
Differences	Place-value groupings must be understood, to align columns correctly.	Like terms are easy to see.
	You "carry" from one column to the next.	No carrying from one variable to another.
	You add from right to left.	You can add from either direction.

Subtracting Polynomials

The procedure for subtraction of polynomials is similar to that used to subtract like terms.

_____ **P**ROCEDURE

To subtract polynomials:

1. Add the opposite of the subtrahend to the minuend.
2. Check by adding the subtrahend and the difference. The result should equal the minuend.

MODEL PROBLEM

Subtract and check: $(5x^2 - 6x + 3) - (2x^2 - 9x - 6)$

Solution 1: _Horizontal arrangement_

$(5x^2 - 6x + 3) - (2x^2 - 9x - 6)$

$= (5x^2 - 6x + 3) + (-2x^2 + 9x + 6)$ Add the opposite.

$= 5x^2 - 6x + 3 - 2x^2 + 9x + 6$ Remove parentheses.

$= 5x^2 - 2x^2 - 6x + 9x + 3 + 6$ Collect like terms.

$= 3x^2 + 3x + 9$ Combine like terms.

Check: by addition

$2x^2 - 9x - 6$ subtrahend

$\underline{3x^2 + 3x + 9}$ difference

$5x^2 - 6x + 3$ minuend

Solution 2: _Vertical arrangement_

Change signs in the subtrahend and add.

$5x^2 - 6x + 3$ minuend

$\overline{\oplus 2x^2 \ominus 9x \ominus 6}$ subtrahend

$3x^2 + 3x + 9$ difference

ANSWER: $3x^2 + 3x + 9$

Now it's your turn to . . . **TRY IT!**   (_See page 317 for solutions._)

1. Subtract and check: $(4a^2 - 5a - 3) - (-7a^2 + 3a - 4)$
2. Simplify the expression: $3w - [4 - (3 - w)]$

In 1–3: **a.** Arrange the polynomial in descending order.
 b. State the degree of the polynomial.
 c. Tell how many terms there are.
 d. Evaluate the polynomial when the replacement value of the variable is:
  (1) 3 **(2)** -2 **(3)** 1.6 **(4)** -3.1

1. $5 + 2x^2 - 3x$ **2.** $6 + x^3 - \frac{1}{2}x^4$ **3.** $2a^2 - 3a - a^3 + a^4$

In 4–17, add and check the result.

4. $5x + 3y$
 $6x + 9y$

5. $4a - 6b$
 $9a + 3b$

6. $-6m + n$
 $-4m - 5n$

7. $-9ab + 8cd$
 $3ab - 8cd$

8. $8r - 3t$
 $-2r + 3t$
 $-6r + 5t$

9. $y + 8z$
 $5y - z$
 $-8y - 5z$

10. $9x^2 + 5$
 $-2x^2 - 8$
 $x^2 - 3$

11. $-4x^2y^2 + 2r^2s^2$
 $-6x^2y^2 - 5r^2s^2$
 $8x^2y^2 + 3r^2s^2$

12. $15x - 26y + 8z$
 $3x - 14y - 3z$

13. $x^2 - 33x + 15$
 $-4x^2 + 18x - 36$

14. $-5a^2 - 6ab - 4b^2$
 $7a^2 + 6ab - 3b^2$

15. $x^2 + 3x + 5$
 $2x^2 - 4x - 1$
 $-5x^2 + 2x + 4$

16. $5c^2 - 4cd + 6d^2$
 $-c^2 + 3cd + 2d^2$
 $-3c^2 + cd - 8d^2$

17. $2.1 + .9z + z^2$
 $- .7z - .2z^2$
 $-.9 + .2z$

In 18–27, simplify the expression.

18. $4a + (9a + 3)$ **19.** $(-6x - 4) + 6x$ **20.** $-5x^3 + (4 - x^3)$

21. $(5x + 3) + (6x - 5)$ **22.** $(5 - 6y) + (-9y + 2)$

23. $(5a + 3b) + (-2a + 4d)$ **24.** $8 + [5 + (6 + x)]$

25. $[-4x + (10 - 5x)] + 5x$ **26.** $(x^2 + 5x - 24) + (-x^2 - 4x + 9)$

27. $(x^3 + 9x - 5) + (-4x^2 - 12x + 5)$

In 28–36, find the sum of the polynomials.

28. $3x - 5, -2x + 3, 2 - x$ **29.** $9a - 4b + c, -5a + 3c + 4b$

30. $3c - 7d, -2c + 5d, -c + 8d, 4c - 6d$ **31.** $4x^2 - 6x - 3, 3x^2 - 5x + 7$

32. $x^2 - 7xy + 3y^2, -2y^2 + 3x^2 - 4xy, xy - 2x^2 - 4y^2$

33. $6ab - 3a^2 + 5b^2, -4b^2 - 4ab, -6a^2 + 3b^2$

34. $x^3 - 4x^2 + 5x, 3x^2 - 5 + 2x^3, -2x + x^2, -4x^3 + 3$

35. $7b - 3b^3 + 5b^2, -8 + 2b^2 - 4b^3, 7 - 5b^3 - 9b, -6b^2 + b^3 - 7$

36. $x^3 - 4x^2y + 5xy^2 - 4y^3, -2xy^2 - 4x^3 + x^2y, -3y^3 + 7xy^2 - 9x^2y$

37. Represent the perimeter of each figure as a polynomial.

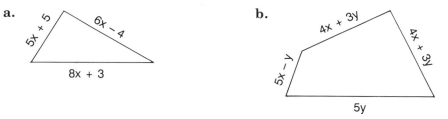

a.

$5x + 5$ $6x - 4$ $8x + 3$

b.

$4x + 3y$ $4x + 3y$ $5x - y$ $5y$

38. Represent the perimeter of a square each of whose sides is:
   a. $4x - 1$ **b.** $x^2 + 2xy + y^2$

39. Represent the perimeter of a rectangle whose width is represented by x and whose length is represented by: **a.** $2x + 1$ **b.** $5 - 2x$

40. Write the opposite (additive inverse) of the expression.
   a. $9x + 6$ **b.** $-5x + 3$ **c.** $-y^2 + 5y - 4$

In 41–54, subtract and check the result.

41. $10a + 8b$
        $\underline{4a + 5b}$

42. $5b + 3c$
        $\underline{4b + c}$

43. $6d + 6e$
        $\underline{9d - 8e}$

44. $8x - 3y$
        $\underline{-4x + 8y}$

45. $4r - 7s$
        $\underline{5r - 7s}$

46. 0
        $\underline{8a - 6b}$

47. $6rs - 7bc$
        $\underline{9rs - 7bc}$

48. $5xy - 9cd$
        $\underline{-3xy + cd}$

49. $x^2 - 6x + 5$
        $\underline{3x^2 - 2x - 2}$

50. $3y^2 - 2y - 1$
        $\underline{-5y^2 - 2y + 6}$

51. $3a^2 - 2ab + 3b^2$
        $\underline{-a^2 - 5ab + 3b^2}$

52. $7a + 6b - 9c$
        $\underline{3a - 6c}$

53. $x^2 - 9$
        $\underline{-2x^2 + 5x - 3}$

54. $5 - 6d - d^2$
        $\underline{ - 4d - d^2}$

In 55–75, simplify the expression.

55. $5c - (4c - 6c^2)$

56. $8r - (-6s - 8r)$

57. $-(5x + 8) - 2x$

58. $(3y - 6) - (8 - 9y)$

59. $(-4x + 7) - (3x - 7)$

60. $(4a - 3b) - (5a - 2b)$

61. $(2c + 3d) - (-6d - 5c)$

62. $(5x^2 + 6x - 9) - (x^2 - 3x + 7)$

63. $(2x^2 - 3x - 1) - (2x^2 + 5x)$

64. $-9d - (2c - 4d) + 4c$

65. $(3y + z) + (z - 5y) - (2z - 2y)$

66. $(a - b) - (a + b) - (-a - b)$

67. $(x^2 - 3x) + (5 - 9x) - (5x^2 - 7)$

68. $5c - [8c - (6 - 3c)]$

69. $12 - [-3 + (6x - 9)]$

70. $10x + [3x - (5x - 4)]$

71. $x^2 - [-3x + (4 - 7x)]$

72. $3x^2 - [7x - (4x - x^2) + 3]$

73. $9a - [5a^2 - (7 + 9a - 2a^2)]$

74. $4y^2 - \{4y + [3y^2 - (6y + 2) + 6]\}$

75. $7m - \{-7m - [-7m - (-7m - 7m)]\}$

76. From $m^2 + 5m - 7$, subtract $m^2 - 3m - 4$.

77. Subtract $2x^2 - 3x + 7$ from $x^2 + 6x - 12$.

78. Subtract $2c^2 + 3c - 4$ from 0.   **79.** By how much does $7x + 5$ exceed $4x - 3$?

80. How much greater than $a^2 + 3ab$ is $4a^2 + 9ab$?

81. How much less than $5x + 3y$ is $2x + y$?

82. What expression must be added to $2x^2 + 5x + 7$ to give the result $8x^2 - 4x - 5$?

83. What expression must be added to $4x^2 - 8$ to give the result 0?

84. From the sum of $y^2 + 2y - 7$ and $2y^2 - 4y + 3$, subtract $3y^2 - 8y - 10$.

85. Subtract the sum of $c^2 - 5$ and $-2c^2 + 3c$ from $4c^2 - 6c + 7$.

86. Use grouping symbols to write an algebraic expression that represents the verbal phrase. Then simplify the expression.

    a. $9x + 2y$ decreased by $-3x + 5y$ **b.** $5x - 7y$ more than $9y - 7x$

    c. $3x^2 - 1$ less than $5x^2 + 7$ **d.** the excess of $3x - 4$ over $9x + 5$

87. Write an expression in simplified form to represent:

    a. the length of arc AC **b.** the length of segment RS

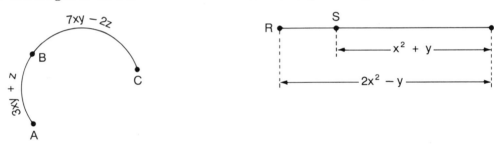

88. Add the matrices.

    a. $\begin{bmatrix} x + y & (-2)^3 \\ -a - b & 4z \end{bmatrix} + \begin{bmatrix} 3x - 2y & -2^3 \\ 2a - b & -4z \end{bmatrix}$

    b. $\begin{bmatrix} x^2 + y^2 & 1 \\ 1.8 & m - mn \\ 9 - y & a + b + c \end{bmatrix} + \begin{bmatrix} 2x^2 - z^2 & r \\ .2 & mn - m \\ -3y + 2 & c - b \end{bmatrix}$

In 89–102, write the answer in simplest algebraic form.

89. James saved xy dollars in June and he saved three times this amount in July. In August, James saved twice as much as in July. How many dollars did James save in the three months?

90. Tina, a jockey, weighed $14xyz$ pounds and had to take off some weight. If she lost $3xyz$ pounds, what was her new weight?

91. Cindy jogged $7(a + b)$ miles in the morning and $5(a + b)$ miles in the afternoon. How many miles did Cindy jog the entire day?

92. Clyde works Monday through Friday, and earns $80(c + d)$ dollars a day. Travel expenses are $(c + d)$ dollars a day and lunch costs 4 times as much as travel. How much money does Clyde have left at the end of the week if he has no other expenses?

93. Mike and Ed went on a fishing trip and shared the expenses. The rod cost $14c^2d$ dollars and the tackle cost $\frac{1}{7}$ as much as the rod. If the bait cost $3c^2d$ dollars, how much did they spend for these items?

94. Rita is $36m^2n$ years of age. How old was she $13m^2n$ years ago?

95. Monday, Steve drove $(c^2 + 2cd - d^2)$ miles. Tuesday, he drove $(3c^2 - 5cd + 8d^2)$ miles. Wednesday, he drove $(9c^2 - 6cd - 5d^2)$ miles. How far did Steve travel during the three days?

96. A projection TV set usually sells for $(x^2 - 3x + 9)$ dollars. If the set is offered at a discount of $(5x + 4)$ dollars, how much will a customer pay for the set?

97. The perimeter of a rectangle is represented by $(3x + 8y - 4z)$ meters. The perimeter of a square is represented by $6x$ meters. By how many meters does the perimeter of the square exceed the perimeter of the rectangle?

98. A merchant's revenue for the month of January was $(9c^2 - 3cd + 8d^2)$ dollars. If his expenses totaled $(8c^2 - 5cd - 2d^2)$ dollars, represent his profit for the month.

99. The thermometer now reads $(2a - 3b + c)$ degrees. By how many degrees should the temperature rise in order for the thermometer to read $(7a - 9b - 4c)$ degrees?

100. Mrs. Barton just remodeled the kitchen, den, and bathrooms in her house. If the kitchen cost $1,400xyz$ dollars, the den cost half as much as the kitchen, and the combined cost for the bathrooms was three times the cost of the kitchen, how much did Mrs. Barton spend in remodeling these rooms?

101. Tad and Ross pooled their money to buy a gift for Avi. Tad had $(2y^2 + 8xy - x^2)$ dollars, Ross had $(3y^2 - 3xy + 2x^2)$ dollars, and the gift cost $(y^2 - 2xy + x^2)$ dollars. How much money was left?

102. Jilly had $(20x + 5y - 4z)$ dollars in her pocket. If she spent $(3x + 2y - z)$ dollars on carfare, $(9x + 8y - 3z)$ dollars on a karate lesson, and $(x - y + 5z)$ dollars on a T-shirt, how much money did she have left?

Karate, an Oriental form of unarmed combat that literally translates as *empty hand,* is believed to have been refined as a martial art by people who were forbidden to carry weapons. Karate blows are struck by hands, feet, knees, and elbows.

6-2 PROPERTIES OF EXPONENTS

Under addition and subtraction, only like terms can be combined. Recall that addition and subtraction make use of the distributive property. For example:

$$2x^2 + 5x^2 = (2 + 5)x^2 = 7x^2$$

Thus, the exponent of the sum is the same as the exponent of the addends.

Under multiplication and division, however, exponents behave differently. For example:

$$(2x^2)(5x^2) = 2 \cdot 5 \cdot x \cdot x \cdot x \cdot x = 10x^4$$

$$\frac{2x^2}{5x^2} = \frac{2 \cdot \overset{1}{\cancel{x}} \cdot \overset{1}{\cancel{x}}}{5 \cdot \underset{1}{\cancel{x}} \cdot \underset{1}{\cancel{x}}} = \frac{2}{5}$$

Multiplication Properties of Exponents

The behavior of exponents under certain multiplications illustrates some general properties and procedures.

Product	Expansion	Result	Property of Exponents
$x^2 \cdot x^3$	$x \cdot x \cdot x \cdot x \cdot x$	$x^2 \cdot x^3 = x^5$	$x^a \cdot x^b = x^{a+b}$
			Multiplication Property
Procedure: To multiply powers of the same base, keep the base and add the exponents.			
$(x^3)^2$	$x^3 \cdot x^3$		$(x^a)^b = x^{ab}$
	$x \cdot x \cdot x \cdot x \cdot x \cdot x$	$(x^3)^2 = x^6$	*Power of a Power Property*
Procedure: To raise a power to a power, keep the base and multiply the exponents.			
$(xy)^2$	$xy \cdot xy$		$(xy)^n = x^n y^n$
	$x \cdot x \cdot y \cdot y$	$(xy)^2 = x^2 y^2$	*Distributive Property Over Multiplication*
Procedure: To raise a product to a power, raise each factor to the power.			

MODEL PROBLEMS

In 1–4, simplify the expression by applying a property of exponents.

1. $m^6 \cdot m = m^{6+1} = m^7$

2. $10^3 \cdot 10^2 = 10^{3+2} = 10^5$ or 100,000

3. $(a^2)^3 = a^{2 \cdot 3} = a^6$

4. $(x^2 y^3)^4 = x^{2 \cdot 4} y^{3 \cdot 4} = x^8 y^{12}$

Division Properties of Exponents

The behavior of exponents under certain divisions illustrates some general properties and procedures. Remember that a denominator may not have a value of 0.

Quotient	Expansion	Result	Property of Exponents
$\dfrac{x^5}{x^3}$	$\dfrac{\overset{1}{\cancel{x}} \cdot \overset{1}{\cancel{x}} \cdot \overset{1}{\cancel{x}} \cdot x \cdot x}{\underset{1}{\cancel{x}} \cdot \underset{1}{\cancel{x}} \cdot \underset{1}{\cancel{x}}}$	$\dfrac{x^5}{x^3} = x^2$	$\dfrac{x^a}{x^b} = x^{a-b}$ *Division Property*
	Procedure: To divide powers of the same base, keep the base and subtract the exponents.		
$\left(\dfrac{x}{y}\right)^2$	$\dfrac{x}{y} \cdot \dfrac{x}{y} = \dfrac{x \cdot x}{y \cdot y}$	$\left(\dfrac{x}{y}\right)^2 = \dfrac{x^2}{y^2}$	$\left(\dfrac{x}{y}\right)^n = \dfrac{x^n}{y^n}$ *Distributive Property Over Division*
	Procedure: To raise a quotient to a power, raise the numerator and the denominator to the power.		

Zero Exponent

Observe how dividing when the exponents are equal leads to two equivalent results.

By expansion: $x^3 \div x^3 = \dfrac{\overset{1}{\cancel{x}} \cdot \overset{1}{\cancel{x}} \cdot \overset{1}{\cancel{x}}}{\underset{1}{\cancel{x}} \cdot \underset{1}{\cancel{x}} \cdot \underset{1}{\cancel{x}}} = 1$

By property: $x^3 \div x^3 = x^{3-3} = x^0$

Conclusion: $x^0 = 1$ (for all real x except 0)

Any number, except 0, raised to the power 0 is 1.

Negative Exponents

Dividing when the exponent of the denominator is greater than the exponent of the numerator leads to two equivalent results.

By expansion: $x^3 \div x^5 = \dfrac{\overset{1}{\cancel{x}} \cdot \overset{1}{\cancel{x}} \cdot \overset{1}{\cancel{x}}}{\underset{1}{\cancel{x}} \cdot \underset{1}{\cancel{x}} \cdot \underset{1}{\cancel{x}} \cdot x \cdot x} = \dfrac{1}{x^2}$

By property: $x^3 \div x^5 = x^{3-5} = x^{-2}$

Conclusion: $x^{-2} = \dfrac{1}{x^2}$ $(x \neq 0)$

In general: $x^{-n} = \dfrac{1}{x^n}$ $(x \neq 0)$

MODEL PROBLEMS

In 1–4, divide. Assume no variable has a value of 0.

1. $x^7 \div x^4 = x^{7-4} = x^3$

2. $10^5 \div 10^3 = 10^{5-3} = 10^2$ or 100

3. $\dfrac{c^5}{c^5} = c^{5-5} = c^0$ or 1

4. $y \div y^5 = y^{1-5} = y^{-4}$ or $\dfrac{1}{y^4}$

5. Rewrite $\dfrac{3}{a^{-4}}$ (where $a \neq 0$) without a negative exponent, and simplify.

 Solution: $\dfrac{3}{a^{-4}} = \dfrac{3}{\dfrac{1}{a^4}}$ Replace the negative exponent.

 $= 3a^4$ Change division to multiplication by the reciprocal.

Observe that division by a base with a negative exponent is equivalent to multiplication by the base with the exponent now positive.

$$\text{In general: } \frac{1}{x^{-n}} = x^n \quad (x \neq 0)$$

6. Rewrite $\dfrac{5^{-3}}{5^{-6}}$ without negative exponents. Simplify.

 Solution: Changing the position between numerator and denominator is equivalent to changing the exponent from negative to positive.

 $\dfrac{5^{-3}}{5^{-6}} = \dfrac{5^6}{5^3} = 5^{6-3} = 5^3$ or 125

 Note. Immediately applying the division property of exponents will also result in a positive exponent.

 $\dfrac{5^{-3}}{5^{-6}} = 5^{-3-(-6)} = 5^{-3+6} = 5^3$

EXERCISES

In 1–25, apply a multiplication property of exponents.

1. $a^2 \cdot a^3$

2. $b^3 \cdot b^4$

3. $c \cdot c^5$

4. $d^4 \cdot d^4$

5. $r^2 \cdot r^4 \cdot r^5$

6. $2^3 \cdot 2^2$

7. $3^4 \cdot 3$

8. $5^4 \cdot 5^4$

9. $4^3 \cdot 4$

10. $2^4 \cdot 2^5 \cdot 2$

11. $(x^3)^2$

12. $(a^4)^2$

13. $(y^3)^2$

14. $(y^5)^2$

15. $(z^3)^2 \cdot (z^4)^2$

16. $(x^2 y^3)^2$

17. $(ab^2)^4$

18. $(rs)^3$

19. $(2^2 \cdot 3^2)^3$

20. $(5 \cdot 2^3)^4$

21. $x^a \cdot x^{2a}$

22. $y^c \cdot y^2$

23. $c^r \cdot c$

24. $x^m \cdot x^n$

25. $(3y)^a \cdot (3y)^b$

26. Explain why each sentence is false.

    a. $2^4 \cdot 2^2 = 2^8$ **b.** $3^3 \cdot 2^2 = 6^5$ **c.** $2^2 + 2^2 = 2^4$ **d.** $(2^2)^3 = 2^5$

In 27–43, divide. Assume all variables are positive integers. Write answers without zero or negative exponents.

27. $a^{10} \div a^5$ **28.** $x \div x^5$ **29.** $c^6 \div c^6$ **30.** $d^2 \div d^6$

31. $\dfrac{d^4}{d^3}$ **32.** $\dfrac{e^6}{e^{11}}$ **33.** $\dfrac{m}{m^4}$ **34.** $\dfrac{r^8}{r^8}$

35. $8^7 \div 8^6$ **36** $y \div y^7$ **37.** $3^2 \div 3^4$ **38.** $m \div m$

39. $5 \div 5^5$ **40.** $x^{5a} \div x^{2a}$ **41.** $r^{2c} \div r^{6c}$ **42.** $s^x \div s^{2x}$ **43.** $a^b \div a^b$

44. Explain why each sentence is false.

    a. $5^6 \div 5^2 = 5^3$ **b.** $3^8 \div 3^4 = 1^4$ **c.** $4^5 \div 2^3 = 2^2$ **d.** $2^0 = 0$

45. a. Which sentences are true?   (1) $(3 + 4)^2 = 3^2 + 4^2$ (2) $(3 - 4)^2 = 3^2 - 4^2$

                                             (3) $(3 \cdot 4)^2 = 3^2 \cdot 4^2$ (4) $(3 \div 4)^2 = 3^2 \div 4^2$

    b. Over which operation(s) is exponentiation distributive?

In 46–77, simplify the expression. Write answers without zero or negative exponents. Assume no variable is equal to zero.

46. $\dfrac{2^3 \cdot 2^4}{2^2}$ **47.** $\dfrac{5^8}{5^4 \cdot 5^6}$ **48.** $\dfrac{10^2 \cdot 10^3}{10^4}$ **49.** $\dfrac{10^8 \cdot 10^2}{(10^5)^2}$ **50.** $\left(\dfrac{x}{y}\right)^3$

51. $\left(\dfrac{a^3}{b^2}\right)^2$ **52.** $\left(\dfrac{m}{n^3}\right)^2$ **53.** $\left(\dfrac{q^3}{r^3}\right)^2$ **54.** a^{-2} **55.** $\dfrac{x^{-4}}{a^{-3}}$

56. $\dfrac{w^{-2}}{w^4}$ **57.** $\dfrac{1}{a^{-3}}$ **58.** $\dfrac{a^{-5}b^3}{x^{-4}y^7}$ **59.** $\dfrac{p^3}{p^{-5}}$ **60.** a^3b^{-2}

61. a^0b^4 **62.** $\dfrac{1}{w^0}$ **63.** $\dfrac{x^0}{y^{-2}}$ **64.** $(3a^2)^0$ **65.** $\dfrac{n^{-4}}{n^{-7}}$

66. $\dfrac{a^4b^{-3}}{a^{-7}b^2}$ **67.** $\dfrac{w^{-4}x^{-6}}{w^0x^{-3}}$ **68.** $\dfrac{mn^5}{m^{-2}n^{-1}}$ **69.** $\dfrac{x^{-3}y^4}{(xy^2)^{-3}}$

70. $(a^4b^{-3})^{-4}$ **71.** $\dfrac{(x^{-5}y^6)^{-3}}{x^2y^{-7}}$ **72.** $\dfrac{a^3}{a^{-5}}$ **73.** $\dfrac{(b^2)^3}{b^4}$

74. $\dfrac{p^{-5}}{p^2}$ **75.** $\dfrac{(q^{-3})^2}{(q^{-2})^3}$ **76.** $\dfrac{(z^2)(z^{-4})}{(z^3)^{-4}}$ **77.** $\dfrac{(w^2)(w^5)^{-3}}{(w^3)^{-1}(w^2)^{-5}}$

In 78–80, multiply the matrices and express the result as a polynomial in simplest form.

78. $[x \ \ y]\begin{bmatrix} x & y \\ x & -x \end{bmatrix}$ **79.** $\begin{bmatrix} a & 2b^2 \\ b & a \end{bmatrix}\begin{bmatrix} b^2 & -b^2 \\ a & -a \end{bmatrix}$ **80.** $[r - s \ \ \ 2r]\begin{bmatrix} -r & 3s & -s \\ s^2 & r^2 & -r \\ r & s & -s \end{bmatrix}$

In 81–90, write the answer as an algebraic expression in simplest form.

81. The Middletown Theater has b^8 rows and b^5 seats in each row. If the theater is full, how many seats are occupied?

82. The Lees drove c^{10} miles to visit their son in camp. If it took them c hours to arrive at the camp, what was their average rate of speed for the trip?

83. Fred paid the government y^9 dollars in income tax last year. This year, he expects his tax to be y times as much as last year's. How many dollars should Fred expect to pay in taxes this year?

84. Every seat is filled in Mrs. Diaz's Math class. There are y^5 rows, each with the same number of seats. If there are y^8 students in the class, how many are seated in each row?

85. The area of a rectangle is x^{11} square feet. If the length is represented by x^9 feet, represent the number of feet in the width of the rectangle.

86. A rectangular solid has length $2x$ feet, width $3y$ feet, and height $4xy$ feet. How many cubic feet are in its volume?

87. Ms. Samson owns 7^{2x} shares of Harley Company stock. If each share is worth 7^x dollars, how much is her stock worth?

If you own a share of stock in a corporation, it means that you are a part owner of the business. If you own a bond, it means that you have loaned money to a business or government.

A *stock exchange* is a marketplace where stocks and bonds are bought and sold. In the dozen major stock exchanges in the United States, millions of shares of stock are traded in a single day.

88. June has 5^{3a} dollars in bills in her wallet. If each bill is worth 5^{2a} dollars, how many bills does she have?

89. The Surrey Theater grossed $(a + b)^5$ dollars in receipts for their Saturday night performance. If the price per ticket was $(a + b)^2$ dollars, how many people attended the theater?

90. The ABC Food Market has $(x + y)^m$ shelves filled with cereal. If there are $(x + y)^n$ boxes on each shelf, how many boxes of cereal are there?

6-3 SCIENTIFIC NOTATION

A number in *scientific notation* is expressed as a product of two factors:

(first factor between 1 and 10) × (second factor an integral power of 10)

Scientific notation is used to express:

1. very large numbers, in which case the power of 10 is a positive integer.
   Example: $430{,}000{,}000 = 4.3 \times 10^8$
2. very small numbers, in which case the power of 10 is a negative integer.
   Example: $.0000057 = 5.7 \times 10^{-6}$

PROCEDURE

To express a number in scientific notation:

1. Determine the first factor, the number between 1 and 10.
2. Determine the power of 10 by counting from where the decimal point *will be* to where the decimal point *is*.

MODEL PROBLEMS

Express in scientific notation: **1.** 32,000,000   **2.** .0000000712

Solutions:

1. $3\,2\,0\,0\,0\,0\,0\,0. = 3.2 \times 10^7$
will be   *1 2 3 4 5 6 7* is

2. $.0\,0\,0\,0\,0\,0\,7\,1\,2 = 7.12 \times 10^{-8}$
is   *8 7 6 5 4 3 2 1* will be

PROCEDURE

To convert *from* scientific notation, look at the power of 10 to tell how many places and which way to move the decimal point.

MODEL PROBLEMS

Express in standard decimal notation: **1.** 4.12×10^6 **2.** 3.4×10^{-5}

Solutions:

1. 10^6 tells you to move the decimal point of the first factor 6 places to the right.

$4.12 \times 10^6 = 4.1\,2\,0\,0\,0\,0 = 4{,}120{,}000$
1 2 3 4 5 6

2. 10^{-5} tells you to move the decimal point of the first factor 5 places to the left.

$3.4 \times 10^{-5} = 0\,0\,0\,0\,3.4 = .000034$
5 4 3 2 1

Calculating in Scientific Notation

PROCEDURE

To multiply or divide numbers in scientific notation:

1. Rearrange the factors of all the numbers so that all of the powers of 10 are grouped together and all of the other quantities are grouped together.
2. Perform the indicated operations in each group of numbers.
3. Rewrite the result in scientific notation.

MODEL PROBLEMS

In 1 and 2, perform the operation, and express the answer in scientific notation.

1. $(2.3 \times 10^3)(4.7 \times 10^4)$

2. $\dfrac{62.9 \times 10^{-3}}{3.7 \times 10^2}$

Solution:

$$(2.3 \times 10^3)(4.7 \times 10^4)$$
$$= (2.3 \times 4.7)(10^3 \times 10^4)$$
$$= 10.81 \times 10^7$$
$$= 1.081 \times 10^8$$

Solution:

$$\frac{62.9 \times 10^{-3}}{3.7 \times 10^2}$$
$$= \frac{62.9}{3.7} \times \frac{10^{-3}}{10^2}$$
$$= 17 \times 10^{-5}$$
$$= 1.7 \times 10^{-4}$$

Now it's your turn to . . . **TRY IT!**

1. Write in scientific notation:
   a. 3,560,000   **b.** .00034   **c.** 856×10^{-4}

2. Write in standard decimal notation:
   a. 5.4×10^3 **b.** 4×10^{-7}

In 3 and 4, perform the operation, and express the answer in scientific notation.

3. $(8 \times 10^{-6})(9.32 \times 10^3)$

4. $\dfrac{6.8 \times 10^7}{1.7 \times 10^2}$

(See page 317 for solutions.)

Scientific notation on the calculator uses only the exponent, not the base. The exponent entry key, usually labeled EXP or EE, automatically includes base 10. Calculate $1 \div 80{,}000{,}000$ by entering **1** ÷ **8** EXP **7** =. The display of

$\boxed{\mathbf{1.25 \; -08}}$ represents 1.25×10^{-8}.

Whether you enter calculations in standard decimal notation or in scientific notation, most scientific calculators will show the result in scientific notation when there are too many digits for the display to accommodate. A word of caution: If any nonzero digits will not fit on the display, a calculator may truncate (cut off) an answer. Change the preceding example by entering **1** ÷ **8** EXP **6** =. If your calculator displays

$\boxed{\mathbf{0.0000001}}$, the digits 25 have been cut off.

For negative exponents, use the +/− key.

Example	Enter
1. $(2.3 \times 10^{-5})(4.7 \times 10^{-4})$	**2.3** EXP **5** +/− × **4.7** EXP **4** +/− =
	Display: $\boxed{\mathbf{1.081 \; -08}}$
2. $\dfrac{5.25 \times 10^{7}}{1.5 \times 10^{-4}}$	**5.25** EXP **7** ÷ (**1.5** EXP **4** +/−) =
	Display: $\boxed{\mathbf{3.5 \quad 11}}$

In 1–9, write the expression in scientific notation.

1. 12,000   **2.** .64   **3.** .000792   **4.** 43.7   **5.** 96,400,000

6. .0314   **7.** 18×10^{4} **8.** $2{,}300 \times 10^{-2}$ **9.** $.0909 \times 10^{4}$

In 10–18, write the expression in standard decimal notation.

10. 2.3×10^{7} **11.** 5×10^{4} **12.** 7.8×10^{-2} **13.** 3.823×10^{4} **14.** 9.612×10^{-1}

15. 6.02×10^{3} **16.** 1.9×10^{2} **17.** 8×10^{-5} **18.** 3.7826×10^{6}

In 19–26, write the given value in scientific notation.

19. A point on Jupiter's equator moves at a speed of 22,000 miles per hour.

20. The 29,000-foot-high peak of Mount Everest was first scaled by Edmund Hillary and Tenzing Norgay, in 1953.

21. In the early 19th century, there were about 60,000,000 bison in the United States. (Protective measures were enacted in 1889, when their numbers had declined to only 551. Today, there are about 15,000 bison.)

22. Neptune lies at an average distance of 2,800,000,000 miles from the sun.

23. Coal has an average heating value of 21,400,000 British thermal units per metric ton.

24. The solubility of marble (calcium carbonate) is .00153 gram per hundred milliliters of water.

25. A germ may measure about .000004 inch in diameter.

26. The diameter of a helium molecule is .000000019 centimeter.

In 27–50, perform the indicated operation, and express the answer in scientific notation.

27. $(3 \times 10^{-5})(4.9 \times 10^7)$

28. $(6 \times 10^4)(6.8 \times 10^{-7})$

29. $(5.72 \times 10^3)(6.8 \times 10^2)$

30. $(9.03 \times 10^{-7})(2.5 \times 10^2)$

31. $(5.01 \times 10^{-6})(4.9 \times 10^{-5})$

32. $(8.7 \times 10^4)(3.2 \times 10^7)$

33. $(1.41 \times 10^2)(5.2 \times 10^{-6})$

34. $(7.2 \times 10^{-6})(4.9 \times 10^4)$

35. $(8.9 \times 10^{-1})(4.9 \times 10^{-1})$

36. $(4.23 \times 10^{-2})(8.83 \times 10^5)$

37. $(3.8 \times 10^{-5})(3.7 \times 10^{-1})$

38. $(3.665 \times 10^5)(6.3 \times 10)$

39. $\dfrac{5.6 \times 10^2}{6.4 \times 10^5}$

40. $\dfrac{5.6 \times 10^2}{7 \times 10^4}$

41. $\dfrac{6 \times 10^{-3}}{1.2 \times 10^{-7}}$

42. $\dfrac{9.1 \times 10^6}{1.3 \times 10^7}$

43. $\dfrac{1.196 \times 10^{11}}{2.3 \times 10^3}$

44. $\dfrac{1.15 \times 10^{-3}}{3.2 \times 10^2}$

45. $\dfrac{7.8 \times 10^4}{7.5 \times 10^5}$

46. $\dfrac{3.62 \times 10}{2.5 \times 10^5}$

47. $\dfrac{7 \times 10^{-4}}{1.6 \times 10^{-2}}$

48. $\dfrac{1.326 \times 10^3}{5.1 \times 10^4}$

49. $\dfrac{1.5438 \times 10^{-2}}{4.65 \times 10^{-7}}$

50. $\dfrac{9 \times 10^5}{1.125 \times 10^6}$

51. Nuclear energy is a practical application of the interchange of mass and energy stated in the relativity formula $E = mc^2$. If $E = 1.49 \times 10^{-10}$ joules and $c = 3.0 \times 10^8$ meters per second, find the number of kilograms in m.

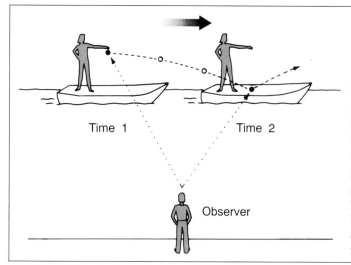

Time 1

Time 2

Observer

Albert Einstein's *Theory of Relativity* states that a natural phenomenon such as energy must be evaluated in association with related phenomena. Thus, the relativity formula $E = mc^2$ relates energy, mass, and motion.

Suppose a person is bouncing a ball on the deck of a moving boat. This person sees the ball go up and down. However, since the boat is moving in time and space, an onshore observer would see the ball move forward as well, tracing the path of an arc.

52. The mass of the sun is 2.4×10^{30} kilograms. The mass of the Earth is 6.0×10^{24} kilograms. How many times the mass of the Earth is the mass of the sun?

53. A single grain of sand has a mass of 2.366×10^{-4} kilograms. The sand is composed of tiny particles of matter, each having a mass of 9.1×10^{-31} kilograms. How many particles are contained in the grain of sand?

In 54–56, give answers in scientific notation written with 2 decimal places.

54. Lake Erie has a surface area of approximately 276,275,000,000 square feet and an average depth of approximately 150 feet.
  a. Find the volume of Lake Erie in cubic feet.
  b. If a teaspoonful of water has a volume of .0020833 cubic feet, how many teaspoonsful of water are in Lake Erie?

55. A quantity of energy consumed can be expressed as the product of the number of watts w and the time t (in seconds). Apply the formula $E = wt$ to find the energy E, in joules, used by a 1,200-watt electric iron in 30 minutes.

56. To convert electrical energy measured in joules to heat measured in calories, use the energy formula $E = wt$ and multiply by a factor of 0.024. Find the heat, in calories, given off by a 1,500-watt hair dryer used for 5 minutes.

57. A huge flock of passenger pigeons, flying at 500 yd./min., took 3 hours to pass over an outdoorsman's camp. He estimated that the flock was 500 yards across, with an average of 10 pigeons per square yard. How many pigeons were in the flock?

The *passenger pigeon* was so named because it traveled great distances, as much as 100 miles a day, in search of food.

These beautiful birds, which by the early 20th century had been hunted to extinction, once existed in vast numbers, traveling in flocks so thick that they blotted out the sun.

6-4 MULTIPLYING POLYNOMIALS

Multiplying a Monomial by a Monomial

Recall that the commutative property of multiplication makes it possible to rearrange the factors of a product and that the associative property makes it possible to multiply the factors in any order. Also, under multiplication with like bases, the exponents are added. Therefore:

$$(-2x^2)(5x^4) = (-2)(5)(x^2)(x^4) = (-2 \cdot 5)(x^2 \cdot x^4) = -10x^6$$

The factors may be rearranged and grouped mentally.

PROCEDURE

To multiply monomials:

1. Mentally rearrange and group factors, using commutative and associative properties.
2. Multiply the numerical coefficients, applying the rules of signs.
3. For variable factors that are powers with the same base, add the exponents.

MODEL PROBLEMS

In 1–4, multiply.

1. $(-4a^3)(-5a^5) = 20a^8$

2. $(3a^2b^3)(4a^3b^4) = 12a^5b^7$

3. $(6c^2d^3)\left(-\frac{1}{2}d\right) = -3c^2d^4$

4. $(-3x^2)^3 = (-3)^3(x^2)^3 = -27x^6$

Multiplying a Polynomial by a Monomial

To multiply a polynomial by a monomial, apply the distributive property:

$$x(4x + 3) = (x)(4x) + (x)(3) = 4x^2 + 3x$$

This result can be illustrated geometrically in terms of areas.

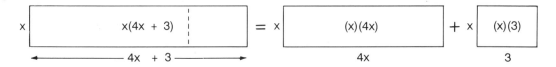

The diagram shows that the area of the largest rectangle is equal to the sum of the areas of the two smaller rectangles.

In 1 and 2, multiply.

1. $-5x(x^2 - 2x + 4) = -5x^3 + 10x^2 - 20x$

2. $-3a^2b^2(4ab^2 - 3b^2) = -12a^3b^4 + 9a^2b^4$

Multiplying a Polynomial by a Polynomial

In arithmetic, each term of a multiplier multiplies each term of a multiplicand. Similarly, in algebraic multiplication, each term of the first polynomial distributes to each term of the second. Note that in algebraic multiplication, you may begin from the left.

$$
\begin{array}{r}
x + 4 \\
x + 3 \\
\hline
\end{array}
$$

$(x + 4)x \longrightarrow x^2 + 4x$

$(x + 4)3 \longrightarrow \quad + 3x + 12$

Add like terms: $x^2 + 7x + 12$

This result can also be illustrated geometrically.

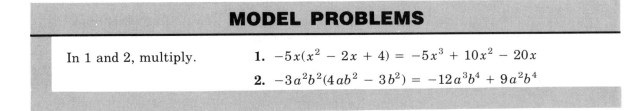

_____ **P**ROCEDURE

To multiply a polynomial by a polynomial.

1. Arrange both polynomials in descending or ascending order of the same variable.
2. Multiply each term of the multiplicand by each term of the multiplier.
3. Combine like terms.

1. Multiply: $(3x - 4)(4x + 5)$

Solution:

$$
\begin{array}{ll}
3x \;-\; 4 & \text{multiplicand} \\
4x \;+\; 5 & \text{multiplier} \\
\hline
12x^2 - 16x & \text{partial product} \\
\quad\;\; + 15x - 20 & \text{partial product} \\
\hline
12x^2 - \;\; x - 20 & \text{product}
\end{array}
$$

ANSWER: $12x^2 - x - 20$

2. Multiply: $(x^2 + 3xy + 9y^2)(x - 3y)$

Solution:

$$
\begin{array}{l}
x^2 + 3xy + 9y^2 \\
x \;-\; 3y \\
\hline
x^3 + 3x^2y + 9xy^2 \\
\quad\;\; - 3x^2y - 9xy^2 - 27y^3 \\
\hline
x^3 + 0 \quad\;\; + 0 \quad\;\; - 27y^3
\end{array}
$$

ANSWER: $x^3 - 27y^3$

3. The length of a rectangle exceeds its width by 7 in. If the length of the rectangle is decreased by 2 in. and the width is increased by 3 in., a new rectangle is formed whose area is 20 sq. in. more than the area of the original. Find the original dimensions.

Solution: Let w = the width of the original rectangle.
Then $w + 7$ = the length of the original rectangle.

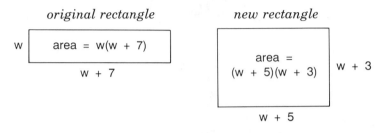

The area of the new rectangle is 20 sq. in. more than the area of the original.

$$(w + 5)(w + 3) = w(w + 7) + 20$$
$$w^2 + 8w + 15 = w^2 + 7w + 20$$

$$8w + 15 = 7w + 20 \qquad \text{Subtract } w^2.$$
$$w + 15 = 20 \qquad \text{Subtract } 7w.$$
$$w = 5 \qquad \text{Subtract } 15.$$
$$w + 7 = 12$$

ANSWER: The original width was 5 in.; the length, 12 in.

Now it's your turn to . . . **TRY IT!** *(See page 317 for solutions.)*

Multiply and check: **1.** $(2a - 7)(3a - 2)$ **2.** $(3a^2 - 2ab - 5b^2)(2a - b)$

3. The Crunchy Oats Corp. is promoting two new cereals. The Nuttee O's box is 8 cm deep, and its height is 5 cm more than its width. The Fruitee O's box has the same depth and width as the other, but is 4 cm shorter, and has 640 cubic cm less volume. Find the width of the boxes.

Shortcuts for Multiplying Binomials

Examine the results of a vertical arrangement for multiplying binomials. Observe how you can compact the arrangement and do the calculations mentally.

$$
\begin{array}{r}
3x - 5 \\
4x + 6 \\
\hline
12x^2 - 20x \\
+ 18x - 30 \\
\hline
12x^2 - 2x - 30
\end{array}
$$

First
Inner
$(3x - 5)(4x + 6)$
Last
Outer

1. Write the first product.

2. Get the outer product and the inner product, and write their sum.

3. Write the last product.

Squaring a binomial, which is finding the product of two identical binomials, is a special case of binomial multiplication. In the product, the first and last terms are respective squares, and the middle term is twice the inner product.

$$(a + b)^2 = (a + b)(a + b) = a^2 + ab + ab + b^2$$

$$\boldsymbol{(a + b)^2 = a^2 + 2ab + b^2}$$

When the first terms of two binomials are identical and the second terms are opposites, the product is a binomial.

$$(a + b)(a - b) = a^2 + ab - ab + b^2$$

$$\boldsymbol{(a + b)(a - b) = a^2 - b^2}$$

EXERCISES

In 1–6, multiply.

1. $4x^2$
    $3x^3$

2. $-4x^3$
    $-7x^4$

3. $-3t^2$
    $-t$

4. $3y^4$
    $-6y^2$

5. $-5d^3$
    $5d^3$

6. $6y$
    $-7y^4$

In 7–27, find the product.

7. $(6)(-2a)$

8. $(4a)(5b)$

9. $(7x)(-2y)(3z)$

10. $(5a^2)(-4a^2)$

11. $(-6x^4)(-3x^3)$

12. $(-7y^2)(5y^5)(-2y^3)$

13. $(20y^3)(-7y^2)$

14. $(18r^5)(-5r^2)$

15. $\left(-\frac{1}{2}s^4\right)\left(-\frac{1}{4}s^2\right)(8s^3)$

16. $(-8y^5)(5y)$

17. $(-9z)(8z^4)(z^3)$

18. $(6x^2y^3)(-4x^4y^2)$

19. $(2r^2s^3)(3r^3s^2)(-r^4s^5)$

20. $(4ab^2)(-2a^2b^3)$

21. $(-2r^4s)(8rs)$

22. $(3ab^3)(-4a^4b)(8ab)$

23. $(-9c)(8cd^2)$

24. $(-3y)(5xy)(15xy^2)$

25. $\left(\frac{2}{3}x^2\right)(-6x)$

26. $(-15ab^2)\left(-\frac{3}{5}a^2b\right)$

27. $\left(\frac{1}{3}xy\right)\left(\frac{1}{2}x\right)(-12x^2y^2)$

In 28–33, write the simplest expression that is equivalent to the given expression.

28. $(7a)^2$

29. $(-3abc)^2$

30. $(-4c^2d)^3$

31. $(-4x)^2(-y)^2$

32. $\left(\frac{1}{2}x^2\right)^3(-4y^3)^2$

33. $10(2x)^2(-y^2)^3$

34. Express each shaded area as a monomial in x.

a.

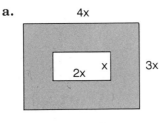

2 rectangles

b.

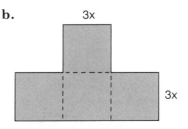

4 congruent squares

35. Express the volume of each figure as a monomial in x.
   a. a cube each of whose edges measures $2x$ inches.
   b. a cylinder whose radius is $4x$ inches and whose altitude is $10x$ inches

In 36–39, write the answer as an algebraic expression in simplest form.

36. Frank rode his bicycle for $2t$ hours. If he was riding at $3rs$ miles per hour, how many miles did Frank cover for the given period of time?

37. Andrea bought $4xy^2$ shares of stock from the ABC Corporation. If each share paid $8y^3$ dollars in dividends, how much money did Andrea earn from her stock purchase?

38. A rectangular table top measures $49c^2d$ cm by $\frac{2}{7}cd^2$ cm. What is its area?

39. The population of Jovial City is decreasing by $(2k)^3$ people each year. How many people will leave Jovial City in the next 6 years?

When the framers of the U.S. Constitution called for a population count every ten years, their chief purpose was to ensure fair representation in the House of Representatives.

Two hundred years later, the census portrait of America's 250 million people has many uses, from setting guidelines for school lunch programs to tracking social and economic change and determining how billions of dollars in government funds are allocated.

Here are some questions from the detailed census form that was distributed to a random one out of six households.

At any time since February 1, 1990, has this person attended regular school or college?
Include only nursery school, kindergarten, elementary school, and schooling which leads to a high school diploma or a college degree.

Do you have COMPLETE plumbing facilities in this house or apartment; that is, 1) hot and cold piped water, 2) a flush toilet, and 3) a bathtub or shower?

Answer ONLY if this is a MOBILE HOME—
What was the total cost for personal property taxes, site rent, registration fees, and license fees on this mobile home and its site last year?

Occupation
What kind of work was this person doing?
(For example: registered nurse, personnel manager, supervisor of order department, gasoline engine assembler, cake icer)

What is this person's ancestry or ethnic origin?
(For example: German, Italian, Afro-Amer., Croatian, Cape Verdean, Dominican, Ecuadoran, Haitian, Cajun, French Canadian, Jamaican, Korean, Lebanese, Mexican, Nigerian, Irish, Polish, Slovak, Taiwanese, Thai, etc.)

Has this person ever been on active-duty military service in the Armed Forces of the United States or ever been in the United States military Reserves or the National Guard?

In 40–53, multiply.

40. $-5(4m - 6n)$

41. $-27\left(\frac{2}{9}x - y\right)$

42. $-16\left(\frac{3}{4}c - \frac{5}{8}d\right)$

43. $-5c^2(15c - 4c^2)$

44. $mn(m + n)$

45. $-ab(a - b)$

46. $3ab(5a^2 - 7b^2)$

47. $-5c^3d^2(9cd^2 - 4c^3d)$

48. $10m^4n(-5n^3 + 3m^2)$

49. $-a^4(10b^2 - a)$

50. $3d(d^2 - 2d + 8)$

51. $\frac{3}{4}(12 - 8x + 4x^2)$

52. $3xy(x^2 + xy + y^2)$

53. $5r^2s^2(-2r^2 + 3rs - 4s^2)$

In 54–60, write the answer as an algebraic expression in simplest form.

54. A car travels $(2x + 5)$ miles per hour. Express the distance it travels in:
   a. 4 hours
   b. h hours
   c. $3x$ hours
   d. $2xy$ hours

55. A hat costs $(2x - 1)$ dollars. Express the cost of:
   a. 10 hats
   b. h hats
   c. $4x$ hats
   d. $3x^2$ hats

56. A building is $(5h + 3)$ yards high. Express its height in feet.

57. If the length of one side of an equilateral triangle is represented by $(5x - 4y)$ feet, represent the perimeter of the triangle.

58. A video rental store has $(x^2 + 3x - 1)$ titles. If each title has $5x$ copies, how many tapes are included in the store's inventory?

59. The price of a subway token is going up by $\frac{1}{4}n$ cents. If Mr. Smith uses $3n^2 + 2n - 8$ of these tokens each year, how much more will Mr. Smith have to pay for his transportation?

60. At the Alamo airport, $(p + q + t^2)$ planes leave runway H every 5 minutes. How many planes will depart from runway H between 10:00 A.M. and 2:00 P.M. inclusive?

In 61–81, multiply.

61. $(a + 2)(a + 3)$

62. $(x - 5)(x - 3)$

63. $(c + 8)(c - 6)$

64. $(x - 7)(x + 2)$

65. $(m + 3)(m - 7)$

66. $(s + 9)(s + 5)$

67. $(6 + y)(5 + y)$

68. $(8 - e)(6 - e)$

69. $(12 - r)(6 + r)$

70. $(2x + 1)(x - 6)$

71. $(2y - 3)(y + 2)$

72. $(c - 5)(2c - 4)$

73. $(2a + 9)(3a + 1)$

74. $(3x - 4)(4x + 3)$

75. $(5y - 2)(3y - 1)$

76. $(2x + 3)(2x - 3)$

77. $(x + y)(x + y)$

78. $(a - b)(a - b)$

79. $(a + 2b)(a + 3b)$

80. $(2c - d)(3c + d)$

81. $(2z + 5w)(3z - 4w)$

In 82–99, find the product mentally.

82. $(a + b)(a - b)$

83. $(x + 5)(x - 5)$

84. $(y + 7)(y - 7)$

85. $(a + 9)(a - 9)$

86. $(10 + a)(10 - a)$

87. $(12 - b)(12 + b)$

88. $(c + d)(c - d)$

89. $(r - s)(r + s)$

90. $(3x + 1)(3x - 1)$

91. $(5c + 4)(5c - 4)$

92. $(5r - 7s)(5r + 7s)$

93. $(x^2 + 8)(x^2 - 8)$

94. $(3 - 5y^2)(3 + 5y^2)$

95. $\left(a + \frac{1}{2}\right)\left(a - \frac{1}{2}\right)$

96. $\left(\frac{3}{4}c - d\right)\left(\frac{3}{4}c + d\right)$

97. $(r + .5)(r - .5)$

98. $(.3 + m)(.3 - m)$

99. $(ab + 8)(ab - 8)$

In 100–107, first express the factors as the sum and difference of the same two numbers, then multiply mentally. For example: $22 \times 18 = (20 + 2)(20 - 2)$

100. 22×18

101. 39×41

102. 53×47

103. 66×74

104. 38×42

105. 55×65

106. 88×92

107. 94×106

In 108–133, multiply.

108. $(x^2 + 3x + 5)(x + 2)$

109. $(y^2 - 2y + 6)(y - 2)$

110. $(2c^2 - 3c - 1)(2c + 1)$

111. $(3 - 2d - d^2)(5 - 2d)$

112. $(c^2 - 2c + 4)(c + 2)$

113. $(d^2 - 3d + 9)(d + 3)$

114. $(2x^2 - 3x + 1)(3x - 2)$

115. $(3y^2 - 9y + 4)(4y + 5)$

116. $(3x^2 - 4xy + y^2)(4x + 3y)$

117. $(4a^2 - 3ab - 2b^2)(2a - 5b)$

118. $(x^3 - 3x^2 + 2x - 4)(3x - 1)$

119. $(2x + 1)(3x - 4)(x + 3)$

120. $(x^2 - 4x + 1)(x^2 + 5x - 2)$

121. $(x + 4)(x + 4)(x + 4)$

122. $(a + 5)^3$

123. $(x - y)^3$

124. $(5 + x^2 - 2x)(2x - 3)$

125. $(5x - 4 + 2x^2)(3 + 4x)$

126. $(2xy + x^2 + y^2)(x + y)$

127. $(3b^2 - 2c^2 - bc)(3b - 2c)$

128. $(r^3 - 2s^4)(r^3 + 2s^4)$

129. $(5c^2d^3 + 7e^5)(5c^2d^3 - 7e^5)$

130. $(a + 5)(a - 5)(a^2 + 25)$

131. $(x - 3)(x + 3)(x^2 + 9)$

132. $(a + b)(a - b)(a^2 + b^2)$

133. $(m^2 + n^2)(m^2 - n^2)(m^4 + n^4)$

In 134–141, simplify the expression.

134. $(x + 7)(x - 2) - x^2$

135. $2(3x + 1)(2x - 3) + 14x$

136. $8x^2 - (4x + 3)(2x - 1)$

137. $(x + 4)(x + 3) - (x - 2)(x - 5)$

138. $(3y + 5)(2y - 3) - (y + 7)(5y - 1)$

139. $(y + 4)^2 - (y - 3)^2$

140. $(x + y)^2 + x(x + 3y)$

141. $r(r - 2s) - (r - s)$

In 142–146, write the answer as a polynomial in simplest form.

142. A school librarian checked out an average of $(b + 3)$ books per person on Tuesday. If $(5b - 2)$ people borrowed books on Tuesday, how many books were checked out?

143. A coliseum has $(3p^2 - 2q + 5)$ rows with $(2p + q)$ seats in each row. How many seats are there in this coliseum?

144. The box in which Ed stores his videotapes is in the shape of a cube. If each edge measures $(x + 2)$ inches, what is the volume of the box?

145. In the AAA basketball tournament, there were $(b^2 + b + 2)$ games played. If there were $(2b - 3)$ players per game on the average, how many played in the tournament?

146. A $9'' \times 12''$ photo is enclosed in a frame of uniform width. If the width of the frame is x inches, express:
a. the area of the frame including the photo
b. the area of the frame only

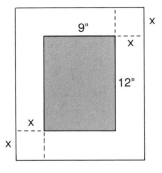

In 147–159, solve algebraically.

147. The length of a rectangle is 8 centimeters more than its width. If the length is increased by 4 centimeters and the width is decreased by 1 centimeter, the area is unchanged. Find the dimensions of the original rectangle.

148. A square and a rectangle have the same area. The length of the rectangle is 3 meters more than the length of a side of the square. The width of the rectangle is 2 meters less than the length of a side of the square. Find the length of each side of the square.

149. The length of each side of a square is increased by 3 feet. The area of the new square is 39 square feet more than the area of the original square. Find the length of each side of the original square.

150. The length of a rectangular garden exceeds its width by 8 feet. If each side of the garden is increased by 2 feet, the area of the garden will be increased by 60 square feet. Find the dimensions of the original garden.

151. The area of a square exceeds the area of a rectangle by 3 square inches. The width of the rectangle is 3 inches shorter and the length of the rectangle 4 inches longer than the side of the square. Find the length of a side of the square.

152. The length of a rectangle is 3 meters more than its width. The length of a side of a square is equal to the length of the rectangle. The area of the square exceeds the area of the rectangle by 24 square meters. Find the dimensions of the rectangle.

153. The length of a rectangle exceeds twice its width by 2 centimeters. If the length is increased by 5 centimeters and the width is decreased by 1 centimeter, a new rectangle is formed whose area exceeds the area of the original rectangle by 20 square centimeters. Find the dimensions of the original rectangle.

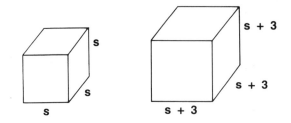

154. If the length of the side, s, of a cube is increased by 3 inches, the volume of the new cube is $(9s^2 + 30s)$ cubic inches more than the volume of the original cube. Find the length of a side of the original cube.

155. If the length of the side, s, of a cube is decreased by 1 mm, the volume of the new cube is $(3s^2 - 5)$ cubic mm less than the volume of the original cube. Find the length of a side of the original cube.

156. The length of a rectangular solid is 3 cm more than the width and the height is 4 cm less than the width. If the length is decreased by 2 cm and the height is increased by 2 cm, the volume of the new rectangular solid is increased by 140 cubic cm. Find the dimensions of the original rectangular solid.

157. The length of a rectangular solid is 2 inches less than the height, and the width is 5 inches less than the height. If the length is decreased by 1 inch and the width is increased by 1 inch, the volume of the new rectangular solid is increased by 16 cubic inches. Find the dimensions of the original rectangular solid.

158. The Olsens and the Sierras have backyard pools of the same uniform depth. The width of the Olsens' pool is 5 feet more than the depth, and the length is 8 feet more than the depth. The width of the Sierras' pool is 6 feet more than the depth, and the length is 7 feet more than the depth. The volume of the Sierras' pool is 8 cubic feet more than the volume of the Olsens' pool. Find the dimensions of each pool.

159. The Cooke Manufacturing Co. uses a model D packing box, whose length is 2 inches more than its width, and whose height is 2 inches less than its width. A new box is needed that has twice the volume. The height is to be the same as that of the model D box, and the length 2 inches less than that of the model D. How must the width of the new box compare with that of the model D?

6-5 DIVIDING POLYNOMIALS

Dividing a Monomial by a Monomial

PROCEDURE

To divide monomials:

1. Divide their numerical coefficients, applying the rules of signs.
2. For variable factors that are powers with the same base, subtract the exponents.
3. Multiply the quotients previously obtained.

MODEL PROBLEMS

1. $\dfrac{24\,a^5}{3\,a^2} = \dfrac{24}{3} \cdot \dfrac{a^5}{a^2} = 8\,a^3$

2. $\dfrac{-15\,x^6 y^5}{-3\,x^3 y^2} = \dfrac{-15}{-3} \cdot \dfrac{x^6}{x^3} \cdot \dfrac{y^5}{y^2} = 5\,x^3 y^3$

3. $\dfrac{3(p + r)^2}{15(p + r)} = \dfrac{3}{15} \cdot \dfrac{(p + r)^2}{(p + r)} = \dfrac{1}{5} \cdot (p + r) = \dfrac{p + r}{5}$

Dividing a Polynomial by a Monomial

PROCEDURE

To divide a polynomial by a monomial, divide each term of the polynomial by the monomial.

MODEL PROBLEMS

In 1 and 2, divide.

1. $(8\,a^5 - 6\,a^4) \div 2\,a^2 = 4\,a^3 - 3\,a^2$

2. $\dfrac{24\,x^3 y^4 - 18\,x^2 y^2 - 6\,xy}{-6\,xy} = -4\,x^2 y^3 + 3\,xy + 1$

Dividing a Polynomial by a Polynomial

To divide one polynomial by another, you use a procedure similar to the one used when dividing whole numbers. See how dividing $x^2 + 6x + 8$ by $x + 2$ follows the same pattern as dividing 736 by 32:

	How to Proceed	*Arithmetic*	*Algebraic*
(1)	Write the usual division form.	$32\overline{)736}$	$x + 2\overline{)x^2 + 6x + 8}$
(2)	Divide the left number of the dividend by the left number of the divisor to obtain the first number of the quotient.	$\dfrac{2}{32\overline{)736}}$	$\dfrac{x}{x + 2\overline{)x^2 + 6x + 8}}$
(3)	Multiply the whole divisor by the first number of the quotient.	$\begin{array}{r} 2 \\ 32\overline{)736} \\ 64 \end{array}$	$\begin{array}{r} x \\ x + 2\overline{)x^2 + 6x + 8} \\ x^2 + 2x \end{array}$
(4)	Subtract this product from the dividend and bring down the next number of the dividend to obtain the new dividend.	$\begin{array}{r} 2 \\ 32\overline{)736} \\ 64 \\ \hline 96 \end{array}$	$\begin{array}{r} x \\ x + 2\overline{)x^2 + 6x + 8} \\ x^2 + 2x \\ \hline 4x + 8 \end{array}$
(5)	Divide the left number of the new dividend by the left number of the divisor to obtain the next number of the quotient.	$\begin{array}{r} 23 \\ 32\overline{)736} \\ 64 \\ \hline 96 \end{array}$	$\begin{array}{r} x + 4 \\ x + 2\overline{)x^2 + 6x + 8} \\ x^2 + 2x \\ \hline 4x + 8 \end{array}$
(6)	Repeat steps 3 and 4, multiplying the whole divisor by the second number of the quotient. Subtract the result from the new dividend. The last remainder is 0.	$\begin{array}{r} 23 \\ 32\overline{)736} \\ 64 \\ \hline 96 \\ 96 \\ \hline 0 \end{array}$	$\begin{array}{r} x + 4 \\ x + 2\overline{)x^2 + 6x + 8} \\ x^2 + 2x \\ \hline 4x + 8 \\ 4x + 8 \\ \hline 0 \end{array}$
		ANSWER: 23	**ANSWER:** $x + 4$

The division process comes to an end when the remainder is 0, or the degree of the remainder is less than the degree of the divisor.

When the remainder is 0, the divisor and the quotient are factors of the dividend.

To check a division: ***quotient $\times$ divisor $+$ remainder $=$ dividend***

From the problem above:　$23 \times 32 = 736$　　$(x + 4)(x + 2) = x^2 + 6x + 8$

Division problems should be written with both divisor and dividend arranged in descending or ascending powers of one variable. For example, if $3x - 1 + x^3 - 3x^2$ is to be divided by $x - 1$, write:

$$x - 1\overline{)x^3 - 3x^2 + 3x - 1}$$

In polynomial division, the dividend must include every power of the variable up to the degree of the dividend. If $x^3 + 8$ is to be divided by $x + 2$, write:

$$x + 2\overline{)x^3 + 0x^2 + 0x + 8}$$

The accompanying flowchart summarizes the procedure, and the decisions you must make, when dividing a polynomial by a polynomial.

POLYNOMIAL LONG DIVISION

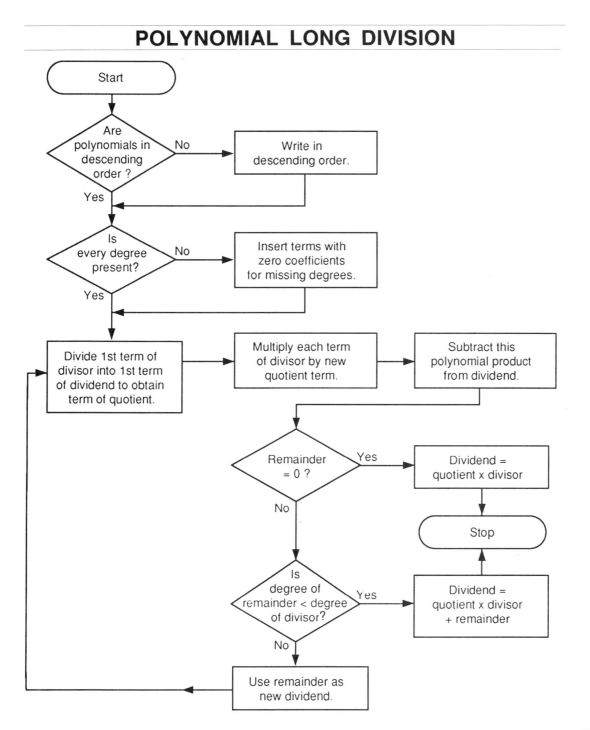

1. Divide $4a - 5 + 3a^2$ by $a - 2$ and check.

2. Divide $8x^3 - 27$ by $2x - 3$.

(See page 318 for solutions.)

EXERCISES

In 1–23, divide.

1. $18x$ by 2

2. $24y^2$ by $-3y$

3. $14x^2y^2$ by $-7x^3$

4. $-12ab$ by $6a$

5. $-2c^2d$ by $-8c^2d^4$

6. $4a^2b^2$ by $-8b^3$

7. $36a^4b^3$ by $9a^2b^3$

8. $-16a^3b^2$ by $-12a^3b$

9. $15c^4d$ by $-5c^4d$

10. $8cd \div 10c^2$

11. $50x^2y \div (-5x^2)$

12. $-14xy^3 \div (-8x^2y^3)$

13. $6a^3b^4 \div 2a^2b^2$

14. $9m^3n^2 \div 9m^3n^2$

15. $8a^2b \div 8ab$

16. $\dfrac{-xyz}{6xy^2z}$

17. $\dfrac{18c^{11}}{27c^5d^2}$

18. $\dfrac{-24a^3b^2}{-8a^2b}$

19. $\dfrac{-12a^4x^6}{5a^3x^5}$

20. $\dfrac{8x^4y^3z}{-2xy^3}$

21. $\dfrac{8(a+b)^5}{2(a+b)^2}$

22. $\dfrac{15(x+y)^2}{6(x+y)^6}$

23. $\dfrac{4(a-2x)}{12(a-2x)^3}$

In 24–26, write the answer as an algebraic expression in simplest form.

24. Lynne purchased $24bcd$ tickets for a football game and used $18bcd$ of them. What fractional part of the tickets purchased was used?

25. Juan bought $4xy$ electric trains for $28x^2y^3$ dollars. What was the cost of each train?

26. If a child puts $(x + y)$ dimes into his piggy bank each week, in how many weeks will he have a total of $7(x + y)^3$ dimes?

27. Using the representations in each diagram, represent the required dimension.

a. the width w of the rectangle

b. the height h of the solid

In 28–43, divide.

28. $(10x + 20y) \div 5$

29. $(cm + cn) \div c$

30. $(12a - 6b) \div -2$

31. $\dfrac{8c^2 - 12d^2}{-4}$

32. $\dfrac{p + prt}{p}$

33. $\dfrac{y^2 - 5y}{-y}$

34. $\dfrac{18d^3 + 12d^2}{6d}$

35. $\dfrac{18r^5 + 12r^3}{6r^2}$

36. $\dfrac{-40x^6 + 16x^4}{8x^4}$

37. $\dfrac{3ab^2 - 4a^2b}{ab}$

38. $\dfrac{2\pi r^2 + 2\pi rh}{2\pi r}$

39. $\dfrac{-6a^2b - 12ab^2}{-2ab}$

40. $\dfrac{1.6a^6x^2 - .8a^5y^2 + 1.2a^4z^2}{.4a^2}$

41. $\dfrac{2.4y^5 + 1.2y^4 - .6y^3}{-.6y^3}$

42. $\dfrac{15r^4s^4 + 20r^3s^3 - 5r^2s^2}{-5r^2s^2}$

43. $\dfrac{x^3y^3 - x^2y^2 + xy}{xy}$

In 44–46, write the answer in simplest form.

44. Represent the number of weeks in $(14x + 7)$ days.

45. The area of a rectangle is $(75r^2 + 15r)$ sq. in. Represent its width if its length is $5r$ in.

46. Tomatoes were planted on $(60x^2 + 20x)$ acres of the Williams farm. If $20x$ bushels of tomatoes were harvested, what was the average number of bushels per acre?

In 47–78, divide and check.

47. $b^2 + 5b + 6$ by $b + 3$

48. $c^2 - 8c + 7$ by $c - 1$

49. $x^2 - 15x - 54$ by $x + 3$

50. $y^2 + 22y + 85$ by $y + 17$

51. $66 + 17x + x^2$ by $6 + x$

52. $30 - t - t^2$ by $5 - t$

53. $3t^2 - 8t + 4$ by $3t - 2$

54. $21a^2 - 10a + 1$ by $3a - 1$

55. $15x^2 - 19x - 56$ by $5x + 7$

56. $16y^2 - 46y + 15$ by $8y - 3$

57. $2x^2 - xy - 6y^2$ by $x - 2y$

58. $21x^2 - 72xy - 165y^2$ by $3x - 15y$

59. $45x^2 + 69xy - 10y^2$ by $3x + 5y$

60. $40x^2 + 11xy - 63y^2$ by $8x - 9y$

61. $56x^2 - 15 - 11x$ by $7x + 3$

62. $15 + 4a^2 - 16a$ by $2a - 3$

63. $a^2 - 8b^2 + 7ab$ by $a + 8b$

64. $cd + c^2 - 30d^2$ by $c - 5d$

65. $15ab + 9b^2 + 6a^2$ by $2a + 3b$

66. $5cd - 3d^2 + 2c^2$ by $2c - d$

67. $x^2 - 64$ by $x - 8$

68. $y^2 - 100$ by $y + 10$

69. $4m^2 - 49n^2$ by $2m + 7n$

70. $64a^2 - 81b^2$ by $8a - 9b$

71. $x^3 - 8x^2 + 17x - 10$ by $x - 5$

72. $6y^3 + y^2 - 28y - 30$ by $2y - 5$

73. $6b^3 - 8b^2 - 17b - 6$ by $3b + 2$

74. $2c^3 - 4c - c^2 + 3$ by $2c + 3$

75. $6y^3 + 11y^2 - 1$ by $3y + 1$

76. $d^3 - 64$ by $d - 4$

77. $8x^3 + 27$ by $2x + 3$

78. $a^3 - 8b^3$ by $a - 2b$

In 79–92, find the quotient and the remainder. Check the answer.

79. $(x^2 - 9x + 7) \div (x - 2)$

80. $(4x^2 + 6x + 9) \div (2x - 5)$

81. $(3x^2 + 9x - 4) \div (3x + 3)$

82. $(12x^2 - 9 + 24x) \div (6x - 3)$

83. $(c^3 - 8c^2 - 6c + 9) \div (c - 2)$

84. $(3c^3 + 14c^2 + 4c - 4) \div (c + 4)$

85. $(2 - 8a + 2a^3 - 5a^2) \div (2a + 3)$

86. $(6y^3 - 10 + 11y^2) \div (3y + 1)$

87. $(10x^2 - 3xy + 9y^2) \div (2x + y)$

88. $(6a^2 + 5ab - 4b^2) \div (3a - 2b)$

89. $(a^2 - 28b^2 + 3ab) \div (a - 6b)$

90. $(10x^2 - 5y^2 + 38xy) \div (2x + 8y)$

91. $(x^2 + 25) \div (x + 5)$

92. $(x^3 - 27) \div (x + 3)$

In 93–98, divide and check.

93. $\dfrac{x^3 + 2x^2 - 2x - 12}{x^2 + 4x + 6}$

94. $\dfrac{6r^3 - 30r + 14r^2 + 12}{2r^2 + 6r - 6}$

95. $\dfrac{2x^3 - x^2 - 4x + 3}{x^2 + 1 - 2x}$

96. $\dfrac{y^4 - 6y^2 + 8}{y^2 - 4}$

97. $\dfrac{4a^4 - 4a^2b^2 - 15b^4}{2a^2 + 3b^2}$

98. $\dfrac{4x^4 + 1}{2x^2 + 2x + 1}$

99. One factor of $x^2 - 8x - 9$ is $x + 1$. Find the other factor.

100. One factor of $3y^2 + 8y + 4$ is $3y + 2$. Find the other factor.

101. Is $x - 2$ a factor of $x^3 - 2x^2 + 4x - 6$? Why?

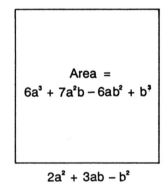

Area =
$6a^3 + 7a^2b - 6ab^2 + b^3$

$2a^2 + 3ab - b^2$

102. a. Given the base and the area of a rectangle as shown in the diagram, represent the height.

   b. Use $a = 3$ and $b = -1$ to find the numerical values of the dimensions. Check that the area is the product of the base and the height.

6-6 FACTORING

Multiplication of **factors** combines expressions; **factoring** breaks a product into its components. It is often helpful in mathematics to look at a numerical or algebraic expression in its factored form.

Prime Factors

To find all the positive integral factors of a number, consider the possible pairs of factors. In finding factors of 24, as shown, do not include pairs beyond 4 · 6. Other pairs, such as 6 · 4, would merely repeat pairs already listed. Thus, the positive integral factors of 24 are 1, 2, 3, 4, 6, 8, 12, and 24.

$$24 = 1 \cdot 24$$
$$24 = 2 \cdot 12$$
$$24 = 3 \cdot 8$$
$$24 = 4 \cdot 6$$

When you factor, you will often use the set of **prime numbers**, which are integers greater than 1 that have no integral factors other than themselves and 1. Some examples of prime numbers are 2, 3, 5, 7, 11, 13, 17. Integers greater than 1 that are not prime are **composite**.

Every composite number can be expressed as the product of prime factors. Although the factors may be written in any order, there is one and only one combination of prime factors whose product is a given integer. A prime factor may appear in a product more than once.

$$21 = 3 \cdot 7$$
$$20 = 2 \cdot 2 \cdot 5 \text{ or } 2^2 \cdot 5$$

MODEL PROBLEM

Express 700 as a product of prime factors.

Solution: $700 = 2 \cdot 350$
$700 = 2 \cdot 2 \cdot 175$
$700 = 2 \cdot 2 \cdot 5 \cdot 35$
$700 = 2 \cdot 2 \cdot 5 \cdot 5 \cdot 7 \text{ or } 2^2 \cdot 5^2 \cdot 7$

Greatest Common Factor

By expressing each of two integers as the product of prime factors, you can find the greatest integer that is a factor of both, their **greatest common factor**.

To find the greatest common factor of 180 and 54, first factor both numbers.

$$180 = 2 \cdot 2 \cdot 3 \cdot 3 \cdot 5 \text{ or } 2^2 \cdot 3^2 \cdot 5 \qquad 54 = 2 \cdot 3 \cdot 3 \cdot 3 \text{ or } 2 \cdot 3^3$$

Note that 2 appears once as a factor in both 180 and 54, and 3 appears twice as a factor in both. Therefore, the greatest common factor of 180 and 54 is $2 \cdot 3^2$, or 18.

The greatest common factor of two or more monomials is the product of the greatest common factor of their numerical coefficients and the highest power of every variable that is a factor of each monomial. For example:

$$\text{Since } 24a^3b^2 = 2 \cdot 2 \cdot 2 \cdot 3 \cdot a \cdot a \cdot a \cdot b \cdot b$$
$$\text{and } 18a^2b = 2 \cdot 3 \cdot 3 \cdot a \cdot a \cdot b,$$

the greatest common factor of the numerical coefficients is $2 \cdot 3$ or 6 and the highest powers of the variables that are factors of each monomial are a^2 and b. Therefore, the greatest common factor of $24a^3b^2$ and $18a^2b$ is $6a^2b$.

Common Monomial Factor

Sometimes, all the terms of a polynomial have a factor in common. When writing a polynomial in factored form, you should find the greatest common factor.

MODEL PROBLEMS

In 1 and 2, factor the polynomial.

1. $\frac{1}{2}na + \frac{1}{2}nl = \frac{1}{2}n(a + l)$ **2.** $6c^3d - 12c^2d^2 + 3cd = 3cd(2c^2 - 4cd + 1)$

3. Use factoring to evaluate the expression.

$87 \times 64 + 87 \times 36 = 87(64 + 36) = 87(100) = 8{,}700$

Factoring by Grouping

Sometimes, when there is no common factor for *all* the terms of a polynomial, there may be common factors for groupings within the polynomial.

$$ax + ay + 5x + 5y$$
$$= (ax + ay) + (5x + 5y) \quad \text{Group within the polynomial.}$$
$$= a(x + y) + 5(x + y) \quad \text{Remove a common factor for each group.}$$
$$= (x + y)(a + 5) \quad \text{Remove } (x + y) \text{ as the common factor.}$$

You may see different ways in which pairs can be grouped. The results of different approaches will be equivalent.

$$ax + ay + 5x + 5y$$
$$= (ax + 5x) + (ay + 5y)$$
$$= x(a + 5) + y(a + 5)$$
$$= (a`+ 5)(x + y)$$

To factor polynomials having four terms and no factor common to all four terms:

1. If possible, group the terms in pairs in such a way that each pair of terms has a greatest common factor.

2. Use the distributive property to factor the greatest common factor in each pair of terms.

3. Use the distributive property again to factor the greatest common factor from the remaining two terms.

MODEL PROBLEMS

In 1 and 2, factor the polynomial.

1. $3x^3 - 5x^2 + 3x - 5$

$= (3x^3 - 5x^2) + (3x - 5)$

$= x^2(3x - 5) + 1(3x - 5)$

$= (3x - 5)(x^2 + 1)$

2. $3a + 5b - 9ax - 15bx$

$= (3a - 9ax) + (5b - 15bx)$

$= 3a(1 - 3x) + 5b(1 - 3x)$

$= (1 - 3x)(3a + 5b)$

Factoring the Difference of Two Squares

An expression of the form $a^2 - b^2$ is called a **difference of two squares**. Factoring an expression that is the difference of two squares is the reverse of multiplying the sum of two terms by the difference of the same two terms.

$$a^2 - b^2 = (a + b)(a - b)$$

NOTE. For a monomial to be a square, its numerical coefficient must be a square and the exponent of each of its variables must be an even number.

MODEL PROBLEMS

In 1–4, factor the polynomial.

	Think	*Write*
1. $r^2 - 9$	$= (r)^2 - (3)^2$	$= (r + 3)(r - 3)$
2. $9y^2 - 16$	$= (3y)^2 - (4)^2$	$= (3y + 2)(3y - 2)$
3. $25x^2 - \frac{1}{49}y^2$	$= (5x)^2 - \left(\frac{1}{7}y\right)^2$	$= \left(5x + \frac{1}{7}y\right)\left(5x - \frac{1}{7}y\right)$
4. $.04 - c^6d^4$	$= (.2)^2 - (c^3d^2)^2$	$= (.2 + c^3d^2)(.2 - c^3d^2)$

1. Tell whether the number is prime.
   a. 5   **b.** 8   **c.** 13   **d.** 23   **e.** 41

2. Write all the prime numbers between the given numbers.
   a. 1 and 10   **b.** 10 and 20   **c.** 20 and 30   **d.** 30 and 40

3. Express the integer as a product of prime numbers.
   a. 35   **b.** 18   **c.** 144   **d.** 400   **e.** 590

Treasure map code

   Mathematicians can become deeply involved with factoring. The success of computer scientists in finding the prime factorization of a 155-digit number caused international excitement. (Try writing a 155-digit number, to get an idea of its enormous size.) The prime factors were themselves large numbers, containing 7, 49, and 99 digits, respectively.

Numbers are often used in writing secret messages. Some secret codes used by U.S. military and naval forces are based on factors of large numbers.

---

4. Determine whether the statement is always true, sometimes true, or never true. If the statement is always true, give an example. If the statement is sometimes true, give two examples, one that is true and one that is false.
   a. The sum of odd numbers is odd.
   b. Prime numbers are odd.
   c. Odd numbers are prime.
   d. The product of two odd prime numbers is prime.
   e. The sum of two prime numbers is prime.
   f. The sum of two odd prime numbers is even.
   g. The factors of an odd number are odd.
   h. The factors of an even number are even.

5. Write all the positive integral factors of the number.
   a. 26   **b.** 50   **c.** 36   **d.** 88   **e.** 100   **f.** 242

6. The product of two monomials is $36x^3y^4$. Find the second factor if the first factor is:
   a. $3x^2y^3$ **b.** $6x^3y^2$ **c.** $12xy^2$ **d.** $9x^3y$ **e.** $18x^3y^2$

7. Find the greatest common factor of the given integers.
   a. 10; 15   **b.** 12; 28   **c.** 14; 35   **d.** 18; 24; 36
   e. 75; 50   **f.** 72; 108   **g.** 144; 200   **h.** 96; 156; 175

8. Find the greatest common factor of the given monomials.

a. $4x$; $4y$
b. 6; $12a$
c. $4r$; $6r^2$; $18r^3$
d. $8xy$; $6xz$
e. $10x^2$; $15xy^2$
f. $7c^3d^3$; $-14c^2d$; $6cd$
g. $36xy^2z$; $-27xy^2z^2$
h. $50m^3n^2$; $75m^3n$
i. $24ab^2c^3$; $18ac^2$; $12abc^2$
j. $12(m + n)$; $18(m + n)$
k. $30(x - y)$; $18(x - y)$

In 9–57, write the expression in factored form.

9. $2a + 2b$
10. $6R - 6r$
11. $bx + by$

12. $xc - xd$
13. $4x + 8y$
14. $12x - 18y$

15. $18c - 27d$
16. $8x + 16$
17. $6x - 18$

18. $8x - 12$
19. $7y - 7$
20. $8 - 4y$

21. $6 - 18c$
22. $y^2 - 3y$
23. $2x^2 + 5x$

24. $3x^2 - 6x$
25. $32x + x^2$
26. $rs^2 - 2r$

27. $ax - 5ab$
28. $3y^4 + 3y^2$
29. $10x - 15x^3$

30. $2x - 4x^3$
31. $p + prt$
32. $s - sr$

33. $\frac{1}{2}hb + \frac{1}{2}hc$
34. $\pi r^2 + \pi R^2$
35. $\pi r^2 + \pi rl$

36. $\pi r^2 + 2\pi rh$
37. $4x^2 + 4y^2$
38. $3a^2 - 9$

39. $5x^2 + 5$
40. $3ab^2 - 6a^2b$
41. $10xy - 15x^2y^2$

42. $21r^3s^2 - 14r^2s$
43. $2x^2 + 8x + 4$
44. $3x^2 - 6x - 30$

45. $ay - 4aw - 12a$
46. $c^3 - c^2 + 2c$
47. $\frac{1}{4}ma + \frac{1}{4}mb + \frac{1}{4}mc$

48. $9ab^2 - 6ab - 3a$
49. $15x^3y^3z^3 - 5xyz$
50. $8a^4b^2c^3 + 12a^2b^2c^2$

51. $28m^4n^3 - 70m^2n^4$
52. $a(x + y) + b(x + y)$
53. $c(t + s) - d(t + s)$

54. $c(m - n) + d(m - n)$
55. $r(y + z) - s(y + z)$

56. $m(x + y) + n(x + y)$
57. $a(v - w) + b(v - w)$

In 58–61, use factoring to evaluate the expression.

58. $35 \times 49 + 35 \times 51$
59. $\frac{1}{2} \times 153 + \frac{1}{2} \times 47$

60. $\frac{22}{7} \times 1,600 - \frac{22}{7} \times 900$
61. $\frac{1}{2} \times 7 \times 6.3 + \frac{1}{2} \times 7 \times 1.7$

In 62–91, factor.

62. $ab + 3a - 2b - 6$
63. $10cf - 4cg + 5df - 2dg$

64. $xy - 3x + 2y^2 - 6y$
65. $4a^3 - 3a^2 + 20a - 15$

66. $10 + 5y + 6x + 3xy$
67. $5aw - 5w + a - 1$

68. $8ab - 2bc + 12ad - 3cd$
69. $a^3 - 2a^2 + 3a - 6$

70. $4mp - np + 4mt - nt$
71. $12x + 2ax + 42 + 7a$

72. $rw + 3rx + 3tw + 9tx$
73. $4m + 20 - 3mt - 15t$

74. $2ah - 10bh - 3a + 15b$

75. $pn - pt - mn + mt$

76. $3ax + 2bx - 15ay - 10by$

77. $15d - 10f - 21cd + 14cf$

78. $4b + 3ac - 12c - ab$

79. $w^7 - 3w^4 - 2w^3 + 6$

80. $6x^4 + 4x^3 - 15x - 10$

81. $4ab + 15 - 20a - 3b$

82. $rt + mw + mt + rw$

83. $3ax - 12x + 4 - a$

84. $rw + arx + acx + cw$

85. $6a^2b - 18a + 5ab^2 - 15b$

86. $3dh + 3y + gy + dgh$

87. $12ax^3 + 5ab - 3bx - 20a^2x^2$

88. $4abx - 4a^2 + 3a - 3bx$

89. $abx + 6dy - 2dx - 3aby$

90. $4c + ab - ac - 4b$

91. $2w^4 - py - 2w^2p + w^2y$

92. Express the term in the form of $(a)^2$.
   a. 144 **b.** $\frac{9}{25}$ **c.** $.81$ **d.** $49d^2$ **e.** $100r^4s^6$

93. Express the binomial as the difference of squares of monomials. If not possible, tell why.
   a. $y^2 - 64$ **b.** $t^2 - 7$ **c.** $25n^2 - 16m^2$
   d. $c^2 - .09d^2$ **e.** $p^2 - \frac{9}{25}q^2$ **f.** $16a^4 - 25b^6$
   g. $9y^9 - 16y^{16}$ **h.** $-9 + m^2$ **i.** $-16 - x^2$

   In 94–129, factor the binomial.

94. $a^2 - 4$ **95.** $c^2 - 100$ **96.** $t^2 - 81$

97. $9 - x^2$ **98.** $144 - c^2$ **99.** $121 - m^2$

100. $16a^2 - b^2$ **101.** $25m^2 - n^2$ **102.** $d^2 - 4c^2$

103. $r^4 - 9$ **104.** $x^4 - 64$ **105.** $25 - s^4$

106. $100x^2 - 81y^2$ **107.** $64e^2 - 9f^2$ **108.** $r^2s^2 - 144$

109. $w^2 - \frac{1}{64}$ **110.** $s^2 - \frac{1}{100}$ **111.** $\frac{1}{81} - t^2$

112. $49x^2 - \frac{1}{9}$ **113.** $\frac{4}{25} - \frac{49}{81}d^2$ **114.** $\frac{1}{9}r^2 - \frac{64}{121}s^2$

115. $x^2 - .64$ **116.** $y^2 - 1.44$ **117.** $.04 - 49r^2$

118. $.16m^2 - 9$ **119.** $81n^2 - .01$ **120.** $.81x^2 - y^2$

121. $64a^2b^2 - c^2d^2$ **122.** $25r^2s^2 - 9t^2u^2$ **123.** $81m^2n^2 - 49x^2y^2$

124. $49m^4 - 64n^4$ **125.** $25x^6 - 121y^{10}$ **126.** $x^4y^8 - 144a^6b^{10}$

127. $(a + b)^2 - c^2$ **128.** $(x - y)^2 - 4$ **129.** $25 - (m + n)^2$

6-7 MORE ABOUT FACTORING

Factoring Trinomials of the Form $x^2 + bx + c$

Factoring a trinomial $x^2 + bx + c$ is the reverse of multiplying binomials.

Consider $(x + 3)(x + 5) = x^2 + 8x + 15$. The three terms of the trinomial product are computed from four individual multiplications, with two terms combined.

$$\text{first term:} \quad (x)(x) = x^2$$
$$\text{middle term:} \quad 3x + 5x = 8x$$
$$\text{last term:} \quad (3)(5) = 15$$

To factor $x^2 + 8x + 15$, choose factors that give the first and last terms of the trinomial and then test these factors to see if they also give the correct middle term. Note that this is a guess-and-check strategy. With practice, you will need fewer guesses.

Except for order, that is $(x + 3)(x + 5)$ or $(x + 5)(x + 3)$, the binomial factors of a trinomial are unique.

PROCEDURE

To factor a trinomial of the form $x^2 + bx + c$:

1. Factor the first term of the trinomial to find the first term of each binomial.
2. Determine the sign of the last term of each binomial by applying the following rules:
   a. If the constant c is positive, both signs are the same as the sign of b.
   b. If the constant c is negative, the last term of one binomial will be positive, the last term of the other will be negative.
3. List the possible pairs of factors of c to determine the last term of each binomial. Use these to write the possible binomial factors of the trinomial.
4. Test the pairs of binomial factors to see which combination will give the correct middle term, bx, of the trinomial.

MODEL PROBLEMS

1. Factor: $x^2 + 7x + 10$

   Solution:

   (1) The product of the first terms of the binomials must be x^2. Therefore, each first term must be x. Write:
   $$x^2 + 7x + 10 = (x \quad)(x \quad)$$

   (2) Since the product of the last terms of the binomials must be $+10$, these last terms must be either both positive or both negative. Since the middle term of the given trinomial, $+7x$, is positive, the last terms of the binomials must both be positive.

(3) The pairs of positive integers whose product is $+10$ are $(+10), (+1)$ and $(+5), (+2)$. The binomial factors are either $(x + 10)(x + 1)$ or $(x + 5)(x + 2)$.

(4) Test each pair of factors.

$(x + 10)(x + 1)$ is not correct because the middle term is not $+7x$.

$$\overset{+10x}{\overbrace{(x + 10)(x + 1)}}$$
$$\underset{+1x}{\underbrace{}}$$

$(x + 5)(x + 2)$ is correct because the middle term is $+7x$.

$$\overset{+5x}{\overbrace{(x + 5)(x + 2)}}$$
$$\underset{+2x}{\underbrace{}}$$

ANSWER: $x^2 + 7x + 10 = (x + 5)(x + 2)$

2. Factor: $y^2 - 8y + 12$

Solution:

(1) The product of the first terms of the binomials must be y^2. Write:
$$y^2 - 8y + 12 = (y \quad)(y \quad)$$

(2) Since the product of the last terms of the binomials must be $+12$, these last terms must be either both positive or both negative. Since the middle term of the given trinomial, $-8y$, is negative, the last terms of the binomials must both be negative.

(3) The pairs of negative integers whose product is $+12$ are:
$$(-1), (-12) \quad (-2), (-6) \quad (-3), (-4)$$
Thus, the possible factors of the trinomial are:
$$(y - 1)(y - 12) \quad (y - 2)(y - 6) \quad (y - 3)(y - 4)$$

(4) Only $(y - 6)(y - 2)$ yields a middle term of $-8y$.

ANSWER: $y^2 - 8y + 12 = (y - 6)(y - 2)$

3. Factor: $c^2 + 5c - 6$

Solution:

(1) The product of the first terms of the binomials must be c^2. Write:
$$c^2 + 5c - 6 = (c \quad)(c \quad)$$

(2) Since the product of the last terms of the binomials must be -6, one of these last terms must be positive, the other negative.

(3) The pairs of integers whose product is -6 are:
$$(+1), (-6) \quad (-1), (+6) \quad (+2), (-3) \quad (-2), (+3)$$
Thus, the possible factors of the trinomial are:
$$(c + 1)(c - 6) \quad (c - 1)(c + 6) \quad (c + 2)(c - 3) \quad (c - 2)(c + 3)$$

(4) Only $(c - 1)(c + 6)$ yields a middle term of $+5c$.

ANSWER: $c^2 + 5c - 6 = (c - 1)(c + 6)$

Factoring Trinomials of the Form *ax*² + *bx* + *c*

The preceding trinomials were of the form $ax^2 + bx + c$, where $a = 1$. If $a \neq 1$, the first terms of the binomials include factors of a, and these factors of a must be considered in combination with the factors of c when checking the middle term.

MODEL PROBLEMS

1. Factor: $2x^2 - 7x - 15$

Solution:

(1) Since the product of the first terms of the binomials must be $2x^2$, one of these terms must be $2x$, the other x. Write:
$$2x^2 - 7x - 15 = (2x \quad)(x \quad)$$

(2) Since the product of the last terms of the binomials must be -15, one of these last terms must be positive, the other negative.

(3) The pairs of integers whose product is -15 are:
  $(+1), (-15)$ $(-1), (+15)$ $(+3), (-5)$ $(-3), (+5)$
The possible pairs of factors are:
$(2x + 1)(x - 15)$ $(2x - 1)(x + 15)$ $(2x + 3)(x - 5)$ $(2x - 3)(x + 5)$
$(2x + 15)(x - 1)$ $(2x - 15)(x + 1)$ $(2x + 5)(x - 3)$ $(2x - 5)(x + 3)$

(4) Only $(2x + 3)(x - 5)$ yields a middle term of $-7x$.

ANSWER: $2x^2 - 7x - 15 = (2x + 3)(x - 5)$

2. Factor: $4x^2 - 4x - 35$

Solution:

(1) Since the product of the first terms of the binomials must be $4x^2$, the first terms must be either $2x$ and $2x$ or $4x$ and x.

(2) Since the product of the last terms of the binomials must be -35, one of these last terms must be positive, the other negative.

(3) The pairs of integers whose product is -35 are:

$(+1), (-35) \quad (-1), (+35) \quad (+5), (-7) \quad (-5), (+7)$

The possible pairs of factors are:

$(4x + 1)(x - 35)$	$(4x + 5)(x - 7)$	$(2x + 1)(2x - 35)$
$(4x - 1)(x + 35)$	$(4x - 5)(x + 7)$	$(2x - 1)(2x + 35)$
$(4x + 35)(x - 1)$	$(4x + 7)(x - 5)$	$(2x + 5)(2x - 7)$
$(4x - 35)(x + 1)$	$(4x - 7)(x + 5)$	$(2x - 5)(2x + 7)$

(4) Only $(2x + 5)(2x - 7)$ yields a middle term of $-4x$.

ANSWER: $4x^2 - 4x - 35 = (2x + 5)(2x - 7)$

NOTE. If you find the middle term of the trinomial product each time you write a trial pair of binomial factors, you will probably find the correct pair of binomial factors without having to list all the possibilities.

Now it's your turn to . . . **TRY IT!**

Factor: **1.** $5x^2 + 3x - 8$ **2.** $6x^2 - 11x - 10$

(See page 318 for solutions.)

Factoring Completely

When factoring a polynomial, it is usual to continue the process until all factors other than monomial factors are prime factors.

_____ **P**ROCEDURE

To factor a polynomial completely:

1. Factor out a greatest common factor, and examine the remaining factor.

2. a. If the remaining factor is a difference of two squares, factor it. If a resulting factor is again a difference of squares, factor again.

   b. If the remaining factor is a trinomial that can be factored, factor into binomials.

3. Write the answer as the product of all the factors. Make certain that in the answer all factors other than monomial factors are prime factors.

The accompanying flowchart summarizes the procedure, and the decisions you must make, when factoring a polynomial as completely as possible.

FACTORING POLYNOMIALS

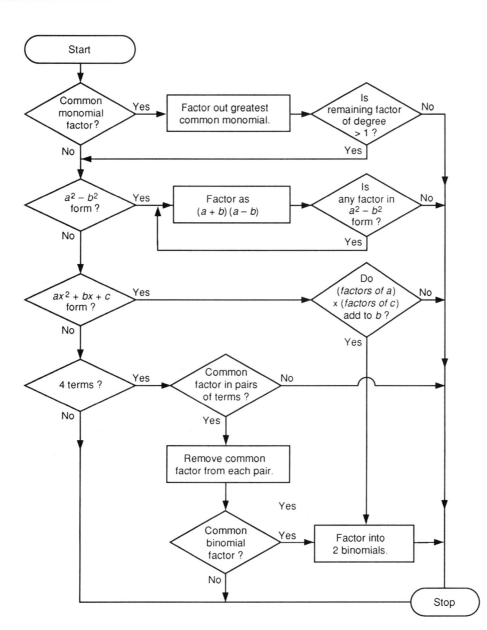

MODEL PROBLEMS

1. Factor: $5x^2 - 45$

How to Proceed	**Solution**
(1) Find the greatest common factor.	$5x^2 - 45 = 5(x^2 - 9)$
(2) Factor the difference of 2 squares.	$5x^2 - 45 = 5(x + 3)(x - 3)$

2. Factor: $3x^2 - 6x - 24$

(1) Find the greatest common factor.	$3x^2 - 6x - 24 = 3(x^2 - 2x - 8)$
(2) Factor the trinomial.	$3x^2 - 6x - 24 = 3(x - 4)(x + 2)$

3. Factor: $x^4 - 16$

(1) Factor the difference of 2 squares.	$x^4 - 16 = (x^2 + 4)(x^2 - 4)$
(2) Factor the difference of 2 squares.	$x^4 - 16 = (x^2 + 4)(x + 2)(x - 2)$

Now it's your turn to . . . **TRY IT!**   *(See page 319 for solutions.)*

In 1–3, factor completely.

1. $6a^2 - 54b^2$ **2.** $5x^4 - 80$ **3.** $3a^5 - 3a^3 - 36a$

EXERCISES

In 1–63, factor and check.

1. $a^2 + 3a + 2$

2. $b^2 + 4b + 3$

3. $c^2 + 6c + 5$

4. $x^2 + 8x + 7$

5. $y^2 + 10y + 9$

6. $r^2 + 12r + 11$

7. $m^2 + 5m + 4$

8. $t^2 + 7t + 10$

9. $x^2 + 9x + 18$

10. $x^2 + 12x + 27$

11. $c^2 + 8c + 15$

12. $y^2 + 12y + 35$

13. $x^2 + 11x + 24$

14. $y^2 + 15y + 36$

15. $z^2 + 13z + 40$

16. $a^2 + 11a + 18$

17. $b^2 + 13b + 30$

18. $16 + 17c + c^2$

19. $x^2 + 2x + 1$

20. $y^2 + 8y + 16$

21. $z^2 + 10z + 25$

22. $a^2 - 8a + 7$

23. $x^2 - 12x + 11$

24. $a^2 - 6a + 5$

25. $x^2 - 5x + 6$

26. $x^2 - 9x + 14$

27. $x^2 - 11x + 10$

28. $y^2 - 6y + 8$

29. $z^2 - 10z + 21$

30. $r^2 - 11r + 18$

31. $a^2 - 9a + 8$

32. $x^2 - 12x + 35$

33. $15 - 8y + y^2$

34. $x^2 - 10x + 24$

35. $y^2 - 13y + 36$

36. $c^2 - 14c + 40$

37. $t^2 - 18t + 72$

38. $r^2 - 16r + 60$

39. $x^2 - 16x + 48$

40. $x^2 - 14x + 49$

41. $y^2 - 16y + 64$

42. $z^2 - 20z + 100$

43. $x^2 - x - 2$

44. $x^2 - 3x - 4$

45. $x^2 - 6x - 7$

46. $y^2 + 4y - 5$

47. $y^2 + 8y - 9$

48. $z^2 - 12z - 13$

49. $a^2 - 3a - 10$

50. $b^2 - 2b - 8$

51. $c^2 - 2c - 15$

52. $c^2 + 2c - 35$

53. $r^2 + 4r - 21$

54. $t^2 + t - 6$

55. $x^2 - 7x - 18$

56. $y^2 - 5y - 24$

57. $z^2 + 9z - 36$

58. $m^2 - 6m - 27$

59. $x^2 + 3x - 40$

60. $x^2 - 13x - 48$

61. $x^2 - 6x - 72$

62. $x^2 + 11x - 60$

63. $x^2 - 2x - 80$

In 64–99, factor and check.

64. $3x^2 - 5x + 2$

65. $5x^2 - 13x - 6$

66. $3x^2 + 10x + 8$

67. $7x^2 + 3x - 10$

68. $2x^2 + 11x + 15$

69. $5a^2 - a - 4$

70. $2x^2 - 3x - 9$

71. $7c^2 - 17c + 6$

72. $11x^2 + 6x - 5$

73. $2x^2 + 5x + 2$

74. $3x^2 + 10x + 3$

75. $2x^2 + 11x + 5$

76. $2x^2 + 7x + 6$

77. $2x^2 + 11x + 12$

78. $3x^2 + 14x + 8$

79. $2y^2 - 3y + 1$

80. $3y^2 - 8y + 4$

81. $14 - 13y + 3y^2$

82. $2x^2 + x - 3$

83. $3x^2 - 5x - 2$

84. $3x^2 + 2x - 5$

85. $2x^2 + x - 6$

86. $5x^2 - 3x - 8$

87. $3x^2 - 5x - 12$

88. $4x^2 - 12x + 5$

89. $6x^2 + 5x - 6$

90. $6x^2 + 5x - 4$

91. $10a^2 - 9a + 2$

92. $10x^2 + 49x - 5$

93. $18y^2 - 23y - 6$

94. $x^2 + 3xy + 2y^2$

95. $c^2 + 4cd - 5d^2$

96. $r^2 - 3rs - 10s^2$

97. $3a^2 - 7ab + 2b^2$

98. $5m^2 + 3mn - 2n^2$

99. $4x^2 - 5xy - 6y^2$

In 100–133, factor completely.

100. $2a^2 - 2b^2$

101. $6x^2 - 6y^2$

102. $4x^2 - 4$

103. $ax^2 - ay^2$

104. $cm^2 - cn^2$

105. $st^2 - s$

106. $2x^2 - 18$

107. $2x^2 - 32$

108. $3x^2 - 27y^2$

109. $18m^2 - 8$

110. $12a^2 - 27b^2$

111. $63c^2 - 7$

112. $x^3 - 4x$

113. $y^3 - 25y$

114. $z^3 - z$

115. $4a^3 - ab^2$

116. $4c^3 - 49c$

117. $9db^2 - d$

118. $4a^2 - 36$

119. $x^4 - 1$

120. $y^4 - 81$

121. $\pi R^2 - \pi r^2$

122. $\pi c^2 - \pi d^2$

123. $100x^2 - 36y^2$

124. $ax^2 + 3ax + 2a$

125. $4r^2 - 4r - 48$

126. $4x^2 - 6x - 4$

127. $2ax^2 - 2ax - 12a$

128. $a^4 - 10a^2 + 9$

129. $y^4 - 13y^2 + 36$

130. $x^2y + 3x^2 - a^2y - 3a^2$

131. $4ax - 8x + 12a - 24$

132. $a^3x^2 - 9a^3 - 4ax^2 + 36a$

133. $3a^2 + 3x^2 - 3a^2x^2 - 3$

6-8 SOLVING A QUADRATIC EQUATION BY FACTORING

Factoring procedures can be used in solving equations. A second-degree equation that can be written in the form $ax^2 + bx + c = 0$, where a, b, and c represent constants, and $a \neq 0$, is called a **quadratic equation**. Examples are:

$$x^2 + 5x - 6 = 0 \qquad x^2 + x = 0 \qquad 3x^2 = 12 \qquad 4x^2 = 4x - 1$$

By the multiplication property of 0, the product of 0 and any other real number is 0. Similarly, if the product of two real numbers is 0, then at least one of the numbers is 0.

In general, if a and b are real numbers, then:

$ab = 0$ if and only if $a = 0$ or $b = 0$

This principle can be applied to solving quadratic equations.

For example, solve the equation:

$$x^2 - 3x + 2 = 0$$

Factor the left side, obtaining the equivalent equation:

$$(x - 1)(x - 2) = 0$$

By the above principle, the product on the left side can equal 0 if and only if:

$$x - 1 = 0 \quad \text{or} \quad x - 2 = 0$$

If $x - 1 = 0$, then $x = 1$. If $x - 2 = 0$, then $x = 2$.

Therefore, the solution set of $x^2 - 3x + 2 = 0$ is $\{1, 2\}$, or the **roots** of the equation are 1 and 2.

The equation $x^2 - 10x + 25 = 0$ can be written $(x - 5)(x - 5) = 0$. Since both factors give the result $x = 5$, the number 5 is called a *double root*. The solution set is simply $\{5\}$.

If a quadratic expression cannot be factored over the integers, the quadratic equation does not have rational roots.

PROCEDURE

To solve an equation by using factoring:

1. If necessary, transform the equation. Do this by removing parentheses, combining like terms on the left side, and making the right side 0.
2. Factor the left side of the equation.
3. Set each factor containing the variable equal to 0.
4. Solve each of the resulting equations.
5. Check by substituting each value of the variable in the original equation.

MODEL PROBLEMS

1. Solve and check: $x^2 - 7x = -10$

How to Proceed   **Solution**

$$x^2 - 7x = -10$$

(1) Transform the equation. $x^2 - 7x + 10 = 0$

(2) Factor the left side. $(x - 2)(x - 5) = 0$

(3) Let each factor $= 0$. $x - 2 = 0 \mid x - 5 = 0$

(4) Solve each first-degree equation. $x = 2 \mid x = 5$

Check:

$$x^2 - 7x = -10 \qquad\qquad x^2 - 7x = -10$$

$$\text{If } x = 2,\ (2)^2 - 7(2) \overset{?}{=} -10 \qquad \text{If } x = 5,\ (5)^2 - 7(5) \overset{?}{=} -10$$

$$4 - 14 \overset{?}{=} -10 \qquad\qquad 25 - 35 \overset{?}{=} -10$$

$$-10 = -10 \checkmark \qquad\qquad -10 = -10 \checkmark$$

ANSWER: $x = 2$ or $x = 5$

2. Solve for x, and check: $3x^2 + 18x = 0$

How to Proceed   **Solution**

$$3x^2 + 18x = 0$$

(1) Factor the left side. $3x(x + 6) = 0$

(2) Let each factor $= 0$. $3x = 0 \mid x + 6 = 0$

(3) Solve each first-degree equation. $x = 0 \mid x = -6$

Check:

$$3x^2 + 18x = 0 \qquad\qquad 3x^2 + 18x = 0$$

$$3(0)^2 + 18(0) \overset{?}{=} 0 \qquad 3(-6)^2 + 18(-6) \overset{?}{=} 0$$

$$0 + 0 = 0 \qquad\qquad 108 - 108 \overset{?}{=} 0$$

$$0 = 0 \checkmark \qquad\qquad 0 = 0 \checkmark$$

ANSWER: $x = -6$ or $x = 0$

3. Find the roots of $2x^2 = 3x$, and check.

Check by substituting both values obtained into the original equation.

ANSWER: The roots are 0 and $\frac{3}{2}$.

Solution

$$2x^2 = 3x$$
$$2x^2 - 3x = 0$$
$$x(2x - 3) = 0$$
$$x = 0 \quad | \quad 2x - 3 = 0$$
$$ \quad 2x = 3$$
$$ \quad x = \frac{3}{2}$$

4. Find the solution set and check:
$$x(x - 6) = -9$$

Solution

$$x(x - 6) = -9$$
$$x^2 - 6x = -9$$
$$x^2 - 6x + 9 = 0$$
$$(x - 3)(x - 3) = 0$$
$$x - 3 = 0 \quad | \quad x - 3 = 0$$
$$x = 3 \quad | \quad x = 3$$

5. Solve for x in terms of a and b:
$$ax^2 + bx = 0$$

Solution

$$ax^2 + bx = 0$$
$$x(ax + b) = 0$$
$$x = 0 \quad | \quad ax + b = 0$$
$$ \quad ax = -b$$
$$ \quad x = -\frac{b}{a}$$

Check by substituting the value(s) obtained into the original equation.

ANSWER: The solution set is $\{3\}$.

ANSWER: $x = 0$ or $x = -\dfrac{b}{a}$

6. Mr. Sloan paved his patio with 50 slate squares. His neighbor, Mr. Stark, using slate squares 3 feet longer on a side, needed 8 slate squares to pave an equal area. How large were Mr. Sloan's slate squares?

Solution: Let $x = $ the length of a side of Mr. Sloan's square.
Then $x + 3 = $ the length of a side of Mr. Stark's square.
And $x^2 = $ the area of a smaller square.
And $(x + 3)^2 = $ the area of a larger square.

50 smaller squares are equal in area to 8 of the larger squares.

$$50x^2 = 8(x + 3)^2 \qquad$$
$$50x^2 = 8(x^2 + 6x + 9) \qquad \text{Square the binomial.}$$
$$50x^2 = 8x^2 + 48x + 72 \qquad \text{Multiply.}$$
$$0 = -42x^2 + 48x + 72 \qquad \text{Combine like terms.}$$
$$0 = 7x^2 - 8x - 12 \qquad \text{Simplify (divide by } -6\text{).}$$
$$0 = (7x + 6)(x - 2) \qquad \text{Factor.}$$
$$7x + 6 = 0 \quad | \quad x - 2 = 0 \qquad \text{Solve for } x.$$
$$x = -\tfrac{6}{7} \quad | \quad x = 2 \qquad \text{Reject the negative value.}$$

Check: Mr. Sloan Mr. Stark
$$50 \cdot 2^2 = 50 \cdot 4 = 200 \text{ sq. ft.} \qquad 8 \cdot 5^2 = 8 \cdot 25 = 200 \text{ sq. ft.} \checkmark$$

ANSWER: Mr. Sloan's slates were 2 feet square.

EXERCISES

In 1–60, find the solution set and check.

1. $x^2 - 3x + 2 = 0$

2. $y^2 - 7y + 6 = 0$

3. $z^2 - 5z + 4 = 0$

4. $x^2 - 8x + 16 = 0$

5. $p^2 - 8p + 12 = 0$

6. $r^2 - 12r + 35 = 0$

7. $c^2 + 6c + 5 = 0$

8. $m^2 + 8m + 7 = 0$

9. $m^2 + 10m + 9 = 0$

10. $x^2 + 2x + 1 = 0$

11. $x^2 + 8x + 15 = 0$

12. $y^2 + 11y + 24 = 0$

13. $x^2 - 4x - 5 = 0$

14. $x^2 - 5x - 6 = 0$

15. $x^2 - 2x - 35 = 0$

16. $x^2 + x - 6 = 0$

17. $q^2 + q - 72 = 0$

18. $x^2 + 2x - 15 = 0$

19. $r^2 - r - 72 = 0$

20. $x^2 - x - 12 = 0$

21. $y^2 - 3y - 10 = 0$

22. $x^2 - 49 = 0$

23. $y^2 - 81 = 0$

24. $z^2 - 4 = 0$

25. $m^2 - 64 = 0$

26. $2r^2 - 18 = 0$

27. $3x^2 - 12 = 0$

28. $d^2 - 2d = 0$

29. $m^2 - 5m = 0$

30. $s^2 - s = 0$

31. $x^2 + 3x = 0$

32. $y^2 + 7y = 0$

33. $z^2 + 8z = 0$

34. $2x^2 - 5x + 2 = 0$

35. $3x^2 - 10x + 3 = 0$

36. $2x^2 + x - 3 = 0$

37. $3x^2 - 8x + 4 = 0$

38. $2x^2 + x - 10 = 0$

39. $5x^2 + 11x + 2 = 0$

40. $x^2 - x = 6$

41. $y^2 - 4y = 12$

42. $y^2 - 3y = 28$

43. $c^2 - 8c = -15$

44. $d^2 + 5d = 14$

45. $2m^2 + 7m = -6$

46. $r^2 = 4$

47. $x^2 = 25$

48. $3x^2 = 12$

49. $y^2 = 6y$

50. $t^2 = 8t$

51. $s^2 = -4s$

52. $y^2 = 8y + 20$

53. $z^2 = 15 - 2z$

54. $2x^2 - x = 15$

55. $x^2 = 9x - 20$

56. $30 + x = x^2$

57. $2y^2 = 7y - 3$

58. $x^2 + 3x - 4 = 50$

59. $x^2 - 8x + 28 = 3x$

60. $2x^2 + 7 = 5 - 5x$

In 61–67, solve for x.

61. $x^2 - b^2 = 0$

62. $4x^2 = c^2$

63. $x^2 + ax = 0$

64. $x^2 = cx$

65. $rx^2 = sx$

66. $x^2 - 5bx + 6b^2 = 0$

67. $x^2 + 4ax = 21a^2$

In 68–88, solve algebraically.

68. When a certain number is added to its square, the result is 30. Find the number.

69. The square of a number exceeds the number by 72. Find the number.

70. In the balcony of a theater, there are 240 seats. The number of seats in each row is 14 more than the number of rows. Find the number of rows.

71. The larger of two positive numbers is 5 more than the smaller. The product of the numbers is 36. Find the numbers.

72. The product of two consecutive integers is 56. Find the integers.

73. The product of two consecutive odd integers is 99. Find the integers.

74. At Langley's Labrador Retriever Kennels during a recent period, the average number of pups in a litter was 5 less than the number of litters. If there were 84 pups altogether, what was the average number in a litter?

Intelligence, steadiness, and gentleness make the Labrador retriever the most favored breed of guide dog today. A large part of the pup's first year is spent being socialized in the home of a "puppy walker." Then follows a 3- to 6-month training period to learn to respond to commands and to ignore distractions. Finally, after the Lab and the new owner train together for about a month, the dog makes it possible for an unsighted person to enjoy independence.

Labrador retrievers are also used in police work, such as bomb, drug, and fire detection.

75. The sum of the squares of two positive consecutive even integers is 100. Find the integers.

76. Find three consecutive odd integers such that the square of the first increased by the product of the other two is 224.

77. If the length of one side of a square is increased by 2 inches and the length of an adjacent side is decreased by 2 inches, the area of the resulting rectangle is 32 square inches. Find the length of one side of the square.

78. The number of square units in the area of a square is 21 more than the number of units in its perimeter. Find the length of a side of the square.

79. The length of a rectangle is three times its width. If the width is diminished by 1 inch and the length is increased by 3 inches, the area of the new rectangle is 72 square inches. Find the dimensions of the original rectangle.

80. A rectangle is 6 cm long and 4 cm wide. If each dimension is increased by the same number of centimeters, a new rectangle is formed whose area is 39 square centimeters more than the area of the original rectangle. By how many centimeters was each dimension increased?

81. Seth's garden is 6 meters long and 4 meters wide. He wishes to double the area of his garden by increasing its length and width by the same amount. Find the number of meters by which each dimension must be increased.

82. The square of Clara's age 2 years from now is equal to 20 times her age 3 years ago. Find Clara's present age.

83. A picture 6 in. by 12 in. is surrounded by a frame of uniform width. If the area of the frame is twice the area of the picture, find the width of the frame.

84. On the Model RV-Q trailer built by a manufacturer of recreational vehicles, the height is 1 foot less than the width, and the length is 3 feet more than the width. A new model, with the same height, will be 1 foot wider and 1 foot longer, with a volume 140 cubic feet greater than that of the RV-Q. What is the width of Model RV-Q?

85. At the end of a bicycle trip, Derek found that his average rate, in miles per hour, was equal to the number of hours he had traveled. He calculated that by increasing his rate by 4 miles per hour and his time by 6 hours, he could have covered twice the distance. What was his average rate?

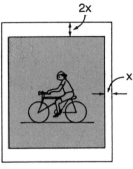

86. An 8″ × 10″ photograph is surrounded by a frame made of 4 strips of wood. The top and bottom strips are twice the width of the side strips. If the total area of the framed picture is 176 square inches, what is the width of a side strip?

Ex. 86

In 87 and 88, the formula $h = k + rt - 16t^2$ gives the altitude h of an object, t seconds after it is thrown upward with an initial velocity r from a height of k feet above the ground.

87. A baseball is thrown upward from the roof of a 96-foot building with an initial velocity of 80 feet per second. In how many seconds will the ball pass the top of the building on the way down?

88. I shot an arrow into the air;
It fell to earth, I knew not where.
If I shot from 5 feet high,
Directly up into the sky
An arrow at 79 feet per second,
Tell me when you would have reckoned
The arrow could again be found:
How long until it hit the ground?

Algebraic operations are in some ways similar to those of arithmetic. In arithmetic, you started with one-digit numbers. In algebra, you started with **monomials**, single-term expressions such as $5x^2$ and $3ab$. In arithmetic, you progressed to numbers with two or more digits. In algebra, two-term expressions such as $x - 7$ and $5a + 2b$ are called **binomials**; expressions with any number of terms are **polynomials**. The largest exponent in a polynomial gives its **degree**.

When you add numbers, you combine digits with the same place value. In algebra, you look for like terms: $7x^2y^3$ is like $-4x^2y^3$, but is unlike $7x^2$ or $7y^3$ or $7xy$. Rules of operations with algebraic expressions include:

$$x^a \cdot x^b = x^{a+b} \qquad (x^a)^c = x^{a \cdot c} \qquad x^a \div x^b = x^{a-b} \qquad x^0 = 1 \qquad x^{-a} = \frac{1}{x^a}$$

The **scientific notation** $a \times 10^n$, where $1 < a < 10$, is a convenient form for writing numbers like 8,300,000,000 or .00000025.

Factoring, rewriting a polynomial as a product, is used in the following techniques:

1. *Finding a common factor* when the same factor appears in all the terms of a polynomial.
2. *Factoring by grouping* with polynomials of four terms when there is no common factor for all terms, but there is a common factor for *pairs* of terms.
3. *Factoring the difference of two squares* if a polynomial is a difference of two perfect squares. Only the difference of two squares can be factored, not the sum.
4. *Factoring the general trinomial $ax^2 + bx + c$* into the product of two binomials. Apply the strategies of (1) guess and check and (2) make a list.

Points to remember regarding factoring: (1) Factoring an expression does not change its value, only its appearance. (2) Some polynomials need more than one kind of factoring in order to be completely factored. Always look for a common factor first. (3) Some polynomials cannot be factored.

An application of factoring is in solving a **quadratic equation** $ax^2 + bx + c = 0$, where $a \neq 0$. If the polynomial can be factored into first-degree factors, solve by setting each factor equal to 0. This makes use of the rule that if a product $a \cdot b = 0$, then $a = 0$ or $b = 0$ or both a and b are 0.

VOCABULARY CHECKUP

SECTION		SECTION	
6-1	*polynomial / monomial / binomial / trinomial / degree of a polynomial*	6-6	*factor / prime number / greatest common factor / difference of two squares*
6-3	*scientific notation*	6-8	*quadratic equation / root*

1. Add and check. (Section 6-1)

   a. $-5m - 3n$ **b.** $8a - 5b - c$
       $4m + 8n$ $6a + b + 3c$
       $\underline{6m - 2n}$ $\underline{-7a + 3b - c}$

2. Subtract and check. (Section 6-1)

   a. $-5m - 3n$ **b.** $8a - 5b - c$
     $\underline{-4m + 8n}$ $\underline{6a + b - 3c}$

In 3–8, perform the indicated operations and simplify. (Section 6-1)

3. $(3a - 5b) + (2b - 7a) + (4a - b)$

4. $(9x^2 - 3y - 2z) + (5y - 4z - 7x^2)$

5. $(3a - 5b) - (2b - 7a)$

6. $(9x^2 - 3y - 2z) - (5y - 4z - 7x^2)$

7. $3a + [4 + (7a - 2) - a]$

8. $2y - [y - (5z - 3y)] - z$

In 9–16, apply a property of exponents. (Section 6-2)

9. $a^3 \cdot a^4$

10. $b^a \cdot b^3$

11. $3^2 \cdot 3^4$

12. $(3y^5)^3$

13. $x \div x^5$

14. $m^7 \div m^4$

15. $4^7 \div 4^3$

16. $10^4 \div 10$

In 17–20, rewrite each expression without using negative exponents. (Section 6-2)

17. $\dfrac{a^{-5}}{b^{-3}}$

18. $\dfrac{m^4}{m^{-7}}$

19. x^{-4}

20. $(a^0 b^5)^{-5}$

In 21–24, write in scientific notation. (Section 6-3)

21. $4{,}375{,}000$

22. $.0064$

23. 289×10^2

24. $1{,}200 \times 10^{-5}$

In 25–28, perform the indicated operation. Express the answer in **(a)** scientific notation and **(b)** standard decimal notation. (Section 6-3)

25. $(2 \times 10^3)(8 \times 10^2)$

26. $(5.3 \times 10^{-2})(1.4 \times 10)$

27. $\dfrac{8.4 \times 10^3}{2.1 \times 10^{-1}}$

28. $\dfrac{1.125 \times 10^{-1}}{3.75 \times 10^{-4}}$

In 29–43, multiply. (Section 6-4)

29. $(-3a^4 b^3)(-5a^7 b)(2b^2)$

30. $(5ab^4)^3$

31. $-4(3a - 2)$

32. $-2a^2(5a^3 - 3a)$

33. $-4x^2 y^3(7x^4 y^3 + 9xy^4)$

34. $3ab^2 c^3(a^2 b^3 c - 2abc)$

35. $(a - 3)(a + 7)$

36. $(2b - 5)(3b - 4)$

37. $(a - b)(a + b)$

38. $(3x + 4)(3x - 4)$

39. $(5a - 11)(5a + 11)$

40. $(6ab + 5c^2)(6ab - 5c^2)$

41. $(x + 2)(x - 2)(x^2 + 4)$

42. $(2x - 1)(3x + 2)(3x - 4)$

43. $(6y^2 - 5y - 3)(2y - 5)$

In 44 and 45, simplify. (Section 6-4)

44. $(3a - 4)(2a + 1) + 5a$

45. $(t - 5)(t - 4) - (t - 3)(t - 2)$

In 46–48, solve algebraically. (Section 6-4)

46. The length of a rectangle is one inch more than its width. If the length is decreased by 3 inches and the width is increased by 4 inches, the area is unchanged. Find the dimensions of the original rectangle.

47. The area of a square exceeds the area of a rectangle by 2 square meters. The length of the rectangle is 3 meters more than a side of the square. The width of the rectangle is 2 meters less than a side of the square. Find the dimensions of the rectangle.

48. The width of a box is 12 cm, and its height is 3 cm more than the width. If the volume is 1,440 cubic cm, what is the depth of the box?

In 49–54, divide. (Section 6-5)

49. $\dfrac{15c^5 y^2}{18c^3 y^7}$

50. $\dfrac{-8a^2 w^3 x}{-2a^2 w}$

51. $\dfrac{-8x^2 - 20x}{-4x}$

52. $\dfrac{6a^2 b^7 - 15a^4 b^5}{3ab^5}$

53. $\dfrac{16a^5 - 40a^4 + 12a^3}{4a^3}$

54. $\dfrac{x^7 y^5 + x^4 y^4 - 2x^3 y^6}{x^3 y^2}$

In 55–58, divide and check. (Section 6-5)

55. $w^2 - 5w - 6$ by $w + 1$

56. $27a^3 - 8$ by $3a - 2$

57. $3a - 4 + 2a^2$ by $2a - 1$

58. $5x^2 - 3xy - 2y^2$ by $x - y$

59. Express as a product of primes. (Section 6-6)
  a. 12   **b.** 100   **c.** 45

60. Find the greatest common factor of: (Section 6-6)
  a. 40 and 72   **b.** $12x^2$ and $8xy$

In 61–69, factor. (Section 6-6)

61. $8x - 16$

62. $3ab + 10ax$

63. $8a^4 b^5 - 20a^6 b^3$

64. $ab + ac + b^2 + bc$

65. $a^3 - 2a^2 + 3a - 6$

66. $x^4 + ax^3 - a^2 x - a^3$

67. $4a^2 - 121$

68. $16 - 81x^2$

69. $c^2 d^2 - h^2$

In 70–78, factor. (Section 6-7)

70. $x^2 - 8x + 15$

71. $y^2 + 8y - 20$

72. $a^2 - 17a + 30$

73. $m^2 - m - 30$

74. $r^2 - 12r + 36$

75. $x^2 + 15x + 36$

76. $2n^2 - 5n - 3$

77. $3c^2 + 13c + 12$

78. $7x^2 - 2x - 9$

In 79–81, factor completely. (Section 6-7)

79. $6a^3 - 2a^2 - 8a$

80. $4c^3 - 100c$

81. $32 - 2w^4$

In 82–87, solve by factoring. (Section 6-8)

82. $x^2 - 9x + 20 = 0$

83. $y^2 - 100 = 0$

84. $5x^2 - 3x = 0$

85. $a^2 = 3a + 4$

86. $9n^2 = 6n - 1$

87. $4x^2 - 3x = 7$

In 88–90, solve algebraically. (Section 6-8)

88. When the square of a number is increased by 3 times the number, the result is 28. Find the number.

89. The length of a rectangular garden is 4 yards more than its width. The area of the garden is 60 square yards. Find the dimensions of the garden.

90. A rectangle is 8 ft. long and 6 ft. wide. If each dimension is increased by the same number of feet, a new rectangle is formed whose area exceeds the area of the original rectangle by 72 sq. ft. Find the dimensions of the new rectangle.

PROBLEMS FOR PLEASURE

1. Simplify the expression as much as possible.

  a. $(x + y)(x - y)(x^2 - xy + y^2)(x^2 + xy + y^2) - (x^2 + y^2)(x^4 - x^2y^2 + y^4)$

  b. $(a - x)(b - x)(c - x)(d - x) \ldots (z - x)$

2. Factor completely.

  a. $(a^2 - 17)^2 - 64$

  b. $(a^4 - 5)^2 - 16$

  c. $(13x^2 - 22.5)^2 - (12x^2 - 13.5)^2$

  d. $(20a^2)^2 - (29a^2 - 16)^2$

3. You have just arrived in town. In a week, you will receive your inheritance, but meanwhile, you have no money. Your only possession of value is a gold chain consisting of seven links, that has been in your family for generations. An innkeeper will let you have food and lodging in exchange for exactly one gold link per day (no advance payment). He has also agreed that you may buy back the chain at the end of seven days.

You want to make the fewest possible cuts in the chain. Tell which links (1 to 7) you would cut and explain how you would arrange the correct payment each day.

CALCULATOR CHALLENGE

1. What natural number less than 50 has the most natural numbers as factors?

2. What is the smallest natural number that has 7 different natural numbers less than 11 as factors?

3. Find the prime factors of 17,329.

4. Find all natural numbers that are factors of 5,043.

In 1–7, choose the correct answer.

1. If $x + y^2 = 6$ and $y = -2$, then $x^2 - y$ is
(A) 2   (B) -2 (C) 6
(D) -6 (E) 102

2. If $x + 3$ is an even integer, then which of the following is an odd integer?
(A) $(x + 3)(x + 4)$
(B) $(x + 3)^2$
(C) $(x + 4)(x + 5)$
(D) $(x + 4)^2$
(E) $(x + 3)(x + 5)$

3. The product of $4x$ and $5x$ gives the same result as the sum of $10x^2$ and
(A) $2x^2$ (B) $2x$ (C) $-x^2$
(D) $10x^2$ (E) $10x$

4. If x is between -1 and 2, and y is between 3 and 4, then x^2y is between
(A) 0 and 16   (B) 3 and 16
(C) 3 and 8   (D) 0 and 8
(E) -4 and 8

5. The expression $(x - (x - (x - (x - 1))))$ equals
(A) 1   (B) -1 (C) $-2x - 1$
(D) $2x + 1$ (E) $2x - 1$

6. If a person drives x^2y miles at a rate of xy^2 miles per hour, then the number of hours required is represented by
(A) $\dfrac{x}{y}$ (B) $\dfrac{x^2}{y^2}$ (C) $\dfrac{y}{x}$
(D) x^2y^2 (E) x^3y^3

7. If x is 10, then $2(x^3 + x)$ has the same value as
(A) 2.2×10^2
(B) 2.02×10^2
(C) 2.2×10^3
(D) 2.02×10^3
(E) 2.02×10^{-3}

Questions 8–21 each consist of two quantities, one in Column A and one in Column B. You are to compare the two quantities and choose:

A   if the quantity in Column A is greater;
B   if the quantity in Column B is greater;
C   if the two quantities are equal;
D   if the relationship cannot be determined from the information given.

1. In certain questions, information concerning one or both of the quantities to be compared is centered above the two columns.

2. In a given question, a symbol that appears in both columns represents the same thing in Column A as it does in Column B.

3. x, n, k, etc. stand for real numbers.

Column A	Column B

$x > 0$

8. $(x + 3)(x - 4)$ $(x - 3)(x + 4)$

$y = x + 1$

9. $(y + 3)^3$ $(x + 4)^3$

$k \neq 0$

10. $\dfrac{9k^3 + 6k^2}{3k^2}$ $\dfrac{12k^2 + 12k}{4k}$

$k \neq 0$

11. $\dfrac{12k + 6k^2}{6k}$ $\dfrac{16k - 8k^2}{8k}$

---

x is negative.

12. $(-3x^3)(-2x^2)$ $12x^6 \div 2x^2$

$y < 6$

13. $6y - (y - 6)$ $6y - (6 - y)$

---

14. $(a - b)^2$ $(b - a)^2$

---

	x is an integer.			$x > y > 1$	

15. $x(x + 2)$ $\qquad$ $(x + 1)(x - 1)$

19. $(x - y)^2$ $\qquad$ $x^2 - y^2$

	$y = x + 2$

16. $y^2 - 8y + 15$ $\qquad$ $x^2 - 4x + 3$

$(x + 1)(x + 2) = 0$

20. $x(x + 1)$ $\qquad$ $x(x + 2)$

---

17. $x^2 + y^2$ $\qquad$ $(x + y)(x + y)$

$x^2 + x + k$

21. The value of k if one factor of the polynomial is $x + 5$ $\qquad$ The value of k if one factor of the polynomial is $x + 4$

	$x > 0$

18. $(x + 3)(x - 4)$ $\qquad$ $(x - 3)(x + 4)$

SPIRAL REVIEW EXERCISES

1. Evaluate $(3x)^2 - 3x^2$ when $x = 4$.

2. Give the name of the property illustrated by $(3)(-8) = (-8)(3)$.

3. Which has the least value? (1) -0.167 (2) -0.0176 (3) -0.176 (4) -0.00167

4. Find the sum of the measures of the acute angles in a right triangle.

5. If d jackets cost $42j$ dollars, represent the cost of m jackets.

6. If $A = \{6, 8, 10, \ldots, 22\}$ and $B = \{2, 5, 8, \ldots, 23\}$, find:
   a. $A \cap B$ $\qquad$ **b.** $A \cup B$

7. Let n represent a whole number less than 30. If n is divided by 7, the remainder is 4. If n is divided by 4, the remainder is 1. Find n.

8. Let $a * b = \dfrac{(a + b)}{2}$. Then the value $5 * 7$ is *not* equal to:

   (1) $7 * 5$ $\qquad$ (2) $(6 * 6) * 6$ $\qquad$ (3) $\dfrac{(10 * 14)}{2}$ $\qquad$ (4) $2 * (5 * 7)$

9. If the perimeter of a rectangle with sides of lengths 5 and x is equal to the perimeter of an equilateral triangle with sides of length 6, what is the value of x?

10. Find the measure of the supplement of a $144°$ angle.

11. If the larger square in the diagram has sides of length $2s$, and the smaller square has sides of length s, the perimeter of the shaded region can be represented by:
   (1) $5s$ $\qquad$ (2) $6s$ $\qquad$ (3) $7s$ $\qquad$ (4) $8s$

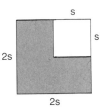

12. Is the set $\{1, 3, 5, 7\}$ closed under addition? Explain.

13. Solve the formula $K = \frac{1}{2}ap$ for the variable a.

14. Solve for x: $9(2 - x) > 2x - 59$

15. One number is 8 less than twice the other, and their sum is 124. Find the numbers.

16. Jane scored 14, 20, 19, 17, and 10 points, respectively, in her first 5 basketball games. How many points must she score on her next game to average 16 points per game?

17. A cube has edges 3 inches long. By how much does the surface area of the cube exceed the area of a right triangle whose legs measure 6 inches and 8 inches?

18. A and n are related by a growth pattern shown in the table. Find x and y.

A	1	5	25	x	625
n	1	4	y	64	256

19. Mr. and Mrs. North make the round trip between Eatonville and Quisley several times each year. The bus fare is $26 round trip per person. If they travel by car, the interstate bridge toll of $4 is paid in one direction only, the thruway toll is $5.60 each way, a city bridge toll is $2.50 each way, and gasoline costs $25. How much do they save by making k round trips by car?

20. A two-part exam was to have been graded as shown in the table. Instead, the test preparer decided to give each question in both parts the same number of points, keeping the total number of points the same. Find the number of points given for each question under the revised grading.

Exam Section	Number of Questions	Points per Question
Part I	30	1.4
Part II	36	$1.\overline{3}$

21. To use a 7-foot board as a seesaw, two friends want to know where to place the fulcrum. If Caia weighs 120 pounds and Meghan weighs 90 pounds, how far from Caia should the fulcrum be placed so that they balance?

22. Jackie walked up a trail at the rate of 5 km per hour and jogged back on the same trail at the rate of 8 km per hour. If she traveled a total of 20 km, how many hours did it take her to complete the entire trip?

23. The Environmental Food Mart offers a 5-cent refund per sack when shoppers bring their own carry-home sacks. Judy bought two nylon tote sacks for $2.39 each (including tax) that she uses on each weekly marketing trip. In how many weeks will the amount of the refund exceed the amount she paid for the sacks?

24. A designer created a repeating 6-step pattern in which each of the steps is different. The first four steps are shown. Step 104 could look like:

(1) Step 1   (2) Step 2
(3) Step 3   (4) Step 4

Step 1 △ ▽ ▢

Step 2 △ ▢ ▽

Step 3 ▢ △ ▽

Step 4 ▢ ▽ △

6-1 ADDING AND SUBTRACTING POLYNOMIALS

TRY IT! *Problems 1 and 2 on page 261*

1. $\quad 4a^2 - 5a - 3 \quad$ minuend
    $\quad \underline{-7a^2 + 3a - 4} \quad$ subtrahend
    $\quad 11a^2 - 8a + 1 \quad$ difference

    Check: *Add*

    $\quad -7a^2 + 3a - 4 \quad$ subtrahend
    $\quad \underline{11a^2 - 8a + 1} \quad$ difference
    $\quad 4a^2 - 5a - 3 \quad$ minuend

    ANSWER: $11a^2 - 8a + 1$

2. $3w - [4 - (3 - w)] = 3w - [4 + (-3 + w)]$
    $\qquad\qquad\qquad\quad = 3w - [4 - 3 + w]$
    $\qquad\qquad\qquad\quad = 3w - [1 + w]$
    $\qquad\qquad\qquad\quad = 3w + [-1 - w]$
    $\qquad\qquad\qquad\quad = 3w - 1 - w$
    $\qquad\qquad\qquad\quad = 2w - 1$

    ANSWER: $2w - 1$

6-3 SCIENTIFIC NOTATION

TRY IT! *Problems 1–4 on page 272*

1. **a.** $3,560,000 = 3.56 \times 10^6$
    b. $.00034 = 3.4 \times 10^{-4}$
    c. $856 \times 10^{-4} = 8.56 \times 10^2 \times 10^{-4}$
    $\qquad\qquad\qquad = 8.56 \times 10^{-2}$

2. **a.** $5.4 \times 10^3 = 5,400$
    b. $4 \times 10^{-7} = .0000004$

3. $\quad (8 \times 10^{-6})(9.32 \times 10^3)$
    $= (8 \times 9.32)(10^{-6} \times 10^3)$
    $= (74.56)(10^{-3})$
    $= 7.456 \times 10^{-2}$

4. $\dfrac{6.8 \times 10^7}{1.7 \times 10^2} = \dfrac{6.8}{1.7} \times \dfrac{10^7}{10^2} = 4 \times 10^5$

6-4 MULTIPLYING POLYNOMIALS

TRY IT! *Problems 1–3 on page 278*

1. *Solution*
    $\quad 2a - 7$
    $\quad \underline{3a - 2}$
    $\quad 6a^2 - 21a$
    $\quad \underline{\quad\;\; - 4a + 14}$
    $\quad 6a^2 - 25a + 14 \quad$ **ANSWER**

2. $3a^2 - 2ab - 5b^2$
    $\underline{2a - b}$
    $6a^3 - 4a^2b - 10ab^2$
    $\underline{\quad\; - 3a^2b + 2ab^2 + 5b^3}$
    $6a^3 - 7a^2b - 8ab^2 + 5b^3 \quad$ **ANSWER**

3.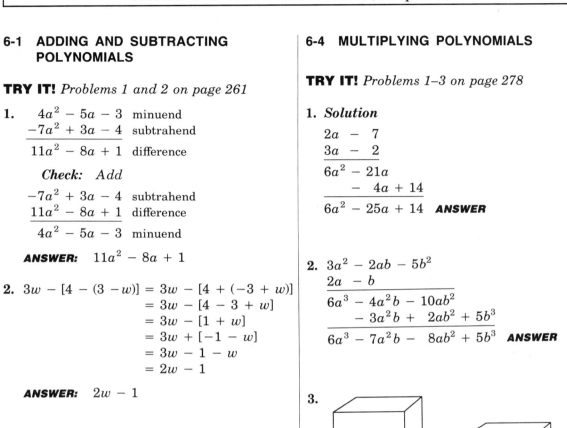

    The volume of the Nuttee O's box is 640 cu. cm more than the volume of the Fruitee O's box.

    $$8w(w + 5) = 8w(w + 1) + 640$$
    $$8w^2 + 40w = 8w^2 + 8w + 640$$
    $$32w = 640$$
    $$w = 20$$

    ANSWER: The width of the boxes is 20 cm.

6-5 DIVIDING POLYNOMIALS

TRY IT! *Problems 1 and 2 on page 288*

1. Arrange terms in descending powers of a.

$$
\begin{array}{r}
3a + 10 \\
a - 2 \overline{\smash{)}\ 3a^2 + 4a - 5} \\
\underline{3a^2 - 6a} \\
10a - 5 \\
\underline{10a - 20} \\
15
\end{array}
$$

Check

$$
\begin{array}{ll}
a - 2 & \text{divisor} \\
\underline{3a + 10} & \text{quotient} \\
3a^2 - 6a & \\
\underline{10a - 20} & \\
3a^2 + 4a - 20 & \\
\underline{+ 15} & \text{remainder} \\
3a^2 + 4a - 5 & \text{dividend}
\end{array}
$$

ANSWER: $3a + 10 + \dfrac{15}{a - 2}$

2. Use zeros for the missing terms.

$$
\begin{array}{r}
4x^2 + 6x + 9 \\
2x - 3 \overline{\smash{)}\ 8x^3 + 0x^2 + 0x - 27} \\
\underline{8x^3 - 12x^2} \\
12x^2 + 0x \\
\underline{12x^2 - 18x} \\
18x - 27 \\
\underline{18x - 27} \\
0
\end{array}
$$

ANSWER: $4x^2 + 6x + 9$

6-7 MORE ABOUT FACTORING

TRY IT! *Problems 1–3 on page 299*

1. The possible factors of $a^2 - 5a + 6$ are $(a - 1)(a - 6)$ and $(a - 2)(a - 3)$.

$$
\overset{-a}{\overbrace{(a - 1)(a - 6)}}
$$
$$
\underbrace{}_{-6a}
$$

$(-a) + (-6a) = -7a$ no

$$
\overset{-2a}{\overbrace{(a - 2)(a - 3)}}
$$
$$
\underbrace{}_{-3a}
$$

$(-2a) + (-3a) = -5a$ yes

ANSWER: $a^2 - 5a + 6 = (a - 2)(a - 3)$

2. The possible factors of $x^2 - 9x - 10$ are:

$(x + 1)(x - 10)$ $\qquad$ $(x + 2)(x - 5)$
$(x - 1)(x + 10)$ $\qquad$ $(x - 2)(x + 5)$

Only $(x + 1)(x - 10)$ yields a middle term of $-9x$.

ANSWER: $x^2 - 9x - 10 = (x + 1)(x - 10)$

3. The possible factors of $c^2 + 3c - 18$ are:

$(c + 1)(c - 18)$ $\qquad$ $(c - 1)(c + 18)$
$(c + 2)(c - 9)$ $\qquad$ $(c - 2)(c + 9)$
$(c + 3)(c - 6)$ $\qquad$ $(c - 3)(c + 6)$

Only $(c - 3)(c + 6)$ yields a middle term of $+3c$.

ANSWER: $c^2 + 3c - 18 = (c - 3)(c + 6)$

TRY IT! *Problems 1 and 2 on page 300*

1. The possible factors of $5x^2 + 3x - 8$ are:

$(5x + 1)(x - 8)$ $\qquad$ $(5x - 1)(x + 8)$
$(5x + 8)(x - 1)$ $\qquad$ $(5x - 8)(x + 1)$
$(5x + 2)(x - 4)$ $\qquad$ $(5x - 2)(x + 4)$
$(5x + 4)(x - 2)$ $\qquad$ $(5x - 4)(x + 2)$

Only $(5x + 8)(x - 1)$ yields a middle term of $+3x$.

ANSWER: $5x^2 + 3x - 8 = (5x + 8)(x - 1)$

2. The possible factors of $6x^2 - 11x - 10$ are:

$$(6x + 1)(x - 10)$$
$$(6x - 1)(x + 10)$$
$$(6x + 10)(x - 1)$$
$$(6x - 10)(x + 1)$$
$$(6x + 2)(x - 5)$$
$$(6x - 2)(x + 5)$$
$$(6x + 5)(x - 2)$$
$$(6x - 5)(x + 2)$$

$$(3x + 1)(2x - 10)$$
$$(3x - 1)(2x + 10)$$
$$(3x + 10)(2x - 1)$$
$$(3x - 10)(2x + 1)$$
$$(3x + 2)(2x - 5)$$
$$(3x - 2)(2x + 5)$$
$$(3x + 5)(2x - 2)$$
$$(3x - 5)(2x + 2)$$

Only $(3x + 2)(2x - 5)$ yields a middle term of $-11x$.

ANSWER:
$$6x^2 - 11x - 10 = (3x + 2)(2x - 5)$$

TRY IT! *Problems 1–3 on page 302*

1. $6a^2 - 54b^2 = 6(a^2 - 9b^2)$
$$= 6(a + 3b)(a - 3b) \quad \textbf{\textit{ANSWER}}$$

2. $\quad 5x^4 - 80$
$$= 5(x^4 - 16)$$
$$= 5(x^2 + 4)(x^2 - 4)$$
$$= 5(x^2 + 4)(x + 2)(x - 2) \quad \textbf{\textit{ANSWER}}$$

3. $\quad 3a^5 - 3a^3 - 36a$
$$= 3a(a^4 - a^2 - 12)$$
$$= 3a(a^2 + 3)(a^2 - 4)$$
$$= 3a(a^2 + 3)(a + 2)(a - 2) \quad \textbf{\textit{ANSWER}}$$

6-8 SOLVING AN EQUATION BY FACTORING

TRY IT! *Problems on page 307*

1.
$$x^2 + 2x = 3$$
$$x^2 + 2x - 3 = 0$$
$$(x + 3)(x - 1) = 0$$
$$x + 3 = 0 \quad | \quad x - 1 = 0$$
$$x = -3 \quad | \quad x = 1$$

Check by substituting both values in the original equation.

ANSWER: $x = -3$ or $x = 1$

2.
$$3x^2 = 6x$$
$$3x^2 - 6x = 0$$
$$3x(x - 2) = 0$$
$$3x = 0 \quad | \quad x - 2 = 0$$
$$x = 0 \quad | \quad x = 2$$

Check by substituting both values in the original equation.

ANSWER: $x = 0$ or $x = 2$

3. $\qquad$ Let x = the first even integer.
Then $x + 2$ = the second even integer.

The product of two consecutive even integers is 80.

$$x(x + 2) = 80$$
$$x^2 + 2x = 80$$
$$x^2 + 2x - 80 = 0$$
$$(x - 8)(x + 10) = 0$$

$$x - 8 = 0 \quad | \quad x + 10 = 0$$
$$x = 8 \quad | \quad x = -10$$
$$x + 2 = 10 \quad | \quad x + 2 = -8$$

Check

$8 \times 10 = 80$; also $-10 \times -8 = 80$

ANSWER: The integers are 8 and 10, or -10 and -8.

Part I

1. If $I = \dfrac{E}{R}$, find R when $I = 5$ and $E = 15$.

2. From $6x^2 - 8x + 6$, subtract $x^2 - 3x - 6$.

3. If three times a certain number is increased by 4, the result is 19. What is the number?

4. Simplify: $|-3| + |7| - |-7|$

5. If $n + 1$ represents an odd integer, express the next larger odd integer in terms of n.

6. Solve for x: $9x + 12 = x - 4$

7. Evaluate $a(b^2 - 5)$ when $a = 4$ and $b = -2$.

8. Find the solution set:

   $w^2 + 4w - 21 = 0$

In 9–20, choose the correct answer.

9. A telephone call costs c cents for the first 3 minutes and m cents for each additional minute. What is the cost, in cents, of a 6-minute call?
   (1) $c + m$ (2) $2c$
   (3) $c + 3m$ (4) $c + 6m$

10. A scout troop left on a bike trip, averaging 8 mph along the way. If one member of the troop started out a half-hour late, averaging 10 mph, how long did it take him to catch up?

    Which equation can be used to solve the problem?
    (1) $10x = 8\left(x + \frac{1}{2}\right)$ (2) $8x = 10\left(x - \frac{1}{2}\right)$
    (3) $10x - 8x = \frac{1}{2}$ (4) $\frac{1}{2}(10x) = 8x$

11. The expression $(y - 2)^2$ is equivalent to
    (1) $y^2 - 4$ (2) $y^2 + 4y + 4$
    (3) $y^2 - 4y + 4$ (4) $y^2 - 4y - 4$

12. The sum of two numbers is 16. If the smaller number is represented by x, then the larger number can be represented by
    (1) $16 - x$ (2) $x - 16$
    (3) $16 + x$ (4) $16x$

13. Which is an illustration of the associative property?
    (1) $a + (b + c) = (a + b) + c$
    (2) $a(b + c) = ab + ac$
    (3) $a + b = b + a$
    (4) $ab + 0 = ab$

14. The expression $7 + 2 - \frac{4}{2}(6 - 4)$ is equal to
    (1) 5 (2) 7 (3) 14 (4) -14

15. The product of $3x^2$ and $6x^4$ is
    (1) $2x^2$ (2) $9x^6$ (3) $18x^6$ (4) $18x^8$

16. If $ab = 0$ and $a > 0$, then
(1) $b < 0$ (2) $b > 0$
(3) $b = 0$ (4) $b = a$

17. Which of the following products is equal to $2x^2 + 5x - 18$?
(1) $(2x - 2)(x + 9)$
(2) $(2x - 9)(x + 9)$
(3) $(x - 9)(2x + 2)$
(4) $(x - 2)(2x + 9)$

18. If the replacement set for x is $\{-1, 0, 1, 2, 3\}$, what is the solution set for the inequality $3x + 5 < 8$?
(1) $\{-1\}$ (2) $\{-1, 0\}$
(3) $\{\ \ \}$ (4) $\{0\}$

19. When $16x^3 - 8x^2 + 4x$ is divided by $4x$, the quotient is
(1) $4x^3 - 2x^2 + x$
(2) $4x^2 - 2x + 1$
(3) $4x^2 - 2x$
(4) $4x^2 - 2x + x$

20. Which solution set is represented by the graph below?

(1) $\{x \mid -3 < x < 2\}$
(2) $\{x \mid -3 \le x < 2\}$
(3) $\{x \mid -3 < x \le 2\}$
(4) $\{x \mid -3 \le x \le 2\}$

Part II

21. Solve algebraically.
Find three consecutive positive integers such that the product of the first and third integers is 15.

22. Use your knowledge of the properties of real numbers to answer the questions.
a. What number is the multiplicative identity element?
b. What is the sum of -2 and its additive inverse?
c. If $\dfrac{a}{b} = 1$, what is the value of $a - b$?
d. What positive number is equal to its reciprocal?
e. If $a + b = c$, and $a = c$, what is the numerical value of b?

23. Translate the phrase into algebraic symbols.
a. 3 more than twice x
b. 4 less than x
c. the product of x and y divided by 3
d. the square of the sum of x and y
e. the sum of the squares of x and y

24. Find the solution set if the replacement set for x is $\{-3, -2, -1, 0, 1, 2\}$.
a. $5(x - 1) = 10$
b. $|x| = 2$
c. $-4x > 4$
d. $2x^2 = 2$
e. $x^2 = x$

CHAPTER 7

RATIONAL EXPRESSIONS AND OPEN SENTENCES

Eighth notes   Quarter notes   Half note

Like all music, American melodies—as in country, jazz, and rock music—have fractional note values.

African tribal cultures have produced a rich array of rhythmic beats and harmonies.

7-1 Simplifying a Rational Expression . . **323**

7-2 Multiplying and Dividing Rational Expressions **328**

7-3 Adding and Subtracting Rational Expressions **334**

7-4 Solving Open Sentences That Contain Fractional Coefficients **342**

7-5 Applying Open Sentences That Contain Fractional Coefficients **346**

7-6 Solving Equations That Contain Rational Expressions **358**

7-7 Applying Equations That Contain Rational Expressions **364**

Chapter Summary **374**

Chapter Review Exercises **375**

Problems for Pleasure **377**

Calculator Challenge **377**

College Test Preparation **378**

Spiral Review Exercises **380**

Solutions to TRY IT! Problems **381**

In this chapter, you will learn how to operate with algebraic fractions by applying properties of real numbers. Factoring plays a major role.

Introducing fractions into open sentences allows a variety of verbal situations that can be modeled algebraically; many will be new to you, and some are extensions of previous experiences.

7-1 SIMPLIFYING A RATIONAL EXPRESSION

The Meaning of a Rational Expression

A *fraction* is a symbol that indicates the quotient of two numbers. For example, the symbol $\frac{3}{4}$ indicates the quotient of the numbers 3 and 4.

An algebraic fraction, or *rational expression,* is a quotient of polynomials.

$$\text{Examples:} \quad \frac{x}{2}, \ \frac{2}{x}, \ \frac{a}{b}, \ \frac{4c}{3d}, \ \frac{x+5}{x-2}, \ \frac{x^2+4x+3}{x+1}$$

Since division by 0 is not possible, a rational expression is not defined for any values of its variable that lead to a denominator of 0. For example:

$$\frac{12}{x-9} \text{ is not defined when } x - 9 = 0, \text{ or } x = 9.$$

Simplifying a Rational Expression

A rational expression is simplified to lowest terms when its numerator and denominator have no common factor other than 1. There are different ways to simplify.

Consider reducing the fraction $\frac{5}{10}$ to $\frac{1}{2}$.

Begin by factoring the numerator and denominator: $\frac{5}{10} = \frac{1 \cdot 5}{2 \cdot 5}$.

Then, either apply the multiplication property of 1 or *cancel* common factors.

$$\frac{1 \cdot 5}{2 \cdot 5} = \frac{1}{2} \cdot \frac{5}{5} = \frac{1}{2} \cdot 1 = \frac{1}{2} \quad \text{or} \quad \frac{1 \cdot 5}{2 \cdot 5} = \frac{1 \cdot \overset{1}{\cancel{5}}}{2 \cdot \underset{1}{\cancel{5}}} = \frac{1}{2}$$

To reduce a rational expression to lowest terms:

1. Factor both the numerator and denominator.

2. Either apply the multiplication property of 1 or cancel common factors.

MODEL PROBLEMS

In 1–4, reduce to lowest terms.

1. $\dfrac{8x^3y^2}{12x^2y^4} = \dfrac{2x}{3y^2} \cdot \dfrac{4x^2y^2}{4x^2y^2}$

$= \dfrac{2x}{3y^2} \cdot 1$

$= \dfrac{2x}{3y^2}$ *ANSWER*

2. $\dfrac{4x - 4y}{x^2 - y^2} = \dfrac{4(x - y)}{(x + y)(x - y)}$

$= \dfrac{4(\overset{1}{\cancel{x - y}})}{(x + y)(\underset{1}{\cancel{x - y}})}$

$= \dfrac{4}{x + y}$ *ANSWER*

3. $\dfrac{(x - 4)^2}{x^2 - 5x + 4} = \dfrac{(x - 4)(x - 4)}{(x - 1)(x - 4)}$

$= \dfrac{(x - 4)}{(x - 1)} \cdot 1$

$= \dfrac{x - 4}{x - 1}$ *ANSWER*

4. $\dfrac{2 - x}{4x - 8} = \dfrac{2 - x}{4(x - 2)}$

$= \dfrac{-x + 2}{4(x - 2)} = \dfrac{-1(\cancel{x - 2})}{4(\cancel{x - 2})}$

$= -\dfrac{1}{4}$ *ANSWER*

5. The $4rs^2t$ pens on a farm held a total of $18r^2t^2$ pigs. Express, in lowest terms, the average number of pigs per pen.

Solution: Write the number of pigs per pen as a fraction.

$\dfrac{\text{pigs}}{\text{pen}} = \dfrac{18r^2t^2}{4rs^2t}$

$= \dfrac{\overset{9}{\cancel{18}}}{\underset{2}{\cancel{4}}} \cdot \dfrac{\overset{r}{\cancel{r^2}}}{\underset{1}{\cancel{r}}} \cdot \dfrac{1}{s^2} \cdot \dfrac{\overset{t}{\cancel{t^2}}}{\underset{1}{\cancel{t}}}$

$= \dfrac{9rt}{2s^2}$ *ANSWER*

Mistakes to Avoid in Reducing a Rational Expression

Students sometimes make mistakes when reducing fractions because they cancel the same quantity in some *part* of the numerator and denominator. Thus, $\dfrac{\cancel{3}x}{\cancel{3}+y} = \dfrac{x}{y}$ is wrong, because 3 is not a factor of the denominator.

Also, $\dfrac{x+\cancel{4}}{y+\cancel{4}} = \dfrac{x}{y}$ is wrong, because 4 is not a factor of the numerator or denominator.

_____ KEEP IN MIND

Cancellation may be used to reduce a rational expression provided that both the numerator and denominator of the fraction are divided by the same factor.

Now it's your turn to . . . **TRY IT!**　　*(See page 381 for solutions.)*

In 1–3, reduce the expression to lowest terms.

1. $\dfrac{a^2 - 4}{5a - 10}$　　2. $\dfrac{2x^2 - 3x - 5}{x + 1}$　　3. $\dfrac{4 - 12x}{3x - 1}$

EXERCISES

In 1–9, represent the symbol as a fraction and give the value of the variable, if any, for which the expression has no meaning.

1. $x \div 7$　　　　　　　　2. $9 \div x$　　　　　　　　3. $(-8) \div 3x$

4. $(x + 6) \div x^2$　　　　5. $15 \div (y - 3)$　　　　6. $(-7b) \div (b + 8)$

7. $(x + 4) \div (2x - 6)$　　8. $(5y + 3) \div (3y + 1)$　　9. $(x - 5) \div (x^2 - 25)$

In 10–37, reduce the expression to lowest terms.

10. $\dfrac{4}{12}$　　11. $\dfrac{6}{8}$　　12. $\dfrac{27}{36}$　　13. $\dfrac{24}{32}$　　14. $\dfrac{6x}{6y}$

15. $\dfrac{24c}{36d}$　　16. $\dfrac{9r}{10r}$　　17. $\dfrac{3m^2}{5m^2}$　　18. $\dfrac{ab}{cb}$　　19. $\dfrac{rs^2}{ts^2}$

20. $\dfrac{x^2y}{x^2z}$　　21. $\dfrac{3ay^2}{6by^2}$　　22. $\dfrac{5xy}{9xy}$　　23. $\dfrac{4xyz}{7xyz}$　　24. $\dfrac{2abc}{4abc}$

25. $\dfrac{6cde}{12cde}$　　26. $\dfrac{15x^3}{5x}$　　27. $\dfrac{18y^4}{6y^2}$　　28. $\dfrac{5x^2}{25x^3}$　　29. $\dfrac{27a}{36a^2}$

30. $\dfrac{8xy^2}{24x^2y}$ **31.** $\dfrac{36a^4y^2}{48ay^3}$ **32.** $\dfrac{18rs^3}{45r^2s}$ **33.** $\dfrac{64a^2b^2c^2}{24ab^2c^2}$ **34.** $\dfrac{12a^2b}{-8ac}$

35. $\dfrac{-20x^2y^2}{-90xy^2}$ **36.** $\dfrac{-32a^3b^3}{48a^3b^3}$ **37.** $\dfrac{5xy}{45x^2y^2}$

In 38–41, express the answer in lowest terms.

38. If a plane travels $15x^2y$ miles in $100xy^3$ hours, represent the rate at which it travels.

39. The senior class sold $24ab^3$ theater tickets for $72a^2$ dollars. Represent the average price for a ticket.

40. In an adult education program, the $64y^3z$ courses offered meet for a total of $80yz$ hours. Represent the average number of hours each course meets.

41. The office copier machine jammed $96mn$ times over a period of $128m^2n^2$ days. Represent the average number of times the copier jammed in one day.

In 42–77, reduce the expression to lowest terms.

42. $\dfrac{5(x+2)}{7(x+2)}$ **43.** $\dfrac{4(m+1)}{8(m+1)}$ **44.** $\dfrac{15(y-3)}{20(y-3)}$ **45.** $\dfrac{18(5-x)}{27(5-x)}$

46. $\dfrac{m(a+b)}{n(a+b)}$ **47.** $\dfrac{x(r-s)}{x(r-s)}$ **48.** $\dfrac{m^2(x+y)}{m(x+y)}$ **49.** $\dfrac{6x^2(r-2s)}{3x(r-2s)}$

50. $\dfrac{5(x-7)}{5x}$ **51.** $\dfrac{8m}{8(m+5)}$ **52.** $\dfrac{2x(a+b)}{8x^2}$ **53.** $\dfrac{9x(x+2)}{9x}$

54. $\dfrac{5x+5y}{7x+7y}$ **55.** $\dfrac{9a-18}{4a-8}$ **56.** $\dfrac{6(x+2y)}{9x+18y}$ **57.** $\dfrac{(x+4)^2}{x+4}$

58. $\dfrac{ab+ac}{db+dc}$ **59.** $\dfrac{rx-ry}{sx-sy}$ **60.** $\dfrac{3x-3}{x^2-1}$ **61.** $\dfrac{y^2-25}{5y+25}$

62. $\dfrac{3m-9}{m^2-9}$ **63.** $\dfrac{64-r^2}{16+2r}$ **64.** $\dfrac{(x+y)^2}{x^2-y^2}$ **65.** $\dfrac{x^2-4}{(x-2)^2}$

66. $\dfrac{9y-18}{3y^2-12}$ **67.** $\dfrac{6y^2-6}{27y+27}$ **68.** $\dfrac{75-12x^2}{15-6x}$ **69.** $\dfrac{5a^2-20}{(a-2)^2}$

70. $\dfrac{1-x}{x-1}$ **71.** $\dfrac{3-b}{b^2-9}$ **72.** $\dfrac{2s-2r}{s^2-r^2}$ **73.** $\dfrac{x^2-y^2}{3y-3x}$

74. $\dfrac{x^2-3x}{2x}$ **75.** $\dfrac{6y-12y^2}{9y}$ **76.** $\dfrac{x}{x^2+x}$ **77.** $\dfrac{3m}{6m-9m^2}$

In 78–81, express the answer in lowest terms.

78. Profit-and-loss statements of Cy Corp. showed that the net change over a period of $(6-2c)$ years was $(2c-6)$ million dollars. Represent the average net change per year.

79. If the telephone company received $(9x-27)$ customer-assistance requests during the last $(3x^2-9x)$ days, represent the average number received in one day.

80. International Airport reported $(4y - 20)$ air disturbances in the past $(y^2 - 25)$ hours. Represent the average number of air disturbances occurring in one hour.

81. A tollbooth attendant at a tunnel estimated that a total of $(3x^3 - 12x)$ riders had passed through during her shift. If there were altogether $(x^2 - 5x - 14)$ buses, automobiles, and other vehicles, represent the average number of riders per vehicle.

The earliest major underwater tunnel was probably one built by the ancient Babylonians under the Euphrates River. The newest is the awesome Channel Tunnel, or Chunnel, extending for 24 miles under the English Channel.

In 82–93, reduce the expression to lowest terms.

82. $\dfrac{3y - 3}{y^2 - 2y + 1}$

83. $\dfrac{x^2 - 3x}{x^2 - 4x + 3}$

84. $\dfrac{x^2 - 25}{x^2 - 2x - 15}$

85. $\dfrac{a^2 - a - 6}{a^2 - 9}$

86. $\dfrac{a^2 - 6a}{a^2 - 7a + 6}$

87. $\dfrac{2x^2 - 50}{x^2 + 8x + 15}$

88. $\dfrac{r^2 - 4r - 5}{r^2 - 2r - 15}$

89. $\dfrac{48 + 8x - x^2}{x^2 + x - 12}$

90. $\dfrac{2x^2 - 7x + 3}{(x - 3)^2}$

91. $\dfrac{3x^2 - 15x + 18}{x^2 - x - 6}$

92. $\dfrac{x^2 - 7xy + 12y^2}{x^2 + xy - 20y^2}$

93. $\dfrac{18c^2 - 32d^2}{6c^2 - cd - 12d^2}$

In 94–96, express the answer in lowest terms.

94. The $(2k^2 + 3k - 2)$ salespeople of Ronco Corp. opened $(2k^2 + 13k - 7)$ new accounts. What was the average number of new accounts per sales representative?

95. Of $(x + 6)^2(1 - x)$ home-mortgage applications received by the Central State Bank, $(36 - x^2)(x - 1)$ were granted. Represent the average number of applications received for each mortgage that was granted.

96. The Crossways Clarion reported that $(m^2 - 10m + 16)$ local couples of the Baby Boom generation had a total of $(6 - 5m + m^2)$ children. Represent the average number of children per couple.

In 97–100, tell whether the solution is correct. State the reason for your answer.

97. $\dfrac{a + \cancel{x}}{b + \cancel{x}} = \dfrac{a}{b}$

98. $\dfrac{\overset{1}{\cancel{b}}\overset{1}{\cancel{(x + 2)}}}{2\cancel{b}\cancel{(x + 2)}} = \dfrac{1}{2}$
${}_{1}{}_{1}$

99. $\dfrac{3x + \overset{1}{\cancel{y}}}{x + \underset{1}{\cancel{y}}} = 3$

100. $\dfrac{\overset{x}{\cancel{x^2}} + \overset{y}{\cancel{y^2}}}{\underset{1}{\cancel{x}} + \underset{1}{\cancel{y}}} = \dfrac{x + y}{2}$

7-2 MULTIPLYING AND DIVIDING RATIONAL EXPRESSIONS

Multiplying Rational Expressions

There are different ways to obtain a product in lowest terms. You can first multiply numerators and multiply denominators, and then reduce the result. Or, you can cancel common factors before you multiply numerators and multiply denominators.

Example: $\dfrac{7}{27} \cdot \dfrac{9}{4} = \dfrac{63}{108} = \dfrac{7 \cdot \overset{1}{\cancel{9}}}{12 \cdot \underset{1}{\cancel{9}}} = \dfrac{7}{12}$ or $\dfrac{7}{27} \cdot \dfrac{9}{4} = \dfrac{7 \cdot \overset{1}{\cancel{9}}}{\underset{3}{\cancel{27}} \cdot 4} = \dfrac{7}{12}$

PROCEDURE

To find the product of two rational expressions and express the result in lowest terms:

1. Factor, when possible, the numerators and denominators.

2. Divide both the numerator and denominator by common factors.

3. Multiply the factors remaining in the numerators and in the denominators.

MODEL PROBLEM

Express the product in reduced form: $\dfrac{a^2 - b^2}{10x^3} \cdot \dfrac{5x^2}{2a + 2b}$

How to Proceed	*Solution*
(1) Factor the numerators and the denominators.	$\dfrac{a^2 - b^2}{10x^3} \cdot \dfrac{5x^2}{2a + 2b} = \dfrac{(a + b)(a - b)}{10x^3} \cdot \dfrac{5x^2}{2(a + b)}$
(2) Divide the numerators and denominators by the common factors, $5x^2$ and $(a + b)$.	$= \dfrac{\overset{1}{\cancel{(a + b)}}(a - b)}{\underset{2x}{\cancel{10x^3}}} \cdot \dfrac{\overset{1}{\cancel{5x^2}}}{2\underset{1}{\cancel{(a + b)}}}$
(3) Multiply the remaining factors.	$= \dfrac{a - b}{4x}$ **ANSWER**

Dividing Rational Expressions

Since multiplication and division are inverse operations, and division is defined as multiplying by the reciprocal, a quotient can be expressed as the product of the dividend and the reciprocal of the divisor.

$$\text{Example:} \quad \frac{8}{7} \div \frac{16}{3} = \frac{8}{7} \cdot \frac{3}{16} = \frac{\overset{1}{\cancel{8}} \cdot 3}{7 \cdot \underset{2}{\cancel{16}}} = \frac{3}{14}$$

___

PROCEDURE

To divide by an algebraic fraction, multiply the dividend by the reciprocal of the divisor.

Note that the reciprocal of $\dfrac{1}{n}$ is n, and the reciprocal of $\dfrac{a + b}{c + d}$ is $\dfrac{c + d}{a + b}$.

MODEL PROBLEM

Divide: $\dfrac{8x^2}{x^2 - 25} \div \dfrac{4x}{3x + 15}$

How to Proceed	*Solution*
(1) Multiply the dividend by the reciprocal of the divisor.	$\dfrac{8x^2}{x^2 - 25} \div \dfrac{4x}{3x + 15} = \dfrac{8x^2}{x^2 - 25} \cdot \dfrac{3x + 15}{4x}$
(2) Factor the numerators and denominators. Divide by the common factors.	$= \dfrac{\overset{2x}{\cancel{8x^2}}}{\underset{1}{(\cancel{x+5})(x - 5)}} \cdot \dfrac{\overset{1}{3(\cancel{x+5})}}{\underset{1}{\cancel{4x}}}$
(3) Multiply the remaining factors.	$= \dfrac{6x}{x - 5}$ **ANSWER**

Now it's your turn to . . . **TRY IT!**   (See page **381** for solutions.)

In 1 and 2, perform the indicated operation and express the result in lowest terms.

1. $\dfrac{12xy^3}{5a^2b^2} \div \dfrac{9y^2}{15a^2b}$

2. $\dfrac{x^2 - 5x + 6}{x^2 - x - 6} \div \dfrac{2x - 4}{4x + 8}$

In 1–20, find the product in lowest terms.

1. $\dfrac{3}{5} \cdot \dfrac{7}{8}$

2. $\dfrac{8}{12} \cdot \dfrac{30}{36}$

3. $\dfrac{3}{8} \cdot 32$

4. $36 \cdot \dfrac{5}{9}$

5. $\dfrac{1}{2} \cdot 20x$

6. $40 \cdot \dfrac{b}{8}$

7. $\dfrac{5}{d} \cdot d^2$

8. $cd \cdot \dfrac{5}{c}$

9. $\dfrac{x^2}{36} \cdot 20$

10. $\dfrac{18}{a^2} \cdot 3a$

11. $6y^2 \cdot \dfrac{4}{3y}$

12. $mn \cdot \dfrac{8}{m^2 n^2}$

13. $\dfrac{3c}{4d} \cdot \dfrac{5r}{3s}$

14. $\dfrac{24x}{35y} \cdot \dfrac{14y}{8x}$

15. $\dfrac{ab}{c} \cdot \dfrac{c}{a}$

16. $\dfrac{12x}{5y} \cdot \dfrac{15y^2}{36x^2}$

17. $\dfrac{m^2}{8} \cdot \dfrac{32}{3m}$

18. $\dfrac{6r^2}{5s^2} \cdot \dfrac{10rs}{6r^3}$

19. $\dfrac{30m^2}{18n} \cdot \dfrac{6n}{5m}$

20. $\dfrac{24a^3 b^2}{7c^3} \cdot \dfrac{21c^2}{12ab}$

In 21–25, represent the answer in lowest terms.

21. Michael's book contains $\dfrac{24n^3}{25m^2}$ chapters, with an average of $\dfrac{15m}{12n^2}$ illustrations per chapter. What is the total number of illustrations in the book?

22. In Tanya's turn at jump rope, she did $\dfrac{64}{120rs^2}$ jumps without stopping, but then Erica succeeded in doing $\dfrac{48s^2}{80rst}$ times as many jumps. How many times did Erica jump?

23. Every week, $\dfrac{180}{w^2}$ dollars were deducted from Mr. Morato's paycheck. Find the total deductions over a period of $9w$ weeks.

24. Marla is a data-processing operator for the Seaview Canning Company, and averages $\dfrac{14d^3}{120e}$ entries per hour. How many entries can she do in $\dfrac{40e}{7d}$ hours?

25. A home economics class baked $\dfrac{2bc^2}{d}$ chocolate layer cakes for the PTA bake sale. If the recipe called for $\dfrac{c}{2bd}$ teaspoonsful of vanilla for each cake, how much vanilla was used altogether?

In 26 and 27, write the answer in lowest terms.

26. $\begin{bmatrix} a & b \\ 5 & 10 \end{bmatrix} \cdot \begin{bmatrix} \dfrac{20}{a} & \dfrac{5}{a} \\ \dfrac{40}{b} & 10b \end{bmatrix}$

27. $\begin{bmatrix} \dfrac{4}{x} & \dfrac{x^2}{2} \\ \dfrac{8}{x} & \dfrac{x^2}{4} \end{bmatrix} \cdot \begin{bmatrix} \dfrac{x^2}{2} & x \\ \dfrac{8}{x^2} & \dfrac{8}{x} \end{bmatrix}$

In 28–47, find the product in lowest terms.

28. $\dfrac{7}{x^2 - 4} \cdot \dfrac{2x + 4}{21}$

29. $\dfrac{3a + 9}{15a} \cdot \dfrac{a^2}{a^2 - 9}$

30. $\dfrac{5x - 5y}{x^2 y} \cdot \dfrac{xy^2}{x^2 - y^2}$

31. $\dfrac{x^2 - 1}{14} \cdot \dfrac{2}{x + 1}$

32. $\dfrac{x^2 - 9}{5c^5} \cdot \dfrac{10c^4}{x - 3}$

33. $\dfrac{a}{x^2 - 4} \cdot (2x + 4)$

34. $\dfrac{(a + 3)^2}{x^2} \cdot \dfrac{4x^2}{4a + 12}$

35. $\dfrac{6k^5}{(x - y)^2} \cdot \dfrac{x^2 - y^2}{9k}$

36. $\dfrac{a(a - b)^2}{4b} \cdot \dfrac{4b}{a(a^2 - b^2)}$

37. $\dfrac{(a - 2)^2}{4b} \cdot \dfrac{16b^3}{4 - a^2}$

38. $\dfrac{a^2 - 7a - 8}{2a + 2} \cdot \dfrac{5}{a - 8}$

39. $\dfrac{x^2 + 6x + 5}{9y^2} \cdot \dfrac{3y}{x + 1}$

40. $\dfrac{y^2 - 2y - 3}{2c^3} \cdot \dfrac{4c^2}{2y + 2}$

41. $\dfrac{4a - 6}{4a + 8} \cdot \dfrac{6a + 12}{5a - 15}$

42. $\dfrac{x^2 - 25}{4x^2 - 9} \cdot \dfrac{2x + 3}{x - 5}$

43. $\dfrac{4x + 8}{6x + 18} \cdot \dfrac{5x + 15}{x^2 - 4}$

44. $\dfrac{y^2 - 81}{(y + 9)^2} \cdot \dfrac{10y + 90}{5y - 45}$

45. $\dfrac{8x}{2x^2 - 8} \cdot \dfrac{8x + 16}{32x^2}$

46. $\dfrac{x + 2d}{5x^2 - 20d^2} \cdot \dfrac{25x - 50d}{25x}$

47. $\dfrac{(x - 4)^2}{2x^2 - 32} \cdot \dfrac{4x + 16}{20x}$

In 48–52, represent the answer in lowest terms.

48. Mr. Silvers swims $\dfrac{(b - 2)^2}{6c^2}$ laps each day at the health club pool. Find the number of laps he can do in $\dfrac{18c^4}{3b - 6}$ days.

49. In one day, $\dfrac{y^2 - 7y + 10}{y - 2}$ bags of popcorn were sold. If each bag cost $\dfrac{18y}{3y - 15}$ cents, find the total receipts for the popcorn.

50. Striking workers at the PDQ Machine Shop were told they would lose $\dfrac{n^2 - 9}{(n - 3)^2}$ days' pay for each day they were off the job. If the strike were to last $\dfrac{2n - 6}{n + 3}$ days, how many days' pay would they lose?

51. During one unusual cold spell, $\dfrac{y^2 - 6y - 16}{3y - 24}$ orange trees were damaged by frost. If the average number of oranges on a tree was $\dfrac{24y}{y^2 - 4}$, how many oranges were affected?

52. Ricardo is building a doghouse, and has completed $\dfrac{1}{(x + y)^2}$ of the job in an hour. If he works at a steady rate, how much of the job can he do in $(x^2 - y^2)$ hours?

In 53 and 54, write the answer in simplest form.

53. $\begin{bmatrix} \dfrac{6}{3r - 3} & 2 \\ \dfrac{1}{(r - 1)^2} & 2r - 2 \end{bmatrix} \cdot \begin{bmatrix} (r - 1)^2 & (r - 1)^2 \\ \dfrac{r - 1}{2} & \dfrac{1}{2} \end{bmatrix}$

54. $2 \cdot \begin{bmatrix} 3a + 3 & \dfrac{2b}{c + 5} \\ (a + 1)^2 & \dfrac{b}{(c + 5)^2} \end{bmatrix} \cdot \begin{bmatrix} \dfrac{1}{a + 1} \\ \dfrac{c^2 + 10c + 25}{b} \end{bmatrix}$

In 55–62, find the product in lowest terms.

55. $\dfrac{x^2 - 7x + 12}{x^2 - 4} \cdot \dfrac{2x + 4}{x + 3}$

56. $\dfrac{x^2 - 3x - 18}{x - 6} \cdot \dfrac{6 - 2x}{x^2 - 9}$

57. $\dfrac{(x - 3)^2}{x^2 - x - 6} \cdot \dfrac{x + 2}{x - 3}$

58. $\dfrac{a - b}{a + b} \cdot \dfrac{a^2 - b^2}{a^2 - 2ab + b^2}$

59. $\dfrac{y^2 - 3y - 10}{y^2 + 3y + 2} \cdot \dfrac{y^2 + 8y + 7}{y^2 - 6y + 5}$

60. $\dfrac{a^2 + 8a + 12}{a^2 - 8a + 16} \cdot \dfrac{20 - a - a^2}{a^2 + 11a + 30}$

61. $\dfrac{2x - 2}{30x^2} \cdot \dfrac{9x^2 + 27x}{x^2 - 9} \cdot \dfrac{x^2 + 2x - 15}{x^2 + 4x - 5}$

62. $\dfrac{i^2 + 6ij + 9j^2}{i^2 - 4j^2} \cdot \dfrac{15i + 30j}{3i^2 + 9ij} \cdot \dfrac{i^2 - ij - 2j^2}{i^2 + 2ij - 3j^2}$

In 63–74, express the quotient in lowest terms.

63. $\frac{7}{10} \div \frac{21}{5}$

64. $\frac{12}{35} \div \frac{4}{7}$

65. $8 \div \frac{1}{2}$

66. $\frac{3}{4} \div 6$

67. $\dfrac{x}{9} \div \dfrac{x}{3}$

68. $\dfrac{2}{b^2} \div \dfrac{2}{b}$

69. $\dfrac{3y}{5y} \div \dfrac{21y}{20y}$

70. $\dfrac{7ab^2}{10cd} \div \dfrac{14b^3}{5c^2d^3}$

71. $\dfrac{xy^2}{x^2y} \div \dfrac{x}{y^3}$

72. $\dfrac{8x^2}{3y^2} \div \dfrac{4x}{6y^3}$

73. $8rs \div \dfrac{24r}{s}$

74. $\dfrac{6a^2b^2}{8c} \div 3ab$

In 75–78, represent the answer in lowest terms.

75. Millicent typed $\dfrac{80cd^2}{70d}$ words in $\dfrac{16d}{56}$ minutes. At that rate, how many words did she type in a minute?

76. If a cruise missile travels at a speed of $36y^2$ miles per minute, how long will it take the missile to travel $\dfrac{108x^3y}{144x^2}$ miles?

77. Mr. and Mrs. Graziele bought $120pq$ tulip bulbs in the garden supply store. If the checkout clerk packed $\dfrac{72p}{q^2}$ bulbs in a sack, how many sacks were needed?

78. Roy received $\dfrac{52x^2y^2}{16y}$ dollars for working $\dfrac{x^2y}{4}$ hours. Find his hourly earnings.

In 79–96, express the quotient in lowest terms.

79. $\dfrac{9}{x^2-1} \div \dfrac{3}{x+1}$

80. $\dfrac{y^2-25}{18} \div \dfrac{y-5}{27}$

81. $\dfrac{3x-3y}{xy^2} \div \dfrac{x^2-y^2}{x^2y}$

82. $\dfrac{3a+12}{18} \div \dfrac{a-2}{2}$

83. $\dfrac{x^2-36}{7y^3} \div \dfrac{x-6}{14y^4}$

84. $\dfrac{b}{a^2-49} \div \dfrac{4b^3}{2a+14}$

85. $\dfrac{(m+1)^2}{n^2} \div \dfrac{6m+6}{9n^2}$

86. $\dfrac{8r^4}{(s-7)^2} \div \dfrac{20r}{s^2-49}$

87. $\dfrac{y^2-3y-10}{8y^2} \div \dfrac{2y-10}{16y^2}$

88. $\dfrac{a^2-1}{2a+2} \div \dfrac{1-a}{3}$

89. $\dfrac{y^2-16}{9y^2-25} \div \dfrac{y-4}{3y+5}$

90. $\dfrac{6x+12}{2x+6} \div \dfrac{x^2-4}{7x+21}$

91. $\dfrac{y^2-49}{(y+7)^2} \div \dfrac{3y-21}{2y+14}$

92. $\dfrac{2x^2-2y^2}{10x} \div \dfrac{6y-6x}{15x^2}$

93. $\dfrac{(x-2)^2}{4x^2-16} \div \dfrac{21x}{3x+6}$

94. $\dfrac{x^2-4x+4}{3x-6} \div (x-2)$

95. $(y^2-9) \div \dfrac{y^2+8y+15}{2y+10}$

96. $\dfrac{x^2-2xy-8y^2}{x^2-16y^2} \div \dfrac{5x+10y}{3x+12y}$

In 97–100, represent the answer in lowest terms.

97. A rock concert raised $\dfrac{x^2-10x+16}{3x-6}$ dollars, to be shared equally by $\dfrac{x^2-64}{9}$ organizations that helped the homeless. How much money was to be given to each organization?

98. During the past summer, D&E Air Conditioning made $\dfrac{5x-15}{20x-20}$ service calls. If the average was $\dfrac{x^2-9}{8(x-1)^2}$ service calls per repairman, how many repairmen were employed?

99. The Ace shopping mall has $\dfrac{m^2-100}{3m-6}$ parking spaces, of which $\dfrac{m^2-8m-20}{12(m+2)^2}$ spaces are reserved for the handicapped. Find the fractional part of the parking spaces that is reserved for the handicapped.

100. It is estimated that the $(20m+20n)$ fruit flies in a genetics research laboratory will increase in one week to $\dfrac{20(m^2+2mn+n^2)}{3n}$. How many times the present number is the projected number?

In 101–104, perform the indicated operations. Express the result in lowest terms.

101. $\dfrac{x-1}{x+1} \cdot \dfrac{2x+2}{x+2} \div \dfrac{4x-4}{x+2}$

102. $\dfrac{x+y}{x^2+y^2} \cdot \dfrac{x}{x-y} \div \dfrac{(x+y)^2}{x^4-y^4}$

103. $\dfrac{3y+6}{y^2-y-6} \div \dfrac{2y+6}{6} \cdot \dfrac{y^2-9}{3y-9}$

104. $\dfrac{w^3-w}{w^2-w-2} \div \dfrac{w^2-w}{w^2+w} \cdot \dfrac{2w-4}{2w+2}$

7-3 ADDING AND SUBTRACTING RATIONAL EXPRESSIONS

When the Denominators Are the Same

The sum of two fractions that have the same denominator is a fraction whose numerator is the sum of the numerators and whose denominator is the common denominator of the given fractions.

$$\frac{5}{7} + \frac{1}{7} = \frac{5+1}{7} = \frac{6}{7}$$

__P__ROCEDURE

To add (or subtract) fractions that have the same denominator:

1. Write a fraction whose numerator is the sum (or difference) of the numerators and whose denominator is the common denominator of the given fractions.
2. Reduce the resulting fraction to lowest terms.

MODEL PROBLEMS

Add or subtract as indicated, reducing the answer to lowest terms.

1. $\dfrac{5}{4x} + \dfrac{9}{4x} - \dfrac{8}{4x}$

$= \dfrac{5+9-8}{4x}$

$= \dfrac{6}{4x} = \dfrac{3}{2x}$

2. $\dfrac{2x}{x+2} + \dfrac{4}{x+2}$

$= \dfrac{2x+4}{x+2}$

$= \dfrac{2(\cancel{x+2})}{\cancel{x+2}} = 2$

3. $\dfrac{4x+7}{x-3} - \dfrac{2x-5}{x-3}$

$= \dfrac{(4x+7) - (2x-5)}{x-3}$

$= \dfrac{4x+7-2x+5}{x-3} = \dfrac{2x+12}{x-3}$

NOTE. As shown in Model Problem 3, you can avoid errors of sign by using parentheses to group numerators that have more than one term.

When the Denominators are Different

To add two fractions with different denominators, you must first transform them into fractions with the same denominator.

To add $\frac{5}{4}$ and $\frac{7}{6}$, rewrite them with the same denominator, 12.

$$\frac{5}{4} = \frac{5 \cdot 3}{4 \cdot 3} = \frac{15}{12} \qquad \frac{7}{6} = \frac{7 \cdot 2}{6 \cdot 2} = \frac{14}{12}$$

Now, add the fractions that have the same denominator.

$$\frac{15}{12} + \frac{14}{12} = \frac{15+14}{12} = \frac{29}{12}$$

Note that using the *lowest common denominator* (L.C.D.) simplifies the work. In this case, the L.C.D. is 12, obtained by inspection or as follows:

(1) Express each denominator as a product of primes: $4 = 2 \cdot 2 = 2^2$ $6 = 2 \cdot 3$
(2) Take the product of the highest power of each of the prime factors:
   L.C.D. $= 2^2 \cdot 3 = 12$

PROCEDURE

To add (or subtract) fractions that have different denominators:

1. Factor each denominator in order to find the lowest common denominator, L.C.D.
2. Transform each fraction to an equivalent fraction whose denominator is the L.C.D.
3. Combine the fractions, reducing if possible.

MODEL PROBLEMS

1. Subtract: $\dfrac{5}{a^2b} - \dfrac{2}{ab^2}$

How to Proceed   *Solution*

(1) Find the L.C.D. Since $a^2b = a^2 \cdot b$ and $ab^2 = a \cdot b^2$, the L.C.D. is $a^2 \cdot b^2$. Transform each fraction into an equivalent fraction with denominator a^2b^2.

$$\frac{5}{a^2b} \cdot \left(\frac{b}{b}\right) - \frac{2}{ab^2} \cdot \left(\frac{a}{a}\right)$$

(2) Multiply each numerator by the same multiplier that was used in its denominator.

$$= \frac{5b}{a^2b^2} - \frac{2a}{a^2b^2}$$

(3) Combine the fractions.

$$= \frac{5b - 2a}{a^2b^2} \quad \textbf{ANSWER}$$

2. Subtract: $\dfrac{2x + 5}{3} - \dfrac{x - 2}{4}$

3. Add: $\dfrac{7}{3x - 15} + \dfrac{2}{5x - 25}$

Solution: $3 = 3 \cdot 1$
$4 = 2 \cdot 2 = 2^2$
L.C.D. $= 3 \cdot 2^2$

$$\frac{2x + 5}{3} - \frac{x - 2}{4}$$

$$= \frac{2x + 5}{3} \cdot \left(\frac{4}{4}\right) - \frac{x - 2}{4} \cdot \left(\frac{3}{3}\right)$$

$$= \frac{8x + 20}{12} - \frac{3x - 6}{12}$$

$$= \frac{(8x + 20) - (3x - 6)}{12}$$

$$= \frac{8x + 20 - 3x + 6}{12} = \frac{5x + 26}{12}$$

Solution: $3x - 15 = 3(x - 5)$
$5x - 25 = 5(x - 5)$
L.C.D. $= 3 \cdot 5 \cdot (x - 5)$

$$\frac{7}{3x - 15} + \frac{2}{5x - 25}$$

$$= \frac{7}{3(x - 5)} + \frac{2}{5(x - 5)}$$

$$= \frac{7}{3(x - 5)} \cdot \left(\frac{5}{5}\right) + \frac{2}{5(x - 5)} \cdot \left(\frac{3}{3}\right)$$

$$= \frac{35}{15(x - 5)} + \frac{6}{15(x - 5)}$$

$$= \frac{35 + 6}{15(x - 5)} = \frac{41}{15(x - 5)} \text{ or } \frac{41}{15x - 75}$$

Now it's your turn to . . . **TRY IT!**   *(See page 381 for solutions.)*

1. Add: $\dfrac{3}{ab^4} + \dfrac{2}{a^3b^3}$

2. Subtract: $\dfrac{3x - 4}{5} - \dfrac{5 - x}{3}$

3. Subtract: $\dfrac{4a - 3}{12a} - \dfrac{2a - 5}{20a}$

4. Add: $\dfrac{4x - 1}{x^2 - 3x + 2} + \dfrac{x - 5}{x^2 + x - 6}$

Mixed Expressions

The mixed number $3\frac{1}{2}$, which means the sum of the integer 3 and the fraction $\frac{1}{2}$, can be expressed as a fraction. To do this, express the integer 3 as the fraction $\frac{3}{1}$ and then add $\frac{3}{1}$ and $\frac{1}{2}$.

$$3\frac{1}{2} = \frac{3}{1} + \frac{1}{2} = \frac{3 \cdot 2}{1 \cdot 2} + \frac{1}{2} = \frac{6}{2} + \frac{1}{2} = \frac{6 + 1}{2} = \frac{7}{2}$$

PROCEDURE

To rewrite a mixed expression as a single fraction:

1. Write the polynomial of the expression as a fraction whose denominator is 1.
2. Add the two fractions.

MODEL PROBLEMS

In 1 and 2, rewrite the mixed expression as a single fraction.

1. $y + \dfrac{5}{y} = \dfrac{y}{1} + \dfrac{5}{y}$

$= \dfrac{y \cdot y}{1 \cdot y} + \dfrac{5}{y}$

$= \dfrac{y^2}{y} + \dfrac{5}{y}$

$= \dfrac{y^2 + 5}{y}$ **ANSWER**

2. $y + 1 - \dfrac{1}{y - 1} = \dfrac{y + 1}{1} - \dfrac{1}{y - 1}$

$= \dfrac{(y + 1)(y - 1)}{1(y - 1)} - \dfrac{1}{y - 1}$

$= \dfrac{y^2 - 1}{y - 1} - \dfrac{1}{y - 1}$

$= \dfrac{y^2 - 1 - 1}{y - 1} = \dfrac{y^2 - 2}{y - 1}$ **ANSWER**

The accompanying flowchart summarizes the steps you must take when adding or subtracting rational expressions.

ADDING AND SUBTRACTING RATIONAL EXPRESSIONS

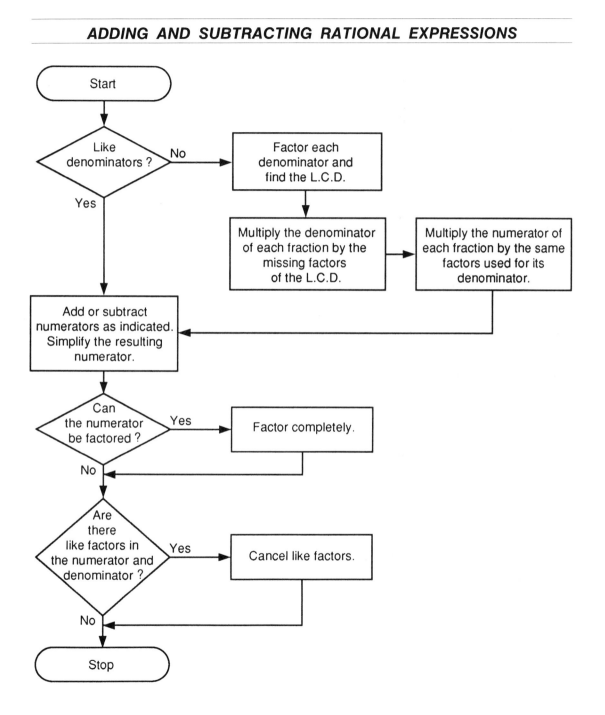

In 1–12, add or subtract as indicated, and reduce the answer to lowest terms.

1. $\dfrac{2}{x} + \dfrac{3}{x}$

2. $\dfrac{9}{15} - \dfrac{6}{15}$

3. $\dfrac{7}{3b} - \dfrac{2}{3b}$

4. $\dfrac{9}{16} - \dfrac{6}{16} + \dfrac{5}{16}$

5. $\dfrac{11a}{9} - \dfrac{5a}{9}$

6. $\dfrac{x}{a} + \dfrac{y}{a}$

7. $\dfrac{6}{12x} - \dfrac{2}{12x}$

8. $\dfrac{7b}{8} + \dfrac{5b}{8} - \dfrac{6b}{8}$

9. $\dfrac{6a}{4x} + \dfrac{5a}{4x}$

10. $\dfrac{11b}{3y} - \dfrac{4b}{3y}$

11. $\dfrac{19c}{12d} + \dfrac{9c}{12d}$

12. $\dfrac{6}{10c} + \dfrac{9}{10c} - \dfrac{3}{10c}$

In 13–15, represent the answer as a single fraction in lowest terms.

13. Among the supplies at camp, the Bennetts had $\dfrac{3b}{2d}$ pounds of quinoa and $\dfrac{5b}{2d}$ pounds of rice. What was the total weight of these grains?

Quinoa (pronounced keen-wah) is a grain that has flourished in the Andes Mountains of South America for at least 5,000 years. A staple food of Andean villagers, it is high in protein and has an appealing nutty flavor. This nutritious grain, which has been described as the best single food on which to survive, is now being grown in the United States.

14. Of $\dfrac{19}{5k}$ Central Airways flights in one day, all but $\dfrac{4}{5k}$ arrived on time. How many flights arrived on schedule?

15. Gem Photo offers a set of studio portraits for $\dfrac{20x}{3y}$ dollars, with a discount of $\dfrac{4x}{3y}$ dollars if you buy the proofs for $\dfrac{5x}{3y}$ dollars. Mrs. Keene decided to buy the entire set with the proofs. Find the cost of her purchase.

In 16–41, add or subtract as indicated, and reduce the answer to lowest terms.

16. $\dfrac{5}{x+2} + \dfrac{3}{x+2}$

17. $\dfrac{2}{a-b} - \dfrac{1}{a-b}$

18. $\dfrac{r}{y-2} + \dfrac{x}{y-2}$

19. $\dfrac{x}{x+1} + \dfrac{1}{x+1}$

20. $\dfrac{2x}{x+3} + \dfrac{6}{x+3}$

21. $\dfrac{y}{y^2-4} - \dfrac{2}{y^2-4}$

22. $\dfrac{2x+1}{2} + \dfrac{3x+6}{2}$

23. $\dfrac{9x-3}{6} + \dfrac{5x+5}{6}$

24. $\dfrac{4x+12}{16x} + \dfrac{8x+4}{16x}$

25. $\dfrac{5x-4}{3} - \dfrac{2x+1}{3}$

26. $\dfrac{6x-9}{6} - \dfrac{4x-8}{6}$

27. $\dfrac{12a-15}{12a} - \dfrac{9a-6}{12a}$

28. $\dfrac{4x+1}{3x+2} + \dfrac{6x-3}{3x+2}$

29. $\dfrac{3c-7}{2c-3} + \dfrac{c+9}{2c-3}$

30. $\dfrac{6y-4}{4y+3} + \dfrac{7-2y}{4y+3}$

31. $\dfrac{9d+6}{2d+1} - \dfrac{7d+5}{2d+1}$

32. $\dfrac{8x-4}{2x+6} - \dfrac{4x-6}{2x+6}$

33. $\dfrac{6x-5}{x^2-1} - \dfrac{5x-6}{x^2-1}$

34. $\dfrac{a^2+3ab}{a+b} + \dfrac{b^2-ab}{a+b}$

35. $\dfrac{x^2-2xy}{x-2y} - \dfrac{xy-2y^2}{x-2y}$

36. $\dfrac{8x-8}{6x-5} - \dfrac{2x-7}{6x-5} + \dfrac{6x-9}{6x-5}$

37. $\dfrac{a+4b}{a^2-b^2} + \dfrac{4a-7b}{a^2-b^2} - \dfrac{3a-b}{a^2-b^2}$

38. $\dfrac{x}{(x+y)(x-y)} - \dfrac{y}{(x+y)(x-y)}$

39. $\dfrac{5a+b}{(2a+3b)(a-b)} - \dfrac{a-5b}{(2a+3b)(a-b)}$

40. $\dfrac{r^2+4r}{r^2-r-6} + \dfrac{8-r^2}{r^2-r-6}$

41. $\dfrac{4m^2+7m}{2m^2+5m+2} - \dfrac{1+7m}{2m^2+5m+2}$

In 42 and 43, find the sum in simplest form.

42. $\begin{bmatrix} \dfrac{a}{2b} & \dfrac{2b}{a} \\[2ex] \dfrac{a}{a+1} & -b \end{bmatrix} + \begin{bmatrix} \dfrac{3a}{2b} & -\dfrac{b}{a} \\[2ex] \dfrac{1}{a+1} & b+2 \end{bmatrix}$

43. $\begin{bmatrix} \dfrac{x}{x+y} & \dfrac{x^2-xy}{x-y} \\[2ex] \dfrac{x-y}{x-2y} & \dfrac{x}{x^2-y^2} \end{bmatrix} + \begin{bmatrix} \dfrac{y}{x+y} & \dfrac{y^2-xy}{x-y} \\[2ex] \dfrac{x-3y}{x-2y} & \dfrac{y}{x^2-y^2} \end{bmatrix}$

In 44–63, rewrite the mixed expression as a single fraction in lowest terms.

44. $5\frac{2}{3}$

45. $9\frac{3}{4}$

46. $5 + \dfrac{1}{x}$

47. $9 - \dfrac{7}{s}$

48. $m + \dfrac{1}{m}$

49. $d - \dfrac{7}{5d}$

50. $\dfrac{a}{b} + c$

51. $3 + \dfrac{5}{x+1}$

52. $6 - \dfrac{4}{x-y}$

53. $7 + \dfrac{2a}{b+c}$

54. $t + \dfrac{1}{t+1}$

55. $s - \dfrac{1}{s-1}$

56. $5 - \dfrac{2x}{x+y}$

57. $\dfrac{4}{y-2} + 4$

58. $8 + \dfrac{c+2}{c-3}$

59. $7 - \dfrac{x+y}{x-y}$

60. $a + 1 + \dfrac{1}{a+1}$

61. $x - 5 - \dfrac{x}{x+3}$

62. $\dfrac{2x-1}{x+2} + 2x - 3$

63. $\dfrac{x^2+3x+2}{x+2} - 2$

64. Find the lowest common denominator for two fractions whose denominators are:

    a. 10; 5   **b.** 8; 12   **c.** $6a$; $2a$ **d.** 3; x **e.** r; s

    f. $15a$; $6b$ **g.** xy; yz **h.** m^2n; mn^2 **i.** $12x^2$; $18y^2$ **j.** $6c^2d^2$; $10cd$

    k. $x+3$; $x-3$ **l.** $4(c+1)$; $6(c+1)$ **m.** $3c+9$; $4c+12$

    n. x; $x+5$ **o.** x^2-1; $3x+3$ **p.** $3x-4$; $4-3x$

65. Transform the given fractions into equivalent fractions that have the L.C.D. as their denominators.

a. $\dfrac{3}{5}$; $\dfrac{4}{3}$

b. $\dfrac{3x}{12}$; $\dfrac{7x}{90}$

c. $\dfrac{5y}{3c}$; $\dfrac{7y}{6c}$

d. $\dfrac{8}{x^2}$; $\dfrac{2}{x}$

e. $\dfrac{7}{4c^2}$; $\dfrac{5}{18d^2}$

f. $\dfrac{a-6}{2}$; $\dfrac{2a+5}{4}$

g. $\dfrac{3c+1}{18d}$; $\dfrac{5c-3}{24d}$

h. $\dfrac{5x-4}{20y}$; $\dfrac{3x-7}{72y}$

i. $\dfrac{1}{x+2}$; $\dfrac{3}{x-2}$

j. $\dfrac{2t-1}{8t-8}$; $\dfrac{4t+1}{6t-6}$

k. $\dfrac{4}{y}$; $\dfrac{y-1}{y+2}$

l. $\dfrac{6x+1}{x^2-9}$; $\dfrac{-3}{x-3}$

m. $\dfrac{7}{1-3a}$; $\dfrac{2}{3a-1}$

In 66–88, add or subtract as indicated, and reduce the answer to lowest terms.

66. $\dfrac{3x}{10}+\dfrac{7x}{5}$

67. $\dfrac{10y}{7}-\dfrac{3y}{4}$

68. $\dfrac{x}{18}-\dfrac{y}{4}$

69. $\dfrac{8x}{5}-\dfrac{3x}{4}+\dfrac{7x}{10}$

70. $\dfrac{9}{4x}+\dfrac{3}{2x}$

71. $\dfrac{7}{8y}-\dfrac{3}{4y}$

72. $\dfrac{8}{5c}-\dfrac{1}{4c}$

73. $\dfrac{1}{2x}-\dfrac{1}{x}+\dfrac{3}{8x}$

74. $\dfrac{5x}{2d}+\dfrac{4x}{3d}$

75. $\dfrac{9a}{8b}-\dfrac{3a}{4b}$

76. $\dfrac{7x}{4y}-\dfrac{3x}{5y}$

77. $\dfrac{3r}{2s}-\dfrac{5r}{4s}-\dfrac{2r}{3s}$

78. $\dfrac{7}{x}-\dfrac{3}{y}$

79. $\dfrac{1}{r^2}-\dfrac{3}{r}$

80. $\dfrac{c}{a}-\dfrac{a}{b}$

81. $\dfrac{1}{x}+\dfrac{1}{y}+\dfrac{1}{z}$

82. $\dfrac{1}{xy}+\dfrac{1}{yz}$

83. $\dfrac{2}{ab}-\dfrac{3}{bc}$

84. $\dfrac{5}{rs}+\dfrac{9}{st}$

85. $\dfrac{x}{a^2b}+\dfrac{y}{ab^2}$

86. $\dfrac{9}{ab}+\dfrac{2}{bc}-\dfrac{3}{ac}$

87. $\dfrac{2}{y^3}-\dfrac{3}{y^2}+\dfrac{7}{y}$

88. $\dfrac{1}{x^2}+\dfrac{3}{xy}-\dfrac{5}{y^2}$

In 89 and 90, represent the answer as a single fraction in lowest terms.

89. Crickets chirp faster as temperature goes up. On a cool day, a cricket might chirp $\dfrac{s}{2rt}$ times a minute. If there are likely to be $\dfrac{3}{r^2t}$ more chirps per minute on a warm day, how many would there be on a warm day?

90. Radioactive elements continually undergo a process of disintegration called radioactive decay. The half-life of Polonium-210 is $\dfrac{5}{6w}$ days. If this is $\dfrac{1}{2w^2}$ days more than the half-life of one of its isotopes, find the half-life of the isotope.

In 91 and 92, find the sum in simplest form.

91. $\begin{bmatrix} \dfrac{1}{x} & x \\[2ex] \dfrac{3y}{4} & \dfrac{x}{y} \end{bmatrix} + \begin{bmatrix} \dfrac{1}{y} & \dfrac{x}{y} \\[2ex] \dfrac{y}{2} & \dfrac{y}{x} \end{bmatrix}$

92. $\begin{bmatrix} \dfrac{3}{p^2q} & \dfrac{p}{q^2} \\[2ex] \dfrac{1}{pq^2} & \dfrac{p}{q^2r} \end{bmatrix} + \begin{bmatrix} \dfrac{5}{pqr} & \dfrac{q}{p} \\[2ex] \dfrac{1}{pq} & \dfrac{p}{qr^2} \end{bmatrix}$

In 93–143, add or subtract as indicated, and reduce the answer to lowest terms.

93. $\dfrac{a-3}{3}+\dfrac{a+1}{6}$

94. $\dfrac{y-1}{2}-\dfrac{y-5}{8}$

95. $\dfrac{x+7}{3}-\dfrac{2x-3}{5}$

96. $\dfrac{3x-5}{4}+\dfrac{2x+3}{6}$

97. $\dfrac{3a-2}{5}-\dfrac{2a+3}{15}$

98. $\dfrac{x+y}{2}-\dfrac{2x-y}{6}$

99. $\dfrac{6a+b}{30}-\dfrac{2a-b}{10}$

100. $\dfrac{x+y}{8}-\dfrac{x+y}{8}$

101. $\dfrac{a-b}{4}-\dfrac{a+b}{6}$

102. $\dfrac{x+5}{4x}+\dfrac{2x-1}{4x}$

103. $\dfrac{9y-2}{12y}-\dfrac{4y+1}{6y}$

104. $\dfrac{3x-4}{2x}-\dfrac{2x-3}{2x}+\dfrac{5x}{2}$

105. $\dfrac{3b+1}{5b}-\dfrac{4b-3}{4b}$

106. $\dfrac{y-4}{4y^2}+\dfrac{3y-5}{3y}$

107. $\dfrac{2x+3}{12x}-\dfrac{3x-6}{8x}-\dfrac{5}{x}$

108. $\dfrac{a+9}{2}+\dfrac{6a-4}{4}$

109. $\dfrac{c+1}{2}-\dfrac{7c+9}{4}$

110. $\dfrac{7x+3}{3}-\dfrac{3x+2}{5}$

111. $\dfrac{3y-4x}{6x}-\dfrac{4x^2-5y}{4x}$

112. $\dfrac{x+y}{x}-\dfrac{y-z}{y}-\dfrac{z-x}{z}$

113. $\dfrac{3x+5}{3}+\dfrac{2-5x}{4}-\dfrac{x-8}{5}$

114. $\dfrac{6}{ab}+\dfrac{a-3}{a^2}-\dfrac{7+b}{b^2}$

115. $\dfrac{2}{3y}-\dfrac{4y-7}{6y^2}+\dfrac{3y-2y^2}{4y^3}$

116. $\dfrac{4a+1}{6a^2b}-\dfrac{3b-5}{4ab^2}-\dfrac{2b+1}{9ab}$

117. $\dfrac{5}{x-3}+\dfrac{7}{2x-6}$

118. $\dfrac{9}{y+1}-\dfrac{3}{4y+4}$

119. $\dfrac{2}{3a-1}+\dfrac{7}{15a-5}$

120. $\dfrac{10}{3x-6}+\dfrac{3}{2x-4}$

121. $\dfrac{11x}{8x-8}-\dfrac{3x}{4x-4}$

122. $\dfrac{3}{2x-3y}+\dfrac{5}{3y-2x}$

123. $\dfrac{2a}{4a-8b}+\dfrac{3b}{3a-6b}$

124. $\dfrac{3x-2}{2x+2}+\dfrac{4x-1}{3x+3}$

125. $\dfrac{5x+2}{6x-3}-\dfrac{3x-5}{8x-4}$

126. $\dfrac{1}{x-5}+\dfrac{1}{x+5}$

127. $\dfrac{9}{y+4}-\dfrac{6}{y-4}$

128. $\dfrac{7}{a+3}+\dfrac{4}{2-a}$

129. $\dfrac{7}{x-2}+\dfrac{3}{x}$

130. $\dfrac{9}{c+8}-\dfrac{2}{c}$

131. $\dfrac{2a+b}{a-b}+\dfrac{a}{b}$

132. $\dfrac{5}{y^2-9}+\dfrac{3}{y-3}$

133. $\dfrac{6}{y^2-16}-\dfrac{5}{y+4}$

134. $\dfrac{9}{a^2-b^2}+\dfrac{3}{b-a}$

135. $\dfrac{3y}{y^2-4}-\dfrac{4}{2y-4}$

136. $\dfrac{x}{x^2-36}-\dfrac{4}{3x+18}$

137. $\dfrac{9}{a^2-ab}+\dfrac{3}{ab-b^2}$

138. $\dfrac{1}{y-3}+\dfrac{2}{y+4}+\dfrac{2}{3}$

139. $\dfrac{1}{(x+2)^3}-\dfrac{1}{(x+2)^2}+\dfrac{1}{x+2}$

140. $\dfrac{7a}{(a-1)(a+3)}+\dfrac{2a-5}{(a+3)(a+2)}$

141. $\dfrac{5}{r^2-4}-\dfrac{3}{r^2+3r-10}$

142. $\dfrac{x+2y}{3x+12y}-\dfrac{6x-y}{x^2+3xy-4y^2}$

143. $\dfrac{2a+7}{a^2-2a-15}-\dfrac{3a-4}{a^2-7a+10}$

7-4 SOLVING OPEN SENTENCES THAT CONTAIN FRACTIONAL COEFFICIENTS

Examples of open sentences that contain fractional coefficients are:

$$\frac{1}{2}x = 10 \qquad \frac{x}{3} + 60 > \frac{5x}{6} \qquad .05x + .25x = 1.90$$

Such an open sentence can be solved by transforming it into an equivalent sentence that does not contain fractional coefficients.

PROCEDURE

To solve an open sentence that contains fractional coefficients:

1. Find the L.C.D.
2. Multiply both sides of the open sentence by the L.C.D.
3. Solve the resulting sentence using the usual methods.
4. Check in the original sentence.

MODEL PROBLEMS

1. Solve and check: $\frac{1}{3}x + \frac{1}{5}x = 8$

How to Proceed	*Solution*
(1) Write the equation, expressing all terms as fractions.	$\dfrac{x}{3} + \dfrac{x}{5} = \dfrac{8}{1}$
(2) Find the L.C.D.	$\text{L.C.D.} = 3 \cdot 5 \cdot 1 = 15$
(3) Multiply both sides of the equation by the L.C.D.	$15\left(\dfrac{x}{3} + \dfrac{x}{5}\right) = 15\left(\dfrac{8}{1}\right)$ $\overset{5}{\cancel{15}}\left(\dfrac{x}{\cancel{3}}\right) + \overset{3}{\cancel{15}}\left(\dfrac{x}{\cancel{5}}\right) = 15\left(\dfrac{8}{1}\right)$
(4) Solve the resulting equation.	$5x + 3x = 120$ $8x = 120$ $x = 15$

Check: Substitute $x = 15$ into the original equation, and carry out the arithmetic for each side of the equation.

ANSWER: $x = 15$

In 2 and 3, solve and check the open sentence.

2.
$$.05x + .4(500 - x) > 25$$
$$\text{L.C.D.} = 100$$
$$100[.05x + .4(500 - x)] > 100(25)$$
$$5x + 40(500 - x) > 2{,}500$$
$$5x + 20{,}000 - 40x > 2{,}500$$
$$-35x + 20{,}000 > 2{,}500$$
$$-35x > -17{,}500$$
$$\text{reverse order} \quad x < 500$$

Check: Choose a value less than 500, say 100, to substitute into the original sentence.

ANSWER: $x < 500$

3.
$$\frac{2x + 7}{6} - \frac{2x - 9}{10} = \frac{3}{1}$$
$$\text{L.C.D.} = 30$$
$$30\left(\frac{2x + 7}{6} - \frac{2x - 9}{10}\right) = 30\left(\frac{3}{1}\right)$$
$$\overset{5}{\cancel{30}}\left(\frac{2x + 7}{\cancel{6}}\right) - \overset{3}{\cancel{30}}\left(\frac{2x - 9}{\cancel{10}}\right) = 30\left(\frac{3}{1}\right)$$
$$\underset{1}{} \qquad \underset{1}{}$$
$$5(2x + 7) - 3(2x - 9) = 90$$
$$10x + 35 - 6x + 27 = 90$$
$$4x + 62 = 90$$
$$4x = 28$$
$$x = 7$$

Check: Substitute $x = 7$ in the original.

ANSWER: $x = 7$

Now it's your turn to . . . **TRY IT!**   (See page 382 for solutions.)

In 1 and 2, solve and check the open sentence.

1. $\frac{1}{3}a - \frac{1}{5}a = 3$ **2.** $\frac{w + 3}{5} + \frac{w - 1}{4} \geq \frac{1}{2}$

CALCULATOR CONNECTION

When checking a solution, it is often helpful to use Memory.

Example: $.02x + .04(2{,}000 - x) = 67$ has the solution $x = 650$. To check, store 650.

Enter: **.02** ⊠ **MR** ⊞ **.04** ⊠ **(** **2000** ⊟ **MR** **)** **=**

The display, 67, checks with the constant side of the equation.

Example: $2.2x + 8 = 12.2 + 3.4x$ has the solution $x = -3.5$. To check, store -3.5 and do a separate calculation for each side of the equation.

Enter: **2.2** ⊠ **MR** ⊞ **8** **=**   Enter: **12.2** ⊞ **3.4** ⊠ **MR** **=**

Since both sides give the same result, 0.3, the solution checks.

In 1–100, solve and check.

1. $\dfrac{x}{7} = 3$

2. $\dfrac{y}{4} = 16$

3. $\dfrac{t}{6} = 18$

4. $\dfrac{3x}{5} = 15$

5. $\dfrac{2m}{3} = 24$

6. $\dfrac{5n}{7} = 35$

7. $\dfrac{x + 8}{4} \geq 6$

8. $\dfrac{m - 2}{9} < 3$

9. $\dfrac{3x - 1}{5} \leq 7$

10. $\dfrac{2r + 6}{5} = -4$

11. $\dfrac{5y - 30}{7} = 0$

12. $\dfrac{7 - 2x}{5} = -1$

13. $\dfrac{x}{5} < \dfrac{8}{10}$

14. $\dfrac{y}{21} \geq \dfrac{3}{7}$

15. $\dfrac{5x}{2} > \dfrac{15}{4}$

16. $.01x = 4$

17. $.02y = 18$

18. $.03n = 12$

19. $54 = .06y$

20. $.08t = 3.36$

21. $.45c = 9$

22. $.5x = 1.5$

23. $8.7 = 3u$

24. $.06a = 1.2$

25. $1.68 > .08b$

26. $.4t \leq .012$

27. $.09d \geq .018$

28. $\dfrac{y + 2}{4} = \dfrac{5}{2}$

29. $\dfrac{m - 5}{35} = \dfrac{5}{7}$

30. $\dfrac{2c + 8}{28} = \dfrac{12}{7}$

31. $\dfrac{2x + 1}{3} = \dfrac{6x - 9}{5}$

32. $\dfrac{3y + 1}{4} = \dfrac{44 - y}{5}$

33. $\dfrac{2m}{3} = \dfrac{3m + 9}{4}$

34. $\dfrac{x}{5} + \dfrac{x}{3} = \dfrac{8}{15}$

35. $\dfrac{y}{3} + \dfrac{y}{2} = 40$

36. $10 = \dfrac{x}{3} + \dfrac{x}{7}$

37. $\dfrac{r}{3} - \dfrac{r}{6} = 2$

38. $\dfrac{2}{5}t - \dfrac{1}{4}t = 3$

39. $1 = \dfrac{3}{4}r - \dfrac{2}{3}r$

40. $\dfrac{3t}{4} - 6 = \dfrac{t}{12}$

41. $\dfrac{2s}{3} = \dfrac{s}{4} + 10$

42. $\dfrac{y}{4} = \dfrac{3y}{5} - 2\dfrac{1}{10}$

43. $.7x - .4 = 1$

44. $.03y - 1.2 = 8.7$

45. $.4x + .08 = 4.24$

46. $.02x - 5 = 7$

47. $.5y - 3 = .5$

48. $2.3m + 4.5 = 16$

49. $.5x - .3x < 8$

50. $2c + .5c \geq 50$

51. $.08y - .9 > .02y$

52. $1.7x = 30 + .2x$

53. $1.5y - 1.7 = y$

54. $.08c = 1.5 + .07c$

55. $\dfrac{1}{4}x - \dfrac{1}{5}x > \dfrac{9}{20}$

56. $y - \dfrac{2}{3}y < 5$

57. $\dfrac{5}{6}c > \dfrac{1}{3}c + 3$

58. $\dfrac{x}{4} - \dfrac{x}{8} \leq \dfrac{5}{8}$

59. $\dfrac{y}{6} \geq \dfrac{y}{12} + 1$

60. $\dfrac{y}{9} - \dfrac{y}{4} > \dfrac{5}{36}$

61. $\dfrac{t}{10} \le 4 + \dfrac{t}{5}$

62. $1 + \dfrac{2x}{3} \ge \dfrac{x}{2}$

63. $2.5x - 1.6x > 4$

64. $2y + 3 \ge .2y$

65. $\dfrac{3x - 1}{7} > 5$

66. $\dfrac{5y - 30}{7} \le 0$

67. $\dfrac{a}{2} + \dfrac{a}{3} + \dfrac{a}{4} = 26$

68. $\dfrac{5c}{8} - \dfrac{c}{3} = \dfrac{5c}{6} - 13$

69. $\dfrac{s}{3} + 7 = \dfrac{s}{5} - 3$

70. $\dfrac{3y}{2} - \dfrac{17}{3} = \dfrac{2y}{3} - \dfrac{3}{2}$

71. $\dfrac{7}{12}y - \dfrac{1}{4} = 2y - \dfrac{5}{3}$

72. $\dfrac{5c}{4} - \dfrac{1}{2} = \dfrac{2c}{3} + 6\tfrac{1}{2}$

73. $\dfrac{x}{3} - 2 = \dfrac{3x - 30}{6}$

74. $\dfrac{2y}{3} - \dfrac{7 - y}{4} = 1$

75. $\dfrac{y + 4}{4} + \dfrac{y - 2}{2} = 3$

76. $\dfrac{t + 1}{2} + \dfrac{2t - 3}{3} = 10$

77. $\dfrac{y + 2}{4} - \dfrac{y - 3}{3} = \dfrac{1}{2}$

78. $\dfrac{7s + 5}{8} - \dfrac{3s + 15}{10} = 2$

79. $2d + \dfrac{1}{4} < \dfrac{7d}{12} + \dfrac{5}{3}$

80. $\dfrac{4c}{3} - \dfrac{7}{9} \ge \dfrac{c}{2} + \dfrac{7}{6}$

81. $\dfrac{2m}{3} \ge \dfrac{7 - m}{4} + 1$

82. $\dfrac{3x - 30}{6} < \dfrac{x}{3} - 2$

83. $\dfrac{6x - 3}{2} > \dfrac{37}{10} + \dfrac{x + 2}{5}$

84. $\dfrac{2y - 3}{3} + \dfrac{y + 1}{2} < 10$

85. $\dfrac{2r - 3}{5} - \dfrac{r - 3}{3} \le 2$

86. $\dfrac{3t - 4}{3} \ge \dfrac{2t + 4}{6} + \dfrac{5t - 1}{9}$

87. $.8m + 2.6 = .2m + 9.8$

88. $.05x - .25 = .02x + .44$

89. $.13x - 1.4 = .08x + 7.6$

90. $.06y + 40 - .03y = 70$

91. $.02(x + 5) < 8$

92. $.05(x - 8) \ge .07x$

93. $.4(x - 9) = .3(x + 4)$

94. $.06(x - 5) = .04(x + 8)$

95. $.04x + .03(2{,}000 - x) = 75$

96. $.02x + .04(1{,}500 - x) = 48$

97. $.05x + 10 \le .06(x + 50)$

98. $.08x > .03(x + 200) - 4$

99. $.07x + .04(9{,}000 - x) > 450$

100. $.06x - .04(3{,}500 - x) < 160$

7-5 APPLYING OPEN SENTENCES THAT CONTAIN FRACTIONAL COEFFICIENTS

MODEL PROBLEMS

1. Ray has marks of 75 in English, 82 in French, and 90 in Algebra. What mark must he get in Social Studies to have an average of 85 for all four subjects?

Solution:
Let x = Ray's mark in Social Studies.

The average of the four marks is 85.

$$\frac{75 + 82 + 90 + x}{4} = 85$$

$$\frac{247 + x}{4} = 85$$

$$247 + x = 340$$
$$x = 93$$

Check your result.

ANSWER: Ray must get a mark of 93.

2. A dealer sold a radio for $39.20, which was 40% above its cost to him. Find the base cost of the radio to the dealer.

Solution:
Let b = the base cost of the radio.
Then $.40b$ = the amount of the markup.

Base plus markup equals selling price.

$$b + .40b = 39.20$$
$$100(b + .40b) = 100(39.20)$$
$$100b + 40b = 3{,}920$$
$$140b = 3{,}920$$
$$b = 28$$

Check your result.

ANSWER: The dealer paid $28.

3. Two companies, Acme Sales and Bronson Sales, offer you a job as a salesperson. Acme pays a straight 7% commission, and Bronson pays $200 per week plus a 3% commission.

   a. For what amount of sales will Acme pay more money?
   b. In trying to decide between jobs, what other information would you like to have?

   Solution: Let x = the dollar amount of weekly sales at each company.
   Then $.07x$ = the weekly amount earned at Acme.
   And $200 + .03x$ = the weekly amount earned at Bronson.

   For what amount of sales will Acme pay more money?

   $$.07x > 200 + .03x$$
   $$100(.07x) > 100(200 + .03x)$$
   $$7x > 20{,}000 + 3x$$
   $$4x > 20{,}000$$
   $$x > 5{,}000$$

   ANSWERS: **a.** Acme pays more for weekly sales greater than $5,000.
   b. To decide between jobs, it would be wise to know the weekly sales of an average salesperson at each company, and the fringe benefits that each company offers.

1. 10% of the students in Mr. Lamont's 5th-period math class earned A's, while 25% of the students in his 8th-period class earned A's. Explain what you would need to know in order to answer the question, "In which class did more students earn A?"

2. Many business situations treat percent as a *rate*, such as the rate of tax charged on an item. The dollar amount on which the tax is calculated is called the *base*, and the dollar amount of tax paid is the *percentage*.

   a. Calculate the percentage when $7\frac{1}{2}\%$ tax is collected on a $15 item.

   b. Using r to represent *rate*, b for *base*, and p for *percentage*, write a rule that can be used for calculations.

3. Use the percentage rule you formulated in Exercise **2b** to do the following calculations on your calculator.

   a. Find: **(1)** 8% of 306 **(2)** $1\frac{1}{4}\%$ of 144 **(3)** 105% of 50 **(4)** $166\frac{2}{3}\%$ of 99

   b. 20 is 10% of what number? **c.** 3% of what number is 3.86?

   d. 125% of what number is 45? **e.** 64 is 80% of what number?

   f. What percent of 10 is 6? **g.** 18 is what percent of 12?

4. **a.** Find the percent of increase when 60 is increased to 75.

   b. Find the percent of decrease when 180 is decreased to 162.

5. Find each number algebraically.

   a. When one-half a number is increased by one-third of that number, the result is 25.

   b. When one-fifth of a number is decreased by one-tenth of that number, the result is 10.

   c. If one-half of a number is increased by 20, the result is 35.

   d. Seven-hundredths of a number increased by 2.5 equals eight-hundredths of the number.

   e. If seventeen-hundredths of a number is decreased by 1.4, the result is the same as when twelve-hundredths of the number is increased by 7.6.

   f. If 5 is added to one-half of a number, the result is the same as when three-fifths of the number is decreased by 3.

   g. If 3 more than 3 times a number is divided by 15, the result is the same as when 18 less than twice the number is divided by 6.

   h. 25% of a number equals 32.

   i. 15% of a number decreased by 40 is equal to 7% of the number.

   In 6–38, solve algebraically.

6. Ari received marks of 85%, 95%, and 80% in his first three spelling tests. What grade must he receive on his next test to obtain an average of 90% for all four tests?

7. What number added to 8% of itself is 64.8?

8. There were 120 planes on an airfield. If 75% of the planes took off for a flight, how many planes took off?

9. The average of the weights of Sari, Vi, and Becky is 110 pounds. How much does Agnes weigh if the average of the weights of the four girls is 112 pounds?

10. How much silver is in 75 pounds of an alloy that is 8% silver?

Liberty Head Nickel   Buffalo Nickel

The content of alloys used in minting U.S. coins is established by law. Nickels have usually been 75% copper and 25% nickel. In the 1980's, the cent was changed from 95% copper and 5% zinc to a copper-plated coin 97.6% zinc and 2.4% copper.

11. The price of a new boat is $3,450. Mr. Klein made a down payment of 15% of the price of the boat when he bought it. How much was his down payment?

12. The average of the heights of three boys is 5 feet 4 inches. What is the height of a fourth boy if the average of the heights of the four boys is $5\frac{1}{2}$ feet?

13. The average of three consecutive even numbers is 20. Find the numbers.

14. Mr. Brink is 24 years older than his son, Stanley. Four years from now, Stanley will be $\frac{1}{5}$ of his father's age at that time. Find Mr. Brink's present age.

15. Helen bought a coat at a "20% off" sale and saved $12.
  a. What was the marked price of the coat?
  b. At what times during the year is clothing most likely to be on sale?

16. The average of three numbers is 31. The second is 1 more than twice the first. The third is 4 less than three times the first. Find the numbers.

17. A businessman is required to collect a 5% sales tax. One day he collected $281 in taxes. Find the total amount of sales he made that day.

18. It is estimated that in ten years the population of Keysport will increase 75% to 2,800.
  a. Find the present population of Keysport.
  b. What might cause the sharp increase?

19. After Mr. Sims lost 15% of his investment, he had less than $2,550 left. How much did he invest originally?

20. The Bluejays lost 15 of their basketball games thus far this season. They have played 38 games. How many games must they win in succession to raise their winning percentage to at least 70%?

21. Mrs. Pryce sells dresses in a department store. She is paid a weekly salary of $240 plus a 3% commission on the dollar volume of her sales. What is the smallest even-dollar amount she must sell if her weekly income is to exceed $492?

STUDENT PASS

STUDENT _____ RM ____

ISSUED BY _____ DATE _____

DESTINATION

GUIDANCE _____   OFFICE _____

MEDIA CENTER _____   RESTROOM _____

NURSE _____   OTHER _____

TIME ISSUED: _____   TIME ARRIVED: _____

RETURN ISSUED: _____   RETURN ARRIVED: _____

22. A music teacher has 30 students in her class, and 40% of them can play an instrument. How many instrument players must be added to the class in order to achieve a class in which at least 50% can play an instrument?

23. In examining the blueprints of a school building, an architect saw that a storage room was half as wide as it was long. A locker room was 2 feet wider and 3 feet shorter than the storage room, and a passageway was $\frac{2}{3}$ the width of the storage room and 6 feet longer. If the locker room and the passageway had the same area, find the length of the storage room.

24. Mrs. Taylor was entitled to take a 2% discount on a bill. She paid the balance with a check for $76.44.
 a. What was the original amount of the bill?
 b. For what reason might the discount be given?

25. Of 54,650 machine parts made by BMK Tool Works, 4% were found to be defective.
 a. How many parts were good?
 b. How might the number of defective parts be determined without testing every item?

26. Ryan bought a used car and arranged to pay for it in equal monthly installments over a two-year period. His annual automobile insurance costs $\frac{1}{5}$ the price of the car. His monthly payments to the finance company and the insurance company total $350. For what price did Ryan buy the car?

27. Most rubber used to come from trees growing wild in the rain forest. In 1988, 43% of Brazilian rubber production was from plantations. By 1991, it had risen to 60%. Find the percent of increase.

The South American rain forests in the Amazon River basin cover an area almost the size of Australia. Scientists think it important to keep this land in its natural state, since it helps to balance the environment of the entire planet.

But, rubber plantations, cattle ranches, and other developments are invading the forests. To some extent, these industries help to improve the living conditions of the native inhabitants, most of whom live in poverty. Researchers are studying the forest's medicinal plants and other resources, hoping to find ways to help the economy without destroying the forests.

28. Bananas supplied by Tropical Fruits are sold by the Big M Market for 36¢ a pound, and those supplied by S.A. Foods are sold for 41¢ a pound. One day, both kinds were delivered, and the manager decided to sell them at a single price. If he wanted to realize at least $39 for 100 pounds of bananas, what is the greatest number of pounds of 36-cent bananas that should be included?

29. As part of his estate, Farmer McGee left $\frac{1}{4}$ of his cows to his daughter and $\frac{1}{3}$ of his cows to his son. His wife received the remaining cows. If he left at least 28 cows for his children, what was the least number of cows there could have been on the farm?

30. Ron got a job at a department store. He can work a 40-hour week at a rate of $6 an hour or he can work on a salary-plus-commission basis, at $160 per week plus a 2% commission on all sales over $1,000. How much merchandise will Ron have to sell on the commission basis in order for his earnings to exceed the flat rate salary?

31. In his frozen yogurt shop, Mr. Cone has a nut topping and a fudge topping that cost him $2.41 and $2.53 a pound, respectively. He decided to make a mixture of the two toppings. In order for 10 pounds of the mixture to cost him under $25, what is the greatest amount of the fudge topping he should use?

32. If the cost of food in Safe Harbor rose by 5.6% between November 1991 and November 1992, how much did a family pay in November 1992 for the same items that cost $175 in 1991?

33. The Public Service Commission is considering a telephone rate increase that would raise $250,320,000 in additional revenues per year, with $37\frac{1}{2}$% of the additional amount coming from a 7% increase in residential rates. How much is now collected from residences?

34. A 1%-milk-fat cheese selling at $1.38 a pound is combined with a 4%-milk-fat cheese selling at $1.32 a pound to make 12 pounds of the mixture. To get the same total receipts by selling the mixture at $1.36 a pound, how many pounds of each type of cheese should be used?

35. a. Find the reduced cost of a computer with a marked price of $1,299, which is on sale at 20% off, with an extra discount of 5% for paying cash.
  b. Are the two successive discounts of 20% and 5% equal to a single discount of 25%? Explain.

36. The Grandview Public Library sold 100 books, charging 25¢ for books less than 10 years old and 10¢ for books more than 10 years old. If more than $19 was received from the sale, how many of the books were less than 10 years old?

37. During rush hour, 600 people ride the Short Line commuter bus to get home from work. The bus line currently charges $.75 for a Zone 1 fare and $1.20 for a Zone 2 fare. If the bus line could collect the same amount of money with a fare of $.90 for all passengers, how many riders live in Zone 2?

38. The 1990 Census counted the population of the United States as approximately 250 million. This was an increase of about 10% over the number counted by the 1980 Census. The 1980 Census showed a population increase of about 12% over the number counted by the 1970 Census. What was the approximate population, to the nearest million, of the United States in 1970?

While all problems are solved within the general framework, some problems are easier to understand if you group them according to situation. Recognizing the pattern of the overall situation will help you deal with variations in conditions.

Percent-Mixture Problems

When a chemist dilutes pure acid with some other substance, the resulting mixture is no longer pure acid. Consistent with the words, *pure acid* is 100% acid. Thus, there are 15 grams of pure acid in 15 grams of a pure-acid solution. However, how many grams of pure acid are there in 15 grams of a solution that is only 40% acid? The answer is no longer 15 grams, but 40% of 15 grams.

_____ **K**EEP **IN** **M**IND

$$\begin{pmatrix}\text{Total amount}\\ \text{of mixture}\end{pmatrix} \times \begin{pmatrix}\text{Percent of}\\ \text{pure substance}\end{pmatrix} = \begin{pmatrix}\text{Amount of}\\ \text{pure substance}\end{pmatrix}$$

MODEL PROBLEM

How much pure acid must be added to 15 grams of an acid solution that is 40% acid in order to produce a solution that is 50% acid?

Solution: Use a table to organize the data. Enter the percents as decimal equivalents. Let n = the number of grams of pure acid to be added.

Kind of solution	Number of grams	Percent pure acid	Number of grams of pure acid
Original solution	15	.40	6
Pure acid to be added	n	1.00	n
New mixture	$15 + n$	.50	$.50(15 + n)$

The amount of pure acid in the original solution plus the amount of pure acid added equals the amount of pure acid in the new solution.

$$6 + n = .50(15 + n)$$
$$6 + n = 7.50 + .50n$$
$$100(6 + n) = 100(7.50 + .50n)$$
$$600 + 100n = 750 + 50n$$
$$600 + 50n = 750$$
$$50n = 150$$
$$n = 3$$

Check:

$$40\% \text{ of } 15 = .40(15) = 6$$
$$6 + 3 = 9$$
$$50\% \text{ of } (15 + 3) = .50(18) = 9✔$$

ANSWER: 3 grams of pure acid must be added.

A chemist has two solutions, one 55% pure acid and the other 30% pure acid. How many grams of each solution must be mixed to produce 60 grams of a solution that is 35% pure acid?

(See page 382 for solution.)

EXERCISES

1. A solution is 40% pure acid. Represent the number of pounds of pure acid in this solution if it weighs:
   a. 100 lb.   **b.** 12 lb.   **c.** x lb.   **d.** $(x - 2)$ lb.

2. Represent the total amount of pure acid in a solution if the solution contains x grams of acid that is 50% pure acid and $(20 - x)$ grams of acid that is 30% pure acid.

3. A solution that is 20% pure iodine weighs 60 ounces.
   a. Find the number of ounces of pure iodine in the solution.
   b. If x ounces of pure iodine are added to this solution, represent:
     (1) the amount of pure iodine in the resulting solution and
     (2) the number of ounces in the resulting solution.

4. 120 grams of a solution of salt and water contains 25% pure salt.
   a. Find the number of grams of pure salt in the solution.
   b. If x grams of water are evaporated from the solution, represent the number of grams in the resulting solution.
   c. If the resulting solution is 30% pure salt, represent in terms of x the number of grams of salt it contains.

   In 5–19, solve algebraically:

5. A chemist has one solution that is 30% pure acid and another solution that is 60% pure acid. How many pounds of each solution must be used to produce 60 pounds of a solution that is 50% pure acid?

6. A farmer has some cream that is 24% butterfat and some cream that is 18% butterfat. How many quarts of each must she use to produce 90 quarts of cream that is 22% butterfat?

7. How many pounds of a solution that is 75% pure acid must be mixed with 16 pounds of a solution that is 30% pure acid to produce a solution that is 55% pure acid?

8. How much pure acid must be added to 25 ounces of a solution of acid and water that is 20% pure acid in order to make a solution that is 50% pure acid?

9. How much salt must be added to 80 pounds of a 5% salt solution to make a 24% salt solution?

10. A chemist has 40 ounces of a solution of iodine and alcohol that is 15% iodine. How much pure iodine must be added to make a solution that is 20% iodine?

11. A solution of iodine and alcohol contains 3 ounces of iodine and 21 ounces of alcohol. How much pure iodine must be added to produce a solution that is 25% iodine?

12. A solution of alcohol and water that is 20% alcohol weighs 60 pounds. How much water must be added to make a solution that is 5% alcohol?

13. A solution contains 8 pounds of acid and 32 pounds of water. How many pounds of water must be evaporated to produce a solution that will be 40% acid?

14. How much water must be added to 30 pounds of a solution of salt water that contains 20% salt so that the resulting solution will be 15% salt?

15. The results of a marketing study by the Delicious Candy Company showed that their hard-candy mixture was most attractive to consumers when 19% of the candy was yellow. The current inventory of 600 pounds of the mixture is 10% yellow. How many pounds of yellow candy must they add to the existing inventory to get the desired mixture?

16. A garden center sold a grass-seed mixture consisting of 60% rye and 40% bluegrass. They also sold pure bluegrass seed. How much bluegrass seed must be added to 5 pounds of the mixture in order to reverse the percentages?

17. Adobe consists primarily of a mixture of clay and soil, with straw added to strengthen the mixture and prevent cracking. An acceptable amount of straw is 4% by weight. If 153 pounds of the clay and soil have already been mixed, what is the weight of the straw that should be added?

Considering that adobe is simply dried mud, adobe houses are astonishingly solid and long-lasting. Almost a thousand years ago, native Americans used adobe to make buildings up to 5 stories high. The Taos Pueblo, a 900-year-old adobe structure, is still lived in.

Adobe houses can be formed into interesting rounded and irregular shapes that blend into the landscape. In the sunshine of the American Southwest, adobe provides solar heating, absorbing heat by day and radiating it by night.

18. Gasohol, a fuel sometimes used in automobiles as a means of conserving energy, is 90% gasoline. The remaining 10% is ethyl alcohol, which can be produced from potatoes, sugar cane, or grains. If a 60-gallon mixture of ethyl alcohol and gasoline in a storage tank is 32% ethyl alcohol, how many gallons of gasoline must be added to obtain gasohol?

19. A food technician is testing mixtures for a new tropical punch made up of pineapple, banana, and papaya juices. If 24 ounces of the mixture contains 25% papaya juice, how much papaya juice must be added to make 55% of the mixture?

Investment Problems

People invest money at varying rates of interest to earn income. The initial amount of money is called the *principal*, which earns income on a yearly, or *annual*, basis. The income earned depends upon the size of the principal and the rate of interest paid.

$$\left(\begin{array}{c}\textbf{Principal in}\\ \textbf{dollars}\end{array}\right) \times \left(\begin{array}{c}\textbf{Annual rate}\\ \textbf{of interest}\end{array}\right) = \left(\begin{array}{c}\textbf{Annual income}\\ \textbf{in dollars}\end{array}\right)$$

MODEL PROBLEM

Mr. Parsons invested a sum of money at 6% and a second sum, $500 more than the first, at 8%. If the total annual income was $180, how much did he invest at each rate?

Solution: Use a table to organize the data, entering the percents as decimal equivalents.

Let p = the number of dollars invested at 6%.
Then $p + 500$ = the number of dollars invested at 8%.

Principal in dollars	Annual rate of interest	Annual income in dollars
p	.06	$.06p$
$p + 500$	.08	$.08(p + 500)$

The total annual income is $180.

$$.06p + .08(p + 500) = 180$$
$$6p + 8(p + 500) = 18{,}000$$
$$6p + 8p + 4{,}000 = 18{,}000$$
$$14p + 4{,}000 = 18{,}000$$
$$14p = 14{,}000$$
$$p = 1{,}000$$
$$p + 500 = 1{,}500$$

Check:

$$\$1{,}000 + \$500 = \$1{,}500$$
$$.06(\$1{,}000) = \$\ 60$$
$$.08(\$1{,}500) = \underline{\$120}$$
$$\text{Total} = \$180 \checkmark$$

ANSWER: $1,000 was invested at 6% and $1,500 was invested at 8%.

Now it's your turn to . . . **TRY IT!**   *(See page 383 for solution.)*

Mrs. Fernandez invested $10,000, part at 9% and the rest at 6%. The annual incomes from these investments were equal. How much was invested at each rate?

1. Represent the annual income when the annual rate is 10% and the amount invested is:
   a. $600   **b.** $2,500   **c.** $x   **d.** $3x   **e.** $(x + 500)   **f.** $(5,000 − x)

2. Represent the total annual income in each of the following.
   a. $4,000 invested at 5% and $6,500 invested at 9%
   b. $3,500 invested at $3\frac{1}{2}$% and $4,200 invested at $7\frac{1}{2}$%
   c. $8,000 invested at 6% and $4x invested at 6%
   d. $x invested at 5% and $4x invested at 6%
   e. $x invested at 6% and $(x + 2,000) invested at 7%
   f. $x invested at 4% and $(8,000 − x) invested at 10%

3. Mr. Walker invested $x at 8%. He also invested $500 more than this sum at 10%.
   a. Represent the amount he invested at 10%.
   b. Represent the annual income from **(1)** the 8% investment **(2)** the 10% investment
   c. Represent the total annual income from both investments.
   d. Write an open sentence to indicate that Mr. Walker's total annual income from these investments was $113.

4. Mrs. Collins invested a portion of $9,000 at 5% and the remainder at 10%. If x represents the amount she invested at 5%:
   a. Represent the amount she invested at 10%.
   b. Represent the annual income from **(1)** the 5% investment **(2)** the 10% investment
   c. Write an open sentence to indicate that the two annual incomes are equal.

   In 5–20, solve algebraically.

5. Mr. Sanchez invested a sum of money at 8%. He invested twice as much at 5%. The total annual income from these investments was $180. Find the amount he invested at each rate.

6. Ms. Todd invested a sum of money at 5%. She invested a second sum, $250 more than the first, at 6%. If her total annual income was $59, how much did she invest at each rate?

7. Mr. Fox invested a sum of money at 6%. He invested a second sum, $2,000 less than the first, at 5%. He invested a third sum, which was $3,000 less than the first, at $7\frac{1}{2}$%. His total annual income was $304. Find the amount he invested at each rate.

8. Mrs. Ryan invested a sum of money at 6%. She invested a second sum, which exceeded twice the first sum by $1,000, at 10%. Her total annual income was $620. Find the amount she invested at each rate.

9. Mr. Carlson has invested $7,500 in two parts, one part at 6% and the other at 10%. Find the amount invested at each rate if the total yearly income is $590.

10. Miss Daniels had a sum of money to invest. She invested $\frac{1}{2}$ of the sum at 5%, $\frac{1}{3}$ of the sum at 6%, and the remainder at 8%. Her total annual income was $350. What was the total sum that Miss Daniels invested?

11. Mr. West invested $7,200, part at 8% and the remainder at 10%. If the annual incomes from both investments were equal, find the amount invested at each rate.

12. Mrs. Chan invested $7,500 in two business enterprises. In one enterprise, she made a 5% profit; in the other, she had a 2% loss. Her net profit for the year was $130. Find the amount invested at each rate.

13. Mrs. Lamb has invested $18,000 in two parts. One part is invested at 8% and the other at 10%. The annual income from the 10% investment is $360 more than the annual income from the 8% investment. Find the amount invested at each rate.

14. Miss Ramirez has invested $6,000 at 5%. How much additional money must she invest at 8% so that her annual income from the two investments will be 6% of the total amount invested?

15. Mr. Peterson invested in a growth stock and made a 20% return on his investment. His annual statement shows that his shareholdings now total $60,000. How much money did Mr. Peterson originally invest?

16. The Reid Corporation invests $\frac{1}{3}$ of its holdings in municipal bonds paying 8% interest, $\frac{1}{4}$ of its holdings in treasury bills paying 10% interest, and the balance in savings institutions paying 6% interest. How much must the Reid Corporation invest if its annual income is to be at least $55,200?

17. During a recent bull market period, Jehara invested $11,000 of her savings in the stock market, part in a high-yield investment that paid 12%, and the rest in a 7% bond. If her investments earned $1,090 in the first year, how much did she invest in each?

In stock-market slang, investors who expect prices to rise are called *bulls*, and those who anticipate falling prices are *bears*. The expressions may come from the animals' styles of attack: the bull thrusts upward with its horns; the bear strikes downward with its paw.

18. Fredda invested $4,000 in a 6% municipal bond and $6,000 in a mutual fund. If her investments earned $780 in a year, what was the rate of return from the mutual fund?

19. In a bear market, the Clancy brothers lost a total of $900 on two investments. They lost 3% on their $6,000 investment, and lost 8% on the other. What was the amount on which they lost 8%?

20. After Mr. Wright earned 6% on a stock investment, he made a second investment, $1,800 more than the first, in another stock. However, the second investment resulted in a 3% loss. If he gained a total of $369 from both investments, what was the amount invested in each?

CALCULATOR CONNECTION

A calculator gives you the ability to perform certain types of calculations that would be almost impossible without a calculator. An example of this is *compound interest*.

The formula used to calculate the future value of an investment is:

$$\textbf{Future value} = \textbf{\textit{P}}(1 + \textbf{\textit{i}})^n$$

where P is the principal, i the interest rate, and n the number of compounding periods.

Example: $1,000 is invested in a 5-year certificate of deposit that pays 8% per year, compounded monthly. At maturity, how much will the CD be worth?

$$P = \$1{,}000$$

$$i = \frac{8\%}{12} \text{ (compounded 12 times each year)}$$

$$n = 5 \times 12, \text{ or } 60 \text{ (months in 5 years)}$$

$$\text{Future value} = 1{,}000 \left(1 + \frac{8\%}{12}\right)^{60}$$

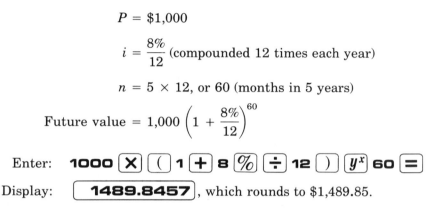

Enter: **1000** $\boxed{\times}$ $\boxed{(}$ **1** $\boxed{+}$ **8** $\boxed{\%}$ $\boxed{\div}$ **12** $\boxed{)}$ $\boxed{y^x}$ **60** $\boxed{=}$

Display: $\boxed{\textbf{1489.8457}}$, which rounds to $1,489.85.

This formula can be rewritten to show how much must be invested to achieve a savings goal. Using the investment circumstances of the previous example, how much should be invested in order to end up with $2,000?

Solve the formula for P and substitute values: $P = \dfrac{\text{Future value}}{(1 + i)^n} = \dfrac{2{,}000}{\left(1 + \dfrac{8\%}{12}\right)^{60}}$

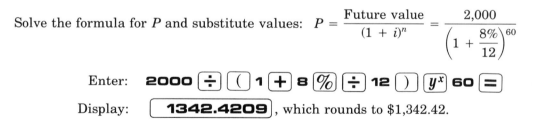

Enter: **2000** $\boxed{\div}$ $\boxed{(}$ **1** $\boxed{+}$ **8** $\boxed{\%}$ $\boxed{\div}$ **12** $\boxed{)}$ $\boxed{y^x}$ **60** $\boxed{=}$

Display: $\boxed{\textbf{1342.4209}}$, which rounds to $1,342.42.

EXERCISES

1. What amounts of interest would be earned on a 90-day investment of $2,500, at an annual rate of 7.2%, compounded daily?   ***ANSWER:*** $44.78

2. The Horgans are investing for their 12-year-old son's college tuition. If they want to have $10,000 when he is 18 years old, how much should they invest at an annual rate of 7.5%, compounded monthly?   ***ANSWER:*** $6,385.22

7-6 SOLVING EQUATIONS THAT CONTAIN RATIONAL EXPRESSIONS

Fractional Equations

An equation that contains rational expressions has a variable in the denominator of one or more of its terms. Such equations are also called **fractional equations**.

$$\text{Examples:} \quad \frac{1}{3} + \frac{1}{x} = \frac{1}{2} \qquad \frac{6}{x} = \frac{7}{x+2}$$

Fractional equations are solved in the same way as are equations with fractional coefficients.

PROCEDURE

To solve a fractional equation:

Clear of fractions by multiplying both sides of the equation by the L.C.D. Remember that the variable cannot represent a number that would make any denominator, and thus the L.C.D., equal to zero.

MODEL PROBLEMS

1. Solve, stating restrictions on the variable, and check: $\dfrac{1}{3} + \dfrac{1}{x} = \dfrac{1}{2}$

Solution: Multiply both members of the equation by the L.C.D., $6x$.
x cannot be equal to 0.

$$\frac{1}{3} + \frac{1}{x} = \frac{1}{2}$$

$$6x\left(\frac{1}{3} + \frac{1}{x}\right) = 6x\left(\frac{1}{2}\right)$$

$$\overset{2x}{\cancel{6x}}\left(\frac{1}{\cancel{3}}\right) + \overset{6}{\cancel{6x}}\left(\frac{1}{\cancel{x}}\right) = \overset{3x}{\cancel{6x}}\left(\frac{1}{\cancel{2}}\right)$$
$$\qquad 1 \qquad\qquad 1 \qquad\qquad 1$$

$$2x + 6 = 3x$$

$$6 = x$$

Check:

$$\frac{1}{3} + \frac{1}{x} = \frac{1}{2}$$

$$\frac{1}{3} + \frac{1}{6} \overset{?}{=} \frac{1}{2}$$

$$\frac{3}{6} \overset{?}{=} \frac{1}{2}$$

$$\frac{1}{2} = \frac{1}{2} ✔$$

ANSWER: $x = 6$

---

2. Solve, stating the restrictions on the variable, and check: $\dfrac{1}{w} + \dfrac{w}{w-2} = 1$

Solution: Multiply both sides by the L.C.D., $w(w-2)$, where $w \ne 0, 2$.

Check

$$\frac{1}{w} + \frac{w}{w-2} = 1$$

$$w(w-2)\left(\frac{1}{w} + \frac{w}{w-2}\right) = 1w(w-2)$$

$$\overset{1}{\cancel{w}}(w-2)\left(\frac{1}{\cancel{w}}\right) + w(\cancel{w-2})\overset{1}{}\left(\frac{w}{\cancel{w-2}}\right) = 1w(w-2)$$

$$w - 2 + w^2 = w^2 - 2w$$
$$w - 2 = -2w$$
$$3w = 2$$
$$w = \tfrac{2}{3}$$

$$\frac{1}{w} + \frac{w}{w-2} = 1$$

$$\frac{1}{\frac{2}{3}} + \frac{\frac{2}{3}}{\frac{2}{3} - 2} \overset{?}{=} 1$$

$$\frac{1}{\frac{2}{3}} + \frac{\frac{2}{3}}{-\frac{4}{3}} \overset{?}{=} 1$$

$$1\left(\tfrac{3}{2}\right) + \left(\tfrac{2}{3}\right)\left(-\tfrac{3}{4}\right) \overset{?}{=} 1$$

$$\tfrac{3}{2} - \tfrac{1}{2} \overset{?}{=} 1$$

$$1 = 1 ✔$$

ANSWER: $w = \tfrac{2}{3}$

3. Solve, stating the restriction on the variable, and check: $\dfrac{10}{x-1} = 12 + \dfrac{10}{x-1}$

Solution: Multiply both sides of the equation by the L.C.D., $x - 1$, where $x \ne 1$.

$$\frac{10}{x-1} = \frac{12}{1} + \frac{10}{x-1}$$

$$(x-1)\frac{10}{x-1} = (x-1)\left(\frac{12}{1} + \frac{10}{x-1}\right)$$

$$(\overset{1}{\cancel{x-1}})\frac{10}{(\cancel{x-1})} = (x-1)\left(\frac{12}{1}\right) + (\overset{1}{\cancel{x-1}})\left(\frac{10}{\cancel{x-1}}\right)$$

$$10 = 12x - 12 + 10$$
$$10 = 12x - 2$$
$$12 = 12x$$
$$1 = x$$

(continued on next page)

Note that $x = 1$ is a solution only of the *transformed* equation:

$$10 \overset{?}{=} 12(x - 1) + 10$$
$$10 \overset{?}{=} 12(1 - 1) + 10$$
$$10 \overset{?}{=} 12(0) + 10$$
$$10 \overset{?}{=} 0 + 10$$
$$10 = 10 \checkmark$$

Check: $x = 1$ does not satisfy the *original* equation:

$$\frac{10}{x - 1} = 12 + \frac{10}{x - 1}$$
$$\frac{10}{1 - 1} \overset{?}{=} 12 + \frac{10}{1 - 1}$$
$$\frac{10}{0} \overset{?}{=} 12 + \frac{10}{0} \quad \text{(undefined)}$$

When the value of the multiplier is 0, the transformed equation is not equivalent to the original equation. The value 1 is *not* a solution of the original equation, and is called ***extraneous***.

ANSWER: $\varnothing$

KEEP IN MIND

When both sides of an equation are multiplied by a variable expression, the resulting equation may not be equivalent to the given equation. Therefore, as with all checks, each candidate for solution must be checked in the given equation.

Now it's your turn to . . . **TRY IT!** (*See page 383 for solutions.*)

In 1–3, solve, stating the restriction on the variable, and check.

1. $\dfrac{5}{3a} - \dfrac{7}{a} = \dfrac{1}{6}$ **2.** $\dfrac{2}{3d} + \dfrac{1}{3} = \dfrac{11}{6d} - \dfrac{1}{4}$ **3.** $\dfrac{3x - 5}{3x + 5} = \dfrac{1}{2}$

Equations Involving Several Variables

The solution of an equation that contains more than one variable is not usually a numerical value. The variable for which you are solving is made to stand alone on one side of the equation, with the remaining term(s) on the other.

PROCEDURE

To solve an equation that involves several variables for one of those variables:
1. Clear of fractions and parentheses.
2. Collect variable terms on one side, obtaining a single coefficient.
3. Divide both sides of the equation by that coefficient.
4. Simplify the answer if necessary.

MODEL PROBLEM

Solve for y: $\dfrac{y}{b} + \dfrac{y}{a} = a + b$

How to Proceed	*Solution*
(1) The variable terms are all on one side.	$\dfrac{y}{b} + \dfrac{y}{a} = a + b$
(2) To clear of fractions, multiply by the L.C.D., ab.	$ab\left(\dfrac{y}{b} + \dfrac{y}{a}\right) = ab(a + b)$
(3) To clear of parentheses, distribute. Cancel.	$\overset{a}{\cancel{ab}}\left(\dfrac{y}{\underset{1}{\cancel{b}}}\right) + \overset{b}{\cancel{ab}}\left(\dfrac{y}{\underset{1}{\cancel{a}}}\right) = ab(a) + ab(b)$
(4) Multiply.	$ay + by = a^2b + ab^2$
(5) Factor out the variable y.	$y(a + b) = a^2b + ab^2$
(6) Divide by $a + b$, the coefficient of the variable y.	$\dfrac{y\overset{1}{\cancel{(a + b)}}}{\underset{1}{\cancel{(a + b)}}} = \dfrac{a^2b + ab^2}{(a + b)}$
(7) Simplify by factoring and cancelling.	$y = \dfrac{ab\overset{1}{\cancel{(a + b)}}}{\underset{1}{\cancel{(a + b)}}}$
	$y = ab$ **ANSWER**

Now it's your turn to . . . **TRY IT!**   (*See page 384 for solutions.*)

1. Solve for x: $cx + d^2 = c^2 + dx$ **2.** Solve the formula $R = \dfrac{g}{g - s}$ for g. $(g \neq s)$

EXERCISES

In 1–40, solve, stating restrictions on the variable, and check.

1. $\dfrac{10}{x} = 5$ **2.** $\dfrac{15}{y} = 3$ **3.** $\dfrac{6}{x} = 12$ **4.** $\dfrac{8x}{x} = 0$

5. $\dfrac{3}{2x} = \dfrac{1}{2}$ **6.** $\dfrac{15}{4x} = \dfrac{1}{8}$ **7.** $\dfrac{7}{3y} = -\dfrac{1}{3}$ **8.** $\dfrac{4}{5y} = -\dfrac{1}{10}$

9. $\dfrac{10}{x} + \dfrac{8}{x} = 9$ **10.** $\dfrac{15}{y} - \dfrac{3}{y} = 4$ **11.** $\dfrac{7}{c} + \dfrac{1}{c} = 16$ **12.** $\dfrac{9}{2x} = \dfrac{7}{2x} + 2$

13. $\dfrac{30}{x} = 7 + \dfrac{18}{2x}$ **14.** $\dfrac{3}{a} = \dfrac{19}{3a} - \dfrac{5}{3}$ **15.** $\dfrac{4}{c} - \dfrac{1}{2} = \dfrac{5}{12} - \dfrac{3}{2c}$

16. $\dfrac{y+9}{2y} + 3 = \dfrac{15}{y}$ **17.** $\dfrac{3}{2x} - 1 = \dfrac{x+1}{x}$ **18.** $\dfrac{2+x}{6x} = \dfrac{3}{5x} + \dfrac{1}{30}$

19. $\dfrac{15}{a+1} = 3$ **20.** $\dfrac{12}{x-2} = 4$ **21.** $\dfrac{10}{x+2} = 2$

22. $\dfrac{9}{2x+1} = 3$ **23.** $\dfrac{10}{2t-1} = 2$ **24.** $\dfrac{16}{1-3t} = 4$

25. $\dfrac{6}{3x-1} = \dfrac{3}{4}$ **26.** $\dfrac{2}{3x-4} = \dfrac{1}{4}$ **27.** $\dfrac{5x}{x+1} = 4$

28. $\dfrac{3}{5-3a} = \dfrac{1}{2}$ **29.** $\dfrac{4z}{7+5z} = \dfrac{1}{3}$ **30.** $\dfrac{1-r}{1+r} = \dfrac{2}{3}$

31. $\dfrac{3}{y} = \dfrac{2}{5-y}$ **32.** $\dfrac{5}{a} = \dfrac{7}{a-4}$ **33.** $\dfrac{2}{m} = \dfrac{5}{3m-1}$

34. $\dfrac{b+1}{b-3} = \dfrac{b-3}{b+1}$ **35.** $\dfrac{y-2}{y+4} = \dfrac{y-3}{y+1}$ **36.** $\dfrac{x-2}{x+1} = \dfrac{x+1}{x-2}$

37. $\dfrac{3}{d-4} + \dfrac{2}{d+4} = \dfrac{24}{d^2-16}$ **38.** $\dfrac{5}{x^2-4} - \dfrac{x+12}{x+2} + \dfrac{x-1}{x-2} = 0$

39. $\dfrac{y}{y+1} - \dfrac{1}{y} = 1$ **40.** $\dfrac{x-3}{x-1} - \dfrac{x-1}{x} = \dfrac{5}{x^2-x}$

In 41–86, solve for x or y and check. Assume that no denominator is 0.

41. $\dfrac{y}{r} = s$ **42.** $\dfrac{x}{3a} = b$ **43.** $s = \dfrac{y}{5t}$ **44.** $\dfrac{x}{3} + b = 4b$

45. $\dfrac{x}{2} - c = d$ **46.** $2m = \dfrac{y}{5} + n$ **47.** $\dfrac{y}{2} - \dfrac{b}{3} = 0$ **48.** $\dfrac{x}{a} = \dfrac{b}{c}$

49. $\dfrac{x}{n} = \dfrac{s}{n^2}$ **50.** $\dfrac{mx}{r} + c = d$ **51.** $a = \dfrac{by}{c} - d$ **52.** $t = s - \dfrac{nx}{d}$

53. $\dfrac{5}{x} = a$ **54.** $\dfrac{r}{x} = t$ **55.** $\dfrac{t}{y} - r = 0$

56. $\dfrac{x-4b}{5} = 8b$ **57.** $\dfrac{2x+c}{3} = 9c$ **58.** $\dfrac{3y-2a}{4} - 7a = 0$

59. $\dfrac{a+b}{x} = c$ **60.** $\dfrac{m+n}{y} - a = 0$ **61.** $\dfrac{a}{b} = \dfrac{c+d}{x}$

62. $\dfrac{y}{4} + \dfrac{y}{6} = 5b$ **63.** $\dfrac{x}{3a} + \dfrac{x}{5a} = 8$ **64.** $\dfrac{y}{2a} - \dfrac{y}{3a} = 3a^2$

65. $\dfrac{8}{x} - \dfrac{7}{x} = e$　　　　　　**66.** $\dfrac{r}{y} + s = \dfrac{t}{y}$　　　　　　**67.** $\dfrac{c}{y} = \dfrac{d}{y} + h$

68. $ax + bx = 4a + 4b$　　　　**69.** $cy + dy = r$　　　　**70.** $3x - a = 3 - ax$

71. $ax - b = bx$　　　　**72.** $ax + b^2 = a^2 - bx$　　　　**73.** $cx - c^2 = dx - d^2$

74. $a(x + b) = 6ax - 9ab$　　　**75.** $8x = 2x - 4(x - 5c)$　　　**76.** $(x - a)(x - b) = x^2$

77. $c(c - y) = d(d + y)$　　　**78.** $\dfrac{x}{a} + \dfrac{x}{b} = 1$　　　**79.** $\dfrac{1}{x} = \dfrac{1}{c} + \dfrac{1}{d}$

80. $\dfrac{x}{d} - \dfrac{x}{c} = c - d$　　　**81.** $\dfrac{1}{a} + \dfrac{1}{x} = \dfrac{1}{b} - \dfrac{1}{x}$　　　**82.** $x = \dfrac{a - x}{b}$

83. $\dfrac{x + s}{r} = \dfrac{r - x}{s}$　　　**84.** $\dfrac{y - 1}{y - 3} = a$　　　**85.** $\dfrac{y + m}{y + n} = \dfrac{y + n}{y + m}$

86. $\dfrac{x - a}{x - b} = \dfrac{b - x}{a - x}$

In 87–101, solve the formula for the indicated variable.

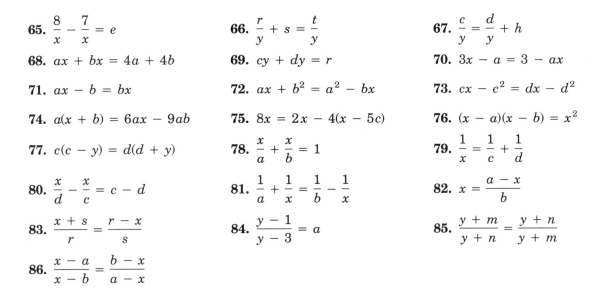

87. $F = \dfrac{mv^2}{r}$ for m

88. $V^2 = \dfrac{L}{KA}$ for A

89. $\dfrac{P}{N} = \dfrac{p}{n}$ for N

90. $\dfrac{E}{R + r} = I$ for r

When a stone tied to a string is whirled around, the string holds the stone in its path by *centripetal force*, according to the formula given in Exercise 87.

By contrast, *centrifugal force* causes the stone, or any object rounding a curve at high speed, to shoot outward. For safety on the highway, roads are "banked," that is, built higher on the outer edges of curves.

91. $F = 32 + \tfrac{9}{5}C$ for C

92. $\dfrac{D}{d} = q + \dfrac{r}{d}$ for d

93. $S = \dfrac{n}{2}(a + l)$ for a　　　**94.** $A = \tfrac{1}{2}h(b + c)$ for b　　　**95.** $A = P(1 + rt)$ for P

96. $A = p + prt$ for p　　　**97.** $n = \dfrac{a - W}{6W}$ for W　　　**98.** $\dfrac{1}{f} = \dfrac{1}{p} + \dfrac{1}{q}$ for p

99. $\dfrac{1}{f} - \dfrac{1}{q} = \dfrac{1}{p}$ for q　　　**100.** $S = \dfrac{rl - a}{r - l}$ for r　　　**101.** $I = \dfrac{E}{r_1 + r_2}$ for r_2

Number Problems

MODEL PROBLEMS

---

1. The denominator of a fraction exceeds the numerator by 7. If 3 is subtracted from the numerator, and the denominator is unchanged, the value of the resulting fraction is $\frac{1}{3}$. Find the original fraction.

Solution:

Let x = numerator of original fraction.
Then $x + 7$ = denominator of original fraction.

And $\dfrac{x}{x + 7}$ = original fraction.

And $\dfrac{x - 3}{x + 7}$ = new fraction.

Value of new fraction is $\frac{1}{3}$.

$$\frac{x - 3}{x + 7} = \frac{1}{3}$$

$$3\overset{1}{\cancel{(x + 7)}} \left(\frac{x - 3}{\cancel{x + 7}}\right) = \overset{1}{\cancel{3}}(x + 7)\left(\frac{1}{\cancel{3}}\right)$$

$$3(x - 3) = 1(x + 7)$$
$$3x - 9 = x + 7$$
$$3x - x = 7 + 9$$
$$2x = 16$$
$$x = 8$$
$$x + 7 = 15$$

Check: The original fraction was $\dfrac{8}{15}$.

The new fraction is $\dfrac{8 - 3}{15} = \dfrac{5}{15} = \dfrac{1}{3}$. ✔

ANSWER: The original fraction was $\dfrac{8}{15}$.

2. The larger of two numbers is 2 less than 4 times the smaller. When the larger number is divided by the smaller, the quotient is 3 and the remainder is 5. Find the numbers.

Solution: Apply the strategy of examining a simpler related problem to see how to write an equation with a quotient and remainder.

$$23 \div 6: \quad \frac{23}{6} = 3 + \frac{5}{6}$$

Thus, in this problem:

Let x = the smaller number.
Then $4x - 2$ = the larger number.

When the larger number is divided by the smaller, the quotient is 3 and the remainder is 5.

$$\frac{4x - 2}{x} = 3 + \frac{5}{x}$$

$$\overset{1}{\cancel{x}}\left(\frac{4x - 2}{\cancel{x}}\right) = x\left(3 + \frac{5}{x}\right)$$

$$4x - 2 = 3x + 5$$
$$4x - 3x = 5 + 2$$
$$x = 7$$
$$4x - 2 = 26$$

Check: The larger number, 26, is 2 less than 4 times the smaller, 7. ✔ If 26 is divided by 7, the quotient is 3 and the remainder is 5. ✔

ANSWER: Smaller 7; larger 26.

| Now it's your turn to . . . **TRY IT!** | *(See page 384 for solution.)* |

The larger of two numbers is 5 less than 4 times the smaller. When the larger number is divided by the smaller, the quotient is 3 and the remainder is 3. Find the numbers.

EXERCISES

In 1–15, solve algebraically.

1. If one-half of a number is 8 more than one-third of the number, find the number.

2. Separate 150 into two parts such that one part is two-thirds of the other part.

3. One-third of the result obtained by adding 5 to a certain number is equal to one-half of the result obtained when 5 is subtracted from the number. Find the number.

4. The numerator of a fraction is 8 less than the denominator of the fraction. The value of the fraction is $\frac{3}{5}$. Find the fraction.

5. The denominator of a fraction exceeds twice the numerator of the fraction by 10. The value of the fraction is $\frac{5}{12}$. Find the fraction.

6. The denominator of a fraction is 30 more than the numerator of the fraction. If 10 is added to the numerator of the fraction and the denominator is unchanged, the value of the resulting fraction becomes $\frac{3}{5}$. Find the original fraction.

7. The numerator of a fraction is three times the denominator. If the numerator is decreased by 1 and the denominator is increased by 2, the value of the resulting fraction is $\frac{5}{2}$. Find the original fraction.

8. What number must be added to both the numerator and denominator of the fraction $\frac{7}{19}$ to make the value of the resulting fraction $\frac{3}{4}$?

9. The numerator of a fraction exceeds the denominator by 3. If 3 is added to the numerator and 3 is subtracted from the denominator, the resulting fraction is equal to $\frac{5}{2}$. Find the original fraction.

10. The numerator of a fraction is 7 less than the denominator. If 3 is added to the numerator and 9 is subtracted from the denominator, the resulting fraction is equal to $\frac{3}{2}$. Find the original fraction.

11. If 1 less than 4 times a certain number is divided by 9, the quotient is 3. Find the number.

12. Separate 96 into two parts such that when the larger part is divided by the smaller part, the quotient is 7.

13. The larger of two numbers is 25 more than the smaller. If the larger is divided by the smaller, the quotient is 5 and the remainder is 1. Find the numbers.

14. The larger of two numbers exceeds twice the smaller by 9. The larger divided by the smaller gives a quotient of 3 and a remainder of 4. Find the numbers.

15. When the reciprocal of a number is decreased by 2, the result is 5. Find the number.

Motion Problems

1. A dogsled racer stopped while crossing some difficult terrain to tend one of his dogs. He had traveled 32 kilometers before stopping, and then traveled 96 kilometers afterward, at twice the earlier rate. If the actual running time was 5 hours, find his average rates before and after stopping.

   Solution: Use a table to organize the data.
   Let r = average rate before stop.
   Then $2r$ = average rate after stop.

	Distance (km)	Rate (km/h)	Time (h)
Before stop	32	r	$\dfrac{32}{r}$
After stop	96	$2r$	$\dfrac{96}{2r}$

   Total travel time was 5 hours.

   $$\frac{32}{r} + \frac{96}{2r} = 5$$

   $$2r\left(\frac{32}{r} + \frac{96}{2r}\right) = 2r(5)$$

   $$\overset{2}{\cancel{2r}}\left(\frac{32}{\underset{1}{\cancel{r}}}\right) + \overset{1}{\cancel{2r}}\left(\frac{96}{\underset{1}{\cancel{2r}}}\right) = 2r(5)$$

   $$64 + 96 = 10r$$
   $$160 = 10r$$
   $$16 = r$$
   $$32 = 2r$$

   Check:
   Travel time before stop:

   $$\frac{32}{r} = \frac{32}{16} = 2 \text{ hours}$$

   Travel time after stop:

   $$\frac{96}{2r} = \frac{96}{32} = 3 \text{ hours}$$

   Total time: 3 + 2, or 5 hours ✓

   ANSWER: The average rate before stopping was 16 km/h; after, 32 km/h.

The 1990 Iditarod Trail Sled Dog Race was run under unusually harsh conditions—the deepest snow in a quarter of a century, a fall of volcanic ash, and aggressive moose along the trail. Nevertheless, Susan Butcher completed the 1,158-mile race from Anchorage, Alaska to Nome in the record time of 11 days, 1 hour, 53 minutes, and 23 seconds, arriving first in a field of 70 mushers for her 4th win in 5 years.

2. A boat can travel 30 mph in still water. It can travel 80 miles downstream in the same time that it requires to travel 40 miles upstream. Find the rate of the stream.

Solution: Use a table to organize the data.

Let c = rate of stream.

Then $30 + c$ = rate of boat downstream.

And $30 - c$ = rate of boat upstream.

	Distance (miles)	Rate (mph)	Time (h)
Down-stream	80	$30 + c$	$\dfrac{80}{30 + c}$
Up-stream	40	$30 - c$	$\dfrac{40}{30 - c}$

Travel time upstream is the same as travel time downstream.

$$\frac{40}{30 - c} = \frac{80}{30 + c}$$

$$\text{L.C.D.} = (30 - c)(30 + c)$$

$$(30 - c)(30 + c)\left(\frac{40}{30 - c}\right) = (30 - c)(30 + c)\left(\frac{80}{30 + c}\right)$$

$$40(30 + c) = 80(30 - c)$$

$$1{,}200 + 40c = 2{,}400 - 80c$$

$$1{,}200 + 120c = 2{,}400$$

$$120c = 1{,}200$$

$$c = 10$$

Check:
Travel time downstream:

$$\frac{80}{30 + c} = \frac{80}{40} = 2 \text{ hours}$$

Travel time upstream:

$$\frac{40}{30 - c} = \frac{40}{20} = 2 \text{ hours}$$

The times are the same. ✔

ANSWER: The rate of the stream is 10 mph.

Now it's your turn to . . . **TRY IT!**

1. Henri traveled 120 miles at a certain rate. By increasing his average rate by 10 mph, he traveled 160 miles on a second trip in the same time that he spent on the 120-mile trip. Find his average rate on the first trip.

2. A plane flying at maximum speed can fly 660 kilometers with an 80 km/h tailwind in the same amount of time that it can fly 540 kilometers against the same wind. What is the maximum speed of the plane when there is no wind?

(See page 384 for solutions.)

In 1–12, solve algebraically.

1. Mr. Sakai biked 12 miles and returned on foot. His rate on the bicycle was 4 times his rate on foot. He spent 5 hours on the entire trip. Find his rate of walking.

2. Ms. Carlsen drove 1,000 miles and returned by plane. She spent 22 hours traveling. If the rate of the plane was 10 times the rate at which she drove, find each rate.

3. Reese traveled 640 miles by ship and returned by plane. He spent 42 hours on the trip. If the rate of the plane was 20 times the rate of the ship, find each rate.

4. Robin drove 70 miles at a certain average rate. If she increased her average rate by 5 miles per hour, she could travel 80 miles in the same time that she spent on the 70-mile trip. Find her average rate on each trip.

5. James drives 15 miles per hour slower than Kenton. If James can travel 150 miles in the same time that Kenton can travel 225 miles, find each rate.

6. One automobile averages 8 mph less than another. The first travels 160 miles in the same time that the other travels 200 miles. Find their rates.

7. The rate of a jet plane exceeds twice the rate of a cargo plane by 100 miles per hour. If the jet can fly 1,800 miles in the same time that the cargo plane requires to fly 750 miles, find each rate.

8. The rate of a train is 35 mph more than the rate of a bus. If it takes the train $\frac{1}{2}$ as much time to travel 350 miles as it does the bus, find each rate.

9. A plane can fly 320 mph in still air. Flying with the wind, the plane flies 1,400 miles in the same time that it requires to fly 1,160 miles against the wind. Find the rate of the wind.

10. A small plane can fly 120 mph in still air. With the wind, it can fly 640 miles in a certain time. Against the wind, it can fly only half this distance in the same time. Find the rate of the wind.

11. Carly planned a trip from Walla Walla, Washington to Pensacola, Florida, a distance of about 3,240 miles. She found travel by train would take 66 hours more than by plane. If the plane travels at 12 times the rate of the train, find the rate of the train.

12. The colorful marine life in a stream in Hawaii can be viewed from a slow-moving glass-bottomed boat. The boat travels 9 miles upstream in the same time it takes to go 15 miles downstream. If the rate of the boat in still water is 8 mph, what is the rate of the stream?

Glass-bottomed boats make underwater sightseeing easy. A sightseer's paradise is found in the coral reefs off the coast of Florida. The reefs are produced by colonies of billions of coral polyps, tiny sea animals that convert the calcium from the sea into limestone. The living polyps that cover the limestone formations give the coral reefs their beautiful shades of orange, rose, purple, and green.

Work Problems

If Harry can paint a wall in 40 minutes, then he will complete $\frac{1}{40}$ of the job in 1 minute. The part of a job that can be completed in 1 unit of time is called the *rate of work*. Thus, $\frac{1}{40}$ is Harry's rate of work. In 2 minutes, Harry will complete $2\left(\frac{1}{40}\right)$, or $\frac{2}{40}$, of the job; in x minutes, he will complete $x\left(\frac{1}{40}\right)$, or $\frac{x}{40}$, of the job. Therefore:

Rate of work × Time of work = Part of the work done

If Sam can paint the same wall in 60 minutes, then his rate of work is $\frac{1}{60}$, and in x minutes he will do $x\left(\frac{1}{60}\right)$, or $\frac{x}{60}$, of the job.

If Harry and Sam both start painting the wall at the same time, for each minute working together, they will do $\frac{1}{40} + \frac{1}{60}$, or $\frac{5}{120}$, of the job.

If Harry and Sam each worked for 24 minutes, Harry did $\frac{24}{40}$, or $\frac{72}{120}$, of the job and Sam did $\frac{24}{60}$, or $\frac{48}{120}$, of the job. Together, they did $\frac{72}{120} + \frac{48}{120}$, or $\frac{120}{120}$, of the job, that is, the whole job.

> **To represent a complete job, the fractions representing parts of the job must have a total value of 1.**

Now it's your turn to . . . **TRY IT!**

a. If May can mow a lawn in 80 minutes, what part of the lawn can she mow in
   (1) 1 minute?   **(2)** x minutes?

b. If Jean can mow the same lawn in 2 hours, what part of the lawn can she mow in
   (1) 1 minute?   **(2)** x minutes?

c. Represent the part of the job that May and Jean, working together, completed in x minutes.

d. If May and Jean mowed the entire lawn in x minutes, what must be the value of the expression given in part **c**?

e. Write an equation whose solution would be the number of minutes that May and Jean, working together, would require to mow the entire lawn.

(See page 385 for solutions.)

1. Rosa, using a power mower, can mow a lawn in 20 minutes. Her brother, Fidel, using a hand mower, can mow the same lawn in 30 minutes. If they work together, how long will it take them to complete the job?

Solution: Use a table to organize the data.

Let x = the number of minutes required to complete the job when Rosa and Fidel work together.

Worker	Rate of work (part of job per min.)	Time worked (min.)	Work done (part of job)
Rosa	$\dfrac{1}{20}$	x	$x\left(\dfrac{1}{20}\right)$ or $\dfrac{x}{20}$
Fidel	$\dfrac{1}{30}$	x	$x\left(\dfrac{1}{30}\right)$ or $\dfrac{x}{30}$

The sum of the fractional parts done by each must be 1.

$$\frac{x}{20} + \frac{x}{30} = 1$$

$$60\left(\frac{x}{20} + \frac{x}{30}\right) = 60(1)$$

$$\overset{3}{\cancel{60}}\left(\frac{x}{\cancel{20}}\right) + \overset{2}{\cancel{60}}\left(\frac{x}{\cancel{30}}\right) = 60(1)$$
$$\underset{1}{} \qquad \underset{1}{}$$

$$3x + 2x = 60$$

$$5x = 60$$

$$x = 12$$

Check:

In 12 minutes, Rosa will mow $\frac{12}{20}$, or $\frac{36}{60}$, of the lawn.

In 12 minutes, Fidel will mow $\frac{12}{30}$, or $\frac{24}{60}$, of the lawn.

In 12 minutes, together, they will mow
$\frac{36}{60} + \frac{24}{60} = \frac{60}{60}$, or the whole lawn. ✔

A gasoline engine, invented in England in 1820, was first used in an automobile in France in 1863. Gasoline-powered lawn mowers were introduced in the United States in the 1940's.

ANSWER: Working together, they take 12 minutes to complete the job.

2. The larger of two pipes can fill a tank twice as fast as the smaller. Together, the two pipes require 20 minutes to fill the tank. Find the number of minutes required for the larger pipe to fill the tank.

Solution:

Let x = the number of minutes required for the larger pipe to fill the tank.

Then $2x$ = the number of minutes required for the smaller pipe to fill the tank.

Size of pipe	Rate of work (part of job per min.)	Time worked (min.)	Work done (part of job)
Larger pipe	$\dfrac{1}{x}$	20	$20\left(\dfrac{1}{x}\right)$ or $\dfrac{20}{x}$
Smaller pipe	$\dfrac{1}{2x}$	20	$20\left(\dfrac{1}{2x}\right)$ or $\dfrac{20}{2x}$

If the tank is filled, the sum of the fractional parts filled by each pipe must be 1.

$$\frac{20}{x} + \frac{20}{2x} = 1$$

$$2x\left(\frac{20}{x} + \frac{20}{2x}\right) = 2x(1)$$

$$\overset{2}{\cancel{2x}}\left(\frac{20}{\cancel{x}}\right) + \overset{1}{\cancel{2x}}\left(\frac{20}{\cancel{2x}}\right) = 2x(1)$$

$$\underset{1}{} \quad \underset{1}{}$$

Check:

$$40 + 20 = 2x$$

In 20 minutes, the larger pipe fills $\frac{20}{30}$, or $\frac{2}{3}$, of the tank.

$$60 = 2x$$

In 20 minutes, the smaller pipe fills $\frac{20}{60}$, or $\frac{1}{3}$, of the tank.

$$30 = x$$

In 20 minutes, the two pipes together fill $\frac{2}{3} + \frac{1}{3} = 1$, or the whole tank. ✔

ANSWER: The larger pipe can fill the tank in 30 minutes.

Now it's your turn to . . . **TRY IT!**

Mr. Cooper can paint a fence in 2 hours. His son Bill can paint the fence in 6 hours. Mr. Cooper painted alone for 1 hour and stopped working. How many hours would Bill require to finish the job?

(See page 385 for solution.)

1. **a.** A programmer can complete a job in 2 hours. What part of the job can she complete in **(1)** one hour? **(2)** x hours?
   b. If her assistant can complete the same job in 3 hours, what part of the job can the assistant complete in **(1)** one hour? **(2)** x hours?
   c. Represent the part of the job that is completed in x hours when both work together.
   d. Write an equation whose solution would be the number of hours that both, working together, would require to complete the job.

2. An oil storage tank can be filled in 4 hours by one pipe and emptied by another pipe in 6 hours. The tank is empty and the valves of both pipes are opened.
   a. Represent the part of the tank filled in 1 hour.
   b. Represent the part of the tank filled in x hours.
   c. Write an equation whose solution would be the number of hours required to fill the tank.

3. Scott and Alyce share a plot in a community vegetable garden. If Scott works alone, he can weed the plot in 30 minutes. If Alyce works alone, she can do the weeding in 20 minutes. If they work together, should they finish the job in
   (a) more than 30 minutes,   (b) 25 minutes, or   (c) less than 20 minutes? Explain.

   In 4–21, solve algebraically.

4. Mrs. Saunders can clean the windows of her house in 3 hours. Her son can clean the windows in 6 hours. How long will it take them to clean the windows if they work together?

5. Mr. Ford can paint the fence around his house in 6 hours. His daughter needs 12 hours to do the job. How many hours would it take them to do the job if they worked together?

6. One pipe can fill an empty tank in 8 minutes, a second can fill it in 12 minutes, and a third can fill it in 24 minutes. How long will it take the three pipes, operating together, to fill it?

7. A farmer, working together with her son, needs 3 hours to plow a field. Working alone, the farmer can plow the field in 4 hours. How long would it take the son, working alone?

8. Mr. Dix can brick a wall in 9 hours. His son Carl can brick the same wall in 18 hours. Mr. Dix started work on the wall, worked for 3 hours, and then stopped working. How many hours would Carl require to complete the wall?

9. An old machine requires three times as many hours to complete a job as a new machine. When both machines work together, they require 9 hours to complete a job. How many hours would it take the new machine to finish the job operating alone?

10. A tank is used to hold molten paraffin for making crayons. An inlet pipe can fill the tank in 3 hours. An outlet pipe can empty it in 6 hours. The tank is empty and the inlet pipe is opened. If the outlet pipe is accidentally opened at the same time, how many hours will it take to fill the tank?

11. A printing press can print an edition of a newspaper in 4 hours. After the press has been at work for 1 hour, another press also starts to print the edition and, together, both presses require 1 more hour to finish the job. How long would it take the second press to print the edition alone?

12. Helen and Dawn, working together, started a job that they could complete in 12 hours, but after they had worked for 4 hours, Helen had to leave. If Dawn could have done the job alone in 18 hours, how long will it take her to finish it?

13. A new printing machine can do a job in 6 hours. An old machine can complete the same job in 16 hours. If 3 new machines and 4 old machines are used to do the job, how many hours will be required to finish it?

14. Tom can paint Mrs. Green's house in 3 days. Jerry can paint Mrs. Green's house in d days. They plan to work together to paint the house.
   a. If $d = 5$, how long will it take them to paint Mrs. Green's house?
   b. If $d > 3$ and is increasing, then the number of days it will take them to paint Mrs. Green's house will always be less than how many days?

15. Diane can weave baskets twice as fast as Sheena. Working together, they can fill a large gift shop's order for baskets in 12 hours. How long would it take Diane, working alone?

Hopis and Navahos have traditionally been among the most skillful native American weavers of baskets and rugs.

Many tribes used a basket cradle or cradleboard for carrying papooses on their mothers' backs. Similar devices for carrying a baby snuggled close to the parent's body have become popular in much of modern America.

16. It takes one day to fill a vat
   With this large pipe,
         two days with that.
   The third pipe needs but one day more;
   The fourth pipe fills the vat in four.
   If all four pipes together run,
   How long before the task is done?

17. A school computer system can receive and process student grades from two scanners simultaneously. One scanner requires 45 minutes to read all of the grades. The other can do the same job in 30 minutes. Together, how long will it take them to read all of the grades?

18. A school computer system can print report cards using two printers simultaneously. The old printer can do the job in 96 minutes. The old printer and the new printer working together can print the report cards in 24 minutes. How long would it take the new printer, working alone, to print these cards?

19. A small document shredder in an ambassador's office requires $2\frac{1}{2}$ hours to shred all of his classified documents. The larger shredder can shred all of the documents in 30 minutes. If needed, how long would it take to shred all of the documents, using both shredders?

20. Mrs. MacDonald's large herd of cattle would take 28 days to eat all of the cattle feed in the silo. Mr. Deere's smaller herd would empty the silo in 70 days. If the herds were combined, how long would the feed in the silo last?

21. Sensor A started inspecting carry-on luggage and, 15 seconds later, was joined by sensor B. Together, A and B scanned the remaining luggage in 30 seconds. If, working alone, A could have done the job in 2 minutes, how long would it have taken B alone?

You work with **rational expressions**, quotients of polynomials, in much the same way that you work with numerical fractions; that is:

- To multiply rational expressions, first cancel factors that appear both in the numerators and in the denominators.
- To divide rational expressions, multiply by the reciprocal of the divisor.
- To be added (or subtracted), rational expressions must first be rewritten with a common denominator.

Other important points to remember about rational expressions:

- $\dfrac{x + n}{n + x} = 1$ 　　　　　　　 $\bullet$ $\dfrac{x - n}{n - x} = -1$

- Cancel with care! When reducing a rational expression, cancel only common factors of the *entire* numerator and the *entire* denominator.
- In subtracting, if the numerator of the subtrahend has more than one term, watch the + and − signs. Using parentheses helps avoid errors.

Earlier open sentences all had integral coefficients. In this chapter, you have seen that when coefficients are fractions or decimals, the open sentences should be transformed so that the coefficients are integers and, thus, the same methods of solution you have already learned can be applied.

When any denominator in an equation contains a variable, the equation is called a **fractional equation**. The method of solution is the same as for an equation with fractional coefficients, except that the replacement set may not contain any number that would make a denominator equal to zero. Be careful to check solutions to fractional equations to be sure that you do not have an **extraneous** value.

Many different real-life situations can be modeled by open sentences that contain fractions or decimals. Some will appear later in this course, as you study such topics as probability and trigonometry. Fractions and decimals appear in formulas with a vast variety of applications, from science to retailing and economics.

VOCABULARY CHECKUP

SECTION

7-1 *fraction / rational expression*

7-3 *lowest common denominator (L.C.D.)*

7-6 *fractional equation / extraneous value*

In 1–4, for which value of the variable is the expression undefined? (Section 7-1)

1. $3 \div x$

2. $5 \div b - 2$

3. $7 \div 3 - d$

4. $4a \div (a - 7)$

In 5–8, reduce the fraction to lowest terms. (Section 7-1)

5. $\dfrac{45}{81}$

6. $\dfrac{12a^2}{9ab}$

7. $\dfrac{8r - 4}{2r - 1}$

8. $\dfrac{x^2 - 3x - 4}{3x - 12}$

In 9–13, find the product in lowest terms. (Section 7-2)

9. $\dfrac{3}{8} \cdot \dfrac{4}{9}$

10. $\dfrac{6x}{5y^2} \cdot \dfrac{15y^4}{4x}$

11. $\dfrac{x^2 - 9}{2x} \cdot \dfrac{6x^2}{3x - 9}$

12. $\dfrac{4a - 12}{a^2 - 9} \cdot \dfrac{a^2 + 4a + 3}{a^2 - 1}$

13. $\dfrac{2a^2 - 32}{6a - 24} \cdot \dfrac{9a - 18}{a^2 + 2a - 8}$

In 14–17, divide. Express the quotient in lowest terms. (Section 7-2)

14. $\dfrac{4}{5} \div \dfrac{3}{2}$

15. $\dfrac{3}{8} \div 12$

16. $\dfrac{4x}{3a} \div \dfrac{a^2x}{6}$

17. $\dfrac{y^2 - 4}{15} \div \dfrac{y + 2}{10}$

In 18–21, add or subtract, as indicated. Reduce answers to lowest terms. (Section 7-3)

18. $\dfrac{8a}{3} - \dfrac{5a}{3}$

19. $\dfrac{3}{5y} + \dfrac{6}{5y} - \dfrac{4}{5y}$

20. $\dfrac{a}{a^2 - 1} + \dfrac{1}{a^2 - 1}$

21. $\dfrac{3n + 11}{2n + 5} - \dfrac{1 - n}{2n + 5}$

In 22–27, add or subtract, as indicated. Reduce answers to lowest terms. (Section 7-3)

22. $\dfrac{2a}{3} - \dfrac{3a}{5}$

23. $\dfrac{3a}{2b} - \dfrac{5a}{4b}$

24. $\dfrac{2}{bc^2} + \dfrac{7}{ab^2}$

25. $\dfrac{3a - 4}{9} - \dfrac{a - 1}{5}$

26. $\dfrac{3x + 2y}{3y} + \dfrac{3x - 2y}{x + y}$

27. $\dfrac{3}{x^2 - 4} + \dfrac{5}{(x + 2)^2}$

In 28–31, express the mixed expression as a fraction in lowest terms. (Section 7-3)

28. $4\frac{5}{8}$

29. $7 - \dfrac{3}{a}$

30. $r + \dfrac{r - 4}{r - 1}$

31. $\dfrac{5}{t - 2} + t + 3$

In 32–37, solve and check. (Section 7-4)

32. $\dfrac{3a}{2} = 5$

33. $\dfrac{b + 4}{5} = 7$

34. $\dfrac{2c - 3}{4} = \dfrac{7}{8}$

35. $\dfrac{y}{3} - \dfrac{y}{4} = 2$

36. $\dfrac{2d}{3} - \dfrac{d}{6} - \dfrac{2d}{9} = \dfrac{d - 1}{4}$

37. $\dfrac{x - 5}{4} - \dfrac{x + 3}{6} = 2$

In 38–40, solve and check. (Section 7-4)

38. $.05a = 7$

39. $.04t - 1.5 = 3$

40. $.06w + 7 = .02w - 1$

In 41–43, solve the inequality. (Section 7-4)

41. $\dfrac{x}{3} - \dfrac{x}{2} < \dfrac{3}{4}$

42. $\dfrac{2a - 5}{7} \geq 6$

43. $\dfrac{b - 4}{2} < \dfrac{b + 4}{3} + 2$

In 44–47, solve algebraically. (Section 7-5)

44. Three families want to pool their resources to rent a vacation home for the summer. The rental cost to each family is expected to be $380. If they invited a fourth family to share rental expenses, by how much would this decrease the cost to each family?

45. A local basketball team plays a 20-game season. In order to win their league championship, they must win at least 65% of their games. At least how many games must they win?

46. Rachel and Rosa each invested the same amount of money in bonds. Rachel got an 8% return on her money and Rosa got a 9% return. If Rosa's annual interest was $500 more than Rachel's, how much did each invest?

47. For Claudia's Cosmetic Compound to have its usual healing effect, it must contain 10% of Claudia's secret ingredient. By mistake, Claudia mixed 60 ounces of her compound containing only 4% of her secret ingredient. How many ounces of her secret ingredient must Claudia add to bring the mixture up to 10%?

In 48–51, solve and check. (Section 7-6)

48. $\dfrac{8}{y} = 2$

49. $\dfrac{11}{r} - \dfrac{7}{r} = \dfrac{2}{3}$

50. $\dfrac{18}{x + 3} = \dfrac{3}{x}$

51. $\dfrac{3}{a + 1} = \dfrac{5}{a + 7}$

In 52–57, solve for x. (Section 7-6)

52. $w = \dfrac{x}{a}$

53. $\dfrac{3x - 2y}{5} = 3y$

54. $ax - bx = c$

55. $\dfrac{x}{a} + \dfrac{x}{b} = a + b$

56. $\dfrac{a}{x + y} = b$

57. $P = \dfrac{bx}{b + x}$

In 58–60, solve algebraically. (Section 7-7)

58. Teresa is flying her single-engine plane against the wind on a 90-mile trip. Flying with the same wind, Teresa could travel 135 miles in the same amount of time. If Teresa's plane can fly at 110 mph in still air, what is the speed of the wind?

59. Tillie tiles a table in 20 hours. Toby can tile the table in 12 hours. How long will it take Tillie and Toby to tile the table together?

60. The numerator of a fraction is 7 less than twice the denominator. If the denominator is increased by 4, the value of the fraction will be $\frac{1}{2}$. Find the original fraction.

1. Perform the indicated operation. Reduce answers to lowest terms.

   a. $\dfrac{a^2 - b^2}{a^3 + b^3} \cdot \dfrac{a^4 + a^2b^2 + b^4}{a^3 - b^3}$

   b. $2a - 3 - \dfrac{a^2 - a - 12}{a + 3}$

   c. $\dfrac{a^3}{a + b} - \dfrac{b(b^3 - a^3)}{a^2 - b^2}$

   d. $\dfrac{(a + b)^2}{a + b + c} - \dfrac{c^2}{a + b + c}$

2. You have more than a dollar in U.S. coins in your pocket, yet you cannot give exact change for a nickel, a dime, a quarter, a half-dollar, or a dollar. What is the most money that you can have and what are the coins?

3. Jackie has grades of 73, 85, 73, 87, and 86. Jenny has grades of 81, 76, and 82. If both of them get the same grade on their next test, what grade would they have to get to have the same average?

4. Suppose you want to travel two miles at an average speed of 60 mph. If you travel 30 mph the first mile, how fast must you go the second mile?

5. Two bicyclists are $\frac{1}{3}$ mile apart. They begin riding toward each other at the same instant at speeds of 8 mph and 12 mph.

   At the same time, a fly that had landed on the front wheel of one of the bicycles begins flying directly toward the other bicycle at a speed of 20 mph. When the fly reaches the front wheel of the second bicycle, it immediately flies back to the front wheel of the first bicycle. The fly then continues back and forth between the front wheels of the two bicycles until the bicycles meet.

   How far does the fly travel before the bikes meet?

1. Find a fraction whose value is $\frac{1}{3}$ and the product of whose numerator and denominator is 507.

2. Find a fraction whose value is $\frac{3}{4}$ and the product of whose numerator and denominator is 1,452.

3. The number 135 is exactly divisible by 15, the product of 1, 3, and 5. Find all the other numbers between 100 and 199 where this exact division can occur.

Questions 1–11 each consist of two quantities, one in Column A and one in Column B. You are to compare the two quantities and choose:

A   if the quantity in Column A is greater;
B   if the quantity in Column B is greater;
C   if the two quantities are equal;
D   if the relationship cannot be determined from the information given.

1. In certain questions, information concerning one or both of the quantities to be compared is centered above the two columns.

2. In a given question, a symbol that appears in both columns represents the same thing in Column A as it does in Column B.

3. x, n, and k, etc. stand for real numbers.

Column A	**Column B**

$$1 < \frac{x}{y} < 2$$

1. x $\qquad\qquad$ y

$$0 < x < y < 1$$

2. $x + y$ $\qquad$ xy

$$x + y = 2$$

3. $\dfrac{1}{x + y}$ $\qquad$ $\dfrac{x}{x + y} + \dfrac{y}{x + y}$

$$a > b > 0$$

4. $a + \dfrac{1}{b}$ $\qquad$ $b + \dfrac{1}{a}$

$$a > b > 0$$

5. $\dfrac{1}{a} + \dfrac{1}{b}$ $\qquad$ $\dfrac{1}{a + b}$

$$x + y \neq 0$$

6. $\dfrac{x^2}{x + y} - \dfrac{y^2}{x + y}$ $\qquad$ $x - y$

$$a > b > 0$$

7. $\dfrac{a + b}{a} - \dfrac{a + b}{b}$ $\qquad$ $\dfrac{a^2}{ab} + \dfrac{b^2}{ab}$

$$n > 0$$

8. $\dfrac{n}{2} + \dfrac{n}{5}$ $\qquad$ $\dfrac{n}{3} + \dfrac{n}{4}$

$$x \neq -2$$

9. $\dfrac{x^2 + 3x + 2}{x + 2}$ $\qquad$ x

$$a \neq 1, a \neq 2$$

10. $\dfrac{a}{a - 1}$ $\qquad$ $\dfrac{a - 1}{a - 2}$

$$a \neq 2$$

11. $\dfrac{a - 3}{a - 2}$ $\qquad$ 1

In 12–25, select the letter of the correct answer.

12. If $x > 4$, which is greatest?

(A) $\dfrac{x + 2}{x + 3}$ (B) $\dfrac{x + 1}{x + 2}$ (C) $\dfrac{x}{x - 1}$

(D) $\dfrac{x - 1}{x - 2}$ (E) $\dfrac{x - 2}{x - 3}$

13. Which of the following *cannot be* a positive integer?

(A) $\dfrac{a - b}{b + a}$ (B) $\dfrac{a + b}{b + a}$ (C) $\dfrac{a - b}{b - a}$

(D) $\dfrac{a + 1}{1 + b}$ (E) $\dfrac{ab}{a + b}$

---

14. Twelve students buy a gift for d dollars and agree to share the expense equally. If 4 students drop out, by how many dollars does each remaining student's share of the expense increase?

(A) $\dfrac{d}{12}$ (B) $\dfrac{d}{24}$ (C) $\dfrac{d}{36}$

(D) $\dfrac{3d}{40}$ (E) $\dfrac{d}{48}$

15. If a man buys some items for n cents a dozen and sells them for $\dfrac{n}{9}$ cents per item, what is his profit, in cents, on each item?

(A) $\dfrac{n}{36}$ (B) $\dfrac{n}{12}$ (C) $\dfrac{3n}{4}$

(D) $\dfrac{4n}{3}$ (E) $\dfrac{n}{18}$

16. If $\frac{4}{5}$ of x is y and $\frac{1}{2}$ of y is z, then z is what percent of x?

(A) 30   (B) 40   (C) 130
(D) 250   (E) 300

17. If $r + s = 100$ and $\dfrac{r}{s} = \frac{1}{4}$, then $s - r =$

(A) -100 (B) 30   (C) 50
(D) -60 (E) 60

18. A cave has a 10-foot ceiling. A stalactite grows downward .004 inch per year, and a stalagmite directly below it grows upward .006 inch per year. How many years does it take them to meet?
(A) 4,000   (B) 8,000   (C) 10,000
(D) 12,000   (E) 14,000

19. One tractor can plow a field in 16 hours and another in 20 hours. In how many hours can they plow it together?
(A) $7\frac{3}{4}$ (B) 8   (C) $8\frac{8}{9}$ (D) 9   (E) $9\frac{8}{9}$

20. If $\dfrac{x}{3} + \dfrac{y}{4} = 10$ and y is twice x, then the average of $\dfrac{x}{3}$ and $\dfrac{y}{4}$ is

(A) 18   (B) 12   (C) 10   (D) 9   (E) 5

21. If $\dfrac{1}{xy} + \dfrac{1}{yz} + \dfrac{1}{xz} = P$, and twice the reciprocal of P is xyz, then $x + y + z$ is

(A) 0   (B) $\frac{1}{2}$ (C) 1   (D) 2   (E) 3

22. If $\dfrac{k - 1}{3} + \dfrac{k - 2}{4} = 5$, then $\dfrac{2k - 5}{3} =$

(A) $2\frac{1}{2}$ (B) $3\frac{2}{3}$ (C) 5   (D) $6\frac{1}{4}$ (E) $7\frac{1}{2}$

23. If $I = \dfrac{V}{r + s}$, then $r =$

(A) $\dfrac{V - Is}{I}$ (B) $\dfrac{V + Is}{I}$ (C) $\dfrac{I}{V - Is}$

(D) $\dfrac{Vs}{I}$ (E) $\dfrac{I}{Vs}$

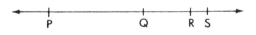

24. In the diagram, PQ is $\frac{2}{3}$ of PR and QR is $\frac{3}{4}$ of QS. PR is what fraction of PS?

(A) $\dfrac{5}{6}$ (B) $\dfrac{7}{8}$ (C) $\dfrac{8}{9}$ (D) $\dfrac{9}{10}$ (E) $\dfrac{11}{12}$

25. If $\dfrac{(2)(3)(4)}{(5)(6)(7)} = \dfrac{x}{y}$ and $\dfrac{(3)(4)(5)}{(6)(7)(8)} = \dfrac{ax}{by}$, then $\dfrac{a}{b}$ is

(A) $\dfrac{5}{8}$ (B) $\dfrac{25}{16}$ (C) $\dfrac{49}{16}$ (D) $\dfrac{5}{49}$ (E) $\dfrac{25}{49}$

1. Simplify the expression: $d(2d - 5) - 3d(d + 2) + d^2$

2. Factor: **a.** $y^2 - 36$ **b.** $x^2 - 11x + 30$

3. Write the next two terms in the sequence 1, 4, 9, 16, 25, _____, _____.

4. Find the measure of the third angle of a triangle if the other two angles measure $78°$ and $84°$.

5. Graph the solution set on the number line. **a.** $3|s| = 15$ **b.** $4(2x + 1) > 12$

6. The area of the shaded region in the figure is what fraction of the area of rectangle $ABCD$?

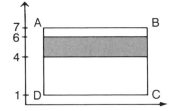

7. Using the formula $V = \pi r^2 h$, find the volume V of a cylinder when $r = 14$ inches and $h = \frac{1}{7}$ inch. (Use $\pi = \frac{22}{7}$.)

8. $2^{12} + 2^{12}$ is equal to:
   (1) 2^{13} (2) 2^{24}
   (3) 4^{12} (4) 4^{24}

9. Ms. Tip bought $(20x - 18y + z)$ shares of stock in the CEO Corp. One year later, she sold $(12x - 8y + 10z)$ shares of the stock. How many shares did she still own?

10. There are y more boys on a committee than there are girls. If there is a total of x students on the committee, write an expression to represent the number of girls.

11. If the sum of two consecutive multiples of 7 is 49, find the smaller of these multiples.

12. Find each product. **a.** $(3.4 \times 10^7)(6.5 \times 10^{-5})$ **b.** $\begin{bmatrix} 2 & -4 \\ 3 & 0 \end{bmatrix} \begin{bmatrix} 3 & 2 \\ -5 & 4 \end{bmatrix}$

13. If $x = 2k$ and $y = 2k - 1$, where k is an integer, then $x^2 - y^2$ is always
    (1) even   (2) odd   (3) a perfect square   (4) a negative number

14. If the area of a square is $16x^2$, what is its perimeter?

15. Kevin spent $(x^2 - 3x - 10)$ cents on movie tickets. If each ticket cost $(x - 5)$ cents, how many tickets did he buy?

16. Two automobiles start from the same place and travel in opposite directions. If their rates are 45 mph and 55 mph, in how many hours will they be 600 miles apart?

17. An operation is defined as $a * b = a^2 - b$.
    a. Is the set of whole numbers closed under operation $*$?
    b. Evaluate: $(1 * 3) * 7$

18. In Carrie's piggy bank, the number of dimes is twice the number of nickels and the number of quarters is two less than the number of nickels. If the total value of the money in the bank is $2.50, find the number of each type of coin.

7-1 SIMPLIFYING A RATIONAL EXPRESSION

TRY IT! *Problems 1–3 on page 325*

1. $\dfrac{a^2 - 4}{5a - 10} = \dfrac{(a + 2)(a - 2)}{5(a - 2)}$

$$= \dfrac{(a + 2)\overset{1}{\cancel{(a - 2)}}}{5\underset{1}{\cancel{(a - 2)}}} = \dfrac{a + 2}{5}$$

2. $\dfrac{2x^2 - 3x - 5}{x + 1} = \dfrac{(2x - 5)\overset{1}{\cancel{(x + 1)}}}{\underset{1}{\cancel{x + 1}}} = 2x - 5$

3. $\dfrac{4 - 12x}{3x - 1} = \dfrac{4(1 - 3x)}{3x - 1} = \dfrac{4(1 - 3x)}{-1 + 3x}$

$$= \dfrac{4(1 - 3x)}{-1(1 - 3x)} = \dfrac{4\overset{1}{\cancel{(1 - 3x)}}}{-1\underset{1}{\cancel{(1 - 3x)}}} = -4$$

7-2 MULTIPLYING AND DIVIDING RATIONAL EXPRESSIONS

TRY IT! *Problems 1 and 2 on page 329*

1. $\dfrac{12xy^3}{5a^2b^2} \div \dfrac{9y^2}{15a^2b} = \dfrac{12xy^3}{5a^2b^2} \cdot \dfrac{15a^2b}{9y^2}$

$$= \dfrac{\overset{4}{\cancel{12}} x \overset{y}{\cancel{y^3}}}{\cancel{5} a^2 b^2} \cdot \dfrac{\overset{}{\cancel{15}} \overset{}{\cancel{a^2}} b}{\cancel{9} \cancel{y^2}} = \dfrac{4xy}{b}$$

2. $\dfrac{x^2 - 5x + 6}{x^2 - x - 6} \div \dfrac{2x - 4}{4x + 8}$

$$= \dfrac{x^2 - 5x + 6}{x^2 - x - 6} \cdot \dfrac{4x + 8}{2x - 4}$$

$$= \dfrac{\cancel{(x - 2)}(x - 3)}{(x - 3)(x + 2)} \cdot \dfrac{\overset{2}{\cancel{4}}(x + 2)}{\cancel{2}(x - 2)} = 2$$

7-3 ADDING AND SUBTRACTING RATIONAL EXPRESSIONS

TRY IT! *Problems 1–4 on page 336*

1. $ab^4 = a \cdot b^4$
   $a^3b^3 = a^3 \cdot b^3$ L.C.D. $= a^3 \cdot b^4$

$$\dfrac{3}{ab^4} + \dfrac{2}{a^3b^3}$$

$$= \dfrac{3}{ab^4} \cdot \left(\dfrac{a^2}{a^2}\right) + \dfrac{2}{a^3b^3} \cdot \left(\dfrac{b}{b}\right)$$

$$= \dfrac{3a^2}{a^3b^4} + \dfrac{2b}{a^3b^4} = \dfrac{3a^2 + 2b}{a^3b^4}$$

2. $5 = 5 \cdot 1$ $3 = 3 \cdot 1$ L.C.D. $= 5 \cdot 3$

$$\dfrac{3x - 4}{5} - \dfrac{5 - x}{3}$$

$$= \dfrac{3x - 4}{5} \cdot \left(\dfrac{3}{3}\right) - \dfrac{5 - x}{3} \cdot \left(\dfrac{5}{5}\right)$$

$$= \dfrac{9x - 12}{15} - \dfrac{25 - 5x}{15}$$

$$= \dfrac{(9x - 12) - (25 - 5x)}{15}$$

$$= \dfrac{9x - 12 - 25 + 5x}{15} = \dfrac{14x - 37}{15}$$

3. $12a = 2^2 \cdot 3 \cdot a$
   $20a = 2^2 \cdot 5 \cdot a$ L.C.D. $= 2^2 \cdot 3 \cdot 5 \cdot a$

$$\dfrac{4a - 3}{12a} - \dfrac{2a - 5}{20a}$$

$$= \dfrac{4a - 3}{12a} \cdot \left(\dfrac{5}{5}\right) - \dfrac{2a - 5}{20a} \cdot \left(\dfrac{3}{3}\right)$$

$$= \dfrac{20a - 15}{60a} - \dfrac{6a - 15}{60a}$$

$$= \dfrac{(20a - 15) - (6a - 15)}{60a}$$

$$= \dfrac{20a - 15 - 6a + 15}{60a} = \dfrac{14a}{60a} = \dfrac{7}{30}$$

4. $x^2 - 3x + 2 = (x - 1)(x - 2)$
$x^2 + x - 6 = (x + 3)(x - 2)$
L.C.D. $= (x - 1)(x - 2)(x + 3)$

$$\frac{4x - 1}{x^2 - 3x + 2} + \frac{x - 5}{x^2 + x - 6}$$

$$= \frac{4x - 1}{(x - 1)(x - 2)} + \frac{x - 5}{(x + 3)(x - 2)}$$

$$= \frac{4x - 1}{(x - 1)(x - 2)} \cdot \left(\frac{x + 3}{x + 3}\right)$$

$$+ \frac{x - 5}{(x + 3)(x - 2)} \cdot \left(\frac{x - 1}{x - 1}\right)$$

$$= \frac{(4x^2 + 11x - 3) + (x^2 - 6x + 5)}{(x - 1)(x - 2)(x + 3)}$$

$$= \frac{5x^2 + 5x + 2}{(x - 1)(x - 2)(x + 3)}$$

7-4 SOLVING OPEN SENTENCES THAT CONTAIN FRACTIONAL COEFFICIENTS

TRY IT! *Problems 1 and 2 on page 343*

1. $\quad \frac{1}{3}a - \frac{1}{5}a = 3$

$$\frac{a}{3} - \frac{a}{5} = \frac{3}{1} \quad \text{L.C.D.} = 15$$

$$15\left(\frac{a}{3} - \frac{a}{5}\right) = 15\left(\frac{3}{1}\right)$$

$$\overset{5}{\cancel{15}}\left(\frac{a}{\cancel{3}}\right) - \overset{3}{\cancel{15}}\left(\frac{a}{\cancel{5}}\right) = 15\left(\frac{3}{1}\right)$$

$$5a - 3a = 45$$
$$2a = 45$$
$$a = 22\tfrac{1}{2}$$

Check:

Substitute $\frac{45}{2}$ in the original sentence.

$$\frac{1}{\cancel{3}}\left(\frac{\overset{15}{\cancel{45}}}{2}\right) - \frac{1}{\cancel{5}}\left(\frac{\overset{9}{\cancel{45}}}{2}\right) \overset{?}{=} 3$$

$$\frac{15}{2} - \frac{9}{2} \overset{?}{=} 3$$

$$\frac{6}{2} \overset{?}{=} 3$$

$$3 = 3 \checkmark$$

2. $\quad \dfrac{w + 3}{5} + \dfrac{w - 1}{4} \geq \dfrac{1}{2} \quad$ L.C.D. $= 20$

$$20\left(\frac{w + 3}{5} + \frac{w - 1}{4}\right) \geq 20\left(\frac{1}{2}\right)$$

$$\overset{4}{\cancel{20}}\left(\frac{w + 3}{\cancel{5}}\right) + \overset{5}{\cancel{20}}\left(\frac{w - 1}{\cancel{4}}\right) \geq \overset{10}{\cancel{20}}\left(\frac{1}{\cancel{2}}\right)$$

$$4(w + 3) + 5(w - 1) \geq 10$$
$$4w + 12 + 5w - 5 \geq 10$$
$$9w + 7 \geq 10$$
$$9w \geq 3$$
$$w \geq \tfrac{1}{3}$$

Check: Substitute a value $\geq \frac{1}{3}$, say 1, in the original sentence.

$$\frac{1 + 3}{5} + \frac{1 - 1}{4} \overset{?}{\geq} \frac{1}{2}$$

$$\frac{4}{5} + 0 \overset{?}{\geq} \frac{1}{2}$$

$$\frac{8}{10} \geq \frac{5}{10} \checkmark$$

7-5 APPLYING OPEN SENTENCES THAT CONTAIN FRACTIONAL COEFFICIENTS

TRY IT! *Problem on page 352*

Let $n =$ no. of grams of 55% solution.
Then $60 - n =$ no. of grams of 30% solution.

Kind of solution	No. of grams	Percent pure acid	No. of grams of pure acid
55% solution	n	.55	$.55n$
30% solution	$60 - n$	.30	$.30(60 - n)$
35% mixture	60	.35	$.35(60)$

The total pure acid in the 55% solution and in the 30% solution is equal to the amount of pure acid in the final 35% solution.

$$.55n + .30(60 - n) = .35(60)$$
$$55n + 30(60 - n) = 35(60)$$
$$55n + 1{,}800 - 30n = 2{,}100$$
$$25n + 1{,}800 = 2{,}100$$
$$25n = 300$$
$$n = 12$$
$$60 - n = 48$$

Check: 55% of 12 = .55(12) = 6.60

30% of 48 = .30(48) = 14.40

35% of 60 = .35(60) = 21.00

6.60 + 14.40 = 21.00

ANSWER: 12 grams of the 55% solution and 48 grams of the 30% solution should be used.

TRY IT! *Problem on page 354*

Let p = amount at 9%.
Then $10,000 - p$ = amount at 6%.

Principal in dollars	Rate of interest	Annual income in dollars
p	.09	$.09p$
$10,000 - p$	.06	$.06(10,000 - p)$

The income from the 9% investment is equal to the income from the 6% investment.

$$.09p = .06(10,000 - p)$$
$$9p = 6(10,000 - p)$$
$$9p = 60,000 - 6p$$
$$15p = 60,000$$
$$p = 4,000$$
$$10,000 - p = 6,000$$

Check: $4,000 + $6,000 = $10,000

.09($4,000) = $360

.06($6,000) = $360

The incomes are equal. ✔

ANSWER: $4,000 at 9%, $6,000 at 6%.

7-6 SOLVING EQUATIONS THAT CONTAIN RATIONAL EXPRESSIONS

TRY IT! *Problems 1–3 on page 360*

1. $\dfrac{5}{3a} - \dfrac{7}{a} = \dfrac{1}{6}$ L.C.D. $= 6a$; $a \neq 0$

$$6a\left(\frac{5}{3a} - \frac{7}{a}\right) = 6a\left(\frac{1}{6}\right)$$

$$\overset{2}{\cancel{6a}}\left(\frac{5}{\cancel{3a}}\right) - \overset{6}{\cancel{6a}}\left(\frac{7}{\cancel{a}}\right) = \overset{a}{\cancel{6a}}\left(\frac{1}{\cancel{6}}\right)$$

$$10 - 42 = a$$
$$-32 = a$$

Check: $\dfrac{5}{3(-32)} - \dfrac{7}{-32} \overset{?}{=} \dfrac{1}{6}$

$$\frac{5}{-96} - \frac{7}{-32} \overset{?}{=} \frac{1}{6}$$

$$-\frac{5}{96} + \frac{21}{96} \overset{?}{=} \frac{1}{6}$$

$$\frac{16}{96} \overset{?}{=} \frac{1}{6}$$

$$\frac{1}{6} = \frac{1}{6} ✔$$

ANSWER: $a = -32$

2. $\dfrac{2}{3d} + \dfrac{1}{3} = \dfrac{11}{6d} - \dfrac{1}{4}$

L.C.D. $= 12d$; $d \neq 0$

$$12d\left(\frac{2}{3d} + \frac{1}{3}\right) = 12d\left(\frac{11}{6d} - \frac{1}{4}\right)$$

$$\overset{4}{\cancel{12d}}\left(\frac{2}{\cancel{3d}}\right) + \overset{4d}{\cancel{12d}}\left(\frac{1}{\cancel{3}}\right) = \overset{2}{\cancel{12d}}\left(\frac{11}{\cancel{6d}}\right) - \overset{3d}{\cancel{12d}}\left(\frac{1}{\cancel{4}}\right)$$

$$8 + 4d = 22 - 3d$$
$$4d = 14 - 3d$$
$$7d = 14$$
$$d = 2$$

Check: $\dfrac{2}{3(2)} + \dfrac{1}{3} \overset{?}{=} \dfrac{11}{6(2)} - \dfrac{1}{4}$

$$\frac{2}{6} + \frac{1}{3} \overset{?}{=} \frac{11}{12} - \frac{1}{4}$$

$$\frac{4}{12} + \frac{4}{12} \overset{?}{=} \frac{11}{12} - \frac{3}{12}$$

$$\frac{8}{12} = \frac{8}{12} ✔$$

ANSWER: $d = 2$

3. $\dfrac{3x - 5}{3x + 5} = \dfrac{1}{2}$ L.C.D. $= 2(3x + 5)$

$$x \neq -\frac{5}{3}$$

$$2\overset{1}{\cancel{(3x + 5)}}\left(\frac{3x - 5}{\cancel{3x + 5}}\right) = \overset{1}{\cancel{2}}(3x + 5)\left(\frac{1}{\cancel{2}}\right)$$

$$6x - 10 = 3x + 5$$
$$3x - 10 = 5$$
$$3x = 15$$
$$x = 5$$

Check:

$$\frac{3(5) - 5}{3(5) + 5} \overset{?}{=} \frac{1}{2}$$

$$\frac{15 - 5}{15 + 5} \overset{?}{=} \frac{1}{2}$$

$$\frac{10}{20} \overset{?}{=} \frac{1}{2}$$

ANSWER: $x = 5$ $\quad \dfrac{1}{2} = \dfrac{1}{2}$ ✔

TRY IT! *Problems 1 and 2 on page 361*

1.
$$cx + d^2 = c^2 + dx$$
$$cx + d^2 - dx = c^2$$
$$cx - dx = c^2 - d^2$$
$$x(c - d) = c^2 - d^2$$
$$x = \frac{c^2 - d^2}{c - d} \quad \text{where } c \neq d$$

$$x = \frac{(c + d)\cancel{(c - d)}}{\cancel{(c - d)}}$$

$$x = c + d$$

2.
$$R = \frac{g}{g - s} \quad \text{L.C.D.} = g - s; \ g \neq s$$

$$(g - s)(R) = \cancel{(g - s)}\left(\frac{g}{\cancel{g - s}}\right)$$

$$gR - sR = g$$
$$gR = g + sR$$
$$gR - g = sR$$
$$g(R - 1) = sR$$

$$g = \frac{sR}{R - 1} \quad \text{where } R \neq 1$$

7-7 APPLYING EQUATIONS THAT CONTAIN RATIONAL EXPRESSIONS

TRY IT! *Problem on page 365*

Let $x =$ the smaller number.
Then $4x - 5 =$ the larger number.

When the larger number is divided by the smaller, the quotient is 3 and the remainder is 3.

$$\frac{4x - 5}{x} = 3 + \frac{3}{x}$$

$$\cancel{x}\left(\frac{4x - 5}{\cancel{x}}\right) = x\left(3 + \frac{3}{x}\right)$$

$$4x - 5 = 3x + 3$$
$$4x - 3x = 3 + 5$$
$$x = 8$$
$$4x - 5 = 27$$

Check: The larger number, 27, is 5 less than 4 times the smaller number, 8. ✔

When 27 is divided by 8, the quotient is 3 and the remainder is 3. ✔

ANSWER: Smaller is 8; larger is 27.

TRY IT! *Problems 1 and 2 on page 367*

1. Let $r =$ the average rate on 1st trip.
Then $r + 10 =$ the average rate on 2nd trip.

	Distance (m)	Rate (mph)	Time (h)
1st trip	120	r	$\dfrac{120}{r}$
2nd trip	160	$r + 10$	$\dfrac{160}{r + 10}$

The time for the first trip was the same as the time for the second trip.

$$\frac{120}{r} = \frac{160}{r + 10}$$

$$\text{L.C.D.} = r(r + 10)$$

$$\cancel{r}(r + 10)\left(\frac{120}{\cancel{r}}\right) = r\cancel{(r + 10)}\left(\frac{160}{\cancel{r + 10}}\right)$$

$$120(r + 10) = 160r$$
$$120r + 1{,}200 = 160r$$
$$1{,}200 = 40r$$
$$30 = r$$
$$40 = r + 10$$

Check:

  Time, 1st trip   Time, 2nd trip

$$\frac{120}{r} = \frac{120}{30} = 4 \text{ hr.} \qquad \frac{160}{r+10} = \frac{160}{40} = 4 \text{ hr.}$$

The times are equal. ✔

    ANSWER: Rate on 1st trip was 30 mph.

2.   Let p = speed of plane with no wind.
Then $p + 80$ = speed of plane with wind.
And $p - 80$ = speed of plane against wind.

	Distance (km)	Rate (km/h)	Time (h)
With wind	660	$p + 80$	$\dfrac{660}{p+80}$
Against wind	540	$p - 80$	$\dfrac{540}{p-80}$

Flying time with the wind is the same as flying time against the wind.

$$\frac{660}{p+80} = \frac{540}{p-80}$$

$$\text{L.C.D.} = (p+80)(p-80)$$

$$\overset{1}{(p+80)}(p-80)\left(\frac{660}{p+80}\right) = (p+80)\overset{1}{(p-80)}\left(\frac{540}{p-80}\right)$$

$$660(p - 80) = 540(p + 80)$$
$$660p - 52{,}800 = 540p + 43{,}200$$
$$120p - 52{,}800 = 43{,}200$$
$$120p = 96{,}000$$
$$p = 800$$

Check: Speed of plane with wind:
$$p + 80 = 800 + 80 = 880 \text{ km/h}$$

    Speed of plane against wind:
$$p - 80 = 800 - 80 = 720 \text{ km/h}$$

Time with wind   Time against wind

$$\frac{660}{880} = \frac{3}{4} \text{ hr.} \qquad \frac{540}{720} = \frac{3}{4} \text{ hr.}$$

The times are equal. ✔

    ANSWER: Speed in still air is 80 km/h.

TRY IT! *Problem on page 369*

a. (1) $\dfrac{1}{80}$ **(2)** $\dfrac{x}{80}$

b. 2 hr. = 120 min.   **(1)** $\dfrac{1}{120}$ **(2)** $\dfrac{x}{120}$

c. $\dfrac{x}{80} + \dfrac{x}{120}$ **d.** 1   **e.** $\dfrac{x}{80} + \dfrac{x}{120} = 1$

TRY IT! *Problem on page 371*

Let x = no. of hrs. for Bill to finish the job.

Worker	Rate of work (part of job per hr.)	Time worked (hr.)	Work done (part of job)
Mr. Cooper	$\dfrac{1}{2}$	1	$1\left(\dfrac{1}{2}\right)$ or $\dfrac{1}{2}$
Bill	$\dfrac{1}{6}$	x	$x\left(\dfrac{1}{6}\right)$ or $\dfrac{x}{6}$

If the job is finished, the sum of the fractional parts done by each must be 1.

$$\frac{1}{2} + \frac{x}{6} = 1$$

$$6\left(\frac{1}{2} + \frac{x}{6}\right) = 6(1)$$

$$\overset{3}{6}\left(\frac{1}{2}\right) + \overset{1}{6}\left(\frac{x}{6}\right) = 6(1)$$

$$3 + x = 6$$

$$x = 3$$

Check: Mr. Cooper did $\frac{1}{2}$ the job.

    In 3 hours, Bill can do $\frac{3}{6}$, or $\frac{1}{2}$, the job.

    $\frac{1}{2} + \frac{1}{2} = 1$, or the whole job.

    ANSWER: Bill needs 3 hrs. to finish the job.

CHAPTER 8

APPLYING RATIO AND PROPORTION

Interesting photographic illusions can be accomplished with mirrors.

In this photo, the figures, which are shown receding infinitely into the distance, remain in proportion.

8-1 Ratio 387

8-2 Proportion 394

8-3 Variation 400

8-4 Similarity 414

8-5 Trigonometric Ratios 420

8-6 Applying the Trigonometric Ratios . . 427

8-7 Probability 439

8-8 Compound Events 448

Chapter Summary 458

Chapter Review Exercises 459

Problems for Pleasure 463

Calculator Challenge 464

College Test Preparation 464

Spiral Review Exercises 467

Solutions to TRY IT! Problems 468

Using mathematics empowers us. For example, we:
(1) Compare numbers to interpret numerical data and to find values we need to know.
(2) Observe how quantities vary to make accurate predictions.
(3) Use mathematical methods of measure to determine lengths we would otherwise be unable to measure.

8-1 RATIO

The Meaning of Ratio

You can compare two numbers, such as 18 and 6,

by subtraction: 18 is 12 more than 6   *by division:* 18 is 3 times 6

When you compare two numbers by division, you are finding their **ratio**, the quotient of the first number divided by the second. Note that the order of comparison is important; $18 \div 6$ is different from $6 \div 18$. Another way to write the terms of a ratio is with the symbol :, as in 18:6.

In general, the ratio of *a* to *b* is written as $\dfrac{a}{b}$, $a \div b$, or $a:b$

where *a* and *b* are an *ordered pair* of numbers and $b \neq 0$.

A ratio compares two quantities of the same type, where the units cancel, and the ratio itself has no unit. For example: $\dfrac{12 \text{ feet}}{2 \text{ feet}} = \dfrac{6}{1}$

If the two quantities in a ratio are expressed in different units, you must convert the expression so that both quantities use the same unit. For example:

$$\frac{5 \text{ minutes}}{1 \text{ hour}} = \frac{5 \text{ minutes}}{60 \text{ minutes}} = \frac{1}{12}$$

A convenient way to achieve the same result is to multiply the original fraction by a **conversion fraction** that is equal to 1, whose units will cancel those in the original fraction. Since 1 hour = 60 minutes, the fraction $\dfrac{1 \text{ hour}}{60 \text{ minutes}}$ is equal to 1. Thus:

$$\frac{5 \text{ minutes}}{1 \text{ hour}} \cdot \frac{1 \text{ hour}}{60 \text{ minutes}} = \frac{5}{60} = \frac{1}{12}$$

For any given ratio, you can use the multiplication property of 1 to find many equivalent ratios. For example:

$$\frac{3}{10} = \frac{3}{10} \cdot \frac{2}{2} = \frac{6}{20} \qquad \frac{3}{10} = \frac{3}{10} \cdot \frac{3}{3} = \frac{9}{30} \qquad \frac{3}{10} = \frac{3}{10} \cdot \frac{x}{x} = \frac{3x}{10x}$$

You see that $3x$ and $10x$ can represent any two numbers whose ratio is $3:10$. In such a representation, x is called the **common ratio factor**.

Also, you can use the division property of a fraction to find equivalent ratios.

$$\frac{24}{16} = \frac{24 \div 2}{16 \div 2} = \frac{12}{8} \qquad \frac{24}{16} = \frac{24 \div 4}{16 \div 4} = \frac{6}{4} \qquad \frac{24}{16} = \frac{24 \div 8}{16 \div 8} = \frac{3}{2}$$

Note that a ratio is in simplest form when there is no common divisor of the terms. Thus, to express the ratio $\frac{24}{16}$ in simplest form, divide both terms by their greatest common divisor, 8, to obtain $\frac{3}{2}$.

Three numbers can be compared in a **continued ratio**, such as $5:4:3$.

MODEL PROBLEMS

1. An oil tank has a capacity of 2,000 liters. There are 500 liters of oil in the tank. What part of the tank is full?

   Solution: $\text{ratio} = \dfrac{\text{number of liters of oil in the tank}}{\text{capacity of the tank in liters}} = \dfrac{500}{2,000} = \dfrac{1}{4}$

   ANSWER: $\frac{1}{4}$ of the tank is full.

2. Express the ratio $1\frac{3}{4}$ to $1\frac{1}{2}$ in simplest form.

   Solution: $1\frac{3}{4} \div 1\frac{1}{2} = \frac{7}{4} \div \frac{3}{2} = \frac{7}{4} \times \frac{2}{3} = \frac{14}{12} = \frac{7}{6}$

3. Compute the ratio of 9 inches to 2 feet.

   Solution 1: Express both quantities in the same unit of measure.

   $$\text{ratio} = \frac{9 \text{ inches}}{2 \text{ feet}} = \frac{9 \text{ inches}}{24 \text{ inches}} = \frac{3}{8}$$

   Solution 2: Multiply by a conversion fraction.

   $$\text{ratio} = \frac{\overset{3}{\cancel{9} \text{ inches}}}{2 \text{ feet}} \cdot \frac{1 \text{ foot}}{\underset{4}{\cancel{12} \text{ inches}}} = \frac{3}{8}$$

   ANSWER: The ratio is $3:8$.

Now it's your turn to . . . **TRY IT!** *(See page 468 for solutions.)*

1. Compute the ratio of 1 minute to 45 seconds.
2. Phil's arm is $1\frac{1}{2}$ ft. long. In a photograph, his arm is 1 in. long. What is the ratio of the actual length of Phil's arm to the length of his arm in the picture?

Using Ratio to Express a Rate

Not only is ratio used to compare two quantities of the same type, it is also possible to use ratio to compare two quantities of different types. For example:

If a plane flies 1,920 kilometers in 3 hours, the ratio of kilometers to hours—or the **rate** at which the plane flies—is

$$\frac{1{,}920 \text{ km}}{3 \text{ hr.}}, \text{ or } \frac{640 \text{ km}}{1 \text{ hr.}}, \text{ read "640 kilometers per hour."}$$

MODEL PROBLEM

There are 5 grams of salt in 100 cc (cubic centimeters) of a solution of salt and water. Express as a rate the ratio of the amount of salt to the amount of solution.

Solution: $\text{ratio} = \dfrac{5 \text{ grams}}{100 \text{ cc}} = .05 \text{ gram per cc}$

Note. The answer could have been written as 1 gram per 20 cc. In general, however, a rate is reduced to a fraction with denominator 1 and written as an integer or as a decimal.

Problems Involving Ratio

Many problems that can be solved by an algebraic model involve the concept of ratio.

MODEL PROBLEM

The measures of the angles of a triangle are in the ratio $1:2:3$.
a. Find the measure of each angle.   **b.** What kind of triangle is it?

Solution: Let x = the common ratio factor.

Then $1x$, $2x$, and $3x$ = the measures of the 3 angles.

The sum of the measures of the 3 angles of a triangle is 180°.

$$1x + 2x + 3x = 180$$
$$6x = 180$$
$$x = 30$$
$$2x = 60$$
$$3x = 90$$

Check:

$$30 + 60 + 90 \overset{?}{=} 180 \qquad 90{:}60{:}30 \overset{?}{=} 3{:}2{:}1$$
$$180 = 180 \checkmark \qquad 3{:}2{:}1 = 3{:}2{:}1 \checkmark$$

ANSWER: **a.** The measures of the 3 angles are 30°, 60°, and 90°.
   b. The triangle is a right triangle.

Now it's your turn to . . . **TRY IT!**   (*See page 468 for solution.*)

Two numbers are in the ratio 2:3. The larger is 30 more than $\frac{1}{2}$ of the smaller. Find the numbers.

EXERCISES

1. a. Express each ratio as a fraction.
   (1) 36 to 12   **(2)** 48:24   **(3)** 40 ÷ 25

   b. Express each ratio in simplest form.
   (1) $\frac{8}{32}$ **(2)** 20:10   **(3)** $3x:2x$

2. a. If the ratio of two numbers is 4:1, how many times the smaller number is the larger?
   b. If the ratio of two numbers is 8:1, the smaller is what fractional part of the larger?

3. a. In each part, tell whether the ratio is equal to $\frac{3}{2}$.
   (1) $\frac{30}{20}$ **(2)** $\frac{9}{4}$ **(3)** $\frac{8}{12}$ **(4)** 9:6   **(5)** $\frac{45}{30}$ **(6)** 18:6

   b. In each part, name the ratios that are equal.
   (1) $\frac{2}{3}, \frac{6}{9}, \frac{10}{30}, \frac{28}{36}, \frac{50}{75}$ **(2)** 10:8, 20:16, 15:13, 4:5, 50:40

   c. In each part, find the number to replace ? making the resulting statement true.
   (1) $\frac{1}{2} = \frac{?}{24}$ **(2)** $\frac{2}{3} = \frac{10}{?}$ **(3)** $\frac{4}{5} = \frac{?}{100}$ **(4)** $\frac{5}{8} = \frac{25}{?}$

   d. Find four pairs of numbers, the ratio of each pair of numbers being:
   (1) $\frac{1}{2}$ **(2)** 3:1   **(3)** $\frac{3}{4}$ **(4)** 2:3

4. Express each ratio in simplest form.
   a. 36 cm to 72 cm   **b.** $1\frac{1}{2}$ hr. to $\frac{1}{2}$ hr.   **c.** 1 ft. to 1 in.
   d. 12 oz. to 3 lb.   **e.** $6 to 50 cents   **f.** 500 g to 2 kg

5. In a class, there are 20 boys and 10 girls. Find the ratio of:
   a. boys to girls   **b.** girls to boys
   c. boys to total number of pupils   **d.** girls to total number of pupils

6. a. A baseball team played 144 games and won 96.
   (1) What is the ratio of the number of games won to the number of games played?
   (2) For every 3 games played, how many games were won?
   b. A student did 6 out of 10 problems correctly.
   (1) What is the ratio of the number correct to the number incorrect?
   (2) For every two answers that were incorrect, how many answers were correct?

7. Express the rate in lowest terms.
   a. 48 patients to 6 nurses
   b. $1.50 to 3 liters
   c. 13.2 pounds to 6 persons
   d. 241.65 miles to 4.5 hours

8. a. If there are 240 tennis balls in 80 cans, how many balls are there in each can?
   b. If an 11-ounce can of shaving cream costs 88 cents, what is the cost of each ounce?
   c. A can of beans is marked 16 ounces net weight and also 454 grams net weight. Find, correct to the nearest gram, the number of grams in 1 ounce.

9. a. In a supermarket, the regular size of Cleanright Cleanser contains 14 ounces and costs 60 cents. The giant size of Cleanright Cleanser, which contains 20 ounces, costs 78 cents. Which is the better buy? Explain.
   b. Bob types 1,800 words in 30 minutes. David types 1,000 words in 20 minutes. Which boy is the faster typist?
   c. Rosa runs 300 meters in 40 seconds. Pearl runs 200 meters in 30 seconds. Who is faster?

10. The table of gravity constants of the planets relative to Earth shows that a weight of 100 pounds on Earth is equivalent to 36 pounds on Mercury, and 261 pounds on Jupiter.

Planet	Gravity
Mercury	0.36
Venus	0.87
Earth	1.00
Mars	0.38
Jupiter	2.61
Saturn	0.90
Uranus	1.07
Neptune	1.41
Pluto	0.30

   a. Deirdre weighs 120 pounds on Earth. To the nearest pound:
   (1) how much would she weigh on Venus?
   (2) how much more would she weigh on Saturn than on Mars?
   b. Philip's weight on Jupiter is 300 pounds. To the nearest pound:
   (1) find his weight on Earth.
   (2) how much more would he weigh on Neptune than on Uranus?
   c. Mimi weighs 102 pounds on Earth.
   (1) How many times greater would her weight be on Saturn than on Pluto?
   (2) How many pounds lighter would she be on Mercury?
   (3) Her weight on Venus would be closest to her weight on what other planet?

11. Octagon Office Supplies sold 18 wastepaper baskets for $107.82. What was the cost per dozen baskets?

12. Gilbert's Losodium Tuna Fish contains 1,680 mg of sodium for every dozen cans. How many milligrams of sodium are there in 9 cans?

13. Write the answer in terms of the given variable(s).
   a. A warehouse worker transports m crates of goods in k hours. How many crates does he transport in 1 hour?
   b. Annie's Knittery orders b balls of wool for making s sweaters. What is the average number of balls of wool needed for each sweater?
   c. If a restaurant uses p pounds of salmon for z servings, how much salmon is used per k servings?

14. A shoe company manufactures 1,400 pairs of a standard leather dress shoe in a 5-day work week, operating 8 hours a day. At the same production rate, how many pairs of these dress shoes does the company manufacture each hour?

15. After 3 days at sea, a shrimp trawler out of Galveston, Texas, had caught 175 pounds of shrimp. At the same rate, how many pounds of shrimp could the trawler expect after 7 days?

Sea turtles, which have been around since the time of the dinosaurs, are now threatened with extinction. The turtles need to breathe air at intervals, and drown if they are caught in nets at the bottom of the sea.

To save this endangered species, the U.S. Government is requiring shrimp trawlers to use turtle excluder devices, or TEDs, specially equipped nets with panels that admit only small sea animals, and let the turtles out an escape hatch.

In 16–44, solve algebraically.

16. Find two numbers in the ratio 1:3 whose sum is 24.

17. Find two numbers whose ratio is 4:1 and whose difference is 36.

18. A line segment 32 inches in length is divided into two parts that are in the ratio 3:5. Find the length of each part.

19. Find the number of degrees in each angle of a triangle if the measures of the three angles are in the ratio: **a.** 2:3:4   **b.** 2:5:8   **c.** 2:2:5   **d.** 1:5:3

20. The measures of the two acute angles of a right triangle are in the ratio 8:1. Find the number of degrees in each acute angle.

21. A triangular flight path covers a total of 4,800 miles. The lengths of the legs of the trip are in the ratio 3:4:5. Find the three distances.

22. The ratio of the number of boys in a school to the number of girls is 11 to 10. If there are 525 pupils in the school, how many of them are boys?

23. The ratio of the number of students in seventh grade to the number of students in eighth grade is $\frac{4}{5}$. If there are 125 students in eighth grade, find the number in seventh grade.

24. Two business partners agree to share their profits in the ratio 4:3. One year the profits were $35,000. How much money did each partner receive?

25. The numbers of history books and math books at Pottstown Library are in the ratio 9:5. If the total number of books for the two subjects is 98, how many are math books?

26. Three fishermen share their 42-pound catch of sea bass in the ratio 2:3:7. Find the number of pounds of sea bass each receives.

27. The measures of two supplementary angles are in the ratio 3:7. Find the number of degrees in each angle.

28. Mrs. Corliss is following a budget in which the ratio of the amount paid for rent to the total monthly income is 1:4. If Mrs. Corliss earns $2,000 a month, what is her monthly rent?

29. The weight of dried apples to the weight of the original fresh apples is in the ratio 2:5. How many pounds of fresh apples are needed to produce at least 98 pounds of dried apples?

30. Carl and Don agreed to share their caddying tips equally. One weekend, the ratio of Carl's tips to Don's was 7:3. Carl gave Don $20, and the two then had equal amounts. Find the amount of Carl's tips.

31. Two numbers are in the ratio 4:15. If 25 is added to the smaller and the larger is decreased by 30, the resulting numbers are equal. Find the original numbers.

32. The weights of two watermelons are in the ratio 3:4. If a 10-pound melon is included, the combined weights will be over 31 pounds. Find the weight of the smallest melon.

33. In a basketball foul shooting contest, the points made by Sam and Wilbur were in the ratio 7:9. Wilbur made 6 more points than Sam. Find the number of points made by each.

34. A chemist wishes to make $12\frac{1}{2}$ quarts of an acid solution by using water and acid in the ratio 3:2. How many quarts of each should she use?

35. The numerator and denominator of a fraction are in the ratio 3:5. If 2 is subtracted from the numerator and 6 is added to the denominator, the resulting fraction has the value $\frac{1}{3}$. Find the original fraction.

36. The amount of snowfall on two days was in the ratio 3:7. If 2 more inches of snow had fallen on each day, the ratio would have been 1:2. How much snow fell on the first day?

37. Carla has $2.50 in her bank in nickels and dimes. The number of nickels and the number of dimes are in the ratio 3:1. How many coins of each type does she have?

38. Two boats start at the same time and place, and travel in opposite directions. The ratio of their speeds is 2:3. In 3 hours, they are 60 miles apart. Find the speed of each boat.

39. The perimeter of a rectangle is 360 feet. If the ratio of its length to its width is 11:4, find the dimensions of the rectangle.

40. The lengths of one leg and the base of an isosceles triangle are in the ratio 4:3. If the perimeter of the triangle is 66 cm, find the lengths of the three sides of the triangle.

41. A ski tram at Snowbird, Utah, takes passengers from 8,200 feet above sea level to 11,000 feet. If the tram ascends at the rate of 21 ft./sec., how long does it take to reach the top?

42. The ratio of the ages of Donna and her older sister Emily is 2:5. In one more year, the ratio of their ages will be 3:7. How old is each now?

43. In a triangle, the ratio of the measures of two angles is 5:1. The measure of the third angle is equal to the difference between the measures of the other two. Find the number of degrees in each angle.

44. The distance between Albertsville and Beckinridge is 450 miles. On an auto club's road map, these cities are 3 inches apart. On the same map, Beckinridge is $5\frac{1}{2}$ inches from Cobb Corner. What is the actual distance between Beckinridge and Cobb Corner?

8-2 PROPORTION

The Meaning of Proportion

An equation stating that two ratios are equal is called a **proportion**. For example, the proportion stating that the ratio $\frac{4}{20}$ is equal to the ratio $\frac{1}{5}$

is written $\frac{4}{20} = \frac{1}{5}$ or $4:20 = 1:5$, and is read "4 is to 20 as 1 is to 5."

In general, a proportion has four terms. The first and fourth terms are the **extremes**, and the second and third terms are the **means**.

In any proportion, the product of the means is equal to the product of the extremes.

$$\text{If } \frac{a}{b} = \frac{c}{d}, \text{ then } ad = bc.$$

You can verify this property of proportion

in particular:

$$\frac{4}{20} = \frac{1}{5}$$

$$20 \times 1 \overset{?}{=} 4 \times 5$$

$$20 = 20 \checkmark$$

in general:

$$\frac{a}{b} = \frac{c}{d}$$

$$\overset{1}{\cancel{b}}d\left(\frac{a}{\cancel{b}}\right) = b\overset{1}{\cancel{d}}\left(\frac{c}{\cancel{d}}\right) \quad \text{Multiply by the L.C.D.}$$

$$ad = bc$$

There are other useful properties of proportion.

In a proportion, the means and extremes can be interchanged.

$$\frac{a}{b} = \frac{c}{d} \text{ is equivalent to } \frac{b}{a} = \frac{d}{c}$$

If the product of two nonzero numbers is equal to the product of two other nonzero numbers, then a proportion can be formed.

$$\text{If } a \cdot d = b \cdot c, \text{ then: } \frac{a}{b} = \frac{c}{d} \text{ or } \frac{b}{a} = \frac{d}{c}$$

Solve for x: $\dfrac{12}{x-2} = \dfrac{32}{x+8}$

Solution: In a proportion, the product of the means is equal to the product of the extremes.

$$\dfrac{12}{x-2} = \dfrac{32}{x+8}$$
$$32(x-2) = 12(x+8)$$
$$32x - 64 = 12x + 96$$
$$32x - 12x = 96 + 64$$
$$20x = 160$$
$$x = 8$$

ANSWER: $x = 8$

Check:

$$\dfrac{12}{x-2} = \dfrac{32}{x+8}$$
$$\dfrac{12}{8-2} \stackrel{?}{=} \dfrac{32}{8+8}$$
$$\dfrac{12}{6} \stackrel{?}{=} \dfrac{32}{16}$$
$$2 = 2 \checkmark$$

NOTE. Do not repeat the cross-multiplication of the solution as a check. Instead, evaluate each ratio separately.

Now it's your turn to . . . **TRY IT!**   (See page 468 for solution.)

Solve for x and check: $\dfrac{x-2}{x+2} = \dfrac{x-1}{x+4}$

Using Proportions to Solve Problems

Proportions can be used to solve problems that involve quantities of different types, and also problems that involve quantities measured in the same units. The following model problems illustrate this distinction.

MODEL PROBLEMS

1. There are about 90 calories in 20 grams of a cheese. Reggie ate 70 grams of this cheese. About how many calories was this?

Solution: Let x = the number of calories in the cheese Reggie ate.

$$\dfrac{x}{70} = \dfrac{90}{20} \quad \begin{array}{l} \leftarrow \text{ number of calories} \\ \leftarrow \text{ number of grams of cheese} \end{array}$$

$$90(70) = 20x$$
$$6{,}300 = 20x$$
$$315 = x$$

Check: $\dfrac{315}{70} \stackrel{?}{=} \dfrac{90}{20}$

$$\dfrac{9}{2} = \dfrac{9}{2} \checkmark$$

ANSWER: There were about 315 calories in the 70 grams of cheese.

2. A board 12 feet long is cut into two pieces whose lengths are in the ratio $3:1$. Find the length of each piece.

Solution 1:

Let x = length of longer piece.
Then $12 - x$ = length of shorter piece.

The lengths of the two pieces are in the ratio 3:1.

$$\frac{x}{12 - x} = \frac{3}{1}$$

$$3(12 - x) = 1x$$
$$36 - 3x = x$$
$$36 = x + 3x$$
$$36 = 4x$$
$$x = 9$$
$$12 - x = 3$$

Solution 2:

Let x = the common ratio factor.
Then $3x$ = length of longer piece.
And $1x$ = length of shorter piece.

The sum of the two lengths is 12 feet.

$$3x + 1x = 12$$
$$4x = 12$$
$$x = 3$$
$$3x = 9$$

Check: $9 + 3 \overset{?}{=} 12$ $9:3 \overset{?}{=} 3:1$

$12 = 12 \checkmark$ $3:1 = 3:1 \checkmark$

ANSWER: The lengths of the pieces are 9 feet and 3 feet.

Now it's your turn to . . . **TRY IT!**   (See page 468 for solution.)

Miss Bjornsen received $40 for working 8 hours. At the same rate, how much would she receive for working 14 hours?

In the Fibonacci sequence 1, 1, 2, 3, 5, 8, 13, $\cdots$, the ratio between any two adjacent numbers after 3 is about 1.6 to 1. This **Golden Ratio** makes many geometric figures appealing to the eye.

The Golden Rectangle, whose length and width are in the ratio $1.6:1$, is often seen in architecture. Today, we see only the ruins of the Parthenon in Athens, Greece, but the original dimensions of this temple were in the Golden Ratio.

This ratio is also the positive solution to the proportion $\frac{1}{x} = \frac{x}{1 + x}$.

1. Can the given ratios form a proportion? If so, write the proportion.

   a. $\dfrac{3}{4}, \dfrac{30}{40}$ **b.** $\dfrac{4}{5}, \dfrac{16}{25}$ **c.** $\dfrac{5x}{9x}, \dfrac{10}{18}$ **d.** $\dfrac{x}{2x}, \dfrac{10}{20}$ **e.** $\dfrac{5a}{6b}, \dfrac{10b}{12a}$

2. Use the given numbers to form a proportion.

   a. 1, 3, 30, 10   **b.** 15, 40, 8, 3   **c.** 28, 6, 24, 7   **d.** 24, 36, 9, 6

3. Find the number to replace ? and make the result a proportion.

   a. $\dfrac{3}{5} = \dfrac{18}{?}$ **b.** $4{:}6 = {?}{:}42$ **c.** $\dfrac{?}{9} = \dfrac{35}{63}$ **d.** $16{:}? = 12{:}9$

4. Rewrite the equation as a proportion.

   a. $4 \cdot 6 = 8 \cdot 3$ **b.** $2 \cdot 10 = 5 \cdot 4$ **c.** $36 \cdot 4 = 12^2$

In 5–16, solve the proportion and check.

5. $\dfrac{30}{4x} = \dfrac{10}{24}$ 6. $\dfrac{5}{15} = \dfrac{x}{x + 8}$ 7. $\dfrac{x + 10}{x} = \dfrac{18}{12}$

8. $\dfrac{x}{12 - x} = \dfrac{10}{30}$ 9. $\dfrac{16}{8} = \dfrac{21 - x}{x}$ 10. $\dfrac{5}{x + 2} = \dfrac{4}{x}$

11. $12{:}15 = x{:}45$ 12. $19{:}x = 57{:}15$ 13. $8{:}2x = 15{:}60$

14. $\dfrac{2x - 1}{21} = \dfrac{3x - 7}{15}$ 15. $\dfrac{x}{x + 4} = \dfrac{x + 1}{x + 6}$ 16. $\dfrac{3x + 1}{5x - 7} = \dfrac{3x + 6}{5x - 3}$

17. Solve for x in terms of the other variables.

    a. $a{:}b = c{:}x$ **b.** $2r{:}s = x{:}t$ **c.** $2x{:}m = 4r{:}s$

In 18–28, solve by using a proportion.

18. If 3 tickets to an event cost \$26.40, find the cost of 7 such tickets.

19. How much would you pay for 5 pencils at the rate of \$3.84 a dozen?

20. Henry scores an average of 7 foul shots out of every 10 attempts. At the same rate, how many shots would he score in 200 attempts?

21. The owner of a plant nursery knows that the color red will occur in a certain plant once out of about 15 seedlings. If the nursery wants to have about 200 red plants of this variety to sell, how many seedlings should be planted?

22. A train traveled 90 miles in $1\frac{1}{2}$ hours. How many miles will the train go in 6 hours, traveling at the same rate?

23. Fifty feet of copper wire weighs 2 pounds. Find the weight of 325 feet of the same wire.

24. A worker received \$28 for working 2 hours. At the same rate of pay, how many hours must she work to earn \$49?

25. A recipe calls for $1\frac{1}{2}$ cups of sugar for a 3-pound cake. How many cups of sugar should be used for a 5-pound cake?

26. The scale on a map is 1 cm = 500 km. If two cities are 875 km apart, how far apart are they on this map?

27. The scale on a blueprint is 1 inch = 20 feet. If on the blueprint the length of a room is $1\frac{1}{4}$ inches, what is the actual length of the room?

28. A picture $3\frac{1}{4}$ inches long and $2\frac{1}{8}$ inches wide is to be enlarged so that its length will become $6\frac{1}{2}$ inches. What will be the width of the enlarged picture?

In 29–36, express the answer in terms of the given variables.

29. If a car travels m miles in n hours, how long will it take to go c miles at the same rate?

30. If 8 pounds of meat costs w dollars, how many pounds can be bought for $3y$ dollars?

31. If the Stars scored p points in $2z$ games, how many points might they score in $3t$ games?

32. If the ABC Corporation mails out g payroll checks every k months, how many of these checks are mailed out in m months?

33. The Read-More Book Corporation binds b books every s hours. At this same rate, how long will it take to bind d books?

34. On the average, f tax returns are processed every k days. If the same rate is applied, how many returns are processed in t days?

35. A theme park collected d dollars in admissions for c park entrants. If everyone paid the same admission price, how much was collected for g entrants?

36. If a cruise ship can go k knots in h hours, how many knots can the ship cover in n hours?

In 37–48, solve algebraically.

37. Two numbers are in the ratio $3:2$. The smaller number is 36. Find the larger number.

38. In a school, the ratio of the number of boys to the number of girls is $5:4$. If 1,008 girls attend the school, what is the number of boys?

39. The ratio of the length of a rectangular field to its width is $10:7$. If the width of the field is 70 meters, find its perimeter.

40. The increase in the price of instant film is expected to be proportional to the increase in the price of silver, which is an essential ingredient. If 24-exposure film costs $9, and silver increases from $8.40 to $11.20 per ounce, what should be the new price of 24-exposure film?

41. The list price of any Salvo calculator is directly proportional to the number of functions it performs. Their 4-function calculator sells for $4.98.
    a. What should be the price of the 10-function calculator?
    b. One retailer is selling the Salvo 24-function calculator for $33.49. Is this price a bargain or an overcharge? By how much does it differ from the list price?

42. The ratio of the number of microwave ovens sold to the number of compact disc players sold by HJS Appliances last year was $3:5$. If the total number of microwaves and CD players sold was less than 1,500, how many microwaves were sold?

43. The length of runway recommended for a commercial passenger plane to land safely is directly proportional to its airspeed in knots when it touches down. A 727 jet touching down at 168 knots requires a 6,000-foot runway. If, at a rush hour period, an air traffic controller must direct the 727 jet to a 5,000-foot runway, what is the jet's maximum permissible touchdown speed?

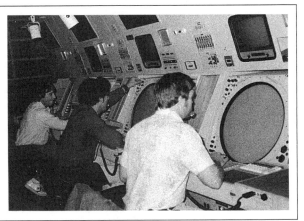

The intensive air traffic control program at the Federal Aviation Administration Academy in Oklahoma City tests to see whether a student can visualize 20 imaginary planes flying in the same airspace at different speeds, directions, and altitudes, keep them from colliding, and bring them down safely.

Controlling is a job for young people, with 22 an ideal age at which to start the two months of schooling and three years of on-the-job training.

44. Assume that the ratio of the number of S.A.D.D. (Students Against Drunk Driving) chapters that existed 5 years ago to the current number of chapters is 5:8, with 1,251 new chapters having been formed during the 5-year period.
   a. How many S.A.D.D. chapters are there now?
   b. If the number of chapters 3 years from now is expected to be in the ratio of 5:3 to the present number, how many new chapters are expected to form in the next 3 years?

45. Suburban Senior High School now has 1,518 students and 66 teachers. If 130 new students are admitted, what is the minimum number of additional teachers that will be needed in order to avoid increasing the student–teacher ratio?

46. Between the hours of 6:00 P.M. Friday and 6:00 P.M. Saturday on a long holiday weekend, 138 people died on U.S. highways. If the death rate remained constant till midnight Monday, estimate the highway death toll for the entire weekend.

47. To estimate the number of fur seal pups in a breeding spot during one summer breeding season, National Marine Fisheries Service workers tagged 2,734 of the pups. Several weeks later, 600 pups in the same rookery were inspected, and of these, 163 had already been tagged. Estimate the number of fur seals in the rookery.

48. On the average, out of every 5 students that apply for admission to Lake Valley College, 3 are accepted. In this year's class, if the number of applicants exceeded the number accepted by 1,250, how many were accepted?

8-3 VARIATION

In general, when two variables are related to each other in an open sentence or a formula, changes in the first variable require corresponding changes in the second variable. For certain relations, you will be able to see patterns of change.

Direct Variation

Consider the formula $p = 4s$ in which the perimeter p of a square is related to the length of the side s. Allow s to change, or *vary*, by taking on selected values. The tables below show selected sets of values for s and the corresponding values of p, which are determined by substitution into the formula.

s	1	2	3	7	10.5
p	4	8	12	28	42

s	9	6	4.1
p	36	24	16.4

Observe that in the first table, the numbers selected for s increase in value, and the corresponding values of p then also increase. How do the s-values in the second table change? How do the corresponding p-values change? (The s-values decrease and the corresponding p-values then also decrease.)

When the formula $p = 4s$ is rewritten as $\frac{p}{s} = 4$, you see that the ratio $\frac{p}{s}$ is always 4, a constant. Look at the ordered pairs of numbers in the tables to verify that they all are in the ratio $4:1$.

In general, when two variables are related so that their *ratio is constant*, that relation is a ***direct variation***. Algebraically:

If *y* varies directly as *x*, the ratio $\frac{y}{x}$ is constant, written:

$$\frac{y}{x} = k \quad \text{or} \quad y = kx, \quad \text{where } k \neq 0 \text{ is the } \textit{constant of variation.}$$

Some consequences of this statement are:

1. All ordered pairs of values for the variables are in the ratio $k:1$.
2. The ratios of any two specific ordered pairs of values for the variables form a proportion.

    If (x_1, y_1), read "x sub 1, y sub 1," represents one specific ordered pair and (x_2, y_2) represents a second specific ordered pair, then: $\dfrac{y_1}{x_1} = \dfrac{y_2}{x_2}$

3. a. If x increases, y also increases; if x decreases, y also decreases.
    b. If x is multiplied by a specific number, then y is multiplied by the same number.

1. If d varies directly as t and if $d = 60$ when $t = 2$, find the value of d when $t = 7$.

   Solution 1: First find the constant of variation.

   Since d varies directly as t: $d = kt$

   If $d = 60$ when $t = 2$: $60 = 2k$
   $$30 = k$$
   Thus, $d = 30t$.

   If $t = 7$, then $d = 30(7)$ or 210.

   Check: $\dfrac{60}{2} \stackrel{?}{=} \dfrac{210}{7}$

   $30 = 30 \checkmark$

   Solution 2: Substitute the two pairs of values into the proportion form.

   Since d varies directly as t,

   $$\frac{d_1}{t_1} = \frac{d_2}{t_2}$$

   $$\frac{60}{2} = \frac{d_2}{7}$$

   $$2d_2 = 420$$

   $$d_2 = 210$$

d	60	?
t	2	7

   ANSWER: $d = 210$ when $t = 7$.

2. At an entrance ramp to the Riveredge Highway, a driver must come to a full stop before proceeding. Emily comes to a stop, then accelerates at a constant rate, reaching 24 miles per hour after 3 seconds. At the same rate of acceleration, how long will it take her to reach a highway speed of 52 miles per hour?

   Solution: At a constant rate of acceleration, the speed varies directly with time.

   Substitute $s = 24$ and $t = 3$, and solve for k.

   Substitute $k = 8$ and $s = 52$ into the original equation, and solve for t.

   $$s = kt$$
   $$24 = k \cdot 3$$
   $$k = 8$$
   $$52 = 8t$$
   $$t = 6\tfrac{1}{2}$$

   ANSWER: It will take Emily $6\frac{1}{2}$ seconds.

Now it's your turn to . . . **TRY IT!**   (See page 468 for solutions.)

1. Examine the table of values for m and n.
   a. Explain why m and n vary directly.
   b. Write a formula to express the relation.

m	−1	1	2
n	−3	3	6

2. If p varies directly as s and $p = 12$ when $s = 3$, find the value of p when $s = 10$.

In 1–3, tell whether one variable varies directly as the other. If it does, express the relation between the variables by means of a formula. If it does not, explain why not.

1.

p	3	6	9
s	1	2	3

2.

n	3	4	5
c	6	8	10

3.

x	4	5	6
y	6	8	10

In 4–6, one variable varies directly as the other. Find the missing numbers and write a formula that relates the variables.

4.

h	1	2	?
A	5	?	25

5.

h	4	8	?
S	6	?	15

6.

L	2	8	?
W	1	?	7

In 7–9, write the relation as a formula, using k as the constant of variation.

7. The perimeter p of an equilateral triangle varies directly as a side s.

8. The circumference c of a circle varies directly as the radius r.

9. The resistance r of a copper wire varies directly as its length ℓ.

In 10–12, write the relation as a formula, defining the variables you choose.

10. At a given time, the length of the shadow of an object varies directly as its height.

11. At a fixed hourly wage, income is directly proportional to the number of hours worked.

12. For uniform motion, distance varies directly as time.

To cushion a driver's head and chest in a front-end collision, cars are equipped with air bags. At an impact at more than about 10 miles per hour, the bag inflates in a split second, about half the blink of an eye.

Another safety device is an electronic control system that senses the speed of each wheel during deceleration. By regulating the pressure applied to each brake, the system reduces the risk of lockup and skidding.

In 13–16, is the relation a direct variation? Explain.

13. $R + T = 80$ **14.** $15T = D$ **15.** $\dfrac{e}{i} = 20$ **16.** $bh = 36$

17. For each of the relations (1)–(4) given below, where k is constant, describe the change:

    a. in y if x is doubled

    b. in x if y is multiplied by 4

        (1) y varies directly as x
        (2) x varies directly as y
        (3) y varies directly as x^2
        (4) x varies directly as y^2

18. The circumference c of a circle varies directly as the diameter d.
    a. If $c = 44$ cm when $d = 14$ cm, find c when $d = 21$ cm.
    b. If $c = 6.28$ ft. when $d = 2$ ft., find d when $c = 62.8$ ft.

19. A salesperson's commission c varies directly as her sales s.
    a. If $c = \$100$ when $s = \$1,000$, find c when $s = \$1,250$.
    b. If $c = \$240$ when $s = \$4,000$, find s when $c = \$300$.

20. The horsepower p developed by a steam engine varies directly as the number of revolutions per minute, r, at which it is operated.
    a. If $p = 1,280$ when $r = 160$, find p when $r = 220$.
    b. If $p = 2,400$ when $r = 250$, find r when $p = 1,920$.

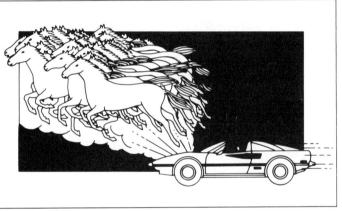

When the Englishman James Watt designed his steam engine in the 18th century, he wanted to describe its power in an easily understandable way. He tested the loads that strong horses could pull, and settled on one horsepower as the power needed to lift a weight of 33,000 pounds one foot in one minute.

The power rating is in inverse proportion to the time it takes to do the job. An engine that can overcome a resistance of 33,000 pounds and move it a foot in $\frac{1}{4}$ minute is a 4-horsepower engine.

21. If 3 men earn \$420 in a day, what will 21 men working at the same wage earn in a day?

22. If 5 caps cost \$35, how much will 9 caps of the same kind cost?

23. If a train travels 240 miles in 4 hours, how far will it go in 7 hours at the same speed?

24. x varies directly as $y + 1$. If $y = 3$ when $x = 2$, find y when $x = 6$.

25. $R + 2$ varies directly as $2S - 3$. If $S = 4$ when $R = 3$, find S when $R = 9$.

26. y varies directly as x^2. If $y = 2$ when $x = 1$, find y when $x = 2$.

Inverse Variation

Consider the relation $\ell \cdot w = 12$ to describe a rectangle of constant area 12 square units and variable length ℓ and width w. The tables show selected ℓ-values and the corresponding w-values.

ℓ	1	2	5	12	16
w	12	6	$\frac{12}{5}$	1	$\frac{3}{4}$

ℓ	24	18	12	6	3	2	1	$\frac{1}{2}$
w	$\frac{1}{2}$	$\frac{2}{3}$	1	2	4	6	12	24

Observe, from the first table, that as the ℓ-values increase, the corresponding w-values decrease. What does the second table show? (The ℓ-values decrease while the corresponding w-values increase.) Verify that in both tables all of the ordered pairs have the same product, 12.

In general, when two variables are related so that their *product is constant*, that relation is an ***inverse variation***. Algebraically:

If *y* varies inversely as *x*, the product *xy* is constant, written:

$$xy = k, \quad \text{where none of the values is 0}$$

Some consequences of this statement are:
1. All ordered pairs of values for the variables have a constant product of k.
2. The products of any two specific ordered pairs of values for the variables are equal. If (x_1, y_1) and (x_2, y_2) represent two specific ordered pairs of values, then:

$$x_1 y_1 = x_2 y_2 \quad \text{or} \quad \frac{x_1}{x_2} = \frac{y_2}{y_1}$$

3. a. If x increases, then y decreases; if x decreases, then y increases.
   b. If x is multiplied by a specific nonzero number, then y is divided by the same number.

MODEL PROBLEMS

1. If y varies inversely as x and if $y = 5$ when $x = 8$, find y when $x = 4$.

Solution 1: First find the constant of variation.

Since y varies inversely as x: $xy = k$

If $y = 5$ when $x = 8$: $(8)(5) = k$

$$40 = k$$

Thus, $xy = 40$.

If $x = 4$, then $4y = 40$ or $y = 10$.

Check: $8 \times 5 \overset{?}{=} 4 \times 10$

$$40 = 40 ✓$$

Solution 2: Since y varies inversely as x, the products of the two pairs of values are equal.

$$x_1 y_1 = x_2 y_2$$
$$8 \cdot 5 = 4 \cdot y_2$$
$$40 = 4y_2$$
$$10 = y_2$$

x	8	4
y	5	?

ANSWER: $y = 10$ when $x = 4$.

Problems about levers and pulleys (refer to Section 5-4) are applications of inverse variation, each obeying a law of the form $x_1y_1 = x_2y_2$.

2. The weights of two people balanced on a seesaw vary inversely with their distances from the fulcrum of the seesaw. If an 88-pound boy and a woman are balanced on a seesaw, what is the weight of the woman if she is 4 feet from the fulcrum while the boy is 6 feet from it?

Solution: Let w_1 and d_1 = the weight and distance for the woman.

Then w_2 and d_2 = the weight and distance for the boy.

Since w varies inversely with d, the products are constant. $w_1 \cdot d_1 = w_2 \cdot d_2$

Substitute the given values into the relation. $w_1 \cdot 4 = 88 \cdot 6$

For the woman: w_1 is unknown and $d_1 = 4$ $4w_1 = 528$

For the boy: $w_2 = 88$ and $d_2 = 6$ $w_1 = 132$

Check: Does the product for the woman $132(4) \stackrel{?}{=} 88(6)$
equal the product for the boy? $528 = 528 ✔$

ANSWER: The woman weighs 132 pounds.

Now it's your turn to . . . **TRY IT!**

1. Study the table of values for m and n.
   a. Explain why m and n vary inversely.
   b. Write a formula to express the relation.

m	−3	2	6
n	−8	12	4

2. The speeds of two connected pulleys vary inversely as their diameters. If at a certain moment, the 16-inch pulley has a speed of 155 ft./min., what is the speed of the 9-inch pulley?

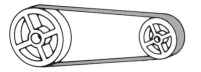

(See page 469 for solutions.)

In 1–3, tell whether one variable varies inversely as the other. If it does, express the relation between the variables by means of a formula. If it does not, explain.

1.

n	2	4	6
c	18	9	6

2.

R	10	20	40
T	4	2	1

3.

x	1	2	3
y	8	7	6

In 4–6, one variable varies inversely as the other. Find the missing numbers and write a formula that relates the variables.

4.

w	2	?	6
d	12	8	?

5.

R	2	6	?
T	72	?	12

6.

ℓ	2	4	?
w	32	?	8

In 7–9, write the relation as a formula, using k as the constant of variation.

7. The time t required to travel a fixed distance varies inversely as the rate of motion, r.

8. The number of days, d, required to complete a job varies inversely as the number of workers, w, on the job.

9. The force f necessary to pry up a rock varies inversely as the length ℓ of the crowbar used.

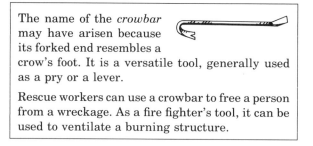

The name of the *crowbar* may have arisen because its forked end resembles a crow's foot. It is a versatile tool, generally used as a pry or a lever.

Rescue workers can use a crowbar to free a person from a wreckage. As a fire fighter's tool, it can be used to ventilate a burning structure.

In 10–12, write the relation as a formula, defining the variables you use.

10. When the temperature of a gas is constant, its volume varies inversely as the pressure.

11. The principal that must be invested to yield a fixed annual income varies inversely as the rate of interest.

12. The temperature at which water boils varies inversely as the number of feet above sea level.

In 13–16, is the relation an inverse variation? Explain.

13. $N - C = 40$ **14.** $NC = 40$ **15.** $t = \dfrac{60}{r}$ **16.** $h = 60b$

17. For each of the relations (1)–(3) given below, where k is constant, describe the change:

    a. in y if x is doubled

    b. in x if y is multiplied by 4

        (1) y varies inversely as x

        (2) x varies inversely as y

        (3) y varies inversely as x^2

18. The number of times, n, a wheel must turn to cover a given distance varies inversely as the radius, r, of the wheel.

    a. If $n = 10$ when $r = 14$, find n when $r = 7$.

    b. If $n = 40$ when $r = 100$, find r when $n = 10$.

19. The measure of the altitude, h, of a parallelogram of constant area varies inversely as the measure of its base, b.

    a. If $h = 4$ in. when $b = 12$ in., find h when $b = 16$ in.

    b. If $h = 9$ cm when $b = 15$ cm, find b when $h = 3$ cm.

20. The number of slices, n, cut from a bread loaf of constant length varies inversely as the uniform thickness, t, of a slice.

    a. If $n = 32$ when $t = \frac{3}{8}$ in., find n when $t = \frac{1}{2}$ in.

    b. If $n = 16$ when $t = 15$ mm, find t when $n = 20$.

21. A man invested \$10,000 at 6% per year. At what rate would he have to invest \$7,500 to have the same annual income?

22. Ten printing presses, all alike, can do a job in 3 hours. How many hours would it take 6 of these printing presses to do the same job?

23. The arms of a lever are 8 feet and 12 feet long, respectively. What weights that differ by 10 pounds will balance the lever?

24. If x varies inversely as $y + 5$, and if $y = 1$ when $x = 2$, find y when $x = 1$.

25. If $M + 1$ varies inversely as $2N - 1$, and if $N = 13$ when $M = 3$, find N when $M = 24$.

26. y varies inversely as x^2. If $y = 4$ when $x = 3$, find y when $x = 6$.

Variations, Direct and Inverse

In 27–32, write a formula, if possible, expressing either direct or inverse variation.

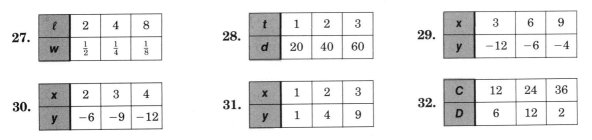

27.

ℓ	2	4	8
w	$\frac{1}{2}$	$\frac{1}{4}$	$\frac{1}{8}$

28.

t	1	2	3
d	20	40	60

29.

x	3	6	9
y	-12	-6	-4

30.

x	2	3	4
y	-6	-9	-12

31.

x	1	2	3
y	1	4	9

32.

C	12	24	36
D	6	12	2

33. Y varies directly as x. If $Y = 35$ when $x = -5$, find Y when $x = -20$.

34. If x varies inversely as y, and if $x = 8$ when $y = 9$, find x when $y = 18$.

35. If n varies inversely as c, and if $n = \frac{1}{3}$ when $c = 60$, find n when $c = 40$.

36. A varies directly as h. If $A = 4.8$ when $h = .4$, find h when $A = 3.6$.

37. N varies directly as d. If $N = 10$ when $d = 8$, find N when $d = 12$.

38. If R varies inversely as T, and if $R = 80$ when $T = \frac{1}{4}$, find R when $T = 2$.

39. If m varies directly as n^2, and if $m = 20$ when $n = 2$, find m when $n = 3$.

40. If y varies inversely as x^3, and if $y = 8$ when $x = 3$, find y when $x = 2$.

41. When two gears are meshed, the number of revolutions per minute varies inversely as the number of teeth in the gears. If a driver gear with 18 teeth is meshed with a driven gear with 12 teeth, and if the driver gear makes 100 revolutions per minute, what is the speed of the driven gear?

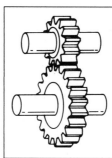

Gears generally operate in pairs, to make it possible for different parts of a machine to rotate at different speeds.

The speeds of rotation, and the power transmitted, are determined by the ratio of the number of teeth in the smaller gear to the number of teeth in the larger.

42. The weight of an object varies directly as its volume. When weight is measured in pounds and volume in cubic feet, the constant of variation for limestone is 170 and for sandstone is 140. Which of the illustrated blocks of stone is heavier, and by how much?

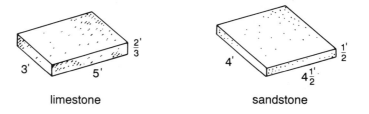

limestone   sandstone

43. If a train travels at a uniform speed, the distance d that it travels varies directly as the time t. If $d = 300$ when $t = 60$, find d when $t = 48$.

44. The time required to do a job varies inversely as the number of workers. If it takes 3 men 4 days to paint a house, how long would it take 2 men, working at the same rate?

45. The speed of a gear varies inversely as the number of teeth. If a gear with 42 teeth makes 18 revolutions per minute, how many revolutions per minute will be made by a gear with 28 teeth?

46. The number of feet a body falls varies directly as the square of the elapsed time. If a stone falls 16 feet the first second, how many feet will it fall in 5 seconds?

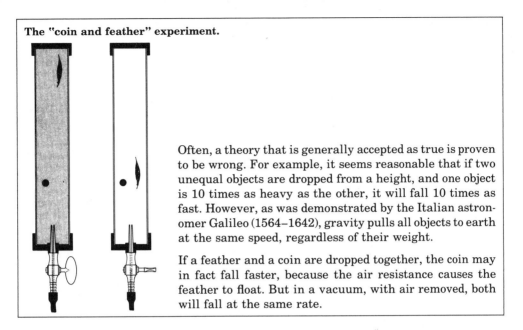

The "coin and feather" experiment.

Often, a theory that is generally accepted as true is proven to be wrong. For example, it seems reasonable that if two unequal objects are dropped from a height, and one object is 10 times as heavy as the other, it will fall 10 times as fast. However, as was demonstrated by the Italian astronomer Galileo (1564–1642), gravity pulls all objects to earth at the same speed, regardless of their weight.

If a feather and a coin are dropped together, the coin may in fact fall faster, because the air resistance causes the feather to float. But in a vacuum, with air removed, both will fall at the same rate.

47. The horsepower required to propel a ship varies directly as the cube of the speed. If it takes 3,000 horsepower to move the ship at a speed of 8 knots, how much horsepower is needed for a speed of 16 knots?

48. The work of the German mathematician Johannes Kepler (1571–1630) established the scientific foundation for astronomy. By his third law of planetary motion, the square of the number of years it takes a planet to revolve around the sun varies directly as the cube of the distance of the planet from the sun.

Let the distance from the earth to the sun be 1 unit. How long would it take a planet whose distance from the sun is 100 units to complete 1 revolution?

49. The length of a spring varies directly as the force applied to the spring. If a 60-pound weight stretches a spring 12 inches, how much will a 17-pound weight stretch the spring?

50. The volume of a sphere is proportional to the cube of the radius. If a beach ball of radius 6 inches has a volume of 288π cubic inches, find the volume of a beach ball with radius 9 inches.

51. The weight of an object varies inversely as its distance from the center of the earth. Assume that the radius of the earth is 4,000 mi. If a man weighs 200 pounds on the surface of the earth, find his weight to the nearest pound in a space station orbiting 210 miles above the earth.

52. The distance traveled by a batted baseball is affected by atmospheric conditions, such as temperature, humidity, and altitude above sea level. In this problem, assume that all balls are batted with an initial velocity of 110 mph, at an angle of 35° from the horizontal. Then the increase in the distance the ball travels is directly proportional to the increase in altitude. A ball that would go 400 feet in Philadelphia (about sea level) would go 40 feet farther in Denver (a mile above sea level).

a. If a ball would travel 405 feet in Chicago, what is the approximate altitude in feet of Chicago's stadium?

b. If Atlanta's stadium is 1,056 feet above sea level, how far would a ball travel in Atlanta?

c. If a baseball were hit in a vacuum, it would travel 750 feet. At what altitude above sea level would a ballfield have to be located in order for altitude to have the same effect?

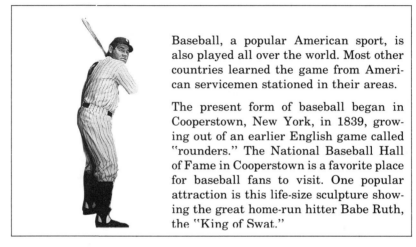

Baseball, a popular American sport, is also played all over the world. Most other countries learned the game from American servicemen stationed in their areas.

The present form of baseball began in Cooperstown, New York, in 1839, growing out of an earlier English game called "rounders." The National Baseball Hall of Fame in Cooperstown is a favorite place for baseball fans to visit. One popular attraction is this life-size sculpture showing the great home-run hitter Babe Ruth, the "King of Swat."

Joint Variation

Consider the formula for the area of a triangle in which the area varies directly as the product of the base and altitude:

$$A = \frac{1}{2}bh \quad \text{or} \quad \frac{A}{bh} = \frac{1}{2}$$

Why is this variation consistent with the definition of direct variation? (The ratio is constant.) This variation, however, does differ from the previous discussion of direct variation. How? (This relation involves more than two variables.)

In general, when three variables are related so that a ratio is constant, that relation is a **_joint variation_**. Algebraically:

If **_y varies jointly as x and z_**, the ratio $\dfrac{y}{xz}$ is constant, written

with a constant of variation, k: $\dfrac{y}{xz} = k$ or $y = kxz$

or as a proportion: $\dfrac{y_1}{x_1 z_1} = \dfrac{y_2}{x_2 z_2}$

y varies jointly as x and the square of z. If $y = 400$ when $x = 4$ and $z = 5$, find y when $x = 3$ and $z = 7$.

Solution 1: First find the constant of variation.

Since y varies jointly as x and z^2:

$$y = kxz^2$$

If $y = 400$ when $x = 4$ and $z = 5$:

$$400 = k(4)(5)^2$$
$$400 = 100k$$
$$4 = k$$

Thus, $y = 4xz^2$.

If $x = 3$ and $z = 7$, $y = 4(3)(7)^2$ or 588.

Solution 2: Substitute the two sets of values into the proportion form.

Since y varies jointly as x and z^2:

$$\frac{y_1}{x_1(z_1)^2} = \frac{y_2}{x_2(z_2)^2}$$

$$\frac{400}{4(5)^2} = \frac{y_2}{3(7)^2}$$

$$\frac{400}{100} = \frac{y_2}{147}$$

$$100y_2 = 400(147)$$

$$y_2 = 588$$

Check: $\dfrac{400}{100} \overset{?}{=} \dfrac{588}{147}$

$4 = 4 \checkmark$ **ANSWER:** $y = 588$ when $x = 3$ and $z = 7$.

Now it's your turn to . . . **TRY IT!** (See page **469** for solutions.)

1. d varies jointly as r and t. If r is doubled and t is tripled, how does d change?
2. i varies jointly as p and r. If $i = 8$ when $p = 100$ and $r = .04$, find i when $p = 400$ and $r = .06$.

EXERCISES

1. If s varies jointly as w and x, and $s = 9$ when $w = 4$ and $x = 2$, find s when $w = 3$ and $x = 5$.

2. If v varies jointly as p and r, and $v = 7$ when $p = 2$ and $r = 3$, find v when $p = 4$ and $r = 5$.

3. If t varies directly as r and the square of s, and $t = 30$ when $r = 4$ and $s = 6$, find t when $r = 3$ and $s = 4$.

4. If r varies jointly as y and z, and $r = 6$ when $y = 3$ and $z = 5$, find r when $y = 2$ and $z = 7$.

5. If m is jointly proportional to v and the square of w, and $m = 6$ when $v = -5$ and $w = 2$, find m when $v = 2$ and $w = 3$.

6. The volume V of a cylinder varies jointly as the altitude h and the square of the radius r. If $V = 3.14$ cu. in. when $h = 1$ in. and $r = 1$ in., find V when $h = 10$ in. and $r = 3$ in.

7. The wind's force on a sail varies jointly as the area of the sail and the square of the wind's velocity. If a 6-mph wind exerts 3 pounds of force on 2 square yards of sail, how much force will a 20-mph wind exert on a 5-square-yard sail?

The path of a sailboat depends on the forward thrust of the boat as well as the force of the wind, with the final direction resulting from a "parallelogram of forces."

To sail to *leeward* (in the same general direction as the wind), all that is needed is a sheet of fabric that will catch the wind to propel the boat forward. It takes more skill to sail to *windward* (against the wind). Since a ship cannot be sailed directly against the wind, forward motion is achieved by zigzagging, or *tacking*.

8. The heat lost through the exterior wall of a house varies jointly as the area of the wall and the difference between indoor and outdoor temperature. If the heat loss through a 200-sq.-ft. wall is 1,200 BTU per hour when the temperature difference is 15° Celsius, find the loss through an identically constructed 160-sq.-ft. wall when there is a 25° Celsius temperature difference.

Combined Variation

Combined variation refers to a situation where both direct and inverse variations occur between 3 or more variables.

9. If t varies inversely as s and directly as v, and $s = 8$ when $t = 30$ and $v = 4$, find s when $t = 24$ and $v = 6$.

10. If w varies inversely as x and directly as y, and $x = 4$ when $w = 12$ and $y = 6$, find x when $w = 20$ and $y = 2$.

11. If p varies directly as r and inversely as the square of q, and $p = 3$ when $q = 6$ and $r = 9$, find p when $q = 5$ and $r = 4$.

12. If b varies jointly as x and y and inversely as z, and $b = 8$ when $x = 4$, $y = 6$, and $z = 3$, find b when $x = -4$, $y = 4$, and $z = 2$.

13. If z varies inversely as c squared and directly as the cube of b, and $z = 4$ when $c = 6$ and $b = -1$, find z when $c = 3$ and $b = 2$.

14. The resistance in an electrical circuit varies directly as the voltage and inversely as the current. If a circuit has a resistance of 20 ohms when the current is 15 amps and voltage is 300 volts, what is the resistance in a 20-amp, 500-volt circuit?

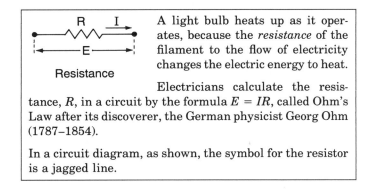

Resistance

A light bulb heats up as it operates, because the *resistance* of the filament to the flow of electricity changes the electric energy to heat.

Electricians calculate the resistance, R, in a circuit by the formula $E = IR$, called Ohm's Law after its discoverer, the German physicist Georg Ohm (1787–1854).

In a circuit diagram, as shown, the symbol for the resistor is a jagged line.

15. A carnival owner found that the number of riders on his merry-go-round varied directly with the duration of the ride and inversely with the price of the ride.
  a. If he attracted an average of 32 riders with a 2-minute ride costing 75 cents, how many riders would he attract with a 3-minute ride for one dollar?
  b. If 3 minutes is required between rides and the carnival is open for 4 hours each night, which price and time structure would be more profitable? By how much?

16. The attraction force F of two bodies varies jointly as their masses m_1 and m_2, and inversely as the square of the distance d between them. If $F = 180$ when $m_1 = 9$, $m_2 = 4$, and $d = 1$, find F when $m_1 = 12$, $m_2 = 15$, and $d = 3$.

17. The safe load of a horizontal beam varies jointly as the width and the square of the depth, and inversely as its length. A beam 15 feet long, 3 inches wide, and 6 inches deep can take a load of 1,800 pounds. What is the safe load for a beam 10 feet long, 4 inches wide, and 2 inches deep?

18. The time it takes for an airplane to circle an airport in a holding pattern varies directly as the distance of the plane from the airport and inversely as the speed of the plane. If a plane circles 14 miles out at a speed relative to the ground of 220 mph, it takes 24 minutes. How long does it take to circle the airport at a speed of 264 mph at 21 miles out?

19. A math class at Horton High is collecting money for a retirement gift for a favorite algebra teacher. The PTA agrees to contribute a fixed fraction of the cost. If all students contribute the same amount of money, that amount will vary directly as the price of the gift and inversely as the number of students contributing. If 40 students contribute $1 each toward a $60 gift, how much would each of 36 students have to contribute toward an $81 gift?

8-4 SIMILARITY

Similar Triangles

The shape of a triangle is determined by its angles. Triangles with the same shape, but not necessarily the same size, are *similar*.

If two triangles are similar, then their corresponding angles are equal in measure and the measures of their corresponding sides are in proportion.

Below are two triangles, ABC and $A'B'C'$ (A' is read A *prime*), that are similar, written $\triangle ABC \sim \triangle A'B'C'$. Since the triangles are known to be similar, their corre-sponding angles are equal in measure, and the mea-sures of their corresponding sides are in proportion.

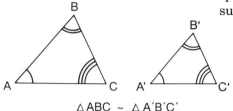

$\triangle ABC \sim \triangle A'B'C'$

Corresponding angles are equal in measure:
$$m\angle A = m\angle A' \quad m\angle B = m\angle B' \quad m\angle C = m\angle C'$$

Corresponding sides are in proportion:
$$\frac{AB}{A'B'} = \frac{BC}{B'C'} = \frac{AC}{A'C'}$$

To say that two triangles are similar, you only need to know that two pairs of angles are equal in measure. Use the numerical values shown in the figure below to explain how applying the property "the sum of the measures of the angles of a triangle is 180°" makes the third pair of angles equal in measure.

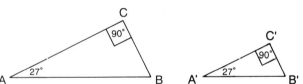

If triangles have two pairs of angles equal in measure, they are similar.

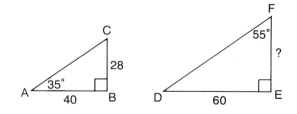

A *pantograph* is an instrument made of 4 bars jointed to form a parallelogram. It is used for copying a diagram, such as a map, on a larger or a smaller scale. The resulting figure is *similar* to the original.

MODEL PROBLEM

In $\triangle ABC$, $m\angle A = 35°$ and $m\angle B = 90°$.

In $\triangle DEF$, $m\angle F = 55°$ and $m\angle E = 90°$.

a. Tell why $\triangle ABC$ is similar to $\triangle DEF$.

b. If $AB = 40$ cm, $BC = 28$ cm, and $DE = 60$ cm, find EF.

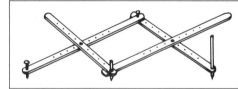

Solution:

a. The triangles are similar since the measures of two pairs of angles are equal.

(1) $m \angle B = m \angle E$ because they are each 90°.

(2) $m \angle C = m \angle F$ because they are each 55°.

In $\triangle ABC$:

$m \angle A + m \angle B = 35 + 90 = 125$

$m \angle C = 180 - 125 = 55$

b. Since the triangles are similar, the measures of their corresponding sides are in proportion.

$$\frac{AB}{DE} = \frac{BC}{EF}$$

$$\frac{40}{60} = \frac{28}{EF}$$

$$40(EF) = 28(60)$$

$$40(EF) = 1{,}680$$

$$EF = 42 \text{ cm} \quad \textbf{ANSWER}$$

Using Similar Triangles in Indirect Measurement

In physical situations that involve measurement, sometimes you can conveniently make a direct measurement by applying a measuring instrument such as a ruler.

In many situations, however, it is inconvenient or impossible to apply the instrument directly to the object. For example, you cannot measure directly the height of a tall building or the distance to the sun. In such cases, methods of indirect measurement are used.

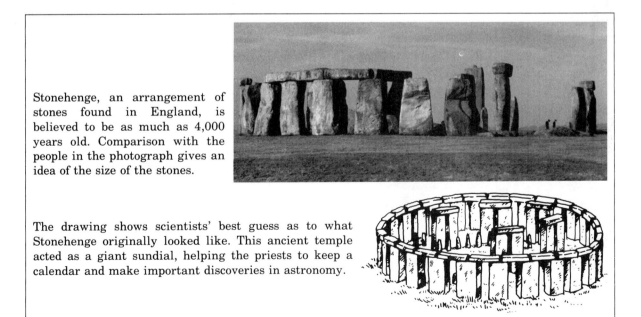

Stonehenge, an arrangement of stones found in England, is believed to be as much as 4,000 years old. Comparison with the people in the photograph gives an idea of the size of the stones.

The drawing shows scientists' best guess as to what Stonehenge originally looked like. This ancient temple acted as a giant sundial, helping the priests to keep a calendar and make important discoveries in astronomy.

MODEL PROBLEM

At the same time that a 6-foot man casts a shadow 5 feet long, a nearby flagpole casts a shadow 10 feet long. Find the height of the flagpole.

Solution: Draw a diagram to model the situation.

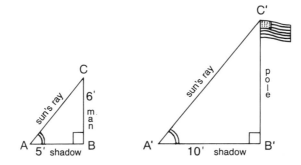

Both the man and the flagpole are vertical, making right angles with the ground. $m \angle B = m \angle B'$

The sun's rays, which cause shadows, can be modeled as parallel lines. When these slanted parallel lines meet the horizontal ground line, angles that are equal in measure are formed. $m \angle A = m \angle A'$

Thus, $\triangle ABC \sim \triangle A'B'C'$, and their corresponding sides are in proportion.

$$\frac{AB}{A'B'} = \frac{BC}{B'C'}$$

$$\frac{5}{10} = \frac{6}{B'C'}$$

$$5(B'C') = 60$$

$$B'C' = 12$$

Check: $\dfrac{\text{man}}{\text{shadow}} \stackrel{?}{=} \dfrac{\text{pole}}{\text{shadow}}$

$$\frac{6}{5} \stackrel{?}{=} \frac{12}{10}$$

$$\frac{6}{5} = \frac{6}{5} \ \checkmark$$

ANSWER: The height of the flagpole is 12 feet.

Now it's your turn to . . . TRY IT!

The town hall, which is 45 feet high, casts a shadow of 20 feet. The adjacent court building casts a 32-foot shadow. How tall is the court building?

(See page 470 for solution.)

In 1 and 2, draw a model and determine whether the triangles are similar. Explain.

1. In $\triangle ABC$, m $\angle A = 40°$ and m $\angle B = 30°$.
   In $\triangle DEF$, m $\angle D = 30°$ and m $\angle E = 40°$.

2. In $\triangle RST$, m $\angle R = 90°$ and m $\angle S = 40°$.
   In $\triangle XYZ$, m $\angle X = 90°$.

3. Select the triangles that are similar and tell why they are similar.

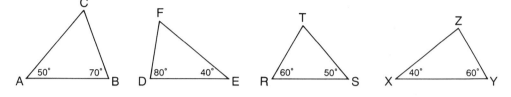

4. a. Are all equiangular triangles similar? Explain.
   b. Are all right triangles similar? Explain.

5. In $\triangle ABC$ and $\triangle RST$, A corresponds to R, B to S, and C to T. If $\triangle ABC \sim \triangle RST$ and if $AB = 9$ in., $AC = 6$ in., and $RS = 3$ in., find RT.

In 6 and 7, the two triangles are similar.

6. Find BC.

7. Find DF and EF.

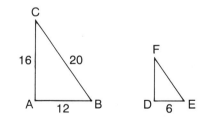

8. Given $\triangle MNO \sim \triangle QRT$.
   a. Complete the proportion:

   $$\frac{MO}{QT} = \frac{?}{QR} = \frac{NO}{?}$$

   b. Find QT and RT.

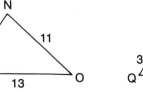

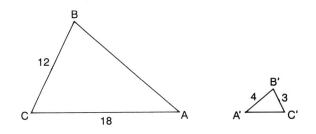

9. Given $\triangle ABC \sim \triangle A'B'C'$, with $m \angle A = m \angle A'$, $m \angle B = m \angle B'$, and $m \angle C = m \angle C'$. If $BC = 12$ cm, $CA = 18$ cm, $A'B' = 4$ cm, and $B'C' = 3$ cm, find:
  a. $A'C'$
  b. the perimeter of $\triangle ABC$

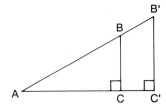

10. a. Why is $\triangle ABC$ similar to $\triangle AB'C'$?
  b. If $BC = 3$ in., $AC = 4$ in., and $AC' = 8$ in., find $B'C'$.
  c. If $AC = 5$ ft., $BC = 2$ ft., and $B'C' = 7$ ft., find AC'.
  d. If $B'C' = 12$ in., $AC = 6$ in., and $CC' = 3$ in., find BC.
  e. If $BC = 10$ ft. and $AC = 8$ ft., find $B'C' : AC'$.

11. A pole 8 ft. high casts a shadow 6 ft. long at the same time that a nearby tree casts a shadow 15 ft. long. Find the height of the tree.

12. A vertical yardstick casts a shadow $2\frac{1}{2}$ ft. long at the same time that a nearby pole casts a shadow 15 ft. long. Find the height of the pole.

13. A tree casts a shadow 40 ft. long at the same time that a nearby boy 5 ft. 6 in. tall casts a shadow 8 ft. long. Find the height of the tree.

14. In the figure, $m \angle AEB = m \angle DEC$ and $m \angle ABE = m \angle DCE$.
  Find AB, the distance across a pond, if:
  a. $CE = 40$ m, $EB = 120$ m, and $CD = 50$ m
  b. $AE = 75$ m, $ED = 30$ m, and $CD = 36$ m

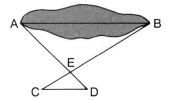

15. The sides of a triangle measure 3, 4, and 5 inches. In a similar triangle, the shortest side is 5 inches long. How long are the other sides?

16. The lengths of the sides of a triangle are r, s, and t. The longest side of a similar triangle measures kt. Represent the perimeter of the second triangle.

17. The sides of a triangle measure 6, 8, and 10 cm. If the shortest side is lengthened 1 cm, what must be the increases in the lengths of the other two sides in order to make the new triangle similar to the old?

18. A detective is following a man as he walks through town late at night. After awhile, the man stops under a streetlight to read what is written on a paper. The streetlight is 16 feet above the ground. If the 6-foot officer casts a 21-foot shadow as he stands watching, how far is he from his suspect?

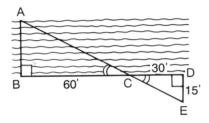

19. Sightings taken from opposite sides of a river enable a surveyor to determine the width of the river. AB represents the width of the river, and the triangles contain two pairs of angles that are equal in measure. Find the width of the river if $BC = 60$ ft., $CD = 30$ ft., and $DE = 15$ ft.

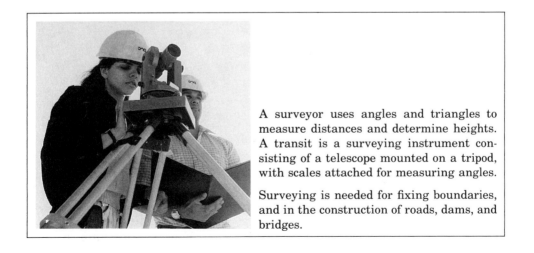

A surveyor uses angles and triangles to measure distances and determine heights. A transit is a surveying instrument consisting of a telescope mounted on a tripod, with scales attached for measuring angles.

Surveying is needed for fixing boundaries, and in the construction of roads, dams, and bridges.

8-5 TRIGONOMETRIC RATIOS

The word **_trigonometry_**, of Greek derivation, means "measurement of triangles." This branch of mathematics, which provides additional methods of indirect measurement, is frequently used by engineers, surveyors, astronomers, and navigators.

Our discussion of trigonometry is limited to the right triangle.

Defining Trigonometric Ratios

For an acute angle of fixed measure, which is located in a right triangle, consider three ratios involving the lengths of the sides of the triangle, and their names.

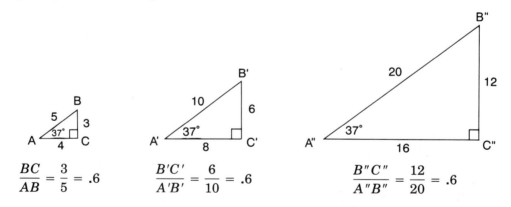

$$\frac{\text{side opposite } \angle A}{\text{hypotenuse}} = \textit{sine A}$$

$$\frac{\text{side adjacent } \angle A}{\text{hypotenuse}} = \textit{cosine A}$$

$$\frac{\text{side opposite } \angle A}{\text{side adjacent } \angle A} = \textit{tangent A}$$

To closely examine these ratios, study the measures of three different right triangles, each of which contains an acute angle of the same measure, 37°. Focus on one ratio, say sine, for the 37° angle.

$$\frac{BC}{AB} = \frac{3}{5} = .6 \qquad\qquad \frac{B'C'}{A'B'} = \frac{6}{10} = .6 \qquad\qquad \frac{B''C''}{A''B''} = \frac{12}{20} = .6$$

Explain why these constant results are not surprising. (All right triangles that have the measure of an acute angle in common are similar, and corresponding sides of similar triangles are in proportion.) Use the diagram to verify that constant results are obtained for the cosine and tangent ratios of a 37° angle. In general:

Each of the trigonometric ratios is constant for an angle of fixed measure.

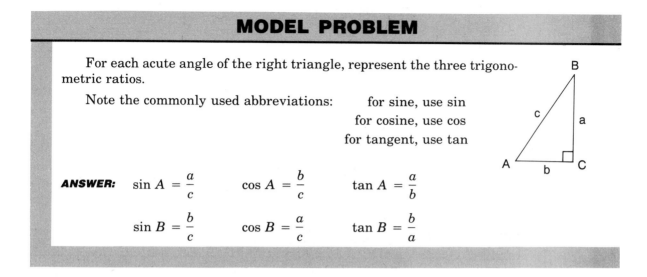

MODEL PROBLEM

For each acute angle of the right triangle, represent the three trigono-metric ratios.

Note the commonly used abbreviations:
for sine, use sin
for cosine, use cos
for tangent, use tan

ANSWER:
$$\sin A = \frac{a}{c} \qquad \cos A = \frac{b}{c} \qquad \tan A = \frac{a}{b}$$

$$\sin B = \frac{b}{c} \qquad \cos B = \frac{a}{c} \qquad \tan B = \frac{b}{a}$$

Using a Table of Trigonometric Ratios

Mathematicians have constructed a table that contains approximations of the trigonometric ratios for angles whose measures are between 0° and 90°. The full table appears on page 423.

To read the value of a trigonometric ratio:

1. Locate the number of degrees in the measure of the angle in the column headed "Angle."

2. Trace right, across to the appropriate ratio heading.

Examples: sin 22° = .3746
cos 51° = .6293

To find the measure of an angle for which the value of a trigonometric ratio is given:

1. Locate the value of the ratio in the column that is so named.

2. Trace left, to the Angle column.

Examples: If tan A = .3249, m∠A = 18°.
If cos A = .3746, m∠A = 68°.

Angle	Sine	Cosine	Tangent	Angle	Sine	Cosine	Tangent
18°			.3249	51°		.6293	
22°	.3746			68°		.3746	

To approximate the measure of an angle for an unlisted trigonometric ratio:

1. Find two consecutive trigonometric values in the table between which the unlisted value fits. Note the corresponding angle measures.

2. Determine the differences between the given value and the listed values.

3. The smaller difference indicates the better approximation for the angle measure.

If sin A = .6497, find m∠A to the nearest degree.

Solution:

m∠A is between 40° and 41°.

Since .0064 is the smaller difference, m∠A is closer to 41°.

Angle	Sine
40°	.6428
A	.6497
41°	.6561

.0069

.0064

ANSWER: If sin A = .6497, m∠A is approximately 41°.

CALCULATOR CONNECTION

Units for Angles

Although this text uses only the degree as the unit of angle measure, calculators can use different units. Scientific calculators generally display angles in degrees when they are first turned on. Check this by looking for the abbreviation *deg* or the letter D.

Finding the Value of a Trigonometric Ratio

To find the value of a trigonometric ratio of an angle of given measure, first enter the measure of the angle, then press the appropriate ratio key.

Example 1: Find tan 31°.

Enter: **31** ⌊tan⌋ ⌊=⌋

Display: .60086061

ANSWER: tan 31° = .6009

Example 2: Find 3 (tan 31°).

Enter: **3** ⌊×⌋ **31** ⌊tan⌋ ⌊=⌋

Display: 1.8025818

ANSWER: 3 (tan 31°) = 1.8026

Finding the Measure of an Angle

To find the measure of an angle for which the value of a trigonometric ratio is given, first enter the value of the trigonometric ratio, and then work with an inverse operation.

Example: Find the measure of the angle whose tangent is .6009.

Enter: **.6009** ⌊tan⁻¹⌋ ⌊=⌋ or **.6009** ⌊inv⌋ ⌊tan⌋ ⌊=⌋

Display: 31.0016

Answer: tan 31° = .6009

VALUES OF THE TRIGONOMETRIC RATIOS

Angle	Sine	Cosine	Tangent	Angle	Sine	Cosine	Tangent
1°	.0175	.9998	.0175	46°	.7193	.6947	1.0355
2°	.0349	.9994	.0349	47°	.7314	.6820	1.0724
3°	.0523	.9986	.0524	48°	.7431	.6691	1.1106
4°	.0698	.9976	.0699	49°	.7547	.6561	1.1504
5°	.0872	.9962	.0875	50°	.7660	.6428	1.1918
6°	.1045	.9945	.1051	51°	.7771	.6293	1.2349
7°	.1219	.9925	.1228	52°	.7880	.6157	1.2799
8°	.1392	.9903	.1405	53°	.7986	.6018	1.3270
9°	.1564	.9877	.1584	54°	.8090	.5878	1.3764
10°	.1736	.9848	.1763	55°	.8192	.5736	1.4281
11°	.1908	.9816	.1944	56°	.8290	.5592	1.4826
12°	.2079	.9781	.2126	57°	.8387	.5446	1.5399
13°	.2250	.9744	.2309	58°	.8480	.5299	1.6003
14°	.2419	.9703	.2493	59°	.8572	.5150	1.6643
15°	.2588	.9659	.2679	60°	.8660	.5000	1.7321
16°	.2756	.9613	.2867	61°	.8746	.4848	1.8040
17°	.2924	.9563	.3057	62°	.8829	.4695	1.8807
18°	.3090	.9511	.3249	63°	.8910	.4540	1.9626
19°	.3256	.9455	.3443	64°	.8988	.4384	2.0503
20°	.3420	.9397	.3640	65°	.9063	.4226	2.1445
21°	.3584	.9336	.3839	66°	.9135	.4067	2.2460
22°	.3746	.9272	.4040	67°	.9205	.3907	2.3559
23°	.3907	.9205	.4245	68°	.9272	.3746	2.4751
24°	.4067	.9135	.4452	69°	.9336	.3584	2.6051
25°	.4226	.9063	.4663	70°	.9397	.3420	2.7475
26°	.4384	.8988	.4877	71°	.9455	.3256	2.9042
27°	.4540	.8910	.5095	72°	.9511	.3090	3.0777
28°	.4695	.8829	.5317	73°	.9563	.2924	3.2709
29°	.4848	.8746	.5543	74°	.9613	.2756	3.4874
30°	.5000	.8660	.5774	75°	.9659	.2588	3.7321
31°	.5150	.8572	.6009	76°	.9703	.2419	4.0108
32°	.5299	.8480	.6249	77°	.9744	.2250	4.3315
33°	.5446	.8387	.6494	78°	.9781	.2079	4.7046
34°	.5592	.8290	.6745	79°	.9816	.1908	5.1446
35°	.5736	.8192	.7002	80°	.9848	.1736	5.6713
36°	.5878	.8090	.7265	81°	.9877	.1564	6.3138
37°	.6018	.7986	.7536	82°	.9903	.1392	7.1154
38°	.6157	.7880	.7813	83°	.9925	.1219	8.1443
39°	.6293	.7771	.8098	84°	.9945	.1045	9.5144
40°	.6428	.7660	.8391	85°	.9962	.0872	11.4301
41°	.6561	.7547	.8693	86°	.9976	.0698	14.3007
42°	.6691	.7431	.9004	87°	.9986	.0523	19.0811
43°	.6820	.7314	.9325	88°	.9994	.0349	28.6363
44°	.6947	.7193	.9657	89°	.9998	.0175	57.2900
45°	.7071	.7071	1.0000	90°	1.0000	.0000	

In 1–3, name the hypotenuse, the leg opposite, and the leg adjacent to each acute angle of the right triangle.

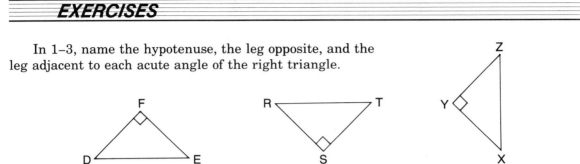

Ex. 1 **Ex. 2** **Ex. 3**

In 4–6, represent the sine ratio for each acute angle of the right triangle.

Ex. 4 **Ex. 5** **Ex. 6**

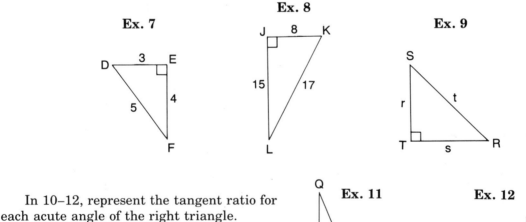

In 7–9, represent the cosine ratio for each acute angle of the right triangle.

Ex. 8

Ex. 7 **Ex. 9**

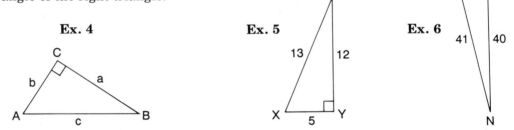

In 10–12, represent the tangent ratio for each acute angle of the right triangle.

Ex. 11 **Ex. 12**

Ex. 10

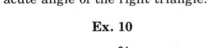

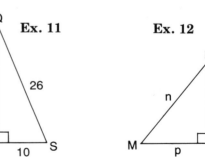

In 13–15, find the value of the indicated ratio.

13. In $\triangle ABC$, $m \angle C = 90°$, $AB = 5$, $BC = 3$, and $AC = 4$. Find $\sin A$.

14. In $\triangle RST$, $m \angle S = 90°$, $RS = 5$, and $RT = 13$, and $ST = 12$. Find $\cos T$.

15. In $\triangle MNO$, $m \angle N = 90°$, $MO = 17$, $MN = 8$, and $NO = 15$. Find $\tan M$.

In 16–31, refer to the figure to complete the ratio.

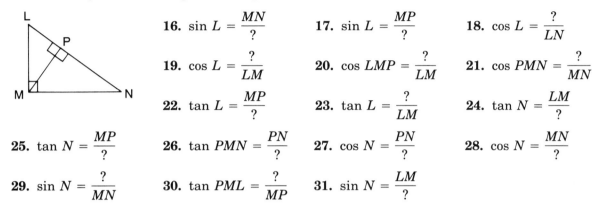

16. $\sin L = \dfrac{MN}{?}$

17. $\sin L = \dfrac{MP}{?}$

18. $\cos L = \dfrac{?}{LN}$

19. $\cos L = \dfrac{?}{LM}$

20. $\cos LMP = \dfrac{?}{LM}$

21. $\cos PMN = \dfrac{?}{MN}$

22. $\tan L = \dfrac{MP}{?}$

23. $\tan L = \dfrac{?}{LM}$

24. $\tan N = \dfrac{LM}{?}$

25. $\tan N = \dfrac{MP}{?}$

26. $\tan PMN = \dfrac{PN}{?}$

27. $\cos N = \dfrac{PN}{?}$

28. $\cos N = \dfrac{MN}{?}$

29. $\sin N = \dfrac{?}{MN}$

30. $\tan PML = \dfrac{?}{MP}$

31. $\sin N = \dfrac{LM}{?}$

32. a. Explain why, in any right triangle, the sine of one of the acute angles has the same value as the cosine of the other acute angle.
   b. Draw a conclusion about the relationship between the tangent values of the two acute angles of any right triangle.

In 33–42, use the table of trigonometric ratios to find the indicated value.

33. $\sin 18°$

34. $\cos 21°$

35. $\tan 45°$

36. $\cos 40°$

37. $\sin 58°$

38. $\tan 89°$

39. $\cos 67°$

40. $\sin 42°$

41. $\tan 74°$

42. $\sin 89°$

In 43–54, use the table to find the measure of angle x.

43. $\cos x = .9397$

44. $\tan x = .1763$

45. $\sin x = .1908$

46. $\tan x = 1.0000$

47. $\sin x = .8387$

48. $\cos x = .0698$

49. $\sin x = .6561$

50. $\tan x = 2.3559$

51. $\cos x = .3584$

52. $\tan x = 4.3315$

53. $\cos x = .0000$

54. $\sin x = .5000$

In 55–66, use the table to find the measure of angle x to the nearest degree.

55. $\tan x = .2281$

56. $\sin x = .2303$

57. $\cos x = .2222$

58. $\sin x = .1900$

59. $\tan x = 3.6231$

60. $\cos x = .9750$

61. $\tan x = .7773$

62. $\sin x = .8740$

63. $\cos x = .5934$

64. $\sin x = .1275$

65. $\tan x = 1.4000$

66. $\cos x = .2968$

In 67–74, use the table to find the measure of angle G to the nearest degree. If there is no possible answer, explain.

67. $\sin G = \dfrac{1}{2}$ **68.** $\cos G = \dfrac{3}{5}$ **69.** $\tan G = \dfrac{1}{8}$ **70.** $\sin G = 2$

71. $\tan G = 2$ **72.** $\tan G = 26$ **73.** $\sin G = \dfrac{9}{10}$ **74.** $\cos G = 1\dfrac{1}{2}$

In 75–89, use the table to determine if the statement is true or false.

75. $\tan 80° \overset{?}{=} 2\,(\tan 40°)$ **76.** $\sin 40° \overset{?}{=} \cos 50°$ **77.** $\cos 60° \overset{?}{=} 3\,(\cos 20°)$

78. $\tan 15° \overset{?}{=} \tan 75°$ **79.** $\cos 45° \overset{?}{=} \dfrac{1}{\sin 45°}$ **80.** $\tan 35° \overset{?}{=} \dfrac{1}{\tan 55°}$

81. $\sin(20° + 30°) \overset{?}{=} \sin 20° + \sin 30°$ **82.** $\cos(50° - 10°) \overset{?}{=} \cos 40°$

83. There is no acute angle for which the values of sine and cosine are equal.

84. As the measure of the angle increases, the value of sine increases.

85. As the measure of the angle increases, the value of tangent increases.

86. As the measure of the angle increases, the values of all the trigonometric ratios increase.

87. The value of the sine of every acute angle is less than 1.

88. The value of the cosine of every acute is less than 1.

89. The values of all the trigonometric ratios of acute angles are less than 1.

In 90–93, determine whether the statement is true or false. Explain your conclusion.

90. $\sin x° \overset{?}{=} \sin(90 - x)°$ **91.** $\sin x° \overset{?}{=} \cos(90 - x)°$

92. $\tan x° \overset{?}{=} \dfrac{1}{\tan(90 - x)°}$ **93.** $\tan x° \overset{?}{=} \tan(90 - x)°$

Many natural phenomena, such as light, sound, and radio transmission waves, can be diagrammed as wave forms.

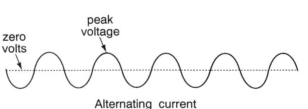

The wave shown, which represents alternating electric current (AC), is in the form of a sine wave. The dotted horizontal line represents a base of zero volts. Above the line, the voltage rises to a maximum, then falls to zero. The portions of the curve below the line represent a flow of current in the opposite direction.

8-6 APPLYING THE TRIGONOMETRIC RATIOS

Using the Trigonometric Ratios to Find Unknown Measures

Since a trigonometric ratio connects the measure of an acute angle of a right triangle to the measures of two of the sides, a trigonometric ratio can be used to find an unknown measure in a right triangle. Of the three measures (angle and two sides), two must be known.

PROCEDURE

To find the length of a side of a right triangle, given the measure of one acute angle and the measure of one side:

1. Identify the placement of the sides of the right triangle with respect to the given angle.
2. Choose the trigonometric ratio, and write an equation.

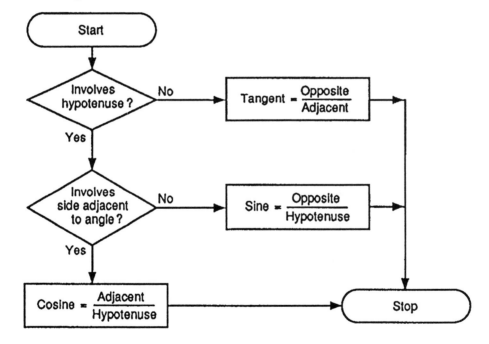

3. Substitute the given values.
4. Substitute the value of the trigonometric ratio, obtained from the table or a calculator.
5. Solve the equation for the remaining unknown value.
6. Round the answer.

Find, to the nearest tenth of an inch, the length of the side of the triangle whose measure is represented by x.

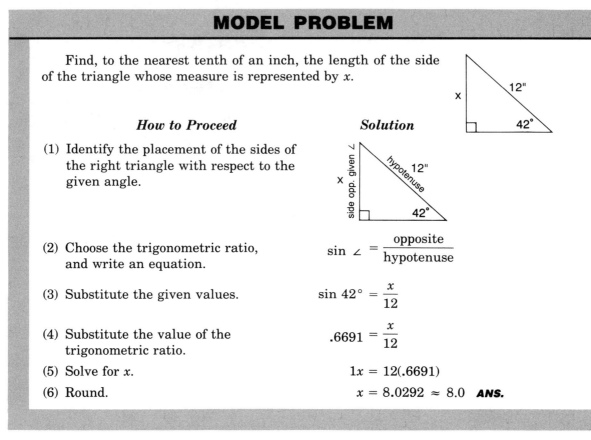

How to Proceed	*Solution*
(1) Identify the placement of the sides of the right triangle with respect to the given angle.	
(2) Choose the trigonometric ratio, and write an equation.	$\sin \angle = \dfrac{\text{opposite}}{\text{hypotenuse}}$
(3) Substitute the given values.	$\sin 42° = \dfrac{x}{12}$
(4) Substitute the value of the trigonometric ratio.	$.6691 = \dfrac{x}{12}$
(5) Solve for x.	$1x = 12(.6691)$
(6) Round.	$x = 8.0292 \approx 8.0$ **ANS.**

NOTE. In some trigonometric equations, the variable appears in the denominator of a ratio. Then, after multiplying the means and the extremes of the proportion, the next step involves dividing by a 4-place decimal.

The long division cannot be avoided in sine and cosine ratios where the length of the hypotenuse is the unknown. However, a tangent ratio can be rewritten by using the other acute angle of the right triangle, thus placing the unknown in the numerator and simplifying the computation. For example, given $\triangle ABC$ with $m \angle C = 90°$, $m \angle A = 35°$, and $BC = 10$. To find AC, use the tangent ratio.

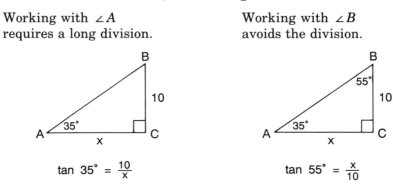

Working with $\angle A$
requires a long division.

Working with $\angle B$
avoids the division.

$\tan 35° = \dfrac{10}{x}$

$\tan 55° = \dfrac{x}{10}$

The following model problem illustrates the procedure for finding the measure of an acute angle of a right triangle, given the lengths of two sides.

MODEL PROBLEM

Find the measure of $\angle x$ to the nearest degree.

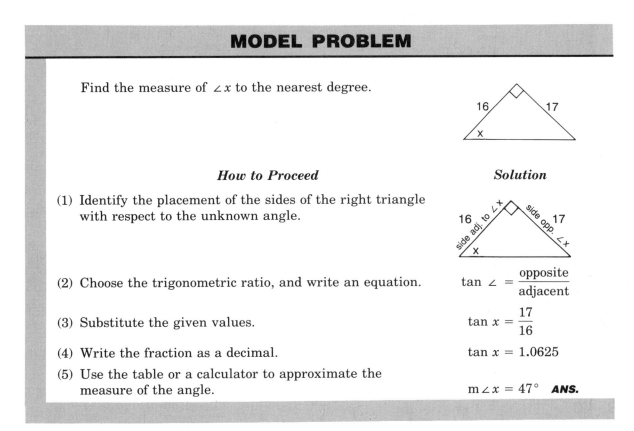

How to Proceed

(1) Identify the placement of the sides of the right triangle with respect to the unknown angle.

(2) Choose the trigonometric ratio, and write an equation.

(3) Substitute the given values.

(4) Write the fraction as a decimal.

(5) Use the table or a calculator to approximate the measure of the angle.

Solution

$$\tan \angle = \frac{\text{opposite}}{\text{adjacent}}$$

$$\tan x = \frac{17}{16}$$

$$\tan x = 1.0625$$

$$m\angle x = 47° \quad \textbf{ANS.}$$

Now it's your turn to . . . **TRY IT!**

1. Find x to the nearest foot.

2. Find x to the nearest degree.

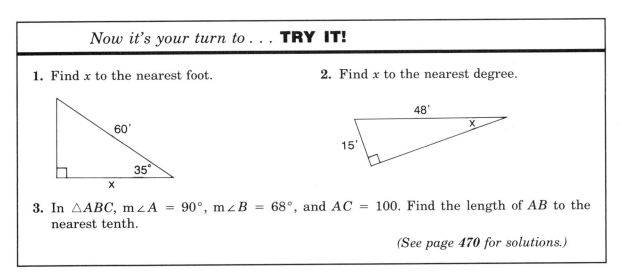

3. In $\triangle ABC$, $m\angle A = 90°$, $m\angle B = 68°$, and $AC = 100$. Find the length of AB to the nearest tenth.

(See page 470 for solutions.)

Using Trigonometry to Solve Problems

When using trigonometry to solve a problem, the angle at which an object is being viewed plays an important role. The observer may be below or above the object.

ANGLE OF ELEVATION

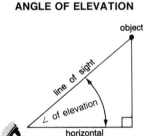

The observer is looking up at the object.

ANGLE OF DEPRESSION

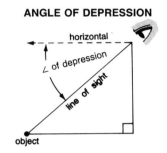

The observer is looking down at the object.

MODEL PROBLEMS

1. From a point on the ground 36 feet from the foot of a tree, the angle of elevation of the top of the tree contains $40°$. Find the height of the tree to the nearest foot.

   Solution: Draw a diagram to represent the situation.

   Let x = the height of the tree.

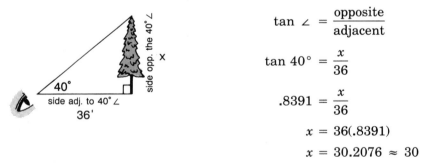

   $$\tan \angle = \frac{\text{opposite}}{\text{adjacent}}$$

   $$\tan 40° = \frac{x}{36}$$

   $$.8391 = \frac{x}{36}$$

   $$x = 36(.8391)$$

   $$x = 30.2076 \approx 30$$

   ANSWER: The height of the tree is 30 feet.

2. From the top of a lighthouse 200 feet high, an observer measures the angle of depression of a boat at sea to be $24°$. Find, to the nearest foot, the distance from the boat to the base of the lighthouse.

   Solution: Since the angle of depression is outside the triangle, use the complement of the angle of depression.

Identify the placement of the two sides of the triangle with respect to the known angle that is *inside* the triangle.

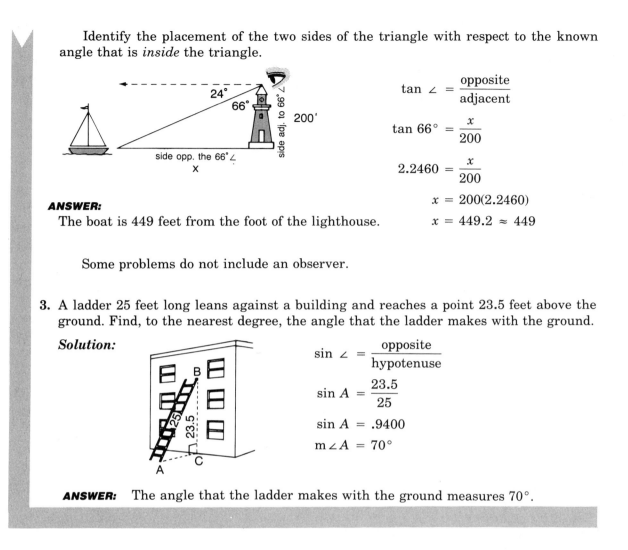

$$\tan \angle = \frac{\text{opposite}}{\text{adjacent}}$$

$$\tan 66° = \frac{x}{200}$$

$$2.2460 = \frac{x}{200}$$

$$x = 200(2.2460)$$

$$x = 449.2 \approx 449$$

ANSWER:
The boat is 449 feet from the foot of the lighthouse.

Some problems do not include an observer.

3. A ladder 25 feet long leans against a building and reaches a point 23.5 feet above the ground. Find, to the nearest degree, the angle that the ladder makes with the ground.

Solution:

$$\sin \angle = \frac{\text{opposite}}{\text{hypotenuse}}$$

$$\sin A = \frac{23.5}{25}$$

$$\sin A = .9400$$

$$m \angle A = 70°$$

ANSWER: The angle that the ladder makes with the ground measures 70°.

Now it's your turn to . . . **TRY IT!**   (See page 470 for solutions.)

1. Find, to the nearest degree, the angle of elevation of the sun when a vertical pole 6 feet high casts a shadow 8 feet long.

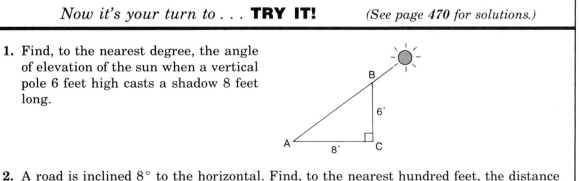

2. A road is inclined 8° to the horizontal. Find, to the nearest hundred feet, the distance a man must drive up this road to increase his altitude 1,000 feet.

Working With More Complicated Diagrams

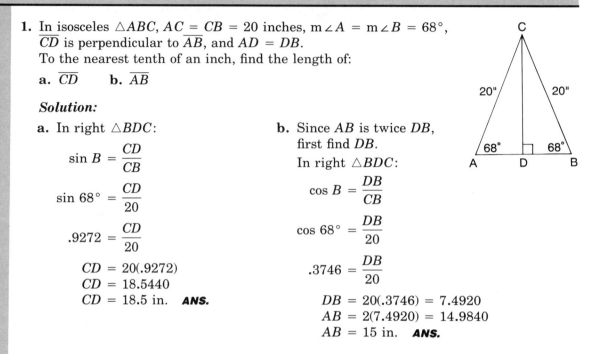

MODEL PROBLEMS

1. In isosceles $\triangle ABC$, $AC = CB = 20$ inches, m $\angle A$ = m $\angle B = 68°$, $\overline{CD}$ is perpendicular to $\overline{AB}$, and $AD = DB$.
   To the nearest tenth of an inch, find the length of:
   a. $\overline{CD}$ **b.** $\overline{AB}$

   Solution:

   a. In right $\triangle BDC$:

   $$\sin B = \frac{CD}{CB}$$

   $$\sin 68° = \frac{CD}{20}$$

   $$.9272 = \frac{CD}{20}$$

   $$CD = 20(.9272)$$
   $$CD = 18.5440$$
   $$CD = 18.5 \text{ in.} \quad \textbf{ANS.}$$

   b. Since AB is twice DB, first find DB.
   In right $\triangle BDC$:

   $$\cos B = \frac{DB}{CB}$$

   $$\cos 68° = \frac{DB}{20}$$

   $$.3746 = \frac{DB}{20}$$

   $$DB = 20(.3746) = 7.4920$$
   $$AB = 2(7.4920) = 14.9840$$
   $$AB = 15 \text{ in.} \quad \textbf{ANS.}$$

2. A pole $\overline{TF}$ rises from the wall $\overline{FB}$ of a shed. From a point P on the ground 27 feet from the foot of the shed, the angle of elevation of the top of the pole contains $38°$ and the angle of elevation of the foot of the pole contains $32°$. Find, to the nearest foot, the height of:
   a. the shed   **b.** the pole

   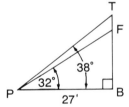

   Solution:

   a. In right $\triangle FPB$:

   $$\tan 32° = \frac{FB}{27}$$

   $$.6249 = \frac{FB}{27}$$

   $$FB = 27(.6249)$$
   $$FB = 16.8723$$
   $$FB = 17' \quad \textbf{ANS.}$$

   b. Since the pole $\overline{TF}$ is not a side of a right triangle, work indirectly.
   First find the height of $\overline{TB}$. In right $\triangle TPB$:

   $$\tan 38° = \frac{TB}{27}$$

   $$.7813 = \frac{TB}{27}$$

   $$TB = 27(.7813)$$
   $$TB = 21.0951$$

   Now, subtract to find TF:

   $$TF = TB - FB$$
   $$TF = 21.0951 - 16.8723$$
   $$TF = 4.2228$$
   $$TF = 4' \quad \textbf{ANS.}$$

$$S \quad O \quad H \quad C \quad A \quad H \quad T \quad O \quad A$$

$$\downarrow \qquad\qquad\qquad \downarrow \qquad\qquad\qquad \downarrow$$

$$\sin \angle = \frac{opposite}{hypotenuse} \qquad \cos \angle = \frac{adjacent}{hypotenuse} \qquad tan \angle = \frac{opposite}{adjacent}$$

EXERCISES

In 1–16, find the length of the side marked x to the nearest foot or the measure of the angle marked x to the nearest degree.

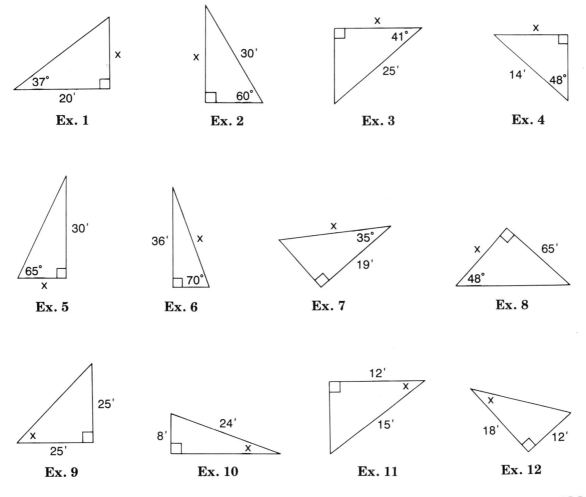

Ex. 1   Ex. 2   Ex. 3   Ex. 4

Ex. 5   Ex. 6   Ex. 7   Ex. 8

Ex. 9   Ex. 10   Ex. 11   Ex. 12

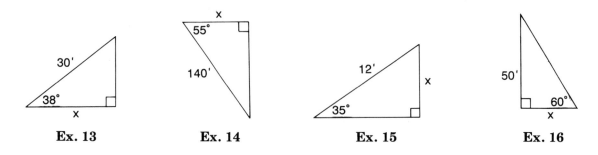

Ex. 13 **Ex. 14** **Ex. 15** **Ex. 16**

17. In △*LMN*, hypotenuse $\overline{LM}$ measures 40 units and m∠*M* = 35°. Find *LN*.

18. △*ABC* has a right angle at *B*. If *AB* = 12 and *BC* = 15, find m∠*A*.

19. In △*FGH*, m∠*F* = 90°, m∠*G* = 40°, and *GH* = 38. Find *FH*.

20. In right triangle *QRS* with the right angle at *Q*, m∠*R* = 27° and *QR* = 15. Find *RS*.

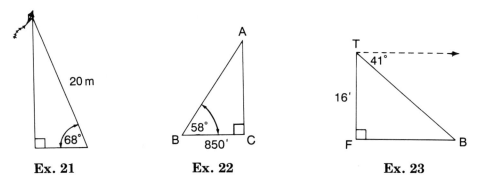

Ex. 21 **Ex. 22** **Ex. 23**

21. As shown, a kite is flying at the end of a 20-meter string. If the string makes an angle of 68° with the ground, how high, to the nearest meter, is the kite?

22. The height of a cloud over an airport at night is determined by projecting a light vertically upward to the cloud. At a point on the ground 850 feet from the light, as shown, the angle of elevation of the spot where the light hits the cloud is found to contain 58°. Find, to the nearest foot, the height of the cloud.

23. As shown, from the top of a tree 16 feet tall, an observer measures the angle of depression of an object on the ground as 41°. Find, to the nearest foot, the distance from the foot of the tree to the object.

24. A straight road is inclined upward at an angle of 16° with the horizontal, as shown. If a horse walked a distance of 2,500 feet up the road, find, to the nearest foot, his increase in altitude.

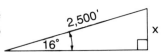

25. For each 12-foot horizontal distance, a wheelchair ramp rises one foot. Find, to the nearest degree, the measure of the angle that the ramp makes with the horizontal.

26. A monument stands on level ground. The angle of elevation of the top of the monument, taken at a point 425 feet from the foot of the monument, is 32°. Find the height of the monument to the nearest foot.

27. A boy flying a kite lets out 150 feet of string that makes an angle of 64° with the ground. If the string is straight, find, to the nearest foot, how high the kite is above the ground.

28. A 25-foot wire attached to the top of a pole makes an angle of 62° with the ground. Find, to the nearest foot, the distance between the point where the wire meets the ground and the foot of the pole.

29. A girl walked 400 feet into a tunnel that slopes downward at an angle of 7° with the horizontal ground. Find, to the nearest ten feet, how far she was beneath the surface.

When the principle of gravity became known, one of its early ingenious applications was in a timepiece. The inclined plane clock, invented in the 1600's, may have been the first designed to show the day as well as the time. The dial remains upright while the clock moves down the incline at an infinitesimal pace, taking a full seven days to complete its journey. The user then returns the clock to the top to start the new week.

30. An airplane A is 1,000 feet above the ground and directly over a church C. The angle of elevation of the plane as seen by a person at a point B on the ground some distance from the church is 22°. Find, to the nearest foot, how far the person is from:
   a. the church   **b.** the plane

31. An observer in a balloon that is 2,000 feet above an airport finds that the angle of depression of a steamer out at sea is 21°. Find, to the nearest hundred feet, the distance between the balloon and the steamer.

32. A plane takes off from a field and climbs at an angle of 12°. Find, to the nearest 100 feet, how far the plane must fly to be at an altitude of 1,200 feet.

33. Find, to the nearest degree, the angle of elevation of the sun when a tree 24 feet high casts a shadow of 36 feet.

34. The foot of a 40-foot ladder leaning against a building is 32 feet from the building. Find, to the nearest degree, the measure of the angle that the ladder makes with the ground.

35. A railroad track slopes upward at an angle of 7° to the horizontal. Find, to the nearest ten feet, the vertical distance it rises in a horizontal distance of 1 mile (5,280 feet).

36. From the top of a cliff 450 feet above sea level, the angle of depression of a boat out at sea is 24°. Find, to the nearest foot, the distance from the top of the cliff to the boat.

37. A television tower is 150 feet high and an observer is 120 feet from the base of the tower. Find, to the nearest degree, the angle of elevation of the top of the tower from the point where the observer is standing.

38. The sliding board is popular at the Parkwood Preschool playground. The bottom of the slide is 1 foot off the ground, and the slide rises at an angle of 50° from the horizontal. If the slide is 10 feet long, how high off the ground is the top of the slide? (Answer to the nearest tenth of a foot.)

39. An 18-foot ladder leans against the side of a house that stands on level ground. The foot of the ladder is 10 feet from the house. Find:
   a. to the nearest degree, the measure of the angle that the foot of the ladder makes with the ground
   b. to the nearest integer, the distance from the point where the top of the ladder touches the house to the ground

40. A telephone pole stands on level ground. A wire, attached to the pole at a point 30 feet above the ground, makes an angle of 61° with the ground. Find, to the nearest foot:
   a. the distance from the base of the pole to the point on the ground where the wire is fastened
   b. the length of the wire

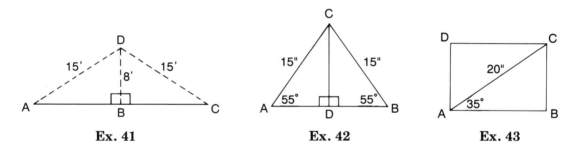

Ex. 41   Ex. 42   Ex. 43

41. The figure represents a view of the roof of a wing of a house. $\overline{DB}$, the altitude of $\triangle ACD$, measures 8 feet. Each of the rafters $\overline{AD}$ and $\overline{CD}$ measures 15 feet. Find, to the nearest degree, the measure of the angle that each rafter makes with $\overline{AC}$, the base of the triangle.

42. In isosceles $\triangle ABC$, $\overline{AC}$ and $\overline{BC}$ each measure 15 inches, m∠A and m∠B are each 55°, and $AD = DB$.
   a. Find CD to the nearest tenth of an inch.
   b. Find AD to the nearest tenth of an inch.
   c. Find the area of triangle ABC to the nearest square inch.

43. In rectangle $ABCD$, diagonal $\overline{AC}$, which measures 20 inches in length, makes a 35° angle with the base $\overline{AB}$.
   a. Find AB to the nearest tenth of an inch.
   b. Find BC to the nearest tenth of an inch.
   c. Find the area of the rectangle to the nearest square inch.

436   Chapter 8   Applying Ratio and Proportion

44. In a rectangle $ABCD$, $AB = 40$ centimeters and $BC = 30$ centimeters.
    a. Find, to the nearest degree, the measure of the angle that diagonal $\overline{AC}$ makes with:
        (1) side $\overline{AB}$ **(2)** side $\overline{AD}$
    b. What is a good way of checking the answers found in part **a**?

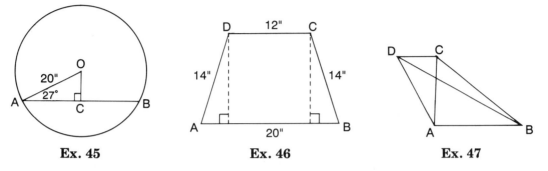

       Ex. 45   **Ex. 46**   **Ex. 47**

45. In the figure, $\overline{AB}$ is a chord in circle O. $\overline{OC}$ is perpendicular to $\overline{AB}$ at its midpoint C. If radius $OA = 20$ in. and $m\angle OAC = 27°$, find, to the nearest tenth of an inch:
    a. the length of $\overline{OC}$ **b.** the length of $\overline{AB}$

46. Each leg of isosceles trapezoid $ABCD$ measures 14 in., and $\overline{AB}$ and $\overline{DC}$ measure 20 in. and 12 in., respectively. For trapezoid $ABCD$, find:
    a. $m\angle A$ to the nearest degree   **b.** the length of the altitude to the nearest inch
    c. the area to the nearest square inch

47. In quadrilateral $ABCD$, $\overline{AB}$ is parallel to $\overline{DC}$, $AB = 28.0$ inches, and $DC = 12.0$ inches. Diagonal $\overline{AC}$ is perpendicular to $\overline{AB}$. The angle between diagonal $\overline{DB}$ and side $\overline{AB}$ measures $24°$. For quadrilateral $ABCD$, find:
    a. AC to the nearest tenth of an inch   **b.** the area to the nearest square inch

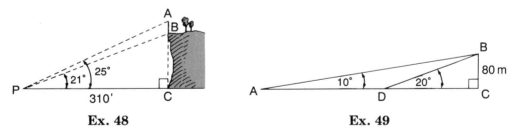

       Ex. 48   **Ex. 49**

48. In the diagram, P represents a point 310 feet from the foot of a vertical cliff $\overline{BC}$. $\overline{AB}$ is a flagpole standing on the edge of the cliff. At P, the angle of elevation of B is $21°$; the angle of elevation of A is $25°$. Find, to the nearest foot:
    a. the distance AC **b.** the height of the flagpole $\overline{AB}$

49. As shown, a ship is headed directly toward a coastline formed by a vertical cliff $\overline{BC}$ that is 80 meters high. At point A, from the ship, the measure of the angle of elevation of B, the top of the cliff, is $10°$. A few minutes later, at point D, the measure of the angle of elevation has increased to $20°$. Find, to the nearest meter:
    a. DC **b.** AC **c.** the distance between the two sightings

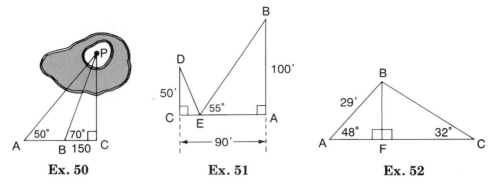

| Ex. 50 | Ex. 51 | Ex. 52 |

50. As shown, point P is on an island in a small lake. To find the distance from P to a point on the shore of the lake, a convenient base line, $\overline{ABC}$, was taken along the shore and the following measurements were made: m∠A = 50°, m∠PBC = 70°, m∠C = 90°, and BC = 150 yards. Find, to the nearest yard: **a.** PC **b.** AP

51. As shown, the heights AB and CD of buildings on opposite sides of an avenue 90 feet wide are, respectively, 100 feet and 50 feet. From a point E on the avenue, the measure of the angle of elevation of B is 55°. Find:
a. to the nearest foot, the distance from E to A
b. to the nearest degree, the measure of the angle of elevation of D from E

52. As shown, from the top of pole $\overline{BF}$ that is standing on level ground, two wires are stretched to the ground and fastened at points A and C that are on opposite sides of the pole. From A, the measure of the angle of elevation of the top of the pole is 48°. From C, the measure of the angle of elevation of the top of the pole is 32°. If AB = 29 feet, find, to the nearest foot:
a. the height of pole $\overline{BF}$ **b.** the length of wire $\overline{BC}$

53. In the diagram, B represents the position of a captive balloon connected by a cable to a ground station at A. Point C is on the ground directly below the balloon, and D is an observation point. Points A, D, and C lie in a straight line on level ground. m∠A = 43°, m∠BDC = 54°, m∠C = 90°, and DC = 170 yards. Find, to the nearest yard:
a. the height, BC, of the balloon **b.** the length, AB, of the cable

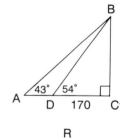

54. $\overline{PQ}$ and $\overline{RS}$ represent buildings on opposite sides of a highway 110 feet wide. From P, the measure of the angle of elevation of R is 27° and the measure of the angle of depression of S is 22°. Find, to the nearest foot:
a. the height of $\overline{PQ}$
b. the height of $\overline{RS}$

55. From the top of a lighthouse 100 feet above sea level, the angles of depression of two boats in line with the foot of the lighthouse are observed to be 18° ad 32° respectively. Find, to the nearest foot, the distance between the boats.

8-7 PROBABILITY

"There is a 30% chance of a thunderstorm today." "A person has a better chance of being hit by lightning than of winning a state lottery." Although there is uncertainty about the final outcome of these statements, it is possible to determine how likely each event is to occur, by using the branch of mathematics called **probability**. This mathematics grew out of a correspondence between two 17th-century French mathematicians, Blaise Pascal and Pierre de Fermat, and is now a widely used branch of mathematics that is part of business, agriculture, medicine, politics, government, and more.

Experimental Probability

An example of a *chance occurrence* is the outcome of flipping a coin. You know from experience that for any particular toss of a coin, you cannot predict heads or tails. Do you know, for 10 or 20 tosses, if heads and tails turn up an equal number of times? Does the number of tosses have any bearing on the likelihood that heads will show? You can find answers to these questions by doing a probability experiment. To get the best estimate of experimental probability, it is helpful to have a large number of trials.

MODEL PROBLEM

Jan tossed a coin 100 times and recorded the number of heads and the number of tails.
Find: **a.** the probability of heads, P(heads)
       b. P(tails)

	Tally	Total
Heads	~~||||~~ ~~||||~~ ~~||||~~ ~~||||~~ ~~||||~~ ~~||||~~ ~~||||~~ ~~||||~~ ~~||||~~		46
Tails	~~||||~~ ~~||||~~ ~~||||~~ ~~||||~~ ~~||||~~ |||| ~~||||~~ ~~||||~~ ~~||||~~ ~~||||~~ ~~||||~~	54

Solution: The experimental results suggest the probable results of any 100 tosses of the coin.

a. Since heads appeared 46 times in

100 tosses, the probability of heads is: $P(\text{heads}) = \dfrac{46}{100}$, or $\dfrac{23}{50}$

b. Since tails appeared 54 times in

100 tosses, the probability of tails is: $P(\text{tails}) = \dfrac{54}{100}$, or $\dfrac{27}{50}$

Now it's your turn to . . . **TRY IT!**

Toss a coin 10 times and record the numbers of heads and tails. Combine your results with those of your classmates. Find $P(H)$ and $P(T)$. Express the probabilities in decimal fractions to the nearest hundredth. Compare your results with the Model Problem.

Predicting Outcomes

A direct application of experimental probability is to predict outcomes. As records are kept for large numbers of people about different events, this accumulated data is used to predict what will happen in similar circumstances in the future.

MODEL PROBLEM

A manufacturer of television sets knows that out of every 1,000 TV's sold, 3 will be repaired under warranty. If the manufacturer sells 200,000 television sets, how many warranty repairs should be expected?

Solution: Write a proportion based on the repair ratio: $\dfrac{\text{number repaired}}{\text{number sold}}$

Let x = the number of repairs expected.

$$\frac{3}{1,000} = \frac{x}{200,000}$$

$$1,000\,x = 3(200,000)$$

$$x = 600$$

ANSWER: 600 repairs should be expected.

Theoretical Probability

There are chance occurrences for which all possible results, or *outcomes*, are known even without doing an experiment. For example, the outcome of tossing a coin can only be heads or tails.

The set of all possible outcomes of a chance occurrence is called a **sample space**. An *event* is a subset of a sample space. For example, the sample space for tossing a coin is {heads, tails}; {heads} is an event, {tails} is a different event, and {heads, tails} is also an event. The numbers of outcomes in the event and in the sample space determine the probability ratio.

> The theoretical probability P of an event E is the ratio of the number of outcomes in the event E to the total number of outcomes in the sample space S.

$$P(E) = \frac{\text{number of outcomes in } E}{\text{number of outcomes in } S}$$

MODEL PROBLEM

When two numbered cubes are tossed, the sum of the dots showing on the tops represents an outcome. The diagram shows the 36 possible outcomes for tossing two differently-colored cubes. Find the probability of tossing:

a. the sum 9

b. the sum 13

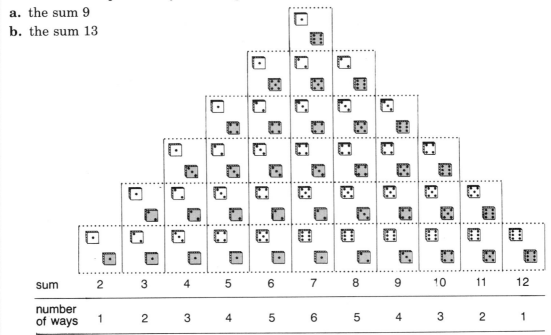

sum	2	3	4	5	6	7	8	9	10	11	12
number of ways	1	2	3	4	5	6	5	4	3	2	1

Solution: The bottom row of numbers in the diagram tells how many ways each sum can be obtained.

a. $P(\text{sum of 9}) = \dfrac{\text{number of ways to get 9}}{\text{total number of outcomes}}$

$P(\text{sum of 9}) = \dfrac{4}{36}$, or $\dfrac{1}{9}$

b. $P(\text{sum of 13}) = \dfrac{\text{ways to get 13}}{\text{total outcomes}}$

$P(\text{sum of 13}) = \dfrac{0}{36} = 0$

Values for Probability

In general, if an event will never occur, it is impossible and its probability is zero.

$$P(\text{impossibility}) = 0$$

If an event always occurs, then it is certain, and its probability is 1.

$$P(\text{certainty}) = 1$$

Every probability ratio has a value between 0 and 1, inclusive. That is:

$$0 \le P(E) \le 1$$

1. Christine had a pouch that contained 4 red, 2 blue, and 3 white marbles.

    a. Without looking, she reached in and pulled out a marble. What is the probability that it was white?

    b. Suppose that Christine had taken a red marble and did not replace it in the pouch. Next, Nancy took a marble. What is the probability that Nancy picked blue?

Solution:

a. $P(\text{white}) = \dfrac{\text{number of white}}{\text{total number}}$

    $P(\text{white}) = \dfrac{3}{9}$, or $\dfrac{1}{3}$

b. Now there are only 8 marbles in the pouch, of which 2 are blue. The probability of Nancy getting a blue marble is:

    $P(\text{blue}) = \dfrac{2}{8}$, or $\dfrac{1}{4}$

2. A fortune cookie producer has 4 different fortunes, A, B, C, and D, one of which is placed in each cookie. There are three times as many A fortunes as B, and twice as many B fortunes as each of C and D. If a cookie is drawn at random from a bag of 30 cookies, find the probability that it contains:

    a. fortune A **b.** fortune B **c.** either fortune A or B

Solution: First, determine the number of fortunes of each type.

    Let x = the number of C's and also the number of D's.

    Then $2x$ = the number of B's, and $6x$ = the number of A's.

The total number of fortune cookies in the bag is 30.

$$x + x + 2x + 6x = 30$$
$$10x = 30$$
$$x = 3$$

Thus, there are 18 A's, 6 B's, 3 C's, and 3 D's.

ANSWERS: **a.** $P(A) = \dfrac{18}{30}$, or $\dfrac{3}{5}$ **b.** $P(B) = \dfrac{6}{30}$, or $\dfrac{1}{5}$ **c.** $P(A \text{ or } B) = \dfrac{18 + 6}{30}$, or $\dfrac{4}{5}$

Now it's your turn to . . . **TRY IT!**

A letter is selected at random from the letters of the word TRIANGLE. Find the probability of selecting: **a.** a vowel   **b.** a consonant

                                      (See page 471 for solution.)

1. Five students each tossed a standard numbered cube 200 times and tallied the number of dots on the top face of the cube. Their results were combined in a table. Find $P(1)$, $P(2)$, $P(3)$, $P(4)$, $P(5)$, and $P(6)$. Express the probability as a decimal fraction to the nearest thousandth.

	Summary of Tosses of a Standard Numbered Cube						
Name	▫	▫	▫	▫	▫	▫	**Total**
Max	26	32	37	36	39	30	200
Sean	28	33	36	36	40	27	200
Becky	34	32	30	34	28	42	200
Caia	36	31	29	33	43	28	200
Billy	33	31	31	30	29	46	200
Total	157	159	163	169	179	173	1,000

2. Toss a standard numbered cube 10 times and record the results. Combine this data with those of your classmates. Find $P(1)$, $P(2)$, $P(3)$, $P(4)$, $P(5)$, and $P(6)$. Express the probabilities as decimal fractions to the nearest thousandth. Compare your results with those in Exercise 1.

3. Ms. Dignan used a spinner in her math class to select students for putting homework problems on the chalkboard. For example, if the spinner pointed to 2, then the students in row 2 went to the chalkboard that day. The class kept a record for one month of how often each row was selected. The results are recorded in the table. What is the probability of each row being selected to put homework on the board?

Spinner Tally					
Row	1	2	3	4	5
Tally	4	2	6	7	4

4. The table shows data, rounded to the nearest thousand, on immigration to the United States in a recent year. What is the probability that an immigrant:
   a. came from Europe
   b. was Asian
   c. came from Africa or Asia
   d. did not come from either Europe or South/Central America

U.S. Immigration in One Year	
Europe	61,000
Asia	258,000
South/Central America	217,000
Africa	18,000
Australia/New Zealand	2,000
Other	3,000
Total	559,000

In 5 and 6, use the following table, which shows the number of successful space launches over a recent 5-year period.

Successful Space Launches						
Year	Russia	U.S.	Japan	Europe	Other	Total
1st	101	18	1	0	1	121
2nd	98	22	3	2	2	127
3rd	97	22	3	4	3	129
4th	98	17	2	3	1	121
5th	91	6	2	2	2	103
Total	485	85	11	11	9	601

5. What was the probability that a successfully launched space vehicle:
   a. in the 1st year, belonged to the United States
   b. in the 5th year, was Japanese
   c. in the 4th year, belonged to Russia

6. **a.** Of the successes by the U.S., what was the probability that one occurred in the 5th year?
   b. Of the successes by Europe, what was the probability that one occurred in the 1st year?
   c. What was the probability that a successfully launched space vehicle during the 5-year period did not belong to the United States or Russia?

7. A popcorn producer knows that 1 out of 4 kernels of his Micro-Pop will not pop when heated. If 1 bag of Micro-Pop completely fills a medium bowl (450 popped pieces = 1 medium bowl), how many kernels should be put in each bag?

8. Weather records show that summer storms occur 3 days out of every 10 at Sunnyside Resort. How many days of stormy weather should guests expect during a 2-week stay?

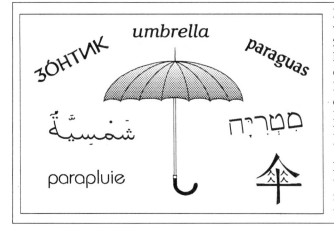

No one knows when or by whom the umbrella was invented, but it was known well before 2,000 B.C. in ancient Egypt. Chinese nobility had many-tiered umbrellas designed like pagodas. A seven-storied umbrella protected the King of Siam. Human jawbones and a skull decorated the umbrella of the King of Dahomey. Because of the uncertainty of weather in the hereafter, princes of India were buried with umbrellas.

Fictional heroes had swords concealed in umbrellas. Robinson Crusoe's umbrella of animal skins kept off rain and sun. Mary Poppins used an umbrella to go sailing across the sky.

9. At the Cleanair Company, the probability of a manufactured filter being defective is $\frac{17}{1,000}$. How many defective filters would be expected in a random sample of 20,000?

10. Mrs. Martinez drives across a drawbridge on her way to and from work every weekday. She has found that she is delayed by a bridge opening 2 days out of every 9. How many days should she expect to be delayed by a bridge opening in the next 21 working days?

11. A hitter has a batting average of .350. If he continues at this rate, he will get 350 hits out of every 1,000 times at bat. How many hits should he get in 80 times at bat?

12. A can of Miracle Graffiti Remover costs $7. The manufacturer advertises "double your money back if you are not completely satisfied with our product." Her previous experience shows that 3 buyers out of 40 will ask for money back. Since 24,240 cans of Miracle Graffiti Remover have been sold, how much money should the manufacturer set aside to pay to dissatisfied customers?

13. A mayoral candidate knows that in an election year, only about 60% of the registered voters turn out to vote. So far, he knows that 2,987 of the registered voters will go to the polls and vote for him. If there are 12,654 registered voters in his town, and voter turnout is as expected, how many additional votes will he need to win the election by getting more than half the votes cast?

14. Quality control figures for the manufacture of Dankin yo-yos show that 96.7% of the 17,500 yo-yos produced in September were satisfactory.
a. If the Toys-R-U stores bought one-fifth of the Dankin yo-yos produced in September, about how many defective yo-yos are they likely to have?
b. If Ian's mom bought one of these yo-yos at Toys-R-U, what is the probability that it is defective?

The yo-yo has been around for centuries. It came to America in the 1920's by way of the Phillipine jungles. As a hunting weapon with a 20-foot thong, it conveniently returned to the hunter, like a boomerang.

In 1991, a group of high school students in Jasonville, Indiana built a record-shattering yo-yo: 6 feet in diameter, weighing 820 pounds—and it works!

In 15 and 16, use the diagram on page 441 for tossing two standard numbered cubes to find the indicated probability. (Recall: 1 is not a prime number.)

15. a. P(sum of 4)   **b.** P(sum of 10)   **c.** P(sum of 12)   **d.** P(sum of 14)

16. a. P(sum that is a prime number)   **b.** P(sum that is a perfect square)
    c. P(sum that is between 1 and 6)   **d.** P(the same number on each cube)
    e. P(not the same number on each cube)   **f.** P(the same prime number on each cube)

17. Lee's bag contained 10 pieces of candy. There were 5 lemon, 2 lime, 1 orange, and 2 mint candies. She shared them with her friends, who reached into the bag without looking.
    a. Connie took a piece of candy. What is the probability that Connie got lemon?
    b. Connie's candy turned out to be lemon. Then Helen took a piece. What is the probability that she got mint?
    c. Helen got lime. If Susan then took a piece, what is the probability that she also got lime?
    d. Susan's piece of candy was lemon. Then Zachary reached into the bag, hoping to get chocolate candy. What is the probability that he got it?

18. Think of a spinner with six equal regions, labeled 1–6. Find the probability that the pointer will stop on:
    a. 4 b. an odd number c. a number greater than 5
    d. a number less than 9 e. a number greater than 3 but less than 7

19. A name is drawn at random from {names of the months}. Find the probability of drawing, in one random selection, a name that:
    a. begins with J b. ends with r c. does not end with r
    d. does not contain an r e. begins with a vowel f. does not begin with a vowel

20. A card is drawn at random from a standard deck of playing cards. Find the probability of drawing: a. a king b. a black king c. a red card d. the ace of hearts
    e. either a red card or a black card f. neither a red card nor a black card

21. Mrs. Kline's third-grade class had a grab bag at the end-of-term party. The students contributed 5 games, 3 decks of cards, 4 jigsaw puzzles, 6 boxes of crayons, and 2 books.
    a. Roy went first. What is the probability that he picked a deck of cards?
    b. If Roy got a jigsaw puzzle, find the probability that Odessa, who was next, also got a jigsaw puzzle.
    c. Odessa picked a box of crayons. Jordana's turn was next. Find the probability that she chose a game.
    d. Jordana got a book. Ian, who went next, wanted a deck of cards. Find the probability that he got the cards.
    e. Ian picked a jigsaw. If Ellen was next, find the probability that she got crayons.

22. A pouch contains 37 marbles, some white, some yellow, and the rest red. There are 5 more yellow marbles than white marbles, and the ratio of red marbles to white marbles is 6:1.
    a. If w represents the number of white marbles, express in terms of w:
       (1) the number of yellow marbles
       (2) the number of red marbles
       (3) the total number of marbles in the pouch
    b. Write and solve an equation to find the number, w, of white marbles.
    c. What is the probability that a randomly-selected marble will be white?

23. There are 24 keys on Brad's calculator. There are 4 more number keys than operation keys, and the ratio of memory keys to operation keys is 1:2. The remaining 5 keys include 2 for clearing, a percent, an equals sign, and a decimal point.
    a. If n represents the number of operation keys, represent the number of number keys in terms of n.
    b. Represent the number of memory keys in terms of n.
    c. Write and solve an equation to find the value of n.
    d. What is the probability that if a key is pressed at random, it will be an operation key?

24. Of the 100 courses offered this semester at Concord Community College, there are:

 16 in social science
 twice as many in music/art as in physical education
 3 fewer in mathematics/computer science than in music/art
 5 more in language arts than in mathematics/computer science
 1 more in foreign languages than in physical education
 2 more in natural science than in foreign languages

 a. If x represents the number of physical education courses, write and solve an equation to find the value of x, and how many of each type of course is offered.

 b. A student selected a course at random. Find the probability that it was:
 (1) a physical education course
 (2) not music/art
 (3) either social science or natural science
 (4) neither language arts nor a foreign language

25. There are 5 types of cartons in a packing plant, A, B, C, D, and E. Of the 360 cartons packed one hour, the number of A cartons exceeded twice the number of B cartons by 80. The number of B cartons was 40 less than the number of C cartons. The number of C cartons was equal to the sum of the D and E cartons, which were equal in number. If an inspector chose a packed carton at random, find the probability that it was:
 a. an E **b.** an A **c.** not a C **d.** either B or D **e.** neither C nor A

26. a. If the probability that it will rain tomorrow is 20%, what is the probability that it will not rain tomorrow?
 b. If the probability that Sam will get a hit his next time at bat is $\frac{2}{5}$, what is the probability that he will not get a hit?
 c. If the probability that an event will occur is p, express, in terms of p, the probability that the event will not occur.

27. A target consists of 2 squares sharing the same center. If the side of the smaller square measures $\frac{1}{2}$ m and the side of the larger square measures 1 m, what is the probability that a dart thrown at the target will land in the inner square?

28. A target consists of two circles sharing the same center. If the diameter of the larger circle is twice the diameter of the smaller circle, what is the probability that a dart thrown at the target will land in the inner circle?

29. A box contains the 48 pieces of a jigsaw puzzle, plus 2 extra pieces that are duplicates of 2 of the puzzle pieces. If a puzzle piece is lost, what is the probability that one of the extra pieces will be the right replacement?

8-8 COMPOUND EVENTS

Whether you determine probability experimentally or theoretically, it is essential that you know the number of elements in sample spaces and events. Sometimes, you can count outcomes one at a time. More often, particularly with more than one event, you need other counting techniques to make sure that all possible outcomes are known.

A *compound event* is composed of two or more single events, such as tossing both a coin and a standard numbered cube. Since the tosses do not depend on each other, they are called *independent events*. You can find the probability of compound events if you can determine the sample spaces, that is, find all possible outcomes.

Tree Diagrams to Illustrate Outcomes

MODEL PROBLEM

Draw a tree diagram to determine all possible outcomes for tossing a coin 3 times.

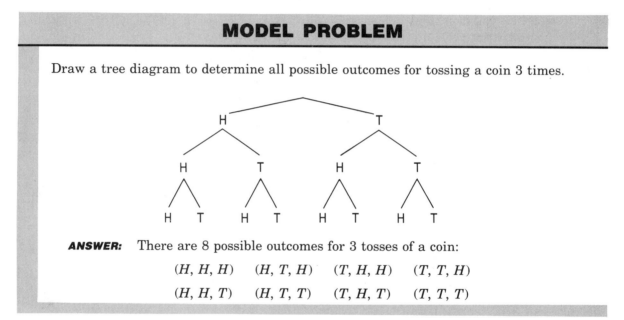

ANSWER: There are 8 possible outcomes for 3 tosses of a coin:

(H, H, H) (H, T, H) (T, H, H) (T, T, H)

(H, H, T) (H, T, T) (T, H, T) (T, T, T)

Now it's your turn to . . . **TRY IT!**

1. Draw a tree diagram to show all possible outcomes for tossing a standard numbered cube and tossing a coin. Give the number of outcomes and write the outcomes as ordered pairs.
2. In his desk, Ali has four colored pencils (yellow, red, blue, and green) and three crayons (purple, white, and orange). Draw a tree diagram to show all possible outcomes for selecting a colored pencil and a crayon. Write the outcomes as ordered pairs.

(See page 471 for solutions.)

The Counting Principle

The tree diagrams suggest a connection between the total number of outcomes and the number of outcomes for each of the single events. For example:

For 3 coins, the total number of outcomes is 8; each coin has two outcomes.
$2 \times 2 \times 2 = 8$

For tossing a standard numbered cube and a coin, the total number of outcomes is 12; the cube has 6 outcomes and the coin has 2. $6 \times 2 = 12$

These examples suggest a pattern that can be used in stating the *Counting Principle:*

> **If one event has *m* outcomes and a second event has *n* outcomes, then both events have *m* × *n* outcomes.**

This principle extends to 3 or more events, that is, if one event has m outcomes, and a second event has n outcomes, and a third event has p outcomes, then the 3 events have $m \times n \times p$ possible outcomes.

MODEL PROBLEM

Joan is buying a car. There are 3 major options she must decide on before choosing the car and determining its cost. Use the Counting Principle to find the number of models from which she can choose.

Body	Engine	Transmission
2 door	4 cylinder	automatic
4 door	6 cylinder	manual

Solution: There are 2 ways to choose the body, 2 ways to choose the engine, and 2 ways to choose the transmission. By the Counting Principle, there are $2 \times 2 \times 2 = 8$ models from which to choose.

EXERCISES

In 1–3, draw a tree diagram to show the possible outcomes. Write the outcomes as ordered pairs or triples.

1. spinning a spinner with equal regions labeled 1, 2, 3, and tossing a standard numbered cube

2. choosing a pants-top outfit from 3 colors each for the pants and the top: red, white, blue

3. choosing a soup-sandwich-fruit lunch from
   3 soups: chicken, beef, tomato
   4 sandwiches: tuna, cheese, ham, salami
   3 fruits: apple, pear, strawberries

In 4–10, use the Counting Principle to determine the number of outcomes.

4. Lawrence is selling ice cream at the school carnival. How many kinds of single-dip cones with one topping can he sell?

Cone	Ice Cream	Topping
Sugar Pretzel	Vanilla Chocolate Pistachio Butter Pecan Orange Sherbert	Nuts Jimmies Marshmallows

5. In a standard deck of cards, how many possibilities are there for:
   a. jack-ten pairs   **b.** heart-club pairs

6. Portia is trying to find the quickest route from her chemistry class on the third floor to her gym class in the basement. If there are 3 ways to the second floor, then 4 ways to the first floor, and 3 ways to the basement, how many different routes are there?

7. A combination lock can be opened by a certain group of 3 letters. There are 3 turns with 8 letters possible on each turn. If Bill knows the order of turns (right, left, right) but does not know the combination, what is the largest number of trials he will have to make?

8. Becky is planning to drive from Los Angeles to New York by way of Denver and Chicago. She has a choice of 3 routes from Los Angeles to Denver, 2 routes from Denver to Chicago and 4 routes from Chicago to New York. In how many different ways can Becky drive from Los Angeles to New York?

9. Sean is layout editor of his school yearbook. He has decided to use 4 arrangements and 3 different colors to make the theme pages. If there is one arrangement and one color per page, how many different theme pages are possible?

10. Jenny is a gymnast who likes to wear a variety of outfits during competitions, each outfit including a leotard, a bodysuit, and a headband. If she has 5 leotards, 4 bodysuits, and 6 headbands, how many possible outfits does she have?

A good gymnast needs a combination of strength, balance, and flexibility. Top gymnasts are usually of high school or college age, with the women even younger than the men.

Competitions involve such events as parallel bars, horse vault, and floor exercise. The power and grace of the participants make gymnastics great fun to watch.

Probability of Compound Events

The Probability of *A and B*

Consider the outcomes of tossing a standard numbered cube and a coin:

$$(1, H) \quad (2, H) \quad (3, H) \quad (4, H) \quad (5, H) \quad (6, H)$$
$$(1, T) \quad (2, T) \quad (3, T) \quad (4, T) \quad (5, T) \quad (6, T)$$

Referring to the sample space, you can determine the probability of getting a 2 on the cube *and* heads on the coin. Since there is 1 such an outcome, $(2, H)$, in the sample space, and there are 12 outcomes in all, the probability is $1:12$.

Another way to find the result is to apply the Counting Principle to probability; that is, multiply the probabilities.

$$P(2 \text{ on cube}) = \frac{1}{6} \quad P(\text{heads on coin}) = \frac{1}{2}$$

$$P(2 \text{ } and \text{ heads}) = \frac{1}{6} \cdot \frac{1}{2} = \frac{1}{12}$$

Set notation can also be used to express this concept.

For the two sets A and B, the compound set A *and* B is the set containing all the elements that belong to both A and B, familiar to you as the intersection of the two sets, written $A \cap B$.

If A and B represent two events, then the probability of the compound event A *and* B is written $P(A \text{ } and \text{ } B)$ or $P(A \cap B)$.

To calculate $P(A \cap B)$, provided A and B are independent events, multiply the probabilities of A and B.

$$P(A \cap B) = P(A) \times P(B)$$

The Probability of *A or B*

Now consider the probability of getting 2 on the cube *or* heads on the coin.

Since there are 2 outcomes with 2 on the cube, $(2, H)$ and $(2, T)$, and 12 outcomes in all, $P(2) = 2:12$. Since there are 6 outcomes with heads on the coin, $(1, H), (2, H), \ldots, (6, H)$, and 12 outcomes in all, $P(\text{heads}) = 6:12$.

The probability of either event is the sum of the probabilities. But since 1 outcome, $(2, H)$, occurs in both events and must not be counted twice, subtract its probability.

$$\text{Thus, } P(2 \text{ } or \text{ heads}) = \frac{2}{12} + \frac{6}{12} - \frac{1}{12} = \frac{7}{12}$$

As before, set notation can be used.

For the two sets A and B, the compound set A *or* B is the set containing all the elements that belong to either A or B or both, familiar to you as the union of the two sets, written $A \cup B$.

To calculate $P(A \cup B)$, add the probabilities of A and B and subtract the probability of the intersection.

$$P(A \cup B) = P(A) + P(B) - P(A \cap B)$$

MODEL PROBLEMS

1. A standard numbered cube is tossed. Find the probability of getting:

   a. an odd number greater than 2

   b. a number that is odd *or* greater than 2

   Solution:

   Let event A be an odd number on the cube: $\qquad A = \{1, 3, 5\}$

   Let event B be a number greater than 2 on the cube: $B = \{3, 4, 5, 6\}$

   The sample space $S = \{1, 2, 3, 4, 5, 6\}$.

   a. $P(A \cap B) = P(A) \times P(B)$

   $\qquad = \dfrac{3}{6} \times \dfrac{4}{6} = \dfrac{12}{36}$, or $\dfrac{1}{3}$ **ANS.**

   A *Venn diagram* can be used to illustrate the result. The intersection of sets A and B, shown as the shaded region, contains 2 elements.

   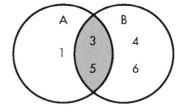

   Since there are 6 outcomes in the sample space:

   P(odd *and* greater than 2) $= \dfrac{2}{6}$, or $\dfrac{1}{3}$

   b. $P(A \cup B) = P(A) + P(B) - P(A \cap B)$

   $\qquad = \dfrac{3}{6} + \dfrac{4}{6} - \dfrac{2}{6} = \dfrac{5}{6}$ **ANS.**

   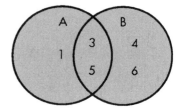

   The shaded region shows the union of sets A and B, which contains 5 elements. Since there are 6 outcomes in the sample space: P(odd *or* greater than 2) $= \dfrac{5}{6}$

2. If a nickel and quarter are tossed, what is the probability of getting heads on the nickel or getting tails on the quarter?

Solution: Let A be heads on the nickel. Then $P(A) = \dfrac{1}{2}$

Let B be tails on the quarter. Then $P(B) = \dfrac{1}{2}$

$$P(A \cap B) = P(A) \times P(B) = \dfrac{1}{2} \times \dfrac{1}{2} = \dfrac{1}{4}$$

$$P(A \cup B) = P(A) + P(B) - P(A \cap B)$$

$$= \dfrac{1}{2} + \dfrac{1}{2} - \dfrac{1}{4} = \dfrac{3}{4} \quad \textbf{ANS.}$$

Now it's your turn to . . . **TRY IT!**   *(See page 471 for solution.)*

If you spin a spinner with equal regions labeled 1, 2, 3, 4, 5, and also toss a coin, find the probability of getting:

a. 3 on the spinner and heads on the coin   **b.** 3 on the spinner or heads on the coin

---

EXERCISES

In 1–3, calculate the probabilities.

1. A 6-sided numbered cube and a coin are tossed. (Recall: 1 is not a prime number.)
   a. P(even number or heads)   **b.** P(odd number and tails)
   c. P(0 or tails)   **d.** P(prime number and heads)
   e. P(6 and tails)   **f.** P(7 and heads)

2. Spinners Y and Z are spun.

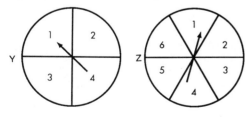

   a. P(3 on Y and 6 on Z)
   b. P(1 on Y or 5 on Z)
   c. P(even numbers on both)
   d. P(even on Y and odd on Z)
   e. P(odd on Y or prime on Z)
   f. P(prime on both)

3. A spinner with equal regions labeled 1–7 is spun and a coin is tossed.
   a. P(5 or tails)   **b.** P(prime or heads)
   c. P(even number and tails)   **d.** P(number $\geq$ 1 and heads)
   e. P(number < 3 or heads)   **f.** P(number > 7 and tails)

4. A silver dollar, a half-dollar, a quarter, a dime, and a nickel are in a box. One coin is drawn and not replaced. A second coin is drawn.
   a. Draw a tree diagram or list the sample space of all possible pairs that are outcomes for this experiment.
   b. What is the probability that the value of the first coin drawn is greater than the value of the second coin drawn?
   c. What is the probability that the value of the 2 coins drawn is greater than $1.00?
   d. What is the probability that the value of the 2 coins drawn is exactly $1.00?

5. A pouch contains only 3 orange marbles numbered 1, 2, and 3, and 2 black marbles numbered 1 and 2. One marble is removed, its color is noted, and the marble is not replaced. A second marble is removed and its color is noted.
   a. Draw a tree diagram or list the sample space of all possible pairs of outcomes.
   b. Find the probability that:
      (1) both marbles drawn are orange
      (2) neither marble drawn is orange
      (3) one of the marbles drawn is green

6. Matt has the following 5 books in his locker: math, history, Latin, science, and English. Without looking at the books, he pulls out one book and then, without replacing the first book, pulls out a second book.
   a. Using the letters *M*, *H*, *L*, *S*, and *E*, draw a tree diagram or write the sample space for pulling 2 books from the locker.
   b. What is the probability that he first pulled the math book and then the English book?
   c. What is the probability that he pulled out 2 different books?

7. Three coins are tossed simultaneously.
   a. Draw a tree diagram or list the sample space for all possible outcomes of heads and tails.
   b. What is the probability that all 3 coins will come up heads?
   c. In how many ways can a head and 2 tails appear?
   d. What is the probability of obtaining a head and 2 tails?

8. A contest offers a first prize of $10,000 or a car. The second prize is $500 or a television set or a set of encyclopedias.
   a. Draw a tree diagram or write the sample space to show all possible pairs of first and second prizes.
   b. If each prize is equally likely to be chosen, find the probability that:
      (1) both prizes are money
      (2) neither prize is money
      (3) one prize is money but the other is not

Technology has made it possible to contain all of the information found in an entire set of encyclopedias on a single computer disk.

9. A game requires spinning a marker and then tossing a single numbered cube. The marker must stop over one of 3 equally likely colors (red, green, blue), and the cube is a standard fair 6-sided type.

    a. Draw a tree diagram or list the sample space of all possible pairs of outcomes for spinning the marker and tossing the cube.

    b. What is the probability of obtaining the result (red, 4)?

    c. What is the probability of obtaining the result (green, 8)?

    d. What is the probability of obtaining the result (blue, even number)?

10. The first step of an experiment is to pick one number from the set {1, 2, 3}. The second step of the experiment is to pick one number from the set {1, 4, 9}.

    a. Draw a tree diagram or list the sample space of all possible pairs of outcomes.

    b. Determine the probability that:

        (1) both numbers are the same

        (2) both numbers are odd

        (3) the second number is the square of the first

11. In Ian's math class, there are 4 students in the first row: 3 girls, Ann, Barbara, and Cathy, and one boy, David. The teacher called one of these students to the board to solve a problem. When the problem was done, the teacher called one of the remaining students in the first row to do a second problem at the board.

    a. Draw a tree diagram or list the sample space of all possible pairs of names for calling two students to the board.

    b. Find the probability that the teacher called Ann first and Barbara second.

    c. Find the probability that the teacher called 2 girls to the board.

    d. Find the probability that David was one of the 2 students called.

Complementary Events

For each event of a sample space, there is a ***complementary event***, or negation. For example, in tossing a coin, if the event is heads, then its negation is *not* heads.

In general, if the probability of an event is written $P(E)$, then the probability of its negation is P (not E).

For example:

On a spinner with 8 equal sections, the probability of getting 5: $P(5) = \dfrac{1}{8}$

             the probability of not getting 5: $P(\text{not } 5) = \dfrac{7}{8}$

Observe that the sum of the probabilities of complementary events is always 1.

$$P(E) + P(\text{not } E) = 1$$

Then, the probability of E not happening is:

$$P(\textbf{not E}) = 1 - P(E)$$

1. What is the probability of not drawing an ace in one random draw of a card from a standard deck?

   Solution: $P(\text{ace}) = \dfrac{\text{number of aces}}{\text{total number of cards}} = \dfrac{4}{52}, \text{ or } \dfrac{1}{13}$

   $P(\text{not ace}) = 1 - P(\text{ace}) = 1 - \dfrac{1}{13} = \dfrac{12}{13}$

2. If the probability that Vera will not win her class election is 0.3, what is the probability that she will win?

$$P(\text{not win}) = 0.3$$
$$P(\text{win}) = 1 - 0.3 = 0.7$$

Calculating Odds

Baseball fans do not usually ask "What is the *probability* that the Chicago Cubs will win the World Series this year?" Fans who want to know the chances of a team winning usually ask, "What are the *odds* that the Chicago Cubs will win?".

Odds that an event will or will not occur are defined as a ratio of probabilities.

$$\textbf{Odds of an event occurring} = \frac{P(E)}{P(\text{not } E)}$$

$$\textbf{Odds of an event not occurring} = \frac{P(\text{not } E)}{P(E)}$$

MODEL PROBLEMS

1. What are the odds in favor of tossing a 4 in one toss of a numbered cube?

   Solution: Since the odds that an event will occur are a ratio of probabilities, first find those probabilities.

   $P(4) = \dfrac{\text{number of successes}}{\text{total number}} = \dfrac{1}{6}$

   $P(\text{not } 4) = 1 - P(4) = 1 - \dfrac{1}{6} = \dfrac{5}{6}$

   Now, form the ratio of the probabilities. $\text{odds}(4) = \dfrac{P(4)}{P(\text{not } 4)} = \dfrac{\frac{1}{6}}{\frac{5}{6}} = \dfrac{1}{5}$

   ANSWER: The odds of tossing a 4 in one toss of a cube are 1 to 5.

2. If there is a 20% chance of rain today, what are the odds against rain?

Solution: You are given the probability that it will rain, that is:

$$P(\text{rain}) = 20\%, \text{ or } \frac{1}{5}$$

$$\text{Thus, } P(\text{not rain}) = 1 - P(\text{rain}) = 1 - \frac{1}{5} = \frac{4}{5}$$

$$\text{And, odds(not rain)} = \frac{P(\text{not rain})}{P(\text{rain})} = \frac{\frac{4}{5}}{\frac{1}{5}} = \frac{4}{1}$$

ANSWER: The odds against rain today are 4 to 1.

EXERCISES

In 1–4, for the event E described, find:

     (1) $P(E)$
     (2) $P(\text{not } E)$
     (3) odds of E occurring
     (4) odds of E not occurring

1. One toss of a standard numbered cube
  a. get 3   **b.** get prime   **c.** get number between 2 and 5
  d. not get 4   **e.** not get 7   **f.** not get 2 or 3

2. One toss of a pair of standard numbered cubes
  a. get sum of 7   **b.** get sum of 2   **c.** get sum of 11 or 12
  d. not get sum of 7 or 11   **e.** not get sum of 1

3. One draw from a standard deck of cards
  a. get an ace   **b.** get a red ace   **c.** not get a red card
  d. get a jack or a king   **e.** not get a queen or an ace
  f. get a club or a red card   **g.** not get a heart or a club

4. A dart is thrown at the inner rectangle of the target shown. The length of the smaller rectangle is twice its width, and the dimensions of the larger rectangle are twice those of the smaller.

CHAPTER SUMMARY

A ***ratio*** is an expression of the form $a:b$ or $\dfrac{a}{b}$ that is used to compare quantities.

1. The types of measures may be the same (as in comparing one length to another), and also the units may be the same (comparing inches to inches), leading to a simplified ratio that has no units.
2. The types of measures may be the same, but the units may be different (inches to feet), leading to a simplified ratio that has no units.
3. The types of measures may be different (as in relating miles to hours). These ratios, called ***rates***, are expressed with units, such as 45 miles per hour.

In solving ratio problems, you can use x as a ***common ratio factor***, and write ax and bx to represent quantities that are in the ratio $a:b$.

When the sides of an equation are equal ratios, the equation is a ***proportion***. A proportion has the property that ***the product of the means equals the product of the extremes***. This property is used in solving a proportion that contains a variable.

Problems using proportions include ***direct variation***, in which the *quotient* of two variables is constant, and ***inverse variation***, in which the *product* of two variables is constant.

Ratios and proportions have applications in mathematics. In geometry, triangles are ***similar*** when the lengths of their corresponding sides are proportional. Ratios in similar triangles are used in ***trigonometry***, a branch of mathematics that allows accurate measurement of lengths that are difficult or impossible to measure directly. Basic trigonometric ratios in the right triangle are ***sine, cosine,*** and ***tangent***.

Another application of ratio and proportion is in ***probability***, the mathematics of chance occurrences. Probability is used daily to make a wide variety of predictions. The value of the probability ratio ranges from 0, for an impossible event, to 1, for a certain event. The closer the value is to 1, the more likely it is that the event will happen; the closer to 0, the less likely. The ***odds*** that an event will occur is the ratio of the number of successful outcomes to the number of unsuccessful outcomes. The number of outcomes in the ***sample space***, that is, the total number of possible outcomes, can be determined by using the ***Counting Principle*** or a tree diagram.

VOCABULARY CHECKUP

SECTION

8-1 *ratio / rate* **8-2** *proportion / means / extremes*
8-3 *direct variation / constant of variation / inverse variation / joint variation*
8-4 *similar triangles / corresponding angles / corresponding sides*
8-5 *trigonometry / sine / cosine / tangent*
8-6 *line of sight / angle of elevation / angle of depression*
8-7 *experimental, and theoretical probability / outcome / sample space / event*
8-8 *compound, independent, and complementary events / Counting Principle / odds*

CHAPTER REVIEW EXERCISES

In 1–6, express each ratio in simplest form. (Section 8-1)

1. $24:36$

2. a day to an hour

3. 10 seconds to 1 hour

4. 2 weeks to 3 days

5. $9x:12x$

6. $6a:6y$

In 7–11, solve by using a ratio. (Section 8-1)

7. Two numbers are in the ratio $5:3$. Their sum is 48. Find the numbers.

8. Agnes and Amtul are in business together. They split profits in the ratio $2:3$. If their business earns a profit of $65,000, how much will Amtul receive?

9. Mrs. Amore gives her three children an allowance each week in a ratio according to their ages. The children's ages are 4, 5, and 7 years old. Each week Mrs. Amore pays out $8 in allowance. How much does her oldest child receive?

10. The ratio of the cost of a hot dog to a can of ginger ale is $2:1$. If $4.05 was spent altogether, how much was spent for the hot dog?

11. Morley runs 3 kilometers in 20 minutes. Marilyn runs 2 kilometers in 15 minutes. Who has the faster average speed?

In 12–20, solve the equation. (Section 8-2)

12. $\dfrac{3}{5} = \dfrac{4}{x}$

13. $\dfrac{x+3}{x} = \dfrac{4}{5}$

14. $\dfrac{2}{x} = \dfrac{7}{x-5}$

15. $\dfrac{3}{2} = \dfrac{x-4}{8}$

16. $3:5 = 6:x$

17. $4x:3 = 7:9$

18. $4:x = 3:12$

19. $\dfrac{x+2}{x-3} = \dfrac{x-6}{x+5}$

20. $\dfrac{3x-4}{x} = \dfrac{6x-1}{2x+3}$

In 21–30, solve by using a proportion. (Section 8-2)

21. If 3 pens cost $2, how much does 1 pen cost?

22. If 7 cans of peas cost $3, how much should 4 cans cost?

23. If Robin drives p miles in q days, how many days will it take her to drive m miles at the same rate?

24. Margo can walk 7 miles in 105 minutes. At the same rate, how long does it take her to walk 2 miles?

25. Cosmo carves a dozen wooden dolls every 18 hours. How many complete dolls can Cosmo carve at the same rate in 50 hours?

26. On a travel map issued by Mr. Keen's auto club, $\frac{1}{4}$ inch represents 200 miles. If the map length between Toronto and Vancouver is $2\frac{1}{2}$ inches, find the actual distance between Toronto and Vancouver.

27. The starting hourly wages of Betty and Carol were in the ratio $3:4$. When both their wages were increased by $3 an hour, their hourly wages were in the ratio $4:5$. If they both work a 35-hour work week, find their weekly wages after the increase.

28. The admissions policy at Clay City University requires that 3 in-city students be admitted for every out-of-city student. If 2,040 freshmen have been admitted in the required ratio, how many are in-city students and how many are from other areas?

29. The U.S. birth rate in a recent year was given as 15.6 per thousand. If the population was 220 million, how many babies were born that year in the U.S.?

30. The table below shows the number of union members among the workers at the Mill City Factory for four successive years.

    a. Find the ratio of union to non-union members for each of the 4 years.

    b. If this ratio remained constant and 200 more workers were hired the next year, how many of these were union members?

Year	Number of Workers	Union Members	Nonunion
1st	600	240	360
2nd	800	320	480
3rd	1,000	400	600
4th	1,200	480	720

In 31–36, tell whether the relation between the variables is direct variation, inverse variation, or neither. If there is a direct or inverse variation, express it by formula. (Section 8-3)

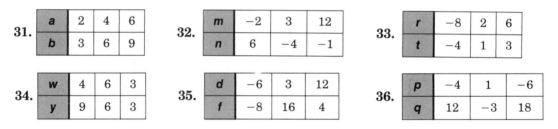

31.
a	2	4	6
b	3	6	9

32.
m	−2	3	12
n	6	−4	−1

33.
r	−8	2	6
t	−4	1	3

34.
w	4	6	3
y	9	6	3

35.
d	−6	3	12
f	−8	16	4

36.
p	−4	1	−6
q	12	−3	18

In 37 and 38, write the relation as a formula. Use k as the constant of variation. (Section 8-3)

37. The cost c of buying personal computers of identical price varies directly as the number n of computers purchased.

38. The time t that it takes to complete a certain job varies inversely as the number of people p working on the job.

In 39–43, find the indicated value. (Section 8-3)

39. If t varies directly as d, and if $t = 5$ when $d = 9$, find t when $d = 6$.

40. If r varies inversely as the square of w, and if $r = 6$ when $w = 4$, find r when $w = -2$.

41. If x varies jointly as y and z, and if $x = 7$ when $y = 3$ and $z = 5$, find x when $y = 4$ and $z = 6$.

42. If m varies directly as p and inversely as q, and if $m = 8$ when $p = 6$ and $q = 3$, find m when $p = -16$ and $q = 4$.

43. A man weighing 150 pounds sits 3 feet from the fulcrum of a seesaw. How far from the fulcrum should his 60-pound daughter sit to balance the seesaw?

In 44 and 45, tell if the triangles are similar. (Section 8-4)

44. In $\triangle ABC$, $m \angle A = 35°$ and $m \angle C = 85°$. In $\triangle PQR$, $m \angle Q = 85°$ and $m \angle P = 60°$.

45. In right triangle PWD, $m \angle D = 19°$. In $\triangle RST$, $m \angle T = 71°$ and $m \angle R = 19°$.

In 46 and 47, use the diagram to find the indicated lengths. (Section 8-4)

46. BD

47. CE

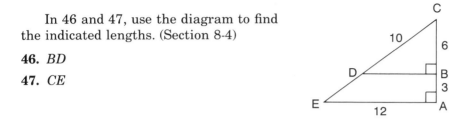

In 48 and 49, find the indicated lengths. (Section 8-4)

48. An 8-foot pole casts a shadow 3 feet long. At the same time, how long a shadow will a 30-foot tree cast?

49. I am 6 ft. tall. My shadow is 27 in. long. My sister casts a 2-ft. shadow. How tall is she?

In 50–52, use the table on page 423 to find: (Section 8-5)

50. tan 9°   **51.** cos 81°   **52.** sin 51°

In 53–55, use the table to find x if: (Section 8-5)

53. $\sin x = .3584$ **54.** $\tan x = 14.3007$ **55.** $\cos x = .7071$

In 56–64, find the indicated value. (Section 8-6)

56. In $\triangle ABC$, $m \angle C = 90°$, $AC = 9$ m, and $BC = 12$ m. Find $m \angle A$ to the nearest degree.

57. A pole casts a 23-foot shadow. The angle of depression from the top of the pole to the end of the shadow is 52°. To the nearest foot, how tall is the pole?

58. In $\triangle DEF$, $m \angle D = 90°$, $DE = 8$ m, and $EF = 12$ m. Find $m \angle F$ to the nearest degree.

59. A tower is held in place by a 95-foot wire that is fastened at the top of the tower. From where the other end of the wire is staked to the ground, the angle of elevation of the top of the tower is 65°. To the nearest foot, how tall is the tower?

60. In $\triangle AEF$, $m \angle F = 90°$, $EF = 9$ in., and $AE = 12$ in. Find $m \angle E$ to the nearest degree.

61. The top of a ladder is placed against the top of a wall. From the base of the ladder, the angle of elevation of the top of the wall is 80°. The base of the ladder is 7 feet from the wall. To the nearest foot, how long is the ladder?

62. A cable car goes 12,000 feet in a straight line from the bottom of a mountain to the top. From the starting point, the angle of elevation to the top of the mountain is 15°. How high is the mountain to the nearest 10 feet?

63. A sliding board is 40 feet long. If the angle the slide makes with the ground measures 30°, how high is the top of the slide?

64. A sailboat is 500 yards from the base of a 20-yard-high cliff. What angle of elevation, to the nearest degree, is required to view the top of the cliff from the sailboat?

65. A spinner has five equal regions, labeled 1–5. Find the probability of the pointer stopping where indicated. (Section 8-7)
  a. on an even number
  b. on a number less than 5
  c. on a number greater than 5

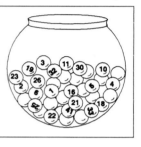

It made a front-page news story when a winner of a state lottery won the lottery again within 4 months. By some calculations, the chance of this happening is 1 in 17 trillion. But mathematicians have demonstrated that the chance is more like 1 in only 30. When very large numbers are involved, statistics show that coincidences happen much more often than you would expect.

In 66 and 67, draw a tree diagram and list the outcomes. (Section 8-8)

66. A coin is tossed 3 times.

67. Two spinners are used, one with 3 equal regions labeled 1, 2, and 3, the other with 4 equal regions labeled *a*, *b*, *c*, and *d*.

Exercises 68–70 refer to a standard deck of cards. (Section 8-8)

68. How many different ace-3 pairs are possible?

69. How many different jack-queen-king combinations are possible?

70. How many different two-card combinations are possible in which one card is higher than 8 (9, 10, jack, queen, king, and ace are all higher than 8) and the other card is lower than 8?

71. Given a coin and a spinner with 3 equal regions, labeled 1–3, find the probability of getting the indicated outcome. (Section 8-8)
  a. tails and an even number
  b. heads or an odd number
  c. heads and a number less than 4

Exercises 72–74 refer to tossing a standard pair of numbered cubes. (Section 8-8)

72. What are the odds of getting a sum of 2?

73. What are the odds against getting a sum of 11?

74. If the probability that a given sum will occur is $\frac{5}{36}$, what is the probability that the sum will not occur?

1. Two numbers are in the ratio $a:b$. The sum of the numbers is 20 and one of the numbers is 4 more than the other. If the ratio is in simplest form and $a > b$, find a and b.

2. A package of balloons contains 4 different colors. The ratio of red to blue balloons is $3:2$. The ratio of blue to yellow is $3:2$. The ratio of yellow to green is $3:2$. If the package contains fewer than 100 balloons, how many balloons does it contain?

3. Four consecutive odd integers can be written as a proportion. What are the integers?

4. Andre met Angela at a dance. He asked for her phone number and she wrote it down for him. The next day, Andre took out the paper with Angela's number on it, and to his dismay, he saw that the number on the paper was 293-516. He assumed that, by mistake, Angela had omitted one of the last 4 digits of her phone number. Andre looked closely at the paper, but he could not tell which digit had been left out. He only remembered that no consecutive digits in the phone number were consecutive whole numbers (increasing or decreasing) and that no digits in the phone number were repeated. Andre calculated the maximum number of calls he would have to make in order to find Angela. What is the probability that he dialed her number on the first try?

5. A builder of fine homes wanted to make a contribution to a hospital that had brought his son through a very serious illness. In order to help support the hospital's research center, the builder decided to raffle off a $200,000 house. Each raffle ticket cost $50. The builder sold only enough tickets to pay for the house and to raise $100,000 for the hospital. If you bought 1 ticket, what is the probability that you will win the house?

6. A school bus route starts at the bus garage and ends at the school parking lot. The bus stops are one mile apart in a square grid, as shown. Find a route that passes each bus stop exactly once.

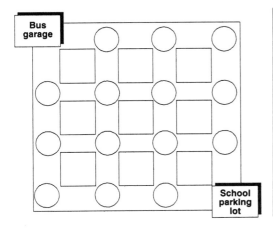

The goal of the so-called traveling salesman problem is to find the shortest possible route for going to a number of locations. Since traveling to only 10 places involves $10 \cdot 9 \cdot 8 \cdots 3 \cdot 2 \cdot 1 = 3,628,800$ possible routes, comparison is not easy.

One industrial application of this problem is in deciding the best order in which to drill the tens of thousands of holes in a circuit board. So far, an imperfect solution to the problem has made it possible to cut the drilling time in half. Mathematicians are still looking for the perfect solution.

1. The ratio of my mother's age to my age is the same as the ratio of my age to my daughter's age. None of us is over 85 or under 20 years old. Neither my mother nor I had a child until we were more than 20 years old. Find our ages. (All ages are whole numbers.)

2. Write a proportion using natural numbers in which the value of each fraction is less than 1 and the product of the means is equal to 989.

3. Find an acute angle x for which:
   a. $\sin x = .5$ **b.** $\cos x = .5$ **c.** $\sin x = \cos x$

4. Pick two acute angles and two obtuse angles. Make a table containing the sine, the cosine, the square of the sine, and the square of the cosine for each angle. See if you can find a rule relating the square of the sine and the square of the cosine for any angle.

COLLEGE TEST PREPARATION

In 1–16, select the letter of the correct answer.

1. If a carton containing a dozen eggs is dropped, which of the following cannot be the ratio of broken eggs to whole eggs?
   (A) 1:2   (B) 1:3   (C) 1:4
   (D) 1:5   (E) 5:7

2. If $1:3 = p:q$, then all of the following equalities are correct except
   (A) $3p = q$ (B) $3q = 9p$
   (C) $2q = 6p$ (D) $-q = -3p$
   (E) $p = 3q$

3. The lengths of three rods are in the ratio of 4:5:6. If the middle rod is 15 cm long, what is the total length of all three rods?
   (A) 36 cm   (B) 38 cm   (C) 40 cm
   (D) 45 cm   (E) 60 cm

4. If $2x + 7 = 12$, then the ratio of $2x$ to $2x + 5$ is
   (A) $\frac{2}{5}$ (B) $\frac{1}{2}$ (C) $\frac{2}{3}$ (D) $\frac{3}{4}$ (E) $\frac{4}{5}$

5. The angle of elevation of the sun when a 30-foot tree casts a 30-foot shadow is
   (A) 30°   (B) 45°   (C) 60°
   (D) 75°   (E) 80°

6. If the probability of event A happening is $\dfrac{x}{x + y}$, then the probability of event A not happening is
   (A) $\dfrac{1}{x + y}$ (B) $\dfrac{y}{x + y}$ (C) $\dfrac{x - y}{x + y}$
   (D) $\dfrac{1 - x}{x + y}$ (E) $\dfrac{x - 1}{x + y}$

7. If $PQ = 1$, then $PR =$
   (A) $\sin P$ (B) $\cos P$ (C) $\tan P$
   (D) $\cos Q$ (E) $\tan Q$

8. If $\triangle PQR$ is similar to $\triangle SQP$, then the ratio of PR to SP is
   (A) $\frac{7}{9}$ (B) $\frac{3}{4}$
   (C) $\frac{7}{16}$ (D) $\frac{9}{16}$ (E) $\frac{7}{12}$

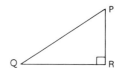

9. A bowl contains only red and black marbles. There are x red marbles, with 2 more black marbles than red. If a marble is selected at random, the probability that it will be black is

(A) $\dfrac{2}{x}$ (B) $\dfrac{x}{x+2}$ (C) $\dfrac{x-2}{x+2}$

(D) $\dfrac{x}{2x+2}$ (E) $\dfrac{x+2}{2x+2}$

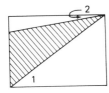

10. If $\tan \angle 1 = \dfrac{3}{4}$ and $\tan \angle 2 = \dfrac{1}{6}$, then the area of the shaded region is what fraction of the area of the rectangle?

(A) $\dfrac{2}{9}$ (B) $\dfrac{1}{3}$ (C) $\dfrac{4}{9}$ (D) $\dfrac{7}{9}$ (E) $\dfrac{7}{18}$

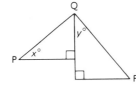

11. If $\tan x° = \tan y°$, then $m\angle PQR$ in degrees is

(A) 30 (B) 45 (C) 60 (D) 90 (E) 120

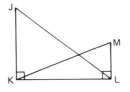

12. If $\tan \angle JLK = \dfrac{3}{4}$ and $\tan \angle MKL = \dfrac{1}{3}$, then the ratio of LM to JK is

(A) $\dfrac{1}{2}$ (B) $\dfrac{1}{3}$ (C) $\dfrac{4}{9}$ (D) $\dfrac{1}{4}$ (E) $\dfrac{5}{9}$

13. Evaluate: $\dfrac{1}{\cos A} - \dfrac{1}{\sin A}$

(A) $\dfrac{1}{5}$ (B) $-\dfrac{1}{5}$ (C) $\dfrac{5}{12}$ (D) $-\dfrac{5}{12}$ (E) 0

14. If two numbers are selected at random from the set $\{1, 2, 3, 4, 5, 6\}$, what is the probability that their product is 12?

(A) $\dfrac{1}{5}$ (B) $\dfrac{1}{3}$ (C) $\dfrac{2}{5}$ (D) $\dfrac{1}{15}$ (E) $\dfrac{2}{15}$

15. Set P is $\{2, 3, 5\}$ and set Q is $\{6, 10, 15\}$. If one element is selected at random from set P and another from set Q, what is the probability that the P element is a divisor of the Q element?

(A) 1 (B) $\dfrac{1}{3}$ (C) $\dfrac{1}{2}$ (D) $\dfrac{2}{3}$ (E) $\dfrac{3}{4}$

16. There are 2^{20} white marbles in a bowl containing 2^{21} marbles. If the rest of the marbles are blue and red, in equal numbers, the probability of drawing a red marble at random is

(A) $\dfrac{1}{3}$ (B) $\dfrac{1}{4}$ (C) $\dfrac{1}{2^{21}}$ (D) $\dfrac{2^{10}}{2^{21}}$ (E) $\dfrac{1}{2^{20}}$

Questions 17–25 each consist of two quantities, one in Column A and one in Column B. You are to compare the two quantities and choose:

A if the quantity in Column A is greater;
B if the quantity in Column B is greater;
C if the two quantities are equal;
D if the relationship cannot be determined from the information given.

1. In certain questions, information concerning one or both of the quantities to be compared is centered above the two columns.

2. In a given question, a symbol that appears in both columns represents the same thing in Column A as it does in Column B.

3. x, n, and k, etc. stand for real numbers.

Column A	Column B

17. $\dfrac{AB}{BC}$ $\dfrac{AD}{DE}$

18. $(PR)(TR)$ $(QR)(SR)$

19. PQ QT

20. $m\angle P + m\angle Q$ $m\angle R + m\angle S$

$$x > y > 0$$

21. $\dfrac{x + y}{x}$ $\dfrac{x + y}{y}$

A standard 6-sided numbered cube

22. If tossed once, probability of obtaining a 4   If tossed twice, probability that the numbers are the same

---

A bank containing only quarters and nickels is shaken until a coin falls out.

23. Probability of a 25¢ coin   Probability of a 5¢ coin

---

24. The probability of tossing 2 numbered cubes and getting a sum of 7 dots   The probability of tossing 2 numbered cubes and getting a sum of 3 or 4 dots

---

Letters are selected at random from the word LETTER

25. The probability of getting L or R on the first draw   The probability of getting T on the first draw

1. If a, b, and c are real numbers, name the property of inequality illustrated by the statement: If $a > b$ and $b > c$, then $a > c$.

2. Simplify.

   a. $64m^2n \div -16m^2$ **b.** $(2x - 1)(x + 3)$ **c.** $\dfrac{9y - 27}{6y - 18}$ **d.** $(x^5)(2x)^2$

3. One factor of $x^2 + 5x - 24$ is $x + 8$. Find the other factor.

4. If the surface area of a cube is 0.24 sq. ft., what is the volume of the cube?

5. Solve.

   a. $\dfrac{2}{3}y = 0.6$ **b.** $.05x + .03(10,000 - x) = 460$ **c.** $\dfrac{x}{4} - 3 = \dfrac{x}{5} + 1$

6. What is the average of $5x + 2$, $11x - 1$, and $2x - 4$?

7. Let $\widehat{n} = n^2 - n + 1$. What is the value of $\widehat{-5}$?

8. If $\begin{bmatrix} 3 & x + 4 & 5 \\ 2 & -7 & x - 8 \end{bmatrix} = \begin{bmatrix} 3 & 9 & 5 \\ 2 & -7 & y \end{bmatrix}$, find the value of y.

9. If $PQ = QR$ on the number line shown, then what is the length of PR?

   P   Q   R

   $\dfrac{4}{n}$ $\dfrac{5}{n}$ $\dfrac{1}{3}$

10. Crunchy Munchy Cereal comes in boxes 12 cm high, 5 cm wide, and 9 cm long. What is the greatest number of cereal boxes that can fit in a shipping carton with inside dimensions 24 cm high, 15 cm wide, and 27 cm long?

11. In a 5-digit palindromic number, the second digit is twice the first digit and the sum of the digits is 6. Find the number.

12. Find the next two numbers in the sequence: 3, 4, 7, 12, 19, __, __

13. A car and a truck are traveling in the same direction. If the truck averages 10 mph more than the car, and the truck traveled 220 miles while the car traveled 180 miles, find their speeds.

14. If $r = 2x$, $s = 3x$ and $t = 6x$, find the value of $\dfrac{r}{s} + \dfrac{s}{t}$.

15. Billy Tell misses the target on his first two shots and hits the bullseye with his next three arrows. If he continues to shoot, what is the least number of consecutive hits he must make in order to hit the target on more than $\frac{4}{5}$ of the total number of arrows shot?

Solutions to **TRY IT!** *Problems (Chapter 8)*

8-1 RATIO

TRY IT! *Problems 1 and 2 on page 388*

1. *Solution 1:* $\dfrac{1 \text{ min.}}{45 \text{ sec.}} = \dfrac{60 \text{ sec.}}{45 \text{ sec.}} = \dfrac{60}{45} = \dfrac{4}{3}$

Solution 2: $\dfrac{1 \text{ min.}}{45 \text{ sec.}} \cdot \dfrac{\overset{4}{\cancel{60}} \text{ sec.}}{1 \text{ min.}} = \dfrac{4}{3}$

2. ratio $= \dfrac{\text{actual length of arm}}{\text{length of arm in picture}}$

$= \dfrac{1\frac{1}{2} \text{ ft.}}{1 \text{ in.}} = \dfrac{18 \text{ in.}}{1 \text{ in.}} = \dfrac{18}{1}$

TRY IT! *Problem on page 390*

Let x = the common ratio factor.
Then $2x$ = the smaller number.
And $3x$ = the larger number.

The larger number is 30 more than $\frac{1}{2}$ of the smaller number.

$$3x = \tfrac{1}{2}(2x) + 30$$
$$3x = x + 30$$
$$2x = 30$$
$$x = 15$$
$$2x = 30$$
$$3x = 45$$

Check: 30:45 is 2:3. ✔
The larger number, 45, is 30 more than 15, which is $\frac{1}{2}$ of the smaller number. ✔

ANSWER: The numbers are 30 and 45.

8-2 PROPORTION

TRY IT! *Problem on page 395*

$$\frac{x-2}{x+2} = \frac{x-1}{x+4}$$
$$(x+2)(x-1) = (x-2)(x+4)$$

$$x^2 + 2x - x - 2 = x^2 - 2x + 4x - 8$$
$$\cancel{x^2} + x - 2 = \cancel{x^2} + 2x - 8$$
$$x - 2 = 2x - 8$$
$$x - 2x = -8 + 2$$
$$-x = -6$$
$$x = 6$$

Check: $\dfrac{x-2}{x+2} = \dfrac{x-1}{x+4}$

$$\frac{6-2}{6+2} \overset{?}{=} \frac{6-1}{6+4}$$

$$\frac{4}{8} \overset{?}{=} \frac{5}{10}$$

$$\frac{1}{2} = \frac{1}{2} ✔$$

ANSWER: $x = 6$

TRY IT! *Problem on page 396*

Let x = the amount she would receive.

$$\frac{x}{14} = \frac{40}{8} \begin{array}{l} \leftarrow \text{ amount of money} \\ \leftarrow \text{ number of hours} \end{array}$$
$$8x = 560$$
$$x = 70$$

Check: $\dfrac{70}{14} \overset{?}{=} \dfrac{40}{8}$

$$5 = 5 ✔$$

ANSWER: $70

8-3 VARIATION

TRY IT! *Problems 1 and 2 on page 401*

1. a. Since the ratio $\dfrac{n}{m} = 3$ for each pair of values, n varies directly as m.

b. The formula is $\dfrac{n}{m} = 3$, or $n = 3m$.

2. *Solution 1:*

$$\text{direct variation:} \quad p = ks$$
$$\text{If } p = 12 \text{ when } s = 3: \quad 12 = 3k$$
$$4 = k$$
$$\text{Thus:} \quad p = 4s$$

If $s = 10$, then $p = 4(10)$, or 40.

Solution 2:

Let x = the unknown value.

p	12	x
s	3	10

$$\frac{p_1}{s_1} = \frac{p_2}{s_2}$$
$$\frac{12}{3} = \frac{x}{10}$$
$$3x = 120$$
$$x = 40$$

Check: $\frac{12}{3} \overset{?}{=} \frac{40}{10}$

$4 = 4 \checkmark$

ANSWER: $p = 40$

TRY IT! *Problems 1 and 2 on page 405*

1. a. Since $mn = 24$ for each pair of values, m varies inversely as n.

b. The formula is $mn = 24$.

2. Let s_1 and d_1 = the speed and diameter of the larger pulley.

Then s_2 and d_2 = the speed and diameter of the smaller pulley.

Given: $s_1 = 155$, $d_1 = 16$, and $d_2 = 9$

$$s_1 \cdot d_1 = s_2 \cdot d_2$$
$$155 \cdot 16 = s_2 \cdot 9$$
$$2{,}480 = 9s_2$$
$$275.\overline{5} = s_2$$

Check: Are the products for the two pulleys equal?

$$155 \times 16 \overset{?}{=} 275.\overline{5} \times 9$$
$$2{,}480 \approx 2{,}479.\overline{9} \checkmark$$

ANSWER: To the nearest foot, the speed of the smaller pulley is 276 ft./min.

TRY IT! *Problems 1 and 2 on page 411*

1. Since d varies jointly as r and t,

$$d = krt$$

Since r is doubled, replace r by $2r$.

Since t is tripled, replace t by $3t$.

$$d = k(2r)(3t)$$
$$d = 6krt$$

Compare the final and initial d-values:

$$6krt \text{ is 6 times } krt$$

ANSWER: d is multiplied by 6.

2. *Solution 1:*

$$i = kpr$$

If $i = 8$ when $p = 100$ and $r = .04$,

$$8 = k(100)(.04)$$
$$8 = 4k$$
$$2 = k$$
$$\text{Thus:} \quad i = 2pr$$

If $p = 400$ and $r = .06$,

$$i = 2(400)(.06) \text{ or } 48$$

Solution 2:

$$\frac{i_1}{p_1 r_1} = \frac{i_2}{p_2 r_2}$$

$$\frac{8}{100(.04)} = \frac{i}{400(.06)}$$

$$\frac{8}{4} = \frac{i}{24}$$

$$4i = 8(24)$$
$$4i = 192$$
$$i = 48$$

ANSWER:

$i = 48$ when $p = 400$ and $r = .06$

8-4 SIMILARITY

TRY IT! *Problem on page 416*

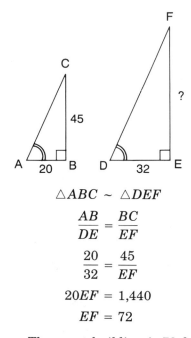

$$\triangle ABC \sim \triangle DEF$$

$$\frac{AB}{DE} = \frac{BC}{EF}$$

$$\frac{20}{32} = \frac{45}{EF}$$

$$20EF = 1{,}440$$

$$EF = 72$$

ANSWER: The court building is 72 feet tall.

8-6 APPLYING THE TRIGONOMETRIC RATIOS

TRY IT! *Problems 1–3 on page 429*

1. $\cos \angle = \dfrac{\text{adjacent}}{\text{hypotenuse}}$

$$\cos 35° = \frac{x}{60}$$

$$.8192 = \frac{x}{60}$$

$$x = 49.152 \approx 49 \text{ ft.} \quad \textbf{ANSWER}$$

2. $\sin \angle = \dfrac{\text{opposite}}{\text{hypotenuse}}$

$$\sin x = \frac{15}{48} = .3125$$

$$x = 18° \quad \textbf{ANSWER}$$

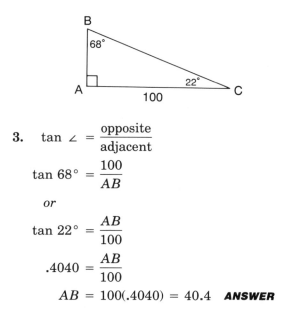

3. $\tan \angle = \dfrac{\text{opposite}}{\text{adjacent}}$

$$\tan 68° = \frac{100}{AB}$$

or

$$\tan 22° = \frac{AB}{100}$$

$$.4040 = \frac{AB}{100}$$

$$AB = 100(.4040) = 40.4 \quad \textbf{ANSWER}$$

TRY IT! *Problems 1 and 2 on page 431*

1. $\tan \angle = \dfrac{\text{opposite}}{\text{adjacent}}$

$$\tan A = \frac{6}{8} = .7500$$

$$m\angle A = 37° \quad \textbf{ANSWER}$$

2.

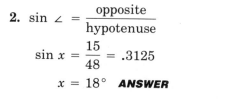

$$\sin \angle = \frac{\text{opposite}}{\text{hypotenuse}}$$

$$\sin A = \frac{BC}{AB}$$

$$\sin 8° = \frac{1{,}000}{AB}$$

$$.1392 = \frac{1{,}000}{AB}$$

$$.1392(AB) = 1{,}000$$

$$AB = \frac{1{,}000}{.1392}$$

$$AB = 7{,}184 \approx 7{,}200 \text{ feet} \quad \textbf{ANSWER}$$

8-7 PROBABILITY

TRY IT! *Problem on page 442*

TRIANGLE has 8 letters, of which 3 are vowels.

a. $P(\text{vowel}) = \dfrac{3}{8}$

b. $P(\text{consonant}) = \dfrac{5}{8}$

8-8 COMPOUND EVENTS

TRY IT! *Problems 1 and 2 on page 448*

1.

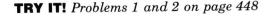

There are 12 outcomes:

$(1, H), (2, H), (3, H), (4, H), (5, H), (6, H)$

$(1, T), (2, T), (3, T), (4, T), (5, T), (6, T)$

2.

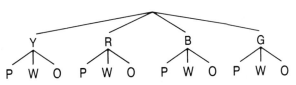

There are 12 outcomes:

$(Y, P), (Y, W), (Y, O)$ $(R, P), (R, W), (R, O)$

$(B, P), (B, W), (B, O)$ $(G, P), (G, W), (G, O)$.

TRY IT! *Problem on page 453*

Solution 1: by sets

Let A be 3 on the spinner. $\quad P(A) = \dfrac{1}{5}$

Let B be heads on the coin. $\quad P(B) = \dfrac{1}{2}$

a. $P(A \cap B) = P(A) \times P(B)$

$$= \frac{1}{5} \times \frac{1}{2} = \frac{1}{10}$$

b. $P(A \cup B) = P(A) + P(B) - P(A \cap B)$

$$= \frac{1}{5} + \frac{1}{2} - \frac{1}{10}$$

$$= \frac{2}{10} + \frac{5}{10} - \frac{1}{10}$$

$$= \frac{6}{10}, \text{ or } \frac{3}{5}$$

Solution 2: by tree diagram

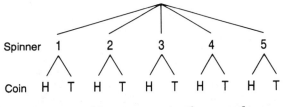

There are 10 outcomes in the sample space:

$(1, H), (2, H), (3, H), (4, H), (5, H)$

$(1, T), (2, T), (3, T), (4, T), (5, T)$

a. Since there is 1 outcome, $(3, H)$, with 3 on the spinner and H on the coin:

$$P(3 \text{ and } H) = \frac{1}{10}$$

b. Since there are 6 outcomes with 3 on the spinner or H on the coin, $(3, T), (1, H)$, $(2, H), (3, H), (4, H)$, and $(5, H)$:

$$P(3 \text{ or } H) = \frac{6}{10}, \text{ or } \frac{3}{5}$$

CHAPTER 9

RELATIONS, FUNCTIONS, AND GRAPHS

In modern play, a coordinate system is used to locate the positions of chess pieces on the board. This popular game originated in India in the 6th century.

9-1 Graphing in a Plane 473

9-2 Relations and Their Graphs 479

9-3 Functions and Their Graphs 483

9-4 Lines and Linear Functions 488

9-5 The Slope of a Line 492

9-6 The Slope-Intercept Form of a
   Linear Equation 499

9-7 Graphing a Linear Inequality
   in the Plane 507

Chapter Summary 511

Chapter Review Exercises 512

Problems for Pleasure 515

Calculator Challenge 515

College Test Preparation 516

Spiral Review Exercises 518

Solutions to TRY IT! Problems 520

SELF-TEST   Chapters 1–9 523

The 17th-century mathematician René Descartes devised a scheme to connect the numbers of Algebra and the points of Geometry. His scheme makes it possible to plot a set of points to display a graph of an open sentence. In this chapter, you will learn about graphing in the plane, with the focus on straight-line graphs and the open sentences that produce them.

9-1 GRAPHING IN A PLANE

Ordered Number Pairs

A part of a map of Watertown shows the tourist information center located at O, the intersection of Main Street and Broadway. From O, a tourist can be directed to any one of four points of interest—A, B, C, or D—by a pair of instructions. For example, to point A, the instructions might be, "travel east 1 block and then north 3 blocks," written (east 1, north 3).

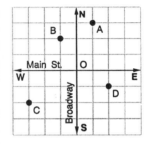

To show the direction of travel, distances east of Broadway can be indicated by positive numbers and distances west of Broadway by negative numbers. Also, distances north of Main Street can be indicated by positive numbers and distances south of Main Street by negative numbers. Therefore, the instructions (east 1, north 3) can be represented by the number pair $(+1, +3)$, or simply $(1, 3)$.

In all such number pairs, the first number always represents a distance east or west of Broadway, and the second a distance north or south of Main. For this reason $(1, 3)$ is called an **ordered number pair**. Be careful not to interchange the numbers because the resulting ordered number pair $(3, 1)$ would represent a different point.

Now it's your turn to . . . **TRY IT!**

For the points B, C, and D on the map of Watertown, write as ordered pairs the directions from point O, using:

a. the words east/west, north/south   **b.** signed numbers

(See page 520 for solutions.)

Points in a Plane

The method used to describe points on a map can be extended to describe points in a plane. Start with two number lines, called **coordinate axes**, drawn at right angles to each other. The horizontal line is the **x-axis,** the vertical line is the **y-axis,** and the point O at which the two axes intersect is the ***origin***. Label each axis with a scale, letting the origin be zero on each scale. Positive numbers are to the right of zero on the x-axis, and above zero on the y-axis. The axes divide the plane into 4 regions called ***quadrants***, which are numbered I, II, III, and IV in a counterclockwise order.

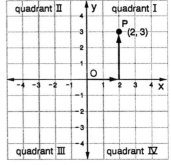

To locate any point in the plane, start at the origin. Point P in the diagram can be located by starting at the origin, moving 2 units to the right along the x-axis, then moving 3 units upward in a direction parallel to the y-axis. Thus, the location of point P can be described by the ordered pair (2, 3).

In general, the location of a point in the plane is described by an ordered pair of numbers (x, y). The first number is the ***x-coordinate***, or ***abscissa***. The second number is the ***y-coordinate***, or ***ordinate***. Together, the abscissa and the ordinate are called the ***coordinates*** of the point.

Now it's your turn to . . . **TRY IT!**   *(See page 520 for solution.)*

On graph paper, draw a set of coordinate axes and graph the point associated with the given ordered number pair. Label each point with its letter.

$A(7, 1)$ $B(3, -2)$ $C(1, 0)$ $D(-5, -3)$ $E(-4, 5)$ $F(0, 7)$ $G(-7, 0)$ $H(0, -2)$

Transformations

A ***transformation*** in a plane changes the positions of points in the plane. The point resulting from a transformation is called the ***image*** of the original point.

The following diagrams show the image of point $P(1, 3)$ under a ***reflection,*** a transformation that produces a "mirror image."

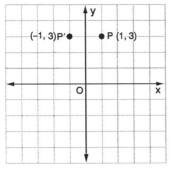

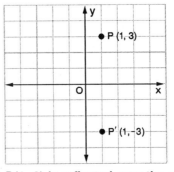

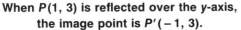

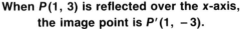

When $P(1, 3)$ is reflected over the y-axis, the image point is $P'(-1, 3)$.   **When $P(1, 3)$ is reflected over the x-axis, the image point is $P'(1, -3)$.**

Another common transformation "slides" a point to a new position in the plane. The diagram at the right shows the image of $P(1, 3)$ under a **translation** that moves P 4 units right and 2 units up to $P'(5, 5)$.

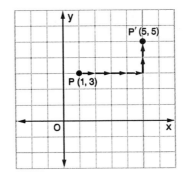

Now it's your turn to . . . **TRY IT!**

1. Write the coordinates of the image point, P', when $P(-2, 5)$ is reflected over:
   a. the y-axis **b.** the x-axis
2. Write the coordinates of the image point, A', when $A(-1, 3)$ is translated:
   a. 3 units right and 4 units up **b.** 2 units left and 3 units down

(See page 520 for solutions.)

Finding Distance and Area

Consider the points $A(1, 3)$ and $B(6, 3)$. Observe that since these points have the same y-value, they lie on a horizontal line segment. To determine the distance from A to B, count boxes or find the absolute value of the difference of the x-values.

Consider the points $M(2, 3)$ and $N(2, -1)$. Observe that since these points have the same x-value, they lie on a vertical line segment. To determine the distance from M to N, count boxes or find the absolute value of the difference of the y-values.

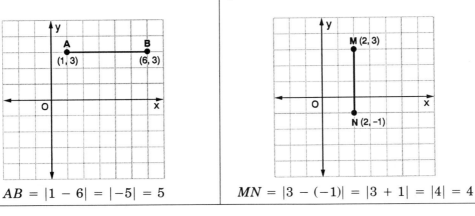

$AB = |1 - 6| = |-5| = 5$

$MN = |3 - (-1)| = |3 + 1| = |4| = 4$

Find the area of $\triangle ABC$ whose vertices are $A(1, 5)$, $B(4, 2)$, and $C(-4, 2)$.

Solution: Use graph paper to draw a diagram.

The formula for the area of a triangle requires that you know the lengths of one side of the triangle and the altitude to that side.

$$A_{\triangle} = \tfrac{1}{2}bh$$

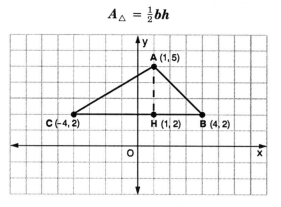

The graph shows that one side of the triangle, $\overline{BC}$, is a horizontal segment whose length can be found by working with the x-values.

$$BC = |4 - (-4)| = |4 + 4| = |8| = 8$$

Since $\overline{BC}$ is horizontal, the altitude to it is vertical. To find the length of the vertical altitude, $\overline{AH}$, first determine the coordinates of H. From the graph, observe $H(1, 2)$.

$$AH = |5 - 2| = |3| = 3$$

Substitute into the area formula.

$$A_{\triangle} = \tfrac{1}{2}bh$$

$$A_{\triangle ABC} = \tfrac{1}{2}(BC)(AH)$$

$$A_{\triangle ABC} = \tfrac{1}{2}(8)(3) = 12 \text{ square units} \quad \textbf{ANS.}$$

Now it's your turn to . . . **TRY IT!**

Find the area of $\triangle ABC$ whose vertices are $A(-2, 2)$, $B(4, 5)$, and $C(4, -4)$.

*(See page **520** for solution.)*

1. Write as ordered number pairs the coordinates of points
   $A, B, C, D, E, F, G, H,$ and O.

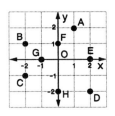

2. Draw a pair of coordinate axes on a sheet of graph paper and graph the point associated with the ordered number pair.

$A(5, 4)$	$B(-3, 2)$	$C(2, -6)$	$D(-4, -5)$
$E\left(2\frac{1}{4}, -3\frac{3}{4}\right)$	$F(-8, 5)$	$G(-1.5, -2.5)$	$H(5, 0)$
$J(-3, 0)$	$K(-10, 0)$	$L\left(3\frac{1}{3}, 0\right)$	$M(0, 4)$
$N(0, -6)$	$P(0, 1)$	$Q(0, -4)$	$R(|2|, |4|)$
$S(|-5|, |3|)$	$T(|1|, |-2|)$	$U(|-3|, |-9|)$	$V(-5, |-2|)$

   In 3–7, name the quadrant in which the graph of the given point appears.

3. $(5, 7)$ 4. $(-3, -2)$ 5. $(-7, 4)$ 6. $(1, -3)$ 7. $(|-2|, |-3|)$

8. Graph several points on the x-axis. What is the value of the ordinate for every point in the set of points on the x-axis?

9. Graph several points on the y-axis. What is the value of the abscissa for every point in the set of points on the y-axis?

10. What are the coordinates of the origin on the coordinate axes?

11. The coordinates of point P are (x, y). Name the set of numbers of which the abscissa must be a member, and the set of which the ordinate must be a member if the graph of P is in:
    a. quadrant I   b. quadrant II   c. quadrant III   d. quadrant IV

    In 12–15, name the quadrant in which the graph of the given point P lies when:
    a. $x > 0$ and $y > 0$ b. $x > 0$ and $y < 0$
    c. $x < 0$ and $y > 0$ d. $x < 0$ and $y < 0$

12. $P(x, y)$ 13. $P(|x|, |y|)$ 14. $P\left(\frac{1}{x}, \frac{1}{y}\right)$ 15. $P\left(\frac{x}{y}, \frac{y}{x}\right)$

    In 16–19, write the coordinates of the image of the given point under a reflection:
    a. in the y-axis   b. in the x-axis

16. $A(4, 3)$ 17. $B(-1, 2)$ 18. $C(-2, -3)$ 19. $D(1, -5)$

20. If $P(x, y)$ represents a point in the coordinate plane, write the coordinates of the image of P under a reflection:
    a. in the y-axis   b. in the x-axis

In 21–24, write the coordinates of the image of the given point under a translation of:
a. 3 units right and 2 units down
b. 2 units left and 5 units up

21. $A(-6, 8)$ **22.** $B(2, 1)$ **23.** $C(0, -5)$ **24.** $D(-4, -1)$

25. If $P(x, y)$ represents a point in the coordinate plane, write the coordinates of the image of P under a translation of a units in the horizontal direction and b units in the vertical direction.

In 26–38: **a.** Graph the given points and connect them, in alphabetical order, returning to the first point, to form a polygon.
          b. Find the area of the polygon formed.

26. $A(3, 0)$ $B(3, 7)$ $C(7, 7)$ $D(7, 0)$

27. $A(0, 5)$ $B(3, 5)$ $C(3, -2)$ $D(0, -2)$

28. $A(-2, -3)$ $B(6, -3)$ $C(6, 4)$ $D(-2, 4)$

29. $A(0, 8)$ $B(6, 0)$ $C(0, 0)$

30. $A(0, 0)$ $B(-9, 0)$ $C(0, -8)$

31. $A(4, 0)$ $B(4, 6)$ $C(-7, 0)$

32. $A(-5, 0)$ $B(-2, 7)$ $C(6, 0)$

33. $A(1, -3)$ $B(4, 5)$ $C(-7, 5)$

34. $A(-4, 3)$ $B(5, 8)$ $C(8, 3)$

35. $A(0, 0)$ $B(7, 0)$ $C(10, 6)$ $D(3, 6)$

36. $A(-8, 0)$ $B(-4, 5)$ $C(9, 5)$ $D(5, 0)$

37. $A(0, 5)$ $B(4, 2)$ $C(4, -7)$ $D(0, -4)$

38. $A(-6, 0)$ $B(6, 4)$ $C(4, 0)$ *Hint:* Draw a vertical altitude from B to the line through $\overline{AC}$.

In 39–42: **a.** Graph the given points and connect them, in alphabetical order, returning to the first point, to form a polygon.
          b. Find the area of the polygon formed.
          Hint: Think of the entire polygon as the sum of simpler polygons.

39. $A(0, 0)$ $B(10, 0)$ $C(5, 8)$ $D(0, 8)$

40. $A(0, 0)$ $B(0, 4)$ $C(3, 8)$ $D(6, 4)$ $E(6, 0)$

41. $A(0, 0)$ $B(12, 0)$ $C(12, 9)$ $D(8, 5)$ $E(4, 5)$ $F(0, 9)$

42. $A(-8, 0)$ $B(-6, 5)$ $C(6, 5)$ $D(8, 0)$ $E(6, -5)$ $F(-6, -5)$

9-2 RELATIONS AND THEIR GRAPHS

A set of ordered pairs is called a *relation*. Often, the rule of the relation is written as an open sentence, for which a solution set of ordered number pairs can be written.

For example, $P = 4s$ relates the perimeter P of a square to the length of a side s. Some ordered number pairs, like (1, 4) and (2, 8), that satisfy the formula are shown in the table below.

s	0	1	2	2.5	3
P	0	4	8	10	12

Observe that since P is a constant multiple of s, this relation is one of direct variation; P varies directly as s.

To show the graph of this relation, begin by plotting the points shown in the table. Then note that other points could be added to the graph, for example, (4, 16) and (.5, 2). In fact, not only can points be added after the points shown, there are infinitely many points that can be added between any two points shown. Thus, to complete the drawing of the graph, connect the points plotted to indicate all the points between, and extend the line with an arrowhead to indicate that it continues without end.

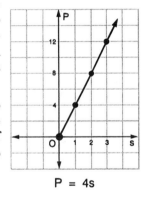

$P = 4s$

Note that the replacement set, or *domain*, for s is limited to nonnegative numbers because s represents a length. Since the domain is restricted, the corresponding values of P—the *range* of the relation—are, in turn, limited. In this case, the range is also the nonnegatives.

MODEL PROBLEMS

1. a. Write a rule to describe the relation that is the set of ordered number pairs shown in the table.
  b. What is the domain of the relation?
  c. What is the range?

x	1	3	0	-2	-10
y	6	18	0	-12	-60

ANSWER: **a.** $y = 6x$
  b. $\{1, 3, 0, -2, -10\}$
  c. $\{6, 18, 0, -12, -60\}$

2. Construct a table of values for the solution set of the equation $y - 2x = 3$ if the replacement set for x is $\{-2, -1, 0, 1\}$, and draw a graph.

Solution: You could begin by substituting each value of x into the equation in its given form to calculate the corresponding y-value.

A more efficient method is to first express y as the subject of the sentence.

$$y - 2x = 3$$
$$y - 2x + 2x = 3 + 2x$$
$$y = 2x + 3$$

ANSWER:

x	2x + 3	y
−2	2(−2) + 3	−1
−1	2(−1) + 3	1
0	2(0) + 3	3
1	2(1) + 3	5

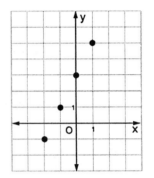

NOTE. Since the replacement set for x has exactly 4 values, the graph consists of exactly 4 distinct points.

3. Construct a table of values for the solution set of the equation $y = |x|$, and draw a graph of the relation.

Solution: Select several values of x, being sure to include some negatives and 0. Calculate the corresponding y-values.

Plot the guide points to discover the pattern of the graph. Connect the points to indicate that infinitely many points lying between also satisfy the rule of the relation. Extend the lines, including arrowheads.

ANSWER:

x	−3	−2	−1	0	1	2	3
y	3	2	1	0	1	2	3

Note that there are no restrictions on the replacement values for x, that is, the domain of this relation is the set of real numbers. However, the range of y is limited to non-negative numbers.

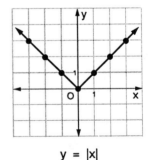

$y = |x|$

1. Write a rule to describe the relation that is the set of ordered number pairs shown in the table.

x	1	3	5	0	-3	-7
y	4	6	8	3	0	-4

2. Construct a table of values for the solution set of the equation $6x + 3y = 12$ if the replacement set for x is $\{-3, -2, -1, 0, 1, 2, 3\}$.

3. For the equation $y = x^2$:
   a. Construct a table of values for the solution set, being sure to include some negatives and 0 for x.
   b. Draw a graph of the relation.
   c. What is the range of y?

   (See page 520 for solutions.)

KEEP IN MIND

1. Every ordered pair of numbers that satisfies an equation represents the coordinates of a point on the graph of the equation.
2. Every point on the graph of an equation has as its coordinates an ordered pair of numbers that satisfies the equation.

EXERCISES

In 1–5, find the missing member in each ordered pair if the second member of the pair is twice the first number.

1. $\left(\frac{1}{2}, ?\right)$ **2.** $(0, ?)$ **3.** $(a, ?)$ **4.** $(?, 10)$ **5.** $(?, a)$

In 6–17, state whether the given ordered pair of numbers is a solution of the sentence.

6. $y = 4x$; $(16, 4)$ **7.** $y = 3x + 1$; $(7, 22)$ **8.** $2x + 3y = 13$; $(5, 1)$

9. $4x - 5y = 18$; $(7, 2)$ **10.** $y > 4x$; $(2, 10)$ **11.** $y < 2x + 3$; $(0, 2)$

12. $2x + 3y \leq 9$; $(0, 3)$ **13.** $3x = y + 4$; $(-7, -1)$ **14.** $x - 2y = 15$; $(1, -7)$

15. $y > 6x$; $(-1, -2)$ **16.** $y \geq 3 - 2x$; $(-1, 6)$ **17.** $5x - 2y \leq 19$; $(3, -2)$

18. Which of the ordered pairs of numbers $(8, -2)$, $(2, -6)$, $(7, 13)$, $(3, 9)$ is a member of the solution set of $x + y = 6$?

19. Which of the ordered pairs of numbers $(1, 8)$, $(5, 2)$, $(3, -1)$, $(0, -4)$ is *not* a member of the solution set of $y < 2x + 1$?

In 20–22, a point is to lie on the graph of the equation. Find its ordinate if its abscissa is the number indicated in the parentheses.

20. $x + 2y = 9$; (3) **21.** $4x - y = 7$; (−1) **22.** $2x + 3y = 5$; (−2)

In 23–30: **a.** Write an equation that describes the relation.
   b. Give four ordered pairs that are members of the relation.

23. y is 5 times x. **24.** y is 1 more than twice x.

25. The sum of x and y is 8. **26.** x decreased by y is 5.

27. Twice x increased by y is 8. **28.** y is twice the square of x.

29. y is less than one-half of x. **30.** y is greater than $3x + 1$.

In 31–38, write a rule to describe the relation.

31.

x	−3	0	1	2.4
y	−3	0	1	2.4

32.

x	−4	−1	0	1
y	4	1	0	−1

33.

a	−2	0	1	3
b	−6	0	3	9

34.

a	−6	0	1	10
b	−3	0	.5	5

35.

s	−3	0	1	2.5
t	4	7	8	9.5

36.

s	−1	0	1	5
t	−5	−4	−3	1

37.

m	−1	0	1	5
n	4	3	2	−2

38.

m	−2	1	2	3
n	−6	12	6	4

In 39–44, construct a table of values for the solution set of the given equation if the replacement set for x is $\{-2, -1, 0, 1, 2\}$, and draw a graph of the points.

39. $y = 3x + 1$ **40.** $x + y = 8$ **41.** $3y = 6x - 18$
42. $y + 2x = 8$ **43.** $4x - y = 6$ **44.** $6x + 2y = 8$

In 45–59, for the given equation:
  a. Construct a table of values for the solution set, being sure to include some negatives and 0 for x.
  b. Draw a graph of the relation.
  c. State any limitations on the domain and range.

45. $y = x$ **46.** $y = -x$ **47.** $y = 2x + 4$ **48.** $x - y = 6$
49. $y - 4x = -2$ **50.** $4y - 2x = -8$ **51.** $y = x^2 + 2$ **52.** $y = -x^2$
53. $y = -x^2 + 2$ **54.** $y = 2|x|$ **55.** $y = |x + 1|$ **56.** $y = -|x|$
57. $y = |x - 1|$ **58.** $xy = 12$ **59.** $xy = -12$

9-3 FUNCTIONS AND THEIR GRAPHS

A set of ordered pairs, or a *relation*, which you have already seen expressed in tabular form, can also be expressed in a **mapping diagram**, which shows the pairings between the elements of the domain and range.

Ordered Pairs

$\{(a, b), (c, d), (e, f)\}$

Mapping Diagram

Tabular Form

Abscissa	a	c	e
Ordinate	b	d	f

When a relation is mapped so that each element of the domain is paired with *exactly one* element of the range, that relation is called a **function.**

Examine each of the following relations to determine which is a function.

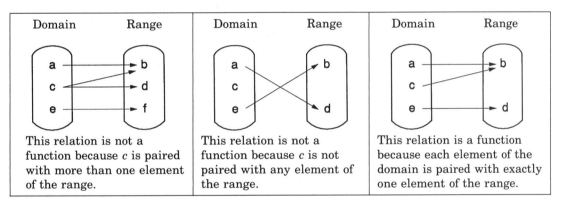

This relation is not a function because c is paired with more than one element of the range.

This relation is not a function because c is not paired with any element of the range.

This relation is a function because each element of the domain is paired with exactly one element of the range.

The graph of a relation can also help you to determine whether or not that relation is a function, by observing how vertical lines pass through the points of the graph.

x	−2	−1	0	1	2
y	2	1	0	1	2

Since no vertical line passes through more than one point on this graph, it is true that each *x*-value is paired with exactly one *y*-value and, thus, this relation is a function.

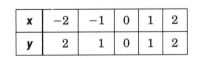

x	0	1	1	2	2
y	0	1	−1	2	−2

Since there are vertical lines that pass through more than one point on this graph, it is not true that each *x*-value is paired with exactly one *y*-value and, thus, this relation is not a function.

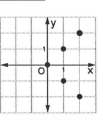

MODEL PROBLEMS

1. Determine whether the given relation is a function:

   a. $\{(-1, 4), (2, 4), (3, 4), (4, 4)\}$ **b.** $\{(2, -1), (2, 0), (2, 1), (2, 2)\}$

   Solution:

   a. The first number of every ordered pair in the relation is associated with a unique second number. Therefore, the relation is a function.

   b. In the ordered pairs, the first number, 2, is associated with different second numbers. Therefore, the relation is not a function.

   In 2–4, for the relation whose graph is shown over the set of real numbers:
   a. State the domain.   **b.** State the range.
   c. Use the vertical-line test to tell whether the relation is a function.

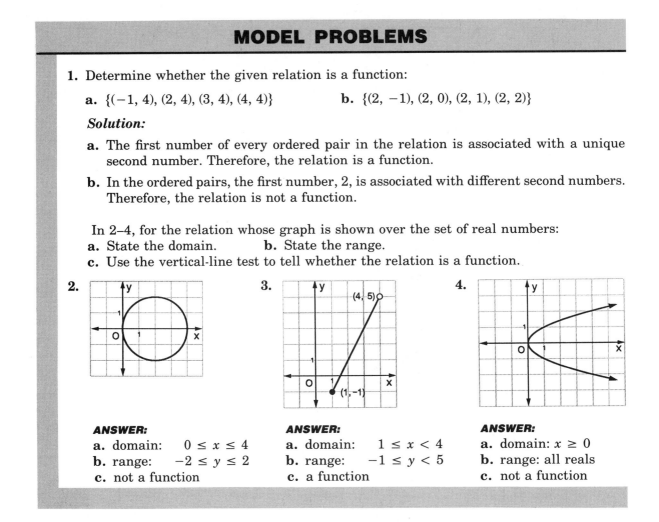

2.

ANSWER:
a. domain: $0 \le x \le 4$
b. range: $-2 \le y \le 2$
c. not a function

3.

ANSWER:
a. domain: $1 \le x < 4$
b. range: $-1 \le y < 5$
c. a function

4.

ANSWER:
a. domain: $x \ge 0$
b. range: all reals
c. not a function

Function Notation

A function can be represented by a letter such as f, g, F, or G. Two forms of notation then specify the pairing, or mapping, between the elements of the domain and range. The first form uses arrow notation to indicate the mapping. For example:

$$f: x \rightarrow 4x \text{ is read "the function } f \text{ that assigns } 4x \text{ to } x\text{"}$$

The second form is useful to indicate specific values. For example:

$$f(x) = 4x \text{ is read "} f \text{ at } x \text{ equals } 4x\text{" or "} f \text{ of } x \text{ equals } 4x\text{"}$$

Then to find the number that is assigned when x is 1, replace x in the notation by 1, and carry out the calculation specified by the rule of the function: $f(1) = 4(1) = 4$
Likewise $f(0) = 4(0) = 0$, and $f\left(\frac{1}{4}\right) = 4\left(\frac{1}{4}\right) = 1$.

1. If g is the function defined by $g(x) = 5x + 3$ for each real number x, find:

   a. $g(-1)$ **b.** $g(|-1|)$ **c.** $g(2b)$ **d.** $g(r + 5)$

   Solution: **ANSWERS:**

   a. $g(-1) = 5(-1) + 3 = -5 + 3 = -2$ $g(-1) = -2$

   b. Since $|-1| = 1$, $g(|-1|) = 5(1) + 3 = 5 + 3 = 8$ $g(|-1|) = 8$

   c. $g(2b) = 5(2b) + 3 = 10b + 3$ $g(2b) = 10b + 3$

   d. $g(r + 5) = 5(r + 5) + 3 = 5r + 25 + 3 = 5r + 28$ $g(r + 5) = 5r + 28$

2. If h is the function defined by $h(x) = x^2 - 2$ for each real number x, find:

   a. $h(2)$ **b.** $h\left(-\frac{1}{2}\right)$ **c.** $h(0)$ **d.** $h(b - 1)$

   Solution: **ANSWERS:**

   a. $h(2) = (2)^2 - 2 = 4 - 2 = 2$ $h(2) = 2$

   b. $h\left(-\frac{1}{2}\right) = \left(-\frac{1}{2}\right)^2 - 2 = \frac{1}{4} - 2 = -1\frac{3}{4}$ $h\left(-\frac{1}{2}\right) = -1\frac{3}{4}$

   c. $h(0) = (0)^2 - 2 = 0 - 2 = -2$ $h(0) = -2$

   d. $h(b - 1) = (b - 1)^2 - 2 = b^2 - 2b + 1 - 2$

                        $= b^2 - 2b - 1$ $h(b - 1) = b^2 - 2b - 1$

Now it's your turn to . . . **TRY IT!**   *(See page 520 for solutions.)*

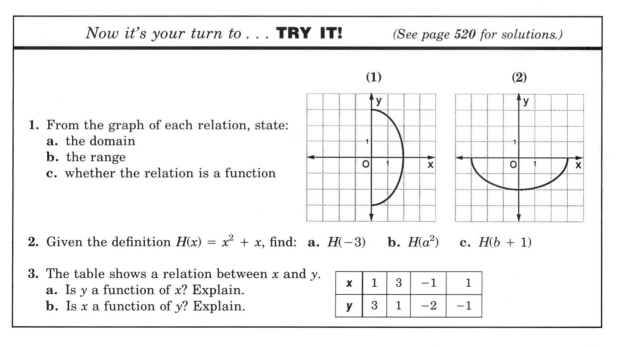

(1)   (2)

1. From the graph of each relation, state:
   a. the domain
   b. the range
   c. whether the relation is a function

2. Given the definition $H(x) = x^2 + x$, find: **a.** $H(-3)$ **b.** $H(a^2)$ **c.** $H(b + 1)$

3. The table shows a relation between x and y.
   a. Is y a function of x? Explain.
   b. Is x a function of y? Explain.

x	1	3	−1	1
y	3	1	−2	−1

In 1–8, state whether the relation is a function. Give the reason for your answer.

1. $\{(1, 3), (2, 6), (3, 9), (4, 12)\}$

2. $\{(4, 9), (5, 11), (6, 13), (7, 15)\}$

3. $\{(\frac{1}{4}, \frac{1}{2}), (\frac{1}{4}, -\frac{1}{2}), (49, 7), (49, -7)\}$

4. $\{(2, 5), (3, 10), (4, 17), (5, 26)\}$

5. $\{(-1, 3), (-2, 4), (-1, 2), (-2, 5)\}$

6. $\{(2, 1), (1, 2), (3, 4), (4, 3)\}$

7. $\{(2, 5), (3, 5), (4, 5), (5, 5)\}$

8. $\{(5, 2), (5, 3), (5, 4), (5, 5)\}$

In 9–16: **a.** Read the mapping diagram to tell if the relation is a function; explain.
  b. Write the coordinates of the set of ordered pairs.

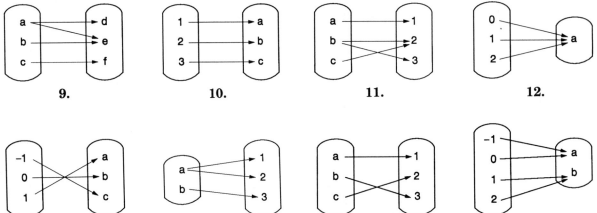

9.   **10.**   **11.**   **12.**

13.   **14.**   **15.**   **16.**

In 17–22, the table describes a relation between the variables x and y.
a. Determine if y is a function of x.
b. Determine if x is a function of y.

17.

x	-2	-1	0	1	2
y	-2	-1	0	1	2

18.

x	-2	-1	0	1	2
y	2	1	0	1	2

19.

x	-2	-1	0	1	2
y	2	1	0	-1	-2

20.

x	0	1	1	2	2
y	0	1	-1	2	-2

21.

x	-2	-1	0	1	2
y	4	1	0	1	4

22.

x	0	1	1	4	4
y	0	1	-1	2	-2

In 23–34, for the relation whose graph is shown over the set of real numbers:
a. State the domain.   **b.** State the range.
c. Use the vertical-line test to tell whether the relation is a function.

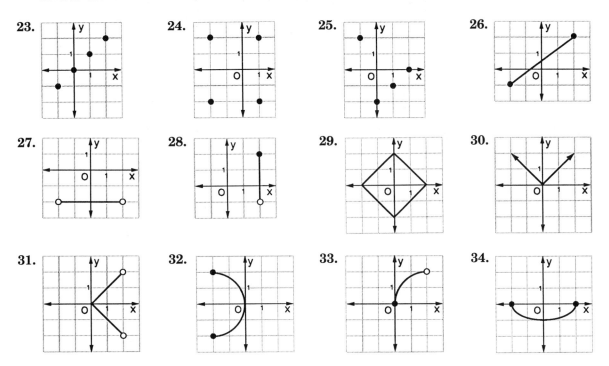

23.   **24.**   **25.**   **26.**

27.   **28.**   **29.**   **30.**

31.   **32.**   **33.**   **34.**

In 35–38, write the coordinates of 3 ordered pairs that are in the set defined by the function over the real numbers.

35. $f: x \rightarrow 2x$ **36.** $g: y \rightarrow \frac{1}{2}y$ **37.** $h: a \rightarrow a + 2$ **38.** $j: x \rightarrow x^2$

In 39–44, the domain of the variable is the set of real numbers. For the function defined, compute the value at: **a.** -2 **b.** 0 **c.** 2 **d.** $|-3|$ **e.** $\frac{1}{3}$

39. $f(y) = 3y - 1$ **40.** $h(x) = x^2 - 2x$ **41.** $F(m) = 2m^2 - 4m + 3$

42. $t(a) = a^3 + 2a^2 - 5$ **43.** $g(r) = -r^2 + 8$ **44.** $f(x) = \dfrac{1}{x}$

45. For the function f defined by $f(x) = x^2 - 2x + 4$ for each real number x, find:
    a. $f(1)$ **b.** $f(-2)$ **c.** $f\left(\frac{3}{5}\right)$ **d.** $f(0)$
    e. $f(.1)$ **f.** $f(|-4|)$ **g.** $2[f(3)]$ **h.** $2[f(2)] - f(1)$
    i. $f(b)$ **j.** $f(3a)$ **k.** $f(b + 3)$ **l.** $f(2b - 1)$

46. For the function F defined by $F(x) = x^2 - 2x$ for each real number x, find:
    a. $F(-2)$ **b.** $-F(-2)$ **c.** $F(-2) - 2$ **d.** $2 - F(-2)$

    e. $F(|-2|)$ **f.** $-[F(|-2|)]$ **g.** $\dfrac{2}{F(|-2|)}$ **h.** $F(-2 + |-2|)$

9-4 LINES AND LINEAR FUNCTIONS

Among the graphs of functions you have seen, some have been straight lines. A function whose graph is a straight line is called a *linear function.* Careful examination of the equations that name straight-line graphs results in the conclusion that the equations of all linear functions can be produced from the following *first-degree* general equation, written here in *standard form:*

$$ax + by = c \text{ where } a, b, \text{ and } c \text{ are real numbers}$$

Depending on whether either of the constants a or b is 0, the line is one of 3 types. That is, the line may be:

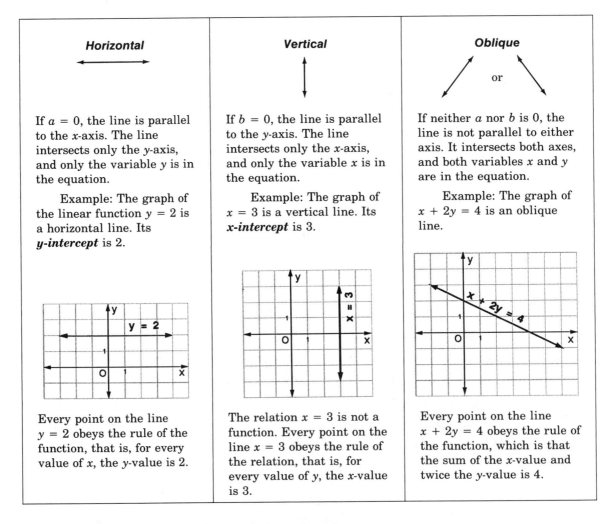

Horizontal	*Vertical*	*Oblique*
If $a = 0$, the line is parallel to the x-axis. The line intersects only the y-axis, and only the variable y is in the equation.	If $b = 0$, the line is parallel to the y-axis. The line intersects only the x-axis, and only the variable x is in the equation.	If neither a nor b is 0, the line is not parallel to either axis. It intersects both axes, and both variables x and y are in the equation.
Example: The graph of the linear function $y = 2$ is a horizontal line. Its *y-intercept* is 2.	Example: The graph of $x = 3$ is a vertical line. Its *x-intercept* is 3.	Example: The graph of $x + 2y = 4$ is an oblique line.
Every point on the line $y = 2$ obeys the rule of the function, that is, for every value of x, the y-value is 2.	The relation $x = 3$ is not a function. Every point on the line $x = 3$ obeys the rule of the relation, that is, for every value of y, the x-value is 3.	Every point on the line $x + 2y = 4$ obeys the rule of the function, which is that the sum of the x-value and twice the y-value is 4.

A line may be graphed using its intercepts.

PROCEDURE

To graph a vertical line $x = a$ or a horizontal line $y = b$:

1. Read the intercept from the equation.
2. Plot the intercept on the appropriate axis.
3. Draw the line through the intercept point, $(a, 0)$ or $(0, b)$.

Include arrowheads to indicate that the line extends without end in both directions, and include the equation of the line.

Now it's your turn to . . . **TRY IT!**   *(See page 521 for solutions.)*

In 1–3, draw the graph of the line.

1. $x = -4$ **2.** $y = 5$ **3.** $3x = 12$

To graph an oblique line from its intercepts, you must first determine the values of the intercepts.

MODEL PROBLEM

Graph the line $2x - y = 4$ by the intercept method.

Solution:

Find the x-intercept by substituting 0 for y; find the y-intercept by substituting 0 for x.

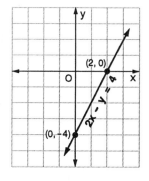

x-intercept	y-intercept
$2x - y = 4$	$2x - y = 4$
$2x - 0 = 4$	$2(0) - y = 4$
$2x = 4$	$-y = 4$
$x = 2$	$y = -4$

Plot the intercepts and draw a line through the intercept points $(2, 0)$ and $(0, -4)$.

Check: Choose any other point on the line, and verify that its coordinates satisfy the equation. Try the point $(1, -2)$.

$$2x - y = 4$$
$$2(1) - (-2) \stackrel{?}{=} 4$$
$$2 + 2 \stackrel{?}{=} 4$$
$$4 = 4 ✓$$

1. The x-intercept of a line is the x-coordinate of the point at which the line intersects the x-axis.

2. An equation for a vertical line, with x-intercept a, is $x = a$.

3. The y-intercept of a line is the y-coordinate of the point at which the line intersects the y-axis.

4. An equation for a horizontal line, with y-intercept b, is $y = b$.

5. For an oblique line:

   to find the x-intercept, let $y = 0$
   to find the y-intercept, let $x = 0$

EXERCISES

In 1–10, draw the graph of the equation.

1. $x = 6$ **2.** $x = \frac{2}{3}$ **3.** $x = 0$ **4.** $x = -3$ **5.** $x = -2$

6. $y = 4$ **7.** $y = 2\frac{1}{4}$ **8.** $y = 0$ **9.** $y = -4$ **10.** $y = -3$

11. Write an equation of the line that is parallel to the x-axis and whose y-intercept is:
   a. 1   **b.** 5   **c.** -4 **d.** -8 **e.** -2.5

12. Write an equation of the line that is parallel to the y-axis and whose x-intercept is:
   a. 3   **b.** 10   **c.** $4\frac{1}{2}$ **d.** -6 **e.** -10

13. Which statement is true about the graph of the equation $y = 6$?
   (1) It is parallel to the y-axis.   (2) It is parallel to the x-axis.
   (3) It passes through the origin.   (4) It has an x-intercept.

14. Which statement is true about the graph of the equation $x = 5$?
   (1) It passes through the origin.   (2) It is parallel to the x-axis.
   (3) It is parallel to the y-axis.   (4) It has a y-intercept.

15. Which statement is true about the graph of the equation $y = x$?
   (1) It is parallel to the x-axis.
   (2) It is parallel to the y-axis.
   (3) It passes through the point $(2, -2)$.
   (4) It passes through the origin.

In 16–21, find the x-intercept and y-intercept of the line that is the graph of the equation.

16. $x + y = 8$ **17.** $x - 5y = 10$ **18.** $y = 4x + 12$

19. $y = 6x$ **20.** $4x - 3y = -12$ **21.** $5x + 3y = 8$

In 22–30, draw the graph of the equation by the intercepts method.

22. $x + y = 3$ **23.** $2x + y = 8$ **24.** $3x - y = -3$

25. $y = 2x + 4$ **26.** $y = -3x + 6$ **27.** $x = 4y - 4$

28. $2x + 3y = 6$ **29.** $5x - 4y = 20$ **30.** $3x - 6y = 9$

31. a. Find the x-intercept and y-intercept for the graph of the equation $y = 3x$.
   b. Explain why you cannot graph the equation $y = 3x$ by the intercepts method.
   c. Draw the graph of $y = 3x$ by constructing a table of values.

32. The linear function $3x + 5y = 300$ models a transportation infrastructure program for Diehl City.

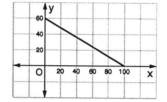

   The scale for the x-axis, measured in miles, represents new subway lines.

   The scale for the y-axis, measured in miles, represents new highways.

   a. How many miles of new subway lines would there be if no money is spent for highways from this budget?
   b. How many miles of new highways would there be if all the money in this budget is used for highways?
   c. Estimate the number of miles of new subway lines and new highways that can be built if the budget is divided about equally between the two projects.

33. The linear function $y = .4x + 90$ models how the resistance in a wire varies with temperature.

   The scale for resistance, measured in ohms, is on the y-axis.

   The scale for temperature, measured in degrees Celsius, is on the x-axis.

   a. Find the resistance when the temperature is $0°C$.
   b. Find the temperature if the resistance were 0 ohms.
   c. Graph the linear function.

34. A dieter, eating only specially prepared foods, has a daily intake of 1,200 calories. The linear function $3x + 5y = 120$ models the diet plan.

   The variable x, measured in ounces, represents liquid food that has 30 calories per ounce.

   The variable y, measured in ounces, represents solid food that has 50 calories per ounce.

   a. Find the value of the x-intercept and interpret its meaning.
   b. Find the value of the y-intercept and interpret its meaning.
   c. If the dieter consumed 15 ounces of liquid food, what is the amount of solid food he can have this day?

Finding the Slope of a Line

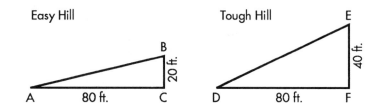

It is more difficult to hike up Tough Hill than it is to hike up Easy Hill since Tough Hill rises 40 ft. vertically over a horizontal distance of 80 ft., whereas Easy Hill rises only 20 ft. vertically over the same horizontal distance of 80 ft. Therefore, Tough Hill is *steeper* than Easy Hill. To compare the steepness of the roads AB and DE that lead up the two hills, compare their **slopes**.

The slope of road AB is the ratio of the change in vertical distance, CB, to the change in horizontal distance, AC:

$$\text{slope of road } AB = \frac{\text{change in vertical distance, } CB}{\text{change in horizontal distance, } AC} = \frac{20 \text{ ft.}}{80 \text{ ft.}} = \frac{1}{4}$$

Also:

$$\text{slope of road } DE = \frac{\text{change in vertical distance, } FE}{\text{change in horizontal distance, } DF} = \frac{40 \text{ ft.}}{80 \text{ ft.}} = \frac{1}{2}$$

This means that road DE rises $\frac{1}{2}$ ft. vertically for each 1 ft. of horizontal distance, whereas road AB rises only $\frac{1}{4}$ ft. vertically for each 1 ft. of horizontal distance and, thus, road DE is steeper than road AB.

In a similar way, you can find the slope of a line in the coordinate plane.

For example, in the figure:

$$\text{slope of } \overleftrightarrow{LM} = \frac{\text{vertical change}}{\text{horizontal change}} = \frac{4}{2} = 2$$

$$\text{slope of } \overleftrightarrow{RS} = \frac{\text{vertical change}}{\text{horizontal change}} = \frac{-2}{3} = -\frac{2}{3}$$

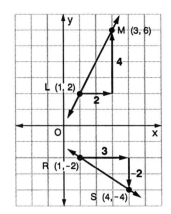

In general, the slope, m, of a line that passes through two points $P_1(x_1, y_1)$ and $P_2(x_2, y_2)$ is the ratio of the difference of the y-values of these points to the difference of the corresponding x-values. Thus:

$$\text{slope of line} = \frac{\text{difference in } y\text{-values}}{\text{difference in } x\text{-values}}$$

$$m = \frac{y_2 - y_1}{x_2 - x_1}$$

The difference in x-values, $x_2 - x_1$, is often represented by Δx, read "delta x." Similarly, the difference in y-values, $y_2 - y_1$, is represented by Δy, read "delta y." Therefore:

$$\text{slope of line} = m = \frac{\Delta y}{\Delta x}$$

MODEL PROBLEM

Find the slope of the line that passes through the two points with coordinates:

a. (2, 3) and (5, 7)   **b.** (2, 7) and (5, 3)
c. (2, 3) and (5, 3)   **d.** (2, 3) and (2, 5)

Solution: Either point can be (x_1, y_1) and the other (x_2, y_2).

a. $(x_1, y_1) = (2, 3)$

$(x_2, y_2) = (5, 7)$

$$m = \frac{y_2 - y_1}{x_2 - x_1}$$

$$m = \frac{7 - 3}{5 - 2} = \frac{4}{3}$$

b. $(x_1, y_1) = (2, 7)$

$(x_2, y_2) = (5, 3)$

$$m = \frac{y_2 - y_1}{x_2 - x_1}$$

$$m = \frac{3 - 7}{5 - 2} = \frac{-4}{3}$$

c. $(x_1, y_1) = (2, 3)$

$(x_2, y_2) = (5, 3)$

$$m = \frac{y_2 - y_1}{x_2 - x_1}$$

$$m = \frac{3 - 3}{5 - 2} = \frac{0}{3} = 0$$

d. $(x_1, y_1) = (2, 3)$

$(x_2, y_2) = (2, 5)$

$$m = \frac{y_2 - y_1}{x_2 - x_1}$$

$$m = \frac{5 - 3}{2 - 2} = \frac{2}{0} \text{ is undefined}$$

Possible Values for Slope

Observe from the preceding Model Problem that it is possible for the value of the slope of a line to be positive, negative, 0, or undefined. Draw a graph for each line of the Model Problem to note the correspondence between the nature of the value of the slope and the nature of the line. In general, the following are true.

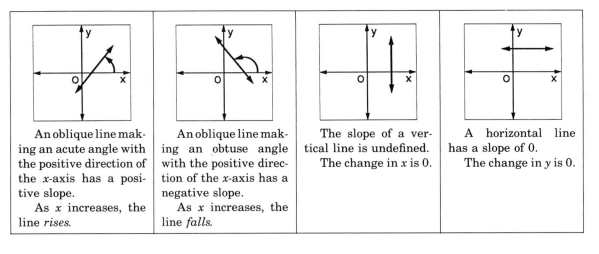

An oblique line making an acute angle with the positive direction of the x-axis has a positive slope. As x increases, the line *rises*.	An oblique line making an obtuse angle with the positive direction of the x-axis has a negative slope. As x increases, the line *falls*.	The slope of a vertical line is undefined. The change in x is 0.	A horizontal line has a slope of 0. The change in y is 0.

Now it's your turn to . . . **TRY IT!**   *(See page 521 for solutions.)*

Find the slope of the line that passes through the two points. Tell from the slope if the line is horizontal, vertical, or oblique.

1. (2, −1) and (2, −6)   **2.** (4, 7) and (−1, −2)

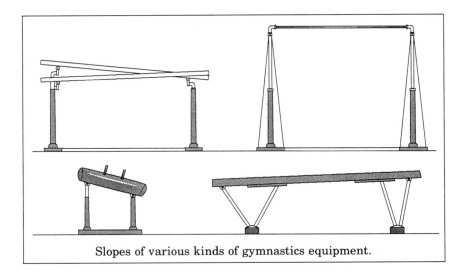

Slopes of various kinds of gymnastics equipment.

Parallel Lines

As you can see in the diagram, lines AB and CD appear to be parallel. For each of these lines, the slope is 2.

$$m = \frac{\text{vertical change}}{\text{horizontal change}} = \frac{2}{1} = 2$$

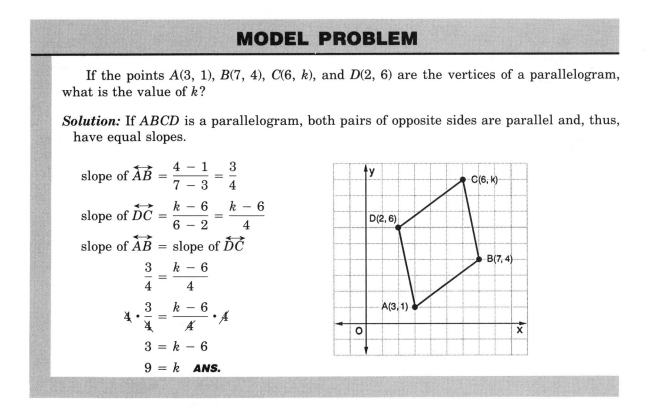

In general:

If two lines are parallel, then they have the same slope.

If two lines have the same slope, then they are parallel.

MODEL PROBLEM

If the points $A(3, 1)$, $B(7, 4)$, $C(6, k)$, and $D(2, 6)$ are the vertices of a parallelogram, what is the value of k?

Solution: If $ABCD$ is a parallelogram, both pairs of opposite sides are parallel and, thus, have equal slopes.

slope of $\overleftrightarrow{AB} = \dfrac{4 - 1}{7 - 3} = \dfrac{3}{4}$

slope of $\overleftrightarrow{DC} = \dfrac{k - 6}{6 - 2} = \dfrac{k - 6}{4}$

slope of $\overleftrightarrow{AB}$ = slope of $\overleftrightarrow{DC}$

$\dfrac{3}{4} = \dfrac{k - 6}{4}$

$4 \cdot \dfrac{3}{4} = \dfrac{k - 6}{4} \cdot 4$

$3 = k - 6$

$9 = k$ **ANS.**

Now it's your turn to . . . **TRY IT!**   *(See page 521 for solution.)*

Find the value of k if the line through $(k, 2)$ and $(4, 0)$ is parallel to the line through $(6, 2)$ and $(5, 4)$.

Collinear Points

Points that lie on the same line are called ***collinear***. Given three points A, B, and C, the points are collinear if the slope of the line through A and B is the same as either the slope of the line through B and C or the slope of the line through A and C.

MODEL PROBLEM

Determine if the points $A(-4, 3)$, $B(-2, 1)$, and $C(1, -3)$ are collinear.

Solution:

$$\text{slope of } \overleftrightarrow{AB} = \frac{1 - 3}{-2 - (-4)} = \frac{-2}{2} = -1$$

$$\text{slope of } \overleftrightarrow{BC} = \frac{-3 - 1}{1 - (-2)} = \frac{-4}{3}$$

Since the slope of $\overleftrightarrow{AB}$ is not equal to the slope of $\overleftrightarrow{BC}$, points A, B, and C are not collinear.

Now it's your turn to . . . **TRY IT!**

Find the value of k if the points $(1, k)$, $(-1, 2)$, and $(-3, -1)$ are collinear.

(See page 521 for solution.)

Falling dominoes show how a small event can trigger a catastrophe, similar to the old saying "the straw that broke the camel's back."

Scientists are working on a theory that may improve predictions about such catastrophes as avalanches, earthquakes, and crashes in the stock market.

This theory proposes that snow piles and other large complex systems—even economies and ecosystems—naturally evolve to a critical state.

Calculations with slope play an important role in this research.

1. In **a–f**, tell whether the line has a positive slope, a negative slope, a slope of zero, or no slope.

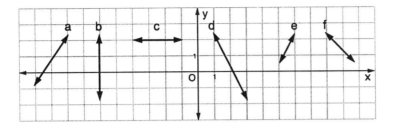

2. In **a–f**, find the slope of the line. If the line has no slope, indicate that fact.

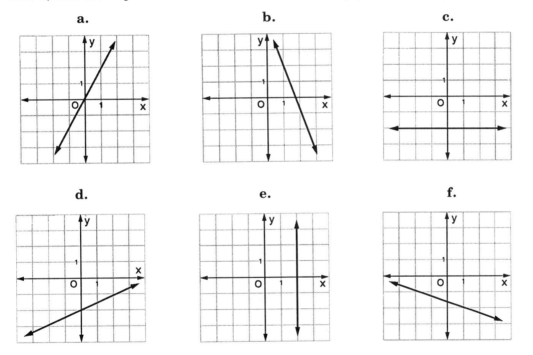

In 3–14, find the slope of the line that passes through the two points. Tell from the slope if the line is oblique, horizontal, or vertical. If the line is oblique, tell whether the angle it makes with the positive direction of the x-axis is acute or obtuse.

3. (0, 0) and (4, 4)

4. (0, 0) and (4, 8)

5. (0, 0) and (9, 3)

6. (0, 0) and (3, −6)

7. (1, 5) and (3, 9)

8. (5, 8) and (4, 3)

9. (7, 3) and (1, −1)

10. (−2, 4) and (0, 2)

11. (5, −2) and (7, −8)

12. (4, 2) and (8, 2)

13. (−1, 3) and (2, 3)

14. (6, −1) and (−2, −1)

15. Find the value of y so that the slope of the line passing through the points $(2, y)$ and $(6, 10)$ will be:

    a. 1   **b.** 2   **c.** $\frac{1}{2}$ **d.** 0

16. Find the value of k so that the line through the two points $(-2, 7)$ and $(3, k)$ is horizontal.

17. Find the value of k so that the line through the two points $(1, -3)$ and $(k, 5)$ is vertical.

18. Which pair of points will determine a line parallel to the x-axis?

    (1) $(1, 1), (2, 3)$ (2) $(1, 1), (3, 3)$ (3) $(2, 3), (2, 5)$ (4) $(2, 5), (4, 5)$

19. Which pair of points will determine an oblique line that makes an obtuse angle with the positive direction of the x-axis?

    (1) $(1, 1), (2, 2)$ (2) $(1, 1), (-2, -2)$ (3) $(1, 1), (-2, 2)$ (4) $(1, -1), (2, 2)$

In 20–25, coordinates are given for points A, B, C, and D. Use the definition of slope to determine whether the line through A and B is parallel to the line through C and D.

20. $A(0, 5), B(2, 6), C(6, 0), D(8, 1)$ **21.** $A(3, -2), B(-1, 0), C(-3, -5), D(1, -3)$

22. $A(4, 3), B(8, 3), C(-9, 7), D(4, 7)$ **23.** $A(-2, -1), B(-5, -4), C(5, 4), D(2, 1)$

24. $A(7, 0), B(4, 1), C(-2, 8), D(4, 6)$ **25.** $A(-6, 7), B(-3, 2), C(4, 1), D(9, -2)$

In 26 and 27, determine the value of k so that the given points are vertices of parallelogram $ABCD$.

26. $A(-2, 3), B(1, 2), C(3, 5), D(0, k)$ **27.** $A(-5, 4), B(1, 6), C(12, 1), D(k, -1)$

In 28–33, use slope to determine whether the points are collinear.

28. $(1, 3), (2, 5), (3, 7)$ **29.** $(-1, 5), (0, 2), (1, -1)$ **30.** $(2, 5), (4, 9), (6, 15)$

31. $(-4, -1), (0, 3), (2, 5)$ **32.** $(-7, -1), (-1, -7), (7, 1)$ **33.** $(3, -4), (1, -4), (-1, 4)$

34. The points $(2, -3)$, $(2, 3)$, and $(k, 0)$ are collinear. The value of k is

    (1) 0   (2) 1   (3) 2   (4) 4

35. The point $(4, -2)$ lies on a line whose slope is $\frac{3}{2}$. The coordinates of another point on this line may be

    (1) $(1, 0)$ (2) $(2, 1)$ (3) $(6, 1)$ (4) $(7, 0)$

9-6 THE SLOPE-INTERCEPT FORM OF A LINEAR EQUATION

For each of the following linear functions, study the graph and the corresponding equation. Observe how some numbers from the graph show up in the equation.

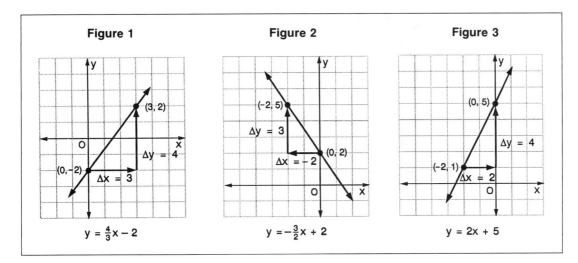

| Figure 1 | Figure 2 | Figure 3 |

$y = \frac{4}{3}x - 2$ $y = -\frac{3}{2}x + 2$ $y = 2x + 5$

Note that the slope of each line is the coefficient of the x-term in the equation and that the y-intercept of each line is the constant that follows the x-term in the equation.

	Equation	Slope $\left(\dfrac{\Delta y}{\Delta x}\right)$	y-intercept
In Figure 1:	$y = \frac{4}{3}x - 2$	$\frac{4}{3}$	-2
In Figure 2:	$y = -\frac{3}{2}x + 2$	$-\frac{3}{2}$	2
In Figure 3:	$y = 2x + 5$	2	5

In general, a linear function can be expressed in the form:

$$y = mx + b$$

where the slope of the line, m, is the coefficient of x, and the y-intercept is the constant b.

MODEL PROBLEM

Find the slope and y-intercept of the line that is the graph of $4x + 2y = 10$.

How to Proceed	*Solution*
(1) Transform the equation into an equivalent equation of the form $y = mx + b$ by solving for y in terms of x.	$4x + 2y = 10$ $2y = -4x + 10$
(2) The coefficient of x is the slope, and the constant term is the y-intercept.	$y = \boxed{-2}x \boxed{+5}$

slope   y-intercept

Writing an Equation of a Line

You can use the slope-intercept form to write an equation of a line when you know the slope and the y-intercept.

MODEL PROBLEMS

1. Write an equation of the line whose slope is $\frac{1}{2}$ and whose y-intercept is -4.

How to Proceed	*Solution*
(1) Write the slope-intercept formula.	$y = mx + b$
(2) Replace m by the numerical value of the slope; replace b by the numerical value of the y-intercept.	slope, $m = \frac{1}{2}$ y-intercept, $b = -4$ $y = \frac{1}{2}x + (-4)$
	or $y = \frac{1}{2}x - 4$ **ANS.**

You can use the slope-intercept form to write an equation of a line when you know the slope and one point.

2. Write an equation of the line whose slope is 4 and that passes through the point $(3, 5)$.

(1) In $y = mx + b$, replace m with the given slope, 4.	$y = mx + b$ $y = 4x + b$
(2) Since the given point $(3, 5)$ is on the line, its coordinates satisfy the equation $y = 4x + b$. Replace x with 3 and y with 5. Solve for b.	$5 = 4(3) + b$ $5 = 12 + b$ $-7 = b$
(3) In $y = 4x + b$, replace b with -7.	$y = 4x - 7$ **ANS.**

You can use the slope-intercept form to write an equation of a line when you know the coordinates of two points.

3. Write an equation of the line that passes through the points (2, 5) and (4, 11).

How to Proceed	***Solution***
(1) Find the slope of the line that passes through the two given points, (2, 5) and (4, 11).	$m = \dfrac{11 - 5}{4 - 2} = \dfrac{6}{2} = 3$
(2) In $y = mx + b$, replace m with the slope, 3.	$y = mx + b$ $y = 3x + b$
(3) Select one point that is on the line, for example (2, 5). Its coordinates must satisfy the equation $y = 3x + b$. Replace x with 2 and y with 5. Solve for b.	$5 = 3(2) + b$ $5 = 6 + b$ $-1 = b$
(4) In $y = 3x + b$, replace b with -1.	$y = 3x - 1$
(5) Check whether the coordinates of the second point (4, 11) satisfy the equation $y = 3x - 1$.	$11 \overset{?}{=} 3(4) - 1$ $11 = 11 ✔$

ANSWER: $y = 3x - 1$

***Now it's your turn to . . .* TRY IT!**

Write an equation of the line:

1. with slope 3 and y-intercept -2

2. with slope -4 and that passes through the point (2, -5)

3. that passes through the two points (5, 4) and (6, 2).

(See page 522 for solutions.)

Being able to write the equation of a line given certain data is important since many real situations can be modeled by a linear function.

MODEL PROBLEM

The 18th-century French scientist Jacques Charles was the first to notice that gases expand when heated and contract when cooled. From his data points, Charles observed that a linear relation exists between volume and temperature.

a. If a gas has a volume of 500 cc at 30°C and a volume of 600 cc at 90°C, write a linear equation to model the relation between volume and temperature of this gas.
b. Draw a graph of the linear function.
c. In theory, the volume of gas could contract until none remained. Estimate the coldest temperature possible to sustain a quantity of this gas.

Solution: To express the volume of the gas as a function of the temperature, $v = f(t)$, consider data points as ordered pairs of the form (t, v); that is, (30, 500) and (90, 600).

a. Apply the slope-intercept form of a line to the given variables.

$$y = mx + b$$
$$v = mt + b$$

Use the coordinates of the two given points to determine the value of the slope m.

$$m = \frac{600 - 500}{90 - 30} = \frac{100}{60} = \frac{5}{3}$$

Substitute the value of m and the coordinates of one point, (30, 500). Solve for b, the y-intercept.

$$v = mt + b$$
$$500 = \tfrac{5}{3}(30) + b$$
$$500 = 50 + b$$
$$450 = b$$

Write the linear function by substituting for m and b.

$$v = mt + b$$
$$v = \tfrac{5}{3}t + 450 \quad \textbf{ANS.}$$

b. Choose appropriate scales. Plot the points (30, 500) and (90, 600), and draw the line through them.

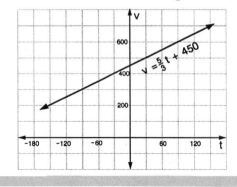

c. Find t when $v = 0$.

$$v = \frac{5}{3}t + 450$$
$$0 = \frac{5}{3}t + 450$$
$$-450 = \frac{5}{3}t$$
$$-450 \cdot \frac{3}{5} = \frac{3}{5} \cdot \frac{5}{3}t$$
$$-270 = t$$

ANS. When temperature is above $-270°C$, there is some quantity of this gas.

Graphing by the Slope-Intercept Method

So far in this chapter, you have seen how to graph a linear function by two methods:

   1. constructing a table of values
   2. plotting the *x*- and *y*-intercepts

Now, consider graphing a linear function by using the slope and *y*-intercept.

MODEL PROBLEM

Draw the graph of $2x + 3y = 9$ using the slope-intercept method.

How to Proceed	*Solution*

(1) Transform the equation into the form $y = mx + b$ to read the slope and *y*-intercept.

$$2x + 3y = 9$$
$$3y = -2x + 9$$
$$y = \boxed{\frac{-2}{3}}\, x \boxed{+\ 3}$$

slope   *y*-intercept

(2) Plot the *y*-intercept, point A.

(3) Use the slope to find two more points on the line. Since slope $= \dfrac{\Delta y}{\Delta x} \xrightarrow{=} \dfrac{-2}{3}$, when *x* changes 3, then *y* changes -2. Start at point A and move 3 units to the right and 2 units down to locate point B. Start at point B and repeat this procedure to locate point C.

(4) Draw the line that passes through the three points.

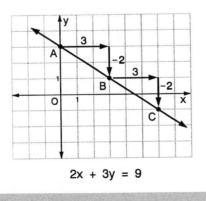

$$2x + 3y = 9$$

In 1–21, find the slope and y-intercept of the line that is the graph of the equation.

1. $y = 3x + 1$ **2.** $y = x - 3$ **3.** $y = 2x$

4. $y = x$ **5.** $y = \frac{1}{2}x + 5$ **6.** $y = \frac{3}{4}x - \frac{1}{2}$

7. $y = -2x + 3$ **8.** $y = -x - 4$ **9.** $y = -3x$

10. $y = -2$ **11.** $y = -\frac{2}{3}x + 4$ **12.** $y = -\frac{1}{2}x - 2$

13. $y - 3x = 7$ **14.** $2x + y = 5$ **15.** $3x - y = 4$

16. $3y = 6x + 9$ **17.** $2y = 5x - 4$ **18.** $2x + 3y = 6$

19. $\frac{1}{2}x + \frac{3}{4} = \frac{1}{3}y$ **20.** $4x - 3y = 0$ **21.** $2y = 5(x + 1)$

In 22–30, write an equation of the line whose slope and y-intercept are respectively:

22. 2 and 7   **23.** 3 and -5 **24.** -4 and 2

25. -1 and -3 **26.** -2 and 4   **27.** -3 and 0

28. $\frac{3}{5}$ and -2 **29.** $\frac{1}{2}$ and 0   **30.** $-\frac{4}{3}$ and $-\frac{1}{3}$

31. If the slope of a line is -3 and the y-intercept is 2, which is an equation for the line?
(1) $y - 3x + 2 = 0$ (2) $y + 3x = 2$ (3) $y = 2x - 3$ (4) $y = 3x - 2$

32. If the slope of a line is 2, which may be an equation for the line?
(1) $2y = 4x + 5$ (2) $\frac{1}{2}y = 4x$ (3) $y + 2x = 6$ (4) $y = 2$

33. If the y-intercept of a line is 4, which may be an equation for the line?
(1) $y = x - 4$ (2) $2y = x + 8$ (3) $4y = x$ (4) $y = 4x$

34. Write an equation of the line whose slope is 3 that has the same y-intercept as the line whose equation is $y = x - 5$.

35. Is the line whose equation is $2y = x + 8$ parallel to the line whose equation is $2y = x - 6$? Explain.

36. Write an equation of the line with y-intercept 4 that is parallel to the line whose equation is $y = 3x + 2$.

In 37–45, write an equation of the line that has the given slope, m, and that passes through the given point.

37. $m = 2, (1, 4)$ **38.** $m = 2, (-3, 4)$ **39.** $m = -1, (0, -2)$

40. $m = -3, (-2, -1)$ **41.** $m = \frac{1}{2}, (4, 2)$ **42.** $m = \frac{2}{3}, (-6, 4)$

43. $m = -\frac{3}{4}, (0, 0)$ **44.** $m = -\frac{5}{3}, (-3, 0)$ **45.** $m = 0, (3, -6)$

46. Write an equation of the line that is:
a. parallel to the line $y = 2x - 4$ and whose y-intercept is 7
b. parallel to the line $y - 3x = 6$ and whose y-intercept is -2
c. parallel to the line $2x + 3y = 12$ and that passes through the origin

47. Write an equation of the line that is:
    a. parallel to the line $y = 4x + 1$ and that passes through the point $(2, 3)$
    b. parallel to the line $y - 3x = 5$ and that passes through the point $(-1, 6)$
    c. parallel to the line $2y - 6x = 9$ and that passes through the point $(-2, 1)$
    d. parallel to the line $y = 4x + 3$ and that has the same y-intercept as the line $y = 5x - 3$

48. Write an equation of the line that is:
    a. parallel to the line $y = 3$ and that passes through the point $(-5, 4)$
    b. parallel to the line $x = 7$ and that passes through the point $(-4, z)$

In 49–57, write an equation of the line that passes through the given points.

49. $(1, 4), (3, 8)$ **50.** $(1, 0), (3, 6)$ **51.** $(3, 1), (9, 7)$

52. $(1, 2), (10, 14)$ **53.** $(0, -1), (6, 8)$ **54.** $(3, 6), (6, 0)$

55. $(-3, 11), (6, 5)$ **56.** $(-2, -5), (-1, -2)$ **57.** $(0, 0), (-3, 5)$

58. Physicists have found that when a weight is attached to a spring to stretch it, the relation between the length ℓ of the spring and the weight w applied to stretch it is linear. Consider a spring of original length 75 mm.

    a. If a weight of 500 grams stretches the spring to 115 mm, write a linear equation relating the extended length of the spring and the weight applied to stretch it, $\ell = f(w)$.
    Hint: When $w = 0, \ell = 75$

    b. What is the length of the spring after a weight of 1 kilogram has been applied?

59. Biologists have found that a linear function can be used to model the relation between the temperature and the number of chirps per minute made by certain crickets.

    a. If these crickets chirp about 100 times per minute at 62°F and 172 times per minute at 80°F, express the temperature t as a linear function of the number of chirps c; that is, $t = f(c)$.

    b. What is the temperature if these crickets chirp about 200 times per minute?

    c. Explain how you can estimate the temperature if you count chirps for only 15 seconds.

In 60–83, graph the linear function using the slope and the y-intercept.

60. $y = 2x + 3$ **61.** $y = 2x - 5$ **62.** $y = 2x$

63. $y = x - 2$ **64.** $y = 2x - 2$ **65.** $y = 3x - 2$

66. $y = 3x$ **67.** $y = 5x$ **68.** $y = -2x$

69. $y = \frac{2}{3}x + 2$ **70.** $y = \frac{1}{2}x - 1$ **71.** $y = \frac{3}{2}x$

72. $y = \frac{1}{3}x$ **73.** $y = -\frac{4}{3}x + 5$ **74.** $y = -\frac{3}{4}x$

75. $y - 2x = 3$ **76.** $3x + y = 4$ **77.** $2y = 4x + 6$

78. $3y = 4x + 9$ **79.** $4x - y = 3$ **80.** $3x + 4y = 12$

81. $2x = 3y + 6$ **82.** $4x + 3y = 0$ **83.** $2x - 3y - 6 = 0$

A *translation* "slides" a figure onto its image without changing the direction of the figure. Changing the y-intercept of a line without changing the slope translates the line up or down along the y-axis. For instance, the line $y = x + 1$ has a y-intercept of 1. If the y-intercept is changed to 4, the line is translated 3 units up in the plane, to the image line $y = x + 4$.

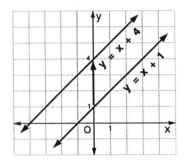

84. Write an equation whose graph is the image of the linear function $y = x + 1$ under a translation of 2 units down in the plane.

85. The linear function $y = \frac{1}{4}x + 5$ models the pedestrian traffic pattern, during peak hours, at Sonia's Sneaker Stop in Sunrise Mall.

   The scale on the y-axis represents the number of people entering the shop, as a percentage of the total number of shoppers at the mall.

   The scale on the x-axis represents the hours from noon to 6 P.M., starting with noon as time 0.

   a. Use the slope-intercept method to draw a graph of this linear function.
   b. In the context of the real situation, interpret the meaning of:
      (1) the slope of the line
      (2) the y-intercept
   c. What percentage of mall shoppers might be expected to enter Sonia's at 8 P.M.?

86. The linear function $y = 4x + 20$ models the average yield of corn on Fortune Farm.

   The scale for production of corn, measured in bushels, is on the y-axis.

   The scale for the years in which the corn was produced, beginning in 1950 as year 0, is on the x-axis.

   a. Use the slope-intercept method to draw a graph of this linear function.
   b. In the context of the real situation, interpret the meaning of:
      (1) the slope of the line
      (2) the y-intercept of the line
   c. How much greater was the yield for 1990 than the yield for 1960?

9-7 GRAPHING A LINEAR INEQUALITY IN THE PLANE

Each point on the graph of a line satisfies the equation of that line. For example, for each point on the graph of $y = 2x$, the ordinate is twice the abscissa. Some points are (1, 2), (0, 0), and (−2, −4). There are, of course, infinitely many such points, all of which lie on that line. (See Figure 1 below.)

Do other points in the plane have a relation to that same line $y = 2x$? Consider the point (1, 4). This point is not on the line and, thus, its coordinates do not satisfy the function $y = 2x$. However, this point does satisfy the relation $y > 2x$. Name other points that satisfy the relation $y > 2x$. Some such points are (1, 5), (1, 6), (1, 25), (0, 3), (0, 4), and (0, 50). Observe from Figure 2 that all points in the region of the plane that is "above" the line $y = 2x$ satisfy the relation $y > 2x$.

Note that the line $y = 2x$ divides the plane into two regions, called **half-planes**. Observe from Figure 3 that all points in the half-plane that is "below" the line $y = 2x$ satisfy the relation $y < 2x$.

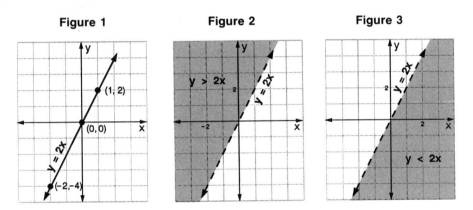

Figure 1   **Figure 2**   **Figure 3**

In general, the graph of $y = mx + b$ is a line that divides the coordinate plane into three sets of points:

1. The set of points *on* the line.
   Each ordered pair is a solution of the equation $y = mx + b$.
   The line is the graph of the function $y = mx + b$.

2. The set of points in the half-plane "above" the line.
   Each ordered pair is a solution of the inequality $y > mx + b$.
   The half-plane is the graph of the relation $y > mx + b$.

3. The set of points in the half-plane "below" the line.
   Each ordered pair is a solution of the inequality $y < mx + b$.
   The half-plane is the graph of the relation $y < mx + b$.

To graph a linear inequality in the plane:

1. Graph the line that is the plane divider.
   a. Draw the line as dotted if the inequality does not also contain an equals sign.
   b. Draw the line as solid if the inequality does also contain an equals sign.
2. Shade the appropriate half-plane.
   a. Shade the half-plane "above" the line if the inequality is of the form $y > mx + b$.
   b. Shade the half-plane "below" the line if the inequality is of the form $y < mx + b$.
3. Check by substituting the coordinates of a point from the shaded region in the original relation.
4. Be sure to include labels on the graph.

MODEL PROBLEMS

1. Graph the inequality $y - 2x \geq 2$.

Solution: To graph the line that is the plane divider, use any of the 3 methods that you learned: (1) table of values (2) x- and y-intercepts (3) slope and y-intercept

Since the inequality includes "equals," the line is solid.
Since the inequality is equivalent to $y \geq 2x + 2$, shade "above" the line.

ANSWER:

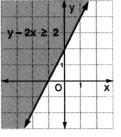

Check: Choose any point from the shaded half-plane. Try $(-2, 2)$.

$$y - 2x \geq 2$$
$$2 - 2(-2) \overset{?}{\geq} 2$$
$$2 + 4 \overset{?}{\geq} 2$$
$$6 \geq 2 ✓$$

2. Graph each of the following linear relations in the coordinate plane:
   a. $x > 1$ **b.** $x \leq 1$ **c.** $y \geq 1$ **d.** $y < 1$

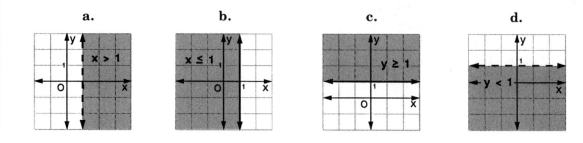

3. The linear relation $\frac{1}{24}x + \frac{1}{32}y \leq 16$ models the gas consumption for Jeremy's car, which has a 16-gallon gas tank.

> The variable x represents city driving, measured in miles, for which the car averages 24 miles per gallon.

> The variable y represents highway driving, measured in miles, for which the car averages 32 miles per gallon.

Jeremy fills his tank near his home in the city and starts out to visit his parents. If Jeremy needs to travel 12 miles to get out of the city and then 400 highway miles to his parents' house, can he make the trip on one tank of gas?

Solution: Graph the relation. Use the x- and y-intercepts to sketch the plane divider.

$$y = 0: \quad x = 24 \times 16 = 384$$
$$x = 0: \quad y = 32 \times 16 = 512$$

Shade the half-plane "below" the line.

The line represents the travel distance at which the car would run out of gas. In the shaded region, there is gas in the tank.

To answer the question, determine whether the point $(12, 400)$ is in the shaded half-plane.

ANSWER: Yes, Jeremy can make the trip on one tank of gas.

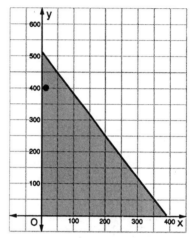

Now it's your turn to . . . **TRY IT!**   *(See page 522 for solutions.)*

Graph the linear relation.

1. $y > 2x - 3$ **2.** $x \leq -2$

EXERCISES

In 1–24, graph the relation in the coordinate plane.

1. $x > 4$ **2.** $x \leq -2$ **3.** $y > 5$

4. $y \leq -3$ **5.** $x \geq 6$ **6.** $y \leq 0$

7. $y > 4x$ **8.** $y \leq 3x$ **9.** $y < x - 2$

10. $y \geq \frac{1}{2}x + 3$ **11.** $x + y < 4$ **12.** $x + y \geq 4$

13. $x + y \leq -3$ **14.** $y - x \geq 5$ **15.** $x - y > 6$

16. $x - y \leq -1$ **17.** $y - 3x > 3$ **18.** $x - 2y \leq 4$

19. $2x + y - 4 \leq 0$ **20.** $y - x + 6 > 0$ **21.** $2y - 6x > 0$

22. $3x + 4y \leq 0$ **23.** $2x - 3y \geq 6$ **24.** $10 \leq 5x - 2y$

In 25–30: **a.** Write the verbal sentence as an open sentence.
   b. Graph the open sentence in the coordinate plane.

25. The ordinate of a point is greater than the abscissa.

26. The ordinate of a point is less than four times the abscissa.

27. The ordinate of a point is greater than or equal to 3 more than the abscissa.

28. The sum of the abscissa and ordinate of a point is less than or equal to 5.

29. The ordinate of a point decreased by three times the abscissa is greater than or equal to 2.

30. The sum of three times the abscissa of a point and twice the ordinate is less than or equal to 12.

31. For breakfast this week, the dietician at New City Elementary School wishes to buy Wheat Squares Cereal and Corn Bites Cereal. The cost of the wheat cereal is 15 cents per ounce and of the corn cereal 18 cents per ounce.

   a. If the budget is $36, write a linear relation, in simplest form, to model the week's cereal costs.
   b. Draw the graph of this relation.
   c. For this budget, what is the maximum number of ounces she can buy of wheat cereal? of corn cereal?
   d. If the dietician buys 150 ounces of each cereal, will she be within the budget?

32. Mr. and Mrs. Dawson are planning some redecorating with carpeting and vinyl floor covering. The carpet they have selected costs $12 per square yard (tax and installation included) and the vinyl $9 per square yard (tax and installation included).

   a. If their total budget for the carpeting and vinyl is $800, write a linear relation, in simplest form, to model the cost.
   b. Draw the graph of this relation.
   c. If the entire budget is used for carpeting, what is the maximum whole number of square yards the Dawsons can buy?
   d. If the Dawsons wish to spend equal amounts on the carpeting and vinyl, estimate the amount of each they can buy.
   e. If the Dawsons reserve $80 for paint, would they be able to get 30 square yards each of carpeting and vinyl? Explain how you would determine this answer from the graph in part **b**.

CHAPTER SUMMARY

Coordinate planes in the form of two-dimensional grids are commonly used as locators, as with road maps. A coordinate plane, which connects Algebra and Geometry, also makes it possible to determine lengths and areas.

A *transformation* changes the positions of points in the plane according to some rule. The study of transformations is especially useful when studying graphs.

Any set of ordered pairs determines a *relation*. A *function* is a special kind of relation in which each member of the domain is paired with exactly one member of the range. A *vertical-line test* on the graph of a relation will determine if it is a function.

A *linear function* is described algebraically by a first-degree equation and graphically by a straight line. A unique characteristic of the line is that its *slope*, the ratio of the vertical and horizontal differences between points on the line, is constant.

Depending upon the nature of the graph of the linear function, which may be *oblique, horizontal,* or *vertical,* the value of the slope is positive, negative, 0, or undefined.

The equation of a linear function may be generalized:

$$\text{\textit{standard form}} \qquad ax + by = c$$
$$\text{\textit{slope-intercept form}} \qquad y = mx + b$$

To draw the graph of a linear function, use any of 3 methods: (1) construct a table of values, (2) plot the x- and y-intercepts, or (3) use the slope and y-intercept.

To draw the graph of a *linear inequality,* first graph the associated line as a *plane divider* and then determine which *half-plane* contains the points whose coordinates satisfy the inequality relation.

VOCABULARY CHECKUP

SECTION

9-1 *ordered number pair / coordinate axes / origin / quadrants / abscissa / ordinate / transformation / reflection / translation*

9-2 *relation / domain / range*

9-3 *function / mapping diagram / f(x)*

9-4 *linear function / standard form / y-intercept / x-intercept / oblique*

9-5 *slope / collinear* **9-6** *slope-intercept form* **9-7** *plane divider / half-plane*

1. Draw a pair of coordinate axes on a sheet of graph paper and graph the point associated with the ordered number pair. (Section 9-1)

   $A(-2, 3)$ $B(-1, -4)$ $C(-3, 0)$ $D(2, 4)$ $E(0, 2)$ $F(4, -3)$

   In 2–4, name the quadrant in which the point $P(x, y)$ lies when: (Section 9-1)

2. $x > 0$ and $y < 0$ 3. $x < 0$ and $y > 0$ 4. P is $(|x|, -4)$

   In 5 and 6, write the coordinates of the image of the given point under a reflection:
   a. in the y-axis   **b.** in the x-axis   (Section 9-1)

5. $A(-5, 2)$ 6. $B(3, -7)$

   In 7 and 8, write the coordinates of the image of the given point under a translation of:
   a. 4 units left and 2 units down   **b.** 8 units right and 5 units up   (Section 9-1)

7. $A(3, 8)$ 8. $B(-3, -11)$

   In 9 and 10:   (Section 9-1)
   a. Graph the given points and connect them, in alphabetical order, returning to the first point to form a polygon.
   b. Find the area of the polygon formed.

9. $A(-2, 5)$ $B(3, 8)$ $C(4, 5)$ 10. $A(-1, 7)$ $B(4, 7)$ $C(4, 1)$ $D(-2, 1)$

11. Which of the ordered pairs of numbers $(-3, 5)$, $(2, -3)$, $(5, -8)$, $(-5, 2)$ is a member of the solution set of $2x + y > 1$? (Section 9-2)

    In 12–15, write an equation that describes the relation. (Section 9-2)

12. x decreased by twice y is 9.   13. y is greater than the square of x.

14. y is 4 more than the square of x.   15. y is 2 less than 4 times x.

16. Write a rule to describe the relation. (Section 9-2)

    a.

x	−1	0	1	2
y	1	3	5	7

    b.

x	−1	0	2	10
y	1	0	−2	−10

    In 17–19, for the given equation: (Section 9-2)
    a. Construct a table of values for the solution set, being sure to include some negatives and 0 for x.
    b. Draw a graph of the relation.
    c. State any limitations on the domain and range.

17. $y = 2x - 1$ 18. $y = |x + 2|$ 19. $y = x^2 - 3$

In 20–24, for the given relation, state: (Section 9-3)

a. the domain   **b.** the range   **c.** whether the relation is a function

20. {(1, 5), (2, 5), (3, 5), (4, 5)}   **21.** {(0, 1), (1, 0)}

22.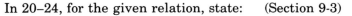

23.

24.

In 25–28, for the function f defined by $f(x) = x^2 - 3x + 2$ for each real number x, find: (Section 9-3)

25. $f(-2)$ **26.** $f(3)$ **27.** $f(2a)$ **28.** $3[f(4)]$

In 29 and 30:   (Section 9-3)

a. Determine if each relation is a function.
b. Write the coordinates of the set of ordered pairs.

29.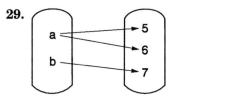

30.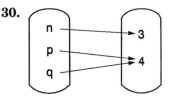

In 31 and 32, the table describes a relation between x and y. (Section 9-3)

a. Determine if y is a function of x.
b. Determine if x is a function of y.

31.

x	3	0	−1	0	3
y	−2	−1	0	1	2

32.

x	−4	−3	−2	−1	0
y	−1	1	3	5	7

In 33–38, draw the graph of the linear function. (Section 9-4)

33. $3x - 2y = 6$ **34.** $y = 4x - 3$ **35.** $x - y = 3$

36. $x = -2y$ **37.** $y = -2x$ **38.** $5x - 4y = 20$

In 39–41, find the x-intercept and the y-intercept of the line that is the graph of the equation. (Section 9-4)

39. $x - y = 12$ **40.** $2x - 5y = -30$ **41.** $y = -3x + 9$

42. Write an equation of the line that is parallel to the x-axis and whose y-intercept is -3. (Section 9-4)

43. Write an equation of the vertical line whose x-intercept is 0. (Section 9-4)

In 44–46, find the slope of the line. (Section 9-5)

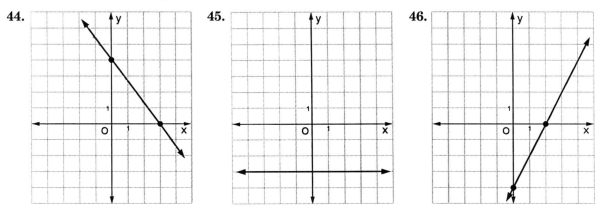

44.

45.

46.

In 47–49, find the slope of the line containing the given points. (Section 9-5)

47. $(5, -2)$ and $(3, 4)$ **48.** $(0, -3)$ and $(4, 8)$ **49.** $(-5, -2)$ and $(0, -2)$

In 50 and 51, given the points $A(0, -2)$, $B(1, 1)$, $C(k, 4)$, and $D(-1, -5)$, find the value of k such that: (Section 9-5)

50. the line through A and B is parallel to the line through C and D

51. the points A, B, and C are collinear

In 52–54, find the slope and y-intercept of the line that is the graph of the equation. (Section 9-6)

52. $y = 7x - 5$ **53.** $y = -\frac{2}{3}x - \frac{2}{5}$ **54.** $3x - 2y = -12$

In 55–57, write an equation of the line whose slope and y-intercept are respectively: (Section 9-6)

55. 3 and -8 **56.** $\frac{3}{4}$ and 0   **57.** $-\frac{5}{6}$ and -1

In 58 and 59, state whether the lines are parallel. (Section 9-6)

58. $y = -2x + 3$ and $x = -2y - 4$ **59.** $y = 5x$ and $y = 5x - 3$

In 60–62, write an equation of the line that has the given slope, m, and passes through the given point. (Section 9-6)

60. $m = -3$, $(5, 2)$ **61.** $m = \frac{2}{5}$, $(0, -4)$ **62.** $m = 0$, $(6, -1)$

In 63–65, write an equation of the line that passes through the given points. (Section 9-6)

63. $(3, -4)$; $(4, 2)$ **64.** $(2, -6)$; $(-2, 1)$ **65.** $(6, 2)$; $(9, 3)$

In 66 and 67, write an equation of the line: (Section 9-6)

66. parallel to $3x - 2y = 7$ and whose y-intercept is 5

67. parallel to $y = 5x - 2$ and that passes through the point $(-2, -5)$

In 68–70, graph the open sentence. (Section 9-7)

68. $y > 3x$ **69.** $y \le x - 2$ **70.** $x < -2$

PROBLEMS FOR PLEASURE

1. The function f is defined by $f(x) = x^2 - 2x + 1$ for every real number x. Find $f(x - 1) - f(x + 3)$.

2. The function f is defined by $f(x) = x + 3$ for every real number x. A second function, g, is defined by $g(x) = 2x - 1$. Find:
   a. $g(x) - f(x)$ **b.** $f(g(x))$ **c.** $g(f(x))$

3. Just as there is a relationship between the slopes of two lines that are *parallel*, there is also a relationship between the slopes of two lines that are *perpendicular*.

    Draw a line through $A(2, 3)$ and $B(4, 6)$, and another line through $C(1, 5)$ and $D(7, 1)$. Line AB is perpendicular to line CD. Find a relationship between their slopes.

    Repeat for line EF through $E(1, 4)$ and $F(-1, 0)$ and line GH through $G(-2, -2)$ and $H(2, -4)$.

    Find a general relationship that applies to the slopes of all pairs of perpendicular lines.

CALCULATOR CHALLENGE

1. The table defines a relation that is a set of ordered number pairs.

x	−3.2	1.25	5.1
y	19.2	−7.5	−30.6

   a. Write a rule to describe the relation.
   b. Find the value of y when $x = -2.5$.
   c. Find the value of x when $y = -9$.

2. Given the linear function $f(x) = ax + b$, write the value of $f(x)$ in scientific notation when $a = 6.2 \times 10^3$, $x = 4.1$, and $b = 4.958 \times 10^4$.

In 1–19, select the letter of the correct answer.

1. The graphs of all of the following equations are parallel except
   (A) $3x = 6 - 2y$ (B) $y = -\frac{3}{2}x + 1$
   (C) $x = -\frac{2}{3}y - 3$ (D) $4y = 8 - 6x$
   (E) $9x - 6y = 7$

Questions 2 and 3 refer to the slope formula $m = \dfrac{y_2 - y_1}{x_2 - x_1}$.

2. A line is vertical when
   (A) $y_1 = y_2$ (B) $y_1 = -y_2$
   (C) $x_1 = x_2$ (D) $x_1 = -x_2$
   (E) $x_1 = y_1$

3. The slope of a line is 1 when
   (A) $x_1 + y_1 = x_2 + y_2$
   (B) $x_1 - y_1 = x_2 - y_2$
   (C) $x_2 + y_1 = y_2 - x_1$
   (D) $x_1 - y_2 = x_2 - y_1$
   (E) $x_1 + x_2 = y_1 + y_2$

4. A pencil lies on graph paper with the eraser at $(-5, 10)$ and the point at $(7, -2)$. If the pencil is moved without changing its direction until the point is at the origin, then the eraser is
   (A) in quadrant I
   (B) in quadrant II
   (C) in quadrant III
   (D) in quadrant IV
   (E) on the y-axis

5. Which is an equation of the line that passes through $(1, p)$ and the origin?
   (A) $y = px$ (B) $y = x + p$
   (C) $x = py$ (D) $y = x - p$
   (E) $y = x - p + 1$

6. If $k > -1$, which number pair does not belong to the solution set of $y \le 2k - 3$?
   (A) $(0, -3)$ (B) $(1, -2)$ (C) $(2, -3)$
   (D) $(1, 0)$ (E) $(3, -1)$

7. If $(0, 1)$ and $(3, 7)$ are both points on the graph of line m, which of the following points also lies on the graph of line m?
   (A) $(3, 8)$ (B) $(2, 6)$ (C) $(1, 3)$
   (D) $(2, 4)$ (E) $(2, 3)$

8. Line k passes through the point $(3, 7)$. If line k has zero slope, then k must pass through which of the following points?
   (A) $(0, 7)$ (B) $(3, 0)$ (C) $(0, 0)$
   (D) $(-3, -7)$ (E) $(0, 3)$

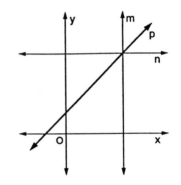

9. If an equation of line m is $x = 3$ and an equation of line n is $y = 4$, then an equation of line p could not be
   (A) $x - y = -1$ (B) $2y = x + 5$
   (C) $y = \frac{2}{3}x + 2$ (D) $x = y - 2$
   (E) $y = \frac{4}{3}x$

10. If $P(2, 3)$ is on the graph of $3x + ky = -9$, then which of the following points is also on the line?
    (A) $(0, -3)$ (B) $(-3, 1)$
    (C) $(4, 6)$ (D) $(7, 6)$
    (E) $(6, 7)$

11. All of the following are solutions of $y = |2 - x|$ except
    (A) $(-3, 5)$ (B) $(4, 6)$
    (C) $(5, 3)$ (D) $(7, 5)$
    (E) $(-6, 8)$

12. If each of the graphs shown below were reflected over the *y*-axis, which of the resulting graphs would not represent a function?

(A)

(B)

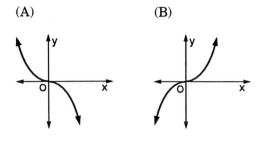

(C)

(D)

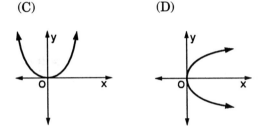

(E)

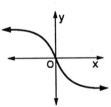

13. A translation slides point A 2 units right and 1 unit down to its image A'. Under a second translation, 6 left and 3 down, the image of A' is $A''(5, 0)$. The original point A has coordinates

(A) $(1, -4)$ (B) $(3, 1)$
(C) $(11, 3)$ (D) $(9, 4)$
(E) $(6, -3)$

14. The point $P(2, -5)$ is reflected in the line $y = x$, then in the line $y = -x$. The coordinates of the resulting image are

(A) $(-2, -5)$ (B) $(2, 5)$
(C) $(-2, 5)$ (D) $(-5, 2)$
(E) $(5, -2)$

15. If $g(x) = 3x$ and $h(x) = \dfrac{3}{3x + 6}$, then $h(x)$ could not be written as

(A) $\dfrac{3}{g(x) + 6}$ (B) $\dfrac{3}{g(x + 2)}$

(C) $g\left(\dfrac{1}{3x + 6}\right)$ (D) $\dfrac{1}{g(x) + 2}$ (E) $\dfrac{1}{x + 2}$

16. If $f(x) = x + 1$, then $f(f(f(x)))$ equals

(A) $x + 3$ (B) $3x + 3$ (C) $x^3 + 3$
(D) $3x + 1$ (E) $x^3 + 1$

17. If $f(x) = \dfrac{x - 1}{2}$ and $g(x) = 2x + 1$, then $f(g(3)) - g(f(3))$ equals

(A) 0   (B) 3   (C) 4   (D) 5   (E) 7

18. If $f(x) = 3x - 5$ and $g(x) = x + 10$, which of the following is equivalent to $3x + 25$?
(A) $f(x) + g(x)$ (B) $f(x) - g(x)$
(C) $6g(x) - f(x)$ (D) $f(g(x))$
(E) $g(f(x))$

19. If the mapping diagrams show the relations f and g, then $f(g(3)) =$
(A) 1   (B) 2   (C) 3   (D) 4   (E) 5

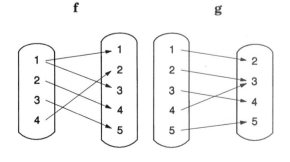

1. For each symbol in **a–f**, choose the matching description.

   a. >
   b. $|a|$
   c. $\dfrac{1}{a}$
   d. $-a$
   e. ∩
   f. ∪

   1. the additive identity element
   2. the multiplicative identity element
   3. the opposite of a
   4. the reciprocal of a
   5. less than
   6. greater than
   7. angle
   8. the absolute value of a
   9. intersection
   10. union
   11. the empty set

2. Solve for y: $y - 3 < 2y$

3. Two numbers are in the ratio 1:4 and their sum is 180. Find the numbers.

4. At the Super Sounds Audio store, $\frac{2}{3}$ of the daily receipts came from tape sales. If today's receipts from tape sales reached $2,280, what were the total receipts?

5. If d varies directly as t, and $d = 240$ when $t = 6$, find d when $t = 5$.

6. If $t = 3$, find the value of: $\dfrac{8}{3t} + \dfrac{1}{3t}$

7. If $\angle C$ is the right angle in isosceles right $\triangle ABC$, which statement is true?
   (1) $\tan A > \sin A$ (2) $\tan A = \cos A$ (3) $\tan A < \sin A$ (4) $\tan A < \cos A$

8. Solve the proportion: $\dfrac{y - 2}{5} = \dfrac{3}{y}$

9. Find the product in lowest terms: $\dfrac{3c - 12}{15} \cdot \dfrac{9}{c^2 - 16}$

10. Given: $A = \begin{bmatrix} 2 & -1 \\ 3 & 0 \end{bmatrix}$ $B = \begin{bmatrix} 1 & x \\ -4 & -2 \end{bmatrix}$ $AB = \begin{bmatrix} 6 & 12 \\ 3 & 15 \end{bmatrix}$

    Find the value of x.

11. Culvert City and Drakesville are 315 miles apart. One car left Culvert City for Drakesville, averaging 55 mph. At the same time, another car left Drakesville for Culvert City, averaging 50 mph. How far from Culvert City did the cars pass each other?

12. If $a \star b = 2b + a$, evaluate: $3 \star (2 \star 5)$

13. Write as a single fraction in lowest terms: $\dfrac{x}{4} - \dfrac{x}{6}$

14. Find three consecutive even integers whose sum is 66.

15. Subtract $3x^2 - 5xy + y^2$ from $x^2 - 8xy - 2y^2$.

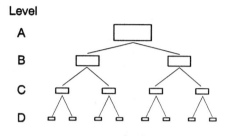

Level

A

B

C

D

16. A weight placed at level A is balanced by equal weights at level B.
The weights at level B are balanced by equal weights at level C.
The weights at level C are balanced by equal weights at level D.

A 200-pound weight is placed at level A, and the balancing weights at level B are each 100 pounds. If the pattern continues, then 4 of the weights at level J weigh:

(1) $\dfrac{25}{64}$ lb. (2) $\dfrac{25}{16}$ lb. (3) $\dfrac{5}{4}$ lb. (4) $\dfrac{25}{8}$ lb. (5) $\dfrac{5}{8}$ lb.

17. A candy store manager sells dark chocolate for \$8 a pound and milk chocolate for \$5.50 a pound. How many pounds of each kind must he use to make 15 pounds of mixed chocolates that can be sold for \$7 a pound?

18. From a point 100 feet from the foot of a cliff, the angle of elevation of the top of the cliff is $60°$. Find the height of the cliff to the nearest foot.

19. If matrix A has 3 rows and 2 columns, and matrix B has 2 rows and 2 columns, then which is true?
(1) $A + B$ has 6 rows and 4 columns.
(2) $A + B$ has 3 rows and 2 columns.
(3) AB has 6 rows and 4 columns.
(4) AB has 3 rows and 2 columns.

Solutions to **TRY IT!** *Problems (Chapter 9)*

9-1 GRAPHING IN A PLANE

TRY IT! *Problem on page 473*

a. B(west 1, north 2)
   C(west 3, south 2)
   D(east 2, south 1)

b. $B(-1, 2)$, $C(-3, -2)$, $D(2, -1)$

TRY IT! *Problem on page 474*

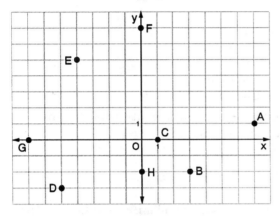

TRY IT! *Problems 1 and 2 on page 475*

1. a. $P'(2, 5)$ **b.** $P'(-2, -5)$
2. a. $A'(2, 7)$ **b.** $A'(-3, 0)$

TRY IT! *Problem on page 476*

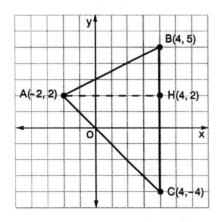

With vertical segment BC as the base, draw horizontal segment AH, with H at $(4, 2)$, as the altitude.

$$BC = |5 - (-4)| = 9$$
$$AH = |4 - (-2)| = 6$$
$$A_{\triangle ABC} = \tfrac{1}{2}(BC)(AH)$$
$$= \tfrac{1}{2}(9)(6) = 27 \text{ sq. units}$$

9-2 RELATIONS AND THEIR GRAPHS

TRY IT! *Problems 1–3 on page 481*

1. $y = x + 3$

2.

x	−3	−2	−1	0	1	2	3
y	10	8	6	4	2	0	−2

3. a.

x	−2	−1	0	1	2
y	4	1	0	1	4

b.

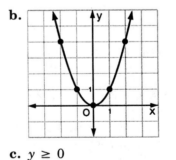

c. $y \geq 0$

9-3 FUNCTIONS AND THEIR GRAPHS

TRY IT! *Problems 1–3 on page 485*

1. (1) a. $0 \leq x \leq 2$ **b.** $-3 \leq y \leq 3$
     c. no
   (2) a. $-3 \leq x \leq 3$ **b.** $-2 \leq y \leq 0$
     c. yes

2. a. 6   **b.** $a^4 + a^2$ **c.** $b^2 + 3b + 2$

3. a. No. The x-value 1 is paired with two different y-values.

   b. Yes. Each y-value is paired with exactly one x-value.

9-4 LINES AND LINEAR FUNCTIONS

TRY IT! *Problems 1–3 on page 489*

1.

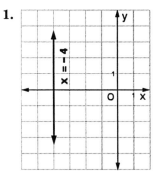

2.

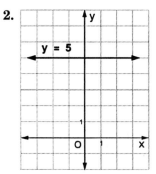

3.

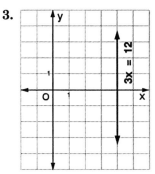

9-5 THE SLOPE OF A LINE

TRY IT! *Problems 1 and 2 on page 494*

1. $m = \dfrac{y_2 - y_1}{x_2 - x_1}$ $P_1(2, -1), P_2(2, -6)$

$m = \dfrac{-6 - (-1)}{2 - 2} = \dfrac{-5}{0}$ undefined

The line is vertical.

2. $P_1(4, 7), P_2(-1, -2)$

$m = \dfrac{-2 - 7}{-1 - 4} = \dfrac{-9}{-5} = \dfrac{9}{5}$

The line is oblique.

TRY IT! *Problem on page 495*

The line through $(k, 2)$ and $(4, 0)$ has the slope $\dfrac{2 - 0}{k - 4}$, or $\dfrac{2}{k - 4}$.

The line through $(6, 2)$ and $(5, 4)$ has the slope $\dfrac{4 - 2}{5 - 6}$, or $\dfrac{2}{-1}$.

Parallel lines have equal slopes.

$$\frac{2}{k - 4} = \frac{2}{-1}$$
$$2k - 8 = -2$$
$$2k = 6$$
$$k = 3$$

TRY IT! *Problem on page 496*

If the points are collinear, the slope of the line through $(1, k)$ and $(-1, 2)$ is equal to the slope of the line through $(-1, 2)$ and $(-3, -1)$.

$$\frac{k - 2}{1 - (-1)} = \frac{2 - (-1)}{-1 - (-3)}$$
$$\frac{k - 2}{2} = \frac{3}{2}$$
$$2k - 4 = 6$$
$$2k = 10$$
$$k = 5$$

9-6 THE SLOPE-INTERCEPT FORM OF A LINEAR EQUATION

TRY IT! *Problems 1–3 on page 501*

1. $y = mx + b$
$y = 3x - 2$

2. Use m and $(2, -5)$ to find b.

$y = mx + b$
$-5 = -4(2) + b$
$3 = b$

$y = -4x + 3$

3. Find the slope.

$$m = \frac{2 - 4}{6 - 5} = \frac{-2}{1} = -2$$

Use m and one point to find b.

$y = mx + b$
$y = -2x + b$
$4 = -2(5) + b$
$4 = -10 + b$
$14 = b$

$y = -2x + 14$

9-7 GRAPHING A LINEAR INEQUALITY IN THE PLANE

TRY IT! *Problems 1 and 2 on page 509*

1.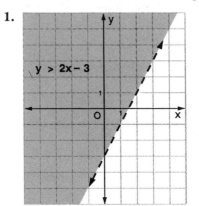

2.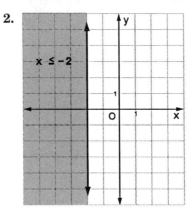

Part I

1. Given the sets $A = \{1, 3, 5\}$, $B = \{1, 3, 4, 5\}$, and $C = \{1, 2, 3, 7\}$, find $A \cup (B \cap C)$.

2. Divide $12x^5 - 18x^3 + 9x$ by $-3x$.

3. The width of a rectangle is $3s$ inches, and its length is $(2s + 1)$ inches. What is the perimeter of the rectangle?

4. Replace the question mark by $>$, $<$, or $=$ to make the sentence true.

$$2 \, |-8| \ \ ? \ \ -4^2$$

5. Find the product in simplest form:

$$\frac{24bc^4}{b - c} \cdot \frac{(b - c)^2}{8b^2c}$$

6. If the domain of b is $\{-3, -2, -1, \ldots 3\}$, graph on a number line the solution set of $-b < 2$.

7. Solve for y: $2(y - 4) = 6(3 - y)$

8. Reduce to lowest terms: $\dfrac{9c - 18}{c^2 - 4}$

9. Solve for w: $\begin{bmatrix} 1 & 5w + 3 \\ 8 & |-9| \end{bmatrix} = \begin{bmatrix} 2^0 & 2w \\ 2^3 & 3^2 \end{bmatrix}$

10. Find three consecutive odd integers whose sum is 63.

11. If r varies inversely as s and if $r = 3$ cm when $s = 8$ cm, find r when $s = 12$ cm.

12. In a right triangle, one acute angle measures $50°$ more than the other. Find the measure of the smallest angle of the triangle.

13. Find 3 prime factors of 105.

In 14–30, choose the correct answer.

14. Solve for x: $\dfrac{x}{3} + \dfrac{x}{2} = 1$
    (1) $x = \frac{5}{6}$ (2) $x = 1\frac{1}{5}$
    (3) $x = 5$ (4) $x = 6$

15. Simplify: $8m - 3m(m - 2) + 6$
    (1) $5m^2 - 10m + 6$
    (2) $-3m^2 - 10m + 6$
    (3) $-3m^2 + 14m + 6$
    (4) $-3m^2 + 8m + 4$

16. Which is the graph of the solution set of $|x| > 3$?

    (1)
    -3 0 3

    (2)
    -3 0 3

    (3)
    -3 0 3

    (4)
    -3 0 3

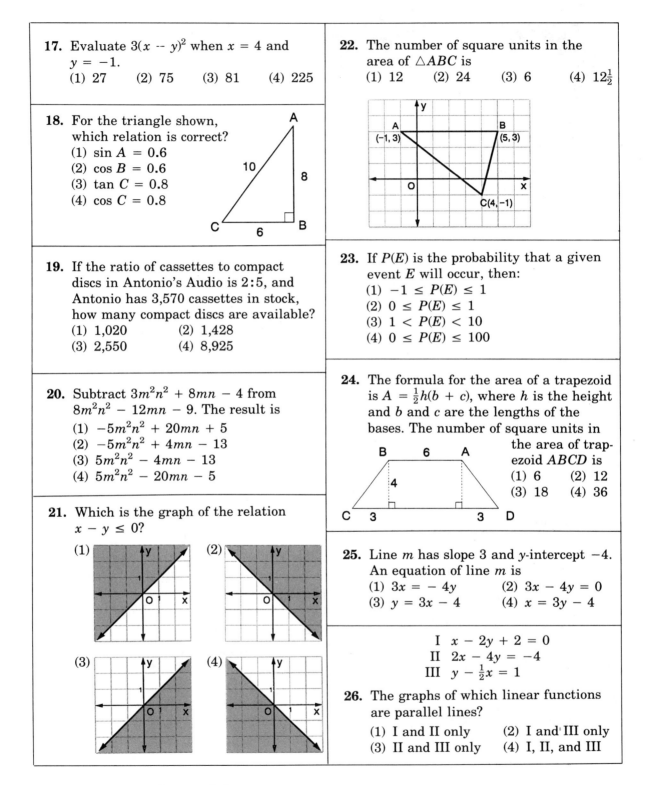

17. Evaluate $3(x - y)^2$ when $x = 4$ and $y = -1$.
   (1) 27   (2) 75   (3) 81   (4) 225

18. For the triangle shown, which relation is correct?
   (1) $\sin A = 0.6$
   (2) $\cos B = 0.6$
   (3) $\tan C = 0.8$
   (4) $\cos C = 0.8$

19. If the ratio of cassettes to compact discs in Antonio's Audio is $2:5$, and Antonio has 3,570 cassettes in stock, how many compact discs are available?
   (1) 1,020   (2) 1,428
   (3) 2,550   (4) 8,925

20. Subtract $3m^2n^2 + 8mn - 4$ from $8m^2n^2 - 12mn - 9$. The result is
   (1) $-5m^2n^2 + 20mn + 5$
   (2) $-5m^2n^2 + 4mn - 13$
   (3) $5m^2n^2 - 4mn - 13$
   (4) $5m^2n^2 - 20mn - 5$

21. Which is the graph of the relation $x - y \le 0$?
   (1)   (2)
   (3)   (4)

22. The number of square units in the area of $\triangle ABC$ is
   (1) 12   (2) 24   (3) 6   (4) $12\frac{1}{2}$

23. If $P(E)$ is the probability that a given event E will occur, then:
   (1) $-1 \le P(E) \le 1$
   (2) $0 \le P(E) \le 1$
   (3) $1 < P(E) < 10$
   (4) $0 \le P(E) \le 100$

24. The formula for the area of a trapezoid is $A = \frac{1}{2}h(b + c)$, where h is the height and b and c are the lengths of the bases. The number of square units in the area of trapezoid $ABCD$ is
   (1) 6   (2) 12
   (3) 18   (4) 36

25. Line m has slope 3 and y-intercept -4. An equation of line m is
   (1) $3x = -4y$ (2) $3x - 4y = 0$
   (3) $y = 3x - 4$ (4) $x = 3y - 4$

      I $x - 2y + 2 = 0$
     II $2x - 4y = -4$
    III $y - \frac{1}{2}x = 1$

26. The graphs of which linear functions are parallel lines?
   (1) I and II only   (2) I and III only
   (3) II and III only   (4) I, II, and III

27. If $a = 2$, find the value of $\dfrac{(a^{-2})(2a^0)(a^3)}{(2a)^0(2a)^2}$

(1) $\frac{1}{4}$ (2) $\frac{1}{2}$ (3) 1   (4) 2

28. By the vertical line test, which could be the graph of a function?
(1) a circle   (2) a rectangle
(3) a vertical line
(4) a horizontal line

29. What is the slope of the line through the points (8, 0) and (6, 8)?
(1) 1   (2) -2 (3) 3   (4) -4

30. If the domain of p is the set of integers, which is the solution set of the equation $(2x - 1)(x + 3) = 0$?
(1) $\{\frac{1}{2}, -3\}$ (2) $\{-3\}$
(3) $\{-\frac{1}{2}, 3\}$ (4) $\varnothing$

Part II

31. The area of rectangle $PQRS$ is 48 sq. cm. Find the length and the width.

32. On a toss of a coin, the outcome can be H (heads) or T (tails). A coin is tossed 3 times.
a. Draw a tree diagram.
b. List the outcomes as ordered triples.
c. What is the probability of tossing $2H$ and $1T$ (in any order)?
d. What is the probability the coin will land the same way on all 3 tosses?

33. **a.** Draw the graph of the linear function $2x - 3y = 9$.
b. What is the slope of the graph?
c. What is the y-intercept?
d. Shade the graph to show the solution set of $2x - 3y \le 9$.
e. Which point is not in the shaded region?   (1) (8, 1)   (2) (0, 0)
(3) $(-4, -4)$ (4) (1, 8)

34. Set $A = \{\ldots -6, -3, 0, 3, 6, \ldots\}$.
a. Is the set closed under addition?
b. Is the set closed under subtraction?
c. Is the set closed under multiplication?
d. Is the set closed under division?
e. Does each element of the set have an additive inverse in the set?

35. In $\triangle ABC$ and $\triangle DEF$, $m\angle A = m\angle D$, $m\angle B = m\angle E$, $AB = 6$ in., and $DE = 9$ in.
a. If $AC = 10$, find DF.
b. If $EF = 12$, find the perimeter of $\triangle ABC$.

36. From an observation point below, the angle of elevation of the top of a vertical cliff is 42°. If the observer is 35 meters from the base of the cliff, find, to the nearest tenth of a meter, the height of the cliff.

CHAPTER 10

SYSTEMS OF EQUATIONS AND INEQUALITIES

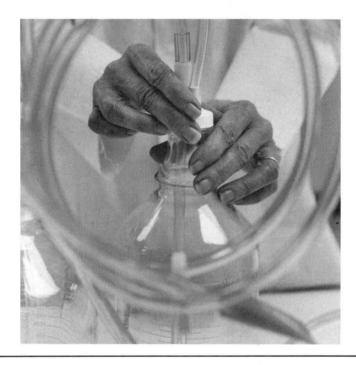

When studying a mixture, a chemist might use systems of linear relations to determine how much of each ingredient to use. Industrial applications of mixtures include petroleum products, paints, and textiles. Other applications are in food preparation, medicine, and agriculture.

10-1 Solving a System of Linear Equations
Graphically 527

10-2 Solving a System of Linear Equations
Algebraically 530

10-3 Solving Verbal Problems by Using
Two Variables 537

10-4 Solving a System of Linear Inequalities
Graphically 554

10-5 Solving a System of
Three Linear Equations 557

10-6 Solving a System of Linear Equations
by Using Matrices 561

Chapter Summary 565

Chapter Review Exercises 566

Problems for Pleasure 567

Calculator Challenge 568

College Test Preparation 568

Spiral Review Exercises 570

Solutions to TRY IT! Problems 571

Working with a set of two or more related equations is important because such *systems* allow more complex problems to be modeled.

In this chapter, focusing first on a system of linear equations in two variables, you will see how this type of system applies to problem solving. You will learn how to solve such a system, both graphically and algebraically, and how to set up a system as a model for a real situation.

You will also consider systems of linear inequalities.

10-1 SOLVING A SYSTEM OF LINEAR EQUATIONS GRAPHICALLY

Consider this problem: Find two numbers whose sum is 4 and whose difference is 2.

Solution of this simple problem illustrates a variety of problem-solving approaches that can be applied to more difficult problems.

For this problem, since the number of possible solutions is limited, you can use the strategy of guessing and checking to determine that the numbers are 3 and 1.

You can also apply an algebraic strategy.

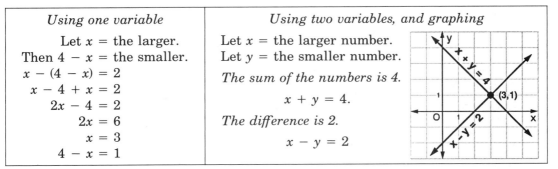

Using one variable	*Using two variables, and graphing*
Let x = the larger.	Let x = the larger number.
Then $4 - x$ = the smaller.	Let y = the smaller number.
$x - (4 - x) = 2$	*The sum of the numbers is 4.*
$x - 4 + x = 2$	$x + y = 4.$
$2x - 4 = 2$	*The difference is 2.*
$2x = 6$	$x - y = 2$
$x = 3$	
$4 - x = 1$	

Observe that by graphing two linear equations on the same set of axes, you can read the coordinates of the *point of intersection*, which is the one point in the plane whose coordinates satisfy both equations.

PROCEDURE

To solve a pair of linear equations graphically:

1. Graph one equation in a coordinate plane.

2. Graph the second equation using the same set of coordinate axes.

3. Find the common solution, which is the ordered number pair associated with the point of intersection of the two graphs.

4. Check the apparent solution by verifying that the ordered pair satisfies both equations.

Solve the linear system graphically, and check: $2x + y = 8$
$$y - x = 2$$

Solution: Use any of the three methods you have learned to draw the graph of each line:
1. x- and y-intercepts   2. slope and y-intercept   3. table of values

How to Proceed

(1) Graph $2x + y = 8$.

x	0	4	1
y	8	0	6

(2) Graph $y - x = 2$.

x	0	-2	1
y	2	0	3

Solution

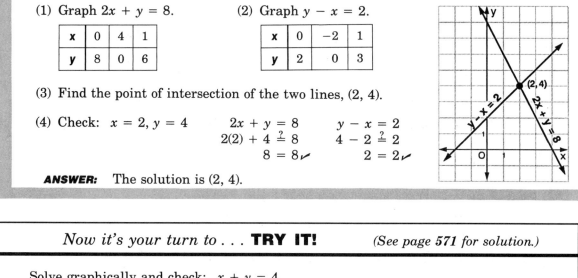

(3) Find the point of intersection of the two lines, $(2, 4)$.

(4) Check: $x = 2$, $y = 4$

$$2x + y = 8 \qquad y - x = 2$$
$$2(2) + 4 \overset{?}{=} 8 \qquad 4 - 2 \overset{?}{=} 2$$
$$8 = 8 ✔ \qquad\qquad 2 = 2 ✔$$

ANSWER: The solution is $(2, 4)$.

Now it's your turn to . . . **TRY IT!**   *(See page 571 for solution.)*

Solve graphically and check: $x + y = 4$
$$y = \tfrac{1}{2}x + 1$$

Will the graphs of a pair of linear equations always intersect when drawn on the same set of axes? Consider the following situations.

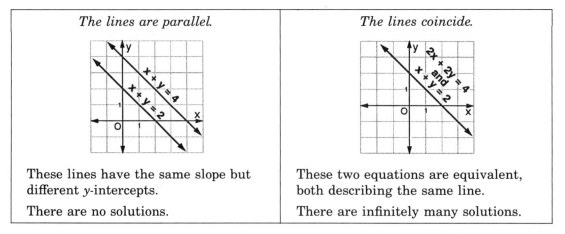

The lines are parallel.	*The lines coincide.*
These lines have the same slope but different y-intercepts.	These two equations are equivalent, both describing the same line.
There are no solutions.	There are infinitely many solutions.

In 1–27, solve the system of equations graphically. Check. When appropriate, write "no solution" or "infinitely many solutions."

1. $y = 2x$
$y = 3x - 3$

2. $y = 2x + 4$
$y = 2x + 5$

3. $y = -2x + 3$
$y = \frac{1}{2}x + 3$

4. $x + y = 7$
$x - y = 1$

5. $x + y = 4$
$x + y = 0$

6. $x + y = -4$
$x - y = 6$

7. $x + 2y = 7$
$y = 2x + 1$

8. $y = 3x$
$2x + y = 5$

9. $x + 3y = 9$
$x = 3$

10. $y - x = -2$
$x - 2y = 4$

11. $3x + y = 6$
$y = 3x + 6$

12. $4x + 2y = 6$
$2x + y = 3$

13. $y = 2x + 4$
$x = y - 5$

14. $x + y = 3$
$2x + y = 4$

15. $y - 3x = 0$
$y = -3x$

16. $2x - y = 1$
$x = y + 1$

17. $3x + y = -1$
$x + 3y = -11$

18. $x = 3$
$y = 4$

19. $y = \frac{1}{3}x - 3$
$2x + y = 4$

20. $3x + y = -4$
$x + 6y = -7$

21. $2y = 8$
$y = -4$

22. $2x = y + 5$
$6x - 3y = 15$

23. $5x - 3y = 9$
$5y = 13 - x$

24. $x = 0$
$y = 0$

25. $x + y - 2 = 0$
$x = y - 8$

26. $y + 2x + 4 = 0$
$y = 2x + 4$

27. $7x - 4y + 7 = 0$
$3x - 5y + 3 = 0$

In 28–33, determine whether the system has no solutions, one solution, or an infinite number of solutions.

28. $x + y = 1$
$x + y = 3$

29. $x + y = 5$
$2x + 2y = 10$

30. $y = 2x + 1$
$y = 3x + 3$

31. $2x - y = 1$
$2y = 4x - 2$

32. $y - 3x = 2$
$y = 3x - 2$

33. $x + 4y = 6$
$x = 2$

In 34–37: **a.** Write a system of two linear equations.
b. Solve the system graphically and check.

34. The sum of two numbers is 3. Their difference is 1. Find the numbers.

35. The sum of two numbers is 3. The larger number is 5 more than the smaller number. Find the numbers.

36. The perimeter of a rectangle is 12 meters. Its length is twice its width. Find the dimensions of the rectangle.

37. The perimeter of a rectangle is 10 centimeters. Its length is 3 centimeters more than its width. Find the dimensions of the rectangle.

10-2 SOLVING A SYSTEM OF LINEAR EQUATIONS ALGEBRAICALLY

You already know how to solve a first-degree equation in one variable. For a system of linear equations in two variables, also called a set of **simultaneous equations,** a strategy for solution is to simplify the system to one equation in one variable.

Eliminating One Variable by Substitution

Consider a linear system. Each equation describes a relation between x and y.

$$x + y = 9$$
$$y = 2x$$

Since y is equal to $2x$ (second sentence), replace the y in the first sentence by $2x$.

$$x + y = 9$$
$$x + 2x = 9$$

Now the situation has been simplified to one equation in one unknown, which you know how to solve.

$$3x = 9$$
$$x = 3$$

Once a value for x is known, return to either relation to determine y.

$$y = 2x$$
$$y = 2(3)$$
$$y = 6$$

Check $x = 3$, $y = 6$ in the original equations. Since each equation is satisfied, $(3, 6)$ is the solution to the system.

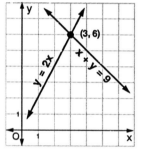

Observe that the algebraic solution of the system is connected to the graphic solution; that is, the answer $(3, 6)$ obtained algebraically is the same as the coordinates of the point of intersection on the graph.

This algebraic method of solving a system of linear equations—the use of substitution to eliminate one variable—is especially useful if one of the equations is already in function form, $y = f(x)$, expressing one of the variables in terms of the other. If not, you must begin by rewriting one of the equations. This technique is summarized as follows.

PROCEDURE

To solve a system of simultaneous linear equations in two variables by using substitution to eliminate one of the variables:

1. One equation must be in function form, $y = f(x)$.
2. Substitute $f(x)$ for y in the other equation.
3. Solve the resulting equation for x.
4. Substitute the value of x into the equation $y = f(x)$ and solve for y.
5. Check in *each* of the *given* equations.

MODEL PROBLEM

Solve the system of equations, and check: $4x + 3y = 27$ [A]

$$2x - y = 1 \quad \text{[B]}$$

How to Proceed	*Solution*
(1) In the second equation, solve for one variable in terms of the other. Solving for y in equation [B] results in equation [C], which is of the form $y = f(x).$*	$2x - y = 1 \qquad \text{[B]}$ $-y = -2x + 1$ $y = 2x - 1 \quad \text{[C]}$
(2) Substitute $f(x)$, or $2x - 1$, for y in equation [A].	$4x + 3y = 27 \qquad \text{[A]}$ $4x + 3(2x - 1) = 27$
(3) Solve the resulting equation for x.	$4x + 6x - 3 = 27$ $10x - 3 = 27$ $10x = 30$ $x = 3$
(4) Substitute 3 for x in equation [C], and solve for y.	$y = 2x - 1 \quad \text{[C]}$ $y = 2(3) - 1$ $y = 6 - 1$ $y = 5$

(5) Check: Substitute 3 for x and 5 for y in each of the original equations.

$$4x + 3y = 27 \qquad 2x - y = 1$$
$$4(3) + 3(5) \overset{?}{=} 27 \qquad 2(3) - 5 \overset{?}{=} 1$$
$$12 + 15 \overset{?}{=} 27 \qquad 6 - 5 \overset{?}{=} 1$$
$$27 = 27 ✓ \qquad 1 = 1 ✓$$

ANSWER: The solution is $(3, 5)$.

*Sometimes, you will need to do $x = f(y)$.

Now it's your turn to . . . TRY IT!

Solve the system of equations, and check: $3x - 4y = 26$

$$x + 2y = 2$$

(See page 571 for solution.)

Eliminating One Variable by Addition

When the equations of a system are both in standard form, $ax + by = c$, one variable can be eliminated by addition (or subtraction).

Consider a linear system in which both equations are in standard form.	$x + y = 7$ $2x - y = 2$
Since the coefficients of y are opposites, adding the equations eliminates y. Solve for x.	$3x + 0y = 9$ $3x = 9$ $x = 3$
Return to either relation to determine y.	$x + y = 7$ $3 + y = 7$ $y = 4$

Check $x = 3$, $y = 4$ in each of the original equations to verify that the solution to the system is (3, 4).

This algebraic method of solving a system of linear equations—using addition to eliminate one of the variables—requires that the coefficients of one of the variables be opposites.

If this is not true of the given equations, you must first apply the multiplication property of equality so that you can work with equivalent equations for which the coefficients of one of the variables are opposites. For example,

Given equations in which the coefficients of y are different:	Multiplying each term of the second equation by 2 makes the coefficients of y opposites:
$x + 2y = 7$ $3x - y = 7$	$x + 2y = 7$ $6x - 2y = 14$

After this transformation, the variable y can be eliminated by addition, yielding $7x = 21$, a single equation in one variable. Complete the solution: $x = 3$ and, by substitution in one of the original equations, $y = 2$. This technique is summarized as follows.

____ PROCEDURE
_____

To solve a system of simultaneous linear equations in two variables by using addition to eliminate one of the variables:

1. Both equations must be in standard form: $ax + by = c$
2. The coefficients of one variable must be opposites. If necessary, apply the multiplication property of equality to one or both equations to make this so.
3. When the coefficients of one variable are opposites, add the two equations to eliminate that variable. Solve the resulting equation.
4. After one value has been determined, substitute in any relation containing both variables to determine the second value.
5. Check in *each* of the *given* equations.

1. Solve the system of equations and check: $5t + v = 13$
$$4t - 3v = 18$$

How to Proceed	*Solution*
(1) The equations are already in standard form: variables lined up on the left side, constants on the right.	$5t + v = 13$ [A] $4t - 3v = 18$ [B]
(2) To make the coefficients of v opposites, multiply equation [A] by 3, resulting in equation [C].	$3(5t + v = 13)$ 3 [A] $4t - 3v = 18$ [B]
(3) Add equations [C] and [B], eliminating v. Solve for t.	$15t + 3v = 39$ [C] $\underline{4t - 3v = 18}$ [B] $19t \qquad = 57$ $t = 3$
(4) Substitute 3 for t in equation [A]. Solve for v.	$5t + v = 13$ $5(3) + v = 13$ $15 + v = 13$ $v = -2$

(5) *Check:* Substitute 3 for t and -2 for v in each of the original equations.

$$5t + v = 13 \qquad\qquad 4t - 3v = 18$$
$$5(3) + (-2) \overset{?}{=} 13 \qquad 4(3) - 3(-2) \overset{?}{=} 18$$
$$15 - 2 \overset{?}{=} 13 \qquad\qquad 12 + 6 \overset{?}{=} 18$$
$$13 = 13 ✔ \qquad\qquad 18 = 18 ✔$$

ANSWER: The solution is $(3, -2)$.

2. Solve the system of equations and check: $7x = 5 - 2y$ [A]
$$3y = 16 - 2x \quad \text{[B]}$$

How to Proceed	*Solution*
(1) Transform the equations to standard form.	$7x + 2y = 5$ [C] $2x + 3y = 16$ [D]
(2) To make the coefficients of y opposites, multiply equation [C] by 3 and equation [D] by -2, obtaining equations [E] and [F].	$3(7x + 2y = 5)$ 3 [C] $-2(2x + 3y = 16)$ -2 [D]
(3) The coefficients of y are opposites. Add equations [E] and [F], eliminating y. Solve for x. Continue, solving for y.	$21x + 6y = 15$ [E] $\underline{-4x - 6y = -32}$ [F] $17x \qquad = -17$ $x = -1$

(4) Substitute -1 for x in equation [B]. Solve for y.

$$3y = 16 - 2x \qquad \text{[B]}$$
$$3y = 16 - 2(-1)$$
$$3y = 16 + 2$$
$$3y = 18$$
$$y = 6$$

(5) *Check:* Substitute -1 for x and 6 for y in each of the original equations.

ANSWER: The solution is $(-1, 6)$.

Now it's your turn to . . . **TRY IT!**   *(See page 571 for solution.)*

Solve the system of equations, and check: $4x + y = 10$
$ x - 2y = 7$

KEEP IN MIND

Sometimes, in the process of solving two simultaneous equations, you may get a statement such as $0 = 5$, which is false. If your work contains no errors, this means that there is no solution for the pair of equations and that their graphs are parallel lines. The system is called *inconsistent*.

Sometimes, you may get a statement such as $0 = 0$, which is always true. If you have made no errors, this means that the lines coincide and that both equations represent the same line. There are infinitely many solutions to the system.

As long as there is a solution, the system is called *consistent*. If there is exactly one solution, the graphs of the lines will intersect in exactly one point, and the equations are *independent*. If there are infinitely many solutions, the graphs of the lines will coincide, and the equations are *dependent*.

EXERCISES

In 1–39, solve the system of equations using substitution to eliminate one variable. Check. If the system has no solution, indicate that the system is inconsistent.
If the system has an infinite number of solutions, indicate that the equations are dependent.

1. $y = x$
$ x + y = 14$

2. $x = y$
$ 2x + 3y = 15$

3. $x = y$
$ 5x - 5y = -2$

4. $y = 2x$
$ x + y = 21$

5. $y = 3x$
$ y - x = 18$

6. $x = 4y$
$ 2x + 3y = 22$

7. $a = -2b$
   $5a - 3b = 13$

8. $r = -3s$
   $3r + 4s = -10$

9. $-2c = d$
   $6c + 5d = -12$

10. $y = x + 1$
   $x + y = 9$

11. $x = y - 2$
   $x + y = 18$

12. $x = 5 - y$
   $x - y = 1$

13. $y = x + 3$
   $3x + 2y = 26$

14. $y = x - 2$
   $3x - y = 16$

15. $x = y + 4$
   $2x - 2y = 8$

16. $y = 2x + 1$
   $x + y = 7$

17. $y = 3x - 1$
   $7x + 2y = 37$

18. $a = 3b + 1$
   $5b - 2a = 1$

19. $a + b = 11$
   $3a - 2b = 8$

20. $r - s = 7$
   $3r - 2s = 18$

21. $x + y = 0$
   $3x + 2y = 5$

22. $3m - 2n = 11$
   $m + 2n = 9$

23. $3c - d = 1$
   $c + 2d = 12$

24. $a - 2b = -2$
   $2a - b = 5$

25. $7x - 3y = 23$
   $x + 2y = 13$

26. $4s - 2t = 4$
   $2s - t = 1$

27. $3x + 2y = 23$
   $x + 3y = 17$

28. $2x = 3y$
   $4x - 3y = 12$

29. $4y = -3x$
   $5x + 8y = 4$

30. $3a = 4b$
   $4a - 5b = 2$

31. $2x + 3y = 7$
   $4x - 5y = 25$

32. $7x + 3y = 3$
   $5x + 6y = 6$

33. $2a + 3b = 3$
   $3a + 4b = 3$

34. $y = 3x$
   $\frac{1}{3}x + \frac{1}{2}y = 11$

35. $t + u = 12$
   $t = \frac{1}{3}u$

36. $10t + u = 24$
   $t + u = \frac{1}{7}(10u + t)$

37. $x + y = 500$
   $y = 1.5x$

38. $x + y = 1,000$
   $.06x = .04y$

39. $x + y = 300$
   $.25x + .25y = 75$

In 40–102, solve the system of equations using addition to eliminate one variable. Check.
If the system has no solutions, indicate that the system is inconsistent.
If the system has an infinite number of solutions, indicate that the equations are dependent.

40. $x + y = 12$
   $x - y = 4$

41. $a + b = 13$
   $a - b = 5$

42. $r + s = -6$
   $r - s = -10$

43. $3x + y = 16$
   $2x + y = 11$

44. $m + 2n = 14$
   $3n + m = 18$

45. $c - 2d = 14$
   $c + 3d = 9$

46. $x + y = 10$
   $x - y = 0$

47. $s + r = 0$
   $r + s = 6$

48. $a - 4b = -8$
   $a - 2b = 0$

49. $x + 2y = 8$
   $x - 2y = 4$

50. $4r + 3s = 29$
   $2r - 3s = 1$

51. $8a + 5b = 9$
   $2a - 5b = -4$

52. $4x + 5y = 23$
   $4x - y = 5$

53. $2a + 3b = 2$
   $5a + 3b = 14$

54. $-2m + 4n = 13$
   $6m + 4n = 9$

55. $3a - b = 3$
$a + 3b = 11$

56. $3r + 9s = 6$
$r + 3s = 10$

57. $4x - y = 10$
$2x + 3y = 12$

58. $5m + 3n = 14$
$2m + n = 0$

59. $5x + 4y = 27$
$x - 2y = 11$

60. $2c - d = -1$
$c + 3d = 17$

61. $2m + 4n = 18$
$m + 2n = 9$

62. $3a - b = 13$
$2a + 3b = 16$

63. $r - 3s = -11$
$3r + s = 17$

64. $a + 3b = 4$
$2a - b = 1$

65. $3x - y = 5$
$5x - 2y = 8$

66. $3x + 4y = 26$
$x - 3y = 0$

67. $5x + 8y = 1$
$3x + 4y = -1$

68. $x - y = -1$
$3x - 2y = 3$

69. $5a - 2b = 3$
$2a - b = 0$

70. $5x - 2y = 20$
$2x + 3y = 27$

71. $x + 4y = 2$
$2x + 8y = 2$

72. $5a + 3b = 17$
$4a - 5b = 21$

73. $2x + 3y = 6$
$3x + 5y = 15$

74. $5r - 2s = 8$
$3r - 7s = -1$

75. $3x + 7y = -2$
$2x + 3y = -3$

76. $4x + 3y = -1$
$5x + 4y = 1$

77. $4a - 6b = 15$
$6a - 4b = 10$

78. $3x + 8y = 16$
$5x + 10y = 25$

79. $2x + y = 17$
$5x = 25 + y$

80. $5r + 3s = 30$
$2r = 12 - 3s$

81. $x - 2y = 8$
$2y = x - 16$

82. $3x + 4y = 16$
$4x = 2y + 14$

83. $6r = s$
$5r = 2s - 14$

84. $3a - 7 = 7b$
$4a = 3b + 22$

85. $3x - 4y = 2$
$x = 2(7 - y)$

86. $6(y + 6) = 2x$
$3x - 9y = 54$

87. $3x + 5(y + 2) = 1$
$8y = -3x$

88. $\frac{1}{3}x + \frac{1}{4}y = 10$
$\frac{1}{3}x - \frac{1}{2}y = 4$

89. $\frac{2}{3}x + \frac{3}{4}y = 2$
$\frac{1}{6}x + \frac{1}{2}y = -2$

90. $\frac{1}{2}a + \frac{1}{3}b = 8$
$\frac{3}{2}a - \frac{4}{3}b = -4$

91. $c - 2d = 1$
$\frac{2}{3}c + 5d = 26$

92. $a - \frac{2}{3}b = 4$
$\frac{3}{5}a + b = 15$

93. $2a = 3b$
$\frac{2}{3}a - \frac{1}{2}b = 2$

94. $.04x + .06y = 26$
$x + y = 500$

95. $.03x + .05y = 17$
$x + y = 400$

96. $.03x = .06y + 9$
$x + y = 600$

97. $x + 2y = 12$
$\dfrac{x}{y} = 1 + \dfrac{6}{y}$

98. $x - 8y = -2$
$\dfrac{3x}{2y} + 1 = \dfrac{10}{y}$

99. $2x + y = 23$
$\dfrac{x - 6}{3y} = \dfrac{1}{5}$

100. $3d = 13 - 2c$
$\dfrac{3c + d}{2} = 8$

101. $3x = 4y$
$\dfrac{3x + 8}{5} = \dfrac{3y - 1}{2}$

102. $\dfrac{a}{3} + \dfrac{a + b}{6} = 3$
$\dfrac{b}{3} - \dfrac{a - b}{2} = 6$

10-3 SOLVING VERBAL PROBLEMS BY USING TWO VARIABLES

You have previously learned how to solve word problems by using one variable. Frequently, a problem can be solved more easily by using two variables rather than one. Recall that using a table or diagram in problem solving is often helpful.

PROCEDURE

To solve word problems by using a system of linear equations involving two variables:

1. Use two different variables to represent the different unknown quantities.
2. Translate two relationships from the problem into a system of two equations.
3. Solve the system of equations to determine the answer(s) to the problem.
4. Check the answer(s) in the original word problem.

MODEL PROBLEMS

1. The sum of two numbers is 10. Three times the larger, decreased by twice the smaller, is 15. Find the numbers.

How to Proceed	*Solution*
(1) Represent the two different unknown quantities by two different variables.	Let ℓ = the larger number. Let s = the smaller number.

(2) Translate two relationships from the problem into a system of two equations.

$$\text{The sum of two numbers is 10.} \qquad \ell + s = 10 \quad [A]$$

$$\begin{array}{l}\textit{Three times the larger, decreased}\\ \textit{by twice the smaller, is 15.}\end{array} \qquad 3\ell - 2s = 15 \quad [B]$$

(3) Solve the system of equations. Multiply equation [A] by 2, obtaining equation [C]. Add equations [C] and [B]. Solve for ℓ.

$$\begin{array}{rl} 2\ell + 2s &= 20 \quad [C]\\ 3\ell - 2s &= 15 \quad [B]\\ \hline 5\ell &= 35\\ \ell &= 7 \end{array}$$

In [A], substitute 7 for ℓ, and solve for s.

$$\begin{array}{rl} \ell + s &= 10\\ 7 + s &= 10\\ s &= 3 \end{array}$$

(4) Check in the original word problem: The sum of the larger number, 7, and the smaller number, 3, is 10. ✔ Three times the larger, decreased by twice the smaller, $(3 \times 7) - (2 \times 3)$, equals $21 - 6$, or 15. ✔

ANSWER: The larger number is 7. The smaller number is 3.

2. A lion cub is running in the direction leading to a deep chasm, and his mother is afraid that he will not see the danger in time. The chasm is 1,800 meters from the lioness. The cub is running at 480 m/min., and the lioness races after him at 660 m/min. If the cub had a 450-meter head start, will his mother catch him in time?

Solution: Set up a system of two equations. Let t = the number of minutes and d = the total distance. Use the formula *distance = rate × time*.

$$\text{lioness:} \quad d = 660t \qquad \text{[A]}$$
$$\text{cub:} \quad d = 480t + 450 \quad \text{[B]}$$

line [A]

t	0	1	2
d	0	660	1,320

line [B]

t	0	1	2
d	450	930	1,410

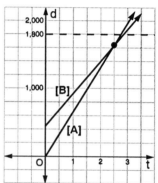

It can be seen from the graph that the lioness catches up with the cub at a distance that is less than 1,800 meters.

ANSWER: The lioness will catch the cub in time.

EXERCISES

In 1–17, solve the problem by using two variables.

1. The sum of two numbers is 105. The smaller number is 5 less than the larger number. Find the numbers.

2. Sue and Carrie together weigh 105 kilograms. Carrie's weight is 15 kilograms less than twice Sue's weight. Find each girl's weight.

3. The perimeter of a rectangle is 50 centimeters. The difference between the length and the width of the rectangle is 9 centimeters. Find the dimensions of the rectangle.

4. One sunny Saturday, Candy challenged Matt to a sales competition. Candy won, selling 15 more than twice as many boxes of candy as Matt sold mats. If the sum of their sales was 87 items, how many boxes of candy did Candy sell?

5. Al has 3 times as much money as Ken. They have a total of $76 between them. How much money does each have?

6. Mrs. Swenson is 10 times as old as her daughter. Eight years from now, she will be 4 times her daughter's age. Find the present age of each.

7. There are 12 more students in Mrs. Reilly's class than in Mr. Horn's class. One-third of Mrs. Reilly's students and half of Mr. Horn's students, a total of 24 students, went on a school trip. How many students remained in each class?

8. Irene has $4.50 in nickels and dimes in her coin bank. The number of dimes is twice the number of nickels. How many coins of each type does Irene have?

9. Jill collects nickels and quarters. Her bank now holds 30 coins amounting to $5.10. How many of each coin does she have?

10. Mrs. Rinaldo changed a $100 bill in a bank. She received $20 bills and $10 bills. The number of $20 bills was 2 more than the number of $10 bills. How many $20 bills did she get?

11. Linda spent $2 for stamps. Some were 20-cent stamps and the rest were 40-cent stamps. The number of 20-cent stamps was 2 less than the number of 40-cent stamps. How many stamps of each kind did Linda buy?

12. It cost the town $350 every time an entrance door to a public office was widened to accommodate a wheelchair. It cost $540 every time a ramp was built to provide wheelchair access to a public building. If the town spent $12,850 to provide larger doorways or ramps at 34 locations, how many ramps were built?

13. The Answer Expert gives wild guesses for $2 and guaranteed correct answers for $3. If he collected $239 from 92 paying customers, how many guaranteed correct answers did he give?

14. The carpentry work in Mrs. White's remodeled den cost $1,000. She paid the carpenter with a personal check for $878 and the balance in $1 and $5 bills. If the number of $1 bills exceeded the number of $5 bills by 14, how many of each kind of bill did she give?

15. Mr. Mix has 6 more quarters than dimes in his pocket. If the number of coins of each type were interchanged, he would have a total of $5.50 in his pocket. How much does he now have in his pocket?

16. Mrs. Sweet bought some bran muffins at 89¢ each and some corn muffins at 79¢ each. If she paid $16.20 for 20 muffins, how many of each type did she buy?

17. Ms. Scarpia buys a $1.65 toll token at the Arapahoe Bridge with two kinds of coins.

   a. How many of each kind of coin does she use if she pays with:
      (1) 13 coins, consisting of quarters and nickels?
      (2) 12 coins, consisting of quarters and dimes?

   b. Using two kinds of coins, can she pay with:
      (1) 17 coins?   (2) 14 coins?   (3) 9 coins?

18. Paul and Hilary are on diet programs. Paul has already lost 7 pounds, and expects to continue to lose weight at the rate of $2\frac{1}{4}$ pounds a week. Hilary is just starting her diet, and hopes that by losing $3\frac{1}{2}$ pounds a week, she can match Paul's total weight loss before their upcoming high school reunion. If each of them diets as planned, can Hilary match Paul's weight loss in less than 2 months? [Use a graph in the solution.]

19. Jaimie bought 100 shares each of stock in three companies, manufacturers of dog food, organic fertilizer, and camping gear, respectively. The dog food shares, which were bought at $2\frac{1}{2}$ (dollars per share), went down by $\frac{1}{4}$ on each successive business day. The fertilizer shares were bought at $1\frac{3}{8}$, and went up by $\frac{1}{8}$ each day. The camping gear shares, bought at 2, remained steady. [Use a graph in the solution.]

    a. On what days was the value of the camping gear stock higher than that of either of the others?

    b. After one week, which stock had the highest value?

20. A plane left Los Angeles, traveling at 525 mph on a transpacific flight. When the plane had covered 1,500 miles, a second plane started along the same route, flying at 600 mph. If the destination of both planes was Tokyo, a distance of about 5,600 miles, which would arrive first? [Use a graph in the solution.]

Mixture Problems

MODEL PROBLEM

    At her gift shop, Patti uses herbs, spices, and dried flowers to make scented mixtures she calls Patti's Potpourri. For a woodsy effect, Patti is mixing dried pine needles priced at $1.20 per ounce with dried rosemary priced at 70 cents per ounce. If she wishes to make 5 pounds of the mixture to sell at $1.00 per ounce, find the number of ounces of each of the dried materials she should use.

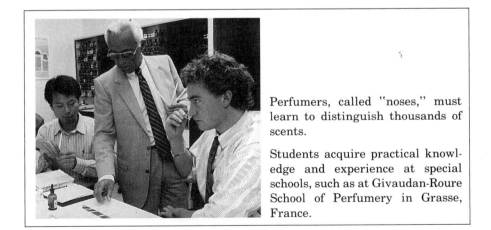

Perfumers, called "noses," must learn to distinguish thousands of scents.

Students acquire practical knowledge and experience at special schools, such as at Givaudan-Roure School of Perfumery in Grasse, France.

Solution: To set up a system of two equations, it is helpful to display the data in a table. Be sure that the units of measure are consistent. Express all prices in terms of cents, and all weights in terms of ounces.

Let p = the number of ounces of pine needles and
r = the number of ounces of rosemary.

Kind of dried material	Number of ounces	Price per ounce in cents	Total value in cents
pine	p	120	$120p$
rosemary	r	70	$70r$
mixture	80	100	$100(80)$

The total number of ounces of the dried mixture is 80.

$$p + r = 80 \qquad \text{[A]}$$

The total value of the pine and rosemary used is 100(80) cents.

$$120p + 70r = 100(80) \quad \text{[B]}$$

To solve the system:

(1) Make the coefficients of r opposites by multiplying equation [A] by -70, resulting in equation [C].

$$-70(p + r = 80) \quad -70 \text{ [A]}$$
$$120p + 70r = 8,000 \qquad \text{[B]}$$

(2) Add equations [C] and [B].

$$\begin{array}{rl} -70p - 70r = -5,600 & \text{[C]} \\ \underline{120p + 70r = 8,000} & \text{[B]} \\ 50p = 2,400 & \\ p = 48 & \end{array}$$

(3) Substitute 48 for p in [A], and solve for r.

$$\begin{array}{rl} p + r = 80 & \text{[A]} \\ 48 + r = 80 & \\ r = 32 & \end{array}$$

(4) *Check:* $48 + 32 = 80$ ✔

Value of 48 oz. @ \$1.20 = \$57.60
Value of 32 oz. @ \$0.70 = <u>\$22.40</u>
Value of 80 oz. @ \$1.00 = \$80.00 ✔

ANSWER: Patti should use 48 oz. of pine and 32 oz. of rosemary.

In 1–8, solve the problem by using two variables.

1. A dealer has some seed worth $2.00 per kilogram and some worth $3.00 per kilogram. He wishes to make a mixture of 80 kilograms that he can sell for $2.20 per kilogram. How many kilograms of each kind should he use?

2. A bake shop has cookies worth $2.70 per pound and cookies worth $4.50 per pound. How many pounds of each kind should the manager of the shop use in order to have 120 pounds of cookies to sell at $3.75 per pound?

3. A merchant mixes candy selling at $1.20 per pound with candy selling at $2.10 per pound to make a 50-pound mixture that he sells at $1.56 per pound. How many pounds of each kind of candy should he mix?

4. How many grams of spice worth $4.20 per gram must be mixed with 30 grams of spice worth $3.60 per gram in order to produce a mixture to sell for $4.00 per gram?

5. Regular gasoline selling at $1.29 per gallon is mixed with premium gasoline selling at $1.49 per gallon to create an intermediate blend selling for $1.43 per gallon. How much premium gasoline should be used in 10 gallons of the intermediate blend?

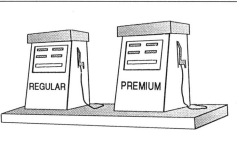

A fuel tends to "knock" during combustion, resulting in a loss of power and a strain on the engine. *Octane*, one of the compounds contained in petroleum, can overcome this tendency.

The quality of a gasoline is described by its octane rating. The higher the octane rating, the less knock, and the more smoothly and efficiently the engine can run.

6. Shady Growth Nursery sells a dozen of a pink blossom variety of mountain laurel shrubs for $18. A dozen shrubs of a hardier, but not as attractive, variety of mountain laurel are sold for $9. If the nursery sells a mix of the two types for $12 a dozen, how many of the hardy laurel should be included?

7. Dulcie's Deli sells ham for $4.89 per pound and cheese spread for $2.49 per pound. A deli special is Savory Spread, made of the cheese spread mixed with ham. If it is sold at $3.39 per pound, how many ounces of ham should be in each pound of Savory Spread?

8. Raisins selling for $3.85 per pound are mixed with nuts selling for $6.25 per pound to make a raisin-nut mixture that sells for $5.00 per pound. How many ounces of nuts will be in 3 pounds of the mixture?

Business Problems

The owner of a men's clothing store bought 6 ties and 8 hats for $140. A week later, at the same prices, he bought 9 ties and 6 hats for $132. Find the price of a tie and the price of a hat.

Solution: Let t = the price of a tie and h = the price of a hat.

$$6 \text{ ties and } 8 \text{ hats cost } \$140. \qquad 6t + 8h = 140 \quad \text{[A]}$$

$$9 \text{ ties and } 6 \text{ hats cost } \$132. \qquad 9t + 6h = 132 \quad \text{[B]}$$

To solve the system:

(1) Make the coefficients of t opposites by multiplying [A] by 3 and [B] by -2, obtaining [C] and [D].

$$3(6t + 8h = 140) \qquad 3\,[A]$$
$$-2(9t + 6h = 132) \quad -2\,[B]$$

(2) Add equations [C] and [D]. Solve for h.

$$
\begin{array}{rl}
18t + 24h = 420 & \text{[C]} \\
-18t - 12h = -264 & \text{[D]} \\
\hline
12h = 156 & \\
h = 13 &
\end{array}
$$

(3) Substitute 13 for h in [A], and solve for t.

$$
\begin{array}{rl}
6t + 8h = 140 & \text{[A]} \\
6t + 8(13) = 140 & \\
6t + 104 = 140 & \\
6t = 36 & \\
t = 6 &
\end{array}
$$

(4) *Check:*

6 ties and 8 hats cost $6(\$6) + 8(\$13)$
$$= \$36 + \$104 = \$140 ✓$$

9 ties and 6 hats cost $9(\$6) + 6(\$13)$
$$= \$54 + \$78 = \$132 ✓$$

ANSWER: A tie cost $6. A hat cost $13.

In 1–11, solve the problem by using two variables.

1. Six boxes of oranges and 5 boxes of grapefruits cost $142. At the same time, 3 boxes of oranges and 2 boxes of grapefruits cost $64. Find the cost of one box of each.

2. On one day, 4 chefs and 5 helpers earned $650. On another day, working the same number of hours and at the same rate of pay, 5 chefs and 6 helpers earned $800. How much does a chef earn and how much does a helper earn each day?

3. A Little League baseball manager bought 4 bats and 9 balls for $191. On another day, she bought 3 bats and 1 dozen balls at the same prices and paid $201. How much did she pay for each bat and each ball?

4. A customer bought 3 cans of string beans and 5 cans of tomatoes for $2.31. The next customer bought 2 cans of string beans and 3 cans of tomatoes for $1.45. Find the cost of 1 can of each.

5. Mrs. Rubero paid $37.40 for 4 pounds of walnuts and 3 pounds of pecans. At the same prices, Mrs. Perez paid $33.00 for 5 pounds of walnuts and 1 pound of pecans. Find the price per pound for each kind of nut.

6. Eight roses and 9 lilies cost $45.60. At the same prices, one dozen roses and 5 lilies cost $48.00. Find the cost of one rose and the cost of one lily.

7. A manufacturer makes two types of electric irons, model R and model S. He sold one customer 6 irons of model R and 7 irons of model S. The bill for this sale was $262. He sold another customer 4 irons of model R and 3 irons of model S. The bill for this sale was $138. What was the price of a model R iron, and what was the price of a model S iron?

8. The Math Club is buying some boxes of stationery and some pens to use as raffle prizes. A box of stationery costs more than a pen. If 3 boxes of stationery and 2 pens cost $18.44, and if the difference between the costs of 5 boxes of stationery and of 3 pens is $20.98, how much is a box of stationery?

9. When Randi bought her word processor, she purchased 3 additional print wheels and 2 memory disks for $21.60. The next customer bought 4 print wheels and 5 memory disks for $34.40. What was the price for each print wheel and each memory disk?

10. At the Storytown Theme Park, a group of 8 adults and 4 children paid $65 in admissions. Another group consisting of 6 adults and 10 children paid $71.50 in admissions. What was the price of admission for each adult and each child?

11. Mr. Jonas is ordering subscriptions to TV Tune-In and Health Watch magazines. He can purchase either 15 issues of TV Tune-In and 12 issues of Health Watch for $13.56 or 24 issues of TV Tune-In and 18 issues of Health Watch for $21.12. What is the price per issue for each magazine?

Investment Problems

MODEL PROBLEM

Mrs. Rich invested some money in 8% bonds and twice as much in 10% bonds. Her annual income from these bonds was $420. Find the amount she invested in each type.

Solution:

Let x = the amount in 8% bonds and y = the amount in 10% bonds.

Investment	Principal in dollars	Annual rate of interest	Annual income in dollars
8% bonds	x	.08	$.08x$
10% bonds	y	.10	$.10y$

The amount invested at 10% is
twice the amount invested at 8%. $y = 2x$ [A]

The total income was $420. $.08x + .10y = 420$ [B]

To solve the system:

(1) Eliminate y by substituting equation [A] into [B].

$$.08x + .10y = 420 \qquad \text{[B]}$$
$$.08x + .10(2x) = 420$$
$$8x + 10(2x) = 42{,}000$$
$$8x + 20x = 42{,}000$$
$$28x = 42{,}000$$
$$x = 1{,}500$$

(2) Substitute 1,500 for x in [A], and solve for y.

$$y = 2x \qquad \text{[A]}$$
$$y = 2(1{,}500)$$
$$y = 3{,}000$$

(3) *Check:* The amount invested at 10%, namely $3,000, is twice as much as the amount invested at 10%, namely $1,500. ✔

The total income is $420: $.08(1{,}500) + .10(3{,}000) = 120 + 300 = \$420$ ✔

ANSWER: The amount invested at 8% is $1,500.
The amount invested at 10% is $3,000.

In 1–10, solve the problem by using two variables.

1. Mrs. Moto invested $1,400, part at 5% and the rest at 8%. Her total annual income from these investments was $100. Find the amount she invested at each rate.

2. Part of $10,400 was invested at 6% and the rest was invested at 7%. The total annual income from these investments is $686. Find the number of dollars invested at each rate.

3. A sum of $7,000 is invested in two parts. One part earns interest at 5% and the other at 8%. The total annual interest is $500. Find the amount invested at each rate.

4. A woman invested some money at 5% and twice as much money at 7%. If the total yearly income from the two investments is $760, how much was invested at each rate?

5. A man invested an amount of money at 10%. He also invested $400 more than the first amount at 8%. The annual incomes from these investments were equal. How much was invested at each rate?

6. Mrs. Washington invested $8,000 at 7%. How much additional money must she invest at 4% in order that her total annual income may equal 5% of her entire investment?

7. Mr. and Mrs. Daniels invested $20,000, part in 6% tax-free bonds and the remainder in 5% tax-free mutual funds. If the total yearly income from these investments was $1,120, how much was invested at each rate?

8. Harold invested $25,500 in two Certificates of Deposit (CDs). One CD was paying 9% annual interest and the other CD was paying 7% interest. If the annual interest on the 7% CD exceeded the annual interest on the 9% CD by $105, how much was invested in each CD?

9. Mr. Duarte invested $240,000 of his company's funds, part in a bond fund paying 6% interest yearly and the remainder in a stock fund paying $6\frac{1}{2}$% yearly. If Mr. Duarte were to reverse the amounts invested in each fund, the annual income from the investments would be increased by $300. How much did he invest in each fund?

10. Mr. Petrov has a sum of money invested at 8% annual interest and another sum, $2,800 greater than the first, in an insured investment paying 6% annual interest. His total annual income from both investments is $756.

   a. How much is invested at each rate?

   b. What is the annual income from the 8% investment?

   c. Mr. Petrov gets a raise and decides to add to the sum he has invested at 8%. How much must he add if he wants the total annual income from both investments to go up to $800?

Digit Problems

In the decimal numeration system, every integer can be written by using only the digits 0, 1, 2, 3, 4, 5, 6, 7, 8, 9.

In a multidigit integer, each place has a value that is 10 times the value of the place immediately at its right. For example, 734 means 7 hundreds plus 3 tens plus 4 ones.

$$734 = 7(100) + 3(10) + 4(1)$$

In general, to represent a two-digit number with tens digit t and units digit u, write $t(10) + u(1)$, or more simply $10t + u$. You may not represent the two-digit number by tu because tu means "t times u."

KEEP IN MIND

If t represents the tens digit and u represents the units digit of a two-digit number:

$10t + u$ represents the original number.

$10u + t$ represents the original number with its digits reversed.

$t + u$ represents the sum of the digits of the original number.

MODEL PROBLEM

In a two-digit number, the sum of the digits is 9. The number is 12 times the tens digit. Find the number.

Solution: Let t = the tens digit.   *The sum of the digits is 9.* $t + u = 9$ [A]
          Let u = the units digit.   *The number is*
          $10t + u$ = the number.   *12 times the tens digit.* $10t + u = 12t$ [B]

To solve the system:

(1) Rewrite equation [B], obtaining [C]. $t + u = 9$ [A]
    Subtract [C] from [A]. $\underline{-2t + u = 0}$ [C]
    Solve for t. $3t = 9$
                                                  $t = 3$

(2) Substitute 3 for t in [A], and solve $t + u = 9$ [A]
    for u. $3 + u = 9$
                                                  $u = 6$

Find the number. $10t + u = 10(3) + 6 = 36$

(3) *Check:* $3 + 6 = 9$ and $36 = 12(3)$

ANSWER: The number is 36.

In 1–16, solve the problem by using two variables.

1. In a two-digit number, the sum of the digits is 10 and the difference of the digits is 4. Find the number if the tens digit is larger than the units digit.

2. The tens digit in a two-digit number is 2 more than twice the units digit. The sum of the digits is 11. Find the number.

3. The units digit in a two-digit number exceeds the tens digit by 6. The sum of the digits is 10. Find the number.

4. The tens digit of a two-digit number is 2 less than 4 times the units digit. The difference between the tens digit and the units digit is 4. Find the number.

5. The units digit of a two-digit number exceeds twice the tens digit by 3. The sum of the digits is 6. Find the number.

6. The units digit of a two-digit number is 4 more than the tens digit. The number is 6 times the units digit. Find the number.

7. The tens digit of a two-digit number is 2 less than the units digit. The number is 4 times the sum of the digits. Find the number.

8. The tens digit of a two-digit number exceeds 3 times the units digit by 1. The number is 8 times the sum of the digits. Find the number.

9. The tens digit of a two-digit number exceeds the units digit by 3. The number is 1 more than 8 times the sum of the digits. Find the number.

10. The units digit of a two-digit number is 11 less than twice the tens digit. The number is 6 less than 7 times the sum of the digits. Find the number.

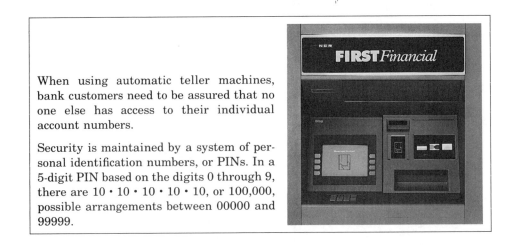

When using automatic teller machines, bank customers need to be assured that no one else has access to their individual account numbers.

Security is maintained by a system of personal identification numbers, or PINs. In a 5-digit PIN based on the digits 0 through 9, there are $10 \cdot 10 \cdot 10 \cdot 10 \cdot 10$, or 100,000, possible arrangements between 00000 and 99999.

11. The sum of the digits of a two-digit number is 11. If 45 is added to the number, the order of the digits is reversed. Find the number.

12. The tens digit of a two-digit number is 1 more than 4 times the units digit. If 63 is subtracted from the number, the order of the digits is reversed. Find the number.

13. In a 3-digit house number, the first two digits are the reverse of the last two digits. The sum of the three digits is 17. The number formed by the last two digits is 4 less than 2 times the number formed by the first two digits. What is the house number?

14. Witnesses agreed that the license number of an antacid salesman's car ended with the letters MUCH and that this was preceded by 2 digits. They further agreed that the digits added up to 10 and that, if reversed, the two-digit number would be decreased by 54. What was the complete license number of the salesman's car?

15. Rita's 3-number locker combination begins with a two-digit number and ends with the reverse of the same number. The middle number, which is the difference of the other two numbers, is 18. If the sum of the first and last numbers is 44 and the first number is smaller than the last, what is Rita's locker combination?

16. In a 4-digit inventory code number, the last two digits are the reverse of the first two digits. The sum of the four digits is 20. The number formed by the first two digits is 2 less than 3 times the number formed by the last two digits. What is the code number?

Motion Problems Involving Currents

Suppose that a boat can travel at the rate of 20 mph in a body of still water, where there is no current. When the boat travels in a body of water where there is a current, the boat will move faster than 20 mph when it is traveling downstream, with the current. It will move slower than 20 mph when it is traveling upstream, against the current. For example, if the boat is traveling in a river that is flowing at the rate of 3 mph, the rate of the boat traveling downstream will be $20 + 3$, or 23 mph. Its rate traveling upstream will be $20 - 3$, or 17 mph.

The speed of an airplane is similarly affected by an air current.

_____ **K**EEP **I**N **M**IND _____

If r = the rate in still water or in still air,

and c = the rate of the water current or air current,

then $r + c$ = the rate traveling with the current,

and $r - c$ = the rate traveling against the current.

MODEL PROBLEM

A motorboat can travel 120 kilometers downstream in 3 hours. It requires 5 hours to make the return trip against the current. Find the rate of the boat in still water and the rate of the current.

Solution:

Let r = the rate of the boat in still water and
c = the rate of the current.

	Rate × (km/h)	Time = (h)	Distance (km)
Downstream	$r + c$	3	$3r + 3c$
Upstream	$r - c$	5	$5r - 5c$

The distance downstream is 120 kilometers. $3r + 3c = 120$ [A]

The distance upstream is 120 kilometers. $5r - 5c = 120$ [B]

To solve the system:

(1) Make the coefficients of c opposites by multiplying [A] by 5, yielding [C], and [B] by 3, yielding [D].

$$5(3r + 3c = 120) \quad 5\,[A]$$
$$3(5r - 5c = 120) \quad 3\,[B]$$

(2) Add equations [C] and [D].
Solve for r.

$$
\begin{array}{ll}
15r + 15c = 600 & [C] \\
\underline{15r - 15c = 360} & [D] \\
30r \qquad\;\; = 960 & \\
r = 32 &
\end{array}
$$

(3) Substitute 32 for r in [A], and solve for c.

$$
\begin{array}{ll}
3r + 3c = 120 & [A] \\
3(32) + 3c = 120 & \\
96 + 3c = 12 & \\
3c = 24 & \\
c = 8 &
\end{array}
$$

(4) *Check:*   Rate downstream = 32 + 8 or 40 km/h

Rate upstream   = 32 − 8 or 24 km/h

Distance downstream = 3(40) = 120 km

Distance upstream   = 5(24) = 120 km

ANSWER: The rate of the boat in still water is 32 kilometers per hour.
The rate of the current is 8 kilometers per hour.

In 1–11, solve the problem by using two variables.

1. A motorboat at maximum speed can travel 180 km in 2 hours going with the current and in 3 hours traveling against the same current. What is the speed of the boat in still water?

2. A plane flies at a constant speed for 6 hours against the wind and travels 1,800 miles. With the same wind, the plane returns in 5 hours. What is the wind speed?

3. A boat is rowed downstream a distance of 16 kilometers in 2 hours and then the same distance upstream in 8 hours. Find the rate of rowing in still water.

4. A boat travels 18 miles downstream in 2 hours, and requires 6 hours to travel back to its starting point. What is the rate of the current?

5. A plane left an airport and flew with the wind for 4 hours, covering 2,000 kilometers. It then returned over the same route against the same wind in 5 hours. Find the rate of the plane in still air and the speed of the wind.

6. Flying with a wind, a plane traveled 2,160 miles in 6 hours. Flying against the same wind, the plane covered only $\frac{2}{3}$ of this distance in the same time. Find the speed of the plane in still air and the speed of the wind.

7. A plane flew a distance of 3,110 kilometers in 5 hours. During the first 3 hours of the flight, it flew with a wind a distance of 1,950 kilometers. During the remainder of the flight, the plane flew against a wind whose average speed was 10 km/h less than what it had been during the first part of the flight. Find the original speed of the wind.

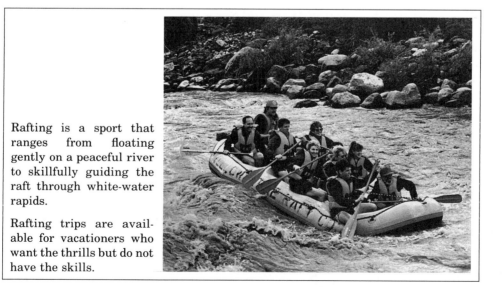

Rafting is a sport that ranges from floating gently on a peaceful river to skillfully guiding the raft through white-water rapids.

Rafting trips are available for vacationers who want the thrills but do not have the skills.

8. Jim and Jodi Evans went on a day cruise to an amusement park 30 miles downriver. The outbound trip took 3 hours. The return trip, going upriver, took 5 hours. Find the speed of the boat in still water and the rate of the current.

9. A plane can travel a distance of 3,300 miles in 6 hours when it is going with the wind. When flying against the same wind, the plane takes 36 minutes longer to return by the same route. Find the speed of the plane in still air and the rate of the wind.

10. A boat can travel 200 miles in 4 hours when it is assisted by the current. However, when the current acts against the boat on the return trip, it can cover only $\frac{3}{5}$ of the distance in the same length of time. Find the rate of the boat in still water and the rate of the current.

11. In a storm, a plane flew against a wind and took 9 hours to cover 3,600 miles. The return trip, traveling with the wind, took 1 hour less. Find the speed of the plane in still air.

Miscellaneous Problems

In 1–16, solve the problem by using two variables.

1. Tickets for a school football game cost $1.00 each if purchased before the day of the game. They cost $1.50 each if bought at the gate. For a particular game, 600 tickets were sold, with receipts of $700. How many tickets were sold at the gate?

2. The perimeter of a rectangle is 28 centimeters. Three times the length increased by 4 times the width is 48 centimeters. Find the dimensions of the rectangle.

3. Mr. Stratton traveled for 7 hours and covered 500 kilometers on the trip. During the first part of the trip, he averaged 80 km/h. During the second part, he averaged 60 km/h. How many hours did he spend on each part of the trip?

4. The sum of Godfrey's age and Marilyn's age is 60 years. Ten years ago, Godfrey was 3 times as old as Marilyn was then. How old is each now?

5. Harriet bought 50 stamps for $14. Some were 20-cent stamps. The rest were 40-cent stamps. How many of each kind did she buy?

6. If 3 is subtracted from the numerator of a fraction, the value of the fraction becomes $\frac{2}{3}$. If 10 is added to the denominator of the original fraction, the value of the fraction becomes $\frac{1}{2}$. Find the original fraction.

7. If the numerator of a fraction is decreased by 1 and the denominator is increased by 5, the value of the fraction becomes $\frac{1}{2}$. If the numerator of the original fraction is increased by 1 and the denominator of the original fraction is decreased by 6, the value of the resulting fraction becomes $\frac{4}{3}$. Find the original fraction.

8. The following lottery data was given for New Jersey and Connecticut:

	Number of Winning Tickets		Sum of Total Winnings in Both States	Difference Between Total Winnings in Each State
Date	N.J.	Conn.		
Jan. 6	5	1	$108,000	
Jan. 13	3	2		$5,000

How much did each winning ticket pay if the individual prize in each state was the same for both weeks?

9. To fill a prescription, a pharmacist can use either 4 size A vials or 3 size B vials of medicine. There are 5 ounces more medicine in 9 B vials than in 2 A vials. How many ounces are in each size vial?

10. The Aces have played more games than the Devils during the basketball season so far. The Aces averaged 52 points a game and the Devils averaged 54 points a game. If the difference between their total points was 70, and the sum of their total points was 1,906, find the number of games played by each team.

11. Robin and Michelle were playing a "nonsense word" game. In this game, the following relationships existed: 3 DUFS equal 4.2 FANTICKS and 1 DUF plus 1 FANTICK equal 12 MORNS. How many MORNS are in a FANTICK?

12. On New Year's Eve, Carrie and Jodi weighed the same amount. Both made resolutions to go on diets beginning January 1st. Their results were as follows: Carrie lost 1 pound every 10 weeks and Jodi gained $\frac{1}{4}$ pound every 5 weeks. After 40 weeks, the average of their weights was 139 pounds. What were their weights on New Year's Eve?

13. Clubs A and B combined have a total of 121 members. If Club A were to increase to 4 times its present membership, and Club B were to decrease its membership by half, their total combined membership would consist of 148 members. How many members are in Club A and Club B at present?

14. Mr. Blanc said to Mr. Adam, "If you give me one dollar, I will have twice as much money as you have." Mr. Adam responded, "Oh, no! If you give me one dollar, we will have the same amount of money." How much money does each have now?

15. Lunch meat at $5.39 per pound is sold with cheese at $2.99 per pound to make a 6-pound deli tray selling for $23.94. How many ounces of lunch meat are on the tray?

16. A pet food manufacturer combines an all-cereal dog food selling at 56 cents for a 14-ounce can with an all-beef dog food selling at 99 cents for a 9-ounce can. How many ounces of each should be in a 1-pound can selling for 99 cents?

10-4 SOLVING A SYSTEM OF LINEAR INEQUALITIES GRAPHICALLY

The solution set of a system of two or more inequalities in two variables consists of all the ordered pairs that satisfy the open sentences of the system.

PROCEDURE

To solve a pair of linear inequalities graphically:

1. Graph one inequality in a coordinate plane.
   Draw its plane divider and shade the appropriate half-plane.

2. Graph the second inequality using the same set of coordinate axes.
   Draw its plane divider and shade the appropriate half-plane.

3. Find the common solution, which is the set of all ordered pairs in the region that has been shaded twice.

4. Check by choosing a point in the twice-shaded region, and verifying that its coordinates satisfy both inequalities.

MODEL PROBLEMS

1. Graph the solution set of the system: $x > 2$
$$y \leq -2$$

Solution:

(1) Graph $x > 2$.
   Draw the plane divider $x = 2$ (dotted, since the inequality does not include =) and shade the half-plane to the right (since the inequality is >).

(2) Using the same set of axes, graph $y \leq -2$.
   Draw the plane divider $y = -2$ (solid, since the inequality does include =) and shade the half-plane below (since the inequality is <).

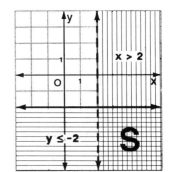

(3) The solution set of the system is in the twice-shaded region labeled S.

(4) Check that, for example, $(4, -3)$ satisfies both inequalities.

2. Graph the solution set of the system: $x + y \geq 4$
$$y < 2x - 3$$

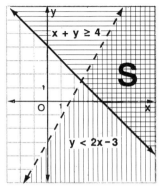

Solution:

(1) Graph $x + y \geq 4$.
   Draw the plane divider $x + y = 4$ (use solid line) and shade the half-plane above.

(2) Using the same set of axes, graph $y < 2x - 3$.
   Draw the plane divider $y = 2x - 3$ (use dotted line) and shade the half-plane below.

(3) The solution set of the system is in the twice-shaded region labeled S.

(4) Check that, for example, (5, 2) satisfies both inequalities.

3. Mr. Lopez is planning a barbecue. He wants to buy no more than 8 pounds of ground beef for making hamburgers and meat loaf. He needs at least 3 pounds for the hamburgers and at least 1 pound for the meat loaf. Show the number of pounds of hamburgers and the number of pounds of meat loaf he can make.

Solution: Set up a system of inequalities, and graph.
   Let h = number of pounds of hamburger and m = number of pounds of meat loaf.

No. of pounds of ground beef is no more than 8.	$h + m \leq 8$	(vertical stripes)
No. of pounds of hamburgers is at least 3.	$h \geq 3$	(gray shading)
No. of pounds of meat loaf is at least 1.	$m \geq 1$	(horizontal stripes)

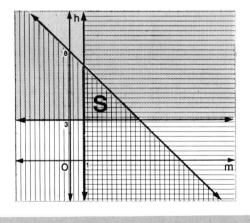

Some possibilities in the solution set S are:

1 pound of meat loaf and
3 to 7 pounds of hamburgers

5 pounds of hamburgers and
1 to 3 pounds of meat loaf

Now it's your turn to . . . **TRY IT!**   (See page 571 for solution.)

Graph the solution set of the system: $2x + y \leq 3$
$$x - y < 1$$

In 1–21, graph the solution set of the system in a coordinate plane.
Label the solution set S.

1. $x > 1$
$y > -2$

2. $x \leq 2$
$y \geq 3$

3. $x > 0$
$y < 0$

4. $x < 0$
$y > 0$

5. $y \geq x$
$x < 2$

6. $y \leq x$
$x \geq -1$

7. $y \geq 2x$
$y > x - 1$

8. $y \leq 2x + 3$
$y \geq -x$

9. $y - x \geq -2$
$y - 2x \leq 1$

10. $x + y > 3$
$x - y < 1$

11. $x - y \leq -2$
$x + y \geq 2$

12. $2x + y > -4$
$x - y \leq 1$

13. $2x + y \leq 4$
$x + y - 2 > 0$

14. $2x + 3y \geq 9$
$x + y - 4 \leq 0$

15. $5x - 3y - 2 \geq 0$
$4x + 2y + 2 < 0$

16. $y \geq x$
$x = 0$

17. $x + y \leq 3$
$y - 2x = 0$

18. $y - 2x - 2 > 0$
$x + y - 2 = 0$

19. $x \leq 2$
$y \leq 2$
$x + y \geq 2$

20. $x + 3y \leq 6$
$2y \geq 2$
$x \geq -2$

21. $y \geq x$
$x + y - 3 \geq 0$
$x - 4 \leq 0$

In 22–29, graph the solution set in a coordinate plane. Label the solution set S.

22. $1 < x < 4$ **23.** $-3 \leq x \leq 2$ **24.** $2 < y \leq 4$ **25.** $-2 \leq y \leq 3$

26. The combined annual sales of the Argo Corporation and the Bilco Corporation were less than 8 million dollars. The Argo Corporation had sales of not less than 3 million dollars and the Bilco Corporation had sales greater than 2 million dollars. What were the possible sales of the Argo and Bilco Corporations for that year?

27. Dr. Fixx prescribes a combination of vitamins and minerals for Mrs. Cass, the total daily intake to be at least 9 units, but no more than 12 units. How many units of vitamins and how many of minerals should she take daily?

28. A gardener uses at most 30 pounds of fertilizer on his daily route. One type that he uses contains nitrogen and potassium in the ratio 4:1. The other type contains nitrogen and potassium in the ratio 3:2. If he has at least 15 pounds of the first type, and at least 6 pounds of the second type, how many pounds of each fertilizer does he use?

29. The Amgas refinery produces at most 30,000 barrels of gasoline and diesel oil in a week. If at least 10,000 barrels must be gasoline and at least 6,000 barrels must be diesel oil, how many barrels of each type of fuel are produced?

10-5 SOLVING A SYSTEM OF THREE LINEAR EQUATIONS

To solve a system in two variables, you know how to apply the strategy of reducing the problem to a simpler situation. This same strategy applies to an algebraic solution for a system in three variables. That is, you will make the situation simpler by first reducing the problem to a system in two variables, and finally to a single equation in one variable.

MODEL PROBLEMS

1. Solve the system of linear equations and check:
$$\begin{aligned} x + 3y + 2z &= 13 \quad [A] \\ x - 2y + 3z &= 6 \quad [B] \\ 2x + 2y - z &= 3 \quad [C] \end{aligned}$$

How to Proceed — *Solution*

(1) Use any pair of the three equations to eliminate one variable.

Work with [A] and [B] to eliminate x. Since the coefficients of x are the same, subtract [B] from [A], obtaining equation [1].

$$\begin{aligned} x + 3y + 2z &= 13 \quad [A] \\ x - 2y + 3z &= 6 \quad [B] \\ \hline 5y - z &= 7 \quad [1] \end{aligned}$$

(2) Use a different pair from the three equations to eliminate the same variable.

Work with [B] and [C] to eliminate x. To get the coefficients of x to be opposites, multiply [B] by -2 and then add the result to [C], obtaining equation [2].

$$\begin{aligned} -2x + 4y - 6z &= -12 \quad -2\,[B] \\ 2x + 2y - z &= 3 \quad [C] \\ \hline 6y - 7z &= -9 \quad [2] \end{aligned}$$

(3) The system has been reduced to two equations in two variables, [1] and [2].

Work with [1] and [2] to eliminate z. To get the coefficients of z to be opposites, multiply [1] by -7 and then add the result to [2]. Solve for y.

$$\begin{aligned} -35y + 7z &= -49 \quad -7\,[1] \\ 6y - 7z &= -9 \quad [2] \\ \hline -29y &= -58 \\ y &= 2 \end{aligned}$$

(4) Replace y by 2 in either equation [1] or [2], and solve for z.

$$\begin{aligned} 5y - z &= 7 \quad [1] \\ 5(2) - z &= 7 \\ 3 &= z \end{aligned}$$

(5) Replace y by 2 and z by 3 in any equation containing the three variables, and solve for x.

$$\begin{aligned} x + 3y + 2z &= 13 \quad [A] \\ x + 3(2) + 2(3) &= 13 \\ x &= 1 \end{aligned}$$

(6) Check $x = 1$, $y = 2$, $z = 3$ in each of the three original equations.

ANSWER: The solution is the ordered triple (1, 2, 3).

2. The sum of the digits of a three-digit number is 14. The units digit is equal to the sum of the hundreds digit and the tens digit. The number with the digits reversed exceeds the original number by 297. Find the original number.

Solution: Use three variables to model the problem.

Let h = the hundreds digit,
t = the tens digit, and
u = the units digit.

Then the number = $100h + 10t + u$ and
the number reversed = $100u + 10t + h$.

The sum of the digits is 14.	$h + t + u = 14$	[A]
The units digit equals the sum of the hundreds and tens digits.	$u = h + t$	[B]
The number reversed exceeds the original number by 297.	$100u + 10t + h = 100h + 10t + u + 297$	[C]

To solve the system of three equations:

(1) Work with [A] and [B] to eliminate u.
In [A], substitute $h + t$ for u.
Simplify to obtain equation [1].

$$h + t + u = 14 \quad [A]$$
$$h + t + (h + t) = 14$$
$$2h + 2t = 14$$
$$h + t = 7 \quad [1]$$

(2) Work with [B] and [C], also to eliminate u.
In [C], substitute for u.
Simplify to obtain [2].
Here, [2] happens to give a value for t.

$$100u + 10t + h = 100h + 10t + u + 297 \quad [C]$$
$$100(h + t) + 10t + h = 100h + 10t + (h + t) + 297$$
$$100h + 100t + 10t + h = 100h + 10t + h + t + 297$$
$$101h + 110t = 101h + 11t + 297$$
$$99t = 297$$
$$t = 3 \quad [2]$$

(3) Replace t by 3 in equation [1] and solve for h.

$$h + t = 7 \quad [1]$$
$$h + 3 = 7$$
$$h = 4$$

(4) Replace t by 3 and h by 4 in any equation containing the three variables, and solve for u.

$$u = h + t \quad [B]$$
$$u = 4 + 3$$
$$u = 7$$

Find values for the number and for the number reversed.

$$100h + 10t + u = 100(4) + 10(3) + 7 = 437$$
$$100u + 10t + h = 100(7) + 10(3) + 4 = 734$$

Check: The sum of the digits is 14.

$$4 + 3 + 7 \overset{?}{=} 14$$
$$14 = 14 \checkmark$$

The units digit equals the sum of the hundreds and tens digits.

$$7 \overset{?}{=} 4 + 3$$
$$7 = 7 \checkmark$$

The number reversed exceeds the original number by 297.

$$734 \overset{?}{=} 437 + 297$$
$$734 = 734 \checkmark$$

ANSWER: The original number is 437.

NOTE. Sometimes, studying the equations closely can lead to a shorter solution. For example:

In [A], replace $(h + t)$ by u. Solve for u.

$$h + t + u = 14 \quad \text{[A]}$$
$$u + u = 14$$
$$2u = 14$$
$$u = 7$$

In [C], subtract $10t$ from each side, substitute 7 for u, and simplify.

$$100u + 10t + h = 100h + 10t + u + 297 \quad \text{[C]}$$
$$100(7) + \cancel{10t} + h = 100h + \cancel{10t} + 7 + 297$$
$$700 + h = 100h + 304$$
$$396 = 99h$$
$$4 = h$$

Replace u by 7 and h by 4 in any equation containing the three variables, and solve for t.

$$h + t + u = 14$$
$$4 + t + 7 = 14$$
$$t + 11 = 14$$
$$t = 3$$

EXERCISES

In 1–12, solve the system of linear equations, and check.

1. $x - 2y + z = 1$
$3x - y + 2z = -3$
$2x + 2y - z = 2$

2. $-3x - y - 2z = -1$
$x - 2y - z = 3$
$2x + y + z = 2$

3. $2x + 2y - 3z = 8$
$4x - y + z = -1$
$y + z = 1$

4. $x + 2y + 3z = -4$
$2x - y - z = 6$
$x + 3y + 2z = -2$

5. $2x + 3y = 1$
$6y + z = 6$
$x + z = 4$

6. $2x + 3y + 5z = 4$
$4x - 6y - 5z = -2$
$2x - 3y - 10z = -4$

7. $2x - y + 3z = 9$
$x + 3y - 4z = -5$
$3x + 2y + 2z = 13$

8. $3x + 2y - z = 3$
$x - 2y + 2z = -7$
$2x + 3y - 3z = 7$

9. $2x + 2y - 4z = -1$
$x - 4y + 8z = 12$
$3x + y + 2z = 4.5$

10. $x - y + 2z = 6$
$2x + 3y - 4z = 3$
$3x - 2y + 2z = 15$

11. $3x - 2y + 4z = -15$
$x + 3y - z = -4$
$2x - y + 2z = -9$

12. $x + y = 2$
$2x + 3z = 22$
$3y - z = -13$

In 13–21, use three variables to solve the problem.

13. The tens digit of a three-digit number is twice the units digit. The sum of the digits is 9. The hundreds digit exceeds the tens digit by 4. Find the number.

14. The units digit of a three-digit number is three times the hundreds digit. The sum of the digits is 12. The difference between the units digit and the tens digit is three times the difference between the hundreds digit and the tens digit. Find the number.

15. The sum of the digits of a three-digit number is 16. The sum of the hundreds digit and the tens digit exceeds the units digit by 2. The number with the digits reversed exceeds the original number by 297. Find the original number.

16. The sum of the digits of a three-digit number is 11. The units digit exceeds the tens digit by 1. If the original number is subtracted from the number obtained when the digits are reversed, the difference exceeds the original number by 52. Find the original number.

17. Mr. Karsh has 23 coins in his pocket consisting of nickels, dimes, and quarters. If he has 3 more dimes than nickels and 2 more quarters than dimes, find the number of coins of each kind that he has.

18. Ms. Gibbs invested a total of $20,000 in three stocks paying, respectively, 4%, 5%, and 6% annual interest. The annual interest from the 4% stock was equal to the annual interest from the 6% stock. If the total interest from the stocks was $980, how much did she invest at each rate?

19.

Item	A	B	C	Total Daily Sales
Cookies	10	12	15	327
Candy Bars	24	18	20	576
Juice	16	0	14	318

Mr. Fontain has 3 types of vending machines (A, B, C) that offer cookies, candy bars, and juice. The chart shows the number of items held by each type of machine, and the total daily sales for each type of machine. How many of each type of machine did he own?

20. For different hourly wages, Alex, Bonita, and Cara work part-time at Jim's Bargain Store. On Saturday, they each work 5 hours and earn a total of $120. On Tuesday, Alex works 4 hours and Bonita works 3 hours, earning between them $54. On Friday, Bonita works 4 hours and Cara works 5 hours, earning between them $80. Find the hourly rate for which each person works.

21. A pool can be filled by three pipes A, B, C. The pool can be filled if pipe C is open for 6 hours, B is open for 4 hours, and A remains closed. It can be filled if A is open for 5 hours, B is open for 2 hours, and C is open for 3 hours. It can be filled if A is open for 5 hours, C is open for 6 hours, and B remains closed. Find how long it would take each pipe alone to fill the pool.

10-6 SOLVING A SYSTEM OF LINEAR EQUATIONS BY USING MATRICES

Consider another algebraic technique for solving a system of two equations in two variables. In this procedure, you will again work with the two equations simultaneously, starting with both of them in standard form, $ax + by = c$. This time, your aim will be to get the equations into a form where $a = 0$ in one of the equations and $b = 0$ in the other. For example: $x + 0y = 15$ If equations are in this form,

$$0x + y = 9 \quad \text{you can read } x = 15 \text{ and } y = 9.$$

Here is a display, a *matrix*, of the constants of the equations above.

$$\begin{bmatrix} 1 & 0 & 15 \\ 0 & 1 & 9 \end{bmatrix}$$

Observe that if the coefficients of the variables are of a certain form,

first equation: $a = 1, b = 0$
second equation: $a = 0, b = 1$

$$\begin{array}{cc} & a \quad b \\ Eq.\ 1 & \begin{bmatrix} 1 & 0 & 15 \\ 0 & 1 & 9 \end{bmatrix} \\ Eq.\ 2 \end{array}$$

then the last column of the matrix displays the solution of the system: $x = 15, y = 9$

In general, the solution to a system of two equations, each in standard form $ax + by = c$, can be read from a matrix. The solution, $x = r, y = s$, is displayed in the last column of the matrix when the other elements of the matrix are 0 and 1, as shown.

$$\begin{bmatrix} 1 & 0 & r \\ 0 & 1 & s \end{bmatrix}$$

You have just seen the end result, what the equations must look like to read the solution. Now, let us start with a system that is not in the final form to see how to achieve that result.

The system in standard form
$$ax + by = c$$
$$x - y = 6$$
$$-x + 2y = 3$$

The system in matrix notation
$$\begin{array}{cc} & a \quad\ \ b \quad c \\ Eq.\ 1 & \begin{bmatrix} 1 & -1 & 6 \\ -1 & 2 & 3 \end{bmatrix} \\ Eq.\ 2 \end{array}$$

<u>Aim:</u> *To get the required value for a in each equation.*

In the first equation, a must be 1. This is already so.

In the second equation, a must be 0.

You must now operate on the equations of the system simultaneously to achieve an equation in which $a = 0$. The equation that results from adding the original equations has $a = 0$; that is, $(x - y = 6) + (-x + 2y = 3)$ yields $0x + y = 9$.

The properties of equality allow replacement of an equation of a system by an equation derived from the system. Thus, $0x + y = 9$ can be used as a replacement.

<u>Do:</u> *Replace the second equation by the sum of the original equations.*

$$x - y = 6$$
$$0x + y = 9$$

$$\begin{bmatrix} 1 & -1 & 6 \\ 0 & 1 & 9 \end{bmatrix}$$

In matrix notation, this operation is: Replace Row 2 by (Row 1 + Row 2).

Aim: To get the required value for b in each equation.

In the first equation, b must be 0.

The equation that results from adding the last pair of equations has $b = 0$; that is, $(x - y = 6) + (0x + y = 9)$ yields $x + 0y = 15$.

Do: Replace the first equation of the last pair by the sum of the equations.

$$\begin{array}{r} x + 0y = 15 \\ 0x + y = 9 \end{array} \qquad \begin{bmatrix} 1 & 0 & 15 \\ 0 & 1 & 9 \end{bmatrix}$$

In matrix notation, this operation is: Replace Row 1 by (Row 1 + Row 2).

In the second equation, b must be 1. This is already so.

Since the system is now in required form, read the solution from the last column of the matrix: x = 15, y = 9

$$\begin{bmatrix} 1 & 0 & 15 \\ 0 & 1 & 9 \end{bmatrix}$$

This algebraic technique of solving a system of linear equations—making the coefficients of the variables conform to a certain pattern so that the solutions to the system can be read from the equations—is more easily carried out using matrix notation.

This technique is summarized as follows:

PROCEDURE

To solve a system of linear equations in two variables by using matrix notation:

1. Both equations must be in standard form: $ax + by = c$

2. Write the system in matrix notation, using the constants of the system.

   The constants of the first equation form the first row of the matrix, and the constants of the second equation form the second row.

$$\begin{array}{c} \\ Eq.\ 1 \\ Eq.\ 2 \end{array} \begin{bmatrix} a & b & c \\ a_1 & b_1 & c_1 \\ a_2 & b_2 & c_2 \end{bmatrix}$$

3. Operate on the rows of the matrix, replacing individual rows by combinations of existing rows, so that the final matrix has the values 1 and 0 in the places shown.

$$\begin{bmatrix} 1 & 0 & r \\ 0 & 1 & s \end{bmatrix}$$

   When the matrix is in the required form, read the solution of the system from the last column: $x = r, y = s$

4. Check in each of the original equations.

1. Using matrix notation, solve the system: $3x + y = 4$
$$2x + y = 2$$

Solution:

(1) Both equations are in standard form.

(2) Write the system in matrix notation.

$$\begin{bmatrix} a_1 & b_1 & c_1 \\ a_2 & b_2 & c_2 \end{bmatrix} = \begin{bmatrix} 3 & 1 & 4 \\ 2 & 1 & 2 \end{bmatrix}$$

(3) To solve the system, use row operations on the matrix:

To get $a_1 = 1$: Replace Row 1 by (Row 1 − Row 2)

$$\begin{bmatrix} 3-2 & 1-1 & 4-2 \\ 2 & 1 & 2 \end{bmatrix} = \begin{bmatrix} 1 & 0 & 2 \\ 2 & 1 & 2 \end{bmatrix}$$

To get $a_2 = 0$: Replace Row 2 by (Row 2 − 2 · Row 1)

$$\begin{bmatrix} 1 & 0 & 2 \\ 2-2(1) & 1-2(0) & 2-2(2) \end{bmatrix} = \begin{bmatrix} 1 & 0 & 2 \\ 0 & 1 & -2 \end{bmatrix}$$

The values of b_1 and b_2 are already as required: $b_1 = 0$ and $b_2 = 1$

Read the solution to the system from the last column of the matrix: $x = 2, y = -2$

(4) Check $x = 2, y = -2$ in each of the original equations.

ANSWER: The solution is $(2, -2)$.

2. Using matrix notation, solve the system: $y = -1 - 4x$
$$3x + 2y = -2$$

Solution:

(1) Rewrite the equations in standard form. $4x + y = -1$
$$3x + 2y = -2$$

(2) Write the system in matrix notation. $\begin{bmatrix} a_1 & b_1 & c_1 \\ a_2 & b_2 & c_2 \end{bmatrix} = \begin{bmatrix} 4 & 1 & -1 \\ 3 & 2 & -2 \end{bmatrix}$

(3) To solve the system, use row operations on the matrix:

To get $a_1 = 1$: Replace Row 1 by (Row 1 − Row 2)

$$\begin{bmatrix} 4-3 & 1-2 & -1-(-2) \\ 3 & 2 & -2 \end{bmatrix} = \begin{bmatrix} 1 & -1 & 1 \\ 3 & 2 & -2 \end{bmatrix}$$

To get $a_2 = 0$: Replace Row 2 by (Row 2 − 3 · Row 1)

$$\begin{bmatrix} 1 & -1 & 1 \\ 3-3(1) & 2-3(-1) & -2-3(1) \end{bmatrix} = \begin{bmatrix} 1 & -1 & 1 \\ 0 & 5 & -5 \end{bmatrix}$$

To get $b_1 = 0$: Replace Row 1 by (Row 1 + $\frac{1}{5}$ Row 2)

$$\begin{bmatrix} 1 + \frac{1}{5}(0) & -1 + \frac{1}{5}(5) & 1 + \frac{1}{5}(-5) \\ 0 & 5 & -5 \end{bmatrix} = \begin{bmatrix} 1 & 0 & 0 \\ 0 & 5 & -5 \end{bmatrix}$$

To get $b_2 = 1$: Replace Row 2 by ($\frac{1}{5}$ Row 2)

$$\begin{bmatrix} 1 & 0 & 0 \\ \frac{1}{5}(0) & \frac{1}{5}(5) & \frac{1}{5}(-5) \end{bmatrix} = \begin{bmatrix} 1 & 0 & 0 \\ 0 & 1 & -1 \end{bmatrix}$$

Read the solution from the last column of the matrix: $x = 0$, $y = -1$

(4) Check $x = 0$, $y = -1$ in each of the original equations.

ANSWER: The solution is $(0, -1)$.

NOTE. When doing row operations to put the matrix in required form, it is sometimes easier to get $b_2 = 1$ before trying to get $b_1 = 0$. For Model Problem 2:

After getting the required values for a_1 and a_2, get $b_2 = 1$ by replacing Row 2 by ($\frac{1}{5} \cdot$ Row 2). Then get $b_1 = 0$ by replacing Row 1 by (Row 1 + Row 2).

EXERCISES

In 1–30, solve the system using matrix notation. Check.

1. $2x + y = 8$
$x + y = 4$

2. $x + 2y = 5$
$x + y = 2$

3. $x - y = 3$
$-x + 2y = 2$

4. $x + y = 3$
$x - y = 5$

5. $3x - 3y = 12$
$x + y = 0$

6. $2x + 4y = -4$
$x + 5y = 1$

7. $2x + y = 8$
$y = x - 7$

8. $3x - 2y = -7$
$y = 2x + 5$

9. $2x + 3y = 18$
$x = 2y - 5$

10. $x - 6y = 13$
$3x + y = 1$

11. $4x - 3y = 2$
$8x + 6y = 8$

12. $2x + 4y = -4$
$y = x - 4$

13. $5x + 5y = 26$
$10x - 10y = 48$

14. $6x - 2y = 12$
$3x + 3y = -10$

15. $x + 3y = 0$
$2x - 5y = 11$

16. $4x - 3y = 29$
$x - 5y = 37$

17. $2x + 4y = 7$
$x = 1 + 8y$

18. $x + 6y = -18$
$4x - y = 3$

19. $3x + 5y = -2$
$y = 9x - 10$

20. $4x + 6y = 3$
$2x - y = \frac{5}{6}$

21. $3x - y = -12$
$x = 10 - 2y$

22. $2x + y = 4$
$x - 3y = -15.5$

23. $3x - 2y = 6$
$x = 2 - 7y$

24. $x + 2y = -5$
$3x - y = 13$

25. $5x - y = -27$
$x + \frac{1}{2}y = -4$

26. $3x + 2y = 0$
$6x - 3y = -7$

27. $3x + 5y = 8$
$x = 4y - 3$

28. $x = 7 - 9y$
$10x - 27y = 5$

29. $5x - 4y = 24$
$x = 1 - 3y$

30. $2x - y = 1$
$x + y = 0.2$

CHAPTER SUMMARY

An equation can describe a relation between two or more variables. A set of related equations, called a *system* or a set of *simultaneous equations*, is used to describe different relations between the same variables.

A solution to a system of two equations is an ordered pair of coordinates that satisfy both equations in the system. The techniques used to solve a system are:

1. *graphic*

   When the equations in a system are graphed on a single set of axes, the lines may:

   a. intersect in one point, giving a single solution.
   b. be parallel, giving no solution.
   c. coincide, giving an infinite number of solutions.

   If a system has a solution, the system is *consistent*.
   An *inconsistent* system has no solution.

2. *algebraic*

   Simplify the system by blending the two equations—either by substitution or addition—to eliminate one variable, thus reducing the problem to one equation in one variable.

   This algebraic technique is extended to a system with three variables by first reducing the problem to an equivalent system in two variables.

   There are other algebraic techniques for solving systems. Some of these involve matrix notation.

A *system of inequalities* also represents different relations between the same variables. Since there are usually infinitely many ordered pairs that satisfy a system of inequalities, the solution set is best represented graphically.

VOCABULARY CHECKUP

SECTION

10-1 *system of linear equations*

10-2 *simultaneous equations / consistent / inconsistent / dependent / independent*

In 1–3, solve the system of equations graphically. (Section 10-1)

1. $y = x - 3$
$y = -2x$

2. $x + y = 3$
$x - y = -1$

3. $4x - 3y = -6$
$2x - y = -4$

In 4–9, solve the system of equations using addition to eliminate one variable. If the system has no solutions, indicate that the system is inconsistent. If the system has an infinite number of solutions, indicate that the equations are dependent. (Section 10-2)

4. $3x + 2y = 5$
$3x - 4y = -19$

5. $x - y = 4$
$3y = 3x - 12$

6. $2x - 3y = 8$
$4x - 6y = 8$

7. $2x + 4y = 7$
$6y = -3x + 2$

8. $x = 2y - 3$
$y = 2x + 3$

9. $3x + 2y = 8$
$2x + 3y = 8$

In 10–12, solve the system of equations using substitution. (Section 10-2)

10. $x = y - 4$
$y = x + 4$

11. $2x + 3y = 7$
$x + 2y = 5$

12. $4x - 3y = -12$
$3x + y = 4$

In 13–20, solve the problem by using two variables. (Section 10-3)

13. A grocer mixed nuts worth $3.40 per pound with nuts worth $2.20 per pound. How many pounds of each kind did he use to make a mixture of 60 pounds to sell at $3.00 per pound?

14. At a dance there were 750 men and women. Two times the number of women who attended was 300 more than the number of men who attended. How many men and how many women attended the dance?

15. At the toll booth on a bridge, a coin machine that takes only quarters and dimes contains $10.50. The number of dimes in the machine is 3 less than twice the number of quarters. How many coins of each type are there in the machine?

16. Miss Valdez invested $10,000, part in a savings account at 6% and the rest in 8% bonds. If the annual interest on the bonds is $240 more than the annual interest paid on the savings account, how much did she invest at each rate?

17. Mr. Amato can travel 36 miles upstream in his motorboat in 3 hours. If he can return in 2 hours, how fast can the motorboat travel in still water?

18. In a two-digit number, the tens digit is 5 greater than the units digit. The number is 63 greater than the sum of its digits. Find the number.

19. A sports manager bought 7 shirts and 4 belts for $118. Another manager, who paid the same prices, paid $77 for 5 shirts and 2 belts. Find the cost of a shirt and the cost of a belt.

20. The perimeter of a rectangle is 78 inches. The length of the rectangle is twice its width. Find the dimensions of the rectangle.

In 21 and 22, graph the solution set of the system of linear inequalities. Label the solution set S. (Section 10-4)

21. $3x - y > 4$
$\quad x + y < 3$

22. $x + 3y \leq 6$
$\quad 2x - 3y \geq 3$

23. At Pete's Fountain Drinks, a large cherry cola consists of 5 to 8 ounces of cherry drink mixed with enough cola to fill a 20-oz. glass. Use a graph in the coordinate plane to show how much of each ingredient is in a large cherry cola. (Section 10-4)

In 24–26, solve the system of linear equations. (Section 10-5)

24. $2x - y + 3z = -1$
$\quad 3x + y - 5z = 15$
$\quad 4x - 3y - 2z = 15$

25. $x - y - z = -9$
$\quad 3x + 2y + 4z = 5$
$\quad 5x - 3y + 3z = -27$

26. $3x - 7y + 2z = -2$
$\quad 7x + 4y + 3z = 7$
$\quad 4x + 5y - 4z = 44$

27. The hundreds digit of a three-digit number exceeds the tens digit by 4. The sum of the three digits is 13. The original number is 99 more than the number formed by reversing the digits. Find the number. (Section 10-5)

In 28–30, solve the system using matrix notation. (Section 10-6)

28. $x - 3y = 5$
$\quad 2x - 3y = 7$

29. $5x + 3y = -3$
$\quad 3x + 2y = -1$

30. $4x - 5y = 8$
$\quad 3x + y = 6$

PROBLEMS FOR PLEASURE

1. Solve the system to find the values of w, x, y, and z.

$$w + x = 2$$
$$x + y = 1$$
$$2x - y + z = 8$$
$$2w + 3y + z = 0$$

2. Al, Brenda, and Carl had a combined bowling score of 509. Brenda's score is 83 pins less than the sum of Al's score and Carl's score. The total of Brenda's score and Carl's score is 3 pins less than 3 times Al's score. What was each person's bowling score?

3. An author was in England born a long, long time ago;
The year in which this man was born you really ought to know.

The sum of digits first and last of that fateful year,
Will equal digit number two I'm sure you're glad to hear.

Now decrease digit number three by one and you will get,
Digit number two again and you're not finished yet.

Three times digit number four will give result the same
As two times digit number three, the last hint in this game.

In what year was this playwright born, and if the year you find,
Give the name of this great man, if it comes to mind.

1. You have two columns of numbers from 5 to 9, as shown below. For each number in Column A, select one number from Column B, multiply the two numbers, and write down the product. The numbers you used may not be selected again. When all numbers have been selected, find the total of all the products. Your object is to find the number combinations that will give the highest possible total.

Col. A	Col. B	Product
5	5	5 × ____ = _____
6	6	6 × ____ = _____
7	7	7 × ____ = _____
8	8	8 × ____ = _____
9	9	9 × ____ = _____

Total _____

2. Nita bought decorations for a Thanksgiving social. Turkey centerpieces sold for $5 each, paper tablecloths were $1.50 each, and balloons were 25¢ each. Nita bought a total of 50 of these items (at least one of each), at a cost of $50. How many of each did she buy?

COLLEGE TEST PREPARATION

In 1–17, select the letter of the correct answer.

1. If the sum of two numbers is 36 and their difference is 12, what is the sum of the reciprocals of the two numbers?
   (A) $\frac{1}{4}$ (B) $\frac{3}{8}$ (C) $\frac{1}{8}$ (D) $\frac{2}{5}$ (E) $\frac{1}{16}$

2. If $x + y = 3 + k$ and $2x + 2y = 10$, then k is (A) 7 (B) 6 (C) 4 (D) 3 (E) 2

3. If $2x + 3y = k$ and $4x = 7y$, then y equals
   (A) $\frac{k}{10}$ (B) $k - 6\frac{1}{2}$ (C) $\frac{2k}{10}$ (D) $\frac{2k}{13}$ (E) $\frac{k}{13}$

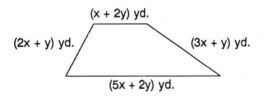

(x + 2y) yd.

(2x + y) yd.   (3x + y) yd.

(5x + 2y) yd.

4. The region above is to be enclosed in fencing that can be purchased in rolls of $(2x + y)$ yards each. How many rolls should be purchased to build the fence?
   (A) 6 (B) 5 (C) 4 (D) 3 (E) 2

5. If $p - q = 3$ and $p + 3q = 9$, what is the average of p and q?
   (A) 5 (B) 4 (C) −3 (D) 3 (E) $2\frac{1}{2}$

6. Which coordinates name a member of the solution set of the system shown?
   $$x > 3$$
   $$x + y < 2$$
   (A) (4, −1) (B) (4, −3)
   (C) (3, −3) (D) (5, 1) (E) (5, −1)

7. If $2x + 5y = 5$ and $x - 2y = 7$, then the average of x and y is
   (A) 12 (B) 6 (C) 4 (D) 3 (E) 2

8. If $2x + 3y = 7$ and $x + 5y = -9$, then $2(3x + 8y)$ equals
   (A) 2 (B) −2 (C) 4 (D) −4 (E) −3

9. If $2x + 3y = 3$ and $x + 2y = 1$, then $x + y$ is what percent of $3x + 5y$?
   (A) 25 (B) $33\frac{1}{3}$ (C) 50
   (D) 75 (E) 80

10. If $2p + 3q = 13$ and $2p + q = 5$, then $4p + 2q =$
    (A) 6 (B) 8 (C) 10 (D) 12 (E) 15

11. One number is 3 more than twice another number. If the difference of the two numbers is 62, the sum of the two numbers is
(A) 180   (B) 171   (C) 164
(D) 135   (E) 121

12. What are the coordinates of the point of intersection of the graphs of the equations $x + y = 1$ and $y = x + 7$?
(A) (4, 1)   (B) (−3, 4)   (C) (1, 7)
(D) (7, 1)   (E) (−3, 7)

13. The perimeter of a rectangle is 56 cm. Three times the length increased by 4 times the width is 96 cm. Three times the width increased by 4 times the length is equal to
(A) 86   (B) 92   (C) 100
(D) 104   (E) 112

14. Part of $8,000 is invested at 7% interest and the rest is put in a bank at 6% interest. The total annual interest from both investments is $530. The difference between the amounts invested is
(A) $1,000   (B) $1,500   (C) $2,000
(D) $2,400   (E) $2,750

15. In a certain fraction, if 1 is subtracted from both the numerator and denominator, the resulting fraction equals $\frac{1}{2}$. If 5 is added to both the numerator and denominator of the original fraction, the resulting fraction equals $\frac{4}{5}$. The sum of the numerator and denominator of the original fraction is
(A) 6   (B) $7\frac{1}{2}$ (C) 8   (D) $8\frac{1}{2}$ (E) 9

16. The grades on a difficult test were adjusted. The top grade of 100 remained 100, but the grade 16 became the passing mark 65 on the new scale. If the old grades x are in a linear relation to the new grades y, and the original average grade was 28, what is the new average?
(A) 40   (B) 50   (C) 60
(D) 70   (E) 80

17. The total of n nickels and d dimes is $2.10. If the numbers of nickels and dimes are interchanged, the value becomes $1.95. How many coins were there?
(A) 15   (B) 18   (C) 20   (D) 27   (E) 30

Questions 18–22 each consist of two quantities, one in Column A and one in Column B. You are to compare the two quantities and choose:

A   if the quantity in Column A is greater;
B   if the quantity in Column B is greater;
C   if the two quantities are equal;
D   if the relationship cannot be determined from the information given.

1. In certain questions, information concerning one or both of the quantities to be compared is centered above the two columns.

2. In a given question, a symbol that appears in both columns represents the same thing in Column A as it does in Column B.

3. x, n, and k, etc. stand for real numbers.

Column A	Column B
$x + y = 5$	
$x - y = 2$	
18. $x^2 + x - y - y^2$	10
$2x + y = 4$	
$3x + 2y = 6$	
19. $x + y$	$x - y$
$3x + 2y = 7$	
$-x - 4y = 2$	
20. $x - y$	4
$x + y + z = 3$	
$x - y - z = -1$	
$2x - y - 2z = 0$	
21. $\dfrac{x - y}{z - x}$	$\dfrac{z - x}{x - y}$

A system of equations of the form $ax + by = c$ is represented by the matrix $\begin{bmatrix} 2 & 1 & 1 \\ -1 & 2 & 12 \end{bmatrix}$.

22. $(x + y)^2$	$x^2 + y^2$

1. Solve for x: $\frac{3}{4}x - 5 = 13$

2. If the measure of an angle is represented by $y°$, give the measure of its complement in terms of y.

3. Find the value of the expression $\frac{1}{2}gt^2$ when $g = -32$ and $t = 3$.

4. Write an equation of the line whose slope is 2 and whose y-intercept is 5.

5. For every integer $m > 1$, let $F(m)$ be defined to be the greatest prime number that is a factor of m. For example, $F(12) = 3$ and $F(20) = 5$. Find $F(63) - F(45)$.

6. Solve for y: $y^2 + 5y - 24 = 0$

7. In the figure, $JKLM$ is a square. If K has coordinates $(3, 4)$, what are the coordinates of L?

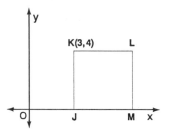

8. Find the slope of the line through the points $(-1, 4)$ and $(7, -2)$.

9. If the perimeter of a square is represented by $12a$ inches, represent the area in terms of a.

10. The measures of the angles of a triangle are in the ratio $4:5:9$. Find the measure of the largest angle of the triangle.

11. Divide, and write the quotient in scientific notation: $\dfrac{60 \times 10^2}{1.2 \times 10^{-3}}$

12. The lengths of the base and the altitude of a rectangle are represented by $2n - 1$ and $n + 2$ respectively. If the perimeter of the rectangle is 44 cm, find the value of n.

13. If the probability that an event will occur is $\frac{7}{12}$, what is the probability that the event will not occur?

14. If a number is multiplied by 3 and added to 9, the result is 8 less than 4 times the original number. Find the original number.

15. Given the matrix product as shown, find the value of x. $\begin{bmatrix} 2 & -1 \\ 3 & 0 \end{bmatrix} \begin{bmatrix} 1 & x \\ -4 & -2 \end{bmatrix} = \begin{bmatrix} 6 & 12 \\ 3 & 15 \end{bmatrix}$

16. Factor completely: $m - m^3$

17. Find the slope of the line through the points $(1, 3)$ and $(3, 7)$.

18. The Plattsville soccer team has won 5 games this season, and lost 3 games. If the team can win the next g games, it will have won 75% of all the games it played. Find the value of g.

10-1 SOLVING A SYSTEM OF LINEAR EQUATIONS GRAPHICALLY

TRY IT! *Problem on page 528*

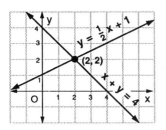

10-2 SOLVING A SYSTEM OF LINEAR EQUATIONS ALGEBRAICALLY

TRY IT! *Problem on page 531*

$$3x - 4y = 26 \qquad [A]$$
$$x + 2y = 2 \qquad [B]$$
$$x = -2y + 2 \qquad [C]$$

$$3x - 4y = 26 \qquad [A]$$
$$3(-2y + 2) - 4y = 26$$
$$-6y + 6 - 4y = 26$$
$$-10y + 6 = 26$$
$$-10y = 20$$
$$y = -2$$

$$x = -2y + 2 \qquad [C]$$
$$x = -2(-2) + 2$$
$$x = 4 + 2$$
$$x = 6$$

Check in both original equations.

ANSWER: The solution is $(6, -2)$.

TRY IT! *Problem on page 534*

$$4x + y = 10 \qquad [A]$$
$$x - 2y = 7 \qquad [B]$$

$$2(4x + y = 10) \qquad 2\,[A]$$
$$x - 2y = 7 \qquad [B]$$

$$8x + 2y = 20 \qquad [C]$$
$$\underline{x - 2y = 7} \qquad [B]$$
$$9x \quad\quad = 27$$
$$x = 3$$

$$4x + y = 10$$
$$4(3) + y = 10$$
$$12 + y = 10$$
$$y = -2$$

Check in both original equations.

ANSWER: The solution is $(3, -2)$.

10-4 SOLVING A SYSTEM OF LINEAR INEQUALITIES GRAPHICALLY

TRY IT! *Problem on page 555*

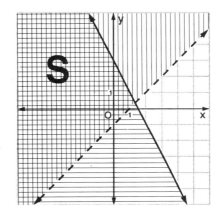

CHAPTER 11

IRRATIONAL NUMBERS

The time required for a pendulum to swing from one side to the other and back is determined by a formula that includes the length of the pendulum and the force of gravity.

$$t = 2\pi\sqrt{\frac{\ell}{g}}$$

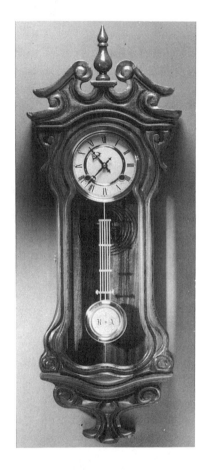

11-1 Finding the Root of a Number **573**

11-2 Simplifying and Combining Radicals **578**

11-3 Multiplying and Dividing Radicals **583**

11-4 Solving a Radical Equation **591**

11-5 Graphing a Quadratic Function ... **596**

11-6 More Methods for Solving a Quadratic Equation **603**

11-7 The Pythagorean Relation **614**

Chapter Summary **621**

Chapter Review Exercises **622**

Problems for Pleasure **624**

Calculator Challenge **624**

College Test Preparation **625**

Spiral Review Exercises **627**

Solutions to TRY IT! Problems **628**

SELF-TEST Chapters 1–11 ...... **633**

The linear equations that you have worked with so far, and the quadratic equations that you have solved by factoring, have had solutions that were rational numbers. Recall that a rational number is one that can be expressed as the ratio of two integers.

Now you will learn how to work with irrational numbers—numbers that cannot be expressed as the ratio of two integers. After learning the basic operations with irrational numbers, you will see these numbers as solutions to quadratic equations that cannot be solved by factoring.

11-1 FINDING THE ROOT OF A NUMBER

Defining the Root of a Number

A familiar operation involving exponents is the operation of *raising to a power*. For example, 25 is the **square** of 5, written $25 = 5^2$. Consider the inverse of this operation, that is, the **square root** of 25 is 5, written $\sqrt{25} = 5$.

The general meaning of these two inverse operations is summarized in the table:

Operation	Symbols	Meaning
raising to a power	$b^n = N$	Use the base b as a factor n times to produce the number N. Read: b raised to the nth power $= N$.
finding a root	$\sqrt[n]{N} = b$	Find the factor b that has been used n times to produce the number N. Read: The nth root of $N = b$.

The names associated with the symbols are shown in the diagram.

Note that, when writing a *square root*, that is, when $n = 2$, it is customary to omit the index, as in $\sqrt{25}$. All other indexes must be written, such as $n = 3$ in $\sqrt[3]{8}$, the **cube root** of 8.

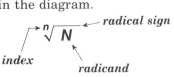

Every rational number can be raised to an integral power (except 0^0), and the result of that operation is unique. For example:

$$4^2 = 16 \qquad 0^3 = 0 \qquad (-4)^2 = 16 \qquad \left(\frac{1}{2}\right)^4 = \frac{1}{16} \qquad (.25)^2 = .0625$$

However, the result of the inverse operation is not necessarily unique. For example, both $+3$ and -3 qualify as square roots of 9 since both $(+3)^2$ and $(-3)^2$ equal 9. To be clear then, the positive square root of a number is called its **principal square root,** and the symbol $\sqrt{9}$ means only the principal square root, 3. To speak of both square roots of 9, we use the symbol $\pm$, read positive *or* negative. Thus, $\pm\sqrt{9}$ means 3 or -3.

Another complication arises when you try to take the square root of a negative number. For example, what is $\sqrt{-4}$? With the numbers you know, there is no answer. Note that $(-2)^2 \neq -4$. What about $\sqrt[3]{-8}$? The answer is -2, since $(-2)^3 = -8$. Thus, in the real number system, when working with roots of negative numbers, you can find values only for cube roots, 5th roots, etc., that is, odd roots.

MODEL PROBLEMS

1. Evaluate each expression.

Expression	*Answer*		*Expression*	*Answer*
a. $\sqrt{64}$	8		**b.** $\pm\sqrt{36}$	± 6
c. $\sqrt[3]{-27}$	-3		**d.** $\sqrt{-16}$	undefined
e. $-\sqrt{49}$	-7		**f.** $\sqrt{0}$	0
g. $(\sqrt{13})^2$	13		**h.** $(.1)^4$	.0001

2. Solve for x: $x^2 = 36$

   Solution: If $x^2 = a$, then $x = \pm\sqrt{a}$ when a is a positive number.

   $$x^2 = 36$$
   $$x = \pm\sqrt{36}$$
   $$x = \pm 6$$

   Check:

   $$x^2 = 36$$
   $$(+6)^2 \overset{?}{=} 36$$
   $$36 = 36 \checkmark$$

   $$x^2 = 36$$
   $$(-6)^2 \overset{?}{=} 36$$
   $$36 = 36 \checkmark$$

   ANSWER: $x = 6$ or $x = -6$

Estimating a Square Root

Some rational numbers are called **perfect squares** because they are squares of rational numbers. Thus, the square root of a perfect square is a rational number. For example:

$$\sqrt{9} = 3 \qquad \sqrt{\tfrac{4}{49}} = \tfrac{2}{7} \qquad \sqrt{.25} = .5 \qquad \sqrt{x^2} = x \quad \text{where } x \text{ is a nonnegative rational number}$$

What about the square root of a number that is not a perfect square, such as $\sqrt{10}$? Since there is no rational number that answers the question, at best you can give an approximation. To estimate, consider the nearest perfect square, namely 9. Thus, $\sqrt{10}$ is somewhat more than $\sqrt{9}$, or $\sqrt{10}$ is just over 3. How exact you need to be depends upon the calculation in which you are using the value. If the requirement is to the nearest tenth, you can guess and check: $\sqrt{10} \approx 3.2$

CALCULATOR CONNECTION

Most calculators have a square-root key. The key is identified by a radical sign, $\sqrt{}$. To use this key, enter the number first, then press the square-root key. For instance, to find $\sqrt{10}$, enter **10**, then press $\boxed{\sqrt{}}$. The result is displayed immediately. You do not have to press $\boxed{=}$.

To find the square root of a longer expression, such as $\sqrt{2 + 3 + 4}$, first perform the addition and display the result, then press $\boxed{\sqrt{}}$.

To evaluate $3 + \sqrt{2}$, enter **3** $\boxed{+}$ **2** $\boxed{\sqrt{}}$ $\boxed{=}$. Note that after you have pressed $\boxed{\sqrt{}}$, the display is **1.41421 . . .** , which is the approximate value of $\sqrt{2}$. You must then press $\boxed{=}$ to perform the addition.

Since there is no real number that is the square root of a negative number, most calculators will display an error message if you try to perform this operation.

MODEL PROBLEMS

1. Between which two consecutive integers is $\sqrt{42}$?

   Solution: Since $6 \times 6 = 36$ and $7 \times 7 = 49$, then $\sqrt{42}$ is between 6 and 7, or $6 < \sqrt{42} < 7$.

2. State whether $\sqrt{56}$ is a rational or an irrational number.

   Solution: Since 56 is not a perfect square, $\sqrt{56}$ is an irrational number.

3. Approximate $\sqrt{38}$ to: **a.** the nearest integer

   b. the nearest tenth

   ANSWER: **a.** $\sqrt{38} \approx 6$

   b. $\sqrt{38} \approx 6.2$

In 1–5, state the index and the radicand of the radical.

1. $\sqrt{36}$ **2.** $\sqrt[3]{125}$ **3.** $\sqrt[4]{81}$ **4.** $\sqrt[5]{32}$ **5.** $\sqrt[n]{1}$

In 6–15, find the principal square root of the number.

6. 81   **7.** 1   **8.** 121   **9.** 225   **10.** 900

11. $\frac{1}{9}$ **12.** $\frac{4}{25}$ **13.** .49   **14.** 1.44   **15.** .04

In 16–54, evaluate the expression.

16. $\sqrt{16}$ **17.** $\sqrt{81}$ **18.** $\sqrt{121}$ **19.** $-\sqrt{64}$ **20.** $-\sqrt{144}$

21. $\sqrt{0}$ **22.** $\pm\sqrt{100}$ **23.** $\pm\sqrt{169}$ **24.** $\sqrt{400}$ **25.** $-\sqrt{625}$

26. $\sqrt{\frac{1}{4}}$ **27.** $-\sqrt{\frac{9}{16}}$ **28.** $\pm\sqrt{\frac{25}{81}}$ **29.** $\sqrt{\frac{49}{100}}$ **30.** $\pm\sqrt{\frac{144}{169}}$

31. $\sqrt{.64}$ **32.** $-\sqrt{1.44}$ **33.** $\pm\sqrt{.09}$ **34.** $-\sqrt{.01}$ **35.** $\pm\sqrt{.0004}$

36. $\sqrt[3]{1}$ **37.** $\sqrt[4]{81}$ **38.** $\sqrt[5]{32}$ **39.** $\sqrt[3]{-8}$ **40.** $-\sqrt[3]{-125}$

41. $\sqrt{(8)^2}$ **42.** $\sqrt{\left(\frac{1}{2}\right)^2}$ **43.** $\sqrt{(.7)^2}$ **44.** $\sqrt{(-4)^2}$ **45.** $\sqrt{(-5)^2}$

46. $(\sqrt{4})^2$ **47.** $(\sqrt{36})^2$ **48.** $(\sqrt{11})^2$ **49.** $(\sqrt{39})^2$ **50.** $(\sqrt{97})(\sqrt{97})$

51. $\sqrt{36} + \sqrt{49}$ **52.** $\sqrt{100} - \sqrt{25}$

53. $(\sqrt{17})^2 + (\sqrt{7})(\sqrt{7})$ **54.** $\sqrt{(-9)^2} - (\sqrt{83})^2$

In 55–57, state which of the four given numbers is a rational number.

55. $\sqrt{7}, \sqrt{14}, \sqrt{25}, \sqrt{30}$ **56.** $\sqrt{\frac{1}{2}}, \sqrt{\frac{3}{25}}, \sqrt{\frac{4}{11}}, \sqrt{\frac{9}{16}}$ **57.** $\sqrt{.5}, \sqrt{.04}, \sqrt{.15}, \sqrt{.1}$

58. a. Compare the given number with its square and state which is greater.

     (1) $\frac{1}{2}$ (2) $\frac{3}{4}$ (3) 1   (4) 2   (5) 4

  b. From the results of part **a**, write a general statement comparing a positive number to its square.

59. a. Compare the principal square root of the given number with the number and state which is greater.

     (1) $\frac{1}{9}$ (2) $\frac{4}{25}$ (3) 1   (4) 4   (5) 9

  b. From the results of part **a**, write a general statement comparing a positive number to its square root.

60. If $n \geq 0$, the expression that is not always equal to n is:

    (1) $\sqrt{n^2}$ (2) $(\sqrt{n})^2$ (3) $(\sqrt{n})^3$ (4) $\sqrt[3]{n^3}$

61. If $n < 0$, the expression that can represent a real number is:

    (1) $\sqrt{n}$ (2) $n\sqrt{n}$ (3) $\sqrt[3]{n}$ (4) $\dfrac{1}{\sqrt{n}}$

In 62–69, solve the equation if the domain is the set of real numbers.

62. $x^2 = 4$ **63.** $y^2 = 100$ **64.** $z^2 = \frac{4}{81}$ **65.** $x^2 = .49$

66. $x^2 - 16 = 0$ **67.** $y^2 - 36 = 0$ **68.** $2x^2 = 50$ **69.** $3x^2 - 75 = 0$

In 70–73, find the length of each side of a square that has the given area.

70. 36 sq. ft.   **71.** 196 sq. yd.   **72.** 1,600 sq. cm   **73.** 441 sq. m

In 74–83, between which two consecutive integers is the given number?

74. $\sqrt{5}$ **75.** $\sqrt{13}$ **76.** $\sqrt{40}$ **77.** $-\sqrt{2}$ **78.** $-\sqrt{20}$

79. $-\sqrt{60}$ **80.** $\sqrt{73}$ **81.** $\sqrt{95}$ **82.** $-\sqrt{125}$ **83.** $\sqrt{143}$

In 84–89, order the given numbers, starting with the smallest.

84. 2, $\sqrt{3}$, -1 **85.** 4, $\sqrt{17}$, 3   **86.** $-\sqrt{15}$, -3, -4

87. 0, $\sqrt{7}$, $-\sqrt{7}$ **88.** 5, $\sqrt{21}$, $\sqrt{30}$ **89.** $-\sqrt{11}$, $-\sqrt{23}$, $-\sqrt{19}$

In 90–99, state whether the number is rational or irrational.

90. $\sqrt{25}$ **91.** $\sqrt{40}$ **92.** $-\sqrt{36}$ **93.** $-\sqrt{54}$ **94.** $\sqrt{144}$

95. $\sqrt{\frac{1}{2}}$ **96.** $-\sqrt{\frac{4}{9}}$ **97.** $\sqrt{\frac{1}{3}}$ **98.** $\sqrt{.36}$ **99.** $\sqrt{.1}$

In 100–109, approximate the value of the expression to the nearest integer.

100. $\sqrt{5}$ **101.** $\sqrt{19}$ **102.** $\sqrt{22}$ **103.** $\sqrt{34}$ **104.** $-\sqrt{55}$

105. $\sqrt{93}$ **106.** $\sqrt{105}$ **107.** $-\sqrt{116}$ **108.** $\sqrt{157}$ **109.** $\sqrt{218}$

In 110–119, approximate the value of the expression to the nearest tenth.

110. $\sqrt{2}$ **111.** $\sqrt{12}$ **112.** $\sqrt{45}$ **113.** $-\sqrt{67}$ **114.** $-\sqrt{86}$

115. $\sqrt{106}$ **116.** $\sqrt{125}$ **117.** $-\sqrt{137}$ **118.** $\sqrt{152}$ **119.** $\sqrt{175}$

11-2 SIMPLIFYING AND COMBINING RADICALS

Simplifying a Radical

Depending on how it is to be used, an irrational number can be approximated, such as $\sqrt{8} \approx 2.8$, or left in *radical form*, $\sqrt{8}$.

Some numbers in radical form can be factored and rewritten.

$$\begin{aligned} \sqrt{8} &= \sqrt{4 \cdot 2} \\ &= \sqrt{4} \cdot \sqrt{2} \\ &= 2 \cdot \sqrt{2} \text{ or } 2\sqrt{2} \end{aligned}$$

When the radicand under a square-root symbol no longer contains a factor that is a perfect square, the number is in *simplest radical form*.

The example above illustrates the following property of square roots:

The square root of a product of nonnegative numbers is equal to the product of the square roots of the numbers.

$$\sqrt{a \cdot b} = \sqrt{a} \cdot \sqrt{b} \quad \text{where } a \text{ and } b \text{ are nonnegative numbers}$$

This property extends to indexes higher than 2.

MODEL PROBLEMS

Write the expression in simplest radical form.

Note

1. $\begin{aligned} \sqrt{108} &= \sqrt{36 \cdot 3} \\ &= \sqrt{36} \cdot \sqrt{3} \\ &= 6\sqrt{3} \end{aligned}$

Choose the largest perfect-square factor at first so that the answer will be in simplest form. If you factor differently, say $\sqrt{108} = \sqrt{4 \cdot 27}$, you must continue:
$$\sqrt{108} = \sqrt{4 \cdot 9 \cdot 3} = 2 \cdot 3\sqrt{3} = 6\sqrt{3}$$

2. $\begin{aligned} 4\sqrt{50} &= 4\sqrt{25 \cdot 2} = 4\sqrt{25} \cdot \sqrt{2} \\ &= 4 \cdot 5\sqrt{2} \\ &= 20\sqrt{2} \end{aligned}$

Once a factor has been removed from the radical sign, multiply it with other coefficients.

3. $\begin{aligned} \sqrt[3]{24} &= \sqrt[3]{8 \cdot 3} \\ &= \sqrt[3]{8} \cdot \sqrt[3]{3} \\ &= 2 \cdot \sqrt[3]{3} \text{ or } 2\sqrt[3]{3} \end{aligned}$

Choose the largest perfect-cube factor.

When a radicand contains a variable, there are other matters to consider. First, you do not know whether the principal square root of x^2 is $+x$ or $-x$ unless you know whether x is a positive or a negative number. For example:

$$\text{If } x = 5, \text{ then } \sqrt{x^2} = \sqrt{(5)^2} = \sqrt{25} = 5 = x.$$
$$\text{If } x = -5, \text{ then } \sqrt{x^2} = \sqrt{(-5)^2} = \sqrt{25} = 5 = -x.$$

In general: $\left.\begin{array}{l}\sqrt{x^2} = x \text{ if } x \text{ is a positive number or zero.}\\ \sqrt{x^2} = -x \text{ if } x \text{ is a negative number.}\end{array}\right\}$ or $\sqrt{x^2} = |x|$

To avoid the need for the absolute value symbol, the domain of the variables that appear under the radical sign will be limited to nonnegative numbers only. Therefore, you may write simply $\sqrt{x^2} = x$.

Note that any even exponent makes the term a perfect square. For example, $\sqrt{x^{10}} = x^5$ since $x^{10} = x^5 \cdot x^5$.

MODEL PROBLEMS

Write the expression in simplest radical form. The domain of the variables is limited to nonnegative numbers.

1. $\sqrt{16m^6} = (\sqrt{16})(\sqrt{m^6}) = 4m^3$ **2.** $\sqrt{\frac{4}{9}a^2b^4} = \left(\sqrt{\frac{4}{9}}\right)(\sqrt{a^2})(\sqrt{b^4}) = \frac{2}{3}ab^2$

3. $\sqrt{.81y^8} = (\sqrt{.81})(\sqrt{y^8}) = .9y^4$ **4.** $\sqrt{75x^2y^3z^7} = \sqrt{25x^2y^2z^6} \cdot \sqrt{3yz} = 5xyz^3\sqrt{3yz}$

Combining Radicals by Addition or Subtraction

For all algebraic terms, only *like terms* can be combined by addition or subtraction.

Like radicals are radicals that have the *same index* and *same radicand*. For example, $3\sqrt{2}$ and $5\sqrt{2}$ are like radicals, as are $4\sqrt[3]{7}$ and $9\sqrt[3]{7}$. However, $3\sqrt{5}$ and $5\sqrt{2}$ are unlike radicals, as are $\sqrt[3]{2}$ and $\sqrt{2}$.

To add or subtract like radicals, use the distributive property as follows:

$$7\sqrt{2} + 3\sqrt{2} \qquad 5\sqrt{3} - \sqrt{3}$$
$$= (7 + 3)\sqrt{2} \qquad = (5 - 1)\sqrt{3}$$
$$= 10\sqrt{2} \qquad = 4\sqrt{3}$$

When it is possible to transform unlike radicals into equivalent radicals that are alike, the resulting like radicals can be added or subtracted. For example:

$$2\sqrt{3} + \sqrt{27} = 2\sqrt{3} + \sqrt{9 \cdot 3}$$
$$= 2\sqrt{3} + 3\sqrt{3}$$
$$= 5\sqrt{3}$$

In 1–3, combine and simplify, as possible.

1. $8\sqrt{3} + 4\sqrt{2} - \sqrt{2} + \sqrt{3}$
$= (8\sqrt{3} + \sqrt{3}) + (4\sqrt{2} - \sqrt{2})$
$= (8 + 1)\sqrt{3} + (4 - 1)\sqrt{2}$
$= 9\sqrt{3} + 3\sqrt{2}$

2. $5\sqrt{3} + 4\sqrt{12} - 2\sqrt{75}$
$= 5\sqrt{3} + 4\sqrt{4 \cdot 3} - 2\sqrt{25 \cdot 3}$
$= 5\sqrt{3} + 4\sqrt{4} \cdot \sqrt{3} - 2\sqrt{25} \cdot \sqrt{3}$
$= 5\sqrt{3} + 4 \cdot 2 \cdot \sqrt{3} - 2 \cdot 5 \cdot \sqrt{3}$
$= 5\sqrt{3} + 8\sqrt{3} - 10\sqrt{3}$
$= (5 + 8 - 10)\sqrt{3} = 3\sqrt{3}$

3. $2\sqrt{2x} + 7\sqrt{8x} - 4\sqrt{50x}$
$= 2\sqrt{2x} + 7\sqrt{4 \cdot 2x} - 4\sqrt{25 \cdot 2x}$
$= 2\sqrt{2x} + 7\sqrt{4} \cdot \sqrt{2x} - 4\sqrt{25} \cdot \sqrt{2x}$
$= 2\sqrt{2x} + 7 \cdot 2 \cdot \sqrt{2x} - 4 \cdot 5 \cdot \sqrt{2x}$
$= 2\sqrt{2x} + 14\sqrt{2x} - 20\sqrt{2x}$
$= (2 + 14 - 20)\sqrt{2x}$
$= -4\sqrt{2x}$

4. Solve:
$x + \sqrt{2} = \sqrt{8}$
$x = \sqrt{8} - \sqrt{2}$
$x = \sqrt{4 \cdot 2} - \sqrt{2}$
$x = 2\sqrt{2} - \sqrt{2}$
$x = \sqrt{2}$

Now it's your turn to . . . **TRY IT!**   *(See page **628** for solutions.)*

Combine: **1.** $3\sqrt{20} - 2\sqrt{5} + 5\sqrt{45}$ **2.** $5\sqrt{3x} + 4\sqrt{27x} - 3\sqrt{75x}$

K<small>EEP IN</small> M<small>IND</small>

The domain of the variables that appear under a radical sign is limited to nonnegative numbers only.

EXERCISES

In 1–36, write the expression in simplest radical form.

1. $\sqrt{8}$ **2.** $\sqrt{12}$ **3.** $\sqrt{20}$ **4.** $-\sqrt{24}$

5. $\sqrt{28}$ **6.** $\sqrt{40}$ **7.** $\sqrt{27}$ **8.** $-\sqrt{45}$

9. $\sqrt{54}$ **10.** $\sqrt{63}$ **11.** $\sqrt{90}$ **12.** $-\sqrt{72}$

13. $\sqrt{98}$ **14.** $\sqrt{99}$ **15.** $\sqrt{108}$ **16.** $-\sqrt{128}$

17. $\sqrt{162}$ **18.** $\sqrt{175}$ **19.** $\sqrt{300}$ **20.** $-\sqrt{500}$

21. $3\sqrt{8}$ **22.** $4\sqrt{12}$ **23.** $2\sqrt{20}$ **24.** $-5\sqrt{24}$

25. $4\sqrt{90}$ **26.** $2\sqrt{45}$ **27.** $3\sqrt{200}$ **28.** $-6\sqrt{98}$

29. $\frac{1}{3}\sqrt{45}$ **30.** $\frac{1}{2}\sqrt{72}$ **31.** $\frac{1}{4}\sqrt{48}$ **32.** $\frac{1}{2}\sqrt{18}$

33. $\frac{3}{4}\sqrt{96}$ **34.** $\frac{2}{3}\sqrt{63}$ **35.** $\frac{3}{8}\sqrt{80}$ **36.** $-\frac{2}{5}\sqrt{125}$

In 37–61, find the square root.

37. $\sqrt{9c^2}$ **38.** $\sqrt{36y^4}$ **39.** $\sqrt{64c^6}$ **40.** $\sqrt{100x^{10}}$ **41.** $\sqrt{81t^8}$

42. $\sqrt{c^2d^2}$ **43.** $\sqrt{x^4y^2}$ **44.** $\sqrt{r^8s^6}$ **45.** $\sqrt{x^{16}y^4}$ **46.** $\sqrt{x^6y^2z^4}$

47. $\sqrt{4x^2y^2}$ **48.** $\sqrt{36a^6b^4}$ **49.** $\sqrt{144a^4b^2}$ **50.** $\sqrt{169x^4y^2}$

51. $\sqrt{25r^8s^{16}t^{12}}$ **52.** $\sqrt{\frac{1}{4}y^2}$ **53.** $\sqrt{\frac{25}{36}x^2y^2}$ **54.** $\sqrt{\frac{49}{9}a^4b^2}$

55. $\sqrt{\frac{81}{121}a^6b^2}$ **56.** $\sqrt{\frac{64}{169}x^8y^{10}z^{16}}$ **57.** $\sqrt{.36m^2}$ **58.** $\sqrt{.49a^2b^2}$

59. $\sqrt{.04x^2y^6}$ **60.** $\sqrt{.01x^4y^2}$ **61.** $\sqrt{1.21a^4b^{16}c^{36}}$

In 62–85, write the expression in simplest radical form.

62. $\sqrt{a^3}$ **63.** $2\sqrt{b^5}$ **64.** $\sqrt{r^2s}$ **65.** $3\sqrt{mn^2}$

66. $\sqrt{x^2y^3}$ **67.** $4\sqrt{x^3y^3}$ **68.** $\sqrt{3x^3y}$ **69.** $-4\sqrt{2a^2b^3}$

70. $\sqrt{49a}$ **71.** $\sqrt{36r^2s}$ **72.** $\sqrt{9x^2y^3}$ **73.** $5\sqrt{4x^3y^3}$

74. $\sqrt{27m^2}$ **75.** $6\sqrt{8r^3}$ **76.** $\sqrt{20x^2y^3}$ **77.** $-7\sqrt{24a^3b^3}$

78. $\sqrt{4x^2y^4z^3}$ **79.** $\sqrt{9a^3b^3c^2}$ **80.** $\sqrt{12r^4s^3t^6}$ **81.** $5\sqrt{18y^3u^6z^8}$

82. $\frac{1}{2}\sqrt{16a^3b^2c^2}$ **83.** $\frac{2}{3}\sqrt{72x^5y^4z^7}$

84. $\frac{3}{5}\sqrt{50r^3s^3t^3}$ **85.** $-\frac{4}{7}\sqrt{147x^2y^6z^{11}}$

In 86–124, combine and simplify, as possible.

86. $8\sqrt{5} + \sqrt{5}$ **87.** $4\sqrt{3} + 2\sqrt{3} - 6\sqrt{3}$

88. $4\sqrt{7} - \sqrt{7} - 5\sqrt{7}$ **89.** $3\sqrt{5} + 6\sqrt{2} - 3\sqrt{2} + \sqrt{5}$

90. $9\sqrt{x} + 3\sqrt{x}$ **91.** $7\sqrt{x} + 3\sqrt{y} - \sqrt{x} - 3\sqrt{y}$

92. $\sqrt{2} + \sqrt{50}$

93. $\sqrt{27} + \sqrt{75}$

94. $\sqrt{80} - \sqrt{5}$

95. $\sqrt{5} + \sqrt{45} + \sqrt{80}$

96. $\sqrt{72} - \sqrt{50}$

97. $\sqrt{12} - \sqrt{48} + \sqrt{3}$

98. $3\sqrt{2} + 2\sqrt{32}$

99. $3\sqrt{32} - 6\sqrt{8}$

100. $5\sqrt{27} - \sqrt{108} + 2\sqrt{75}$

101. $3\sqrt{40} - \sqrt{90}$

102. $3\sqrt{8} - \sqrt{2}$

103. $5\sqrt{8} - 3\sqrt{18} + \sqrt{3}$

104. $3\sqrt{50} - 5\sqrt{18}$

105. $3\sqrt{28} - 2\sqrt{63}$

106. $\sqrt{98} - 4\sqrt{8} + 3\sqrt{128}$

107. $\frac{1}{2}\sqrt{20} + \sqrt{45}$

108. $\frac{2}{3}\sqrt{18} - \sqrt{72}$

109. $4\sqrt{18} - \frac{3}{4}\sqrt{32} - \frac{1}{2}\sqrt{8}$

110. $\sqrt{7a} + \sqrt{28a}$

111. $\sqrt{81x} + \sqrt{25x}$

112. $\sqrt{100b} - \sqrt{64b} + \sqrt{9b}$

113. $3\sqrt{3x} - \sqrt{12x}$

114. $5\sqrt{8b} + 3\sqrt{32b}$

115. $5\sqrt{3y} - \sqrt{27x} + \sqrt{12y}$

116. $\sqrt{3a^2} + \sqrt{12a^2}$

117. $4\sqrt{5y^2} - \sqrt{20y^2}$

118. $4\sqrt{12r^2} + 2\sqrt{75r^2} - 3\sqrt{27r^2}$

119. $\sqrt{7x^3} + \sqrt{28x^3}$

120. $3\sqrt{2y^3} - \sqrt{8y^3}$

121. $5\sqrt{12a^3} - 2\sqrt{3a^3} + \sqrt{27a^3}$

122. $x\sqrt{8y} + 3\sqrt{2x^2y}$

123. $b\sqrt{27a} - 6\sqrt{3ab^2}$

124. $\sqrt{3x^3} + 3\sqrt{12x^3} - x\sqrt{75}$

In 125–130, solve for x.

125. $x - \sqrt{20} = \sqrt{45}$

126. $x = \frac{1}{4}\sqrt{128}$

127. $x + \sqrt{28} = \sqrt{63}$

128. $\sqrt{98} = \sqrt{32} + x$

129. $2\sqrt{11} + x = 4\sqrt{99}$

130. $2x = \sqrt{48}$

131. Find the perimeter of a square whose side measures:

    a. $\sqrt{3}$ cm
    b. $3\sqrt{7}$ in.
    c. $(5 + 7\sqrt{6})$ ft.

132. Find the perimeter of a rectangle whose dimensions are:

    a. length: $3\sqrt{5}$ ft.
       width: $2\sqrt{5}$ ft.

    b. length: $8\sqrt{3}$ yd.
       width: 6 yd.

    c. length: $(4 + 2\sqrt{7})$ m
       width: $(11 - 2\sqrt{7})$ m

133. Find the perimeter of an isosceles triangle whose base measures 10 in. and whose legs each measure $(6 + \sqrt{2})$ in.

11-3 MULTIPLYING AND DIVIDING RADICALS

Multiplication

To simplify radicals, you have used the principle that for nonnegative numbers a and b, $\sqrt{a \cdot b} = \sqrt{a} \cdot \sqrt{b}$. Consider also:

$\sqrt{4} \cdot \sqrt{9} = 2 \cdot 3 = 6$ and $\sqrt{36} = 6$. Thus, $\sqrt{4} \cdot \sqrt{9} = \sqrt{4 \cdot 9} = \sqrt{36}$.

This example illustrates an alternate use of the same principle:

$$\sqrt{a} \cdot \sqrt{b} = \sqrt{a \cdot b} \qquad \text{where } a \text{ and } b \text{ are nonnegative numbers}$$

Apply this principle, as well as the associative and commutative properties of multiplication, to evaluate products.

_____ **P**ROCEDURE _____

To multiply two radicals with rational coefficients:

1. Multiply the coefficients to find the coefficient of the product.
2. Multiply the radicands to find the radicand of the product.
3. If possible, simplify the result.

MODEL PROBLEMS

Multiply and simplify.

1. $3\sqrt{6} \cdot 5\sqrt{2}$

$= 3 \cdot 5 \sqrt{6 \cdot 2}$

$= 15\sqrt{12}$

$= 15\sqrt{4} \cdot \sqrt{3}$

$= 15 \cdot 2 \cdot \sqrt{3}$

$= 30\sqrt{3}$

2. $(2\sqrt{3})^2$

$= 2^2 \cdot (\sqrt{3})^2$

$= 4 \cdot 3$

$= 12$

3. $\sqrt{3x} \cdot \sqrt{6x}$

$= \sqrt{3x \cdot 6x}$

$= \sqrt{18x^2}$

$= \sqrt{9x^2 \cdot 2}$

$= \sqrt{9x^2} \cdot \sqrt{2}$

$= 3x\sqrt{2}$

Observe that the product of two irrational numbers is sometimes an irrational number and sometimes a rational number.

Products involving polynomials require the distributive property. For example:

$$\sqrt{3}(\sqrt{2} + \sqrt{5})$$
$$= (\sqrt{3})(\sqrt{2}) + (\sqrt{3})(\sqrt{5})$$
$$= \sqrt{6} + \sqrt{15}$$

Multiplication can be done

<table>
<tr><td align="center">vertically</td><td align="center">or</td><td align="center">horizontally</td></tr>
</table>

$$
\begin{array}{r}
7 - \sqrt{2} \\
5 + \sqrt{2} \\
\hline
35 - 5\sqrt{2} \\
+ 7\sqrt{2} - \sqrt{4} \\
\hline
35 + 2\sqrt{2} - 2 \\
= 33 + 2\sqrt{2}
\end{array}
$$

$$-5\sqrt{2}$$
$$(7 - \sqrt{2})(5 + \sqrt{2})$$
$$+ 7\sqrt{2}$$
$$= 35 + 2\sqrt{2} - \sqrt{4}$$
$$= 35 + 2\sqrt{2} - 2$$
$$= 33 + 2\sqrt{2}$$

Now it's your turn to . . . **TRY IT!**   *(See page **628** for solutions.)*

Multiply and simplify.

1. $4\sqrt{10} \cdot 3\sqrt{5}$ **2.** $(3\sqrt{5})^2$ **3.** $3\sqrt{2}(3\sqrt{8} - 4\sqrt{3})$ **4.** $(5 - \sqrt{3})(2 - \sqrt{3})$

Division

Since $\sqrt{\dfrac{16}{25}} = \dfrac{4}{5}$ and $\dfrac{\sqrt{16}}{\sqrt{25}} = \dfrac{4}{5}$, then $\sqrt{\dfrac{16}{25}} = \dfrac{\sqrt{16}}{\sqrt{25}}$.

This example illustrates the following property of radicals:

The square root of a quotient is equal to the square root of its numerator divided by the square root of its denominator.

$$\sqrt{\dfrac{a}{b}} = \dfrac{\sqrt{a}}{\sqrt{b}} \qquad \text{where } a \text{ is nonnegative and } b \text{ is positive}$$

Reversing the equality makes it also true that:

The quotient of two square roots is equal to the square root of the quotient of the radicands.

1. When a single radical contains a fraction in which only the denominator is a perfect square, write the radical as a quotient of two radicals and simplify the denominator.

2. To divide two radicals with rational coefficients:

   a. Divide the coefficients to find the coefficient of the quotient.

   b. Divide the radicands to find the radicand of the quotient.

   c. If possible, simplify the result.

MODEL PROBLEMS

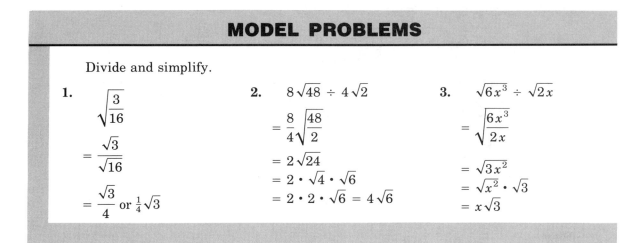

Divide and simplify.

1. $\sqrt{\dfrac{3}{16}}$

$= \dfrac{\sqrt{3}}{\sqrt{16}}$

$= \dfrac{\sqrt{3}}{4}$ or $\frac{1}{4}\sqrt{3}$

2. $8\sqrt{48} \div 4\sqrt{2}$

$= \dfrac{8}{4}\sqrt{\dfrac{48}{2}}$

$= 2\sqrt{24}$

$= 2 \cdot \sqrt{4} \cdot \sqrt{6}$

$= 2 \cdot 2 \cdot \sqrt{6} = 4\sqrt{6}$

3. $\sqrt{6x^3} \div \sqrt{2x}$

$= \sqrt{\dfrac{6x^3}{2x}}$

$= \sqrt{3x^2}$

$= \sqrt{x^2} \cdot \sqrt{3}$

$= x\sqrt{3}$

When the numerator of a quotient is a polynomial, *each term* must be divided by the divisor. For example:

$$\frac{\sqrt{40} + \sqrt{10}}{\sqrt{5}} = \frac{\sqrt{40}}{\sqrt{5}} + \frac{\sqrt{10}}{\sqrt{5}} = \sqrt{8} + \sqrt{2}$$

$$= 2\sqrt{2} + \sqrt{2}$$

$$= 3\sqrt{2}$$

Now it's your turn to . . . **TRY IT!**   (See page **628** for solutions.)

Divide and simplify.

1. $\sqrt{\dfrac{3}{25}}$ **2.** $21\sqrt{96} \div 3\sqrt{3}$ **3.** $\dfrac{\sqrt{12} + \sqrt{27}}{\sqrt{3}}$

Rationalizing the Denominator

A fraction with a radical in the denominator is not in simplest form. Rewriting the fraction so that the denominator is rational makes it easier to approximate its rational value.

For example, to find the value of $\frac{\sqrt{3}}{3}$ given that $\sqrt{3}$ is approximately 1.73, the computation can be done mentally: $\frac{\sqrt{3}}{3} \approx \frac{1.73}{3} \approx .58$ But, evaluating the equivalent fraction $\frac{1}{\sqrt{3}}$, or approximately $\frac{1}{1.73}$, requires long division.

To simplify $\frac{1}{\sqrt{3}}$, you multiply $\frac{1}{\sqrt{3}}$ by 1 in the form $\frac{\sqrt{3}}{\sqrt{3}}$:

$$\frac{1}{\sqrt{3}} = \frac{1}{\sqrt{3}} \cdot \frac{\sqrt{3}}{\sqrt{3}} = \frac{\sqrt{3}}{\sqrt{9}} = \frac{\sqrt{3}}{3}$$

By performing this multiplication, you transformed a fraction with an irrational denominator into an equivalent fraction with a rational denominator. This process is called *rationalizing the denominator* of the fraction.

PROCEDURE

To rationalize the denominator of a fraction:

1. Multiply the fraction by 1 represented as a fraction whose numerator and denominator are the smallest radical that will make the denominator of the resulting fraction a rational number.

2. Simplify the resulting fraction, if possible.

MODEL PROBLEMS

Rationalize the denominator.

1. $\dfrac{12}{\sqrt{2}}$

$= \dfrac{12}{\sqrt{2}} \cdot \dfrac{\sqrt{2}}{\sqrt{2}}$

$= \dfrac{12\sqrt{2}}{\sqrt{4}}$

$= \dfrac{12\sqrt{2}}{2} = 6\sqrt{2}$

2. $\dfrac{4}{\sqrt{12}}$

$= \dfrac{4}{\sqrt{12}} \cdot \dfrac{\sqrt{3}}{\sqrt{3}}$

$= \dfrac{4\sqrt{3}}{\sqrt{36}}$

$= \dfrac{4\sqrt{3}}{6} = \dfrac{2\sqrt{3}}{3}$ or $\frac{2}{3}\sqrt{3}$

3. $6\sqrt{\dfrac{1}{3}}$

$= 6\dfrac{\sqrt{1}}{\sqrt{3}} \cdot \dfrac{\sqrt{3}}{\sqrt{3}}$

$= 6\dfrac{\sqrt{3}}{\sqrt{9}}$

$= 6\dfrac{\sqrt{3}}{3} = 2\sqrt{3}$

SIMPLIFYING RADICALS

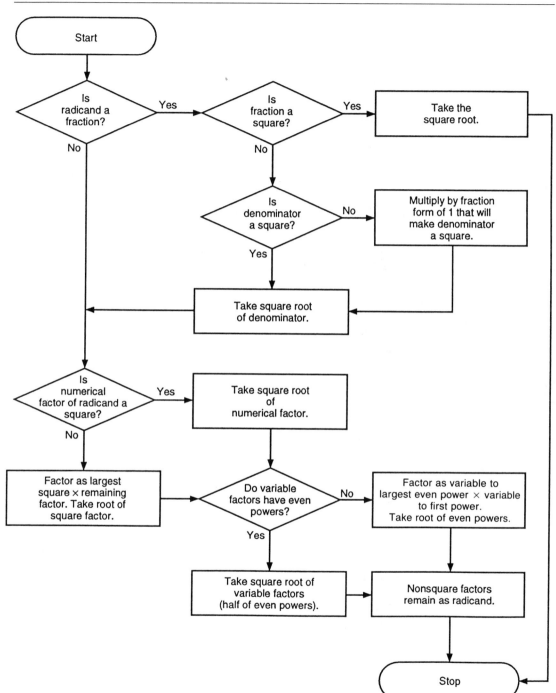

(See page **628** for solutions.)

Now it's your turn to . . . **TRY IT!**

Rationalize the denominator.

1. $\dfrac{4}{\sqrt{5}}$ **2.** $\dfrac{3}{\sqrt{8}}$ **3.** $\sqrt{\dfrac{3}{2}}$

EXERCISES

In 1–60, multiply and simplify.

1. $\sqrt{5} \cdot \sqrt{5}$ **2.** $\sqrt{71} \cdot \sqrt{71}$ **3.** $\sqrt{113} \cdot \sqrt{113}$

4. $\sqrt{a} \cdot \sqrt{a}$ **5.** $\sqrt{r} \cdot \sqrt{r}$ **6.** $\sqrt{2x} \cdot \sqrt{2x}$

7. $\sqrt{12} \cdot \sqrt{3}$ **8.** $\sqrt{32} \cdot \sqrt{2}$ **9.** $2\sqrt{18} \cdot 3\sqrt{8}$

10. $\sqrt{14} \cdot \sqrt{2}$ **11.** $\sqrt{21} \cdot \sqrt{3}$ **12.** $\sqrt{60} \cdot \sqrt{5}$

13. $3\sqrt{6} \cdot \sqrt{3}$ **14.** $5\sqrt{8} \cdot 7\sqrt{3}$ **15.** $3\sqrt{3} \cdot \sqrt{18}$

16. $\frac{2}{3}\sqrt{24} \cdot 9\sqrt{3}$ **17.** $5\sqrt{6} \cdot \frac{2}{3}\sqrt{15}$ **18.** $\frac{1}{3}\sqrt{18} \cdot 12\sqrt{3}$

19. $6\sqrt{\frac{1}{3}} \cdot 4\sqrt{12}$ **20.** $7\sqrt{56} \cdot 3\sqrt{\frac{1}{2}}$ **21.** $8\sqrt{225} \cdot 3\sqrt{\frac{1}{3}}$

22. $(5\sqrt{x})(3\sqrt{x})$ **23.** $(-4\sqrt{a})(3\sqrt{a})$ **24.** $(-\frac{1}{2}\sqrt{y})(-6\sqrt{y})$

25. $(\sqrt{3})^2$ **26.** $(\sqrt{5})^2$ **27.** $(\sqrt{y})^2$

28. $(\sqrt{t})^2$ **29.** $(3\sqrt{6})^2$ **30.** $\left(\frac{1}{3}\sqrt{3}\right)^2$

31. $\sqrt{25x} \cdot \sqrt{4x}$ **32.** $\sqrt{27a} \cdot \sqrt{3a}$ **33.** $\sqrt{15x} \cdot \sqrt{3x}$

34. $\sqrt{9a} \cdot \sqrt{ab}$ **35.** $\sqrt{5x^2y} \cdot \sqrt{10y}$ **36.** $\sqrt{3a^2} \cdot \sqrt{18b^2}$

37. $(\sqrt{3a})^2$ **38.** $(\sqrt{5x})^2$ **39.** $(\sqrt{6z})^2$

40. $(3\sqrt{r})^2$ **41.** $(2\sqrt{t})^2$ **42.** $(5\sqrt{3m})^2$

43. $(\sqrt{2x+1})^2$ **44.** $(\sqrt{3x-2})^2$ **45.** $(\sqrt{5x-3})^2$

46. $5(\sqrt{7} + \sqrt{3})$ **47.** $\sqrt{5}(\sqrt{5} + \sqrt{11})$ **48.** $\sqrt{2}(\sqrt{8} - 2\sqrt{2} + 5)$

49. $\sqrt{2}(8\sqrt{2} + 2\sqrt{8} - 3\sqrt{32})$ **50.** $2\sqrt{3}(3\sqrt{5} - 2\sqrt{20} - \sqrt{45})$

51. $(5 + \sqrt{2})(6 + \sqrt{2})$ **52.** $(3 + \sqrt{5})(7 - \sqrt{5})$ **53.** $(5 + \sqrt{3})(5 - \sqrt{3})$

54. $(\sqrt{7} - 4)(\sqrt{7} + 4)$ **55.** $(\sqrt{a} + \sqrt{b})(\sqrt{a} + \sqrt{b})$ **56.** $(\sqrt{a} - \sqrt{b})(\sqrt{a} + \sqrt{b})$

57. $(\sqrt{3} - 2)^2$ **58.** $(3 + \sqrt{7})^2$

59. $(3\sqrt{2} + 5)^2$ **60.** $(x - \sqrt{y})^2$

61. Find the area of a square whose side measures $5\sqrt{3}$ ft.

62. Find, in simplest form, the volume of a rectangular solid with length $4\sqrt{2}$ in., width $2\sqrt{3}$ in., and height $2\sqrt{6}$ in.

63. Find, in simplest form, the volume of a pyramid with height $3\sqrt{3}$ cm and whose square base has a side of $2\sqrt{3}$ cm.

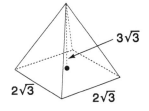

64. Evaluate $x^2 - 3x + 2$ when x is equal to $5 - \sqrt{3}$.

65. Show, by substitution, that if $y = 3\sqrt{2}$, the expression $y^2 - 3y - 6$ equals $3(4 - 3\sqrt{2})$.

66. Show, by substitution, that if $x = 1 + \sqrt{3}$, the expression $x^2 - 2x - 2$ has the value 0.

In 67–69, state whether the product is an irrational number or a rational number.

67. $(5\sqrt{12})(4\sqrt{3})$ **68.** $8\sqrt{2}(3\sqrt{32} - 2\sqrt{18})$ **69.** $(5 + \sqrt{3})(9 - 2\sqrt{3})$

In 70–79, simplify.

70. $\sqrt{\dfrac{49}{81}}$ **71.** $\dfrac{1}{4}\sqrt{\dfrac{36}{121}}$ **72.** $2\sqrt{\dfrac{7}{64}}$ **73.** $6\sqrt{\dfrac{20}{9}}$ **74.** $4\sqrt{\dfrac{75}{100}}$

75. $5\sqrt{\dfrac{40}{49}}$ **76.** $\dfrac{3}{4}\sqrt{\dfrac{48}{169}}$ **77.** $\dfrac{2}{3}\sqrt{\dfrac{45}{5}}$ **78.** $\dfrac{3}{4}\sqrt{\dfrac{120}{10}}$ **79.** $5\sqrt{\dfrac{28}{225}}$

In 80–118, divide and simplify.

80. $\sqrt{72} \div \sqrt{2}$ **81.** $\sqrt{75} \div \sqrt{3}$ **82.** $\sqrt{18} \div \sqrt{3}$ **83.** $\sqrt{70} \div \sqrt{10}$

84. $\sqrt{14} \div \sqrt{2}$ **85.** $8\sqrt{48} \div 2\sqrt{3}$ **86.** $\sqrt{24} \div \sqrt{2}$ **87.** $\sqrt{150} \div \sqrt{3}$

88. $21\sqrt{40} \div \sqrt{5}$ **89.** $9\sqrt{6} \div 3\sqrt{6}$ **90.** $7\sqrt{3} \div 3\sqrt{3}$ **91.** $2\sqrt{2} \div 8\sqrt{2}$

92. $\sqrt{9y} \div \sqrt{y}$ **93.** $8\sqrt{3a} \div 2\sqrt{a}$ **94.** $5\sqrt{24x} \div 10\sqrt{6x}$

95. $\dfrac{12\sqrt{20}}{3\sqrt{5}}$ **96.** $\dfrac{20\sqrt{50}}{4\sqrt{2}}$ **97.** $\dfrac{25\sqrt{24}}{5\sqrt{2}}$ **98.** $\dfrac{14\sqrt{150}}{7\sqrt{2}}$

99. $\dfrac{3\sqrt{54}}{6\sqrt{3}}$ **100.** $\dfrac{\sqrt{y^3}}{\sqrt{y}}$ **101.** $\dfrac{6\sqrt{27a^5}}{2\sqrt{3a^3}}$ **102.** $\dfrac{5\sqrt{48a^3b}}{10\sqrt{3ab}}$

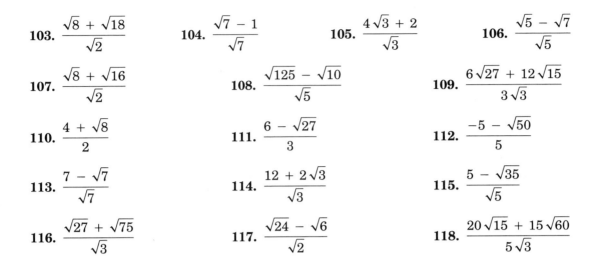

103. $\dfrac{\sqrt{8} + \sqrt{18}}{\sqrt{2}}$

104. $\dfrac{\sqrt{7} - 1}{\sqrt{7}}$

105. $\dfrac{4\sqrt{3} + 2}{\sqrt{3}}$

106. $\dfrac{\sqrt{5} - \sqrt{7}}{\sqrt{5}}$

107. $\dfrac{\sqrt{8} + \sqrt{16}}{\sqrt{2}}$

108. $\dfrac{\sqrt{125} - \sqrt{10}}{\sqrt{5}}$

109. $\dfrac{6\sqrt{27} + 12\sqrt{15}}{3\sqrt{3}}$

110. $\dfrac{4 + \sqrt{8}}{2}$

111. $\dfrac{6 - \sqrt{27}}{3}$

112. $\dfrac{-5 - \sqrt{50}}{5}$

113. $\dfrac{7 - \sqrt{7}}{\sqrt{7}}$

114. $\dfrac{12 + 2\sqrt{3}}{\sqrt{3}}$

115. $\dfrac{5 - \sqrt{35}}{\sqrt{5}}$

116. $\dfrac{\sqrt{27} + \sqrt{75}}{\sqrt{3}}$

117. $\dfrac{\sqrt{24} - \sqrt{6}}{\sqrt{2}}$

118. $\dfrac{20\sqrt{15} + 15\sqrt{60}}{5\sqrt{3}}$

119. Find, in simplest form, the length of a rectangle whose area is $14\sqrt{24}$ square inches and whose width is $2\sqrt{2}$ inches.

120. Find, in simplest form, the height of a rectangular solid whose volume is $28\sqrt{96}$ cm^3, with length $4\sqrt{2}$ cm and width $7\sqrt{3}$ cm.

In 121–150, rationalize the denominator and simplify.

121. $\dfrac{1}{\sqrt{2}}$

122. $\dfrac{1}{\sqrt{5}}$

123. $\dfrac{1}{\sqrt{7}}$

124. $\dfrac{1}{\sqrt{x}}$

125. $\dfrac{1}{\sqrt{a}}$

126. $\dfrac{3}{\sqrt{2}}$

127. $\dfrac{5}{\sqrt{3}}$

128. $\dfrac{7}{\sqrt{5}}$

129. $\dfrac{9}{\sqrt{7}}$

130. $\dfrac{8}{\sqrt{11}}$

131. $\dfrac{8}{\sqrt{2}}$

132. $\dfrac{9}{\sqrt{3}}$

133. $\dfrac{25}{\sqrt{5}}$

134. $\dfrac{6}{\sqrt{6}}$

135. $\dfrac{7}{\sqrt{7}}$

136. $\dfrac{12}{\sqrt{8}}$

137. $\dfrac{6}{\sqrt{12}}$

138. $\dfrac{36}{\sqrt{18}}$

139. $\dfrac{25}{\sqrt{20}}$

140. $\dfrac{30}{\sqrt{50}}$

141. $\dfrac{12}{2\sqrt{3}}$

142. $\dfrac{18}{3\sqrt{2}}$

143. $\dfrac{60}{3\sqrt{8}}$

144. $\dfrac{\sqrt{6}}{\sqrt{5}}$

145. $\dfrac{15\sqrt{3}}{\sqrt{5}}$

146. $\dfrac{1}{\sqrt{y}}$

147. $\dfrac{\sqrt{r}}{\sqrt{s}}$

148. $\dfrac{ab}{\sqrt{b}}$

149. $\dfrac{8}{\sqrt{2a}}$

150. $\dfrac{\sqrt{2s}}{\sqrt{g}}$

11-4 SOLVING A RADICAL EQUATION

You are familiar with the concept of using an inverse operation to solve an equation. For example, for the simple first-degree equation $x + 2 = 7$, you would apply the inverse of the operation shown. That is, since 2 is added to x, you would subtract 2 from each side of the equation, obtaining $x = 5$.

Now you will see how to apply the concept of inverse operation to solve an equation that contains the variable in a radical, called a *radical equation*. For example:

$$\sqrt{x} = 2 \quad \text{This is a radical equation.}$$

$$(\sqrt{x})^2 = 2^2 \quad \text{To solve, apply the inverse of the square-root operation.}$$

$$x = 4$$

As always, it is important to check the apparent answer in the original equation. In this case, $x = 4$ does satisfy the original equation.

Now consider the radical equation $\sqrt{x} = -2$.

Solution	*Check*
$\sqrt{x} = -2$	$\sqrt{x} = -2$
$(\sqrt{x})^2 = (-2)^2$	$\sqrt{4} \overset{?}{=} -2$
$x = 4$	$2 \neq -2$

You see that 4 is not a root of $\sqrt{x} = -2$. Since there is no other possible answer, the solution set of $\sqrt{x} = -2$ is the empty set, $\varnothing$.

Observe why this happened. If $a^2 = b^2$, it is not necessarily true that $a = b$. For example, $(-7)^2 = (+7)^2$, but $-7 \neq +7$. Therefore, when you square both members of an equation, the solution set of the "squared" equation may not be the same as the solution set of the original equation. That is, the "squared" equation and the original equation may not be equivalent.

Thus, whenever you solve an equation by squaring both members, it is essential to check the roots of the "squared" equation in the given equation to see whether these roots also satisfy the given equation.

PROCEDURE

To solve an equation containing a square-root radical:

1. Isolate the radical on one side of the equation.
2. Square both sides of the equation.
3. Solve the "squared" equation.
4. Check the roots of the "squared" equation in the given equation.

1. Solve and check: $\sqrt{3x} = 12$

How to Proceed	*Solution*	*Check*
(1) Write the equation.	$\sqrt{3x} = 12$	$\sqrt{3x} = 12$
(2) Square both sides.	$(\sqrt{3x})^2 = (12)^2$	$\sqrt{3 \times 48} \stackrel{?}{=} 12$
(3) Simplify.	$3x = 144$	$\sqrt{144} \stackrel{?}{=} 12$
	$x = 48$	$12 = 12$ ✔

ANSWER: $x = 48$

2. Solve and check: $\sqrt{2x - 1} = 3$

(1) Write the equation.	$\sqrt{2x - 1} = 3$	$\sqrt{2x - 1} = 3$
(2) Square both sides.	$(\sqrt{2x - 1})^2 = (3)^2$	$\sqrt{(2)(5) - 1} \stackrel{?}{=} 3$
(3) Simplify.	$2x - 1 = 9$	$\sqrt{9} \stackrel{?}{=} 3$
	$2x = 10$	$3 = 3$ ✔
	$x = 5$	

ANSWER: $x = 5$

3. Solve and check: $3\sqrt{x} + 1 = 3$

(1) Write the equation.	$3\sqrt{x} + 1 = 3$	$3\sqrt{x} + 1 = 3$
(2) Isolate the radical term.	$3\sqrt{x} = 2$	$3\sqrt{\frac{4}{9}} + 1 \stackrel{?}{=} 3$
(3) Square both sides.	$(3\sqrt{x})^2 = (2)^2$	$3\left(\frac{2}{3}\right) + 1 \stackrel{?}{=} 3$
(4) Simplify.	$9x = 4$	$2 + 1 \stackrel{?}{=} 3$
	$x = \frac{4}{9}$	$3 = 3$ ✔

ANSWER: $x = \frac{4}{9}$

4. Solve and check: $\sqrt{3x + 4} - 3 = -7$

(1) Write the equation.	$\sqrt{3x + 4} - 3 = -7$	$\sqrt{3x + 4} - 3 = -7$
(2) Isolate the radical.	$\sqrt{3x + 4} = -4$	$\sqrt{(3)(4) + 4} - 3 \stackrel{?}{=} -7$
(3) Square both sides.	$(\sqrt{3x + 4})^2 = (-4)^2$	$\sqrt{16} - 3 \stackrel{?}{=} -7$
(4) Simplify.	$3x + 4 = 16$	$4 - 3 \stackrel{?}{=} -7$
	$3x = 12$	$1 \neq -7$
	$x = 4$	

ANSWER: The solution set is the empty set, $\varnothing$.

5. The time t required for a pendulum to swing from one side to the other and back is given by the formula:

$$t = 2\pi\sqrt{\frac{\ell}{g}} \qquad \text{where } \ell = \text{the length of the pendulum and}$$
$$g = \text{the force of gravity}$$

Given that $g = 32\ \dfrac{\text{ft.}}{\text{sec.}^2}$, find ℓ when $t = 10$ sec.

Solution:

$$t = 2\pi\sqrt{\frac{\ell}{g}}$$

$$10 = 2\pi\sqrt{\frac{\ell}{32}} \qquad \text{Substitute the given values: } t = 10, g = 32$$

$$(10)^2 = \left(2\pi\sqrt{\frac{\ell}{32}}\right)^2 \qquad \text{Square both sides of the equation.}$$

$$100 = 4\pi^2 \cdot \frac{\ell}{32} \qquad \text{Simplify the squares.}$$

$$3{,}200 = 4\pi^2 \cdot \ell \qquad \text{Multiply both sides by 32.}$$

$$\frac{800}{\pi^2} = \ell \qquad \text{Divide both sides by } 4\pi^2.$$

ANSWER: $\ell = \left(\dfrac{800}{\pi^2}\right)$ ft., or $\ell \approx 81$ ft.

Note how the units of the formula work out.

$$t = 2\pi\sqrt{\frac{\ell}{g}}$$

$$\text{sec.} = \sqrt{\frac{\text{ft.}}{\frac{\text{ft.}}{\text{sec.}^2}}} = \sqrt{\cancel{\text{ft.}} \times \frac{\text{sec.}^2}{\cancel{\text{ft.}}}} = \text{sec.}$$

Now it's your turn to . . . **TRY IT!**

Solve and check: **1.** $\sqrt{3x} + 10 = 4$ **2.** $\sqrt{3x - 2} + 3 = 7$

*(See page **629** for solutions.)*

In 1–42, solve the equation and check.

1. $\sqrt{x} = 5$　　　　　　**2.** $\sqrt{y} = 8$　　　　　　**3.** $\sqrt{z} = -4$

4. $\sqrt{a} = 1.2$　　　　　**5.** $\sqrt{b} = \frac{1}{4}$　　　　　**6.** $\sqrt{c} = \frac{2}{9}$

7. $\sqrt{5x} = 5$　　　　　**8.** $\sqrt{2x} = 8$　　　　　**9.** $\sqrt{3y} = 4$

10. $2\sqrt{x} = 6$　　　　　**11.** $5\sqrt{y} = 10$　　　　　**12.** $4\sqrt{a} = 36$

13. $2\sqrt{y} = 1$　　　　　**14.** $3\sqrt{m} = 4$　　　　　**15.** $5\sqrt{r} = -2$

16. $\sqrt{\dfrac{x}{4}} = 6$　　　　　**17.** $\sqrt{\dfrac{y}{3}} = 5$　　　　　**18.** $\sqrt{\dfrac{3z}{2}} = 6$

19. $3\sqrt{2x} = 12$　　　　**20.** $2\sqrt{5x} = 20$　　　　**21.** $4\sqrt{3x} = -9$

22. $\sqrt{x + 1} = 4$　　　　**23.** $\sqrt{b - 1} = 8$　　　　**24.** $\sqrt{r + 3} = 5$

25. $\sqrt{2x - 1} = 7$　　　**26.** $\sqrt{3a + 3} = 6$　　　**27.** $\sqrt{2b - 4} = 10$

28. $\sqrt{x} + 3 = 8$　　　　**29.** $\sqrt{m} - 6 = 2$　　　　**30.** $\sqrt{2r} + 7 = 8$

31. $2\sqrt{x} + 1 = 5$　　　**32.** $3\sqrt{x} - 2 = 10$　　　**33.** $4\sqrt{y} + 1 = 8$

34. $\sqrt{2x + 1} - 1 = 4$　　**35.** $\sqrt{5y - 1} - 3 = 0$　　**36.** $5 + \sqrt{2x - 4} = 1$

37. $\dfrac{10}{\sqrt{x}} = \sqrt{x}$　　　**38.** $\dfrac{3}{\sqrt{y - 5}} = \sqrt{y - 5}$　　　**39.** $\sqrt{2x + 1} = \dfrac{7}{\sqrt{2x + 1}}$

40. $\sqrt{3x - 8} = \sqrt{x}$　　**41.** $\sqrt{5x} = \sqrt{2x + 6}$　　**42.** $\sqrt{4z - 3} = \sqrt{3z + 4}$

In 43–45, solve the equation for the indicated quantity.

43. $s = \sqrt{A}$ for A　　　　**44.** $R = \sqrt{\dfrac{A}{\pi}}$ for A　　　　**45.** $e = \sqrt{\dfrac{s}{6}}$ for s

46. The square root of 3 times a number is 6. Find the number.

47. Five times the square root of a number is 20. Find the number.

48. The square root of the sum of twice a number and 5 is 3. Find the number.

49. When twice a number is decreased by 4, the square root of the result is equal to 10. Find the number.

50. The velocity v of a satellite moving in a circular orbit near the surface of the earth is given by the formula:

$$v = \sqrt{gr} \quad \text{where } g = \text{the force of gravity and}$$
$$r = \text{the radius of the earth}$$

Given that $g = 9.81 \dfrac{\text{m}}{\text{sec.}^2}$ and $v = 7.91 \times 10^3 \dfrac{\text{m}}{\text{sec.}}$, find r.

51. The time t that it takes a freely falling object to hit the ground is given by the formula:

$$t = \sqrt{\dfrac{2s}{g}} \quad \text{where } s = \text{the height of the object above the ground and}$$
$$g = \text{the force of gravity}$$

Given that $g = 32 \dfrac{\text{ft.}}{\text{sec.}^2}$, find s when $t = 9$ sec.

52. In physics, a description of an object in motion is given by the formula:

$$v_f = \sqrt{(v_0)^2 + 2as} \quad \text{where } v_f = \text{the final velocity}$$
$$v_0 = \text{the initial velocity}$$
$$a = \text{the acceleration, and}$$
$$s = \text{the distance traveled}$$

a. If $v_0 = 4 \dfrac{\text{m}}{\text{sec.}}$, $a = 1.6 \dfrac{\text{m}}{\text{sec.}^2}$, and $s = 500$ m, find v_f to the nearest hundredth.

b. If $v_f = 55 \dfrac{\text{m}}{\text{sec.}}$, $v_0 = 5 \dfrac{\text{m}}{\text{sec.}}$, and $s = 1$ km, find a to the nearest tenth.

53. The resonant frequency f of an electrical circuit is given by the formula:

$$f = \dfrac{1}{2\pi \sqrt{LC}} \quad \text{where } L = \text{the inductance and } C = \text{the capacitance}$$

a. If $L = 3 \times 10^{-2}$ henries and $C = 12 \times 10^{-6}$ farads, find the number of cycles per second in f.

b. If $L = 2 \times 10^{-2}$ henries and $f = \dfrac{10^4}{8\pi} \dfrac{\text{cycles}}{\text{sec.}}$, find the number of farads in C.

11-5 GRAPHING A QUADRATIC FUNCTION

Recall that the standard form of a *quadratic equation*, a second-degree equation in one variable, is:

$$ax^2 + bx + c = 0 \quad \text{where } a \neq 0$$

You are familiar with the solution of such an equation when it is possible to factor the polynomial. For example:

$$x^2 - 2x = 3$$

$$x^2 - 2x - 3 = 0 \qquad \text{Rewrite the equation in standard form.}$$

$$(x + 1)(x - 3) = 0 \qquad \text{Factor.}$$

$$
\begin{array}{l|l}
x + 1 = 0 & x - 3 = 0 \\
x = -1 & x = 3
\end{array}
\qquad
\begin{array}{l}
\text{Set each factor} = 0 \text{ and solve the} \\
\text{resulting first-degree equations.}
\end{array}
$$

Checking each of the apparent answers in the original equation establishes that the equation has two roots. The solution set is $\{-1, 3\}$.

It is also possible to see the roots of this equation graphically by graphing the quadratic relation $y = x^2 - 2x - 3$. To draw the graph, construct a table of values. Use integers for x from -2 through 4.

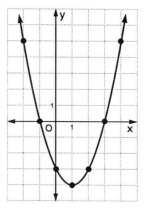

x	$x^2 - 2x - 3$	y
-2	$(-2)^2 - 2(-2) - 3$	5
-1	$(-1)^2 - 2(-1) - 3$	0
0	$0^2 - 2(0) - 3$	-3
1	$1^2 - 2(1) - 3$	-4
2	$2^2 - 2(2) - 3$	-3
3	$3^2 - 2(3) - 3$	0
4	$4^2 - 2(4) - 3$	5

Study this graph to note first that, by the vertical-line test, the quadratic relation $y = x^2 - 2x - 3$ is a function. Observe that this graph, which is called a ***parabola***, has two x-intercepts. Note that the x-intercepts, for which the y-values are 0, are the roots of the equation $x^2 - 2x - 3 = 0$.

In general, the roots of the quadratic equation $ax^2 + bx + c = 0$ are the x-intercepts of the parabola $y = ax^2 + bx + c$.

a. Solve and check: $x^2 + 2x - 8 = 0$

b. Graph the parabola $y = x^2 + 2x - 8$ by constructing a table of values, using integral values for x from -5 through 3.

To learn more about a parabola, study the two graphs drawn from the tables shown.

x	x² + 2x − 3	y
−4	$(-4)^2 + 2(-4) - 3$	5
−3	$(-3)^2 + 2(-3) - 3$	0
−2	$(-2)^2 + 2(-2) - 3$	−3
−1	$(-1)^2 + 2(-1) - 3$	−4
0	$0^2 + 2(0) - 3$	−3
1	$1^2 + 2(1) - 3$	0
2	$2^2 + 2(2) - 3$	5

x	−x² − 2x + 3	y
−4	$-(-4)^2 - 2(-4) + 3$	−5
−3	$-(-3)^2 - 2(-3) + 3$	0
−2	$-(-2)^2 - 2(-2) + 3$	3
−1	$-(-1)^2 - 2(-1) + 3$	4
0	$-(0)^2 - 2(0) + 3$	3
1	$-(1)^2 - 2(1) + 3$	0
2	$-(2)^2 - 2(2) + 3$	−5

Figure 1

Figure 2

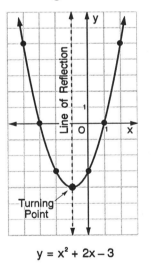

$y = x^2 + 2x - 3$

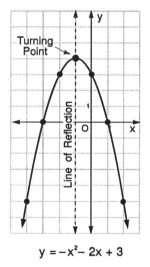

$y = -x^2 - 2x + 3$

It is customary to look at a graph from left to right to observe its nature. Note that in Figure 1, the parabola falls, then turns, then rises, while in Figure 2, the parabola rises, then turns, then falls. Every parabola has a ***turning point***. In the case of parabolas like Figure 1, the turning point is the lowest point on the curve, the *minimum point*. In parabolas like Figure 2, the turning point is the highest point on the curve, the *maximum point*.

Observe, from the tables of values, the pattern formed by the *y*-values for each curve. Note that on opposite sides of the turning point, corresponding *y*-values are the same.

See how this shows up on the graph: a vertical line through the turning point is a *line of reflection;* that is, each point on the parabola has a corresponding image point on the other side of the line of reflection.

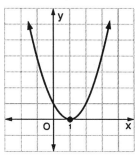

When the parabola $y = ax^2 + bx + c$ touches the x-axis in only one point, the two roots of the equation $ax^2 + bx + c = 0$ have the same value, called a **double root**. For example, 1 is a double root of the equation $x^2 - 2x + 1 = 0$.

Note that the polynomial $x^2 - 2x + 1$ has two identical factors: $(x - 1)(x - 1)$

$$y = x^2 - 2x + 1$$

MODEL PROBLEMS

1. a. Graph the parabola $y = x^2 - 3x + 2$ by constructing a table of values, using integral values for x from -1 through 4.

b. Write the coordinates of the turning point.

c. Write an equation for the line of reflection.

d. Under a reflection in the line in part **c**, what is the image point of $(-1, 6)$?

e. From the graph, read the roots of the equation $x^2 - 3x + 2 = 0$.

Solution

a.

x	$x^2 - 3x + 2$	y
-1	$(-1)^2 - 3(-1) + 2$	6
0	$0^2 - 3(0) + 2$	2
1	$1^2 - 3(1) + 2$	0
2	$2^2 - 3(2) + 2$	0
3	$3^2 - 3(3) + 2$	2
4	$4^2 - 3(4) + 2$	6

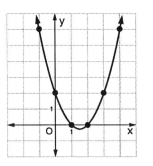

b. Since the turning point did not show up in the table of values, its x-coordinate is not an integer. Note, from the symmetry of the y-values in the table, that the x-value of the turning point is midway between 1 and 2, that is, 1.5. To determine the corresponding y-value, substitute $x = 1.5$ into the function.

$$y = x^2 - 3x + 2$$
$$(1.5)^2 - 3(1.5) + 2 = 2.25 - 4.5 + 2 = -.25$$

ANSWER: The turning point is $(1.5, -.25)$.

c. The line of reflection, also called the **axis of symmetry** of the parabola, is a vertical line through the turning point.

   ANSWER: An equation of the line of reflection is $x = 1.5$.

d. On the parabola, the image of $(-1, 6)$ is $(4, 6)$.

e. The roots of the equation $x^2 - 3x + 2 = 0$ are the x-intercepts of the parabola. That is, when $y = 0$, $x = 1$ or $x = 2$.

2. Connie has 100 feet of wire fencing with which to enclose an outdoor rectangular playpen for her puppy.

   a. Use one variable x to represent the dimensions of the rectangle.

   b. Write the area y as a function of x.

   c. Draw the parabola that is the graph of the area function by constructing a table of values, using values for x in intervals of 10 from 0 through 50.

   d. Find the dimensions of the rectangle that give the maximum area.

   e. What is the maximum area?

Solution:

a. Let each width $= x$.
   Then from the original 100 feet, there are $(100 - 2x)$ feet left for the two lengths.

   Thus, each length $= (50 - x)$ feet.

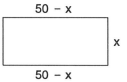

b. The area y of the rectangle $= x(50 - x)$ or $y = -x^2 + 50x$.

c.

x	$-x^2 + 50x$	y
0	$-(0)^2 + 50(0)$	0
10	$-(10)^2 + 50(10)$	400
20	$-(20)^2 + 50(20)$	600
30	$-(30)^2 + 50(30)$	600
40	$-(40)^2 + 50(40)$	400
50	$-(50)^2 + 50(50)$	0

d. The maximum area would occur at the turning point of the parabola. From the table, note that the x-coordinate of the turning point is midway between 20 and 30, that is $x = 25$.

Thus, for maximum area, the width is 25, and the length $= 50 - x = 50 - 25 = 25$.

e. The maximum area occurs when the rectangle is a square of area $(25)^2$, or 625 sq. ft.

1. a. Graph the parabola $y = -x^2 + 5x - 6$ by constructing a table of values, using integral values for x from 0 through 5.
   b. Write the coordinates of the turning point.
   c. Write an equation for the axis of symmetry.
   d. From the graph, read the roots of the equation $-x^2 + 5x - 6 = 0$.

2. Using an existing fence as one side, Connie plans to enclose an outdoor rectangular playpen for her puppy with 100 feet of wire fencing.
   a. Draw a diagram to model the situation using one variable x to represent the dimensions of the playpen.
   b. Write the area y as a function of x.
   c. Draw the parabola that is the graph of the area function by constructing a table of values, using values for x in intervals of 10 from 0 through 50.
   d. What is the maximum area?

Graphing a Quadratic Inequality

The procedure for graphing a quadratic inequality in the plane is the same as for graphing a linear inequality. This time, however, the plane divider is a parabola, which divides the plane into two regions "above" and "below" the parabola.

MODEL PROBLEM

Graph the inequality $y > x^2 - 4x + 4$ by constructing a table of values, choosing integers for x from -1 through 5.

Solution:

x	$x^2 - 4x + 4$	y
−1	$(-1)^2 - 4(-1) + 4$	9
0	$0^2 - 4(0) + 4$	4
1	$1^2 - 4(1) + 4$	1
2	$2^2 - 4(2) + 4$	0
3	$3^2 - 4(3) + 4$	1
4	$4^2 - 4(4) + 4$	4
5	$5^2 - 4(5) + 4$	9

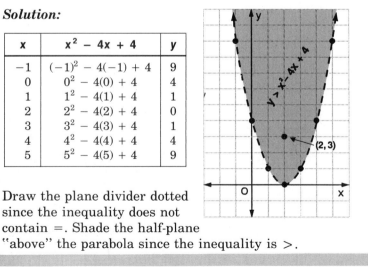

Draw the plane divider dotted since the inequality does not contain =. Shade the half-plane "above" the parabola since the inequality is >.

Choose a check point, say (2, 3), to verify the graph.

$$y > x^2 - 4x + 4$$
$$3 \overset{?}{>} 2^2 - 4(2) + 4$$
$$3 \overset{?}{>} 4 - 8 + 4$$
$$3 > 0 ✓$$

In 1–20:

a. Graph the given quadratic function of the form $y = ax^2 + bx + c$ by constructing a table of values, using the integral values of x in the given interval.

b. Write the coordinates of the turning point of the parabola.

c. Write an equation for the axis of symmetry of the parabola.

d. From the graph, find the roots of the equation $ax^2 + bx + c = 0$. For nonintegral roots, approximate the values between consecutive integers.

1. $y = x^2 - 6x + 8$ $0 \leq x \leq 6$ **2.** $y = x^2 - 4x + 3$ $-1 \leq x \leq 5$

3. $y = -x^2 + 6x - 8$ $0 \leq x \leq 6$ **4.** $y = -x^2 + 4x - 3$ $-1 \leq x \leq 5$

5. $y = x^2 - 2x - 3$ $-2 \leq x \leq 4$ **6.** $y = x^2 + x - 6$ $-4 \leq x \leq 3$

7. $y = x^2$ $-3 \leq x \leq 3$ **8.** $y = 2x^2$ $-2 \leq x \leq 2$

9. $y = -x^2 + 3$ $-3 \leq x \leq 3$ **10.** $y = x^2 - x - 1$ $-3 \leq x \leq 4$

11. $y = x^2 - 4$ $-3 \leq x \leq 3$ **12.** $y = -x^2$ $-3 \leq x \leq 3$

13. $y = x^2 - 2x - 2$ $-2 \leq x \leq 4$ **14.** $y = x^2 + x - 3$ $-3 \leq x \leq 2$

15. $y = -x^2 + x + 3$ $-2 \leq x \leq 3$ **16.** $y = -x^2 - x + 3$ $-3 \leq x \leq 2$

17. $y = 4x^2 - 4x + 1$ $-1 \leq x \leq 2$ **18.** $y = -4x^2 + 12x - 9$ $0 \leq x \leq 3$

19. $y = (x + 3)^2$ $-6 \leq x \leq 0$ **20.** $y = -(x - 2)^2$ $-1 \leq x \leq 5$

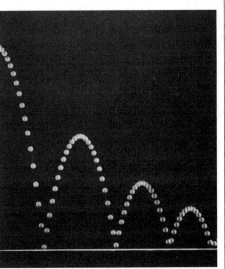

In our physical world, the parabola is the shape naturally assumed by objects in certain types of motion. For example, the path of a basketball aimed at an overhead hoop is a parabola, the path of a fly ball hit by a batter is a parabola, and the path of a bouncing ball is a series of parabolas.

A 3-dimensional parabolic shell is used in situations that require special reflective properties, such as an automobile headlight, or intense amplification, such as a satellite dish antenna.

21. A projectile is fired from the ground so that its height h above the ground is a function of the time t it has traveled: $h = 96t - 16t^2$

    a. Draw the parabola that is the graph of the height of the projectile by constructing a table of values, using integral values for t from 0 through 6.

    b. After how many seconds does the projectile reach its maximum height?

    c. What is its maximum height?

22. The cost c of producing x units of a radio is given by the function:
$c = 10 - 0.048x + 0.0002x^2$

    a. Draw the parabola that is the graph of the cost of production by constructing a table of values, using values for x in intervals of 10 from 90 through 150.

    b. What number of radios will minimize the cost of production?

    c. What is the minimum cost per radio?

23. Using a river as one border and 48 feet of fencing for the other three sides, Tom is planning to mark off a rectangular vegetable plot.

    a. Draw a diagram to model the situation using one variable x to represent the dimensions of the rectangle.

    b. Write the area of the rectangle, y, as a function of x.

    c. Draw the parabola that is the graph of the area function by constructing a table of values, using integral values of x from 9 through 15.

    d. What is the maximum area for the rectangular plot?

24. A cost analysis of fares at the Pleased Passengers Bus Company revealed that when the fare on the commuter express between Stoneville and Rock Harbor was $3.00, the average number of riders was 20. For each fare increase of 25 cents, the average number of riders decreases by 1.

    a. Represent the declining number of riders by x and represent the corresponding fare increase in terms of x.

    b. Write the total receipts of a bus run, r, in terms of x.

    c. Draw the parabola that is the graph of the revenue function by constructing a table of values, using integral values of x from 1 through 7.

    d. What fare should the bus company charge to maximize the revenue per trip?

    In 25–30, graph the quadratic inequality. To construct a table of values for the plane divider, use the integral values of x in the given interval.

25. $y < x^2$ $-3 \le x \le 3$ **26.** $y < -2x^2$ $-2 \le x \le 2$

27. $y \ge x^2 - 2x$ $-2 \le x \le 4$ **28.** $y > -x^2 + 4x - 3$ $-1 \le x \le 5$

29. $y \le x^2 - 5$ $-3 \le x \le 3$ **30.** $y > -x^2 - x + 3$ $-3 \le x \le 2$

11-6 MORE METHODS FOR SOLVING A QUADRATIC EQUATION

Solving a Quadratic Equation by Taking the Square Root of Both Sides

Consider a quadratic equation $ax^2 + bx + c = 0$ for which the value of b is 0. Sometimes such an equation can be solved by factoring over the integers.

$$x^2 - 4 = 0$$
$$(x + 2)(x - 2) = 0$$
$$x + 2 = 0 \mid x - 2 = 0$$
$$x = -2 \mid \quad x = 2$$

Another method for solving a quadratic in which $b = 0$ relies upon taking the square root of both sides of the equation, recalling that a positive number has two square roots that are opposites.

$$x^2 - 4 = 0$$
$$x^2 = 4$$
$$x = \pm\sqrt{4}$$
$$x = \pm 2$$

Note that this method applies to this case of a quadratic even when it cannot be factored over the integers, thus producing irrational roots.

$$x^2 - 5 = 0$$
$$x^2 = 5$$
$$x = \pm\sqrt{5}$$

This method of solving a quadratic equation, by taking the square root of both sides of the equation, is appropriate as long as the variable is contained within a perfect square. For example:

$(x - 2)^2 = 9$ The variable is in a perfect square.

$\sqrt{(x - 2)^2} = \pm\sqrt{9}$ Take the square root of both sides, recalling that a positive number has two square roots that are opposites.

$x - 2 = \pm 3$

$x = 2 \pm 3$ Isolate x by adding 2 to both sides.

$x = 2 + 3 \mid x = 2 - 3$ Calculate the two values of x.
$x = 5 \mid x = -1$

Check both values, 5 and -1, in the original equation.

$$(x - 2)^2 = 9 \qquad\qquad (x - 2)^2 = 9$$
$$(5 - 2)^2 \stackrel{?}{=} 9 \qquad\qquad (-1 - 2)^2 \stackrel{?}{=} 9$$
$$(3)^2 \stackrel{?}{=} 9 \qquad\qquad (-3)^2 \stackrel{?}{=} 9$$
$$9 = 9 \checkmark \qquad\qquad 9 = 9 \checkmark$$

Now it's your turn to . . . TRY IT!

Solve and check: **1.** $3x^2 = 48$ **2.** $2x^2 - 34 = 0$ **3.** $(x - 1)^2 = 16$

(See page 630 for solutions.)

In 1–24, solve the equation and check.

1. $y^2 = 64$ **2.** $\frac{1}{2}z^2 = 8$ **3.** $a^2 - 25 = 0$

4. $b^2 - .49 = 0$ **5.** $5y^2 = 45$ **6.** $27r^2 = 243$

7. $2x^2 - 8 = 0$ **8.** $5y^2 - 5 = 0$ **9.** $3y^2 - 300 = 0$

10. $x^2 + 9 = 25$ **11.** $y^2 + 25 = 169$ **12.** $z^2 + .01 = .37$

13. $4x^2 + 5 = 21$ **14.** $2x^2 - 11 = 39$ **15.** $.05x^2 - 3 = 2$

16. $2x^2 + 3x^2 = 45$ **17.** $6x^2 - 4x^2 = 98$ **18.** $7x^2 = 4x^2 + .75$

19. $4y^2 - 13 = y^2 + 14$ **20.** $2y^2 - 5 = 15 - 3y^2$ **21.** $2z^2 + 7 = 10 - z^2$

22. $\dfrac{y^2}{3} = 12$ **23.** $\dfrac{x}{9} = \dfrac{4}{x}$ **24.** $\dfrac{4x}{25} = \dfrac{4}{x}$

In 25–45, solve for x in simplest radical form.

25. $x^2 = 35$ **26.** $x^2 = 20$ **27.** $x^2 = 27$

28. $3x^2 = 6$ **29.** $2x^2 = 100$ **30.** $\frac{1}{2}x^2 = 6$

31. $2x^2 - 16 = 0$ **32.** $3x^2 - 60 = 0$ **33.** $x^2 + 25 = 100$

34. $\frac{1}{3}x^2 - 3 = 3$ **35.** $3x^2 + 5 = 29$ **36.** $5x^2 - 40 = 100$

37. $3x^2 + 4x^2 = 35$ **38.** $8x^2 - 6x^2 = 54$ **39.** $9x^2 = 4x^2 + 100$

40. $3x^2 - 28 = 2x^2 + 33$ **41.** $3x^2 - 6 = 34 - 2x^2$ **42.** $x^2 - 4 = 80 - 2x^2$

43. $\dfrac{x}{3} = \dfrac{5}{x}$ **44.** $\dfrac{x}{8} = \dfrac{4}{x}$ **45.** $\dfrac{2x}{9} = \dfrac{6}{x}$

In 46–54, find the positive value of x correct to the nearest tenth.

46. $x^2 = 24$ **47.** $\frac{1}{3}x^2 = 31$ **48.** $10x^2 = 264$

49. $4x^2 - 160 = 0$ **50.** $x^2 + 16 = 64$ **51.** $2x^2 + 7 = 67$

52. $7x^2 = x^2 + 198$ **53.** $x^2 - 42 = 82 - 3x^2$ **54.** $\dfrac{2x}{43} = \dfrac{8}{x}$

In 55–63, solve for x, expressing irrational answers in radical form.

55. $(x - 1)^2 = 4$ **56.** $(x - 3)^2 = 49$ **57.** $(x + 2)^2 = 16$

58. $(x - 2)^2 = 3$ **59.** $(x - 4)^2 = 11$ **60.** $(x + 5)^2 = 23$

61. $\dfrac{4}{x - 1} = \dfrac{x - 1}{9}$ **62.** $\dfrac{x + 3}{2} = \dfrac{8}{x + 3}$ **63.** $\dfrac{x - 2}{3} = \dfrac{5}{x - 2}$

In 64–69, solve for x in terms of the other quantities.

64. $x^2 = b^2$ **65.** $x^2 = 25a^2$ **66.** $9x^2 = r^2$

67. $4x^2 - a^2 = 0$ **68.** $x^2 + a^2 = c^2$ **69.** $x^2 + b^2 = c^2$

In 70–77, solve for the indicated quantity.

70. Solve for S: $S^2 = A$ **71.** Solve for e: $S^2 = 6e^2$

72. Solve for r: $A = \pi r^2$ **73.** Solve for r: $S = 4\pi r^2$

74. Solve for I: $W = I^2 R$ **75.** Solve for r: $V = \pi r^2 h$

76. Solve for t: $s = \frac{1}{2}gt^2$ **77.** Solve for v: $F = \dfrac{mv^2}{gr}$

Completing a Trinomial Square

You have seen that when the variable is contained in a perfect square, a quadratic equation can be solved by taking the square root of both sides of the equation. Soon you will see how every quadratic equation can be rewritten so that the variable is contained in a perfect square.

First consider the perfect square that contains the variable.

The expression that is the result of the square of a binomial is called a ***trinomial square.*** $(x + 3)^2 = x^2 + 6x + 9$

Observe from this trinomial square that the constant in the trinomial is the square of the constant in the binomial. Also note that the constant in the binomial is half of the coefficient of x in the trinomial.

$$9 \text{ is the square of } 3$$
$$(x + 3)^2 = x^2 + 6x + 9$$
$$3 \text{ is half of } 6$$

Thus, beginning with only two terms of a trinomial square, you can determine what number to add to *complete the square.*

What number must be added to complete the square? $x^2 + 6x + ?$

To complete the trinomial square, add the square of half the coefficient of x. $x^2 + 6x + \left(\frac{1}{2} \cdot 6\right)^2$

Check to see that the completed trinomial is the square of a binomial. $x^2 + 6x + 9 = (x + 3)^2$

KEEP IN MIND

To complete a trinomial square starting with a binomial of the form $x^2 + px$, add the square of half the coefficient of x.

$$x^2 + px + \left(\tfrac{1}{2}p\right)^2 = \left(x + \tfrac{1}{2}p\right)^2$$

MODEL PROBLEMS

| **Complete the Square** | **Add the square of half the coefficient of x.** | **The completed trinomial is the square of a binomial.** |

1. $x^2 + 8x + ?$ $x^2 + 8x + \left(\frac{1}{2} \cdot 8\right)^2$ $x^2 + 8x + 16 = (x + 4)^2$

2. $x^2 - 10x + ?$ $x^2 - 10x + \left(\frac{1}{2} \cdot (-10)\right)^2$ $x^2 - 10x + 25 = (x - 5)^2$

3. $x^2 + 3x + ?$ $x^2 + 3x + \left(\frac{1}{2} \cdot 3\right)^2$ $x^2 + 3x + \frac{9}{4} = \left(x + \frac{3}{2}\right)^2$

4. $2x^2 + 5x + ?$
$= 2\left(x^2 + \frac{5}{2}x + ?\right)$ $2\left|x^2 + \frac{5}{2}x + \left(\frac{1}{2} \cdot \frac{5}{2}\right)^2\right|$ $2\left(x^2 + \frac{5}{2}x + \frac{25}{16}\right) = 2\left(x + \frac{5}{4}\right)^2$

As in Model Problem 4, when the coefficient of x^2 is other than 1, first factor out that coefficient. Then proceed to complete the square as before.

Now it's your turn to . . . **TRY IT!**   (*See page 631 for solutions.*)

Complete the square, and express the resulting trinomial as the square of a binomial.
   1. $x^2 + 14x + ?$ **2.** $x^2 - 6x + ?$ **3.** $x^2 + 7x + ?$ **4.** $3x^2 + 6x + ?$

EXERCISES

In 1–15, replace the question mark with a number that will complete the square, and express the resulting trinomial as the square of a binomial.

1. $x^2 + 4x + ?$ **2.** $x^2 - 12x + ?$ **3.** $x^2 + 3x + ?$

4. $x^2 - 4x + ?$ **5.** $c^2 + 14c + ?$ **6.** $x^2 - 18x + ?$

7. $r^2 + 5r + ?$ **8.** $x^2 - 9x + ?$ **9.** $x^2 - x + ?$

10. $x^2 + \frac{1}{2}x + ?$ **11.** $y^2 - \frac{1}{3}y + ?$ **12.** $x^2 - \frac{4}{5}x + ?$

13. $2x^2 + 10x + ?$ **14.** $3x^2 + 9x + ?$ **15.** $2x^2 + 3x + ?$

In 16–21, determine whether the trinomial is a perfect square.

16. $x^2 + 4x + 4$ **17.** $x^2 - 14x + 49$ **18.** $x^2 - 6x + 6$

19. $x^2 - 10x - 25$ **20.** $x^2 - 3x + 2.25$ **21.** $x^2 - x + 1$

Solving a Quadratic Equation by Completing the Square

Now that you can complete a trinomial square, you will be able to rewrite any quadratic equation so that the variable is contained in a perfect square. For example:

$x^2 + 6x - 7 = 0$	The trinomial on the left is not a perfect square.
$x^2 + 6x = 7$	Transpose the constant so that only terms containing the variable are on the left side.
$\left(\frac{1}{2} \cdot 6\right)^2$ or 9	Determine the number that will complete the square on the left side of the equation; that is, the square of half the coefficient of the x-term.
$x^2 + 6x + 9 = 7 + 9$	Add the number that completes the trinomial square to both sides of the equation.
$(x + 3)^2 = 16$	The trinomial square is the square of a binomial.
$x + 3 = \pm\sqrt{16}$	Take the square root of both sides of the equation, recalling that a positive number has two square roots.
$x + 3 = \pm 4$	
$x + 3 = 4 \ \vert \ x + 3 = -4$	Calculate the two values of x.
$x = 1 \ \ \vert \ \ \ \ x = -7$	

Checking both values in the original equation verifies that the solution set is $\{-7, 1\}$.

The fact that the roots of the equation are integers indicates that the original equation could have been solved by factoring. You may have noticed this at the outset. Note now that the solution by completing the square is applicable to *any* quadratic equation. Thus, you have a general method for solving a quadratic equation, especially those that are not factorable.

PROCEDURE

To solve a quadratic equation by completing the square:

1. Rewrite the equation in the form $x^2 + px = q$.

   You may have to transpose terms, or divide both sides by the coefficient of x^2.

2. Determine the number that will complete the trinomial square, $\left(\frac{1}{2}p\right)^2$, and add it to both sides of the equation.

3. Express the trinomial as the square of a binomial.

4. Take the square root of both sides of the equation.

5. Calculate the two values of the variable.

6. Check both values in the original equation.

MODEL PROBLEM

Solve the quadratic equation by completing the square: $x^2 + 4 = 6x$

How to Proceed	**Solution**
(1) Rewrite the equation in the form $x^2 + px = q$.	$x^2 - 6x = -4$
(2) Determine the number that will complete the trinomial square, and add it to both sides of the equation.	$x^2 - 6x + 9 = -4 + 9$
(3) Express the trinomial as the square of a binomial.	$(x - 3)^2 = 5$
(4) Take the square root of both sides of the equation.	$x - 3 = \pm\sqrt{5}$
(5) Calculate the two values of the variable.	$\begin{array}{c\|c} x - 3 = \sqrt{5} & x - 3 = -\sqrt{5} \\ x = 3 + \sqrt{5} & x = 3 - \sqrt{5} \end{array}$

(6) Check. When the roots are irrational, they occur in pairs, such as $3 \pm \sqrt{5}$. Thus, it is sufficient to check one root.

$$x^2 - 6x = -4$$

$$(3 + \sqrt{5})^2 - 6(3 + \sqrt{5}) \stackrel{?}{=} -4$$

$$9 + 6\sqrt{5} + 5 - 18 - 6\sqrt{5} \stackrel{?}{=} -4$$

$$-4 = -4 ✔$$

ANSWER: $x = 3 + \sqrt{5}$ or $x = 3 - \sqrt{5}$, making the solution set $\{3 \pm \sqrt{5}\}$.

NOTE. You can also find decimal approximations for irrational roots. In this case, to approximate the roots to the nearest tenth, first find $\sqrt{5}$ to two decimal places.

$$3 + \sqrt{5} \approx 3 + 2.24 \approx 5.24 \approx 5.2 \qquad 3 - \sqrt{5} \approx 3 - 2.24 \approx .76 \approx .8$$

Now it's your turn to . . . **TRY IT!** *(See page 631 for solutions).*

Solve by completing the square. Leave irrational answers in radical form.

1. $x^2 - 6x + 8 = 0$ **2.** $x^2 - 4x = -1$

EXERCISES

In 1–21, solve the equation by completing the square. Express irrational answers in simplest radical form.

1. $x^2 + 2x = 8$ **2.** $z^2 - 4z = 21$ **3.** $x^2 - 4x + 3 = 0$

4. $x^2 = 6x + 40$ **5.** $y^2 - 5y - 14 = 0$ **6.** $z^2 = 6 - z$

7. $x^2 - 4x = -4$ **8.** $y^2 - 9 = 6y$ **9.** $2c^2 - 3c - 2 = 0$

10. $2x^2 + 7x + 3 = 0$ **11.** $3r^2 = 5r + 2$ **12.** $5y^2 + 9y - 2 = 0$

13. $x^2 + 2x = 1$ **14.** $x^2 - 4x + 2 = 0$ **15.** $x^2 + 3x - 5 = 0$

16. $m^2 - 1 = 4m$ **17.** $a^2 = 6a + 19$ **18.** $2x^2 - 4x = 1$

19. $3x^2 - 12x + 5 = 0$ **20.** $3x^2 - 2 = 6x$ **21.** $5x^2 = 3x + 1$

In 22–27, solve the equation by completing the square. Approximate the answers correct to the nearest tenth.

22. $x^2 - 6x + 9 = 7$ **23.** $x^2 + 4x + 4 = 3$ **24.** $x^2 - 2x = 4$

25. $x^2 = 8x + 2$ **26.** $2x^2 - 20x = 5$ **27.** $3x^2 - 4x - 2 = 0$

Solving a Quadratic Equation By Formula

Since any quadratic equation can be solved by the method of completing the square, you can solve the general quadratic equation $ax^2 + bx + c = 0$ for x in terms of a, b, and c. Then the general roots can be used as formulas for finding the roots of any quadratic equation. Study the following parallel solutions to see how the same method is used in solving a particular quadratic equation and the general quadratic equation.

$$3x^2 + 5x + 1 = 0 \qquad\qquad\qquad ax^2 + bx + c = 0 \;\; (a \neq 0)$$

$$3x^2 + 5x = -1 \qquad\qquad\qquad\qquad ax^2 + bx = -c$$

$$x^2 + \frac{5}{3}x = -\frac{1}{3} \qquad\qquad\qquad\qquad x^2 + \frac{b}{a}x = -\frac{c}{a}$$

$$\left[\frac{1}{2}\left(\frac{5}{3}\right)\right]^2 = \left(\frac{5}{6}\right)^2 = \frac{25}{36} \qquad\qquad \left[\frac{1}{2}\left(\frac{b}{a}\right)\right]^2 = \left(\frac{b}{2a}\right)^2 = \frac{b^2}{4a^2}$$

$$x^2 + \frac{5}{3}x + \frac{25}{36} = \frac{25}{36} - \frac{1}{3} \qquad\qquad x^2 + \frac{b}{a}x + \frac{b^2}{4a^2} = \frac{b^2}{4a^2} - \frac{c}{a}$$

$$\left(x + \frac{5}{6}\right)^2 = \frac{13}{36} \qquad\qquad\qquad \left(x + \frac{b}{2a}\right)^2 = \frac{b^2 - 4ac}{4a^2}$$

$$x + \frac{5}{6} = \pm\sqrt{\frac{13}{36}} \qquad\qquad\qquad x + \frac{b}{2a} = \pm\sqrt{\frac{b^2 - 4ac}{4a^2}}$$

$$x = -\frac{5}{6} \pm \frac{\sqrt{13}}{6} \qquad\qquad\qquad x = -\frac{b}{2a} \pm \frac{\sqrt{b^2 - 4ac}}{2a}$$

$$x = \frac{-5 \pm \sqrt{13}}{6} \qquad\qquad\qquad\qquad x = \frac{-b \pm \sqrt{b^2 - 4ac}}{2a}$$

The last step of the general solution is called the ***quadratic formula***. This formula can be used in solving any quadratic equation.

PROCEDURE

To solve a quadratic equation by the quadratic formula:

1. Write the given equation in the standard form $ax^2 + bx + c = 0$.
2. Determine the values of a, b, and c.
3. Substitute the values of a, b, and c into the quadratic formula: $x = \dfrac{-b \pm \sqrt{b^2 - 4ac}}{2a}$
4. Calculate the two values of x.

MODEL PROBLEMS

1. Use the quadratic formula to solve: $x^2 = 6x - 1$

Solution: Rewrite the equation in standard form and identify the values of a, b, and c.

$$ax^2 + bx + c = 0$$
$$x^2 - 6x + 1 = 0 \qquad a = 1 \quad b = -6 \quad c = 1$$

Substitute into the formula, and evaluate.

$$x = \frac{-b \pm \sqrt{b^2 - 4ac}}{2a}$$

$$x = \frac{-(-6) \pm \sqrt{(-6)^2 - 4(1)(1)}}{2(1)}$$

$$x = \frac{6 \pm \sqrt{36 - 4}}{2}$$

$$x = \frac{6 \pm \sqrt{32}}{2}$$

$$x = \frac{6 \pm 4\sqrt{2}}{2}$$

$$x = 3 \pm 2\sqrt{2}$$

Check one of the pair of irrational values in the original equation.

$$x^2 = 6x - 1$$
$$(3 + 2\sqrt{2})(3 + 2\sqrt{2}) \overset{?}{=} 6(3 + 2\sqrt{2}) - 1$$
$$9 + 12\sqrt{2} + 8 \overset{?}{=} 18 + 12\sqrt{2} - 1$$
$$17 + 12\sqrt{2} = 17 + 12\sqrt{2} \checkmark$$

ANSWER: $x = 3 + 2\sqrt{2}$ or $x = 3 - 2\sqrt{2}$, making the solution set $\{3 \pm 2\sqrt{2}\}$.

2. A rectangular sheet of tin is to be made into an open box by cutting congruent squares from the four corners, and turning up the edges. If the dimensions of the tin sheet are 8 inches by 10 inches and the area of the base of the open box is to be 60 square inches, find, to the nearest tenth of an inch, the length of the side of the square that should be cut from each corner.

Solution: Let x = the length of the side of the square.

    Then $8 - 2x$ = the remaining width of the tin sheet.

    And $10 - 2x$ = the remaining length of the tin sheet.

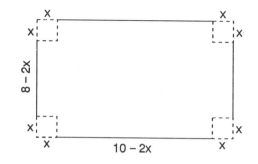

The area of the base of the box is 60.

$$(8 - 2x)(10 - 2x) = 60$$

$$80 - 36x + 4x^2 = 60$$

$$4x^2 - 36x + 20 = 0$$

$$x^2 - 9x + 5 = 0$$

From standard form:
$a = 1, b = -9, c = 5$

$$x = \frac{-(-9) \pm \sqrt{(-9)^2 - 4(1)(5)}}{2(1)}$$

Substitute into the quadratic formula, and evaluate.

$$x = \frac{9 \pm \sqrt{61}}{2}$$

$\sqrt{61} \approx 7.81$

$$x \approx \frac{9 + 7.81}{2} \qquad x \approx \frac{9 - 7.81}{2}$$

$$x \approx \frac{16.81}{2} \qquad x \approx \frac{1.19}{2}$$

$$x \approx 8.40 \approx 8.4 \qquad x \approx .59 \approx .6$$

Reject $x \approx 8.4$ since it leads to negative values for the dimensions of the box.

Check $x \approx .6$:   width = $8 - 2(.6) = 6.8$ in.
               length = $10 - 2(.6) = 8.8$ in.

The area of the base of the box is $6.8 \times 8.8 \approx 60$ sq. in. ✔

> **ANSWER:**   The length of the side of the square that should be cut from each corner is .6 in.

1. Solve and check: $x^2 + 11 = 10x$ Express answers in simplest radical form.

2. A rectangular picture frame of uniform width has outside dimensions of 8 in. by 12 in. If the area of the frame is the same as the area of the picture, how wide is the picture frame? Answer to the nearest tenth.

EXERCISES

In 1–27, solve the equation by using the quadratic formula. Express irrational roots in simplest radical form.

1. $x^2 + 2x - 24 = 0$ **2.** $x^2 - 9x + 20 = 0$ **3.** $x^2 - 6x + 9 = 0$

4. $x^2 + 12 = 7x$ **5.** $x^2 - 30 = x$ **6.** $x^2 = 5x + 14$

7. $2x^2 - 5x + 2 = 0$ **8.** $3x^2 - 10x + 3 = 0$ **9.** $5x^2 = 3x + 2$

10. $x^2 - 8x = 0$ **11.** $x^2 - 9 = 0$ **12.** $4x^2 = 25$

13. $x^2 + 2x - 1 = 0$ **14.** $y^2 - 2y - 2 = 0$ **15.** $z^2 + 3z = 5$

16. $x^2 - 4x + 1 = 0$ **17.** $x^2 + 5x = 2$ **18.** $y^2 = 5y + 2$

19. $2x^2 - 3x - 1 = 0$ **20.** $2x^2 + x - 4 = 0$ **21.** $3x^2 - 2x - 3 = 0$

22. $2y^2 + y = 5$ **23.** $5y^2 - 1 = 2y$ **24.** $2x^2 = 6x - 1$

25. $\frac{1}{4}x^2 - \frac{3}{2}x - 1 = 0$ **26.** $1 + \dfrac{7}{x} + \dfrac{2}{x^2} = 0$ **27.** $2x = 5 + \dfrac{4}{x}$

In 28–42, solve the equation by using the quadratic formula. Approximate irrational roots correct to the nearest tenth.

28. $x^2 + 3x - 3 = 0$ **29.** $y^2 - 4y + 2 = 0$ **30.** $2c^2 - 7c + 1 = 0$

31. $x^2 + 2x - 4 = 0$ **32.** $3x^2 - 2x - 6 = 0$ **33.** $2x^2 - 8x + 1 = 0$

34. $3x^2 + 5x - 1 = 0$ **35.** $2x^2 + 4x = 3$ **36.** $2x^2 - 10x = 9$

37. $2x^2 - 2x = 3$ **38.** $x^2 = 20x + 10$ **39.** $x^2 = 12 - 9x$

40. $x(x - 3) = -2$ **41.** $x(2x + 9) = 3$ **42.** $x = 4 + \dfrac{2}{x}$

In 43–53, solve algebraically. Express irrational answers in simplest radical form unless otherwise indicated.

43. The sum of two numbers is 12. The sum of their squares is 104. Find the numbers.

44. The difference between two numbers is 3. The sum of the squares of the numbers is 89. Find the numbers.

45. The sum of a number and its reciprocal is $\frac{5}{2}$. Find the number.

46. Mr. Perkins drove from his home to Boston, 40 miles away. He then returned home from Boston, traveling over the same road at an average speed that was 30 mph faster than his rate going. He spent a total of 2 hours traveling. Find his speed each way.

47. The rate of a motorboat in still water is 10 mph. The boat traveled 24 miles upstream and returned the same distance downstream. The round trip required 5 hours. Find the rate of the current.

48. A group of sorority sisters went to see a college drama production and spent $30 for tickets. The number of dollars in the price of a ticket exceeded the number of sorority sisters by 7. How many sorority sisters were in the group?

49. If Elizabeth and Huck work together, they can paint a fence in 4 hours. If they work alone, Elizabeth needs 6 hours more than Huck to paint the fence. How many hours would each one working alone need to complete the job?

50. Mr. Perez wishes to build a bin to hold 9 tons of coal. The bin is to be 6 feet deep and 4 feet longer than it is wide. Allowing 40 cubic feet for 1 ton of coal, what must be the dimensions of the bin?

To reach underlying seams of coal, *strip miners* remove the surface of the land, leaving behind barren landscapes. In an effort to preserve our planet, many former mine sites have now been restored, and have been brought to life as apple orchards, camp sites, fishing ponds, and grazing grounds.

51. A square and a rectangle have equal area. The length of the rectangle is 5 inches more than twice the length of a side of the square, and the width of the rectangle is 2 inches less than the length of a side of the square. Find the length of the side of the square to the nearest tenth of an inch.

52. A 5 by 7-inch photo is surrounded by a frame of uniform width. If the area of the frame alone is 24 square inches, find the width of the frame to the nearest tenth of an inch.

53. The equation of a circle with center at the origin and radius of 6 is $x^2 + y^2 = 36$.

   a. Find the coordinates of the two points on the circle whose y-coordinates are 2 more than their x-coordinates.

   b. Name the quadrant in which each of these points lies.

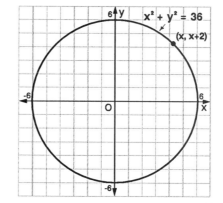

11-7 THE PYTHAGOREAN RELATION

Consider the right triangle whose sides measure 3, 4, and 5 cm.

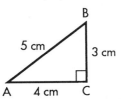

Observe the numerical relationship among the three lengths:

$$5^2 = 3^2 + 4^2$$

This example illustrates a relationship that is true in all right triangles:

In a right triangle, the square of the length of the hypotenuse is equal to the sum of the squares of the lengths of the other two sides.

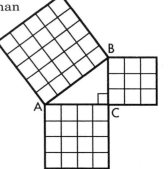

$$c^2 = a^2 + b^2$$

This general property of the right triangle was recorded more than 2,000 years ago by the Greek mathematician Pythagoras.

A geometric representation of the Pythagorean Relation is that the area of the square built on the hypotenuse of a right triangle is equal to the sum of the areas of the squares built on the legs.

One application of the Pythagorean Relation is to find the measure of the third side of a right triangle, given the measures of the other two sides.

MODEL PROBLEMS

1. Find the length of the hypotenuse of a right triangle in which the lengths of the other two sides are 5 and 12 inches.

   Solution: Let the lengths of the sides of the triangle be:

   hypotenuse = c, side $a = 5$, side $b = 12$

   $$c^2 = a^2 + b^2$$
   $$c^2 = 5^2 + 12^2$$
   $$c^2 = 25 + 144$$
   $$c^2 = 169$$
   $$c = \pm\sqrt{169} = \pm13 \quad \text{Reject the negative value as a length.}$$

   ANSWER: 13 inches

2. The hypotenuse of a right triangle measures 20 cm. If one leg measures 16 cm, find the length of the other leg.

   Solution: Let the lengths of the sides of the triangle be:

   hypotenuse $c = 20$, side $a = 16$, side $b = b$

   $$c^2 = a^2 + b^2$$
   $$20^2 = 16^2 + b^2$$
   $$400 = 256 + b^2$$
   $$144 = b^2$$
   $$\pm\sqrt{144} = b$$

   ANSWER: 12 cm

When a right triangle is a part of another geometric figure, the Pythagorean Relation can be used to determine the length of a diagonal or an altitude.

3. Find the length of the diagonal of a square whose side measures 8 cm.

Solution:

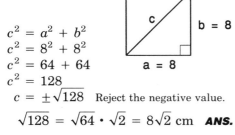

$$c^2 = a^2 + b^2$$
$$c^2 = 8^2 + 8^2$$
$$c^2 = 64 + 64$$
$$c^2 = 128$$
$$c = \pm\sqrt{128} \quad \text{Reject the negative value.}$$

$$\sqrt{128} = \sqrt{64} \cdot \sqrt{2} = 8\sqrt{2} \text{ cm} \quad \textbf{ANS.}$$

4. Find the length of the altitude of an equilateral triangle whose side measures 10 cm.

Solution:

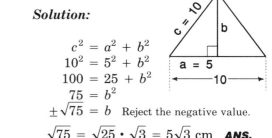

$$c^2 = a^2 + b^2$$
$$10^2 = 5^2 + b^2$$
$$100 = 25 + b^2$$
$$75 = b^2$$
$$\pm\sqrt{75} = b \quad \text{Reject the negative value.}$$

$$\sqrt{75} = \sqrt{25} \cdot \sqrt{3} = 5\sqrt{3} \text{ cm} \quad \textbf{ANS.}$$

5. In isosceles trapezoid $ABCD$, the bases are $\overline{AB}$ and $\overline{CD}$. Leg DA is 5 inches more than CD, and AB is 2 inches more than twice CD.

  a. If CD is represented by x, represent the measures of the other sides.
  b. If the perimeter of the figure is 52 inches, find the value of x.
  c. Find the length of the altitude.

Solution:

a.

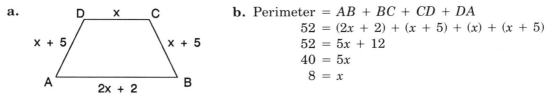

b. Perimeter $= AB + BC + CD + DA$
$$52 = (2x + 2) + (x + 5) + (x) + (x + 5)$$
$$52 = 5x + 12$$
$$40 = 5x$$
$$8 = x$$

c. Using $x = 8$, evaluate the measures of the four sides of the isosceles trapezoid:
$$CD = x = 8 \quad AB = 2x + 2 = 16 + 2 = 18 \quad AD = BC = x + 5 = 8 + 5 = 13$$

The altitudes, $\overline{DP}$ and $\overline{CQ}$, of the isosceles trapezoid divide the figure into a rectangle and two congruent right triangles. In rectangle $PQCD$, since opposite sides have the same measure, $CD = QP = 8$. When QP is subtracted from AB, $18 - 8$, there are 10 inches left for the two congruent lengths AP and QB. Thus, $AP = QB = 5$.

To determine the length of altitude $\overline{DP}$, apply the Pythagorean Relation to right $\triangle APD$.

$$(AD)^2 = (DP)^2 + (AP)^2$$
$$13^2 = (DP)^2 + 5^2$$
$$169 = (DP)^2 + 25$$
$$144 = (DP)^2$$
$$12 = DP$$

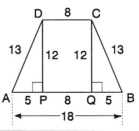

1. Find the length of the hypotenuse of an isosceles right triangle each of whose legs measures 5 inches.

2. Find the length of the base of a rectangle with diagonal 17 cm and height 8 cm.

The Distance Formula

In the coordinate plane, you are familiar with finding lengths of horizontal and vertical line segments. Now you will see how the Pythagorean Relation provides a method for finding the length of an oblique line segment.

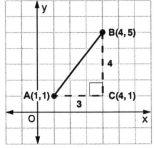

Consider the oblique line segment that joins the two points $A(1, 1)$ and $B(4, 5)$. Note that a right triangle, $\triangle ABC$, can be formed with AB as the hypotenuse, making $C(4, 1)$. In $\triangle ABC$:

$$AC = |4 - 1| = 3 \qquad BC = |5 - 1| = 4$$

Apply the Pythagorean Relation:

$$(AB)^2 = (AC)^2 + (BC)^2$$

$$(AB)^2 = 3^2 + 4^2 = 9 + 16 = 25$$

$$AB = 5 \quad \text{Reject the negative value.}$$

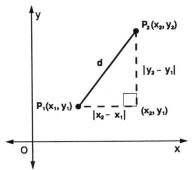

Repeating these same steps using the general points $P_1(x_1, y_1)$ and $P_2(x_2, y_2)$ produces a formula for finding the length P_1P_2, or the general distance d.

$$d^2 = (x_2 - x_1)^2 + (y_2 - y_1)^2$$

$$\boldsymbol{d = \sqrt{(x_2 - x_1)^2 + (y_2 - y_1)^2}}$$

MODEL PROBLEMS

1. Find the distance between $(-2, 3)$ and $(1, 2)$.

   Solution: To apply the distance formula, use either point as P_1 and the other as P_2. Let $(-2, 3) = (x_1, y_1)$ and $(1, 2) = (x_2, y_2)$.

   $$d = \sqrt{(x_2 - x_1)^2 + (y_2 - y_1)^2}$$

   $$d = \sqrt{[1 - (-2)]^2 + (2 - 3)^2}$$

   $$d = \sqrt{(1 + 2)^2 + (-1)^2} = \sqrt{9 + 1} = \sqrt{10} \quad \textbf{ANS.}$$

2. Determine if $\triangle ABC$, with $A(-4, -1)$, $B(-2, 7)$, $C(1, 2)$, is a right triangle.

Solution: To be a right triangle, the lengths of the sides must satisfy the Pythagorean Relation.

First use the distance formula to find the lengths.

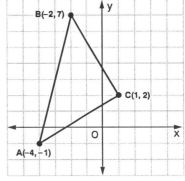

$$AB = \sqrt{[-2 - (-4)]^2 + [7 - (-1)]^2}$$
$$AB = \sqrt{(-2 + 4)^2 + (7 + 1)^2}$$
$$AB = \sqrt{4 + 64} = \sqrt{68}$$

$$BC = \sqrt{[1 - (-2)]^2 + (2 - 7)^2}$$
$$BC = \sqrt{9 + 25} = \sqrt{34}$$

$$AC = \sqrt{[1 - (-4)]^2 + [2 - (-1)]^2}$$
$$AC = \sqrt{25 + 9} = \sqrt{34}$$

Determine if the lengths satisfy the Pythagorean Relation. Note that AB, the longest distance, would have to be the hypotenuse of the triangle.

$$(AB)^2 \stackrel{?}{=} (BC)^2 + (AC)^2$$
$$(\sqrt{68})^2 \stackrel{?}{=} (\sqrt{34})^2 + (\sqrt{34})^2$$
$$68 \stackrel{?}{=} 34 + 34$$
$$68 = 68 \checkmark$$

ANSWER: $\triangle ABC$ is a right triangle.

Now it's your turn to . . . **TRY IT!**

Find the length of the line segment that joins the two points $(-4, -2)$ and $(3, 6)$.

(See page 632 for solution.)

EXERCISES

In 1–9, find the length of the third side of the right triangle, the length of whose hypotenuse is represented by c and the lengths of whose other sides are a and b. Express irrational results in simplest radical form.

1. $c = 10$, $a = 6$ **2.** $c = 17$, $b = 15$ **3.** $c = 25$, $b = 20$

4. $a = \sqrt{2}$, $b = \sqrt{2}$ **5.** $a = 4$, $b = 4\sqrt{3}$ **6.** $a = 5\sqrt{3}$, $c = 10$

7. $a = 3$, $b = 3$ **8.** $b = \sqrt{3}$, $c = \sqrt{15}$ **9.** $a = 4\sqrt{2}$, $c = 8$

In 10–15, approximate to the nearest tenth, the length of the third side of the right triangle, the length of whose hypotenuse is represented by c and the lengths of whose other sides are a and b.

10. $a = 5$, $b = 7$ **11.** $c = 15$, $a = 5$ **12.** $c = 12$, $b = 3$

13. $c = 35$, $b = 25$ **14.** $a = 7$, $b = 7$ **15.** $a = 5\sqrt{2}$, $c = 10$

In 16–19, find the value of x, expressing irrational results in simplest radical form.

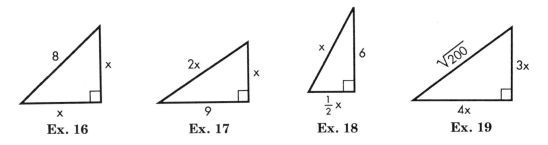

Ex. 16   Ex. 17   Ex. 18   Ex. 19

20. The ratio of the lengths of the legs of a right triangle is $5:12$. What is the ratio of the length of the shorter leg to the hypotenuse?

(1) $13:5$ (2) $\sqrt{119}:5$

(3) $5:13$ (4) $5:\sqrt{119}$

In 21–32, express irrational results in simplest form.

21. Find the length of the diagonal of a square whose side measures 10 cm.

22. The length of a diagonal of a rectangle is 26 in. If the length of the base is 24 in., find the height.

23. Find the length of an altitude of an equilateral triangle whose side measures 8 ft.

24. The length of the diagonal of a square is $6\sqrt{2}$ in. Find the length of a side.

25. A rectangle has a diagonal of length 10 cm and one side of length 6 cm. Find the perimeter of the rectangle.

26. If the length of a diagonal of a square is $8\sqrt{2}$ cm, find the perimeter of the square.

27. Find the area of an isosceles right triangle whose hypotenuse measures 18 inches.

28. In feet, the lengths of the two bases of an isosceles trapezoid are 12 and 28. If the length of an altitude is 6 ft., find the length of a leg.

29. A ladder 39 feet long leans against a building and reaches the ledge of a window. If the foot of the ladder is 15 feet from the foot of the building, how high is the window ledge above the ground?

30. A wire stretches from the top of a pole 24 feet high to a stake in the ground that is 18 feet from the foot of the pole. Find the length of the wire.

31. A man traveled 24 miles north and then 10 miles east. How far was he from his starting point?

32. Tom and Henry started from the same place. Tom traveled west at the rate of 30 miles per hour and Henry traveled south at the rate of 40 miles per hour. How far apart were they at the end of one hour?

33. Approximate, to the nearest foot, the diameter of the largest circular table top that can be taken through a rectangular doorway of base 4 feet and height 7 feet.

34. A baseball diamond has the shape of a square measuring 90 feet on each side. Approximate, to the nearest tenth of a foot, the distance from home plate to second base.

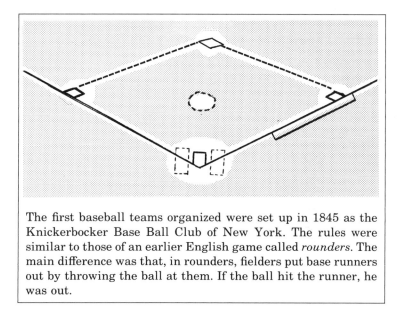

The first baseball teams organized were set up in 1845 as the Knickerbocker Base Ball Club of New York. The rules were similar to those of an earlier English game called *rounders*. The main difference was that, in rounders, fielders put base runners out by throwing the ball at them. If the ball hit the runner, he was out.

35. In cm, the bases of an isosceles trapezoid measure 15 and 9, and each of the legs measures 5 cm. For this trapezoid, find:

   a. the length of an altitude   **b.** the perimeter   **c.** the area

In 36–39, solve algebraically.

36. In right △*ABC*, ∠*C* is the right angle. If *AC* is 1 cm more than *BC* and *AB* is 2 cm more than *BC*, find the perimeter of the triangle.

37. The length of a rectangle is 7 inches more than the width. If the diagonal measures 17 inches, find the area of the rectangle.

38. The length of the longer base of an isosceles trapezoid exceeds the length of the shorter base by 6 cm, and the length of each leg is 5 cm. Find the length of the altitude.

39. In an isosceles trapezoid, the lengths of the bases are in the ratio of 1:2 and the length of the altitude is the same as that of the shorter base. If the area of the trapezoid is 24 square inches, find the length of each base.

In 40–51, find the length of the line segment that joins the given points. Express irrational answers in simplest radical form.

40. (4, 1) and (1, 1)

41. (−3, −3) and (1, 3)

42. (7, 2) and (5, 2)

43. (−6, 3) and (1, 3)

44. (−4, 0) and (0, −3)

45. (5, 0) and (0, −12)

46. (6, 5) and (1, 2)

47. (2, 5) and (−1, 4)

48. (−3, 0) and (1, −2)

49. (−5, 1) and (7, 6)

50. (−3, −6) and (−7, 11)

51. (2, −4) and (−8, 4)

52. A circle whose center is the origin passes through the point (3, 4). What is the length of its radius?

53. The point (2, 4) is on a circle whose center is (6, 1). Find the length of the radius of the circle.

54. Right $\triangle ABC$ has vertices $A(0, 4)$, $B(0, 0)$, and $C(4, 0)$. Find the length of hypotenuse $\overline{AC}$.

55. The coordinates of two opposite vertices of quadrilateral $ABCD$ are $A(5, 8)$ and $C(−3, 2)$. Find the length of diagonal $\overline{AC}$.

56. Which point lies farthest from the origin?
   (1) (0, −9)   (2) (−7, 6)
   (3) (8, 5)   (4) (−2, 9)

57. The triangle with vertices at (0, 0), (5, 0), and (−2, −2) is
   (1) right   (2) isosceles
   (3) scalene   (4) equilateral

58. Which set of numbers cannot be the lengths of the sides of a right triangle?
   (1) {3, 4, 5}   (2) {6, 8, 10}
   (3) {8, 15, 16}   (4) {8, 15, 17}

59. Given: $A(4, 2)$, $B(7, 10)$, $C(10, 2)$
   Determine if $\triangle ABC$ is isosceles.

60. Given: $A(3, 1)$, $B(10, 1)$, $C(8, 4)$, $D(5, 4)$
   Determine if trapezoid $ABCD$ is isosceles.

In 61–64, determine if the triangle formed by the given points is a right triangle.

61. $M(9, 10)$, $N(13, 2)$, $P(−3, −6)$

62. $R(5, −2)$, $S(6, 7)$, $T(1, 6)$

63. $D(−2, 1)$, $E(5, 2)$, $F(6, −5)$

64. $L(−1, 6)$, $M(3, −2)$, $N(−2, −1)$

CHAPTER SUMMARY

Finding a **square root** of a number is the inverse operation to squaring a number. When a radicand is a **perfect square**, the square root is rational. When the radicand is not a perfect square, the square root is irrational, but can be simplified if the radicand has perfect-square factors. For example: $\sqrt{50} = \sqrt{25} \cdot \sqrt{2} = 5\sqrt{2}$

Only **like radicals** can be combined by addition or subtraction. Be careful: $\sqrt{a + b}$ is *not* equal to $\sqrt{a} + \sqrt{b}$. As with other algebraic expressions, use the distributive, commutative, and associative properties to multiply or divide radical expressions. A fraction with a radical in the denominator is not in simplest form.

A **radical equation** contains the variable in a radicand. Square both sides of the equation to remove the radical sign and, since the resulting equation may not be equivalent to the original, check the solutions carefully.

When you first learned about a quadratic equation, the method for solution—factoring—was restricted to those quadratics whose roots were rational numbers. Now you have seen methods that can be used to solve *any* quadratic—completing the square and the quadratic formula. When the roots of a quadratic are irrational, they always occur in pairs such as $2 \pm \sqrt{3}$.

The two real roots of the quadratic equation $ax^2 + bx + c = 0$ are the x-intercepts of the graph of the corresponding quadratic function $y = ax^2 + bx + c$. The graph of this function, called a **parabola**, has special characteristics of symmetry, and is useful in real-world design.

Involving squaring and square roots, the **Pythagorean Relation**, like the trigonometric ratios, allows you to determine the length of a side of a right triangle without actually measuring the side.

VOCABULARY CHECKUP

SECTION

11-1 *square / square root / cube root / radical / radical sign / radicand / index / principal square root / perfect square*

11-2 *like radicals*   **11-3** *rationalizing a denominator*   **11-4** *radical equation*

11-5 *parabola / turning point / axis of symmetry / double root*

11-6 *trinomial square / completing the square / quadratic formula*

11-7 *Pythagorean Relation*

CHAPTER REVIEW EXERCISES

In 1–5, simplify the expression. (Section 11-1)

1. $\sqrt{.04}$ **2.** $-\sqrt{\dfrac{49}{81}}$ **3.** $\sqrt{(64)^2}$ **4.** $(\sqrt{7})^2$ **5.** $\sqrt[3]{-64}$

In 6–8, arrange the given numbers in increasing order. (Section 11-1)

6. $7, \sqrt{7}, 6$ **7.** $-\sqrt{3}, -4, -\sqrt{5}$ **8.** $\sqrt{37}, 6, 7$

In 9–11, approximate the value of the expression to the nearest integer. (Section 11-1)

9. $\sqrt{33}$ **10.** $-\sqrt{50}$ **11.** $\sqrt{88}$

In 12–14, find the square root. (Section 11-2)

12. $\sqrt{144a^6}$ **13.** $\sqrt{100x^4b^{16}}$ **14.** $\sqrt{.81x^{10}}$

In 15–18, simplify the expression. (Section 11-2)

15. $3\sqrt{50}$ **16.** $\dfrac{2}{3}\sqrt{75}$ **17.** $\sqrt{rs^5}$ **18.** $-3\sqrt{12x^6y^3}$

In 19–22, combine and simplify. (Section 11-2)

19. $4\sqrt{5} - 2\sqrt{3} + 8\sqrt{5} - \sqrt{3}$ **20.** $3\sqrt{75} - 4\sqrt{27}$

21. $\dfrac{2}{3}\sqrt{45} + 6\sqrt{20} - 11\sqrt{5}$ **22.** $4\sqrt{6x} - 2\sqrt{54x}$

In 23–25, multiply, then simplify. (Section 11-3)

23. $\sqrt{3r^3} \cdot \sqrt{5r^5}$ **24.** $3\sqrt{12a} \cdot 2\sqrt{3a^3}$ **25.** $(4\sqrt{3y})^2$

In 26–36, simplify. (Section 11-3)

26. $3\sqrt{\dfrac{80}{5}}$ **27.** $15\sqrt{48} \div 5\sqrt{2}$ **28.** $\dfrac{28\sqrt{12x^7}}{14\sqrt{3x^3}}$ **29.** $\dfrac{8}{\sqrt{3}}$ **30.** $\dfrac{24}{4\sqrt{3}}$ **31.** $\sqrt{\dfrac{3}{5}}$ **32.** $\sqrt{\dfrac{x}{y}}$

33. $3\sqrt{2}(\sqrt{18} - \sqrt{5})$ **34.** $(5 - 2\sqrt{3})(2 + 3\sqrt{3})$ **35.** $\dfrac{24\sqrt{60} - 18\sqrt{15}}{6\sqrt{5}}$ **36.** $\dfrac{\sqrt{3} + \sqrt{8}}{\sqrt{2}}$

In 37–40, solve and check. (Section 11-4)

37. $\sqrt{3a} = 6$ **38.** $\sqrt{5b - 4} = 9$ **39.** $2\sqrt{c} + 7 = 3$ **40.** $\sqrt{3d - 2} - 2 = 5$

41. The relationship between the radius and the area of a circle can be expressed by the formula
$r = \sqrt{\dfrac{A}{\pi}}$, where A is the area of the circle and r is the radius. (Section 11-4)

   a. Find A when $r = 7$ m. **b.** Find A when $r = 3\pi$ in.

In 42 and 43:   (Section 11-5)

a. Graph the given quadratic function of the form $y = ax^2 + bx + c$ by constructing a table of values using the integral values of x in the given interval.
b. Write the coordinates of the turning point of the parabola.
c. Write an equation for the axis of symmetry of the parabola.
d. From the graph, find the roots of the equation $ax^2 + bx + c = 0$. For nonintegral roots, approximate the values between consecutive integers.

42. $y = x^2 - 4x$ $-1 \leq x \leq 5$

43. $y = x^2 + 3x + 2$ $-4 \leq x \leq 1$

44. Graph the quadratic inequality $y > x^2 + 1$. To construct a table of values for the plane divider, use the integral values of x in the interval $-3 \leq x \leq 3$. (Section 11-5)

In 45–47, solve the equation and check. (Section 11-6)

45. $6a^2 = 216$

46. $3b^2 + 8 = 200$

47. $3c^2 - 11 = 89 - c^2$

In 48–50, solve. Write irrational answers in simplest radical form. (Section 11-6)

48. $7d^2 = 140$

49. $(f - 3)^2 = 49$

50. $(g - 5)^2 = 40$

In 51–53, replace the question mark with a number that will complete the square, and express the resulting trinomial as the square of a binomial. (Section 11-6)

51. $r^2 - 6r + ?$

52. $p^2 + 11p + ?$

53. $x^2 - \frac{2}{3}x + ?$

In 54–56, solve the equation by completing the square. Express irrational answers in simplest radical form. (Section 11-6)

54. $c^2 + 10c - 24 = 0$

55. $y^2 + 1 = 7y$

56. $x^2 - 3x - 2 = 0$

In 57–59, solve the equation by using the quadratic formula. (Section 11-6)

57. $3h^2 - 5h = 2$

58. $2y^2 + 5y = -3$

59. $x^2 - 6x = 0$

In 60–62, solve the equation. Approximate answers to the nearest tenth. (Section 11-6)

60. $3a^2 = 72$

61. $b^2 - 8b = 5$

62. $2c^2 - 7c - 6 = 0$

In 63–66, the legs of a right triangle are represented by a and b and the hypotenuse is represented by c. (Section 11-7)

63. Find a if $b = 9$ cm and $c = 15$ cm.

64. Find c if $a = 5$ ft. and $b = 12$ ft.

65. Find b if $a = 3\sqrt{7}$ in. and $c = 8$ in.

66. Find c if $a = b = 10$ in.

In 67 and 68, find the indicated lengths. (Section 11-7)

67. A 20-foot ladder is leaning against a wall. If the base of the ladder is 12 feet from the wall, how high up on the wall is the end of the ladder?

68. Mary Jane drove 7 miles north and 24 miles east. How far is she from her starting point?

In 69–71, find the length of the line segment that joins the given points. (Section 11-7)

69. $(5, -2)$ and $(3, -1)$ **70.** $(0, 9)$ and $(-6, 17)$ **71.** $(-7, -3)$ and $(-4, 1)$

PROBLEMS FOR PLEASURE

In 1 and 2, simplify the expression.

1. a. $\sqrt[3]{\sqrt{64}}$ **b.** $\sqrt{\sqrt[3]{64}}$ **2.** $(\sqrt{3} + 2\sqrt{2})(7 - 3\sqrt{5})(3\sqrt{5} + 7)(2\sqrt{2} - \sqrt{3})$

3. Find the value of x if $\sqrt[3]{5 - x} = -2$.

4. Solve: **a.** $x^2 - 2\sqrt{11}x + 7 = 0$ **b.** $\sqrt{2}x^2 + 4\sqrt{2}x + 3\sqrt{2} = 0$

5. One number is 7 more than a second number. The sum of the squares of the two numbers is equal to the square of 4 less than their sum. Find the numbers.

6. Explain how you can fold a standard-size sheet of paper measuring $8\frac{1}{2}''$ by $11''$ so that the length of the fold is very close to $12''$.

CALCULATOR CHALLENGE

1. Paul used his calculator to determine whether 3.7, 3.5, and 1.2 could be the lengths of the sides of a right triangle. He entered the key sequence:

$$3.7 \;\boxed{x^2}\; \boxed{-}\; 3.5\; \boxed{x^2}\; \boxed{-}\; 1.2\; \boxed{x^2}\; \boxed{=}$$

When the resulting display was 0, he concluded that the numbers did satisfy the Pythagorean relation. Do you agree with his conclusion and his method? If so, explain how his method can be generalized. If not, explain why the method would not always work.

2. When laying railroad track, spaces are usually left between sections to allow for expansion of the metal in hot weather. At the Grand Valley Gamepark, a track one mile long, without spaces, was laid over level ground. One hot July day, the track expanded by a total of one inch over its entire length. Since it was firmly secured at both ends, it rose in the middle, forming an isosceles triangle with the ground as the base. How high above the ground did the middle of the track rise?

3. a. The size of a rectangular TV screen refers to the length of the diagonal. A 19-inch TV has a diagonal 19 inches long. The *aspect ratio* of a TV screen is the ratio of its length to its width. If a standard 19-inch TV has an aspect ratio of 4:3, what is the viewing area of the screen?

  b. High-definition TV (HDTV) has a wider screen to accommodate movies on TV. Its aspect ratio is 5:3. What is the smallest whole-number diagonal measure in inches that would provide HDTV with at least the same viewing area as a conventional 19-inch TV?

Questions 1–7 each consist of two quantities, one in Column A and one in Column B. You are to compare the two quantities and choose:

A if the quantity in Column A is greater;
B if the quantity in Column B is greater;
C if the two quantities are equal;
D if the relationship cannot be determined from the information given.

1. In certain questions, information concerning one or both of the quantities to be compared is centered above the two columns.

2. In a given question, a symbol that appears in both columns represents the same thing in Column A as it does in Column B.

3. x, n, and k, etc. stand for real numbers.

	Column A	Column B
1.	$\dfrac{1}{4} + \dfrac{1}{9}$	$\sqrt{\dfrac{1}{4}} + \sqrt{\dfrac{1}{9}}$
2.	2	$\sqrt[3]{9}$
3.	$\dfrac{1}{5}$	$\sqrt{\dfrac{1}{24}}$
4.	$2\sqrt{3}$	$(\sqrt{3})^2$
5.	$\sqrt{\dfrac{16}{25}}$	$\left(\dfrac{16}{25}\right)^2$
6.	$\dfrac{\sqrt{3}+\sqrt{3}}{\sqrt{3}}$	$\dfrac{\sqrt{5}+\sqrt{5}}{\sqrt{5}}$

$$x = \sqrt{5},\ y = \sqrt{3}$$

7.	3	$(x+y)(x-y)$

In 8–22, select the letter of the correct answer.

8. The area of a rectangle is 64 and the base is $4\sqrt{2}$. The perimeter must be
   (A) $32\sqrt{2}$ (B) $24\sqrt{2}$ (C) $16\sqrt{2}$
   (D) $12\sqrt{2}$ (E) $8\sqrt{2}$

9. $\dfrac{\sqrt{x}}{\sqrt{y}} = k$, where k is a whole number.
   Which number pair (x, y) below does not satisfy the preceding condition?
   (A) $(100, 25)$ (B) $(36, 9)$ (C) $(72, 2)$
   (D) $(54, 6)$ (E) $(12, 9)$

10. If $\sqrt{36x^6} = 48$, then $\sqrt{36x^2}$ equals
    (A) 6 (B) 12 (C) 16
    (D) 24 (E) 30

11. If $x + \sqrt{2} = \sqrt{4}$ and $y + \sqrt{8} = \sqrt{16}$, then the ratio of x to y equals
    (A) $\dfrac{\sqrt{2}}{\sqrt{16}}$ (B) $\frac{1}{4}$ (C) $\dfrac{1}{\sqrt{2}}$ (D) $\frac{1}{2}$ (E) $\sqrt{2}$

12. If $\sqrt{8}x - \sqrt{3}y = 1$, which of the following does not lie on the graph of the line?
    (A) $(\sqrt{2}, \sqrt{3})$ (B) $(-\sqrt{8}, -\sqrt{27})$
    (C) $(-1, \sqrt{3})$ (D) $(4\sqrt{2}, 5\sqrt{3})$
    (E) $(5\sqrt{8}, 13\sqrt{3})$

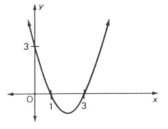

13. The graph of $y = x^2 + ax + b$ is shown. Find the y-intercept of
    $$y + 1 = x^2 + ax + b.$$
    (A) 2 (B) 3 (C) 4 (D) −1 (E) −2

14. If $y = x^2 + kx + 5$ represents a parabola that passes through $(2, 3)$, then k is
(A) -2 (B) 3 (C) -1
(D) -4 (E) -3

15. If $x^2 + 2xy + y^2 = 16$, then the average of x and y could be
(A) 1 (B) 2 (C) 4 (D) 8 (E) 16

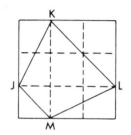

16. If $x^2 + 8x + 16$ is the area of square $ABCD$ and $x^2 + 2x + 1$ is the area of square II, then the area of square I is
(A) $6x + 15$ (B) 15 (C) 9
(D) $3x + 3$ (E) $x^2 + 5x + 4$

17. If $f(x) = x^2 + 3x$, then $f(x + 1) =$
(A) $x^2 + 5x + 4$
(B) $x^2 + 3x + 4$
(C) $x^2 + 4x + 3$
(D) $x^2 + 5x + 5$
(E) $x^3 + 4x^2 + 3x$

18. If $\dfrac{(x - y)^2}{(x + y)^2} = 1$, then
(A) $x = y = 0$
(B) $x = 0$ or $y = 0$
(C) $x = y = 1$
(D) $x = 1$ or $y = 1$
(E) $x = y$

19. If the average of x numbers is $x - 8$, what is the average of the same x numbers and -9?
(A) $x + 1$ (B) $x - 1$ (C) $x - 3$
(D) $x - 8$ (E) $x - 9$

20. If $\dfrac{2x^2 + kx - 5}{x + 1} = 1$ when x is 3, then k is

(A) 1 (B) -1 (C) -3 (D) 3 (E) 5

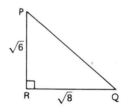

21. The square above is divided by the dashed lines into 9 one-foot squares. What is the perimeter, in feet, of the quadrilateral $JKLM$ inscribed in the square?
(A) $8\sqrt{2}$ (B) $4\sqrt{5} + 3\sqrt{2}$
(C) $2\sqrt{2} + 2\sqrt{5}$ (D) $3\sqrt{2} + 2\sqrt{5}$
(E) $3\sqrt{2} + 3\sqrt{5}$

22. Which is the best estimate for the perimeter of $\triangle PQR$?
(A) 5 (B) 7 (C) 9 (D) 11 (E) 13

1. Simplify: $(x^2 - 5x + 6) - (x^2 - x - 4)$

2. If one factor of $72c^5$ is $-9c^2$, find the other factor.

3. The sides of a right triangle measure 5 cm, 12 cm, and 13 cm. Find the perimeter of a similar triangle if its hypotenuse measures 26 cm.

4. If $\frac{4}{5}x = \frac{1}{2}y$, find $\dfrac{x}{y}$.   5. Multiply and simplify: $\dfrac{x^2 - 9}{x^2 - x - 6} \cdot \dfrac{3x + 6}{2x + 6}$

6. The average of 5 integers is greater than 33. If the average of the first 4 integers is 25, what is the least possible value of the 5th integer?

7. What is the remainder when $(3m - 1)^2$ is divided by 3?

8. The two acute angles of a right triangle have measures in the ratio 2:3. Find the number of degrees in the smaller angle.

In 9 and 10, solve for y.

9. $y^3 - 2y^2 - 15y = 0$

10. $\dfrac{y + 1}{y} + 1 = \dfrac{3}{2y}$

11. If the operation $*$ is as defined in the table, then $x * (y * z) =$
(1) $(z * x) * y$
(2) $y * (z * x)$
(3) $z * (x * y)$
(4) $x * (z * y)$

$*$	x	y	z
x	x	z	y
y	y	x	z
z	z	y	x

12. Graph the solution set on a number line: $|x| \le 3$

13. Solve the system: $3x - y = 10$
    $x + y = 2$

14. A gum ball machine contains gum balls of three different colors, with the colors red, green, and yellow occurring in the ratio 4:3:2. What is the probability that the next gum ball to drop out will be red?

15. The display on Ken's digital watch showed three consecutive even integers, in increasing order. If the sum of the first two digits was equal to the third digit, what time was it?

16. Which of the lines (1)–(4) shown on the graph has the slope with the smallest numerical value?

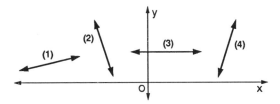

17. In the sequence $x, y, 15, 31, 63$, each term after the first term, x, is formed by doubling the previous term and then adding 1. What is the value of $x + y$?

18. If $p = q - 1 = r + 3 = s + 5 = t - 7$, which variable has the smallest value?
    (1) p (2) q (3) r (4) s (5) t

Solutions to **TRY IT!** Problems (Chapter 11)

11-2 SIMPLIFYING AND COMBINING RADICALS

TRY IT! *Problems 1 and 2 on page 580*

1.
$$3\sqrt{20} - 2\sqrt{5} + 5\sqrt{45}$$
$$= 3\sqrt{4 \cdot 5} - 2\sqrt{5} + 5\sqrt{9 \cdot 5}$$
$$= 3\sqrt{4} \cdot \sqrt{5} - 2\sqrt{5} + 5\sqrt{9} \cdot \sqrt{5}$$
$$= 3 \cdot 2 \cdot \sqrt{5} - 2\sqrt{5} + 5 \cdot 3 \cdot \sqrt{5}$$
$$= 6\sqrt{5} - 2\sqrt{5} + 15\sqrt{5}$$
$$= (6 - 2 + 15)\sqrt{5}$$
$$= 19\sqrt{5}$$

2.
$$5\sqrt{3x} + 4\sqrt{27x} - 3\sqrt{75x}$$
$$= 5\sqrt{3x} + 4\sqrt{9 \cdot 3x} - 3\sqrt{25 \cdot 3x}$$
$$= 5\sqrt{3x} + 4\sqrt{9} \cdot \sqrt{3x} - 3\sqrt{25} \cdot \sqrt{3x}$$
$$= 5\sqrt{3x} + 4 \cdot 3 \cdot \sqrt{3x} - 3 \cdot 5 \cdot \sqrt{3x}$$
$$= 5\sqrt{3x} + 12\sqrt{3x} - 15\sqrt{3x}$$
$$= (5 + 12 - 15)\sqrt{3x}$$
$$= 2\sqrt{3x}$$

11-3 MULTIPLYING AND DIVIDING RADICALS

TRY IT! *Problems 1–4 on page 584*

1.
$$4\sqrt{10} \cdot 3\sqrt{5}$$
$$= 4 \cdot 3\sqrt{10 \cdot 5}$$
$$= 12\sqrt{50}$$
$$= 12\sqrt{25 \cdot 2}$$
$$= 12\sqrt{25} \cdot \sqrt{2}$$
$$= 12 \cdot 5\sqrt{2}$$
$$= 60\sqrt{2}$$

2.
$$(3\sqrt{5})^2$$
$$= 3^2(\sqrt{5})^2$$
$$= 9 \cdot 5$$
$$= 45$$

3.
$$3\sqrt{2}(3\sqrt{8} - 4\sqrt{3})$$
$$= (3\sqrt{2})(3\sqrt{8}) - (3\sqrt{2})(4\sqrt{3})$$
$$= 3 \cdot 3\sqrt{2 \cdot 8} - 3 \cdot 4\sqrt{2 \cdot 3}$$
$$= 9\sqrt{16} - 12\sqrt{6}$$
$$= 9 \cdot 4 - 12\sqrt{6}$$
$$= 36 - 12\sqrt{6}$$

4.
$$\begin{array}{r} 5 - \sqrt{3} \\ 2 - \sqrt{3} \\ \hline 10 - 2\sqrt{3} \\ -5\sqrt{3} + 3 \\ \hline 10 - 7\sqrt{3} + 3 = 13 - 7\sqrt{3} \end{array}$$

TRY IT! *Problems 1–3 on page 585*

1. $\sqrt{\dfrac{3}{25}} = \dfrac{\sqrt{3}}{\sqrt{25}} = \dfrac{\sqrt{3}}{5}$

2. $21\sqrt{96} \div 3\sqrt{3}$

$$= \frac{21}{3}\sqrt{\frac{96}{3}}$$
$$= 7\sqrt{32}$$
$$= 7\sqrt{16 \cdot 2}$$
$$= 7\sqrt{16} \cdot \sqrt{2} = 7 \cdot 4\sqrt{2} = 28\sqrt{2}$$

3.
$$\frac{\sqrt{12} + \sqrt{27}}{\sqrt{3}}$$
$$= \frac{\sqrt{12}}{\sqrt{3}} + \frac{\sqrt{27}}{\sqrt{3}}$$
$$= \sqrt{\frac{12}{3}} + \sqrt{\frac{27}{3}}$$
$$= \sqrt{4} + \sqrt{9} = 2 + 3 = 5$$

TRY IT! *Problems 1–3 on page 588*

1. $\dfrac{4}{\sqrt{5}} = \dfrac{4}{\sqrt{5}} \cdot \dfrac{\sqrt{5}}{\sqrt{5}} = \dfrac{4\sqrt{5}}{5}$

2. $\dfrac{3}{\sqrt{8}} = \dfrac{3}{\sqrt{8}} \cdot \dfrac{\sqrt{2}}{\sqrt{2}} = \dfrac{3\sqrt{2}}{\sqrt{16}} = \dfrac{3\sqrt{2}}{4}$

3. $\sqrt{\dfrac{3}{2}} = \dfrac{\sqrt{3}}{\sqrt{2}} = \dfrac{\sqrt{3}}{\sqrt{2}} \cdot \dfrac{\sqrt{2}}{\sqrt{2}} = \dfrac{\sqrt{6}}{\sqrt{4}} = \dfrac{\sqrt{6}}{2}$

11-4 SOLVING A RADICAL EQUATION

TRY IT! *Problems 1 and 2 on page 593*

1.

Solution	*Check*
$\sqrt{3x} + 10 = 4$	$\sqrt{3x} + 10 = 4$
$\sqrt{3x} = -6$	$\sqrt{(3)(12)} + 10 \stackrel{?}{=} 4$
$3x = 36$	$\sqrt{36} + 10 \stackrel{?}{=} 4$
$x = 12$	$6 + 10 \stackrel{?}{=} 4$
	$16 \neq 4$

ANSWER: $\varnothing$

2.

Solution	*Check*
$\sqrt{3x - 2} + 3 = 7$	$\sqrt{3x - 2} + 3 = 7$
$\sqrt{3x - 2} = 4$	$\sqrt{(3)(6) - 2} + 3 \stackrel{?}{=} 7$
$3x - 2 = 16$	$\sqrt{18 - 2} + 3 \stackrel{?}{=} 7$
$3x = 18$	$\sqrt{16} + 3 \stackrel{?}{=} 7$
$x = 6$	$4 + 3 \stackrel{?}{=} 7$
	$7 = 7 \checkmark$

ANSWER: $x = 6$

11-5 GRAPHING A QUADRATIC FUNCTION

TRY IT! *Problem on page 597*

a.

$$x^2 + 2x - 8 = 0$$
$$(x + 4)(x - 2) = 0$$
$$x + 4 = 0 \quad | \quad x - 2 = 0$$
$$x = -4 \quad | \quad x = 2$$

Check

$x^2 + 2x - 8 = 0$	$x^2 + 2x - 8 = 0$
$(-4)^2 + 2(-4) - 8 \stackrel{?}{=} 0$	$2^2 + 2(2) - 8 \stackrel{?}{=} 0$
$16 - 8 - 8 \stackrel{?}{=} 0$	$4 + 4 - 8 \stackrel{?}{=} 0$
$0 = 0 \checkmark$	$0 = 0 \checkmark$

ANSWER: $x = -4$ or $x = 2$

b.

x	$x^2 + 2x - 8$	y
-5	$(-5)^2 + 2(-5) - 8$	7
-4	$(-4)^2 + 2(-4) - 8$	0
-3	$(-3)^2 + 2(-3) - 8$	-5
-2	$(-2)^2 + 2(-2) - 8$	-8
-1	$(-1)^2 + 2(-1) - 8$	-9
0	$0^2 + 2(0) - 8$	-8
1	$1^2 + 2(1) - 8$	-5
2	$2^2 + 2(2) - 8$	0
3	$3^2 + 2(3) - 8$	7

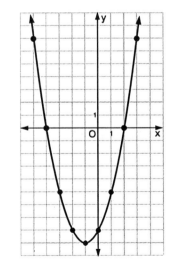

TRY IT! *Problems 1 and 2 on page 600*

1. a.

x	$-x^2 + 5x - 6$	y
0	$-(0)^2 + 5(0) - 6$	-6
1	$-(1)^2 + 5(1) - 6$	-2
2	$-(2)^2 + 5(2) - 6$	0
3	$-(3)^2 + 5(3) - 6$	0
4	$-(4)^2 + 5(4) - 6$	-2
5	$-(5)^2 + 5(5) - 6$	-6

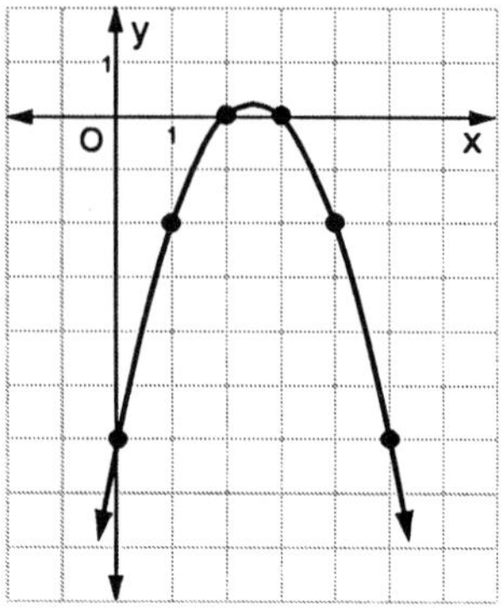

b. Turning point: $(2.5, .25)$
c. Axis of symmetry: $x = 2.5$
d. The roots of $-x^2 + 5x - 6 = 0$ are $x = 2$ or $x = 3$.

2. a.

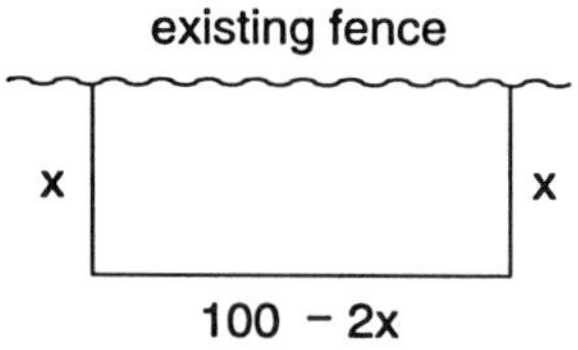

existing fence

x   x

$100 - 2x$

b. $y = x(100 - 2x)$ or $y = -2x^2 + 100x$

c.

x	$-2x^2 + 100x$	y
0	$-2(0)^2 + 100(0)$	0
10	$-2(10)^2 + 100(10)$	800
20	$-2(20)^2 + 100(20)$	$1,200$
30	$-2(30)^2 + 100(30)$	$1,200$
40	$-2(40)^2 + 100(40)$	800
50	$-2(50)^2 + 100(50)$	0

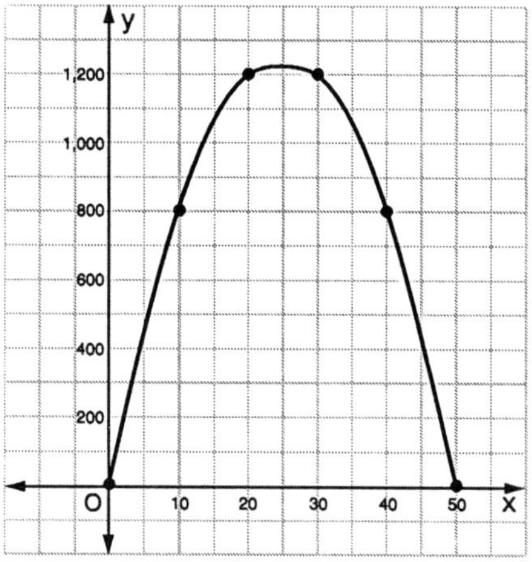

d. 25 ft. × 50 ft. = 1,250 sq. ft.

11-6 MORE METHODS FOR SOLVING A QUADRATIC EQUATION

TRY IT! *Problems 1–3 on page 603*

1. $3x^2 = 48$
$\quad x^2 = 16$
$\quad\quad x = \pm\sqrt{16}$
$\quad\quad x = \pm 4$

Check:

$3x^2 = 48$
$3(4)^2 \overset{?}{=} 48$
$3(16) \overset{?}{=} 48$
$\quad\quad 48 = 48 ✓$

$3x^2 = 48$
$3(-4)^2 \overset{?}{=} 48$
$3(16) \overset{?}{=} 48$
$\quad\quad 48 = 48 ✓$

ANSWER: $x = \pm 4$

2. $2x^2 - 34 = 0$

$$2x^2 = 34$$
$$x^2 = 17$$
$$x = \pm\sqrt{17}$$

Check

$$2x^2 - 34 = 0 \qquad\qquad 2x^2 - 34 = 0$$
$$2(\sqrt{17})^2 - 34 \stackrel{?}{=} 0 \qquad 2(-\sqrt{17})^2 - 34 \stackrel{?}{=} 0$$
$$2(17) - 34 \stackrel{?}{=} 0 \qquad\qquad 2(17) - 34 \stackrel{?}{=} 0$$
$$34 - 34 \stackrel{?}{=} 0 \qquad\qquad 34 - 34 \stackrel{?}{=} 0$$
$$0 = 0 \checkmark \qquad\qquad\qquad 0 = 0 \checkmark$$

ANSWER: $x = \pm\sqrt{17}$

3. $(x - 1)^2 = 16$

$$x - 1 = \pm 4$$
$$x = 1 \pm 4$$
$$x = 1 + 4 \;\big|\; x = 1 - 4$$
$$x = 5 \qquad\big|\; x = -3$$

Check

$$(x - 1)^2 = 16 \qquad\qquad (x - 1)^2 = 16$$
$$(5 - 1)^2 \stackrel{?}{=} 16 \qquad\qquad (-3 - 1)^2 \stackrel{?}{=} 16$$
$$4^2 \stackrel{?}{=} 16 \qquad\qquad\qquad (-4)^2 \stackrel{?}{=} 16$$
$$16 = 16 \checkmark \qquad\qquad\qquad 16 = 16 \checkmark$$

ANSWER: $x = -3$ or $x = 5$

TRY IT! *Problems 1–4 on page 606*

1. $x^2 + 14x + \left(\frac{1}{2} \cdot 14\right)^2$

$$x^2 + 14x + 49 = (x + 7)^2$$

2. $x^2 - 6x + \left(\frac{1}{2} \cdot (-6)\right)^2$

$$x^2 - 6x + 9 = (x - 3)^2$$

3. $x^2 + 7x + \left(\frac{1}{2} \cdot 7\right)^2$

$$x^2 + 7x + \frac{49}{4} = \left(x + \frac{7}{2}\right)^2$$

4. $3\left[x^2 + 2x + \left(\frac{1}{2} \cdot 2\right)^2\right]$

$$3(x^2 + 2x + 1) = 3(x + 1)^2$$

TRY IT! *Problems 1 and 2 on page 608*

1. $x^2 - 6x + 8 = 0$

$$x^2 - 6x = -8$$
$$x^2 - 6x + 9 = -8 + 9$$
$$(x - 3)^2 = 1$$
$$x - 3 = \pm 1$$
$$x - 3 = 1 \;\big|\; x - 3 = -1$$
$$x = 4 \qquad\big|\qquad x = 2$$

Check both values in the original equation.

ANSWER: $x = 2$ or $x = 4$

2.
$$x^2 - 4x = -1$$
$$x^2 - 4x + 4 = -1 + 4$$
$$(x - 2)^2 = 3$$
$$x - 2 = \pm\sqrt{3}$$
$$x - 2 = \sqrt{3} \;\big|\; x - 2 = -\sqrt{3}$$
$$x = 2 + \sqrt{3} \;\big|\; x = 2 - \sqrt{3}$$

Check one of the irrational values.

$$x^2 - 4x = -1$$
$$(2 + \sqrt{3})^2 - 4(2 + \sqrt{3}) \stackrel{?}{=} -1$$
$$4 + 4\sqrt{3} + 3 - 8 - 4\sqrt{3} \stackrel{?}{=} -1$$
$$7 - 8 \stackrel{?}{=} -1$$
$$-1 = -1 \checkmark$$

ANSWER: $x = 2 \pm \sqrt{3}$

TRY IT! *Problems 1 and 2 on page 612*

1. $x^2 + 11 = 10x$

$$x^2 - 10x + 11 = 0$$
$$a = 1 \quad b = -10 \quad c = 11$$

$$x = \frac{-b \pm \sqrt{b^2 - 4ac}}{2a}$$

$$x = \frac{-(-10) \pm \sqrt{(-10)^2 - 4(1)(11)}}{2(1)}$$

$$x = \frac{10 \pm \sqrt{100 - 44}}{2}$$

$$x = \frac{10 \pm \sqrt{56}}{2} = \frac{10 \pm 2\sqrt{14}}{2} = 5 \pm \sqrt{14}$$

Check one of the pair of irrational values in the original equation.

$$x^2 + 11 = 10x$$

$$(5 + \sqrt{14})^2 + 11 \stackrel{?}{=} 10(5 + \sqrt{14})$$

$$25 + 10\sqrt{14} + 14 + 11 \stackrel{?}{=} 50 + 10\sqrt{14}$$

$$50 + 10\sqrt{14} = 50 + 10\sqrt{14} \text{✓}$$

ANSWER: $x = 5 \pm \sqrt{14}$

2. Let x = the width of the frame.
Then $12 - 2x$ = the length of the picture.
And $8 - 2x$ = the width of the picture.

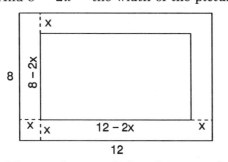

The total area of the picture and the frame is 8×12, or 96 sq. in. Since the picture and the frame have equal areas, each area is 48 sq. in.

The area of the picture is 48.

$$(12 - 2x)(8 - 2x) = 48$$
$$96 - 40x + 4x^2 = 48$$
$$4x^2 - 40x + 48 = 0$$
$$x^2 - 10x + 12 = 0$$

$$x = \frac{-(-10) \pm \sqrt{(-10)^2 - 4(1)(12)}}{2(1)}$$

$$x = \frac{10 \pm \sqrt{52}}{2}$$

$$x \approx \frac{10 \pm 7.21}{2}$$

$$x \approx \frac{10 + 7.21}{2} \qquad x \approx \frac{10 - 7.21}{2}$$

$$x \approx 8.6 \qquad\qquad x \approx 1.4$$

Reject $x \approx 8.6$ since it leads to negative values for the dimensions of the picture.

Check 1.4: length = $12 - 2(1.4) = 9.2$ in.
width = $8 - 2(1.4) = 5.2$ in.
Area of picture = $5.2 \times 9.2 \approx 48$ sq. in. ✓

ANSWER: The width of the frame is 1.4 in.

11-7 THE PYTHAGOREAN RELATION

TRY IT! *Problems 1 and 2 on page 616*

1.
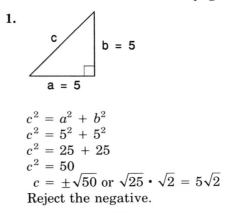

$$c^2 = a^2 + b^2$$
$$c^2 = 5^2 + 5^2$$
$$c^2 = 25 + 25$$
$$c^2 = 50$$
$$c = \pm\sqrt{50} \text{ or } \sqrt{25} \cdot \sqrt{2} = 5\sqrt{2}$$
Reject the negative.

ANSWER: $5\sqrt{2}$ inches

2.
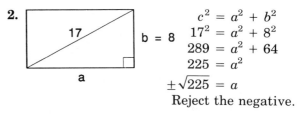

$$c^2 = a^2 + b^2$$
$$17^2 = a^2 + 8^2$$
$$289 = a^2 + 64$$
$$225 = a^2$$
$$\pm\sqrt{225} = a$$
Reject the negative.

ANSWER: 15 cm

TRY IT! *Problem on page 617*

Let $(-4, -2) = (x_1, y_1)$ and $(3, 6) = (x_2, y_2)$.
$$d = \sqrt{(x_2 - x_1)^2 + (y_2 - y_1)^2}$$
$$d = \sqrt{[3 - (-4)]^2 + [6 - (-2)]^2}$$
$$d = \sqrt{(3 + 4)^2 + (6 + 2)^2}$$
$$d = \sqrt{7^2 + 8^2} = \sqrt{49 + 64} = \sqrt{113}$$

Part I

1. If $a + b = 0$ and $a = 5$, what is the value of b?

2. Solve for x: $6x = 2(x + 10)$

3. Factor: $r^2 - 64$

4. A boy 6 feet tall casts a shadow 4 feet long. At the same time, a nearby flagpole casts a shadow 20 feet long. What is the height, in feet, of the flagpole?

5. Find the value of $(ab)^2$ if $a = 3$ and $b = -2$.

6. Express $\dfrac{2x}{3} - \dfrac{x}{5}$ as a single fraction.

7. Solve for c: $\dfrac{c}{6} = \dfrac{c + 1}{8}$

8. Find the value of $\sqrt{45}$ to the nearest tenth.

9. Solve the system of equations for x:
$$3x + y = 7$$
$$x + y = 5$$

10. If n represents an odd integer, express the next smaller odd integer in terms of n.

11. The sum of the measures of two angles is $80°$. If the angles are in the ratio $1:3$, what is the number of degrees in the smaller angle?

12. If 60% of a number is 120, find the number.

13. From $3x^2 - 4x + 8$, subtract $2x^2 - 6x - 3$.

14. Solve for x: $0.1x - 0.02x = 8$

15. In the diagram, the cosine of which angle is equal to $\frac{3}{5}$?

16. The lengths of the sides of a triangle are represented by $y + 6$, $3y - 1$, and $2y - 6$. Express the perimeter of the triangle as a binomial in terms of y.

17. Given $M = \{2, 4, 6\}$ and $N = \{1, 3\}$. If one number is selected at random from set M and one number is selected at random from set N, what is the probability that their sum will be even?

In 18–30, choose the correct answer.

18. Which are the factors of $x^2 + 5x - 24$?
(1) $(x - 2)(x + 12)$
(2) $(x - 3)(x + 8)$
(3) $(x - 4)(x + 6)$
(4) $(x - 1)(x + 24)$

19. The value of $|5| - |2| - |-3|$ is
(1) 0   (2) 8   (3) 10   (4) 4

20. The product of $2x^3$ and $3x^2$ is
(1) $6x^6$ (2) $6x^5$ (3) $5x^6$ (4) $5x^5$

21. A taxi ride costs c cents for the first $\frac{1}{4}$-mile and m cents for each additional $\frac{1}{8}$-mile. What is the cost, in cents, of a $\frac{5}{8}$-mile ride?
(1) $c + m$ (2) $2c$
(3) $c + 3m$ (4) $c + 6m$

22. What is the solution set for n when $3n - 2 \geq n + 6$?
(1) $\{n | n \geq 2\}$ (2) $\{n | n = 2\}$
(3) $\{n | 3 < n < 6\}$ (4) $\{n | n \geq 4\}$

23. Which is not a rational number?
(1) 0   (2) -8 (3) $\sqrt{9}$ (4) $\sqrt{6}$

24. If $c = ax + b$, then x equals
(1) $\dfrac{c + b}{a}$ (2) $\dfrac{c - b}{a}$
(3) $\dfrac{c}{a} - b$ (4) $\dfrac{c}{a} + b$

25. The expression $6\sqrt{2} + \sqrt{32}$ is equivalent to
(1) $7\sqrt{34}$ (2) 20
(3) $10\sqrt{2}$ (4) $6\sqrt{34}$

26. Which statement describes the graph of the equation $x = 3$?
(1) It has a slope of 3.
(2) It passes through the origin.
(3) It is parallel to the x-axis.
(4) It is parallel to the y-axis.

27. Which statement illustrates the associative property of multiplication?
(1) $(a \times b) \times c = a \times (b \times c)$
(2) $a \times b = b \times a$
(3) $a \times 0 = 0$
(4) $a \times \dfrac{1}{a} = 1$

28. When $6x^2 + 7x + 2$ is divided by $2x + 1$, the quotient is
(1) $3x + 5$
(2) $3x + \dfrac{7}{2}$
(3) $3x + 2$
(4) $3x + 7$

29. If the length of each side of a square is multiplied by 2, then the area of the square is multiplied by
(1) $\dfrac{1}{2}$ (2) 2   (3) 3   (4) 4

30.

Which inequality is represented by the graph?
(1) $-2 \leq x < 3$
(2) $-2 < x \leq 3$
(3) $x > 3$ or $x \leq 2$
(4) $x \geq 3$ or $x < -2$

Part II

31. Solve graphically and check:
$$y = x$$
$$x + 2y = -6$$

32. a. Perform the indicated operation and express the result in lowest terms:
$$\frac{x^2 + 12x + 35}{x + 6} \div \frac{2x + 14}{x^2 - 36}$$

b. Solve for y and check:
$$\frac{y}{5} + 7 = \frac{y}{2} - 2$$

33. Write an equation or a system of equations that can be used to solve each of the following problems. In each case, state what the variable or variables represent. (Solution of the equations is not required.)

a. Susan's change purse contained quarters, dimes, and nickels. She had twice as many nickels as quarters and four more nickels than dimes. She had a total amount of $5.10. How many coins of each kind did Susan have in her purse?

b. John rode his bicycle to town at the rate of 15 miles per hour. He left the bicycle in town for minor repairs. He walked home along the same route at the rate of 3 miles per hour. The entire trip took 3 hours. How long did it take John to walk back?

34. The sum of the digits of a two-digit number is 17. If 9 is subtracted from the number, the result is the original number with the digits reversed. Solve algebraically to find the original number.

35. Use any strategy (algebraic, trial and error, making a table, etc.) to solve this problem.

There are two pairs of integers that satisfy both of these conditions:
 The larger integer is 9 more than the smaller integer.
 The sum of the squares of the integers is 41.

a. Find the two pairs of integers.

b. Show that one pair of integers found in part **a** satisfies both given conditions.

36. In right triangle ABC, angle C is a right angle. The length of $\overline{AC}$ is 7 and the length of $\overline{BC}$ is 8.

a. Find, to the nearest degree, the measure of angle A.

b. Find, to the nearest integer, the length of $\overline{AB}$.

37. The replacement set for x is $\{-3, -2, -1, 0, 1, 2\}$. Find the solution set of each open sentence.

a. $5(x - 1) = 10$ **b.** $|x| = 2$

c. $-4x > 4$ **d.** $2x^2 = 2$

e. $x^2 = x$

GLOSSARY

absolute value The (nonnegative) distance of a number from 0 on the number line. The absolute value of a is written $|a|$. Example: $|-5| = 5$.

acute angle An angle whose measure is between $0°$ and $90°$.

addition property of equality If the same quantity is added to two equal quantities, the results are equal. If $a = b$, then $a + c = b + c$.

addition property of inequality If the same quantity is added to two unequal quantities, the results are unequal in the same order. If $a < b$, then $a + c < b + c$.

algebraic model An equation that represents a real situation.

altitude A line segment from a vertex perpendicular to the opposite side, the base, of a figure.

associative property The way that numbers are grouped does not change their sum or product. $(a + b) + c = a + (b + c)$ and $(ab)c = a(bc)$.

base A side of a plane or solid figure, often used in determining area.

base A number that is raised to a power. In the expression 3^2, 3 is the base.

binomial A polynomial with two terms.

closure A set is closed under an operation if the result of performing the operation on members of the set is always in the set. Example: $\{0, 1, 2, \ldots\}$ is closed under multiplication, but not under division.

coefficient The numerical factor in a product of a number and variable(s). In $3ab$, 3 is the coefficient of ab.

commutative property The order in which two numbers are written does not change their sum or product. $a + b = b + a$ and $ab = ba$.

complementary angles A pair of angles whose measures add up to $90°$.

congruent (symbol $\cong$) Having the same size and shape.

constant A term whose value does not change. In $x^2 + 5$, the constant is 5.

coordinate A number associated with a point. On a number line, a point has one coordinate; in the plane, a point has an ordered pair of coordinates.

cosine In a right triangle, the ratio of the length of the side adjacent to a given angle to the length of the hypotenuse.

degree of a polynomial The highest exponent of the variable.

difference of two squares A binomial of the form $a^2 - b^2$.

distributive property Multiplication across a sum: $a(b + c) = ab + ac$. Also used in factoring: $ab + ac = a(b + c)$.

domain of a variable The replacement set; a set of numbers that may be substituted for the variable.

domain of a relation The set consisting of the first number from each ordered pair in a relation.

element A member of a set.

empty set (symbol $\{\ \}$ or $\varnothing$) The set containing no elements.

equation A mathematical statement that two quantities are equal.

equilateral triangle A triangle with 3 sides of equal length.

even integer An integer that is exactly divisible by 2; an element of the set: $\{\ldots, -2, 0, 2, 4, \ldots\}$

event The outcome, or the set of outcomes, for which the probability is to be determined.

exponent A number that indicates how many times another number, the base, is to be multiplied by itself. In 4^3, the exponent is 3; thus $4^3 = 4 \times 4 \times 4 = 64$.

factor A part of a product. Example: Since $6x$ can be written as $1 \cdot 6x, 2 \cdot 3x$, etc., the factors of $6x$ are 1, 2, 3, 6, x, $2x$, $3x$, and $6x$.

factor To write a number or an expression as a product.
Examples: $12 = 2^2 \cdot 3$, $2a + 2b = 2(a + b)$

formula An equation that expresses a relationship between two or more variables.

fractional equation An equation in which at least one fraction has a variable in the denominator.

function A set of ordered pairs (a, b), where each a is matched with exactly one b. The relationship between a and b can often be described by an equation.

general framework A guideline for problem solving, consisting of 4 parts: (1) analyze the problem, (2) solve the problem, (3) check your result, and (4) learn from the problem.

graph A visual representation of a set of numbers, such as the solution set of an open sentence.

greatest common factor The largest expression that can be divided evenly into each of two or more terms.

identity A number that does not change the number with which it is combined. The additive identity is 0, since $a + 0 = a$; the multiplicative identity is 1, since $1 \cdot a = a$.

inequality A mathematical statement that quantities are unequal.
Symbols are $<$, $\leq$, $>$, and $\geq$.

integer An element of the set:
$$\{\ldots, -2, -1, 0, 1, 2, \ldots\}$$

inverse operation An operation that "undoes" another operation. Example: Addition and subtraction are inverse operations.

irrational number A number that cannot be written as a quotient of integers.
Examples: $\sqrt{3}$, π

isosceles triangle A triangle with two sides, called the legs, of equal length. The third side is called the base.

matrix A display of numerical data in a rectangular array of rows and columns.

monomial A product of numbers and/or variables, with no variable in a denominator.
Examples: $\frac{1}{2} ab$, $-3x^2 y^3$

multiplication property of equality If two equal quantities are multiplied by the same number, the results are equal.
If $a = b$, then $ac = bc$.

multiplication property of inequality If two unequal quantities are multiplied by the same nonzero number, the results are:
unequal in the same order if the multiplier is positive; if $a < b$, and $c > 0$, then $ac < bc$.
unequal in reverse order if the multiplier is negative; if $a < b$, and $c < 0$, then $ac > bc$.

natural number An element of the set $\{1, 2, 3, \ldots\}$.

number line A visual representation of the set of real numbers. Every point on the number line represents a real number, and every real number can be represented by a point on the line.

obtuse angle An angle whose measure is between $90°$ and $180°$.

odd integer An integer that is not evenly divisible by 2; a member of the set:
$$\{\ldots, -3, -1, 1, 3, \ldots\}$$

open sentence An equation or an inequality that contains a variable.

opposite An additive inverse; a number that is opposite in sign. Pairs of opposites are 5 and -5, or -3 and 3. Zero is its own opposite.

order of operations The order in which operations are to be performed if two or more operations appear in the same expression.

parallel Having the same direction. Parallel lines do not meet.

perimeter The distance around a figure, obtained by adding the lengths of all its sides.

perpendicular Meeting at right angles.

polynomial A monomial, or the sum or difference of monomials.
Examples: $3cd$, $5x - 4xyz$, $x^2 - 5x + 6$

probability The likelihood that an event will happen; described by a ratio with a value from 0 (for an impossible event) to 1 (for an event certain to occur).

proportion An equation of 2 equal ratios.

Pythagorean relation In a right triangle with a hypotenuse of length c and legs of lengths a and b, $a^2 + b^2 = c^2$.

range of a relation The set containing all the second elements from the ordered pairs in the relation.

ratio a comparison of two quantities by the operation of division.

rational expression A fraction in which the numerator and the denominator are polynomials.

rational number A number that can be written as a fraction in which the numerator and the denominator are integers, and the denominator is not 0. Examples: $-\frac{2}{3}$, 4, 1.25

real numbers The set of all rational and irrational numbers.

reciprocal A multiplicative inverse; for any nonzero number $\frac{a}{b}$, the reciprocal is $\frac{b}{a}$. Pairs of reciprocals are $\frac{2}{3}$ and $\frac{3}{2}$, or -5 and $-\frac{1}{5}$.

reflection A mirror image of a figure.

relation A set of ordered pairs. The relationship between the first and second members of the pairs can often be described by an open sentence.

replacement set Also called the domain; the set of numbers that may be substituted for a variable.

right angle An angle whose measure is 90°.

right triangle A triangle that contains a right angle.

root A solution of an equation.

scalene triangle A triangle in which all three sides have different lengths.

scientific notation A way of writing a number as a product of two factors, the first factor a number between 1 and 10, and the second a power of 10. Example: In scientific notation, 25,000 is 2.5×10^4.

set A collection of numbers or objects. Braces, { }, are used to indicate a set.

similar (symbol ~) Having the same shape, but not necessarily the same size.

sine In a right triangle, the ratio of the length of the side opposite a given angle to the length of the hypotenuse.

solution set The set of all solutions of an open sentence.

straight angle An angle whose measure is 180°.

strategy A problem-solving technique.

substitution property If two expressions are equal, one may be substituted for the other.

supplementary angles A pair of angles whose measures have a sum of 180°.

tangent In a right triangle, the ratio of the lengths of the sides opposite and adjacent to a given angle.

term A numeral, a variable, or a product or quotient of numerals and variables.

transitive property For equality; If $a = b$ and $b = c$, then $a = c$. For inequality: If $a < b$ and $b < c$, then $a < c$.

translation A "slide" of a figure in the plane.

trigonometry A branch of mathematics that can be used to find the measures of sides or angles in a triangle.

trinomial A polynomial with 3 terms.

variable A symbol, usually a letter, used to represent an unknown number.

variation The pattern of change in related number pairs. Examples: In $y = 2x$, x and y vary directly; in $xy = 4$, x and y vary inversely.

Venn diagram A diagram used to show a relationship between sets.

vertical angles Opposite angles formed by pairs of intersecting lines.

whole number An element of the set: $\{0, 1, 2, 3, \ldots\}$

ANSWERS TO ODD-NUMBERED QUESTIONS

CHAPTER 1. PROBLEM SOLVING

1-1 Guess and Check: page 4

1. 10 dimes   **3.** (4)   **5.** $27   **7.** (3)   **9.** (3)

1-2 Draw a Diagram: pages 9–10

1. 4 smaller triangles   **3.** 5 students
5. 6 routes
7. *WTX, WTY, WBX, WBY, DTX, DTY, DBX, DBY, MTX, MTY, MBX, MBY*
9. 17 cones   **11.** 60 feet   **13.** 18 meals

1-3 Make a List or a Table: pages 13–14

1.

1.00	1.50	3.00	**4.50**	7.75	**9.75**
.30	**.45**	**.90**	1.35	**2.33**	2.93

3. 30,000; 70,000; 30,000; 15,000   **5.** Rhonda
7. Ms. P. is a meteorologist. Ms. Q. is an administrator. Ms. R. is a therapist.

9.

1	2	3	4	5	6	8	9	10	12	15	18
360	180	120	90	72	60	45	40	36	30	24	20

11. 2 ways (one $7 token and ten $3 tokens, or four $7 and three $3)

1-4 Find a Pattern: pages 17–18

1. a. The amount saved increases $2 per week. Total amount saved is the square of the week number.
   b. $17   **c.** $144
3. a. 9 P.M. (add 4 days + 7 hours)
   b. 4 P.M. (1 day and 22 hours ago)
5. (3)   **7.** 11
9. yes (the sum is divisible by 5 when you add a number ending in 8 or 0)
11. 29 feet (The path—up 3, down 2—goes from 14 to 17 ft. high in 1 ft. of length. At low points, height is half the length; 14 is half of 28.)

1-5 Work Backward: page 22

1. 11   **3.** 89   **5.** 17 tomatoes
7. $32,000 Start with the $54,000 and work backward. Alphonse just lost an investment.

Since he always invests half of his money, he had twice as much before his last venture. He had $108,000.

	Before Venture	Put Away	Invested	Gained	After Venture
3rd	72,000	36,000	36,000	36,000	108,000
2nd	48,000	24,000	24,000	24,000	72,000
1st	32,000	16,000	16,000	16,000	48,000

9. $C = 1$
$D = 0$
$B = 5$
$A = 8$

$$\begin{array}{r} A\,B\,B \\ +\ C\,B\,D \\ \hline C\,D\,D\,B \end{array}$$

Since $B + D = B$, $D = 0$.

$$\begin{array}{r} A\,B\,B \\ +\ 1\,B\,0 \\ \hline 1\,0\,0\,B \end{array}$$

Since $B \neq 0$, $B + B = 10$.
$B = 5$
Carry 1 to the hundreds column.
$1 + A + 1 = 10$ $A = 8$

1-6 Choose a Strategy: pages 25–27

Note: Usually, there are alternate strategies, not all of them listed here.

1. 4, 6 Guess and check.
3. Doreen has Sunday, Thursday, and Saturday off. Make a table.
5. (4) Guess and check.   **7.** True. Make a list.
9. 12 feet. Draw a diagram and *work backward.*
11. Largest to smallest: *POR, LOR, MOR, KOR, BOR.* Make a table. (a) The numbers in the table show one order in which the entries may be made. (b) When a box is checked, all other boxes in its row and its column are eliminated.

	BOR	KOR	LOR	MOR	POR
Largest	×(8)	×(5)	×(3)	×(10)	✓(13)
	×(8)	×(5)	✓(11)	×(10)	×(12)
	×(8)	×(5)	×(10)	✓(9)	×(10)
	×(5)	✓(4)	×(5)	×(5)	×(5)
Smallest	✓(7)	×(5)	×(1)	×(2)	×(6)

13. 4:30 P.M. Work backward.

15. Yes, the result is divisible by 3. To find a pattern, look at the sums.

Sum divisible by 3?

$2 + 4 + 6 + 8 + 10 + 12 + 14 + 16 + 18 + 20$

2	6	12	20	30	42	56	72	90	110
N	Y	Y	N	Y	Y	N	Y	Y	N

One pattern is No-Yes-Yes.

Continue the pattern.

22 24 26 28 30

Y   Y   N   Y   Y

Chapter Review Exercises: pages 29–30

1. 11 days **3.** 26 seedlings

5. (3) (She got a cheeseburger and a drink)

7. $31\frac{1}{2}$, 40 (the differences are $2\frac{1}{2}$, $3\frac{1}{2}$, $4\frac{1}{2}$, $\cdots$)

9. $101.88

11. 52 salespeople

Start with one arrangement. Eliminate numbers that do not fit other arrangements.

Divisible by 7, + 3 remaining.	94	87	80	73	66	59	52
Divisible by 5, +2 remaining?	×	✔	×	×	×	×	✔
Divisible by 3, +1 remaining?	✔	×	×	✔	×	×	✔

13. (3)

Problems for Pleasure: pages 30–31

1. $F = 2$, $O = 9$, $R = 7$, $T = 8$, $Y = 6$, $E = 5$, $N = 0$, $I = 1$, $S = 3$, $X = 4$

```
  F O R T Y
+   T E N
+   T E N
  S I X T Y
```

From the ones column, $N = 0$ or $N = 5$. From the tens column, $E = 0$ or $E = 5$. However, if $N = 5$, then 1 must be carried to the tens column and no value of E would make $T + E + E + 1 = T$. Therefore, $N = 0$, $E = 5$, and 1 is carried to the hundreds column.

The hundreds column is $R + T + T + 1 = X$. Carry 1 or 2 to the thousands column. The only possible digit to carry from the thousands column is 1. Thus, in the ten-thousands column, $F + 1 = S$, that is, F and S are consecutive integers. Thus, $O = 8$ or $O = 9$ and $I = 0$ or $I = 1$. However, $N = 0$, so $I = 1$. This means that $O = 9$

and 2 was carried into the thousands column. Then $R + T + T + 1 = 20 + X$. For this to be true, $T > 5$. With 2, 3, 4, 6, 7, and 8 available, the following table shows the possibilities for R, T, and X.

2	3	4	6	7	8	$R + T + T + 1$	Comments
				T	R	21	X = 1, not possible
				R	T	21	
		X		T	R	23	But no digits remain
		X	R		T	23	so that F and S can be consecutive
			X	R	T	24	only possible solution

From the table, $X = 4$, $R = 7$, and $T = 8$. Thus, $F = 2$ and $S = 3$, since these are the only possible consecutive numbers. By elimination, $Y = 6$.

```
   29,786
 +    850
 +    850
   31,486
```

3. 50 students

Draw a Venn diagram. Some of the information can be filled in directly (see figure). Other information can be deduced: (1) 28 students are studying Italian; $28 - (11 + 4 + 5) = 8$.

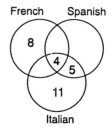

Write 8 in the blank region of the Italian circle. Similarly, fill in the blank regions of (2) the French circle (write 5) and (3) the Spanish circle (write 9). Finally, add all the parts. There are 50 students.

5. 32 *Routes*

AB, AC, BC, AD, BD, CD, DE, DF, DG, DH, GH, FH, EH, FG, EG, EF

16 Routes × 2 = 32

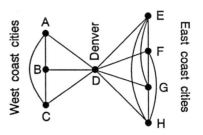

Calculator Challenge: page 31

1. Some possible patterns are described.
   a. The second number appears twice as the product.
   b. The numbers are palindromes. The middle digit is the sum of the first two digits.
   c. After the decimal point, the dividend is repeated in paired digits.
   d. Another 1 appears each time.
3. At least 59 calls

Number of calls	50	55	60	59
Basic service	$18.55	$19.40	$20.25	$20.08

College Test Preparation: pages 32–33

1. (E)   3. (E)   5. (A)   7. (B)
9. (C)   11. (C)   13. (B)

Spiral Review Exercises.: pages 33–34

1. 960,000   3. 5 feet 9 inches   5. 86
7. 18
   $\frac{2}{3}$ won, $\frac{1}{3}$ lost; 6 is $\frac{1}{3}$ of 18
9. $2   11. $\frac{1}{12}$ 13. $\frac{3}{8}$ pound
15. 6 times
   Start at 2:00. Find the numbers of minutes after 2:00 that are multiples of both 4 and 3, that is, multiples of 12. 2:00, 2:12, 2:24, 2:36, 2:48, 2:60 or 3:00
17. 28 eggs

Start with 4; count by 6's	10	16	22		28
Start with 3; count by 5's	8	13	18	23	28

CHAPTER 2. THE LANGUAGE OF ALGEBRA

2-1 Algebraic Expressions: page 47

1. 6   3. 9   5. 16   7. 12   9. 19   11. 0
13. 15   15. 34   17. 50   19. 4   21. 3   23. 8
25. 3   27. 104   29. 60   31. 22   33. 45   35. 2
37. 4   39. 17   41. 16   43. 36   45. 78   47. 5
49. 6   51. 1   53. 24   55. 10   57. 35   59. 9
61. 18   63. 59

2-2 Terms, Factors, and Exponents: pages 52–53

1. a. 5, n, 5n b. 7, m, n, 7m, 7n, mn, 7mn
   c. 13, x, y, 13x, 13y, xy, 13xy
3. a. base, t; exponent, 1
   b. base, 10; exponent, 6
   c. base, 5y; exponent, 4
5. a. 10^4 b. a^4b^2 c. $(6a)^3$ 7. 52   9. 2
11. $3 \cdot n \cdot n \cdot n \cdot n \cdot n$ 13. 144   15. 64
17. 0   19. 100   21. 2,700   23. 169   25. 144
27. 106   29. 24   31. 169   33. 1   35. 8
37. 65   39. 14   41. 85

43.

Start	
Write the expression.	4$(b + c)^2$
Let $b = 3$, $c = 2$.	4$(3 + 2)^2$
In the parentheses, add.	4$(5)^2$
Evaluate the power.	4(25)
Multiply.	100
Stop	

45. 64 **47.** 81 **49.** 6 **51.** 48 **53.** 27,648
55. 100 **57.** 91 **59.** 48 **61.** 78 **63.** 8
65. 116 **67.** 28 **69.** 32 **71.** 11 **73.** 7 **75.** 1
77. 1

2-3 Sets: pages 58–59

1. $\{1, 3, 5, 7, \ldots\}$ **3.** $\{\ldots, -3, -1, 1, 3, 5, 7, 9\}$
5. an infinite set **7.** a finite set

In 9–12, answers will vary.

9. {odd numbers between 14 and 20}
11. {letters in the word *car*}
13. (c) equivalent **15.** (e) equal **17.** (b) equal
19. true **21.** false **23.** true **25.** 0 **27.** .625
29. 2.25 **31.** 5.5 **33.** $.58\overline{3}$ **35.** $\frac{81}{100}$ **37.** $3\frac{1}{5}$
39. $-\frac{233}{1,000}$ **41.** always true **43.** never true
45. never true **47.** sometimes true
49. 1,221 and 2,442
   1,001 and 2,002
   1,331 and 2,662
51. 4   (686, 767, 848, 929)
53. a. 1 5 10 10 5 1
   b. Place a 1 at each end of the new row. For
   the other numbers, add each pair of
   adjacent numbers from the row above.

2-4 The Number Line: pages 64–65

1. A, $1\frac{1}{2}$; R, $2\frac{1}{2}$; T, 3

3.

5.

In 6–9, answers will vary.

7. $\frac{11}{200}$, .056, $\frac{57}{1,000}$ **9.** .4, $\frac{1}{2}$, .52 **11.** D **13.** G
15. E
17. One number is positive and one number is
   negative. They have the same absolute value.
19. .1 **21.** $-\frac{1}{2}$ **23.** 5 **25.** 0 **27.** $-.01$
29. The number on the right is larger.
31. north of the equator
33. -500 **35.** -8 **37.** 0 **39.** $-\frac{3}{4}$
41. false; if $a = -1$, then $-a = 1$
43. false; if $a = 0$, the opposite of a is 0.
45. false; if $a = -1$, the opposite of a is 1, which
   is to the right.
47. a. 3 **b.** -3 **49. a.** $1\frac{1}{2}$ **b.** $-1\frac{1}{2}$
51. a. 2.7 **b.** -2.7 **53.** true **55.** true
57. true **59.** 10 **61.** 24 **63.** -2

2-5 Using an Equation in an Algebraic Model: pages 70–73

1. a. $7 + 3 \times 9$ **b.** $5(8 - 6)$
   c. $20 - (30 \div 5)$ **d.** $3 \times 7 - 8$
3. a. $n - 2$ **b.** $\frac{3}{4}n$
   c. $4n + 3$ **d.** $3n + 8$
5. $(y + 5)$ years old **7.** $(k - 10)$ kilograms
9. $(c - d)$ pounds **11.** $(c + 25)$ dollars
13. a. $5n$ cents **b.** $25q$ cents
   c. $(10d + 50h)$ cents
   d. $[5x + 10(25 - x)]$ cents
15. dn dollars
17. a. hm dollars **b.** $104m$ dollars
19. $36(d - s)$ dollars **21.** $\frac{q}{t}$ dollars
23. $[(f + w) + (g + w)]$ dollars
25. $n - 1$
27. $7n = 70$ **29.** $2n + 7 = 27$ **31.** (2) **33.** (4)
35. (3) **37.** (1)

In 39–42, answers will vary.

39. A number increased by 7 equals 12.
41. The quotient of a number and 4 is 8.
43. positive rational numbers
45. positive real numbers
47. positive rational numbers
49. whole numbers
51. Let $x =$ the number.
   $5x - 9 = 31$
53. Let $x =$ the number of hours the first week.
   $x + 2x = 45$
55. Let $x =$ the weight of the backpack.
   $\frac{1}{2}x + 8 = x - 4$
57. Let $x =$ the number of crystals added.
   $100 + x = 3x$

2-6 Working with an Inequality: pages 80–82

1. true **3.** false **5.** true
7. false **9.** true
11. a. $-4 < 8$ **b.** $8 > -4$
13. a. $-4 < -2 < 3$ **b.** $3 > -2 > -4$
15. a. $-2\frac{1}{2} < -1.5 < 3\frac{1}{2}$ **b.** $3\frac{1}{2} > -1.5 > -2\frac{1}{2}$
17. $9 < 11 < 20$

In 19–22, answers will vary.

19. The sum of 9 and 8 is greater than 16.
21. -2 is less than -1, and -1 is less than 0.

23. a. false **b.** true **c.** false **d.** false
 e. true

25. $\frac{7}{2}$ **27.** $\frac{5}{8}$ **29.** $-3\frac{1}{3}$ **31.** -5.7

33. $\{.3, .31, .313113111\ldots, .333\ldots\}$

35. $\{.27, .272272227, .\overline{27}, \frac{2}{7}\}$

37. C, D, E, F, G, H, I, J **39.** no points

41. L, M, N, P

43.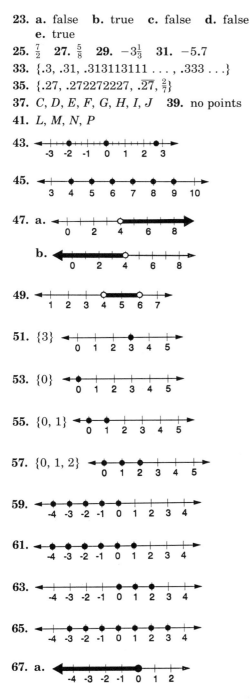

45.

47. a.

b.

49.

51. $\{3\}$

53. $\{0\}$

55. $\{0, 1\}$

57. $\{0, 1, 2\}$

59.

61.

63.

65.

67. a.

b. The darkened circle at 0 shows that 0 is included. The darkened line and arrow to the left represent the infinite set of real numbers less than 0.

69. a.

b. The open circle at 2 shows that 2 is not included. The darkened lines and arrows to the right and left represent the set of real numbers greater than or less than 2.

71. a.

b. The open circle at -3 shows that -3 is not included, and the open circle at 2 shows that 2 is not included. The darkened line represents the infinite set of real numbers between -3 and 2.

73. a.

b. The darkened circle at -3 shows that -3 is included, and the darkened circle at 3 shows that 3 is included. The darkened line represents the infinite set of real numbers between -3 and 3.

75.

77.

79. Let x = the area of one triangle.
 $2x < 250$

81. Let x = the number being hired at Newton.
 $x + (x + 7) \geq 41$

83. Let j = the amount saved in January.
 $j + (j + 50) + 3j \geq 450$

85. Let r = the number of beachfront rooms.
 $r + \frac{1}{2}r + (\frac{1}{2}r - 30) \leq 300$

Chapter Review Exercises: pages 84–86

1. 22 **3.** 8 **5.** 50 **7.** 39 **9.** 27 **11.** 93

13. 9 **15.** 10 **17.** $\{B, A, N\}$

19. $\{3, 5, 7, 9, \ldots\}$

In 21–24, answers will vary.

21. {odd natural numbers}
23. {letters in the word *ear*}
25. I, R **27.** W, I, R **29.** $.625$ **31.** $5.1\overline{6}$
33. $.41\overline{6}$ **35.** $\frac{17}{1,000}$ **37.** $\frac{873}{100}$ or $8\frac{73}{100}$ **39.** B, C, D
41. 8 **43.** 10 **45.** 1 **47.** 9 **49.** $(7 \times 6) - 8$
51. xy **53.** $(p - 7)$ degrees **55.** $10d$ cents

57. $\left(\dfrac{d}{b}\right)$ dollars **59.** $5(p - 7) = 3$

61. $16 - 2p = 9$ **63.** true **65.** true
67. $-4 > -7$ **69.** $-\frac{1}{4} > -\frac{1}{3} > -\frac{1}{2}$
71. $.9$ **73.** $\frac{3}{7}$
75. Let $x =$ the number of yards for the blouse.
$x + 2x > 4$
77. Let $x =$ the number of lambs.
$4x + 2x \leq 150$
79. F, H, K

81.

83.

85.

87.

89.

Problems for Pleasure: *pages 86–87*

1. 3 **3.** 32
5. sometimes irrational, sometimes rational

.565565556. . .	.565565556. . .
$-.212212221. . .$	$-.121121112. . .$
.353353335. . .	.444444444. . .
irrational	rational

Calculator Challenge: *pages 87–88*

1. The primary purpose of the game is to develop skill in estimating. One approach to estimating is to round the given numbers. Some possibilities in Game 1 are:

1. $40,000 \div 20$ 6. $\frac{1}{2}$ of 770
2. 500×90 7. $\frac{4}{5}$ of 60
3. $\frac{1}{2}$ of 84   8. $64 \div .8$
4. $2,100 \div 20$ 9. 70×70
5. $\frac{1}{3}$ of 540   10. $1,400 \div 10$

3. a. $78 = 3^4 - 3$ **b.** $352 = 2^5(3 + 2 \times 4)$

College Test Preparation: *pages 88–89*

1. (B) **3.** (E) **5.** (E) **7.** (D) **9.** (A) **11.** (D)
13. (D) **15.** (B) **17.** (C) **19.** (C) **21.** (A)
23. (A) **25.** (B)

Spiral Review Exercises: *pages 90–91*

1. a. 21,200,000
   b. 41,400,000
   c. 8,600
3. 248.935 **5.** 28, 39
7. 12 *Hint:* draw a diagram
9. 12 **11.** $2.39 **13.** 2 dozen **15.** 52 cents
17. 75 cents
19. a. 840 calories
   b. 495 calories
   c. 363 calories
21. 9 **23.** 48

CHAPTER 3. REAL NUMBERS: PROPERTIES AND OPERATIONS

3-1 Closure and Substitution: *page 97*

1. yes **3.** yes **5.** yes **7.** yes
9. a. yes **b.** no **c.** yes **d.** no
11. yes **13.** no **15.** no **17.** no **19.** yes

21. a. Not closed; $5 + 3$ is not odd.
   b. Not closed; $7 - 3$ is not odd.
   c. Closed; the result is always in the set.
   d. Not closed; $3 \div 7$ is not in the set.

23. a. Not closed; $5 + 5$ is not in the set.
   b. Not closed; $5 - 3$ is not in the set.
   c. Not closed; 5×3 is not in the set.
   d. Not closed; $1 \div 3$ is not in the set.

25. a. Not closed; $1 + 1$ is not in the set.
   b. Not closed; $0 - 1$ is not in the set.
   c. Closed; the result is always in the set.
   d. Not closed; $1 \div 0$ is undefined.
27. a. Not closed; $1 + 1$ is not in the set.
   b. Not closed; $1 - 1$ is not in the set.
   c. Closed; result is always in the set.
   d. Closed; same as part c.

In 29–33, answers will vary.

29. 2×7, $10 + 4$, $\frac{140}{10}$

31. $10x + 8x$, $\dfrac{18x}{1}$, $(2 \times 9)x$

33. $9t$, $\dfrac{27t}{3}$, $12t - 3t$

35. $\dfrac{40}{t}$ m.p.h. **37.** $n + 1$ **39.** $\dfrac{1}{c}$

3-2 Addition of Real Numbers: pages 103–104

1.

Starting from 0, the graph shows a translation of 3 units to the right followed by a translation of 4 units to the right, arriving at 7.

3.

Starting from 0, the graph shows a translation of 7 units to the right followed by a translation of 4 units to the left, arriving at 3.

5.

$4 + (-4) = 0$

7.

$0 + (-6) = -6$

9.

$[8 + (-4)] + (-6) = -2.$

11.

$(-5 + 2) + 3 = 0$

13. -25 **15.** 8 **17.** $18\frac{1}{4}$ **19.** -7.8 **21.** 50
23. 27 **25.** $6\frac{1}{2}$ **27.** 2.3 **29.** -2.3 **31.** -50
33. 25 **35.** -28 **37.** -12 **39.** -17 **41.** 6
43. -26 **45.** 0 **47.** -1.5 **49.** 0 **51.** 0

53. 80 **55.** 2 **57.** 0 **59.** 5 **61.** 15 yards
63. 28 miles **65.** 8 **67.** $-C$ **69.** -5 **71.** y
73. -7 **75.** -4 **77.** 2
79. Addition property of opposites.
81. Property of the opposite of a sum.
83. (1) Substitution property.
    (2) Associative property of addition.
    (3) Addition property of opposites.
    (4) Addition property of zero.

In 85 and 86, answers will vary.

85. $15 + (-8)$
    $= (7 + 8) + (-8)$ Subst. prop.
    $= 7 + [8 + (-8)]$ Assoc. prop. of add.
    $= 7 + 0$ Add. prop. of opp.
    $= 7$ Add. prop. of 0
87. false **89.** true **91.** true
93. False; for example, odd numbers are closed under multiplication; but, $5 + 3 = 8$.

3-3 Multiplication of Real Numbers:
pages 108–109

1. 115 **3.** 120 **5.** 0 **7.** -3.6 **9.** 4 **11.** $\frac{1}{6}$
13. 48 **15.** -77 **17.** 0 **19.** 2 **21.** 1 **23.** -3
25. 24 **27.** -60 **29.** -120 **31.** 11 **33.** 4
35. 8 **37. a.** $\frac{1}{5}$ **b.** $\frac{4}{3}$ **c.** $\frac{2}{5}$ **d.** $-\frac{1}{7}$
39. 16 **41.** 125 **43.** -125 **45.** $\frac{1}{4}$ **47.** $\frac{8}{27}$
49. $-\frac{1}{64}$
51. Commutative property of multiplication.
53. Multiplication property of one.

In 55 and 56, answers will vary.

55. 1. Multiplying a positive number by a positive number: If Rosa saves $10 a month, in 5 months she will have $50 more, represented by: $10 \times 5 = 50$.
    2. Multiplying a negative number by a positive number: If Rosa spends $10 a month, in 5 months she will have $50 less, represented by: $-10 \times 5 = -50$
    3. Multiplying a positive number by a negative number: If Rosa saves $10 a month, five months ago she had $50 less, represented by: $10 \times (-5) = -50$
    4. Multiplying a negative number by a negative number: If Rosa spends $10 a month, five months ago she had $50 more, represented by: $(-10) \times (-5) = 50$

3-4 Subtraction and Division: pages 113–115

1. 8 **3.** 1 **5.** -11 **7.** -54 **9.** -12
11. 14.5 **13.** 1.6 **15.** 10 **17.** -11 **19.** 30
21. 1.5 **23.** 2 **25.** 27 **27.** 9 **29.** -26
31. 60 **33.** 20 **35.** 10.4 **37.** 8 **39.** -28
41. -38
43. a. $3°$ **b.** $28°$ **c.** $-12°$ **d.** $-16°$
45. 18th floor **47.** 110 feet **49.** $230
51. a. false **b.** false
53. a. false **b.** false **c.** false
55. $-\frac{1}{5}$ **57.** -10 **59.** $-\frac{1}{x}$ **61.** -13 **63.** -5
65. -16 **67.** 5 **69.** -13 **71.** $\frac{5}{4}$ **73.** $-\frac{25}{2}$
75. 0 **77.** -3 **79.** 5 **81.** -36 **83.** $\frac{9}{8}$
85. $-.2$ **87.** -18 **89.** 25 **91.** .7 **93.** 1
95. 0 **97.** 112
99. a. $\frac{1}{x-5}$ **b.** 5 **101. a.** $\frac{1}{2x-1}$ **b.** $\frac{1}{2}$
103. a. false **b.** false
105. a. false **b.** false **c.** false

3-5 The Distributive Property; Summary of Properties: page 118

1. yes **3.** yes **5.** yes **7.** $-4p - 4q$
9. $15m - 7m$
11. $128 \times 615 - 28 \times 615$
$= (128 - 28)\,615 = 61{,}500$
13. $937 \times .8 + 937 \times .2 = 937\,(.8 + .2) = 937$
15. $50 \times 8\frac{3}{5} = 50\,(8 + \frac{3}{5})$
$= 50 \times 8 + 50 \times \frac{3}{5} = 400 + 30 = 430$
17. $-14x - 7$ **19.** $-10m + 18n$ **21.** $2(p + g)$
23. $a(t + r)$ **25.** $7(x + 3)$
27. (1) Associative property of addition
(2) Distributive property
(3) Substitution property
29. a. positive integers, negative integers, even integers, rational numbers, real numbers
b. even integers, rational numbers, real numbers
c. positive integers, odd integers, even integers, rational numbers, real numbers
d. rational numbers, real numbers

In 31–36, answers will vary.

31. $(ab)c = a(bc)$ Assoc. prop. of $\times$
$= a(cb)$ Comm. prop. of $\times$

33. $a(b + c) = ab + ac$ Dist. prop.
$= ab + ca$ Comm. prop. of $\times$
35. $m + n + (-m)$
$= (n + m) + (-m)$ Comm. prop. of $+$
$= n + [m + (-m)]$ Assoc. prop. of $+$
$= n + 0$ Add. prop. of opp.
$= n$ Add. prop. of 0

3-6 Using the Properties of the Real Number System: pages 122–123

1. $10x$ **3.** $11c$ **5.** $4.7y$ **7.** $11ab$ **9.** $14m$
11. $-\frac{3}{4}xy$ **13. a.** $15x$ **b.** $24m$ **c.** $23y$ **d.** $13a$
15. a. $5y^2 - 2y^2 = 3y^2$
$5(12)^2 - 2(12)^2 \stackrel{?}{=} 3(12)^2$
$720 - 288 \stackrel{?}{=} 432$
$432 = 432$
b. $5y^2 - 2y^2 = 3y^2$
$5(-2)^2 - 2(-2)^2 \stackrel{?}{=} 3\,(-2)^2$
$20 - 8 \stackrel{?}{=} 12$
$12 = 12$
c. $5y^2 - 2y^2 = 3y^2$
$5(.2)^2 - 2(.2)^2 \stackrel{?}{=} 3(.2)^2$
$.2 - .08 \stackrel{?}{=} .12$
$.12 = .12$
17. $15m + 11$ **19.** $6.6 + 15.8b$
21. $6\frac{1}{3}a + 2b + 3\frac{3}{4}c$
23. $7m + 35 + 8m + 4 = 15m + 39$
25. a. $2a + b$ **b.** $2a + 4b$
c. $4a + 2b + 6c$ **d.** $5x + 9y$
27. $14x - 3$ **29.** $-12e + 25$ **31.** $4b + 4$
33. $2 + 8s$ **35.** $13x - 11$ **37.** $6b$ **39.** $49x - 14$
41. $36(x + 2) + 12(2x - 1) = (60x + 60)$ inches
43. $10 + 7(2x - 1) = (14x + 3)$ days
45. -30 **47.** -40 **49.** -1 **51.** -3 **53.** 36
55. -27 **57.** -9 **59.** 32 **61.** -108 **63.** 2
65. 450 **67.** 54 **69.** -9 **71.** -2 **73.** 12
75. -43 **77.** 9 **79.** 98 **81.** 9 **83.** 34 **85.** 28
87. -96 **89.** -80 **91.** $-\frac{2}{3}$ **93.** 24
95. 31 stamps **97.** $288

3-7 Defining an Operation: pages 126–127

1. $3 * (-2) = 3^2 - (-2) = 9 + 2 = 11$
3. $3 \odot (-2) = 3^2(-2) - (-2) = 9(-2) - (-2)$
$= -18 + 2 = -16$

5. a. 7 **b.** −8 **c.** $-\frac{1}{7}$ **d.** .4

7. a. 1 **b.** 10 **c.** −5 **d.** −8

9. a. {5, 10} **b.** {2, 5, 7, 10, 12}

11. a. ∅ **b.** {integers}

13. a. {multiples of 15} **b.** {multiples of 3 or 5}

15. a. {c, h, m} **b.** {b, c, d, e, f, g, h, i, j, l, m}
   c. {h, m} **d.** {d, f, g, h, l, m}
   e. {c, d, e, f, g, h, l, m}

17. (4) $\boxed{3} + \boxed{4} = 9 + 16 = 25 = \boxed{5}$

19. (3) $.8 \div 4 \div 1 = .2 \div 1 = .2$
   $10 \div 5 \div 10 = 2 \div 10 = .2$

21. $\boxed{\begin{array}{|c|c|} \hline 7 & 8 \\ \hline x & x \\ \hline \end{array}} = 8x - 7x = 12$
   $x = 12$

3-8 Another Arithmetic: Matrices: pages 132–135

1. a. 96 juniors study Spanish.
   b. 309 students study French.
   c. 197 freshman study foreign languages.
   d. 714 students study foreign languages.

3. a.

	pennies	nickels	dimes	quarters
Saturday	15	20	18	12
Sunday	13	22	16	20

b. (1) 20 rolls of quarters deposited Sunday.
   (2) 65 rolls of coins deposited Saturday.
   (3) 34 rolls of dimes deposited during the weekend.
   (4) 136 rolls of coins deposited during the weekend.

c. Find the sum of each column and multiply by the number of coins in a roll by the value of each coin. Then, find the sum of the products.

5. $\begin{bmatrix} -1 \\ -1 \end{bmatrix}$ **7.** $\begin{bmatrix} 8 & -6 & 0 \\ -6 & 4 & 6 \\ -1 & 2 & -4 \end{bmatrix}$

9. When a matrix is multiplied by a constant, each element in the matrix is multiplied by the same constant.

11. [.5 1 0 −1.5]

In 13 and 17, detailed computation is shown.

13. $[4(2) + (-2)(-1) \quad 4(-3) + (-2)(5)]$
   $= [10 \quad -22]$

15. $\begin{bmatrix} -10 \\ -9 \\ 20 \end{bmatrix}$

17. $\begin{bmatrix} -4 & -12 \\ 6 & 8 \\ -2 & -10 \end{bmatrix} \begin{bmatrix} 2 & -3 & 5 \\ -3 & 1 & 0 \end{bmatrix}$

$= \begin{bmatrix} -4(2) + (-12)(-3) & -4(-3) + (-12)(1) & -4(5) + (-12)(0) \\ 6(2) + 8(-3) & 6(-3) + 8(1) & 6(5) + 8(0) \\ -2(2) + (-10)(-3) & -2(-3) + (-10)(1) & -2(5) + (-10)(0) \end{bmatrix}$

$= \begin{bmatrix} 28 & 0 & -20 \\ -12 & -10 & 30 \\ 26 & -4 & -10 \end{bmatrix}$

19. a. The number of elements in a row of the first matrix is not equal to the number of elements in a column of the second matrix.

b. A square matrix, for which the multiplication identity matrix is a square matrix of the same size.

c. $\begin{bmatrix} 1 & 0 & 0 & 0 \\ 0 & 1 & 0 & 0 \\ 0 & 0 & 1 & 0 \\ 0 & 0 & 0 & 1 \end{bmatrix}$

21. $\begin{bmatrix} 15 & -3 \\ -6 & 0 \end{bmatrix}\begin{bmatrix} 3 \\ -4 \end{bmatrix} = \begin{bmatrix} 57 \\ -18 \end{bmatrix}$

23. Associative property of multiplication

25. a. Matrix A

	orch.	mezz.	balc.
requests	[83	55	62]

b. Matrix B

	Astro	Lido	Vox
orchestra	35	40	37
mezzanine	25	30	32
balcony	15	10	20

c.

Astro   Lido   Vox

$AB = [5{,}210 \quad 5{,}590 \quad 6{,}071]$

d. (1) Vox
(2) Astro
(3) $861

Chapter Review Exercises: pages 137–138

1. not closed **3.** not closed **5.** -6 **7.** 8
9. -2 **11.** -7 **13.** -4 **15.** 24 **17.** $-\frac{7}{12}$
19. -36 **21.** 6
23. Commutative property of addition.
25. Addition property of zero.
27. Commutative property of multiplication.
29. $6a + 15$ **31.** $12b + 9bx$ **33.** $hm - hn$
35. $(5 + a)d$ **37.** $3(3g - 4)$ **39.** $a(x + 12)$

In 40–49, answers will vary.

41. $de = ed$ **43.** $g \cdot 1 = g$
45. $m(n \cdot p) = (m \cdot n)p$ **47.** $r \cdot 0 = 0$
49. $v + 0 = v$ **51.** $-2b$ **53.** $7a + w$
55. $4p - 18$ **57.** $16h + 6$ **59.** 20 **61.** 40
63. -60 **65.** -72 **67.** -28
69. a. -4 **b.** 5 **c.** 0
71. a. $\{11, 13\}$
b. $\{1, 3, 5, 7, 9, 11, 13, 15, 17, 19\}$
73. $\{1, 2, 4, 5\}$
75. There are 29 girls in 3rd grade.
77. $a = 4$

79. $\begin{bmatrix} 4 & -4 \\ -1 & 8 \\ 7 & 1 \end{bmatrix}$

Problems for Pleasure: page 139

1. a.

#	P	R	I	M	E
P	P	P	R	R	P
R	P	R	M	E	R
I	R	M	I	I	I
M	R	E	I	M	M
E	P	R	I	M	E

b. $P \mathbin{\#} R \mathbin{\#} I \mathbin{\#} M \mathbin{\#} E = (P) \mathbin{\#} I \mathbin{\#} M \mathbin{\#} E$
$= (R) \mathbin{\#} M \mathbin{\#} E$
$= E \mathbin{\#} E = E$

c. E **d.** M
e. P and I have no inverses. Under operation #, they do not combine with another element to produce identity E.

3. Weighing on a balance scale is done by adding and subtracting weights. To add, the weights are put on the same side of the balance; to subtract, the weights are on opposite sides.

In this explanation, an arrow will represent a weighing. On the left of the arrows are the desired weights. On the right are the ways to achieve the desired weights.

The smallest number of weights is found by avoiding duplications. You should not be able to weigh anything in more than 1 way. Thus, if you start with a 1-ounce weight, the next weight should not be another 1-ounce weight. The second weight should not be 2 ounces because there would be more than 1 way to weigh a 1-ounce item: $1 \rightarrow 1$ and $1 \rightarrow 2 - 1$. The second weight is 3 ounces: then $1 \rightarrow 1$, $2 \rightarrow 3 - 1$, $3 \rightarrow 3$, and $4 \rightarrow 3 + 1$.

The third weight cannot be 4 ounces, since $1 \rightarrow 1$ and $1 \rightarrow 4 - 3$. Nor can it be 5 ounces, since $2 \rightarrow 3 - 1$ and $2 \rightarrow 5 - 3$. Six ounces can't be used because $3 \rightarrow 3$ and $3 \rightarrow 6 - 3$. Nor can you use 7 ounces, since $4 \rightarrow 3 + 1$ and $4 \rightarrow 7 - 3$. Eight ounces also gives a duplication, since $4 \rightarrow 3 + 1$ and $4 \rightarrow 8 - (3 + 1)$. The third weight is 9 ounces, because items from 5 ounces to 13 ounces can then be weighed with no duplications.

The fourth weight, 27 ounces (given in the problem), can then be used with the other weights to weigh items from 14 to 40 ounces.

Therefore, Pat should purchase weights of 1 ounce, 3 ounces, and 9 ounces.

Calculator Challenge: *page 140*

1. **a.** 9, −13 **b.** −6, −27
   c. 94, −19 **d.** 21, 83
3. Start, +15, +18, +26, ×3, −51, ×2, −98, −3, −67, +42, ×4, +34, −45, End; Score: 493

College Test Preparation: *pages 140–141*

1. (C) 3. (C) 5. (D) 7. (E) 9. (E) 11. (A)
13. (C) 15. (A) 17. (C) 19. (B) 21. (A)

Spiral Review Exercises: *pages 142–143*

1. **a.** (3) is less than
   b. (4) is greater than or equal to
   c. (11) is not equal to **d.** (7) null set
   e. (8) absolute value of a **f.** (10) a cubed
3. 3,400,000 5. 42 feet 7. $8n − 3 = 29$
9. $−.43 < −.428 < −.401 < −.4$
11. (2) 13. 1.5 gallons 15. 12 years old
17. 1991 and 2002
19. **a.** 31 **b.** 49 **c.** (1) b (2) a

SELF-TEST   Chapters 1–3: *pages 146–147*

1. 288 2. $\{0, 1, 2\}$ 3. $.\overline{27}$
4. $|3| − |−5| < 5 − |3|$ 5. $1\frac{1}{3}$ 6. $\{5, −5\}$
7. $−8$ 8. $5x − y$ 9. 34 10. $−\frac{2}{3}$ 11. 4
12. $\{3\frac{1}{2}\}$ 13. (1) 14. (1) 15. (3) 16. (3)
17. (2) 18. (1) 19. (4) 20. (2)
21. Let $a = 2$ and $b = 3$.
   $a * b = 2 \cdot 3 − (2 − 3) = 6 − (−1) = 7$
   $b * a = 3 \cdot 2 − (3 − 2) = 6 − 1 = 5$
22. (1) Commutative property of multiplication
    (2) Associative property of multiplication
    (3) Multiplicative inverse
    (4) Multiplicative identity
23. **a.** $2n + 5 = 40$
    b. $n + 12 = 4 − n$
    c. $\dfrac{6n}{4} > 8$
24. **a.** 10 **b.** 16 **c.** 16 **d.** 10 **e.** 52

CHAPTER 4. EQUATIONS AND INEQUALITIES

4-1 Properties of Equality: *pages 151–152*

1. Multiplication property.
3. Addition property.
5. Subtraction property.
7. Multiplication property.
9. Addition property. 11. Addition property.
13. Subtraction property.
15. Multiplication property.
17. Addition property. 19. Division property.
21. Division property.
23. Multiplication property.
25. Given.
    Addition property.
    Division property.
27. Given.
    Subtraction property.
    Division property.
29. Given.
    Addition property.
    Division property.
31. Given.
    Subtraction property.
    Division property.

33. Yes
35. No; 1 is less than 2, but 2 is not less than 1.
37. No; 5 is not greater than 5.
39. No; if John III is the son of John II, and John II is the son of John, then John III is the grandson and not the son of John.
41. Yes

4-2 Solving an Equation by Applying Inverses: *pages 154–155*

1. $x = 21$ 3. $x = 2$ 5. $x = −7$
7. $x = 98$ 9. $y = −9$ 11. $y = \frac{1}{2}$
13. $y = 162$ 15. $y = 162$ 17. $z = −6$
19. $z = −12$ 21. $r = −\frac{1}{5}$ 23. $r = −18$
25. $t = 1\frac{1}{2}$ 27. $t = 4$ 29. $t = 1$
31. $t = 2\frac{1}{2}$ 33. $m = −25$ 35. $n = 0$
37. $b = 0$ 39. $q = 24$ 41. $d = 7.3$
43. $x = .8$ 45. $c = 7\frac{3}{4}$ 47. $x = 2$
49. $m = \frac{4}{3}$ 51. $x = \frac{2}{3}$ 53. $m = 10$
55. $x = 4$ 57. $a = −8$ 59. $x = \frac{1}{3}$
61. $b = \frac{1}{2}$ 63. $t = \frac{2}{3}$ 65. $c = −16$
67. $y = −5$ 69. $r = −3$ 71. $z = −4$
73. $p = 9$ 75. $r = −6$

---

77. a. $\{12\}$ **b.** $\{12\}$ **79. a.** $\varnothing$ **b.** $\{-1\}$
81. a. $\varnothing$ **b.** $\{0\}$ **83. a.** $\{108\}$ **b.** $\{108\}$
85. a. $\varnothing$ **b.** $\{-100\}$
87. $\quad t = 3$ **89.** $\quad r = 24$
$\quad\quad t + 7 = 10$ $\quad\quad\quad 5r = 120$
91. $\quad t = 3$ **93.** $\quad\quad m = 14$
$\quad\quad 2t = 6$ $\quad\quad\quad 5m - 7 = 63$
95. a. $x = -19$ **b.** $y = 8$
97. a. $x = -2$ **b.** $y = 0$

4-3 Using An Equation to Solve a Problem:
pages 157–159

In 1–40, unless otherwise stated, let $x = $ the unknown value.

1. $x - 7 = 46$ **3.** $x - 10 = 42$
$\quad\quad x = 53$ $\quad\quad\quad\quad\quad x = 52$
5. $x - 8 = 7$ **7.** $x + 225 = 2{,}670$
$\quad\quad x = 15$ yr. $\quad\quad\quad\quad x = \$2{,}445$
9. $x + 3.75 = 8.25$ **11.** $x - 78 = 1{,}125$
$\quad\quad\quad x = \$4.50$ $\quad\quad\quad x = 1{,}203$ sophomores
13. $x - 7.25 = 3.50$ **15.** $5x = 150$
$\quad\quad\quad x = \$10.75$ $\quad\quad\quad x = 30$ miles
17. $12x = 360$ **19.** $2\frac{1}{2}x = 90$
$\quad\quad x = 30°$ $\quad\quad\quad\quad x = \$36$
21. $\frac{1}{5}x = 550$ **23.** $\frac{3}{4}x = 18$
$\quad\quad x = \$2{,}750$ $\quad\quad\quad x = 24$ games
25. $22 + x = 41$ **27.** $6x - 4 = 68$
$\quad\quad x = 19$ versions $\quad\quad x = 12$
29. $38 + \frac{5}{9}x = 128$ **31.** $2x - 5 = 115$
$\quad\quad\quad x = 162$ $\quad\quad\quad x = 60$ houses
33. $4x + 10 = 130$ **35.** $3x + 15 = 120.60$
$\quad\quad x = 30$ adult tickets $\quad x = \$35.20$
37. $\quad$ Let $f = $ number on first day.
$\quad\quad \frac{3}{4}f - 30 = 1{,}980$
$\quad\quad\quad\quad f = 2{,}680$
$\quad\quad\text{total} = 2{,}680 + 1{,}980$
$\quad\quad\quad\quad = 4{,}660$ envelopes
39. $\frac{2}{3}x + 3 = 25$
$\quad\quad x = 33$ participants

4-4 Combining Like Terms in an Equation:
pages 162–165

1. $a = 12\frac{1}{2}$ **3.** $x = 13\frac{1}{2}$ **5.** $x = 3$ **7.** $x = 2$
9. $y = 3$ **11.** $y = -10$ **13.** $x = 4$ **15.** $c = 3$
17. $y = 9$ **19.** $y = 3$ **21.** $y = 7$ **23.** $y = \frac{1}{3}$
25. $y = 16$ **27.** $c = 16$ **29.** $x = 9$ **31.** $x = 24$
33. $m = -50$ **35.** $y = 10$ **37.** $x = \frac{3}{5}$

---

39. $t = 6$ **41.** $s = 5$ **43.** $x = -4$ **45.** $b = 9$
47. $x = 2$ **49.** $x = 0$ **51.** $a = 1$ **53.** $y = 4$
55. $c = 4$ **57.** $z = \frac{1}{4}$ **59.** $a = 3$ **61.** $d = -10$
63. $d = -5$ **65.** $m = -\frac{1}{2}$ **67.** $b = 9$
69. $c = 8$ **71.** $t = 15$ **73.** $y = 4$ **75.** $y = -3$
77. $y = 3$ **79.** $x = 3$ **81.** $y = 6$ **83.** $d = 7$
85. $z = \frac{6}{7}$ **87.** $c = -5$ **89.** $a = 2$ **91.** $x = 11$
93. $m = \frac{7}{3}$ **95.** $x = 6$ **97.** $r = 1$ **99.** $z = -1$
101. a. $-2 = -2$
$\quad$ **b.** The variable has dropped out.
$\quad$ **c.** Answers will vary.
$\quad$ **d.** Any number will satisfy the equation.
103. a. $x = 6$ $\quad\quad$ **b.** $y = 4$
105. a. $x = 0$ $\quad\quad$ **b.** $y = 20$
107. $x = 16$
109. a. Let $w = $ width.
$\quad\quad$ Then $w + 2 = $ height.
$\quad$ **b.** $w + 3, w + 3, w$
$\quad$ **c.** $w = 10$
111. Let $x = $ the smaller number.
$\quad\quad 5x - x = 96$
$\quad\quad\quad x = 24, 5x = 120$
113. Let $x = $ amount earned by Bob.
$\quad\quad 3x + x = 24$
$\quad\quad\quad x = 6$
$\quad$ Bob earned \$6; Diane earned \$18.
115. Let $x = $ original number.
$\quad\quad 4x - 9 = 3x - 1$
$\quad\quad\quad x = 8$
117. Let $x = $ Sylvia's height.
$\quad\quad x - \frac{3}{4}x = 15$
$\quad\quad\quad x = 60$
$\quad$ Caia's height, 45 inches; Sylvia's height, 60 inches.
119. Let $x = $ the cost of rug.
$\quad\quad 2(x + 15) + 4x = 300$
$\quad\quad\quad\quad\quad x = 45$
$\quad$ rug, \$45; bed, \$60
121. Let $x = $ Bullets' half-time score.
$\quad\quad x + 24 = 2(x + 6)$
$\quad\quad\quad x = 12$
$\quad$ Bullets, 12; Greenbacks, 15
123. Let $x = $ number of students signed up.
$\quad\quad 6(x + 4) = 8(x - 2)$
$\quad\quad\quad x = 20$ students
125. Let $x = $ original price.
$\quad\quad 8x + 8(x + .04) = 19.20$
$\quad\quad\quad\quad\quad x = \$1.18$

127. Let x = difference of earnings.
$$\tfrac{6}{5}(100) - \tfrac{6}{5}(90) = x$$
$$\$12 = x$$

129. Let x = Team A's score this year.
$$4(x - 9) = x + 6$$
$$x = 14$$
team A, 5; team B, 20

131. Let x = amount of the estate.
$$x - .50x - .40x = 10,000$$
$$x = \$100,000$$

133. Let x = suggested cost of player.
$$x - .1x - 5 = 211$$
$$x = \$240$$

4-5 Working with a Formula: pages 168–172

1. $l = 10f$ **3.** $s = c + p$ **5.** $F = \tfrac{9}{5}C + 32$

7. $R = \dfrac{D}{T}$ **9.** $w = \dfrac{T}{p}$ **11.** $n = \dfrac{lwh}{1.25}$

13. $c = 100m$ **15.** $i = 36y$ **17.** $o = 16p$

19. $n = rt$ **21.** $n = 2t + s$ **23.** $c = x + 6y$

25. $c = a + 8b$ **27. a.** 45   **b.** 30

29. a. 68°   **b.** 95°   **c.** 32°   **d.** 23°   **e.** −4°

31. a. 420   **b.** 264   **c.** 11.2   **d.** −72

33. 100   **35.** 6   **37.** 20   **39.** 81

41. 80   **43.** 35

45. a. $h = 4$ **b.** $b = 14$ **c.** $c = 3.5$

47. $V = 40$ liters   **49.** $d = 150$ meters

51. $x = \dfrac{b}{5}$ **53.** $y = \dfrac{s}{r}$ **55.** $y = \dfrac{5}{c}$

57. $x = r - 5$ **59.** $y = d - c$ **61.** $y = 9 - d$

63. $x = r + 2$ **65.** $x = d + c$ **67.** $x = 9$

69. $x = \dfrac{c + 5}{b}$ **71.** $y = \dfrac{t - s}{r}$ **73.** $x = 1$

75. $x = 6a$ **77.** $x = 2$ **79.** $s = \dfrac{P}{4}$

81. $\pi = \dfrac{C}{D}$ **83.** $r = \dfrac{C}{2\pi}$ **85.** $N = \dfrac{360}{C}$

87. $H = \dfrac{3A}{B}$ **89.** $g = S - c$ **91.** $g = \dfrac{2S}{t^2}$

93. $l = \dfrac{P - 2w}{2}$ **95.** $a = \dfrac{2S}{n} - l$

97. $b = \dfrac{2A}{h} - c$ **99.** $R = \dfrac{E}{I} - r$

101. $b = P - 2a$ **103.** $a = \dfrac{P - b - c}{2}$

105. a. $h = \dfrac{2A}{b}$ **b.** $h = 6$

107. a. $L = \dfrac{P - 2W}{2}$ **b.** $L = 19$

109. a. $r = \dfrac{A - P}{tP}$ **b.** $r = .1$

111. $A = 2b^2$

4-6 Solving an Inequality: pages 177–181

1. Multiplication property.
3. Addition property.
5. Subtraction property.
7. Subtraction property.
9. $>$ **11.** $>$
13. Subtraction property.
15. Multiplication property.
17. Addition property.
19. Division property.
21. Given.
   Addition property.
   Division property.
23. Given.
   Subtraction property.
   Division property.
25. Given.
   Subtraction property.
   Division property.
27. a. $\{-4, -3, -2, -1, 0\}$ **b.** $\{0, 1, 2, 3, 4\}$

29. $y > 2\tfrac{1}{2}$

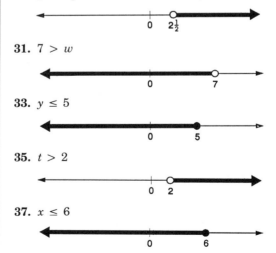

31. $7 > w$

33. $y \le 5$

35. $t > 2$

37. $x \le 6$

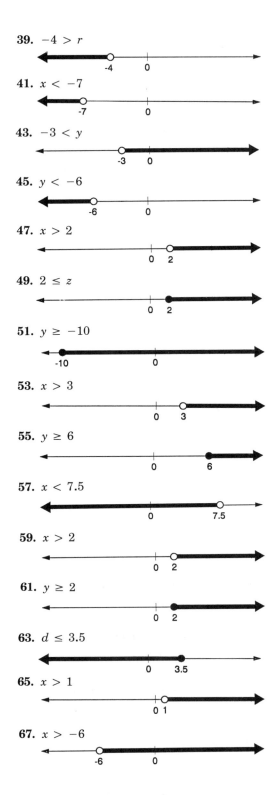

39. $-4 > r$

41. $x < -7$

43. $-3 < y$

45. $y < -6$

47. $x > 2$

49. $2 \leq z$

51. $y \geq -10$

53. $x > 3$

55. $y \geq 6$

57. $x < 7.5$

59. $x > 2$

61. $y \geq 2$

63. $d \leq 3.5$

65. $x > 1$

67. $x > -6$

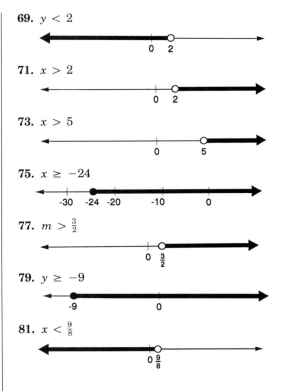

69. $y < 2$

71. $x > 2$

73. $x > 5$

75. $x \geq -24$

77. $m > \frac{3}{2}$

79. $y \geq -9$

81. $x < \frac{9}{8}$

83. a. $5 > 0$
    b. The variable has dropped out.
    c. {all real numbers}
    d. The inequality is true for all replacement
      values from the real numbers.

85. Let $x =$ the number.
$$2x + 6 < 48$$
$$x < 21$$

87. Let $x =$ weight now.
$$x - 7 > 136$$
$$x > 143$$

89. Let $x =$ the cost of the paper.
$$x + \tfrac{1}{2}x \leq 360$$
$$x \leq 240$$

91. Let $x =$ the distance Gladys drives.
$$39 + .35\,(x - 40) < 60$$
$$x < 100$$

93. Let $x =$ the number of times Brian rents the video.
$$2.5x < 29.95$$
$$x < 11.98$$
Brian can rent the video 11 times.

95. a. Let x = the number of couples.
$$5x + 200 < 500$$
$$x < 60$$
Gem Studios will charge less than
Photo Plus for fewer than 60 couples.

b. $5x + 200 > 500$
$$x > 60$$
Photo Plus will charge less for more than
60 couples.

c.

	Gem	Photo Plus
(1)	$500 + 50x + 200 + 5x \le 70x$	$500 + 50x + 500 \le 70x$
	$46\frac{2}{3} \le x$	$50 \le x$

At least 47 couples must attend.

(2)	$500 + 50x + 200 + 5x \le 60x$	$500 + 50x + 500 \le 60x$
	$140 \le x$	$100 \le x$

At least 100 couples must attend.

(3)	$500 + 50x + 200 + 5x \le 50x$	$500 + 50x + 500 \le 50x$
	$x \le -140$	$1{,}000 \le 0$

The cost cannot be as low as $50 per couple.

97. Let x = the width of the envelope.
$$2(x + 1\tfrac{1}{2}x) = 35$$
$$x = 7$$
Yes. The slot is 8 in. wide and the letter is
7 in. wide.

99. $0 < m < 45$

101. Let m = the amount of the purchases.
$$10 < m + 5.92 \le 30$$
$$4.08 < m \le 24.08$$

103. a. Let m = the time in months.
$$10{,}000 - 65m$$
b. For $m \ge 24$, $10{,}000 - 65m \le \$8{,}440$
c. $10{,}000 - 65m < 5{,}000$
$$m > 76\tfrac{12}{13}$$
77 months is the least.

105. a. Let h = the number of hours of labor.
$$48h + 232$$
b. $400 \le 48h + 232 \le 520$
$$3\tfrac{1}{2} \le h \le 6$$

4-7 Absolute Value in an Open Sentence:
pages 184–186

1. $\{-14, 14\}$ **3.** $\{-12, 12\}$ **5.** $\{-1, 1\}$
7. $\{1, 11\}$ **9.** $\{-5, 3\}$ **11.** $\{-10, -8\}$
13. $\{-7\frac{4}{5}, 5\}$ **15.** $\{-20, 20\}$ **17.** $\{-48, 48\}$
19. $\{-2, 2\}$
21. $y + 4 = 10$ *or* $y + 4 = -10$
$$y = 6 \qquad y = -14$$
23. $z - 3 = -24$ *or* $z - 3 = 24$
$$z = -21 \qquad z = 27$$

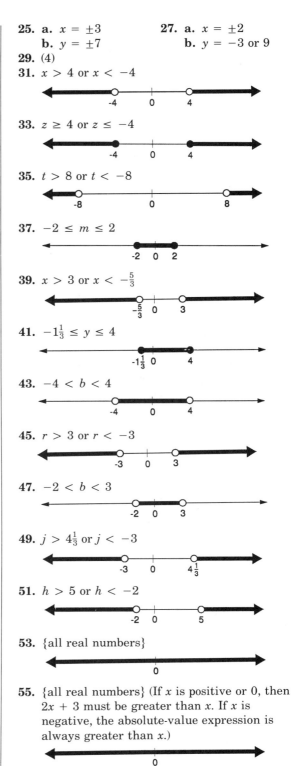

25. a. $x = \pm 3$ **27. a.** $x = \pm 2$
b. $y = \pm 7$ **b.** $y = -3$ or 9
29. (4)
31. $x > 4$ or $x < -4$

33. $z \ge 4$ or $z \le -4$

35. $t > 8$ or $t < -8$

37. $-2 \le m \le 2$

39. $x > 3$ or $x < -\frac{5}{3}$

41. $-1\frac{1}{3} \le y \le 4$

43. $-4 < b < 4$

45. $r > 3$ or $r < -3$

47. $-2 < b < 3$

49. $j > 4\frac{1}{3}$ or $j < -3$

51. $h > 5$ or $h < -2$

53. {all real numbers}

55. {all real numbers} (If x is positive or 0, then
$2x + 3$ must be greater than x. If x is
negative, the absolute-value expression is
always greater than x.)

57. $\{z \mid z \le 2 \text{ or } z \ge 3\}$ (The inequality is true for 0 or negative values of z. Solve for $z > 0$: $5z - 12 \ge z$ gives $z \ge 3$ and $5z - 12 \le -z$ gives $z \le 2$.)

59. Let $x =$ the number.
$$3|x| + 1 > 7$$
$$|x| > 2$$
$$x > 2 \quad \text{or} \quad x < -2$$

61. a. Let $m =$ the number of the milepost.
$$|m - 131|$$
  b. $|m - 131| < 7$
$$-7 < m - 131 < 7$$
$$124 < m < 138$$
He must be between mileposts 124 and 138.

63. a. $|n - 2,384|$
  b. $|n - 2,384| \le 15$
$$-15 \le n - 2,384 \le 15$$
$$2,369 \le n \le 2,399$$

65. a. (1) $|w - 34.5| < .3$
      (2) $|h - 68| < .3$
  b. $34.2 < w < 34.8$
$67.7 < h < 68.3$; yes
  c. no; the width must be greater than 34.2.

67. a. $|w - 6,417| = 2$
  b. $|x - 6,417| \le 10$; $x \in \{\text{odd integers}\}$
  c. $|y - 6,417| \le 12$; $y \in \{\text{odd integers}\}$

Chapter Review Exercises: *pages 188–190*

1. Division property.
3. Addition property.
5. Division property.
7. Multiplication property.
9. Given.
  Subtraction property.
  Division property.
11. a. yes
  b. yes; Sarah is the same age as Sarah.
13. $m = 9$ **15.** $w = \frac{15}{7}$ **17.** $f = 12$
19. $r = 4$ **21. a.** $x = -6$
  $2r - 1 = 7$ **b.** $y = 2$
23. Let $x =$ the remaining distance.
$$x + 37 = 100$$
$$x = 63 \text{ km}$$
25. Let $x =$ the smaller number.
$$2x + 21 + x = 84$$
$$x = 21$$
$$2x + 21 = 63$$

27. $a = 6$ **29.** $t = 10$ **31.** $h = 5$ **33.** $x = 6$
35. a. $x = -\frac{2}{3}$ **b.** $y = -\frac{9}{2}$
37. Let $x =$ the number.
$$3x - 19 = x + 5$$
$$x = 12$$
39. Let $x =$ the average noise level.
$$2x + 40 = 5x - 100$$
$$x = 80 \text{ decibels}$$
41. $d = 90 - m$ **43.** $I = 30$ **45.** $R = 27$
47. $y = 8$ **49.** $y = 2a$ **51.** $h = \dfrac{V}{lw}$
53. Addition property. **55.** $x > y$
57. Division property.
59. Given.
  Subtraction property.
  Division property.
61. $w \le 2.6$

63. $n > -\frac{11}{12}$

65. Let $x =$ the number.
$$3x - 13 \le 14$$
$$x \le 9$$

67. $|3x| = 27$
$$3x = 27 \text{ or } 3x = -27$$
$$x = 9 \qquad x = -9$$

69. $4|x - 3| = 32$
$$|x - 3| = 8$$
$$x - 3 = 8 \text{ or } x - 3 = -8$$
$$x = 11 \qquad x = -5$$

71. a. $x = 11$ or -5
  b. $y = 10$ or -12

73. $-4 < r < 5$

Problems for Pleasure: *page 191*

1. Let $x =$ the first number and $y =$ the second number. Since "their sum is 4 more than their difference," $x + y = (x - y) + 4$. Solving for y, $y = 2$. Note that this is true for any value of x.

　　To find a value for x, use the statement "two numbers have a difference that is twice their sum." Then $x - y = 2(x + y)$. Since $y = 2$, this becomes $x - 2 = 2(x + 2)$. Then $x = -6$. The numbers are -6 and 2. Note that this solution only works if -6 is used as the first number in each subtraction, which is assumed when you "let x = the first number."

3. All are true.

Calculator Challenge: *pages 191–192*

1. **a.** 121   **b.** 70 years old   **c.** 35, 110
   d. -2 **e.** 8

3. **a.**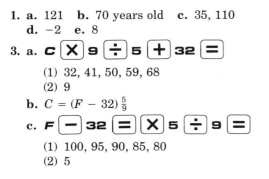
   (1) 32, 41, 50, 59, 68
   (2) 9
   b. $C = (F - 32)\frac{5}{9}$
   c.
   (1) 100, 95, 90, 85, 80
   (2) 5

College Test Preparation: *pages 192–193*

1. (A)   **3.** (B)   **5.** (B)   **7.** (A)   **9.** (A)   **11.** (C)
13. (A)   **15.** (A)   **17.** (A)   **19.** (D)   **21.** (B)
23. (C)   **25.** (C)

Spiral Review Exercises: *pages 193–194*

1. $-\frac{1}{2}$ **3.** $\begin{bmatrix} 8 & -1 \\ -4 & 2 \end{bmatrix}$ **5.** $\frac{1}{2}$ **7.** $\frac{mr}{s}$ cents

9. a. $\{B, R, N\}$
   b. $\{B, A, R, N, U, I, G\}$

11. $\boxed{8}$ = 135

13.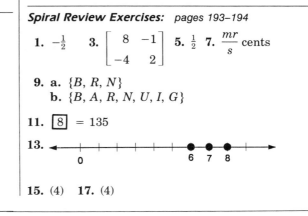

15. (4)   **17.** (4)

CHAPTER 5. MORE APPLICATIONS OF EQUATIONS AND INEQUALITIES

5-1 Consecutive-Integer Problems

Preparing to Solve Consecutive-Integer Problems:
pages 199–200

1. **a.** 15, 16, 17, 18
   b. 0, 1, 2, 3
   c. $-10, -9, -8, -7$
   d. $y, y + 1, y + 2, y + 3$
   e. $2y, 2y + 1, 2y + 2, 2y + 3$
   f. $3y - 2, 3y - 1, 3y, 3y + 1$

3. **a.** 13, 15, 17, 19
   b. $-15, -13, -11, -9$
   c. $-3, -1, 1, 3$
   d. $z, z + 2, z + 4, z + 6$
   e. $2z + 1, 2z + 3, 2z + 5, 2z + 7$
   f. $2z - 1, 2z + 1, 2z + 3, 2z + 5$

5. **a.** **(1)** odd **(2)** odd **(3)** even **(4)** even **(5)** odd
   (6) even
   b. **(1)** even **(2)** even **(3)** even **(4)** odd
   (5) even **(6)** odd

7. **a.** even   **b.** odd   **c.** even

Solving Consecutive-Integer Problems:
pages 201–202

1. **a.** **(1)** Let n = first integer.
   $$n + n + 1 + n + 2 = 99$$
   $$n = 32$$
   32, 33, 34
   (2) $n + n + 1 + n + 2 = -12$
   $$n = -5$$
   $-5, -4, -3$

   b. **(1)** Let n = first integer.
   $$n + n + 1 > 31$$
   $$n > 15$$
   16, 17
   (2) $n + n + 1 > -5$
   $$n > -3$$
   $-2, -1$

   c. **(1)** Let n = first even integer.
   $$n + n + 2 + n + 4 = 48$$
   $$n = 14$$
   14, 16, 18
   (2) $n + n + 2 + n + 4 = -60$
   $$n = -22$$
   $-22, -20, -18$

---

d. (1) Let n = first even integer.

$$n + n + 2 + n + 4 + n + 6 < 60$$
$$n < 12$$
$$10, 12, 14, 16$$

(2) $n + n + 2 + n + 4 + n + 6 < -8$
$$n < -5$$
$$-6, -4, -2, 0$$

e. (1) Let n = first odd integer.

$$n + n + 2 + n + 4 \leq 27$$
$$n \leq 7$$
$$7, 9, 11$$

(2) $n + n + 2 + n + 4 \leq -6$
$$n \leq -4$$
$$-5, -3, -1$$

f. Let x = first integer.

$$x + 1 + x + 3 = 132$$
$$x = 64$$
$$64, 65, 66, 67$$

g. Let n = first integer.

$$n + n + 1 + n + 2 < 0$$
$$n < -1$$
$$-2, -1, 0$$

h. Let x = first even integer.

$$x + x + 2 + x + 4 > 49$$
$$x > 14\tfrac{1}{3}$$
$$16, 18, 20$$

3. Let x = first even integer.

$$2(x + 2 + x + 4) \geq x + 33$$
$$x \geq 7 \qquad 8, 10, 12$$

5. Let x = first integer.

$$x + 2(x + 1) - (x + 2) < 25$$
$$x < 12\tfrac{1}{2} \quad 12, 13, 14$$

7. No. The sum of 3 consecutive odd integers is divisible by 3.

9. Let h = 1st player's hits.

$$h + h + 2 + h + 4 = 246$$
$$h = 80 \qquad 80, 82, 84$$

11. -36 or -34

13. Let x = amount to first son.

$$x + x + 2 + x + 4 \leq 100$$
$$x \leq 31\tfrac{1}{3}$$

a. $30, $32, $34   **b.** $4

15. Let x = length of the shortest side of $\triangle ABC$.

$$x + x + 2 + x + 4 = 24$$
$$x = 6$$

The sides of $\triangle ABC$ are 6, 8, and 10.

$$RS \geq 5$$
$$3RS \geq 15$$

The minimum perimeter is 15 cm.

5-2 Money-Value Problems

Preparing to Solve Money-Value Problems:
page 203

1. a. 16 cents   **b.** p cents   **c.** $2p$ cents
3. a. 30 cents   **b.** $10y$ cents   **c.** $(10y + 20)$ cents
5. a. 400 cents   **b.** 1,500 cents   **c.** $100D$ cents

Solving Money-Value Problems: *pages 205–206*

1. Let x = the number of dimes.
$$10x + 25(4x) = 220$$
$$x = 2 \qquad \text{2 dimes, 8 quarters}$$

3. Let x = the number of nickels.
$$5x + 25(15 - x) \leq 140$$
$$x \geq 11\tfrac{3}{4}$$
at least 12 nickels

5. Let x = the number of children's tickets.
$$2x + 4(x + 200) \geq 1,700$$
$$x \geq 150$$
150 children's tickets, 350 adult's tickets

7. Let x = the number of dimes.
$$10x + 25(2x + 1) > 500$$
$$x > 7\tfrac{11}{12}$$
8 dimes, 17 quarters

9. Let x = the number of nickels.
$$5x + 10(45 - x) = 350$$
$$x = 20$$
20 nickels, 25 dimes

11. Let x = the number of 35¢ stamps.
$$35x + 20(70 - x) = 1,700$$
$$x = 20$$
35¢ stamps, 20; 20¢ stamps, 50

13. Let x = the number of cans of Balanced Meal.
$$60x + 70(x + 8) \leq 5,000$$
$$x \leq 34\tfrac{2}{13}$$
42 cans of All-Beef

15. Let x = the number of Specials.
$$300x + 440(2x) + 40(336 - 3x) = 98,240$$
$$x = 80$$
80 special combination, 160 jumbo,
96 individual

17. Let x = the number of 7-day cruises.
$$2,350x + 1,650(24 - x) = 42,400$$
$$x = 4$$
Two couples bought 7-day package.

19. Let x = the number of 20¢ stamps.
$$20x + 7(100 - x) = 1,000$$
$$13x = 300$$
No. 300 is not divisible by 13.

5-3 Motion Problems

Preparing to Solve Motion Problems:
pages 207–209

1. a. 400 km   **b.** 280 km
c. $80x$ km   **d.** $(160x + 80)$ km
e. $(800 - 80x)$ km

3.
8x miles | 10x miles

5.
28x miles | 28x miles

7. $\frac{200}{40} = 5$ hours; Nadine's driving time
$\frac{250}{60} = 4\frac{1}{6}$ hours; Cyril's driving time
Cyril arrived first.
Cyril arrived 50 minutes earlier.

9. $3(55) = 165$ miles; Lynette's distance from Culver City
$2(85) = 170$ miles; Gerri's distance from Culver City.
They were 5 miles apart at 3:00 P.M.

11. $\frac{1.1}{3} = \frac{11}{30}$ hr = 22 minutes to walk
$\frac{1.1}{33} = \frac{11}{330} = \frac{1}{30}$ hr = 2 minutes by bus
It takes 20 minutes longer to walk.

Solving Motion Problems: pages 211–213

1. Let x = the number of hours.
$40x + 30x = 350$
$x = 5$ hours

3. Let x = the number of hours traveled.
$35x + 40x = 300$
$x = 4$ hours
$8 + 4 = 12$ At 12:00 noon

5. Let x = the number of hours.
$67x + 53x = 800$
$x = 6\frac{2}{3}$ hours

7. Let x = the number of hours traveled.
$44x + 36x = 180$
$x = 2\frac{1}{4}$
first truck, $44(2\frac{1}{4}) = 99$ miles
second truck, $36(2\frac{1}{4}) = 81$ miles

9. Let x = the speed of the northbound train.
$x + x + 20 \leq 100$
$x \leq 40$
Northbound train travels at not more than 40 mph.

11. Let x = the rate of the bus.
$3x + 4(x + 15) = 375$
$x = 45$
bus, 45 mph; train, 60 mph.

13. Let x = the number of hours before noon.
$40x + 25(8 - x) = 275$
$x = 5$
$12 - 5 = 7$
7 A.M. to 3 P.M.

15. Let x = the time Ricky traveled.
$55x - 45(x - 1) \geq 75$
$x \geq 3$
$7 + 3 = 10$ At 10 A.M.

17. Let x = the time traveled by the express train.
$50t = 30(t + 2)$
$t = 3$ hours

19. Let x = the first rate.
$2x + x - 14 \geq 118$
$x \geq 44$
44 mph

21. Let t = the jet plane's time.
$500t = 300(t + 2)$
$t = 3$
$3(500) = 1{,}500$ miles

23. Let x = the rate of the car.
$500 < 5x + 5(x + 10) < 600$
$45 < x < 55$
The rate of the train is between 55 and 65 mph.

25. Let t = time going.
$600t = 400(\frac{5}{2} - t)$
$t = 1$
She can fly $1(600) = 600$ km.

5-4 Lever and Pulley Problems: pages 215–216

1. Let x = Lillian's weight.
$5x = 4(80)$
$x = 64$ pounds

3. Let x = speed of larger pulley.
$28x = 14(140)$
$x = 70$ rpm

5. Let s = speed of larger pulley.
$$12s = 8(1{,}452)$$
$$s = 968 \text{ rpm}$$

7. Let x = the weight of the rock.
$$\tfrac{1}{2}x = 4\tfrac{1}{2}(180)$$
$$x = 1{,}620 \text{ lb.}$$

9. Let x = the distance Ann is from the fulcrum.
$$40x = 60(10 - x)$$
$$x = 6$$
Ann, 6 ft.; Martha, 4

11. Let d = diameter of second pulley.
$$675d = 5(540)$$
$$d = 4 \text{ feet}$$

13. Let s = speed of second pulley
$$3s = 5(18)$$
$$s = 30 \text{ rpm}$$

15. Let x = the distance of the object from the fulcrum.
$$600x = 200(6 - x)$$
$$x = 1\tfrac{1}{2} \text{ feet from the object}$$

17. Let x = the weight of the block.
$$\tfrac{1}{2}x = 5\tfrac{1}{2}(200)$$
$$x = 2{,}200 \text{ pounds}$$

5-5 Angle Problems

Preparing to Solve Angle Problems: page 218

1. a. straight angle
   b. obtuse angle
   c. acute angle
3. acute angle, right angle, obtuse angle, straight angle
5. a. $0° < x < 30°$ **b.** $18° < x < 36°$
7. a. $30°$ **b.** $90°$ **c.** $120°$
   d. $180°$ **e.** $165°$ **f.** $15°$
9. a. $60°$ **b.** $(90 - z)°$ **c.** $z°$
11. $180°$

Solving Angle Problems: pages 220–221

1. $5x = 3x + 10$
   $x = 5$
   m$\angle RTM = 25°$

3. Let a = the measure of the angle.
   $12a = 360$
   $a = 30°$

5. Let x = the measure of the first angle.
$$x + \tfrac{1}{2}x = 180$$
$$x = 120$$
$120°$ and $60°$

7. Let x = the measure of the first angle.
$$5x + x = 180$$
$$x = 30$$
$30°$

9. Let x = the measure of the first angle.
$$1\tfrac{1}{5}x = 90$$
$$x = 75$$
$15°$

11. Let x = the measure of the first angle.
$$x + x + 10 = 180$$
$$x = 85$$
$95°$

13. Let x = the measure of the first angle.
$$x + x + 2 = 90$$
$$x = 44$$
$44°$ and $46°$

15. Let x = the measure of the first angle.
$$x + 2x - 36 = 180$$
$$x = 72$$
$72°$ and $108°$

17. Let x = the measure of the first angle.
$$\tfrac{2}{3}(x) = \tfrac{1}{2}(90 - x - 20)$$
$$x = 30$$
$30°$ and $60°$

19. Let a = the measure of $\angle A$.
$$a + a + 10a = 180$$
$$a = 15°$$

21. Let a = measure of $\angle BAC$.
Then $a - 2$ = measure of $\angle DAE$.
And $2a + 5$ = measure of $\angle CAD$.
$$a + a - 2 + 2a + 5 = 35$$
$$a = 8$$
m$\angle BAC = 8°$

5-6 Triangle Problems

Preparing to Solve Triangle Problems: page 223

1. $80°$ **3.** $90°$ **5.** $60°$
7. a. $90°$ **b.** complementary   **9.** right
11. a. $70°$ **b.** $45°$ **c.** $40°$
13. The sum of the measures of the angles of a quadrilateral is $360°$.
15. (3)   **17.** (3)

Solving Triangle Problems: *pages 225–226*

1. m $\angle S = 40°$, m $\angle R = 20°$, m $\angle T = 120°$

3. 18°, 72°, 90° **5.** 12°, 84°, 84°

7. 20°, 80°, 80° **9.** 30°, 75°, 75°

11. Let x = leg of isosceles triangle.

$x + x > 15$

$\quad x > 7\frac{1}{2}$

8 is the least integral value.

13. a. $x + 2x - 8 > 2x + 1$

$\qquad\qquad x > 9 \qquad x = 10$ feet

b. If $x = 10$, $2x - 8 + x + 2x + 1 = 43$ feet

$\qquad\qquad\qquad 2x + 1 + 8 + 14 = 43$ feet

15. Let x = the measure of the vertex angle.

$x + 4x + 4x = 180$

$\qquad\quad x = 20$

80°

17. Let x = the measure of the angle at the mainland.

Then $(3x + 12)$ = the measure of the angle at the first island.

And $2x$ = the measure of the angle at the second.

$x + (3x + 12) + 2x = 180$

$\qquad\qquad\quad x = 28$

28°, 56°, 96°

5-7 Perimeter and Area Problems

Perimeter: *pages 228–232*

1. a. (1) 26 ft. **b. (1)** 18 cm **c. (1)** 16 ft.

$\quad$ **(2)** 10 m $\quad\quad$ **(2)** 14.4 in. $\quad$ **(2)** 12 in.

$\quad$ **(3)** $19\frac{3}{4}$ ft. $\quad$ **(3)** 28 ft. $\quad\quad$ **(3)** 32.5 in.

$\quad$ **(4)** 12 ft.

3. a. 70 in. **b.** 84 in. **c.** 112 in.

5. a. doubled **b.** tripled

7. Let x = the width.

$2x + 2(5 + x) = 66$

$\qquad\qquad x = 14$

14 cm by 19 cm

9. Let x = the length of the third side.

$x + 4x + 4x = 144$

$\qquad\quad x = 16$

16 cm, 64 cm, 64 cm

11. Let x = the original width.

$2(x - 1) + 2(2x + 4) = 198$

$\qquad\qquad\qquad x = 32$

32 in. by 64 in.

13. Let x = original width.

$2x + 2(x + 4) + 8 = 2(2x) + 2(x + 4 - 2)$

$\qquad\qquad\qquad x = 6$

6 feet by 10 feet

15. Let x = the length of the base.

$2x + 2(12x - 5) = x + 2(4 + 8x)$

$\qquad\qquad\qquad x = 2$

2 feet

17. a. (3) **b.** $10.77

19. Let x = the distance between viewers.

$x + 3x + 3x = 35$

$\qquad\qquad x = 5$

15 feet

21. Divide the maze into 8 vertical borders and 7 vertical paths. The solid borders each contain 22 plants. Subtract 2 from 22 for each break in a border. The 8 borders contain 22, 14, 14, 14, 16, 14, 14, and 22 plants respectively. Add 2 for every barrier across the 7 paths. The paths have 2, 4, 5, 4, 5, 5, and 3 barriers respectively. 186 plants are needed.

23. Let x = the distance between islands.

$x + 2x + (2x + 80) = 1,000$

$\qquad\qquad\qquad x = 184$

368 nautical miles, 184 nautical miles, and 448 nautical miles

25. Let x = the distance from Chicago to St. Louis.

$x + (3x + 37) + (3x + 37 - 119) = 1,978$

$\qquad\qquad\qquad\qquad x = 289$

289 miles

Area: *pages 232–235*

1. a. 80 m^2 **b.** 25.5 sq. yd.

$\quad$ **c.** 51 sq. ft. **d.** 480 sq. in.

3. a. 81 sq. yd. **b.** 625 cm^2

$\quad$ **c.** 1,024 sq. ft. **d.** 6.25 m^2

5. a. 70 sq. in. **b.** 99 sq. in.

$\quad$ **c.** 10 sq. in. **d.** 54 sq. in.

7. a. 112 sq. ft. **b.** 176 m^2

$\quad$ **c.** 30 sq. in. **d.** 32 cm^2

9. 1,728 sq. in. **11.** 416 sq. in.

13. 6.16 cm^2 or 1.96π cm^2

15. rectangle: $3 \times 5 = 15$

semicircle: $\frac{1}{2}\pi r^2 = \frac{1}{2}(3.14)(1.5)^2 = 3.5325$

Area $= 15 + 3.5325 = 18.5$ sq. ft.

17. a. 108 tiles **b.** 192 tiles

19. No, 6 small squares can fill a 36 cm square.
7 small squares can fill a 49 cm square.

21. a. Let x = width of framed painting.
$$38x = 448 + 30(x - 8)$$
$$x = 26 \text{ cm}$$
b. $30(26 - 8) = 540 \text{ cm}^2$

23. a. (1) 40 ft.
(2) 30 ft.
(3) 26 ft.
(4) 24 ft.
b. 8

5-8 Surface Area and Volume Problems

Solid Figures: *pages 237–238*

1. a. There are 6 faces, all rectangles.
b. There are 5 faces, a square and 4 triangles.
c. There are 7 faces, 2 pentagons and 5 rectangles.

3. Answers will vary.

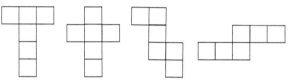

5.

Number of sides on base	3	4	5	6	n
a. Vertices	6	8	10	12	$2n$
b. Faces	5	6	7	8	$n + 2$
c. Edges	9	12	15	18	$3n$

d. Answers will vary.
$$v = e - f + 2$$
or $e = f + v - 2$
or $f + v - e = 2$

Surface Area: *pages 238–240*

1. a. 126 sq. ft.   **b.** 318 sq. in.
c. 43.28 sq. ft.   **d.** $91\frac{3}{4}$ sq. in.

3. a. 2,464 m   **b.** 154 sq. ft.
c. 1,256 m^2 **d.** 5,024 sq. in.
e. 98.56 cm^2

5. quadrupled   **7.** quadrupled

9. a. 154 cm^2 **b.** 660 cm^2 **c.** 968 cm^2

11. $9 \times 16 \times 16 = 2{,}304$ sq. in.

13. a. $2(2 \cdot 2) + 2(2 \cdot h) + 2(2 \cdot h) = 20$
$$h = \frac{3}{2} \text{ ft.}$$
b. $\frac{16}{9}$ sq. yd.; $2(18) = \$36$

Volume: *pages 240–244*

1. a. 140 cu. ft.   **b.** 280 cu. in.
c. 214.2 cu. in.   **d.** 105 cu. in.

3. a. $53\frac{19}{81}$ cm^3 **b.** 113,040 m^3 **c.** $179\frac{2}{3}$ mm^3

5. a. 486 cm^3 **b.** 120 cm^3
c. 270 cm^3 **d.** 420 cm^3

7. a. $V = \underbrace{e \times e}_{\substack{\text{Area of} \\ \text{base, } B}} \times \underbrace{e}_{h} = Bh$

b. $V = \underbrace{\pi r^2}_{\substack{\text{Area of} \\ \text{base, } B}} h = Bh$

9. a. $149\frac{1}{3}$ cu. in.   **b.** 168 cu. in.   **c.** 112 cu. in.

11. Volume of sphere $= \frac{4}{3}\pi r^3$
Volume of cone $= \frac{1}{3}\pi r^2(2r) = \frac{2}{3}\pi r^3$
The volume of the sphere is twice the volume of the cone.

13. Volume of pyramid
$$= \frac{1}{3}s \cdot s \cdot h = \frac{1}{3} \cdot 36 \cdot h = 12h$$
Volume of cone $= \frac{1}{3}\pi 3^2 h = 3\pi h = 9.42h$
The volume of the pyramid is greater.

15. 14 feet   **17.** $19\frac{1}{4}$ cu. in.

19. a. yellow: $\frac{440}{7}$ cu. in.
green: $\frac{396}{7}$ cu. in.
The yellow one.

b. $6\frac{2}{7}$ cu. in.

21. a. 1,152 cu. in.   **b.** 24
23. a. 1,392 sq. in.   **b.** 3,456 cu. in.
25. a. $2(25 + 15)20 + 25 \cdot 15 = 1{,}975$ cm^2
b. $25 \cdot 15 \cdot 20 \cdot 6 = 45{,}000$ g

27. a. $16 \times 16 \times 5 = 1{,}280$ sq. in. or .99 sq. yd.
b. \$12
c. $(16 - 1)^3 = 3{,}375$ cu. in.

29. $\frac{1}{2}(10 \cdot 6)x = 420$
$$x = 14 \text{ ft.}$$

---

31. a. $8(6h) = 240$
$h = 5$ ft.

b. $\dfrac{45 \cdot 8 \cdot 10}{8 \cdot 6 \cdot 5} = \dfrac{45}{6} \cdot \dfrac{8}{8} \cdot \dfrac{10}{5} \approx 7 \cdot 1 \cdot 2$
(need whole numbers)
14 crates

33. a. $\pi r^2(3r) = 81\pi$ **b.** $r = 3$ in.

5-9 A Roundup of Problems: pages 245–246

1. Let x = the measure of the smallest angle.
$x + 4x + 90 = 180$
$x = 18°$

3. Let x = the smaller angle.
$x + x + 68 = 180$
$x = 56$
$56°$ and $124°$

5. Let x = the bus fare.
$10(x + x + 30) = 3,800$
$x = 175$
bus, \$1.75; train, \$2.05

7. Let x = the last digit.
$x + (x + 2) + (x + 4) = 15$
$x = 3$
753

9. Let x = the time Troy walks.
$4x = 12(1 - x)$
$x = \frac{3}{4}$
45 minutes

11. Let x = the number of ounces of chocolates.
$42x + 30(87 - x) < 3,300$
$x < 57\frac{1}{2}$
57 ounces

13. Let x = the rate traveled toward Sacramento.
$7x + 6(x + 4) = 778$
$x = 58$
first rate, 58 mph; second rate, 62 mph

15. Let x = the number of dimes.
$25 - x - (x + 2) + 10(x) + 5(x + 2) = 137$
$x = 8$
8 dimes, 10 nickels, and 7 pennies

17. Let x = Jack's distance from the fulcrum.
$120x = 90(7 - x)$
$x = 3$
Jack sits 3 ft. from the fulcrum, and Jill 4 ft.

19. 000,000; 000,001; 000,002

Chapter Review Exercises: pages 248–251

1. Let x = the first integer.
$x + (x + 1) + (x + 2) = 84$
$x = 27$
27, 28, 29

3. Let x = the smallest integer.
$3(x + 4) = x - 2$
$x = -7$
$-7, -5,$ and -3

5. Let x = the number of hamburgers.
$1.25(x + 31) + 1.50x > 1,000$
$x > 349\frac{2}{7}$
350 hamburgers

7. Let x = the time in hours.
$50x + 55x = 420$
$x = 4$
200 miles

9. Let x = the rate in good weather.
$3x + 2(x - 75) = 1,600$
$x = 350$
1st rate, 350 mph; 2nd rate, 275 mph

11. Let x = time going out.
$4x = 2(6 - x)$
$x = 2$
8 miles distance

13. Let x = the daughter's weight.
$7x = 3(140)$
$x = 60$
60 pounds

15. Let d = the diameter of the first pulley.
$270d = 9(360)$
$d = 12$ in.

17. $150°$

19. Let x = the measure of the first angle.
$x + 15 = 2(180 - x)$
$x = 115$
$115°$ and $65°$

21. $5x - 18 = 2x$ **23.** $75°$
$x = 6$
$m \angle AEC = 12°$

25. Let $15x$ = the measure of the first angle.
$15x + 15(x + 1) + 15(x + 2) = 180$
$x = 3$
$45°, 60°, 75°$

27. Let x = the measure of a base angle.
$x + x + 4(x + x) = 180$
$x = 18$
$144°$

29. $(4 \times 15) + (13 \times 5) + (5 \times 10) = 175$ cm²
or $(5 \times 22) + (4 \times 10) + (5 \times 5) = 175$ cm²
31. Let x = Sarah's distance from either speaker.
$$x + x + (x + 2) = 32$$
$$x = 10$$
10 feet
33. Let x = the length of a side of the square.
$$3(x + 8) - 4x = 20$$
$$x = 4$$
square, 4 inches; triangle, 12 inches
35. 288 sq. feet (Guess and check) **37.** 48 cu. ft.
39. $\frac{1}{3}(3)^2\pi(10) = 30\pi$ or 94.25 cu. in.

Problems for Pleasure: *page 251*

1.
$$
\begin{aligned}
1\ Q &= .25 \\
2\ D &= .20 \\
2\ N &= .10 \\
\underline{45\ P} &= \underline{.45} \\
50 \text{ coins} &= 1.00
\end{aligned}
$$

3. Let x = the shortest side.
$x + y$ = the middle side.
$x + 5$ = the longest side.
Possible values for y are 4, 3, 2, 1
If $y = 4$; $x + x + 4 + x + 5 = 47$
$$x = 12\tfrac{2}{3}$$
not a whole number
If $y = 3$; $x + x + 3 + x + 5 = 47$
$$x = 13$$
The lengths are 13, 16, 18

Calculator Challenge: *page 252*

1. a. $15 \times 25 = 375$, $8 \times 15 = 120$,
$375 + 120 = 495$
15 25¢ stamps, 8 15¢ stamps
b. $15 \times 3 = 45$, $26 \times 1.50 = 39$,
$45 + 39 = 84$
15 \$3 tickets, 26 \$1.50 tickets
c. $-83, -81, -79$
d. \$6.60 **e.** 366

---

3. a.

W	L	H	Area of Front and Back	Area of Sides	Area of Top and Bottom	Total Area	Cost per Box (nearest cent)
1	1	48	2	96	96	194	.25
1	2	24	4	96	48	148	.19
1	3	16	6	96	32	134	.17
1	4	12	8	96	24	128	.16
1	6	8	12	96	16	124	.16
1	8	6	16	96	12	124	.16
1	12	4	24	96	8	128	.16
1	16	3	32	96	6	134	.17
1	24	2	48	96	4	148	.19
1	48	1	96	96	2	194	.25
2	1	24	4	48	96	148	.19
2	2	12	8	48	48	104	.13
2	3	8	12	48	32	92	.12
2	4	6	16	48	24	88	.11
2	6	4	24	48	16	88	.11
2	8	3	32	48	12	92	.12
2	12	2	48	48	8	104	.13
2	24	1	96	48	4	148	.19
3	1	16	6	32	96	134	.17
3	2	8	12	32	48	92	.12
3	4	4	24	32	24	80	.10

$3 \times 4 \times 4$ is least expensive package.

b. $2 \times 4 \times 6$ 11¢

College Test Preparation: *pages 253–254*

1. (C) **3.** (B) **5.** (C) **7.** (D) **9.** (C)
11. (D) **13.** (D) **15.** (A)

Spiral Review Exercises: *pages 254–255*

1. $a + (-a) = 0$ **3.** $24y - 3 = 3(8y - 1)$
5. ps jars **7.** $c = 2$ **9.** $.41\overline{6}$ **11.** 18
13. a. $\frac{16}{17}$ **b.** $\frac{4}{7}$ **15.** (1)
17. area, 10; perimeter, 16 **19.** 2 $(n = 852)$

CHAPTER 6. POLYNOMIALS

6-1 Adding and Subtracting Polynomials:
pages 262–265

1. **a.** $2x^2 - 3x + 5$
   b. degree 2
   c. 3 terms
   d. (1) 14 (2) 19 (3) 5.32 (4) 33.52
3. **a.** $a^4 - a^3 + 2a^2 - 3a$
   b. degree 4
   c. 4 terms
   d. (1) 63 (2) 38 (3) 2.78 (4) 150.06
5. $13a - 3b$ **7.** $-6ab$ **9.** $-2y + 2z$
11. $-2x^2y^2$ **13.** $-3x^2 - 15x - 21$
15. $-2x^2 + x + 8$ **17.** $1.2 + .4z + .8z^2$
19. -4 **21.** $11x - 2$ **23.** $3a + 3b + 4d$
25. $-4x + 10$ **27.** $x^3 - 4x^2 - 3x$
29. $4a + 4c$ **31.** $7x^2 - 11x + 4$
33. $-9a^2 + 2ab + 4b^2$
35. $-11b^3 + b^2 - 2b - 8$
37. **a.** $19x + 4$ **b.** $13x + 10y$
39. **a.** $6x + 2$ **b.** $10 - 2x$
41. $6a + 3b$ **43.** $-3d + 14e$ **45.** $-r$
47. $-3rs$ **49.** $-2x^2 - 4x + 7$ **51.** $4a^2 + 3ab$
53. $3x^2 - 5x - 6$ **55.** $c + 6c^2$ **57.** $-7x - 8$
59. $-7x + 14$ **61.** $7c + 9d$ **63.** $-8x - 1$
65. 0 **67.** $-4x^2 - 12x + 12$ **69.** $24 - 6x$
71. $x^2 + 10x - 4$ **73.** $-7a^2 + 18a + 7$
75. $21m$ **77.** $-x^2 + 9x - 19$ **79.** $3x + 8$
81. $3x + 2y$ **83.** $-4x^2 + 8$ **85.** $5c^2 - 9c + 12$
87. **a.** $10xy - z$ **b.** $x^2 - 2y$
89. $10xy$ dollars **91.** $(12a + 12b)$ miles
93. $19c^2d$ dollars **95.** $(13c^2 - 9cd + 2d^2)$ miles
97. $(3x - 8y + 4z)$ meters
99. $(5a - 6b - 5c)$ degrees
101. $(4y^2 + 7xy)$ dollars

6-2 Properties of Exponents: pages 268–270

1. a^5 **3.** c^6 **5.** r^{11} **7.** 3^5 **9.** 4^4 **11.** x^6
13. y^6 **15.** z^{14} **17.** a^4b^8 **19.** $2^6 \cdot 3^6$ **21.** x^{3a}
23. c^{r+1} **25.** $(3y)^{a+b}$ **27.** a^5 **29.** 1 **31.** d
33. $\dfrac{1}{m^3}$ **35.** 8 **37.** $\dfrac{1}{3^2}$ **39.** $\dfrac{1}{5^4}$ **41.** $\dfrac{1}{r^{4c}}$ **43.** 1
45. **a.** (3) and (4)
    b. multiplication and division
47. $\dfrac{1}{5^2}$ **49.** 1 **51.** $\dfrac{a^6}{b^4}$ **53.** $\dfrac{q^6}{r^6}$ **55.** $\dfrac{a^3}{x^4}$ **57.** a^3
59. p^8 **61.** b^4 **63.** y^2 **65.** n^3 **67.** $\dfrac{1}{w^4x^3}$

69. y^{10} **71.** $\dfrac{x^{13}}{y^{11}}$ **73.** b^2 **75.** 1 **77.** 1

79. $\begin{bmatrix} ab^2 + 2ab^2 & -ab^2 - 2ab^2 \\ b^3 + a^2 & -b^3 - a^2 \end{bmatrix}$

$= \begin{bmatrix} 3ab^2 & -3ab^2 \\ a^2 + b^3 & -a^2 - b^3 \end{bmatrix}$

81. b^{13} seats **83.** y^{10} dollars **85.** x^2 feet
87. 7^{3x} dollars **89.** $(a + b)^3$ people

6-3 Scientific Notation: pages 273–275

1. 1.2×10^4 **3.** 7.92×10^{-4} **5.** 9.64×10^7
7. 1.8×10^5 **9.** 9.09×10^2 **11.** 50,000
13. 38,230 **15.** 6,020 **17.** .00008
19. 2.2×10^4 **21.** 6.0×10^7 **23.** 2.14×10^7
25. 4.0×10^{-6} **27.** 1.47×10^3
29. 3.8896×10^6 **31.** 2.4549×10^{-10}
33. 7.332×10^{-4} **35.** 4.361×10^{-1}
37. 1.406×10^{-5} **39.** 8.75×10^{-4}
41. 5.0×10^4 **43.** 5.2×10^7
45. 1.04×10^{-1} **47.** 4.375×10^{-2}
49. 3.32×10^4 **51.** $1.6\bar{5} \times 10^{-27}$ kg
53. 2.6×10^{26} **55.** 2.16×10^6 **57.** 4.5×10^8

6-4 Multiplying Polynomials: pages 279–284

1. $12x^5$ **3.** $3t^3$ **5.** $-25d^6$ **7.** $-12a$
9. $-42xyz$ **11.** $18x^7$ **13.** $-140y^5$ **15.** s^9
17. $-72z^8$ **19.** $-6r^9s^{10}$ **21.** $-16r^5s^2$
23. $-72c^2d^2$ **25.** $-4x^3$ **27.** $-2x^4y^3$
29. $9a^2b^2c^2$ **31.** $16x^2y^2$ **33.** $-40x^2y^6$
35. **a.** $8x^3$ cu. in.
    b. $160\pi x^3$ cu. in.
37. $32xy^5$ dollars **39.** $48k^3$ **41.** $-6x + 27y$
43. $-75c^3 + 20c^4$ **45.** $-a^2b + ab^2$
47. $-45c^4d^4 + 20c^6d^3$
49. $-10a^4b^2 + a^5$ **51.** $9 - 6x + 3x^2$
53. $-10r^4s^2 + 15r^3s^3 - 20r^3s^4$
55. **a.** $(20x - 10)$ dollars **b.** $(2hx - h)$ dollars.
    c. $(8x^2 - 4x)$ dollars **d.** $(6x^3 - 3x^2)$ dollars
57. $(15x - 12y)$ feet **59.** $(\frac{3}{4}n^3 + \frac{1}{2}n^2 - 2n)$ cents
61. $a^2 + 5a + 6$ **63.** $c^2 + 2c - 48$
65. $m^2 - 4m - 21$ **67.** $30 + 11y + y^2$
69. $72 + 6r - r^2$ **71.** $2y^2 + y - 6$

73. $6a^2 + 29a + 9$ **75.** $15y^2 - 11y + 2$
77. $x^2 + 2xy + y^2$ **79.** $a^2 + 5ab + 6b^2$
81. $6z^2 + 7wz - 20w^2$ **83.** $x^2 - 25$
85. $a^2 - 81$ **87.** $144 - b^2$ **89.** $r^2 - s^2$
91. $25c^2 - 16$ **93.** $x^4 - 64$ **95.** $a^2 - \frac{1}{4}$
97. $r^2 - .25$ **99.** $a^2b^2 - 64$
101. $39 \times 41 = (40 - 1)(40 + 1)$
$$= 1600 - 1 = 1,599$$
103. $66 \times 74 = (70 - 4)(70 + 4)$
$$= 4,900 - 16 = 4,884$$
105. $55 \times 65 = (60 - 5)(60 + 5)$
$$= 3,600 - 25 = 3,575$$
107. $94 \times 106 = (100 - 6)(100 + 6)$
$$= 10,000 - 36 = 9,964$$
109. $y^3 - 4y^2 + 10y - 12$
111. $15 - 16d - d^2 + 2d^3$
113. $d^3 + 27$ **115.** $12y^3 - 21y^2 - 29y + 20$
117. $8a^3 - 26a^2b + 11ab^2 + 10b^3$
119. $6x^3 + 13x^2 - 19x - 12$
121. $x^3 + 12x^2 + 48x + 64$
123. $x^3 - 3x^2y + 3xy^2 - y^3$
125. $8x^3 + 26x^2 - x - 12$
127. $9b^3 - 9b^2c - 4bc^2 + 4c^3$
129. $25c^4d^6 - 49e^{10}$ **131.** $x^4 - 81$
133. $m^8 - n^8$ **135.** $12x^2 - 6$
137. $14x + 2$ **139.** $14y + 7$
141. $r^2 - 2rs - r + s$
143. $6p^3 + 3p^2q - 4pq + 10p + 5q - 2q^2$
145. $2b^3 - b^2 + b - 6$
147. Let x = width.
$$(x + 12)(x - 1) = x(x + 8)$$
$$x = 4$$
4 cm = width, 12 cm = length.
149. Let x = a side of old square.
$$(x + 3)^2 = 39 + x^2$$
$$x = 5 \text{ ft.}$$
151. Let x = side of a square.
$$x^2 - (x - 3)(x + 4) = 3$$
$$x = 9 \text{ in.}$$
153. Let x = width.
$$(x - 1)(2x + 7) = x(2x + 2) + 20$$
$$x = 9 \text{ cm}, 2x + 2 = 20 \text{ cm}$$
155. $(s - 1)^3 + (3s^2 - 5) = s^3$
$$s = 2 \text{ mm}$$
157. Let x = height.
$$x(x - 3)(x - 4) = 16 + x(x - 2)(x - 5)$$
$$x = 8 \text{ in.}, x - 2 = 6 \text{ in.},$$
$$x - 5 = 3 \text{ in.}$$

159. Let x = width.
$$2x(x + 2)(x - 2) = w(x - 2)x$$
$$2x + 4 = w$$
The new width is 4 in. more than twice the original.

6-5 Dividing Polynomials: *pages 288–290*

1. $9x$ **3.** $\dfrac{-2y^2}{x}$ **5.** $\dfrac{1}{4d^3}$ **7.** $4a^2$

9. -3 **11.** $-10y$ **13.** $3ab^2$ **15.** a

17. $\dfrac{2c^6}{3d^2}$ **19.** $-\dfrac{12ax}{5}$ **21.** $4(a + b)^3$

23. $\dfrac{1}{3(a - 2x)^2}$ **25.** $7xy^2$ dollars

27. a. $w = 8xy$ **b.** $h = \dfrac{32a^2b}{3}$

29. $m + n$ **31.** $-2c^2 + 3d^2$ **33.** $-y + 5$
35. $3r^3 + 2r$ **37.** $3b - 4a$ **39.** $3a + 6b$
41. $-4y^2 - 2y + 1$ **43.** $x^2y^2 - xy + 1$
45. $(15r + 3)$ in. **47.** $b + 2$ **49.** $x - 18$
51. $x + 11$ **53.** $t - 2$ **55.** $3x - 8$
57. $2x + 3y$ **59.** $15x - 2y$ **61.** $8x - 5$
63. $a - b$ **65.** $3a + 3b$ **67.** $x + 8$
69. $2m - 7n$ **71.** $x^2 - 3x + 2$
73. $2b^2 - 4b - 3$ **75.** $2y^2 + 3y - 1$

77. $4x^2 - 6x + 9$ **79.** $x - 7 + \dfrac{-7}{x - 2}$

81. $x + 2 + \dfrac{-10}{3x + 3}$ **83.** $c^2 - 6c - 18 + \dfrac{-27}{c - 2}$

85. $a^2 - 4a + 2 + \dfrac{-4}{2a + 3}$

87. $5x - 4y + \dfrac{13y^2}{2x + y}$

89. $a + 9b + \dfrac{26b^2}{a - 6b}$ **91.** $x - 5 + \dfrac{50}{x + 5}$

93. $x - 2$ **95.** $2x + 3$ **97.** $2a^2 - 5b^2$
99. $x - 9$

101.
$$
\begin{array}{r}
x^2 + 4 \\
x - 2 \overline{)\, x^3 - 2x^2 + 4x - 6} \\
\underline{x^3 - 2x^2 } \\
4x - 6 \\
\underline{4x - 8} \\
2 \text{ Remainder}
\end{array}
$$
No, there is a remainder.

6-6 Factoring: pages 294–296

1. a. prime   **b.** not prime
   c. prime   **d.** prime
   e. prime

3. a. $35 = 5 \cdot 7$ **b.** $18 = 2 \cdot 3^2$
   c. $144 = 2^4 \cdot 3^2$ **d.** $400 = 2^4 \cdot 5^2$
   e. $590 = 2 \cdot 5 \cdot 59$

5. a. 26: 1, 2, 13, 26
   b. 50: 1, 2, 5, 10, 25, 50
   c. 36: 1, 2, 3, 4, 6, 9, 12, 18, 36
   d. 88: 1, 2, 4, 8, 11, 22, 44, 88
   e. 100: 1, 2, 4, 5, 10, 20, 25, 50, 100
   f. 242: 1, 2, 11, 22, 121, 242

7. a. 5   **b.** 4   **c.** 7   **d.** 6
   e. 25   **f.** 36   **g.** 8   **h.** 1

9. $2(a + b)$ **11.** $b(x + y)$ **13.** $4(x + 2y)$
15. $9(2c - 3d)$ **17.** $6(x - 3)$ **19.** $7(y - 1)$
21. $6(1 - 3c)$ **23.** $x(2x + 5)$ **25.** $x(32 + x)$
27. $a(x - 5b)$ **29.** $5x(2 - 3x^2)$ **31.** $p(1 + rt)$
33. $\frac{1}{2}h(b + c)$ **35.** $\pi r(r + l)$ **37.** $4(x^2 + y^2)$
39. $5(x^2 + 1)$ **41.** $5xy(2 - 3xy)$
43. $2(x^2 + 4x + 2)$ **45.** $a(y - 4w - 12)$
47. $\frac{1}{4}m(a + b + c)$ **49.** $5xyz(3x^2y^2z^2 - 1)$
51. $14m^2n^3(2m^2 - 5n)$ **53.** $(c - d)(t + s)$
55. $(r - s)(y + z)$ **57.** $(a + b)(v - w)$

59. $\frac{1}{2} \times 153 + \frac{1}{2} \times 47$
     $= \frac{1}{2}(153 + 47) = \frac{1}{2}(200) = 100$

61. $\frac{1}{2} \times 7 \times 6.3 + \frac{1}{2} \times 7 \times 1.7$
     $= \frac{1}{2} \times 7(6.3 + 1.7) = \frac{7}{2}(8) = 28$

63. $(2c + d)(5f - 2g)$ **65.** $(a^2 + 5)(4a - 3)$
67. $(5w + 1)(a - 1)$ **69.** $(a^2 + 3)(a - 2)$
71. $(2x + 7)(6 + a)$ **73.** $(4 - 3t)(m + 5)$
75. $(p - m)(n - t)$ **77.** $(5 - 7c)(3d - 2f)$
79. $(w^4 - 2)(w^3 - 3)$ **81.** $(4a - 3)(b - 5)$
83. $(3x - 1)(a - 4)$ **85.** $(6a + 5b)(ab - 3)$
87. $(b - 4ax^2)(5a - 3x)$ **89.** $(2d - ab)(3y - x)$
91. $(2w^2 + y)(w^2 - p)$

93. a. $y^2 - 8^2$
   b. 7 is not a perfect square, but a prime
     number.
   c. $(5n)^2 - (4m)^2$
   d. $c^2 - (.3d)^2$
   e. $p^2 - (\frac{3}{5}q)^2$
   f. $(4a^2)^2 - (5b^3)^2$
   g. 9 is an odd exponent and $9y^9$ is not a
     square.

h. $m^2 - 3^2$
i. (-16) is a negative number and is not a
   square.

95. $(c + 10)(c - 10)$ **97.** $(3 + x)(3 - x)$
99. $(11 + m)(11 - m)$ **101.** $(5m + n)(5m - n)$
103. $(r^2 + 3)(r^2 - 3)$ **105.** $(5 + s^2)(5 - s^2)$
107. $(8e + 3f)(8e - 3f)$ **109.** $(w + \frac{1}{8})(w - \frac{1}{8})$
111. $(\frac{1}{9} + t)(\frac{1}{9} - t)$ **113.** $\left(\dfrac{2}{5} + \dfrac{7d}{9}\right)\left(\dfrac{2}{5} - \dfrac{7d}{9}\right)$
115. $(x + .8)(x - .8)$ **117.** $(.2 + 7r)(.2 - 7r)$
119. $(9n + .1)(9n - .1)$
121. $(8ab + cd)(8ab - cd)$
123. $(9mn + 7xy)(9mn - 7xy)$
125. $(5x^3 + 11y^5)(5x^3 - 11y^5)$
127. $[(a + b) + c][(a + b) - c]$
129. $[5 + (m + n)][5 - (m + n)]$

6-7 More About Factoring: pages 302–303

1. $(a + 2)(a + 1)$ **3.** $(c + 5)(c + 1)$
5. $(y + 9)(y + 1)$ **7.** $(m + 4)(m + 1)$
9. $(x + 6)(x + 3)$ **11.** $(c + 5)(c + 3)$
13. $(x + 3)(x + 8)$ **15.** $(z + 8)(z + 5)$
17. $(b + 3)(b + 10)$ **19.** $(x + 1)(x + 1)$
21. $(z + 5)(z + 5)$ **23.** $(x - 11)(x - 1)$
25. $(x - 3)(x - 2)$ **27.** $(x - 10)(x - 1)$
29. $(z - 7)(z - 3)$ **31.** $(a - 8)(a - 1)$
33. $(5 - y)(3 - y)$
35. $(y - 9)(y - 4)$ **37.** $(t - 12)(t - 6)$
39. $(x - 12)(x - 4)$ **41.** $(y - 8)(y - 8)$
43. $(x - 2)(x + 1)$ **45.** $(x - 7)(x + 1)$
47. $(y + 9)(y - 1)$ **49.** $(a - 5)(a + 2)$
51. $(c - 5)(c + 3)$ **53.** $(r + 7)(r - 3)$
55. $(x - 9)(x + 2)$ **57.** $(z + 12)(z - 3)$
59. $(x + 8)(x - 5)$ **61.** $(x - 12)(x + 6)$
63. $(x - 10)(x + 8)$ **65.** $(5x + 2)(x - 3)$
67. $(7x + 10)(x - 1)$ **69.** $(5a + 4)(a - 1)$
71. $(7c - 3)(c - 2)$ **73.** $(2x + 1)(x + 2)$
75. $(2x + 1)(x + 5)$ **77.** $(2x + 3)(x + 4)$
79. $(2y - 1)(y - 1)$ **81.** $(7 - 3y)(2 - y)$
83. $(3x + 1)(x - 2)$ **85.** $(2x - 3)(x + 2)$
87. $(3x + 4)(x - 3)$ **89.** $(3x - 2)(2x + 3)$
91. $(5a - 2)(2a - 1)$ **93.** $(9y + 2)(2y - 3)$
95. $(c + 5d)(c - d)$ **97.** $(3a - b)(a - 2b)$
99. $(4x + 3y)(x - 2y)$ **101.** $6(x - y)(x + y)$
103. $a(x - y)(x + y)$ **105.** $s(t - 1)(t + 1)$
107. $2(x - 4)(x + 4)$ **109.** $2(3m - 2)(3m + 2)$

111. $7(3c - 1)(3c + 1)$　　**113.** $y(y - 5)(y + 5)$
115. $a(2a - b)(2a + b)$　　**117.** $d(3b - 1)(3b + 1)$
119. $(x - 1)(x + 1)(x^2 + 1)$
121. $\pi(R - r)(R + r)$
123. $4(5x - 3y)(5x + 3y)$　　**125.** $4(r - 4)(r + 3)$
127. $2a(x - 3)(x + 2)$
129. $(y - 3)(y + 3)(y - 2)(y + 2)$
131. $4(x + 3)(a - 2)$
133. $3(x - 1)(x + 1)(1 - a)(1 + a)$

6-8　Solving a Quadratic Equation by Factoring:
pages 307–309

1. $\{1, 2\}$　**3.** $\{1, 4\}$　**5.** $\{2, 6\}$　**7.** $\{-5, -1\}$
9. $\{-9, -1\}$　**11.** $\{-5, -3\}$　**13.** $\{5, -1\}$
15. $\{7, -5\}$　**17.** $\{-9, 8\}$　**19.** $\{9, -8\}$
21. $\{5, -2\}$　**23.** $\{9, -9\}$　**25.** $\{8, -8\}$
27. $\{2, -2\}$　**29.** $\{0, 5\}$　**31.** $\{0, -3\}$
33. $\{0, -8\}$　**35.** $\{\frac{1}{3}, 3\}$　**37.** $\{\frac{2}{3}, 2\}$
39. $\{-\frac{1}{5}, -2\}$　**41.** $\{6, -2\}$　**43.** $\{5, 3\}$
45. $\{-\frac{3}{2}, -2\}$　**47.** $\{5, -5\}$　**49.** $\{0, 6\}$
51. $\{0, -4\}$　**53.** $\{-5, 3\}$　**55.** $\{4, 5\}$
57. $\{\frac{1}{2}, 3\}$　**59.** $\{4, 7\}$　**61.** $x = b, x = -b$
63. $x = 0, x = -a$

65. $x = 0, x = \dfrac{s}{r}$　**67.** $x = -7a, x = 3a$

69. $x = 9, x = -8$

71. Let x = the smaller number.
$x(x + 5) = 36$
The smaller number is 4; the larger, 9.

73. Let x = the smaller number.
$x(x + 2) = 99$
The integers are 9 and 11, or -11 and -9.

75. Let x = the first integer.
$x^2 + (x + 2)^2 = 100$
The integers are 6 and 8.

77. Let x = the length of a side of a square.
$(x + 2)(x - 2) = 32$
6 in. is the length of the side of the square.

79. Let x = the width of a rectangle.
$(x - 1)(3x + 3) = 72$
The width is 5 inches, and the length 15 inches.

81. Let x = the number of meters change.
$(x + 6)(x + 4) = 2(6 \cdot 4)$
Each dimension should be increased 2 m.

83. Let x = the width of the frame.
$(6 + 2x)(12 + 2x) = 3(6)(12)$
The width is 3 inches.

85. Let x = the rate.
$(x + 4)(x + 6) = 2x^2$
The rate is 12 miles per hour.

87.　$h = k + rt - 16t^2$
$96 = 96 + 80x - 16x^2$
The ball will return is 5 seconds.

Chapter Review Exercises:　pages 311–313

1. a. $5m + 3n$　**b.** $7a - b + c$　**3.** $-4b$
5. $10a - 7b$　**7.** $9a + 2$　**9.** a^7　**11.** 3^6

13. $\dfrac{1}{x^4}$　**15.** 4^4　**17.** $\dfrac{b^3}{a^5}$　**19.** $\dfrac{1}{x^4}$

21. 4.375×10^6　**23.** 2.89×10^4
25. a. 1.6×10^6　**b.** $1,600,000$
27. a. 4.0×10^4　**b.** $40,000$
29. $30a^{11}b^6$　**31.** $-12a + 8$
33. $-28x^6y^6 - 36x^3y^7$　**35.** $a^2 + 4a - 21$
37. $a^2 - b^2$　**39.** $25a^2 - 121$　**41.** $x^4 - 16$
43. $12y^3 - 40y^2 + 19y + 15$　**45.** $-4t + 14$

47. Let x = the side of the square.
$x^2 - 2 = (x + 3)(x - 2)$
$x = 4$
7m by 2 m

49. $\dfrac{5c^2}{6y^5}$　**51.** $2x + 5$　**53.** $4a^2 - 10a + 3$

55. $w - 6$　**57.** $a + 2 + \dfrac{-2}{2a - 1}$

59. a. $12 = 2 \times 2 \times 3$
b. $100 = 2 \times 2 \times 5 \times 5$
c. $45 = 3 \times 3 \times 5$

61. $8(x - 2)$　**63.** $4a^4b^3(2b^2 - 5a^2)$
65. $(a^2 + 3)(a - 2)$　**67.** $(2a - 11)(2a + 11)$
69. $(cd - h)(cd + h)$　**71.** $(y + 10)(y - 2)$
73. $(m - 6)(m + 5)$　**75.** $(x + 12)(x + 3)$
77. $(3c + 4)(c + 3)$　**79.** $2a(3a - 4)(a + 1)$
81. $2(2 - w)(2 + w)(4 + w^2)$
83. $y = 10, y = -10$　**85.** $a = -1, a = 4$
87. $x = -1, x = \frac{7}{4}$

89. Let x = width of the garden.
$x(x + 4) = 60$
6 yards by 10 yards

1. a.
$$(x + y)(x - y)(x^2 - xy + y^2)(x^2 + xy + y^2) - (x^2 - y^2)(x^4 + x^2y^2 + y^4)$$
$$= (x + y)(x^2 - xy + y^2)(x - y)(x^2 + xy + y^2) - (x^2 - y^2)(x^4 + x^2y^2 + y^4)$$
$$= \qquad (x^3 + y^3) \qquad \qquad (x^3 - y^3) \qquad - \qquad (x^6 - y^6)$$
$$= (x^6 - y^6) - (x^6 - y^6)$$
$$= 0$$

b.
$$(a - x)(b - x)(c - x)(d - x) \ldots (z - x)$$
$$= (a - x)(b - x)(c - x)(d - x) \ldots (x - x)(y - x)(z - x)$$
$$= (a - x)(b - x)(c - x)(d - x) \ldots (0)(y - x)(z - x)$$
$$= 0$$

3. Cut only 1 link, either link 3 or link 5. This gives you pieces of the chain containing 2 links, 1 link, and 4 links. Pay the first day with the single link. On the second day, trade the 2-link piece for the single link. On day three, give the innkeeper the single link. On day four, trade the 4-link piece for the three links the innkeeper already has. On day five, add the single link. On day six, trade the 2-link piece for the single link. On day seven, add the single link.

1. 48   **3.** 13, 31, 43

1. (C)   **3.** (D)   **5.** (A)   **7.** (D)   **9.** (C)   **11.** (D)
13. (A)   **15.** (D)   **17.** (D)   **19.** (B)   **21.** (B)

1. 96   **3.** (3)   **5.** $\dfrac{42jm}{d}$ **7.** $n = 25$

9. $x = 4$ **11.** (4)   **13.** $a = \dfrac{2K}{p}$

15. Let x = the first number.
$$x + 2x - 8 = 124$$
$$x = 44$$
$$2x - 8 = 80$$

17. Let x = the difference between the surface areas.
$$6(3)^2 - x = \tfrac{1}{2}(6)(8)$$
$$x = 30 \text{ sq. in.}$$

19. Let x = the amount saved.
$$26(2)(k) - x = 4k + 5.6k(2) + 2.5k(2) + 25k$$
$$x = \$6.80k$$

21. Let x = Caia's distance from the fulcrum.
$$120x = 90(7 - x)$$
$$x = 3 \text{ ft.}$$

23. Let x = the number of weeks.
$$2(2.39) < .05(2x)$$
$$47.8 < x$$
In 48 weeks.

1. 3   **2.** $5x^2 - 5x + 12$ **3.** 5   **4.** 3   **5.** $n + 3$
6. $x = -2$ **7.** -4 **8.** $\{-7, 3\}$ **9.** (3)   **10.** (1)
11. (3)   **12.** (1)   **13.** (1)   **14.** (1)   **15.** (3)
16. (3)   **17.** (4)   **18.** (2)   **19.** (2)   **20.** (3)
21. 3, 4, and 5
22. a. 1   **b.** 0   **c.** 0   **d.** 1   **e.** 0
23. a. $2x + 3$ **b.** $x - 4$ **c.** $\dfrac{xy}{3}$
    d. $(x + y)^2$ **e.** $x^2 + y^2$
24. a. $\varnothing$ **b.** $\{-2, 2\}$ **c.** $\{-3, -2\}$
    d. $\{-1, 1\}$ **e.** $\{0, 1\}$

7-1 Simplifying a Rational Expression:
pages 325–327

1. $\dfrac{x}{7}$ **3.** $\dfrac{-8}{3x}$, $x = 0$ **5.** $\dfrac{15}{y - 3}$, $y = 3$

7. $\dfrac{x + 4}{2x - 6}$, $x = 3$ **9.** $\dfrac{x - 5}{x^2 - 25}$, $x = 5$, $x = -5$

11. $\frac{3}{4}$ **13.** $\frac{3}{4}$ **15.** $\dfrac{2c}{3d}$ **17.** $\frac{3}{5}$ **19.** $\dfrac{r}{t}$ **21.** $\dfrac{a}{2b}$

23. $\frac{4}{7}$ **25.** $\frac{1}{2}$ **27.** $3y^2$ **29.** $\dfrac{3}{4a}$ **31.** $\dfrac{3a^2}{4y}$ **33.** $\dfrac{8a}{3}$

35. $\dfrac{2x}{9}$ **37.** $\dfrac{1}{9xy}$ **39.** $\dfrac{3a}{b^3}$ dollars **41.** $\dfrac{3}{4mn}$ times

43. $\frac{1}{2}$ **45.** $\frac{2}{3}$ **47.** 1 **49.** $2x$ **51.** $\dfrac{m}{m + 5}$

53. $x + 2$ **55.** $\frac{9}{4}$ **57.** $x + 4$ **59.** $\dfrac{r}{s}$ **61.** $\dfrac{y - 5}{5}$

63. $\dfrac{8 - r}{2}$ **65.** $\dfrac{x + 2}{x - 2}$ **67.** $\dfrac{2(y - 1)}{9}$

69. $\dfrac{5(a + 2)}{a - 2}$ **71.** $-\dfrac{1}{b + 3}$ **73.** $-\dfrac{x + y}{3}$

75. $\dfrac{2(1 - 2y)}{3}$ **77.** $\dfrac{1}{2 - 3m}$ **79.** $\dfrac{x}{3}$ requests

81. $\dfrac{3x(x - 2)}{x - 7}$ riders **83.** $\dfrac{x}{x - 1}$ **85.** $\dfrac{a + 2}{a + 3}$

87. $\dfrac{2(x + 5)}{x + 3}$ **89.** $-\dfrac{x - 12}{x - 3}$ **91.** $\dfrac{3(x - 2)}{x + 2}$

93. $\dfrac{2(3c - 4d)}{2c - 3d}$ **95.** $\dfrac{x + 6}{x - 6}$ applications

97. Wrong. 7 is not a factor of the numerator or the denominator.

99. Wrong. $x + y$ is not a factor of the numerator.

7-2 Multiplying and Dividing Rational
Expressions: pages 330–333

1. $\frac{21}{40}$ **3.** 12 **5.** $10x$ **7.** $5d$ **9.** $\dfrac{5x^2}{9}$ **11.** $8y$

13. $\dfrac{5cr}{4ds}$ **15.** b **17.** $\dfrac{4m}{3}$ **19.** $2m$

21. $\dfrac{6n}{5m}$ illustrations **23.** $\dfrac{1,620}{w}$ dollars

25. $\dfrac{c^3}{d^2}$ teaspoons **27.** $\begin{bmatrix} 2x + 4 & 4x + 4 \\ 4x + 2 & 2x + 8 \end{bmatrix}$

29. $\dfrac{a}{5(a - 3)}$ **31.** $\dfrac{x - 1}{7}$ **33.** $\dfrac{2a}{x - 2}$

35. $\dfrac{2k^4(x + y)}{3(x - y)}$ **37.** $-\dfrac{4b^2(a - 2)}{a + 2}$ **39.** $\dfrac{x + 5}{3y}$

41. $\dfrac{3(2a - 3)}{5(a - 3)}$ **43.** $\dfrac{10}{3(x - 2)}$ **45.** $\dfrac{1}{x(x - 2)}$

47. $\dfrac{x - 4}{10x}$ **49.** $6y$ cents **51.** $\dfrac{8y}{y - 2}$ oranges

53. $\begin{bmatrix} 3r - 3 & 2r - 1 \\ r^2 - 2r + 2 & r \end{bmatrix}$

55. $\dfrac{2(x - 3)(x - 4)}{(x - 2)(x + 3)}$ **57.** 1 **59.** $\dfrac{y + 7}{y - 1}$ **61.** $\dfrac{3}{5x}$

63. $\frac{1}{6}$ **65.** 16 **67.** $\frac{1}{3}$ **69.** $\frac{4}{7}$ **71.** $\dfrac{y^4}{x^2}$ **73.** $\dfrac{s^2}{3}$

75. $4c$ words **77.** $\dfrac{5q^3}{3}$ sacks **79.** $\dfrac{3}{x - 1}$

81. $\dfrac{3x}{y(x + y)}$ **83.** $2y(x + 6)$ **85.** $\dfrac{3(m + 1)}{2}$

87. $y + 2$ **89.** $\dfrac{y + 4}{3y - 5}$ **91.** $\frac{2}{3}$ **93.** $\dfrac{x - 2}{28x}$

95. $2(y - 3)$ **97.** $\dfrac{x + 8}{3}$ dollars **99.** $\dfrac{1}{4(m + 10)}$

101. $\frac{1}{2}$ **103.** $\dfrac{3}{y - 3}$

7-3 Adding and Subtracting Rational
Expressions: pages 338–341

1. $\dfrac{5}{x}$ **3.** $\dfrac{5}{3b}$ **5.** $\dfrac{2a}{3}$ **7.** $\dfrac{1}{3x}$ **9.** $\dfrac{11a}{4x}$ **11.** $\dfrac{7c}{3d}$

13. $\dfrac{4b}{d}$ pounds **15.** $\dfrac{7x}{y}$ **17.** $\dfrac{1}{a - b}$ **19.** 1

21. $\dfrac{1}{y + 2}$ **23.** $\dfrac{7x + 1}{3}$ **25.** $\dfrac{3x - 5}{3}$ **27.** $\dfrac{a - 3}{4a}$

29. $\dfrac{4c + 2}{2c - 3}$ **31.** 1 **33.** $\dfrac{1}{x - 1}$ **35.** $x - y$

37. $\dfrac{2}{a + b}$ **39.** $\dfrac{2}{a - b}$ **41.** $\dfrac{2m - 1}{m + 2}$

43. $\begin{bmatrix} 1 & x - y \\ 2 & \dfrac{1}{x - y} \end{bmatrix}$ **45.** $\frac{39}{4}$ **47.** $\dfrac{9s - 7}{s}$

49. $\dfrac{5d^2 - 7}{5d}$ **51.** $\dfrac{3x + 8}{x + 1}$ **53.** $\dfrac{7b + 7c + 2a}{b + c}$

55. $\dfrac{s^2 - s - 1}{s - 1}$ **57.** $\dfrac{4y - 4}{y - 2}$ **59.** $\dfrac{6x - 8y}{x - y}$

61. $\dfrac{x^2 - 3x - 15}{x - 3}$ **63.** $x - 1$

65. a. $\dfrac{9}{15}; \dfrac{20}{15}$ **b.** $\dfrac{45x}{180}; \dfrac{14x}{180}$

c. $\dfrac{10y}{6c}; \dfrac{7y}{6c}$ **d.** $\dfrac{8}{x^2}; \dfrac{2x}{x^2}$

e. $\dfrac{63d^2}{36c^2d^2}; \dfrac{10c^2}{36c^2d^2}$ **f.** $\dfrac{2a - 12}{4}; \dfrac{2a + 5}{4}$

g. $\dfrac{12c + 4}{72d}; \dfrac{15c - 9}{72d}$ **h.** $\dfrac{90x - 72}{360y}; \dfrac{15x - 35}{360y}$

i. $\dfrac{x - 2}{x^2 - 4}; \dfrac{3x + 6}{x^2 - 4}$ **j.** $\dfrac{6t - 3}{24(t - 1)}; \dfrac{16t + 4}{24(t - 1)}$

k. $\dfrac{4y + 8}{y^2 + 2y}; \dfrac{y^2 - y}{y^2 + 2y}$ **l.** $\dfrac{6x + 1}{x^2 - 9}; \dfrac{-3x - 9}{x^2 - 9}$

m. $\dfrac{-7}{3a - 1}; \dfrac{2}{3a - 1}$

67. $\dfrac{19y}{28}$ **69.** $\dfrac{31x}{20}$ **71.** $\dfrac{1}{8y}$ **73.** $\dfrac{-1}{8x}$ **75.** $\dfrac{3a}{8b}$

77. $\dfrac{-5r}{12s}$ **79.** $\dfrac{1 - 3r}{r^2}$ **81.** $\dfrac{yz + xz + xy}{xyz}$

83. $\dfrac{2c - 3a}{abc}$ **85.** $\dfrac{bx + ay}{a^2b^2}$

87. $\dfrac{2 - 3y + 7y^2}{y^3}$ **89.** $\dfrac{rs + 6}{2r^2t}$ chirps

91. $\begin{bmatrix} \dfrac{x + y}{xy} & \dfrac{xy + x}{y} \\ \dfrac{5y}{4} & \dfrac{x^2 + y^2}{xy} \end{bmatrix}$

93. $\dfrac{3a - 5}{6}$ **95.** $\dfrac{-x + 44}{15}$ **97.** $\dfrac{7a - 9}{15}$

99. $\dfrac{2b}{15}$ **101.** $\dfrac{a - 5b}{12}$ **103.** $\dfrac{y - 4}{12y}$

105. $\dfrac{-8b + 19}{20b}$ **107.** $\dfrac{-5x - 96}{24x}$ **109.** $\dfrac{-5c - 7}{4}$

111. $\dfrac{-12x^2 - 8x + 21y}{12x}$ **113.** $\dfrac{-27x + 226}{60}$

115. $\dfrac{-6y + 23}{12y^2}$ **117.** $\dfrac{17}{2x - 6}$ **119.** $\dfrac{17}{15a - 5}$

121. $\dfrac{5x}{8(x - 1)}$ **123.** $\dfrac{a + 2b}{2(a - 2b)}$ **125.** $\dfrac{11x + 23}{12(2x - 1)}$

127. $\dfrac{3y - 60}{y^2 - 16}$ **129.** $\dfrac{10x - 6}{x^2 - 2x}$ **131.** $\dfrac{a^2 + ab + b^2}{b(a - b)}$

133. $\dfrac{-5y + 26}{y^2 - 16}$ **135.** $\dfrac{y - 4}{y^2 - 4}$ **137.** $\dfrac{3a + 9b}{ab(a - b)}$

139. $\dfrac{x^2 + 3x + 3}{(x + 2)^3}$ **141.** $\dfrac{2r + 19}{(r - 2)(r + 2)(r - 5)}$

143. $\dfrac{-a^2 - 2a - 2}{(a - 5)(a + 3)(a - 2)}$

7-4 Solving Open Sentences that Contain Fractional Coefficients: pages 344–345

1. $x = 21$ **3.** $t = 108$ **5.** $m = 36$ **7.** $x \geq 16$
9. $x \leq 12$ **11.** $y = 6$ **13.** $x < 4$ **15.** $x > \frac{3}{2}$
17. $y = 900$ **19.** $y = 900$ **21.** $c = 20$
23. $u = 2.9$ **25.** $b < 21$ **27.** $d \geq .2$
29. $m = 30$ **31.** $x = 4$ **33.** $m = -27$
35. $y = 48$ **37.** $r = 12$ **39.** $r = 12$ **41.** $s = 24$
43. $x = 2$ **45.** $x = \frac{52}{5}$ **47.** $y = 7$ **49.** $x < 40$
51. $y > 15$ **53.** $y = 3.4$ **55.** $x > 9$ **57.** $c > 6$
59. $y \geq 12$ **61.** $t \geq -40$ **63.** $x > \frac{40}{9}$
65. $x > 12$ **67.** $a = 24$ **69.** $s = -75$ **71.** $y = 1$
73. $x = 18$ **75.** $y = 4$ **77.** $y = 12$ **79.** $d < 1$
81. $m \geq 3$ **83.** $x > 2$ **85.** $r \leq 24$ **87.** $m = 12$
89. $x = 180$ **91.** $x < 395$ **93.** $x = 48$
95. $x = 1,500$ **97.** $x \geq 700$ **99.** $x > 3,000$

7-5 Applying Open Sentences that Contain Fractional Coefficients: pages 347–350

1. The number of students in each class.
3. a. (1) 24.48 **(2)** 1.8 **(3)** 52.5 **(4)** 165
  b. 200 **c.** 128.667 **d.** 36
  e. 80 **f.** 60% **g.** 150%
5. Let $x =$ the number.

  a. $\dfrac{x}{2} + \dfrac{x}{3} = 25$ **b.** $\dfrac{x}{5} - \dfrac{x}{10} = 10$
    $x = 30$ $x = 100$

  c. $\dfrac{x}{2} + 20 = 35$ **d.** $\dfrac{7x}{100} + 2.5 = \dfrac{8x}{100}$
    $x = 30$ $x = 250$

  e. $\dfrac{17x}{100} - 1.4 = \dfrac{12x}{100} + 7.6$
    $x = 180$

  f. $\dfrac{x}{2} + 5 = \dfrac{3x}{5} - 3$ **g.** $\dfrac{3x + 3}{15} = \dfrac{2x - 18}{6}$
    $x = 80$ $x = 24$

  h. $.25x = 32$ **i.** $.15x - 40 = .07x$
    $x = 128$ $x = 500$

7. Let x = the number.
$$x + .08x = 64.8$$
$$x = 60$$

9. Let p = Agnes's weight.
$$\frac{3(110) + p}{4} = 112$$
$$p = 118 \text{ pounds}$$

11. Let d = down payment.
$$d = .15(3,450)$$
$$d = \$517.50$$

13. Let n = first consecutive even integer.
$$\frac{n + n + 2 + n + 4}{3} = 20$$
$$n = 18 \qquad 18, 20, 22$$

15. a. Let m = the marked price of the coat.
$$.20m = 12$$
$$m = \$60$$

   b. After Christmas, for winter clothes
After July 4, for summer clothes.

17. Let s = the total sales.
$$.05s = 281$$
$$s = \$5,620$$

19. Let m = original investment.
$$m - .15m < 3000$$
$$m < \$3,000$$

21. Let s = amount of sales.
$$240 + .03s > 492$$
$$s > 8,400 \quad \text{At least } \$8,401$$

23. Let w = width of storage room.
$$(w + 2)(2w - 3) = \tfrac{2}{3}w(2w + 6)$$
$$w = 6$$
$$2w = 12 \quad \text{Length is 12 feet.}$$

25. a. Let g = good parts.
$$g = .96 (54,650)$$
$$g = 52,464 \text{ good parts}$$

   b. Answers will vary; for example, find the percent defective in a sample, then calculate that percent of the total number.

27. Let x = percent of increase.
$$x = \tfrac{17}{60}$$
$$x = 28\tfrac{1}{3}\%$$

29. Let x = the number of cows.
$$\frac{x}{4} + \frac{x}{3} \geq 28$$
$$x \geq 48 \text{ cows}$$

31. Let x = pounds of nut topping.
$(10 - x)$ = pounds of fudge topping.
$$2.41x + 2.53(10 - x) < 25$$
$$x > 2\tfrac{1}{2}$$
$$10 - x < 7\tfrac{1}{2} \text{ pounds of fudge}$$

33. Let m = amount collected from residential customers.
$$.07m = .375 (250,320,000)$$
$$m = \$1,341,000,000$$

35. a. Let c = reduced price of computer.
$$c = .75 (1,299)$$
$$c = \$974.25$$

   b. No. If two successive discounts were taken, price would be \$987.24, because the 5% discount is taken on a smaller amount.
20% of \$1,299 = \$259.80
$$\begin{array}{r} 1,299.00 \\ -259.00 \\ \hline 1,039.20 \end{array}$$
5% of \$1,039.20 = \$51.96
$$\begin{array}{r} -51.96 \\ \hline \$987.24 \end{array}$$

37. Let r = number of riders in Zone 1.
$$.75r + 1.20(600 - r) = .90(600)$$
$$r = 400$$
$$600 - r = 200 \text{ riders in Zone 2}$$

Percent-Mixture Problems: *pages 352–353*

1. a. 40 lbs. **b.** 4.8 lbs. **c.** $.40x$ lbs.
   d. $40(x - 2)$ lbs.

3. a. 12 ounces   **b. (1)** $(12 + x)$ ounces
                     (2) $(60 + x)$ ounces

5. Let x = pounds of 30% acid solution.
$$.30x + .60(60 - x) = .50(60)$$
$$x = 20$$
20 pounds of 30% acid solution
40 pounds of 60% acid solution

7. Let x = pounds of 75% acid solution.
$$.75 + .30(16) = .55(x + 16)$$
$$x = 20 \text{ pounds}$$

9. Let s = pounds of salt to be added.
$$s + .05(80) = .24(s + 80)$$
$$s = 20 \text{ pounds}$$

11. Let x = ounces of iodine to be added.
$$3 + x = .25(24 + x)$$
$$x = 4 \text{ ounces}$$

13. Let w = pounds of water evaporated.
$$8 = .40(40 - w)$$
$$w = 20 \text{ pounds to be evaporated.}$$

15. Let c = pounds of yellow candy.

$$.10(600) + c = .19(600 + c)$$
$$c = 66\tfrac{2}{3} \text{ pounds}$$

17. Let s = pounds of straw to be added.

$$s = .04(153 + s)$$
$$s = 6\tfrac{3}{8} \text{ pounds}$$

19. Let p = ounces of papaya juice to be added.

$$.25(24) + p = .55(24 + p)$$
$$p = 16 \text{ ounces}$$

Investment Problems: pages 355–356

1. a. \$60 **b.** \$250 **c.** \$.10x **d.** \$.10$(3x)$
e. \$.10$(x + 500)$ **f.** \$.10$(5,000 - x)$

3. a. \$$(500 + x)$ **b.** (1) \$.08x (2) \$.10$(500 + x)$
c. \$$(.18x + 50)$ **d.** $.18x + 50 = 113$

5. Let x = dollars invested at 8%.

$$.08x + .05(2x) = 180$$
$$x = 1,000$$
\$1,000 at 8%; \$2,000 at 5%

7. \$3,400 at 6%; \$1,400 at 5%; \$400 at 7.5%
9. \$4,000 at 6%; \$3,500 at 10%
11. \$4,000 at 8%; \$3,200 at 10%

13. Let x = dollars at 8%.

$$.10(18,000 - x) = .08x + 360$$
$$x = 8,000$$
\$8,000 at 8%; \$10,000 at 10%

15. Let m = the original investment.

$$m + .20m = 60,000$$
$$m = \$50,000 \quad \text{original investment.}$$

17. \$4,600 at 7%; \$6,400 at 12% **19.** \$9,000

7-6 Solving Equations that Contain Rational Expressions: pages 361–363

1. $x = 2$ **3.** $x = \tfrac{1}{2}$ **5.** $x = 3$ **7.** $y = -7$
9. $x = 2$ **11.** $c = \tfrac{1}{2}$ **13.** $x = 3$ **15.** $c = 6$
17. $x = \tfrac{1}{4}$ **19.** $a = 4$ **21.** $x = 3$ **23.** $t = 3$
25. $x = 3$ **27.** $x = 4$ **29.** $z = 1$ **31.** $y = 3$
33. $m = 2$ **35.** $y = 5$ **37.** $\varnothing$ **39.** $y = -\tfrac{1}{2}$
41. $y = rs$ **43.** $y = 5st$ **45.** $x = 2(c + d)$

47. $y = \dfrac{2b}{3}$ **49.** $x = \dfrac{s}{n}$ **51.** $y = \dfrac{c}{b}(a + d)$

53. $x = \dfrac{5}{a}$ **55.** $y = \dfrac{t}{r}$ **57.** $x = 13c$

59. $x = \dfrac{a + b}{c}$ **61.** $x = \dfrac{b}{a}(c + d)$ **63.** $x = 15a$

65. $x = \dfrac{1}{e}$ **67.** $y = \dfrac{c - d}{h}$ **69.** $y = \dfrac{r}{c + d}$

71. $x = \dfrac{b}{a - b}$ **73.** $x = c + d$ **75.** $x = 2c$

77. $y = c - d$ **79.** $x = \dfrac{cd}{c + d}$ **81.** $x = \dfrac{2ab}{a - b}$

83. $x = r - s$ **85.** $y = -\dfrac{(m + n)}{2}$ **87.** $m = \dfrac{Fr}{v^2}$

89. $N = \dfrac{nP}{p}$ **91.** $C = \tfrac{5}{9}(F - 32)$

93. $a = \dfrac{2S - nl}{n}$ **95.** $P = \dfrac{A}{1 + rt}$

97. $W = \dfrac{a}{6n + 1}$ **99.** $q = \dfrac{fp}{p - f}$

101. $r_2 = \dfrac{E - Ir_1}{I}$

7-7 Applying Equations that Contain Rational Expressions

Number Problems: page 365

1. Let n = the number.

$$\frac{n}{2} = \frac{n}{3} + 8$$
$$n = 48$$

3. 25 **5.** $\tfrac{25}{60}$ **7.** $\tfrac{36}{12}$ **9.** $\tfrac{12}{9}$ **11.** 7
13. 6, 31 **15.** $\tfrac{1}{7}$

Motion Problems: page 368

1.

	Rate	Time	Distance
Walk	r	$\dfrac{12}{r}$	12
Bike	$4r$	$\dfrac{12}{4r}$	12

$$\frac{12}{r} + \frac{12}{4r} = 5$$
$$r = 3 \text{ mph.}$$

3. ship, 16 mph; plane, 320 mph
5. James, 30 mph; Kenton, 45 mph
7. cargo plane, 250 mph; jet plane, 600 mph
9. 30 mph **11.** 45 mph

---

1. a. (1) $\frac{1}{2}$ (2) $\frac{x}{2}$ **b.** (1) $\frac{1}{3}$ (2) $\frac{x}{3}$

   c. $\frac{x}{2} + \frac{x}{3}$ **d.** $\frac{x}{2} + \frac{x}{3} = 1$

3. (c) It would take less time working together than for either person working alone
5. 4 hours   **7.** 12 hours   **9.** 12 hours
11. 2 hours   **13.** $1\frac{1}{3}$ hours   **15.** 18 hours
17. 18 min.   **19.** 25 min.
21. Let x = the number of seconds needed by B alone.

$$\frac{45}{120} + \frac{30}{x} = 1$$
$$x = 48 \text{ seconds}$$

Chapter Review Exercises: pages 375–376

1. $x = 0$ **3.** none   **5.** $\frac{5}{9}$ **7.** 4   **9.** $\frac{1}{6}$

11. $x(x + 3)$ **13.** 3   **15.** $\frac{1}{32}$ **17.** $\frac{2(y - 2)}{3}$

19. $\frac{1}{y}$ **21.** 2   **23.** $\frac{a}{4b}$ **25.** $\frac{6a - 11}{45}$

27. $\frac{8x - 4}{(x + 2)^2(x - 2)}$ **29.** $\frac{7a - 3}{a}$ **31.** $\frac{t^2 + t - 1}{t - 2}$

33. $b = 31$ **35.** $y = 24$ **37.** $x = 45$
39. $t = 112\frac{1}{2}$ **41.** $x > -4\frac{1}{2}$ **43.** $b < 32$
45. at least 13 games   **47.** 4 ounces   **49.** $r = 6$

51. $a = 8$ **53.** $x = \frac{17y}{3}$ **55.** $x = ab$

57. $x = \frac{bP}{b - P}$

59. Let t = time if they work together.
$$\frac{t}{20} + \frac{t}{12} = 1$$
$$t = 7\frac{1}{2} \text{ hours}$$

Problems for Pleasure: page 377

1. a. $\frac{a^2 - b^2}{a^3 + b^3} \cdot \frac{a^4 + a^2b^2 + b^4}{a^3 - b^3} = \frac{a^6 - b^6}{a^6 - b^6} = 1$

  b. $2a - 3 - \frac{a^2 - a - 12}{a + 3}$

    $= 2a - 3 - \frac{(a + 3)(a - 4)}{a + 3} = a + 1$

---

  c. $\frac{a^3}{a + b} - \frac{b(b^3 - a^3)}{a^2 - b^2}$

    $= \frac{a^3(a - b) - b(b^3 - a^3)}{a^2 - b^2}$

    $= \frac{a^4 - a^3b - b^4 + a^3b}{a^2 - b^2}$

    $= a^2 + b^2$

  d. $\frac{(a + b)^2}{a + b + c} - \frac{c^2}{a + b + c}$

    $= \frac{(a + b - c)(a + b + c)}{a + b + c} = a + b - c$

3. $\frac{73 + 85 + 73 + 87 + 86 + x}{6}$

  $= \frac{81 + 76 + 82 + x}{4}$

  $\frac{404 + x}{6} = \frac{239 + x}{4}$

    $x = 91$

5. The bicyclists are covering a total distance of one-third of a mile at a combined speed of 20 mph. Since the rate and time for the fly is the same as the rate and time for the cyclists, the distance must be the same; $\frac{1}{3}$ mile.

Calculator Challenge: page 377

1. $\frac{13}{39}$ **3.** 111, 112, 115, 128, 132, 144, 175

College Test Preparation: pages 378–379

1. (A)   **3.** (B)   **5.** (A)   **7.** (B)   **9.** (A)   **11.** (D)
13. (C)   **15.** (A)   **17.** (E)   **19.** (C)   **21.** (D)
23. (A)   **25.** (B)

Spiral Review Exercises: page 380

1. $-11d$ **3.** 36, -49

5. a.

  b.

7. 88 cu. in.   **9.** $(8x - 10y - 9z)$ shares   **11.** 21
13. (2)   **15.** $(x + 2)$ tickets   **17. a.** No.   **b.** -3

CHAPTER 8. APPLYING RATIO AND PROPORTION

8-1 Ratio: pages 390–393

1. a. (1) $\frac{3}{1}$ (2) $\frac{2}{1}$ (3) $\frac{8}{5}$ **b.** (1) $\frac{1}{4}$ (2) 2 (3) $\frac{3}{2}$

3. a. (1) yes (2) no (3) no (4) yes
(5) yes (6) no

b. (1) $\frac{2}{3}, \frac{6}{9}, \frac{50}{75}$ (2) 10:8, 20:16, 50:40

c. (1) 12 (2) 15 (3) 80 (4) 40

In part **d,** answers will vary.

d. (1) $\frac{2}{4}, \frac{3}{6}, \frac{4}{8}, \frac{5}{10}$ (2) 6:2, 9:3, 12:4, 15:5
(3) $\frac{6}{8}, \frac{9}{12}, \frac{12}{16}, \frac{15}{20}$ (4) 4:6, 6:9, 8:12, 10:15

5. a. 2 to 1 **b.** 1 to 2 **c.** 2 to 3 **d.** 1 to 3

7. a. $\frac{8}{1}$ **b.** $\frac{\$.50}{1}$ **c.** $\frac{11}{5}$ **d.** $\frac{537}{10}$

9. a. The giant size, costing 3.9 cents per ounce, is a better buy. The regular size costs 4.3 cents per ounce.
b. Bob **c.** Rosa

11. $71.88 **13. a.** $\frac{m}{k}$ **b.** $\frac{b}{s}$ **c.** $\frac{kp}{z}$

15. $408\frac{1}{3}$ pounds **17.** 48 and 12

19. a. 40°, 60°, and 80° **b.** 24°, 60°, and 96°
c. 40°, 40°, and 100° **d.** 20°, 100°, and 60°

21. 1,200 miles, 1,600 miles, and 2,000 miles
23. 100 students **25.** 35 books
27. 54° and 126° **29.** at least 245 pounds
31. 20 and 75 **33.** 21 points and 27 points
35. $\frac{9}{15}$ **37.** 30 nickels and 10 dimes
39. 132 ft. × 48 ft. **41.** 2.2 minutes
43. 90°, 18°, and 72°

8-2 Proportion: pages 397–399

1. a. $\frac{3}{4} = \frac{30}{40}$ **b.** no **c.** $\frac{5x}{9x} = \frac{10}{18}$

d. $\frac{x}{2x} = \frac{10}{20}$ **e.** no

3. a. 30 **b.** 28 **c.** 5 **d.** 12
5. $x = 18$ **7.** $x = 20$ **9.** $x = 7$ **11.** $x = 36$
13. $x = 16$ **15.** $x = 4$

17. a. $x = \dfrac{bc}{a}$ **b.** $x = \dfrac{2rt}{s}$ **c.** $x = \dfrac{2mr}{s}$

19. $1.60 **21.** 3,000 plants **23.** 13 pounds

25. $2\frac{1}{2}$ cups **27.** 25 ft. **29.** $\dfrac{cn}{m}$ hours

31. $\dfrac{3pt}{2z}$ points **33.** $\dfrac{bd}{s}$ books **35.** $\dfrac{dg}{c}$ dollars

37. 54 **39.** 340 m
41. a. $12.45 **b.** overcharge $3.61
43. 140 knots **45.** 6 teachers **47.** 10,064

8-3 Variation

Direct Variation: pages 402–403

1. p varies directly as s.
$p = 3s$

3. x does not vary directly as x.
The ratio of x to y is different for each pair.
5. $h = \frac{2}{3}S$ h, 10; S, 12
7. $p = ks$ **9.** $r = kl$

11. Let $h =$ the number of hours.
Let $i =$ the income.
$i = kh$

13. no The ratio, $\dfrac{R}{T}$, is not constant.

15. yes
17. a. (1) $y = kx$ y is doubled.
(2) $x = ky$ y is doubled.
(3) $y = kx^2$ y is quadrupled.
(4) $x = ky^2$ y is multiplied by $\sqrt{2}$

b. (1) $y = kx$ x is multiplied by 4
(2) $x = ky$ x is multiplied by 4
(3) $y = kx^2$ x is multiplied by 2
(4) $x = ky^2$ x is multiplied by 16

19. a. $c = \$125$ **b.** $s = \$5,000$
21. $2,940 **23.** 420 mi. **25.** $S = 7$

Inverse Variation: pages 406–410

1. yes **3.** no
$nc = 36$ xy is not a constant.
5. $RT = 144$ R, 12; T, 24
7. $tr = k$ **9.** $fl = k$ **11.** $pi = k$
13. No, the product is not constant.
15. Yes, the product is constant.

17. a. (1) y is halved. **b.** (1) x is quartered.
(2) y is halved. (2) x is quartered.
(3) y is quartered. (3) x is halved.

19. a. $h = 3$ in. **b.** $b = 45$ cm
21. 8% **23.** 30 pounds and 20 pounds
25. $N = 2\frac{1}{2}$ **27.** $lw = 1$ **29.** $xy = -36$

31. $y = x^2$ **33.** $Y = 140$ **35.** $n = \frac{1}{2}$
37. $N = 15$ **39.** $m = 45$ **41.** 150 rpm
43. $d = 240$ **45.** 27 rpm   **47.** 24,000 units
49. $3\frac{2}{5}$ in.   **51.** 190 lb.

Joint Variation: pages 411–413

1. $s = 16\frac{7}{8}$ **3.** $t = 10$ **5.** $m = -5.4$ **7.** $83\frac{1}{3}$ lb.
9. $s = 15$ **11.** $z = 1\frac{23}{25}$ **13.** $z = -128$
15. a. 36 riders
    b. $1.00 would be $288 more profitable.
17. 400 lb.   **19.** $1.50

8-4 Similarity: pages 417–419

1. Yes, two pairs of corresponding angles are equal in measure.
3. $\triangle ABC \sim \triangle SRT$, $\triangle DEF \sim \triangle ZXY$
Two pairs of corresponding angles are equal in measure in each pair of triangles.
5. $RT = 2$ in.   **7.** $DF = 8$, $EF = 10$
9. a. $4\frac{1}{2}$ cm   **b.** 46 cm
11. 20 ft.   **13.** $27\frac{1}{2}$ ft.   **15.** $6\frac{2}{3}$ in. and $8\frac{1}{3}$ in.
17. $1\frac{1}{3}$ cm and $1\frac{2}{3}$ cm   **19.** 30 ft.

8-5 Trigonometric Ratios: pages 424–426

	angle	hypotenuse	leg opposite	leg adjacent
1.	D	$\overline{DE}$	$\overline{EF}$	$\overline{DF}$
	E	$\overline{DE}$	$\overline{DF}$	$\overline{EF}$
3.	X	$\overline{XZ}$	$\overline{YZ}$	$\overline{XY}$
	Z	$\overline{XZ}$	$\overline{XY}$	$\overline{YZ}$

5. $\sin X = \frac{12}{13}$, $\sin Z = \frac{5}{13}$
7. $\cos D = \frac{3}{5}$, $\cos F = \frac{4}{5}$
9. $\cos S = \frac{r}{t}$, $\cos R = \frac{s}{t}$
11. $\tan Q = \frac{10}{24}$, $\tan S = \frac{24}{10}$ **13.** $\sin A = \frac{3}{5}$
15. $\tan M = \frac{15}{8}$ **17.** LM **19.** LP **21.** PM
23. MN **25.** NP **27.** MN **29.** MP **31.** LN
33. $\sin 18° = .3090$ **35.** $\tan 45° = 1.000$
37. $\sin 58° = .8480$ **39.** $\cos 67° = .3907$
41. $\tan 74° = 3.4874$ **43.** $m\angle x = 20°$
45. $m\angle x = 11°$ **47.** $m\angle x = 57°$
49. $m\angle x = 41°$ **51.** $m\angle x = 69°$
53. $m\angle x = 90°$ **55.** $m\angle x = 13°$
57. $m\angle x = 77°$ **59.** $m\angle x = 75°$
61. $m\angle x = 38°$ **63.** $m\angle x = 54°$

65. $m\angle x = 54°$ **67.** $m\angle G = 30°$
69. $m\angle G = 7°$ **71.** $m\angle G = 63°$
73. $m\angle G = 64°$ **75.** False   **77.** False
79. False   **81.** False   **83.** False
85. True for acute angles   **87.** True   **89.** False
91. True, $x°$ and $(90 - x)°$ are the measures of complementary angles.
93. False for all values except $45°$.

$$\tan x° = \frac{1}{\tan (90 - x)°}$$

8-6 Applying the Trigonometric Ratios: pages 433–438

1. $x = 15$ ft.   **3.** $x = 19$ ft.   **5.** $x = 14$ ft.
7. $x = 23$ ft.   **9.** $m\angle x = 45°$ **11.** $m\angle x = 37°$
13. $x = 24$ ft.   **15.** $x = 7$ ft.   **17.** $LN = 23$ units
19. $FH = 24$ **21.** $x = 19$ m   **23.** $x = 18$ ft.
25. $x = 5°$ **27.** $x = 135$ ft.   **29.** $x = 50$ ft.
31. 5,600 ft.   **33.** $x = 34°$ **35.** 650 ft.   **37.** $51°$
39. a. $56°$ **b.** 15 ft.   **41.** $32°$
43. a. 16.4 in.   **b.** 11.5 in.   **c.** 189 sq. in.
45. a. 9.1 in.   **b.** 35.6 in.
47. a. $AC = 12.5 + 5.3 = 17.8$ in.
    b. $17.8 \times 20 = 356$ sq. in.
49. a. 220 m   **b.** 454 m   **c.** 234 m
51. a. 70 ft.   **b.** $68°$
53. a. 234 yd.   **b.** 343 yd.   **55.** 148 ft.

8-7 Probability: pages 443–447

1. $P(1) = .157$ **3.** $P(1) = \frac{4}{23}$
    $P(2) = .159$ $P(2) = \frac{2}{23}$
    $P(3) = .163$ $P(3) = \frac{6}{23}$
    $P(4) = .169$ $P(4) = \frac{7}{23}$
    $P(5) = .179$ $P(5) = \frac{4}{23}$
    $P(6) = .173$
5. a. $\frac{18}{121}$ **b.** $\frac{2}{103}$ **c.** $\frac{98}{121}$
7. 600   **9.** 340   **11.** 28.   **13.** 810
15. a. $\frac{1}{12}$ **b.** $\frac{1}{12}$ **c.** $\frac{1}{36}$ **d.** 0
17. a. $\frac{1}{2}$ **b.** $\frac{2}{9}$ **c.** $\frac{1}{8}$ **d.** 0
19. a. $\frac{1}{4}$ **b.** $\frac{1}{3}$ **c.** $\frac{2}{3}$ **d.** $\frac{1}{3}$ **e.** $\frac{1}{4}$ **f.** $\frac{3}{4}$
21. a. $\frac{3}{20}$ **b.** $\frac{3}{19}$ **c.** $\frac{5}{18}$ **d.** $\frac{3}{17}$ **e.** $\frac{5}{16}$
23. a. $n + 4$ **b.** $\frac{1}{2}n$
    c. $2\frac{1}{2}n + 9 = 24$
          $n = 6$
    d. $\frac{1}{4}$

25. $D = E$, $C = D + E$, $B = C - 40$, $A = 80 + 2B$, $A + B + C + D + E = 360$
Let $x = D$.
$D = E = x$, $C = 2x$, $B = 2x - 40$, $A = 4x$, $4x + 2x - 40 + 2x + 2x = 360$
$x = 40$
a. $\frac{1}{9}$ **b.** $\frac{4}{9}$ **c.** $\frac{7}{9}$ **d.** $\frac{2}{9}$ **e.** $\frac{1}{3}$
27. $\frac{1}{4}$ **29.** $\frac{2}{25}$

8-8 Compound Events: pages 449–450

1.

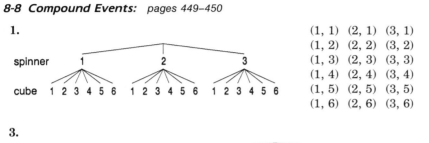

$(1, 1)$ $(2, 1)$ $(3, 1)$
$(1, 2)$ $(2, 2)$ $(3, 2)$
$(1, 3)$ $(2, 3)$ $(3, 3)$
$(1, 4)$ $(2, 4)$ $(3, 4)$
$(1, 5)$ $(2, 5)$ $(3, 5)$
$(1, 6)$ $(2, 6)$ $(3, 6)$

3.

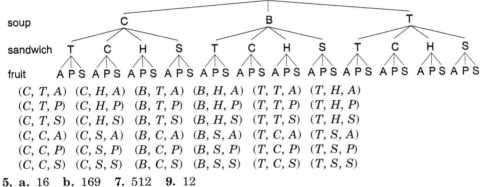

(C, T, A) (C, H, A) (B, T, A) (B, H, A) (T, T, A) (T, H, A)
(C, T, P) (C, H, P) (B, T, P) (B, H, P) (T, T, P) (T, H, P)
(C, T, S) (C, H, S) (B, T, S) (B, H, S) (T, T, S) (T, H, S)
(C, C, A) (C, S, A) (B, C, A) (B, S, A) (T, C, A) (T, S, A)
(C, C, P) (C, S, P) (B, C, P) (B, S, P) (T, C, P) (T, S, P)
(C, C, S) (C, S, S) (B, C, S) (B, S, S) (T, C, S) (T, S, S)

5. a. 16 **b.** 169 **7.** 512 **9.** 12

Probability of Compound Events: pages 453–455

1. a. $\frac{3}{4}$ **b.** $\frac{1}{4}$ **c.** $\frac{1}{2}$ **d.** $\frac{1}{4}$ **e.** $\frac{1}{12}$ **f.** 0
3. a. $\frac{4}{7}$ **b.** $\frac{11}{14}$ **c.** $\frac{3}{14}$ **d.** $\frac{1}{2}$ **e.** $\frac{9}{14}$ **f.** 0
5. a. O_1O_2, O_1O_3, O_1b_1, O_1b_2, O_2O_1, O_2O_3, O_2b_1,
O_2b_2, O_3O_1, O_3O_2, O_3b_1, O_3b_2, b_1O_1, b_1O_2,
b_1O_3, b_1b_2, b_2O_1, b_2O_2, b_2O_3, b_2b_1
b. (1) $\frac{3}{10}$ **(2)** $\frac{1}{10}$ **(3)** 0
7. a. *hhh, hht, hth, htt, thh, tht, tth, ttt.*
b. $\frac{1}{8}$ **c.** 3 **d.** $\frac{3}{8}$
9. a. red, 1; red, 2; red, 3; red, 4; red, 5; red, 6;
green, 1; green, 2;, green, 3; green, 4;
green, 5; green, 6; blue, 1; blue, 2; blue, 3;
blue, 4; blue, 5; blue, 6
b. $\frac{1}{18}$ **c.** 0 **d.** $\frac{1}{6}$
11. a. Using the initials: *A,B*; *A,C*; *A,D*; *B,A*; *B,C*;
B,D; *C,A*; *C,B*; *C,D*; *D,A*; *D,B*; *D,C*
b. $\frac{1}{12}$ **c.** $\frac{1}{2}$ **d.** $\frac{1}{2}$

Complementary Events: page 457

1. a. (1) $\frac{1}{6}$ **(2)** $\frac{5}{6}$ **(3)** $\frac{1}{5}$ **(4)** $\frac{5}{1}$
b. (1) $\frac{1}{2}$ **(2)** $\frac{1}{2}$ **(3)** $\frac{1}{1}$ **(4)** $\frac{1}{1}$
c. (1) $\frac{1}{3}$ **(2)** $\frac{2}{3}$ **(3)** $\frac{1}{2}$ **(4)** $\frac{2}{1}$
d. (1) $\frac{5}{6}$ **(2)** $\frac{1}{6}$ **(3)** $\frac{5}{1}$ **(4)** $\frac{1}{5}$
e. (1) 1 **(2)** 0 **(3)** $\frac{1}{0}$ **(4)** $\frac{0}{1}$
f. (1) $\frac{2}{3}$ **(2)** $\frac{1}{3}$ **(3)** $\frac{2}{1}$ **(4)** $\frac{1}{2}$

3. a. (1) $\frac{1}{13}$ **(2)** $\frac{12}{13}$ **(3)** $\frac{1}{12}$ **(4)** $\frac{12}{1}$
b. (1) $\frac{1}{26}$ **(2)** $\frac{25}{26}$ **(3)** $\frac{1}{25}$ **(4)** $\frac{25}{1}$
c. (1) $\frac{1}{2}$ **(2)** $\frac{1}{2}$ **(3)** $\frac{1}{1}$ **(4)** $\frac{1}{1}$
d. (1) $\frac{2}{13}$ **(2)** $\frac{11}{13}$ **(3)** $\frac{2}{11}$ **(4)** $\frac{11}{2}$
e. (1) $\frac{11}{13}$ **(2)** $\frac{2}{13}$ **(3)** $\frac{11}{2}$ **(4)** $\frac{2}{11}$
f. (1) $\frac{3}{4}$ **(2)** $\frac{1}{4}$ **(3)** $\frac{3}{1}$ **(4)** $\frac{1}{3}$
g. (1) $\frac{1}{2}$ **(2)** $\frac{1}{2}$ **(3)** $\frac{1}{1}$ **(4)** $\frac{1}{1}$

1. 2:3   **3.** 1:360   **5.** 3:4   **7.** 30 and 18
9. $3.50   **11.** Morley   **13.** -15 **15.** 16   **17.** $\frac{7}{12}$

19. $\frac{1}{2}$ **21.** $.67   **23.** $\frac{mq}{p}$ days   **25.** 33 dolls

27. $420 and $525   **29.** 3,432,000 babies
31. direct, $a = \frac{2}{3}b$ **33.** direct, $r = 2t$
35. inverse, $df = 48$ **37.** $c = kn$ **39.** $t = 3\frac{1}{3}$
41. $x = 11\frac{1}{5}$ **43.** $x = 7\frac{1}{2}$ **45.** yes   **47.** 15
49. 5 ft. 4 in.   **51.** .1564   **53.** 21°   **55.** 45°
57. 29 ft.   **59.** 86 ft.   **61.** 40 ft.   **63.** 20 ft.
65. a. $\frac{2}{5}$ **b.** $\frac{4}{5}$ **c.** 0

67.

(1, a)   (2, a)   (3, a)
(1, b)   (2, b)   (3, b)
(1, c)   (2, c)   (3, c)
(1, d)   (2, d)   (3, d)

69. 64   **71. a.** $\frac{1}{6}$ **b.** $\frac{5}{6}$ **c.** $\frac{1}{2}$ **73.** 17 to 1

Problems for Pleasure: page 463

1. $a = 3, b = 2$

3. Try: $\dfrac{x}{x + 6} = \dfrac{x + 2}{x + 4}$ $(x + 4)x \overset{?}{=} (x + 6)(x + 2)$,

$x = -3$

$\dfrac{x}{x + 2} = \dfrac{x + 4}{x + 6}$ $x(x + 6) \overset{?}{=} (x + 2)(x + 4)$,

$8 = 0$ reject

$\dfrac{x}{x + 4} = \dfrac{x + 2}{x + 6}$ $x(x + 6) \overset{?}{=} (x + 2)(x + 4)$,

$8 = 0$ reject

$\dfrac{x + 2}{x} = \dfrac{x + 4}{x + 6}$ $x(x + 4) \overset{?}{=} (x + 6)(x + 2)$,

$x = -3$

$\dfrac{x}{x + 6} = \dfrac{x + 4}{x + 2}$ $x(x + 2) \overset{?}{=} (x + 6)(x + 4)$,

$x = -3$

$\dfrac{x}{x + 4} = \dfrac{x + 6}{x + 2}$ $x(x + 2) \overset{?}{=} (x + 4)(x + 6)$

$x = -3$

$-3, -1, 1, 3$

5. Let $x =$ the number of tickets.
$50x = 200,000 + 100,000$
$x = 6,000$
$\frac{1}{6000}$

Calculator Challenge: page 464

1. $\dfrac{x}{y} = \dfrac{y}{z}$

$85 \geq x > y + 20$
$\quad\quad y > z + 20$
$\quad\quad 20 \leq z$

$85 \geq x > y + 20 > z + 40 \geq 60$
$85 \geq x \geq 60, 65 \geq y \geq 40, 45 \geq z \geq 20$

Try $x \geq 85$; $85z = y^2$
$\quad 85z$ must be a perfect square.
$\quad 85 = 17 \cdot 5$

For z, try multiples of 17. $45 \geq 34 \geq 20$
$x \cdot z = 85 \cdot 34$ is not a perfect square.
Reject 85.

Try $x = 84$; $84z = y^2$
$\quad 84z$ must be a perfect square.
$\quad 84 = 2^2 \cdot 3 \cdot 7$

Look for a z value that is a multiple of 3 and or 7 between 45 and 20: 21, 28, 35, 42.

Try 21
$\quad xz = 84 \cdot 21 = 1,764; \sqrt{1,764} = 42$
$\quad x = 84, y = 42, z = 21$
$\quad \dfrac{84}{42} = \dfrac{42}{21}$

No other values fit the conditions.
3. a. $\sin x = .5, x = 30°$
$\quad$ **b.** $\cos x = .5, x = 60°$
$\quad$ **c.** $\sin x = \cos x, x = 45°$

College Test Preparation: pages 464–466

1. (C) **3.** (D) **5.** (B) **7.** (B) **9.** (E) **11.** (D)
13. (C) **15.** (D) **17.** (C) **19.** (B) **21.** (B)
23. (D) **25.** (C)

Spiral Review Exercises: page 467

1. Transitive property of inequality
3. $x - 3$
5. a. $y = .9$ **b.** $x = 8,000$ **c.** $x = 80$
7. $\left(-5\right) = 31$ **9.** $\frac{1}{9}$ **11.** 12,021
13. car, 45 m/h; truck, 55 m/h   **15.** 8 hits

9-1 Graphing in a Plane: pages 477–478

1. $A(1, 2)$, $B(-2, 1)$, $C(-2, -1)$, $D(2, -2)$, $E(2, 0)$, $F(0, 1)$, $G(-1, 0)$, $H(0, -2)$, $O(0, 0)$

3. quadrant I 5. quadrant II 7. quadrant I

9. 0

11. **a.** abscissa, positive numbers; ordinate, positive numbers
    b. abscissa, negative numbers; ordinate, positive numbers
    c. abscissa, negative numbers; ordinate, negative numbers
    d. abscissa, positive numbers; ordinate, negative numbers

13. **a.** quadrant I   **b.** quadrant I
    c. quadrant I   **d.** quadrant I

15. **a.** quadrant I   **b.** quadrant III
    c. quadrant III   **d.** quadrant I

17. **a.** $(1, 2)$ **b.** $(-1, -2)$
19. **a.** $(-1, -5)$ **b.** $(1, 5)$
21. **a.** $(-3, 6)$ **b.** $(-8, 13)$
23. **a.** $(3, -7)$ **b.** $(-2, 0)$ 25. $P'(x + a, y + b)$

27. **a.**

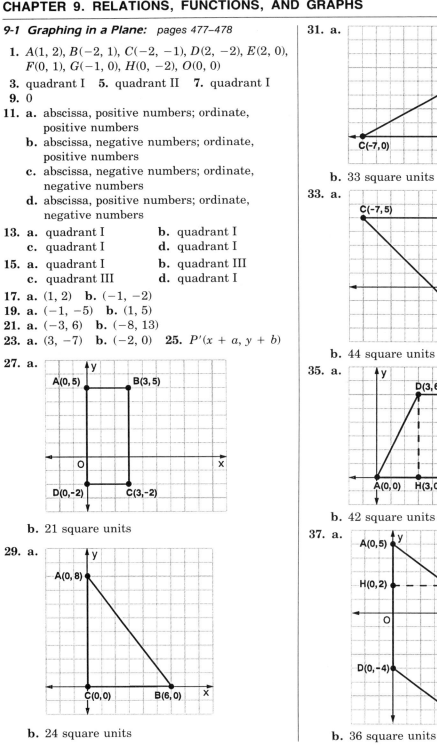

b. 21 square units

29. **a.**

b. 24 square units

31. **a.**

b. 33 square units

33. **a.**

b. 44 square units

35. **a.**

b. 42 square units

37. **a.**

b. 36 square units

39. a.

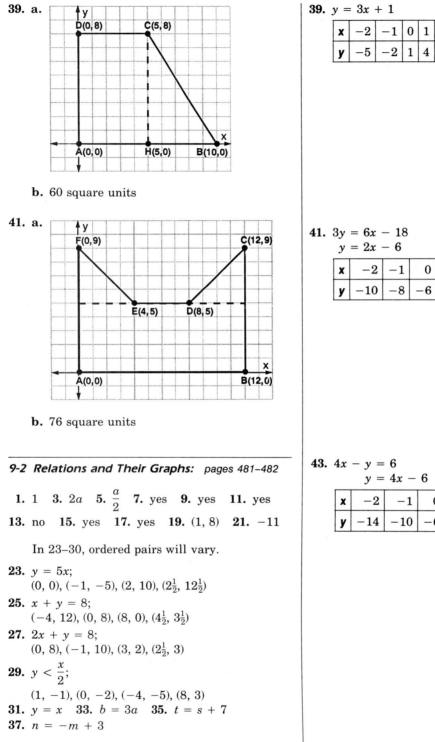

b. 60 square units

41. a.

b. 76 square units

9-2 Relations and Their Graphs: pages 481–482

1. 1 **3.** 2a **5.** $\frac{a}{2}$ **7.** yes **9.** yes **11.** yes

13. no **15.** yes **17.** yes **19.** (1, 8) **21.** −11

In 23–30, ordered pairs will vary.

23. $y = 5x$;
  (0, 0), (−1, −5), (2, 10), ($2\frac{1}{2}$, $12\frac{1}{2}$)

25. $x + y = 8$;
  (−4, 12), (0, 8), (8, 0), ($4\frac{1}{2}$, $3\frac{1}{2}$)

27. $2x + y = 8$;
  (0, 8), (−1, 10), (3, 2), ($2\frac{1}{2}$, 3)

29. $y < \frac{x}{2}$;
  (1, −1), (0, −2), (−4, −5), (8, 3)

31. $y = x$ **33.** $b = 3a$ **35.** $t = s + 7$

37. $n = -m + 3$

39. $y = 3x + 1$

x	−2	−1	0	1	2
y	−5	−2	1	4	7

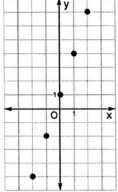

41. $3y = 6x - 18$
  $y = 2x - 6$

x	−2	−1	0	1	2
y	−10	−8	−6	−4	−2

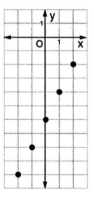

43. $4x - y = 6$
  $y = 4x - 6$

x	−2	−1	0	1	2
y	−14	−10	−6	−2	2

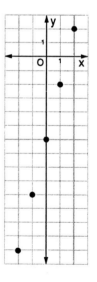

45. a.

x	−2	−1	0	1	2
y	−2	−1	0	1	2

b. 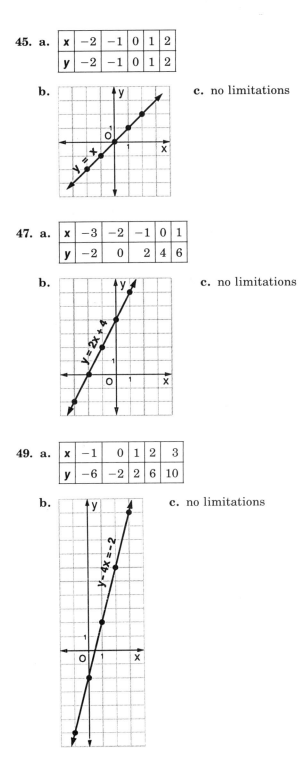   **c.** no limitations

47. a.

x	−3	−2	−1	0	1
y	−2	0	2	4	6

b.   **c.** no limitations

49. a.

x	−1	0	1	2	3
y	−6	−2	2	6	10

b.   **c.** no limitations

51. a.

x	−2	−1	0	1	2
y	6	3	2	3	6

c. no limitations on domain; the range of $y \geq 2$

b.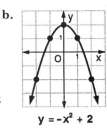

$y = x^2 + 2$

53. a.

x	−2	−1	0	1	2
y	−2	1	2	1	−2

c. no limitations on domain; range of $y \leq 2$

b.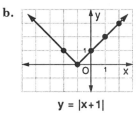

$y = -x^2 + 2$

55. a.

x	−2	−1	0	1	2
y	1	0	1	2	3

b.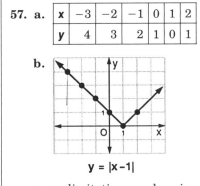

$y = |x+1|$

c. no limitations on domain; range of $y \geq 0$

57. a.

x	−3	−2	−1	0	1	2
y	4	3	2	1	0	1

b.

$y = |x-1|$

c. no limitations on domain; range of $y \geq 0$

59. a.

x	-4	-3	-2	-1	0	1	2	3	4
y	3	4	6	12	–	-12	-6	-4	-3

b.

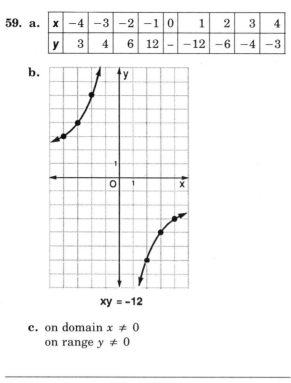

xy = -12

c. on domain $x \neq 0$
on range $y \neq 0$

9-3 *Functions and Their Graphs:* pages 486–487

1. Yes; each first number is associated with a unique second number.

3. No; $\frac{1}{4}$ is associated with $\frac{1}{2}$, $-\frac{1}{2}$; 49 is associated with 7, -7.

5. No; -1 is associated with 3, 2; -2 is associated with 4, 5.

7. Yes; same as **1.**

9. a. No; a is associated with d, e.
   b. $\{(a, d), (a, e), (b, e), (c, f)\}$

11. a. No; b is associated with 2, 3.
   b. $\{(a, 1), (b, 2), (b, 3), (c, 2)\}$

13. a. Yes; same as 10.
   b. $\{(-1, c), (0, b), (1, a)\}$

15. a. Yes; same as 10.
   b. $\{(a, 1), (b, 3), (c, 2)\}$

17. a. yes **b.** yes **19. a.** yes **b.** yes
21. a. yes **b.** no
23. a. $\{-1, 0, 1, 2\}$ **b.** $\{-1, 0, 1, 2\}$ **c.** yes
25. a. $\{-1, 0, 1, 2\}$ **b.** $\{-2, -1, 0, 2\}$ **c.** yes
27. a. $\{x \mid -2 < x < 2\}$ **b.** $\{-2\}$ **c.** yes
29. a. $\{x \mid -2 \leq x \leq 2\}$ **b.** $\{y \mid -2 \leq y \leq 2\}$
   c. no

31. a. $\{x \mid 0 \leq x < 2\}$ **b.** $\{y \mid -2 < y < 2\}$
   c. no
33. a. $\{x \mid 0 \leq x < 2\}$ **b.** $\{y \mid 0 \leq y < 2\}$
   c. yes

In 35–38, answers will vary.

35. $\{(0, 0), (1, 2), (2, 4)\}$
37. $\{(-4, -2), (0, 2), (3, 5)\}$
39. a. -7 **b.** -1 **c.** 5 **d.** 8 **e.** 0
41. a. 19 **b.** 3 **c.** 3. **d.** 9 **e.** $\frac{17}{9}$
43. a. 4 **b.** 8 **c.** 4 **d.** -1 **e.** $\frac{71}{9}$
45. a. 3 **b.** 12 **c.** $\frac{79}{25}$ **d.** 4 **e.** 7 **f.** 12
   g. 14 **h.** 5 **i.** $b^2 - 2b + 4$
   j. $9a^2 - 6a + 4$
   k. $(b + 3)^2 - 2(b + 3) + 4$ or $b^2 + 4b + 7$
   l. $(2b - 1)^2 - 2(2b - 1) + 4$ or $4b^2 - 8b + 7$

9-4 *Lines and Linear Functions:* pages 490–491

1.

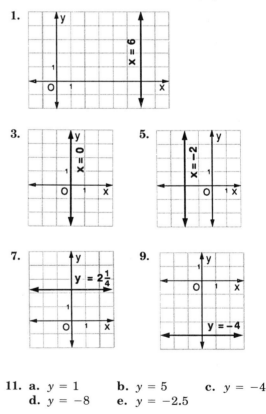

11. a. $y = 1$ **b.** $y = 5$ **c.** $y = -4$
   d. $y = -8$ **e.** $y = -2.5$

13. (2) **15.** (4)

682 Answers/Chapter 9. Relations, Functions, and Graphs

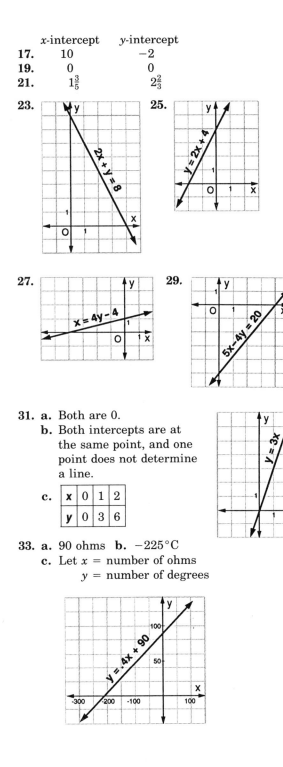

	x-intercept	y-intercept
17.	10	-2
19.	0	0
21.	$1\frac{3}{5}$	$2\frac{2}{3}$

23. [graph: $2x + y = 8$] **25.** [graph: $y = 2x + 4$]

27. [graph: $x = 4y - 4$] **29.** [graph: $5x - 4y = 20$]

31. a. Both are 0.
 b. Both intercepts are at the same point, and one point does not determine a line.
 c.

x	0	1	2
y	0	3	6

[graph: $y = 3x$]

33. a. 90 ohms **b.** $-225°C$
 c. Let x = number of ohms
    y = number of degrees

[graph: $y = .4x + 90$]

9-5 The Slope of a Line: pages 497–498

1. a. positive slope **b.** no slope
  c. slope of zero **d.** negative slope
  e. positive slope **f.** negative slope

3. $m = 1$, oblique, acute
5. $m = \frac{1}{3}$, oblique, acute
7. $m = 2$, oblique, acute
9. $m = \frac{2}{3}$, oblique, acute
11. $m = -3$, oblique, obtuse
13. $m = 0$, horizontal
15. a. $y = 6$ **b.** $y = 2$ **c.** $y = 8$ **d.** $y = 10$
17. 1 **19.** (3) **21.** not parallel **23.** parallel
25. not parallel **27.** 6 **29.** yes **31.** yes
33. no **35.** (3)

9-6 The Slope-Intercept Form of a Linear Equation: pages 504–506

	slope	y-intercept
1.	3	1
3.	2	0
5.	$\frac{1}{2}$	5
7.	-2	3
9.	-3	0
11.	$-\frac{2}{3}$	4
13.	3	7
15.	3	-4
17.	$\frac{5}{2}$	-2
19.	$\frac{3}{2}$	$2\frac{1}{4}$
21.	$\frac{5}{2}$	$2\frac{1}{2}$

23. $y = 3x - 5$ **25.** $y = -x - 3$ **27.** $y = -3x$
29. $y = \frac{1}{2}x$ **31.** (2) **33.** (2)
35. Yes; the slope of both lines is $\frac{1}{2}$.
37. $y = 2x + 2$ **39.** $y = -x - 2$ **41.** $y = \frac{1}{2}x$
43. $y = -\frac{3}{4}x$ **45.** $y = -6$
47. a. $y = 4x - 5$ **b.** $y = 3x + 9$
  c. $y = 3x + 7$ **d.** $y = 4x - 3$
49. $y = 2x + 2$ **51.** $y = x - 2$ **53.** $y = \frac{3}{2}x - 1$
55. $y = -\frac{2}{3}x + 9$ **57.** $y = -\frac{5}{3}x$
59. a. $t = \frac{1}{4}c + 37$ **b.** $87°F$
  c. In the formula, the unit for c is chirps per minute. Thus, chirps counted for 15 seconds need to be multiplied by 4. This multiplication by 4 cancels the value of the slope. Therefore, to approximate the temperature after counting chirps for 15 seconds, multiply by 4 and add 37.

61.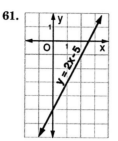

63.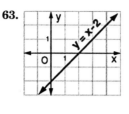

65.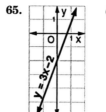

67.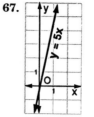

69.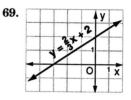

71.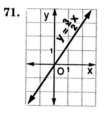

73.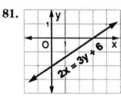

75.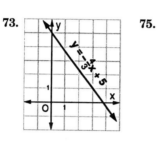

77.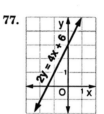

79.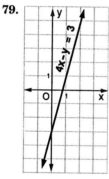

81.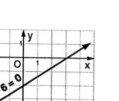

83.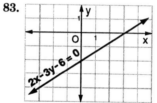

85. a.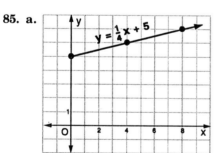

b. **(1)** Slope $= \frac{1}{4}$ means that every 4 hours, the number of Sonia's customers increases by 1% of the total number of mall shoppers.

(2) At noon, 5% of the mall shoppers are at Sonia's.

c. At 8 P.M., 7% of the total number of shoppers might be expected to enter Sonia's Sneaker Stop.

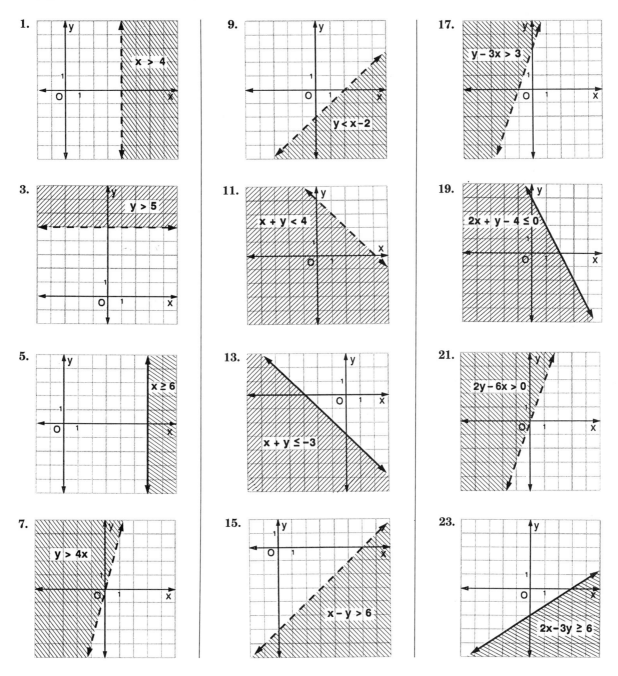

1. $x > 4$

9. $y < x - 2$

17. $y - 3x > 3$

3. $y > 5$

11. $x + y < 4$

19. $2x + y - 4 \leq 0$

5. $x \geq 6$

13. $x + y \leq -3$

21. $2y - 6x > 0$

7. $y > 4x$

15. $x - y > 6$

23. $2x - 3y \geq 6$

25. a. $y > x$

b.
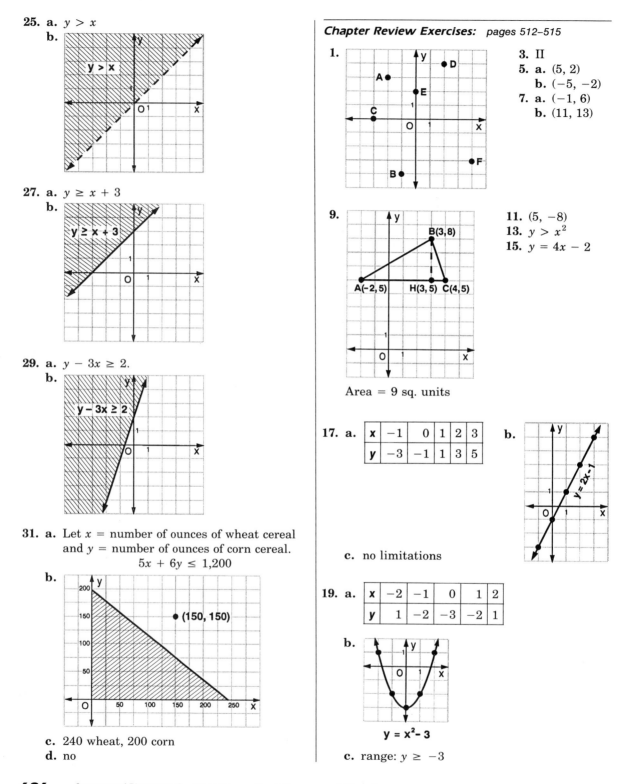

27. a. $y \geq x + 3$

b.

29. a. $y - 3x \geq 2$.

b.

31. a. Let x = number of ounces of wheat cereal and y = number of ounces of corn cereal.
$$5x + 6y \leq 1,200$$

b.

c. 240 wheat, 200 corn
d. no

Chapter Review Exercises: pages 512–515

1.

3. II
5. a. $(5, 2)$
    b. $(-5, -2)$
7. a. $(-1, 6)$
    b. $(11, 13)$

9.

Area = 9 sq. units

11. $(5, -8)$
13. $y > x^2$
15. $y = 4x - 2$

17. a.

x	−1	0	1	2	3
y	−3	−1	1	3	5

b.

c. no limitations

19. a.

x	−2	−1	0	1	2
y	1	−2	−3	−2	1

b.

$y = x^2 - 3$

c. range: $y \geq -3$

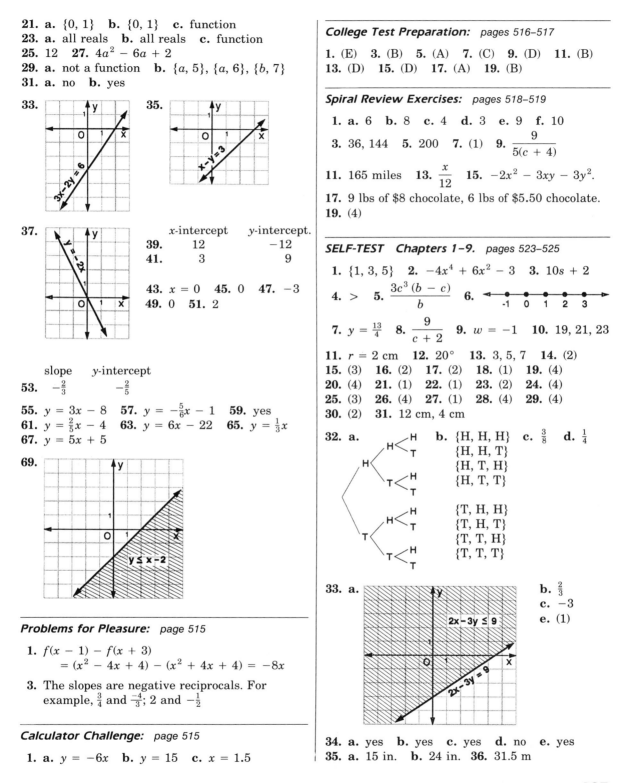

21. a. $\{0, 1\}$ **b.** $\{0, 1\}$ **c.** function
23. a. all reals **b.** all reals **c.** function
25. 12 **27.** $4a^2 - 6a + 2$
29. a. not a function **b.** $\{a, 5\}, \{a, 6\}, \{b, 7\}$
31. a. no **b.** yes

33.

35.

37.

	x-intercept	y-intercept.
39.	12	-12
41.	3	9

43. $x = 0$ **45.** 0 **47.** -3
49. 0 **51.** 2

	slope	y-intercept
53.	$-\frac{2}{3}$	$-\frac{2}{5}$

55. $y = 3x - 8$ **57.** $y = -\frac{5}{6}x - 1$ **59.** yes
61. $y = \frac{2}{5}x - 4$ **63.** $y = 6x - 22$ **65.** $y = \frac{1}{3}x$
67. $y = 5x + 5$

69.

Problems for Pleasure: page 515

1. $f(x - 1) - f(x + 3)$
$= (x^2 - 4x + 4) - (x^2 + 4x + 4) = -8x$

3. The slopes are negative reciprocals. For example, $\frac{3}{4}$ and $\frac{-4}{3}$; 2 and $-\frac{1}{2}$

Calculator Challenge: page 515

1. a. $y = -6x$ **b.** $y = 15$ **c.** $x = 1.5$

College Test Preparation: pages 516–517

1. (E) **3.** (B) **5.** (A) **7.** (C) **9.** (D) **11.** (B)
13. (D) **15.** (D) **17.** (A) **19.** (B)

Spiral Review Exercises: pages 518–519

1. a. 6 **b.** 8 **c.** 4 **d.** 3 **e.** 9 **f.** 10
3. 36, 144 **5.** 200 **7.** (1) **9.** $\dfrac{9}{5(c + 4)}$
11. 165 miles **13.** $\dfrac{x}{12}$ **15.** $-2x^2 - 3xy - 3y^2$.
17. 9 lbs of \$8 chocolate, 6 lbs of \$5.50 chocolate.
19. (4)

SELF-TEST Chapters 1–9. pages 523–525

1. $\{1, 3, 5\}$ **2.** $-4x^4 + 6x^2 - 3$ **3.** $10s + 2$
4. $>$ **5.** $\dfrac{3c^3(b - c)}{b}$ **6.**
7. $y = \frac{13}{4}$ **8.** $\dfrac{9}{c + 2}$ **9.** $w = -1$ **10.** 19, 21, 23
11. $r = 2$ cm **12.** $20°$ **13.** 3, 5, 7 **14.** (2)
15. (3) **16.** (2) **17.** (2) **18.** (1) **19.** (4)
20. (4) **21.** (1) **22.** (1) **23.** (2) **24.** (4)
25. (3) **26.** (4) **27.** (1) **28.** (4) **29.** (4)
30. (2) **31.** 12 cm, 4 cm
32. a. **b.** $\{H, H, H\}$ **c.** $\frac{3}{8}$ **d.** $\frac{1}{4}$
$\{H, H, T\}$
$\{H, T, H\}$
$\{H, T, T\}$

$\{T, H, H\}$
$\{T, H, T\}$
$\{T, T, H\}$
$\{T, T, T\}$

33. a. **b.** $\frac{2}{3}$
c. -3
e. (1)

34. a. yes **b.** yes **c.** yes **d.** no **e.** yes
35. a. 15 in. **b.** 24 in. **36.** 31.5 m

10-1 Solving a System of Linear Equations Graphically: page 529

1.

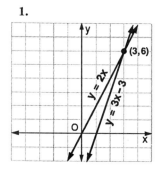

3.

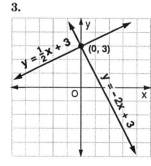

5.

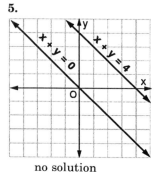

no solution

7.

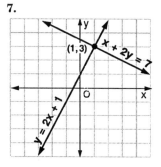

9.

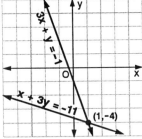

11.

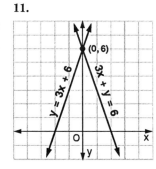

13.

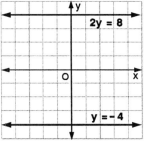

15.

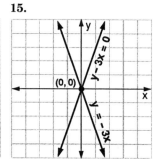

17.

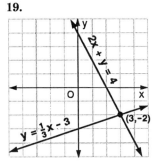

19.

21.

no solution

23.

25.

27.

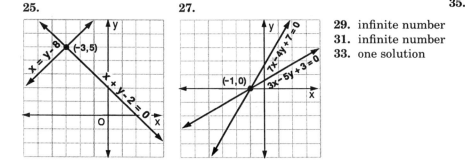

29. infinite number
31. infinite number
33. one solution

35. a. $x + y = 3$
     $x = y + 5$

b.

4 and −1

37. a. $2x + 2y = 10$
$x = y + 3$

b.

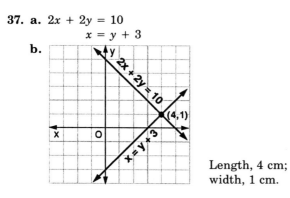

Length, 4 cm; width, 1 cm.

10-2 Solving a System of Linear Equations Algebraically: *pages 534–536*

Note: In 1–102, the coordinates are given in alphabetical order.

1. (7, 7) **3.** inconsistent **5.** (9, 27)
7. (2, −1) **9.** (3, −6) **11.** (8, 10) **13.** (4, 7)
15. dependent **17.** (3, 8) **19.** (6, 5) **21.** (5, −5)
23. (2, 5) **25.** (5, 4) **27.** (5, 4) **29.** (−4, 3)
31. (5, −1) **33.** (−3, 3) **35.** (3, 9)
37. (200, 300) **39.** dependent **41.** (9, 4)
43. (5, 1) **45.** (12, −1) **47.** inconsistent
49. (6, 1) **51.** $(\frac{1}{2}, 1)$ **53.** (4, −2) **55.** (2, 3)
57. (3, 2) **59.** (7, −2) **61.** dependent **63.** (4, 5)
65. (2, 1) **67.** (−3, 2) **69.** (3, 6)
71. inconsistent **73.** (−15, 12) **75.** (−3, 1)
77. $(0, -2\frac{1}{2})$ **79.** (6, 5) **81.** inconsistent
83. (2, 12) **85.** (6, 4) **87.** (−8, 3) **89.** (12, −8)
91. (9, 4) **93.** (6, 4) **95.** (150, 250) **97.** (8, 2)
99. (9, 5) **101.** (4, 3)

10-3 Solving Verbal Problems by Using Two Variables: *pages 538–540*

1. $s + l = 105$
$s = l - 5$
$s = 50; l = 55$

3. $2l + 2w = 50$
$l = w + 9$
$l = 17$ cm; $w = 8$ cm

5. $A = 3K$
$A + K = 76$
Al, $57; Ken, $19

7. $R = H + 12$
$\frac{1}{3}R + \frac{1}{2}H = 24$
In Mrs. Reilly's, 24; in Mr. Horn's, 12

9. $n + q = 30$
$5n + 25q = 510$
12 nickels, 18 quarters

11. Let x = the number of 20¢ stamps
and y = the number of 40¢ stamps.
$x = y - 2$
$20x + 40y = 200$
Two 20¢ stamps, four 40¢ stamps

13. $w + c = 92$
$2w + 3c = 239$
55 correct answers

15. $q = d + 6$
$10q + 25d = 550$
He now has $6.40.

17. a. (1) $q + n = 13$
$25q + 5n = 165$
5 quarters, 8 nickels
(2) $q + d = 12$
$25q + 10d = 165$
3 quarters, 9 dimes

b. Use equations, or use the strategy of making a list.
(1) yes; 16 dimes, 1 nickel, or 4 quarters, 13 nickels
(2) no
(3) yes; 6 quarters, 3 nickels, or 5 quarters, 4 dimes

19. Let p = share price and d = number of days.
Dog food: $p = 2\frac{1}{2} - \frac{1}{4}d$ [D]
Fertilizer: $p = 1\frac{3}{8} + \frac{1}{8}d$ [F]
Camping gear: $p = 2$ [C]

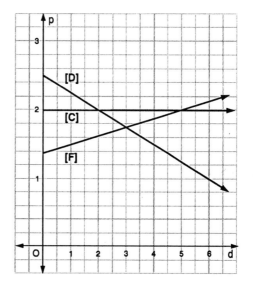

a. On days 3 and 4, the value of the camping gear stock was highest.
b. After day 5, the fertilizer stock had the highest value.

Mixture Problems: page 542

1. Let x = the number of kg of $2 seed
   and y = the number of kg of $3 seed.
   $$x + y = 80$$
   $$2x + 3y = 2.20(80)$$
   64 kg of $2 seed, 16 kg of $3 seed

3. Let x = the number of pounds of $1.20 candy
   and y = the number of pounds of $2.10 candy.
   $$x + y = 50$$
   $$1.20x + 2.10y = 1.56(50)$$
   30 lb. of $1.20 candy, 20 lb. of $2.10 candy

5. $$r + p = 10$$
   $$1.29r + 1.49p = 1.43(10)$$
   7 gal. of premium

7. $$4.89h + 2.49c = 3.39(h + c)$$
   $$h + c = 1$$
   $\frac{3}{8}$ lb., or 6 oz., of ham

Business Problems: page 544

1. Let r = cost per box of oranges
   and g = cost per box of grapefruit.
   $$6r + 5g = 142$$
   $$3r + 2g = 64$$
   oranges, $12; grapefruit, $14

3. Let x = price per bat
   and y = price per ball.
   $$4x + 9y = 191$$
   $$3x + 12y = 201$$
   bat, $23; ball, $11

5. $$4w + 3p = 37.40$$
   $$5w + 1p = 33.00$$
   walnuts, $5.60; pecans, $5

7. $$6R + 7S = 262$$
   $$4R + 3S = 138$$
   model R, $18; model S, $22

9. $$3p + 2m = 21.60$$
   $$4p + 5m = 34.40$$
   print wheel, $5.60; memory disk, $2.40

11. $$15T + 12H = 13.56$$
    $$24T + 18H = 21.12$$
    TV Tune-In, 52¢; Health Watch, 48¢

Investment Problems: page 546

1. $$x + y = 1,400$$
   $$.05x + .08y = 100$$
   $400 at 5%; $1,00⟩ at 8%

3. $$x + y = 7,000$$
   $$.05x + .08y = 500$$
   $2,000 at 5%; $5,000 at 8%

5. $$.10x = .08y$$
   $$y = x + 400$$
   $1,600 at 10%; $2,000 at 8%

7. $$x + y = 20,000$$
   $$.06x + .05y = 1,120$$
   $12,000 at 6%; $8,000 at 5%

9. $$x + y = 240,000$$
   $$.06x + .065y + 300 = .065x + .06y$$
   $150,000 at 6%; $90,000 at $6\frac{1}{2}$%

Digit Problems: pages 548–549

1. $t + u = 10$ 3. $u = t + 6$
   $t - u = 4$ $t + u = 10$
   73   28

5. $$u = 2t + 3$$
   $$t + u = 6$$
   15

7. $$t = u - 2$$
   $$10t + u = 4(t + u)$$
   24

9. $$t = u + 3$$
   $$10t + u = 8(t + u) + 1$$
   41

11. $$t + u = 11$$
    $$10t + u + 45 = 10u + t$$
    38

13. $$10h + t = 10u + t$$
    $$h + t + u = 17$$
    $$10t + u = 2(10h + t) - 4$$
    494

15. $$10u + t - 18 = 10t + u$$
    $$10t + u + 10u + t = 44$$
    1ɜ − 18 − 31

Motion Problems Involving Currents: pages 551–552

1. $2(r + c) = 180$ 3. $2(r + c) = 16$
   $3(r - c) = 180$ $8(r - c) = 16$
   75 km/h   rate, 5 km/h

5. $$4(r + w) = 2,000$$
   $$5(r - w) = 2,000$$
   rate, 450 km/h; wind, 50 km/h

7. $3(r + w) = 1,950$
$2(r - (w - 10)) = 1,160$
wind, 40 km/h

9. $6(r + w) = 3,300$
$6\frac{3}{5}(r - w) = 3,300$
rate, 525 mph; wind, 25 mph

11. $9(r - w) = 3,600$
$8(r - w) = 3,600$
rate, 425 mph

Miscellaneous Problems: pages 552–553

1. Let x = the number of tickets brought before,
and y = the number of tickets bought at the
gate.
$x + y = 600$
$x + 1.50y = 700$
200 sold at the gate

3. $x + y = 7$
$80x + 60y = 500$
4 hr. at 80 km/h; 3 hr at 60 km/h

5. $x + y = 50$
$20x + 40y = 1,400$
thirty 20¢ stamps; twenty 40¢ stamps

7. $\dfrac{n - 1}{d + 5} = \dfrac{1}{2}$ **9.** $4A = 3B$
$9B - 2A = 5$
$A, \frac{1}{2}$ oz.; $B, \frac{2}{3}$ oz.

$\dfrac{n + 1}{d - 6} = \dfrac{4}{3}$

11. $3D = 4.2F$ or $D = 1.4F$
$D + F = 12M$

$\dfrac{n}{d} = \dfrac{11}{15}$ 5 MORNS

13. $A + B = 121$
$4A + \frac{1}{2}B = 148$ Club A, 25; Club B, 96

15. $m + c = 6$
$5.39m + 2.99c = 23.94$ $2\frac{1}{2}$ lb. or 40 oz.

**10-4 Solving a System of Linear Inequalities
Graphically:** page 556

1.
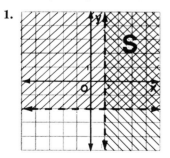

3.
5.
7.
9.
11.
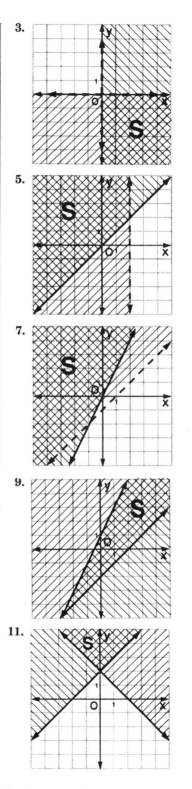

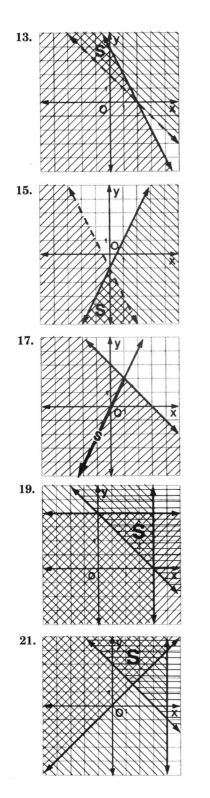

13.

15.

17.

19.

21.

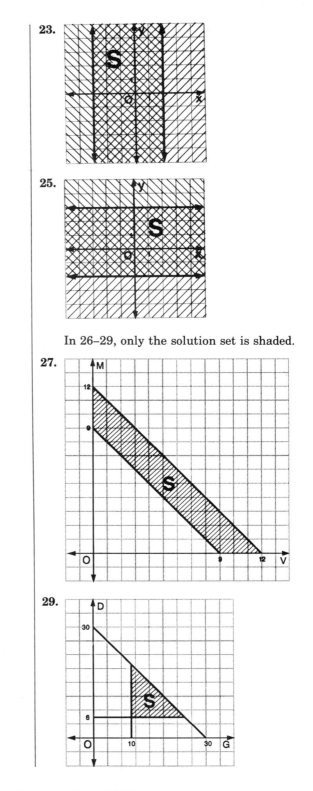

23.

25.

In 26–29, only the solution set is shaded.

27.

29.

1. $x = 1, y = -2, z = -4$
3. $x = \frac{1}{2}, y = 2, z = -1$
5. $x = 0, y = \frac{1}{3}, z = 4$
7. $x = 1, y = 2, z = 3$
9. $x = 2, y = -2, z = \frac{1}{4}$
11. $x = -3, y = -1, z = -2$
13. 621   **15.** 457
17. 5 nickels, 8 dimes, 10 quarters
19. 12 type A, 6 type B, 9 type C
21. A 10 hours, B 8 hours, C 12 hours

10-6 Solving of System of Linear Equations by Using Matrices: page 564

1. $2x + y = 8$
   $x + y = 4$
   To get $a_1 = 1$: Replace Row 1 by
   $\qquad\qquad$ (Row 1 − Row 2)
   To get $a_2 = 0$: Replace Row 2 by
   $\qquad\qquad$ (Row 2 − Row 1)
   $b_1 = 0$ ✔ $b_2 = 1$ ✔
   $x = 4, y = 0$

3. $x - y = 3$
   $-x + 2y = 2$
   $a_1 = 1$ ✔
   To get $a_2 = 0$: Replace Row 2 by
   $\qquad\qquad$ (Row 1 + Row 2)
   To get $b_1 = 0$: Replace Row 1 by
   $\qquad\qquad$ (Row 1 + Row 2)
   $b_2 = 1$ ✔
   $x = 8, y = 5$

5. $3x - 3y = 12$
   $x + y = 0$
   To get $a_1 = 1$: Replace Row 1 by $\frac{1}{3} \cdot$ Row 1
   To get $a_2 = 0$: Replace Row 2 by
   $\qquad\qquad$ (−1 · Row 1 + Row 2)
   To get $b_1 = 0$: Replace Row 1 by
   $\qquad\qquad$ ($\frac{1}{2} \cdot$ Row 2 + Row 1)
   To get $b_2 = 1$: Replace Row 2 by $\frac{1}{2} \cdot$ Row 2
   $x = 2, y = -2$

7. $2x + y = 8$
   $\qquad\quad y = x - 7$
   $\quad x - y = 7$
   To get $a_1 = 1$: Replace Row 1 by $\frac{1}{2} \cdot$ Row 1
   To get $a_2 = 0$: Replace Row 2 by
   $\qquad\qquad$ (−1 · Row 1 + Row 2)

To get $b_1 = 0$: Replace Row 1 by
$\qquad\qquad$ ($\frac{1}{3} \cdot$ Row 2 + Row 1)
To get $b_2 = 1$: Replace Row 2 by $-\frac{2}{3} \cdot$ Row 2
$x = 5, y = -2$

9. $2x + 3y = 18$
   $\quad x - 2y = -5$
   To get $a_1 = 1$: Replace Row 1 by $\frac{1}{2} \cdot$ Row 1.
   To get $a_2 = 0$: Replace Row 2 by
   $\qquad\qquad$ (−1 · Row 1 + Row 2)
   To get $b_2 = 1$: Replace Row 2 by $-\frac{2}{7} \cdot$ Row 2
   To get $b_1 = 0$: Replace Row 1 by
   $\qquad\qquad$ ($-\frac{3}{2} \cdot$ Row 2 + Row 1)
   $x = 3, y = 4$

11. $4x - 3y = 2$
    $8x + 6y = 8$
    To get $a_1 = 1$: Replace Row 1 by $\frac{1}{4} \cdot$ Row 1
    To get $a_2 = 0$: Replace Row 2 by
    $\qquad\qquad$ (−8 · Row 1 + Row 2)
    To get $b_2 = 1$: Replace Row 2 by $\frac{1}{12} \cdot$ Row 2.
    To get $b_1 = 0$: Replace Row 1 by
    $\qquad\qquad$ ($\frac{3}{4} \cdot$ Row 2 + Row 1)
    $x = \frac{3}{4}, y = \frac{1}{3}$

13. $\quad 5x + 5y = 26$
    $10x - 10y = 48$
    To get $a_1 = 1$: Replace Row 1 by $\frac{1}{5} \cdot$ Row 1
    To get $a_2 = 0$: Replace Row 2 by
    $\qquad\qquad$ (−10 · Row 1 + Row 2)
    To get $b_1 = 0$: Replace Row 1 by
    $\qquad\qquad$ ($\frac{1}{20} \cdot$ Row 2 + Row 1)
    To get $b_2 = 1$: Replace Row 2 by $-\frac{1}{20} \cdot$ Row 2
    $x = 5, y = \frac{1}{5}$

15. $\quad x + 3y = 0$
    $2x - 5y = 11$
    $a_1 = 1$ ✔
    To get $a_2 = 0$: Replace Row 2 by
    $\qquad\qquad$ (−2 · Row 1 + Row 2)
    To get $b_2 = 1$: Replace Row 2 by $-\frac{1}{11} \cdot$ Row 2
    To get $b_1 = 0$: Replace Row 1 by
    $\qquad\qquad$ (−3 · Row 2 + Row 1)
    $x = 3, y = -1$

17. $2x + 4y = 7$
    $\quad x - 8y = 1$
    To get $a_1 = 1$: Replace Row 1 by $\frac{1}{2} \cdot$ Row 1.
    To get $a_2 = 0$: Replace Row 2 by
    $\qquad\qquad$ (−1 · Row 1 + Row 2)
    To get $b_1 = 0$: Replace Row 1 by
    $\qquad\qquad$ ($\frac{1}{5} \cdot$ Row 2 + Row 1)
    To get $b_2 = 1$: Replace Row 2 by $-\frac{1}{10} \cdot$ Row 2
    $x = 3, y = \frac{1}{4}$

19. $3x + 5y = -2$
$9x - y = 10$
To get $a_1 = 1$: Replace Row 1 by $\frac{1}{3} \cdot$ Row 1
To get $a_2 = 0$: Replace Row 2 by
$$(-9 \cdot \text{Row 1} + \text{Row 2})$$
To get $b_2 = 1$: Replace Row 2 by $-\frac{1}{16} \cdot$ Row 2
To get $b_1 = 0$: Replace Row 1 by
$$(-\frac{5}{3} \cdot \text{Row 2} + \text{Row 1})$$
$x = 1, y = -1$

21. $3x - y = -12$
$x + 2y = 10$
To get $a_1 = 1$: Replace Row 1 by $\frac{1}{3} \cdot$ Row 1
To get $a_2 = 0$: Replace Row 2 by
$$(-1 \cdot \text{Row 1} + \text{Row 2})$$
To get $b_2 = 1$: Replace Row 2 by $\frac{3}{7} \cdot$ Row 2
To get $b_1 = 0$: Replace Row 1 by
$$(\frac{1}{3} \cdot \text{Row 2} + \text{Row 1})$$
$x = -2, y = 6$

23. $3x - 2y = 6$
$x + 7y = 2$
To get $a_1 = 1$: Replace Row 1 by $\frac{1}{3} \cdot$ Row 1
To get $a_2 = 0$: Replace Row 2 by
$$(-1 \cdot \text{Row 1} + \text{Row 2})$$
To get $b_2 = 1$: Replace Row 2 by $\frac{3}{23} \cdot$ Row 2
To get $b_1 = 0$: Replace Row 1 by
$$\frac{2}{3} \cdot \text{Row 2} + \text{Row 1}$$
$x = 2, y = 0$

25. $5x - y = -27$
$x + \frac{1}{2}y = -4$
To get $a_1 = 1$: Replace Row 1 by $\frac{1}{5} \cdot$ Row 1
To get $a_2 = 0$: Replace Row 2 by
$$(-1 \cdot \text{Row 1} + \text{Row 2})$$
To get $b_2 = 1$: Replace Row 2 by $\frac{10}{7} \cdot$ Row 2
To get $b_1 = 0$: Replace Row 1 by
$$(\frac{1}{5} \cdot \text{Row 2} + \text{Row 1})$$
$x = -5, y = 2$

27. $3x + 5y = 8$
$x - 4y = -3$
To get $a_1 = 1$: Replace Row 1 by $\frac{1}{3} \cdot$ Row 1
To get $a_2 = 0$: Replace Row 2 by
$$(-1 \cdot \text{Row 1} + \text{Row 2})$$
To get $b_2 = 1$: Replace Row 2 by $-\frac{3}{17} \cdot$ Row 2
To get $b_1 = 0$: Replace Row 1 by
$$(-\frac{5}{3} \cdot \text{Row 2} + \text{Row 1})$$
$x = 1, y = 1$

29. $5x - 4y = 24$
$x + 3y = 1$
To get $a_1 = 1$: Replace Row 1 by $\frac{1}{5} \cdot$ Row 1
To get $a_2 = 0$: Replace Row 2 by
$$(-1 \cdot \text{Row 1} + \text{Row 2})$$
To get $b_2 = 1$: Replace Row 2 by $\frac{5}{19} \cdot$ Row 2
To get $b_1 = 0$: Replace Row 1 by
$$(\frac{4}{5} \cdot \text{Row 2} + \text{Row 1})$$
$x = 4, y = -1$

Chapter Review Exercises: *pages 566–567*

1.

3.

5. dependent **7.** inconsistent **9.** $(\frac{8}{5}, \frac{8}{5})$
11. $(-1, 3)$
13. 40 lbs. of $3.40 nuts;
20 lbs. of $2.20 nuts
15. 45 dimes; 24 quarters **17.** 15 mph
19. shirt $12; belt $8.50

21.

23.

25. $(-3, 5, 1)$ **27.** 625

29. $5x + 3y = -3$
    $3x + 2y = -1$
To get $a_1 = 1$: Replace Row 1 by
        ($2 \cdot$ Row 2 $-$ Row 1)
To get $a_2 = 0$: Replace Row 2 by
        ($3 \cdot$ Row 1 $-$ Row 2)
To get $b_1 = 0$: Replace Row 1 by (Row 1 $-$ Row 2)
$b_2 = 1$ ✓
$x = -3, y = 4$

Problems for Pleasure: page 567

 1. $w = 0, x = 2, y = -1, z = 3$
 3. 1564; Shakespeare
    The four-digit year must begin with 1. Thus,
    the year is $1stf$. First hint: $1 + f = s$. Second
    hint: $t - 1 = s$. Since $1 + f$ and $t - 1$ both
    equal s, then $1 + f = t - 1$, or $f - t = -2$.
    Third hint: $3f = 2t$, or $3f - 2t = 0$. Solve
    $f - t = -2$ and $3f - 2t = 0$. Then $f = 4$ and
    $t = 6$. Substitute 4 for f in $1 + f = s$, and
    $s = 5$. The year is 1564. The man is William
    Shakespeare.

Calculator Challenge: page 568

 1. $5 \times 5 = 25$
    $6 \times 6 = 36$
    $7 \times 7 = 49$
    $8 \times 8 = 64$
    $9 \times 9 = 81$
    Total $= 255$

College Test Preparation: pages 568–569

 1. (C) **3.** (D) **5.** (D) **7.** (E) **9.** (C) **11.** (A)
13. (C) **15.** (C) **17.** (D) **19.** (C) **21.** (C)

Spiral Review Exercises: page 570

 1. $x = 24$ **3.** -144 **5.** 2 **7.** (7, 4) **9.** $9a^2$
11. 5.0×10^6 **13.** $\frac{5}{12}$ **15.** $x = 5$ **17.** 2

CHAPTER 11. IRRATIONAL NUMBERS

11-1 Finding the Root of a Number:
   pages 576–577

 1. index, 2; radicand, 36
 3. index, 4; radicand, 81
 5. index, n; radicand, 1 **7.** 1 **9.** 15 **11.** $\frac{1}{3}$
13. .7 **15.** .2 **17.** 9 **19.** -8 **21.** 0
23. ± 13 **25.** -25 **27.** $-\frac{3}{4}$ **29.** $\frac{7}{10}$ **31.** .8
33. $\pm .3$ **35.** $\pm .02$ **37.** 3 **39.** -2 **41.** 8
43. .7 **45.** 5 **47.** 36 **49.** 39 **51.** 13 **53.** 24
55. $\sqrt{25}$ **57.** $\sqrt{.04}$

59. a. (1) principal square root
       (2) principal square root
       (3) equal **(4)** given number
       (5) given number

   b. If the number is less than 1, its principal
      square root is greater; if the number is
      greater than 1, it is greater than its
      principal square root.

61. (3) **63.** $y = \pm 10$ **65.** $x = \pm .7$
67. $y = \pm 6$ **69.** $x = \pm 5$ **71.** 14 yd. **73.** 21 m
75. 3 and 4 **77.** -2 and -1 **79.** -8 and -7
81. 9 and 10 **83.** 11 and 12 **85.** 3, 4, $\sqrt{17}$
87. $-\sqrt{7}, 0, \sqrt{7}$ **89.** $-\sqrt{23}, -\sqrt{19}, -\sqrt{11}$

91. irrational **93.** irrational **95.** irrational
97. irrational **99.** irrational **101.** 4 **103.** 6
105. 10 **107.** -11 **109.** 15 **111.** 3.5
113. -8.2 **115.** 10.3 **117.** -11.7 **119.** 13.2

11-2 Simplifying and Combining Radicals:
   pages 580–582

 1. $2\sqrt{2}$ **3.** $2\sqrt{5}$ **5.** $2\sqrt{7}$ **7.** $3\sqrt{3}$ **9.** $3\sqrt{6}$
11. $3\sqrt{10}$ **13.** $7\sqrt{2}$ **15.** $6\sqrt{3}$ **17.** $9\sqrt{2}$
19. $10\sqrt{3}$ **21.** $6\sqrt{2}$ **23.** $4\sqrt{5}$ **25.** $12\sqrt{10}$
27. $30\sqrt{2}$ **29.** $\sqrt{5}$ **31.** $\sqrt{3}$ **33.** $3\sqrt{6}$ **35.** $\frac{3}{2}\sqrt{5}$
37. $3c$ **39.** $8c^3$ **41.** $9t^4$ **43.** $x^2 y$ **45.** $x^8 y^2$
47. $2xy$ **49.** $12a^2 b$ **51.** $5r^4 s^8 t^6$ **53.** $\frac{5}{6}xy$
55. $\frac{9}{11}a^3 b$ **57.** $.6m$ **59.** $.2xy^3$ **61.** $1.1a^2 b^8 c^{18}$
63. $2b^2\sqrt{b}$ **65.** $3n\sqrt{m}$ **67.** $4xy\sqrt{xy}$
69. $-4ab\sqrt{2b}$ **71.** $6r\sqrt{s}$ **73.** $10xy\sqrt{xy}$
75. $12r\sqrt{2r}$ **77.** $-14ab\sqrt{6ab}$ **79.** $3abc\sqrt{ab}$
81. $15yu^3 z^4\sqrt{2y}$ **83.** $4x^2 y^2 z^3\sqrt{2xz}$
85. $-4xy^3 z^5\sqrt{3z}$ **87.** 0 **89.** $4\sqrt{5} + 3\sqrt{2}$
91. $6\sqrt{x}$ **93.** $8\sqrt{3}$ **95.** $8\sqrt{5}$ **97.** $-\sqrt{3}$ **99.** 0
101. $3\sqrt{10}$ **103.** $\sqrt{2} + \sqrt{3}$ **105.** 0 **107.** $4\sqrt{5}$

109. $8\sqrt{2}$ **111.** $14\sqrt{x}$ **113.** $\sqrt{3x}$
115. $7\sqrt{3y} - 3\sqrt{3x}$ **117.** $2y\sqrt{5}$ **119.** $3x\sqrt{7x}$
121. $11a\sqrt{3a}$ **123.** $-3b\sqrt{3a}$ **125.** $x = 5\sqrt{5}$
127. $x = \sqrt{7}$ **129.** $x = 10\sqrt{11}$
131. a. $4\sqrt{3}$ cm   **b.** $12\sqrt{7}$ in.   **c.** $(20 + 28\sqrt{6})$ ft.
133. $(22 + 2\sqrt{2})$ in.

11-3 Multiplying and Dividing Radicals:
pages 588–590

1. 5   **3.** 113   **5.** r **7.** 6   **9.** 72   **11.** $3\sqrt{7}$
13. $9\sqrt{2}$ **15.** $9\sqrt{6}$ **17.** $10\sqrt{10}$ **19.** 48
21. $120\sqrt{3}$ **23.** $-12a$ **25.** 3   **27.** y **29.** 54
31. $10x$ **33.** $3x\sqrt{5}$ **35.** $5xy\sqrt{2}$ **37.** $3a$ **39.** $6z$
41. $4t$ **43.** $2x + 1$ **45.** $5x - 3$ **47.** $5 + \sqrt{55}$
49. 0   **51.** $32 + 11\sqrt{2}$ **53.** 22
55. $a + b + 2\sqrt{ab}$ **57.** $7 - 4\sqrt{3}$
59. $43 + 30\sqrt{2}$ **61.** 75 sq. ft.   **63.** $12\sqrt{3}$ cu. cm
65. $(3\sqrt{2})^2 - 3(3\sqrt{2}) - 6 = 3(4 - 3\sqrt{2})$
67. rational   **69.** irrational   **71.** $\frac{3}{22}$ **73.** $4\sqrt{5}$
75. $\frac{10}{7}\sqrt{10}$ **77.** 2   **79.** $\frac{2}{3}\sqrt{7}$ **81.** 5   **83.** $\sqrt{7}$
85. 16   **87.** $5\sqrt{2}$ **89.** 3   **91.** $\frac{1}{4}$ **93.** $4\sqrt{3}$
95. 8   **97.** $10\sqrt{3}$ **99.** $\frac{3}{2}\sqrt{2}$ **101.** $9a$ **103.** 5
105. $\dfrac{12 + 2\sqrt{3}}{3}$ **107.** $2 + 2\sqrt{2}$ **109.** $6 + 4\sqrt{5}$
111. $2 - \sqrt{3}$ **113.** $\sqrt{7} - 1$ **115.** $\sqrt{5} - \sqrt{7}$
117. $\sqrt{3}$ **119.** $14\sqrt{3}$ in.   **121.** $\frac{1}{2}\sqrt{2}$ **123.** $\frac{1}{7}\sqrt{7}$
125. $\dfrac{1}{a}\sqrt{a}$ **127.** $\frac{5}{3}\sqrt{3}$ **129.** $\frac{9}{7}\sqrt{7}$ **131.** $4\sqrt{2}$
133. $5\sqrt{5}$ **135.** $\sqrt{7}$ **137.** $\sqrt{3}$ **139.** $\frac{5}{2}\sqrt{5}$
141. $2\sqrt{3}$ **143.** $5\sqrt{2}$ **145.** $3\sqrt{15}$ **147.** $\dfrac{1}{s}\sqrt{rs}$
149. $\dfrac{4}{a}\sqrt{2a}$

11-4 Solving a Radical Equation: pages 594–595

1. $x = 25$ **3.** no root   **5.** $b = \frac{1}{16}$ **7.** $x = 5$
9. $y = \frac{16}{3}$ **11.** $y = 4$ **13.** $y = \frac{1}{4}$ **15.** no root
17. $y = 75$ **19.** $x = 8$ **21.** no root   **23.** $b = 65$
25. $x = 25$ **27.** $b = 52$ **29.** $m = 64$ **31.** $x = 4$
33. $y = \frac{49}{16}$ **35.** $y = 2$ **37.** $x = 10$ **39.** $x = 3$
41. $y = 2$ **43.** $A = s^2$ **45.** $s = 6e^2$ **47.** 16
49. 52   **51.** 1,296 ft.
53. a. $\dfrac{10^4}{12\pi}$ or about 265 cycles per second.
    b. 8×10^{-6} or $\dfrac{8}{10^6}$

11-5 Graphing a Quadratic Function:
pages 601–602

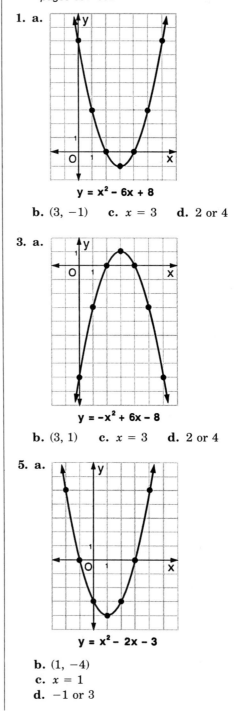

1. a.

$y = x^2 - 6x + 8$

b. $(3, -1)$ **c.** $x = 3$ **d.** 2 or 4

3. a.

$y = -x^2 + 6x - 8$

b. $(3, 1)$ **c.** $x = 3$ **d.** 2 or 4

5. a.

$y = x^2 - 2x - 3$

b. $(1, -4)$
c. $x = 1$
d. -1 or 3

7. a.
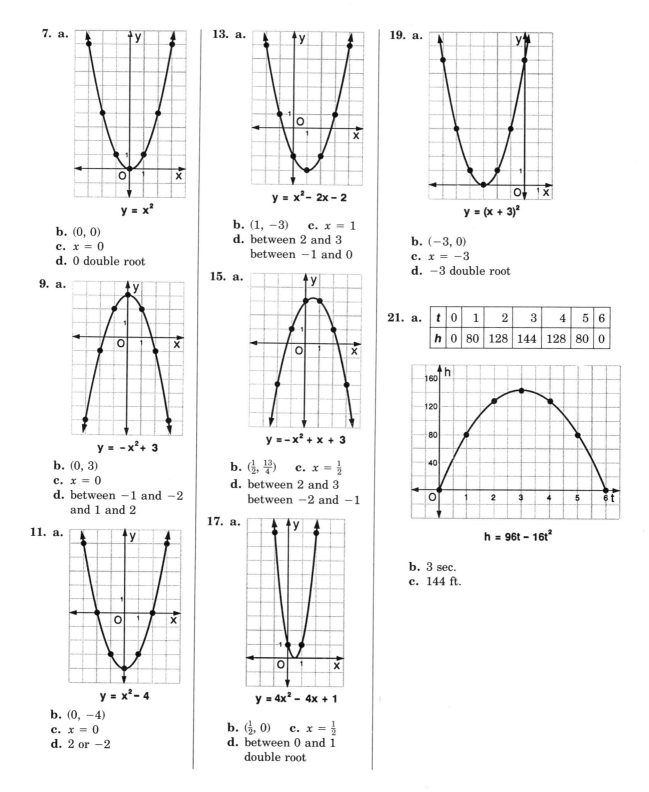
$y = x^2$

b. $(0, 0)$
c. $x = 0$
d. 0 double root

9. a.
$y = -x^2 + 3$

b. $(0, 3)$
c. $x = 0$
d. between -1 and -2
and 1 and 2

11. a.
$y = x^2 - 4$

b. $(0, -4)$
c. $x = 0$
d. 2 or -2

13. a.
$y = x^2 - 2x - 2$

b. $(1, -3)$ **c.** $x = 1$
d. between 2 and 3
between -1 and 0

15. a.
$y = -x^2 + x + 3$

b. $(\frac{1}{2}, \frac{13}{4})$ **c.** $x = \frac{1}{2}$
d. between 2 and 3
between -2 and -1

17. a.
$y = 4x^2 - 4x + 1$

b. $(\frac{1}{2}, 0)$ **c.** $x = \frac{1}{2}$
d. between 0 and 1
double root

19. a.
$y = (x + 3)^2$

b. $(-3, 0)$
c. $x = -3$
d. -3 double root

21. a.

t	0	1	2	3	4	5	6
h	0	80	128	144	128	80	0

$h = 96t - 16t^2$

b. 3 sec.
c. 144 ft.

23. a.

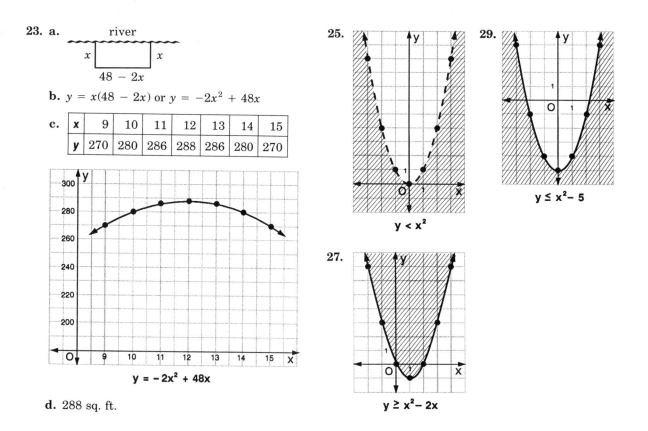

river

x x

$48 - 2x$

b. $y = x(48 - 2x)$ or $y = -2x^2 + 48x$

c.

x	9	10	11	12	13	14	15
y	270	280	286	288	286	280	270

$y = -2x^2 + 48x$

d. 288 sq. ft.

25.

$y < x^2$

29.

$y \leq x^2 - 5$

27.

$y \geq x^2 - 2x$

11-6 More Methods for Solving a Quadratic Equation

Solving a Quadratic Equation by Taking the Square Root of Both Sides: pages 604–605

1. $y = \pm 8$ **3.** $a = \pm 5$ **5.** $y = \pm 3$ **7.** $x = \pm 2$
9. $y = \pm 10$ **11.** $y = \pm 12$ **13.** $x = \pm 2$
15. $x = \pm 10$ **17.** $x = \pm 7$ **19.** $y = \pm 3$
21. $z = \pm 1$ **23.** $x = \pm 6$ **25.** $x = \pm\sqrt{35}$
27. $x = \pm 3\sqrt{3}$ **29.** $x = \pm 5\sqrt{2}$ **31.** $x = \pm 2\sqrt{2}$
33. $x = \pm 5\sqrt{3}$ **35.** $x = \pm 2\sqrt{2}$ **37.** $x = \pm\sqrt{5}$
39. $x = \pm 2\sqrt{5}$ **41.** $x = \pm 2\sqrt{2}$ **43.** $x = \pm\sqrt{15}$
45. $x = \pm 3\sqrt{3}$ **47.** $x = 9.6$ **49.** $x = 6.3$
51. $x = 5.5$ **53.** $x = 5.6$ **55.** $x = 3$ or $x = -1$
57. $x = 2$ or $x = -6$ **59.** $x = 4 \pm \sqrt{11}$
61. $x = 7$ or $x = -5$ **63.** $x = 2 \pm \sqrt{15}$

65. $x = \pm 5a$ **67.** $x = \pm\dfrac{a}{2}$

69. $x = \pm\sqrt{c^2 - b^2}$ **71.** $e = \pm\dfrac{S}{6}\sqrt{6}$

73. $r = \pm\dfrac{1}{2\pi}\sqrt{\pi S}$ **75.** $r = \pm\dfrac{1}{\pi h}\sqrt{\pi h V}$

77. $V = \pm\dfrac{1}{m}\sqrt{grmF}$

Completing a Trinomial Square: page 606

1. $4, (x + 2)^2$ **3.** $\frac{9}{4}, (x + \frac{3}{2})^2$ **5.** $49, (c + 7)^2$
7. $\frac{25}{4}, (r + \frac{5}{2})^2$ **9.** $\frac{1}{4}, (x - \frac{1}{2})^2$ **11.** $\frac{1}{36}, (y - \frac{1}{6})^2$
13. $\frac{25}{2}, 2(x + \frac{5}{2})^2$ **15.** $\frac{9}{8}, 2(x + \frac{3}{4})^2$ **17.** yes
19. no   **21.** no

Solving a Quadratic Equation by Completing the Square: pages 608–609

1. $x = 2$ or $x = -4$ **3.** $x = 1$ or $x = 3$
5. $y = 7$ or $y = -2$ **7.** $x = 2$
9. $c = 2$ or $c = -\frac{1}{2}$ **11.** $r = 2$ or $r = -\frac{1}{3}$

---

13. $x = -1 \pm \sqrt{2}$ **15.** $x = -\dfrac{3}{2} \pm \dfrac{\sqrt{29}}{2}$

17. $a = 3 \pm 2\sqrt{7}$ **19.** $x = 2 \pm \dfrac{\sqrt{21}}{3}$

21. $x = \dfrac{3}{10} \pm \dfrac{\sqrt{29}}{10}$ **23.** $x \approx -.3$ or $x \approx -3.7$

25. $x \approx 8.2$ or $x \approx -.2$ **27.** $x \approx 1.7$ or $x \approx -.4$

Solving a Quadratic Equation by Formula:
pages 612–613

1. $x = 4$ or $x = -6$ **3.** $x = 3$
5. $x = 6$ or $x = -5$ **7.** $x = 2$ or $x = \frac{1}{2}$
9. $x = 1$ or $x = -\frac{2}{5}$ **11.** $x = \pm 3$

13. $x = -1 \pm \sqrt{2}$ **15.** $z = -\dfrac{3}{2} \pm \dfrac{\sqrt{29}}{2}$

17. $x = -\dfrac{5}{2} \pm \dfrac{\sqrt{33}}{2}$ **19.** $x = \dfrac{3}{4} \pm \dfrac{\sqrt{17}}{4}$

21. $x = \dfrac{1}{3} \pm \dfrac{\sqrt{10}}{3}$ **23.** $y = \dfrac{1}{5} \pm \dfrac{\sqrt{6}}{5}$

25. $x = 3 \pm \sqrt{13}$ **27.** $x = \dfrac{5}{4} \pm \dfrac{\sqrt{57}}{4}$

29. $y \approx 3.4$ or $y \approx .6$ **31.** $x \approx 1.2$ or $x \approx -3.2$
33. $x \approx 3.9$ or $x \approx .1$ **35.** $x \approx .6$ or $x \approx -2.6$
37. $x \approx 1.8$ or $x \approx -.8$
39. $x \approx 1.2$ or $x \approx -10.2$
41. $x \approx .3$ or $x \approx -4.8$ **43.** 2 and 10
45. 2 or $\frac{1}{2}$ **47.** 2 mph
49. Huck 6 hr; Elizabeth 12 hr.   **51.** 2.7 in.
53. a. $A(-1 + \sqrt{17}, 1 + \sqrt{17})$
      $B(-1 - \sqrt{17}, 1 - \sqrt{17})$
   b. A is in quadrant I
      B is in quadrant III

11-7 The Pythagorean Relation: pages 617–620

1. $b = 8$ **3.** $a = 15$ **5.** $c = 8$ **7.** $c = 3\sqrt{2}$
9. $b = 4\sqrt{2}$ **11.** $b \approx 14.1$ **13.** $a \approx 24.5$
15. $b \approx 7.1$ **17.** $x = 3\sqrt{3}$ **19.** $x = 2\sqrt{2}$
21. $10\sqrt{2}$ cm.   **23.** $4\sqrt{3}$ ft.   **25.** 28 cm
27. 81 sq. in.   **29.** 36 ft.   **31.** 26 miles   **33.** 8 ft.
35. a. 4 cm   **b.** 34 cm   **c.** 48 cm^2
37. 120 in.^{2} **39.** 4 in., 8 in.   **41.** $2\sqrt{13}$ **43.** 7
45. 13   **47.** $\sqrt{10}$ **49.** 13   **51.** $2\sqrt{41}$ **53.** 5
55. 10   **57.** (3)   **59.** yes   **61.** yes   **63.** yes

---

Chapter Review Exercises: pages 622–624

1. .2   **3.** 64   **5.** -4 **7.** $-4, -\sqrt{5}, -\sqrt{3}$ **9.** 6
11. 9   **13.** $10x^2 b^8$ **15.** $15\sqrt{2}$ **17.** $s^2\sqrt{rs}$
19. $12\sqrt{5} - 3\sqrt{3}$ **21.** $3\sqrt{5}$ **23.** $r^4\sqrt{15}$ **25.** $48y$

27. $6\sqrt{6}$ **29.** $\frac{8}{3}\sqrt{3}$ **31.** $\dfrac{\sqrt{15}}{5}$ **33.** $18 - 3\sqrt{10}$

35. $5\sqrt{3}$ **37.** $a = 12$ **39.** no solution.
41. a. 49π sq. m   **b.** $9\pi^3$ sq. in.

43. a.

$$y = x^2 + 3x + 2$$

   b. $\left(-\frac{3}{2}, -\frac{1}{4}\right)$
   c. $x = -\frac{3}{2}$
   d. -2, or -1

45. $a = \pm 6$ **47.** $c = \pm 5$ **49.** 10 or -4 **51.** 9

53. $\frac{1}{9}$ **55.** $\dfrac{7 \pm 3\sqrt{5}}{2}$ **57.** $-\frac{1}{3}$ or 2   **59.** 0 or 6

61. 8.6 or $-.6$ **63.** $a = 12$ cm   **65.** 1 in.

67. 16 feet   **69.** $\sqrt{5}$ **71.** 5

Problems for Pleasure: page 624

1. a. $\sqrt[3]{\sqrt{64}} = \sqrt[3]{8} = 2$ **b.** $\sqrt{\sqrt[3]{64}} = \sqrt{4} = 2$

3. $x = 13$

5. Let $x =$ the second number. Then $x + 7 =$ the
first number.

$$(x + 7)^2 + x^2 = [(x + 7 + x) - 4]^2$$
$$x^2 + 14x + 49 + x^2 = (2x + 3)^2$$
$$2x^2 + 14x + 49 = 4x^2 + 12x + 9$$
$$0 = 2x^2 - 2x - 40$$
$$2x^2 - 2x - 40 = 0$$
$$x^2 - x - 20 = 0$$
$$(x - 5)(x + 4) = 0$$

$x - 5 = 0$	$x + 4 = 0$
$x = 5$	$x = -4$
$x + 7 = 12$	$x + 7 = 3$

There are two answers: (12, 5) or (3, -4)

1. The conclusion and the method are valid. The key sequence, with a 0 result, corresponds to the equation $c^2 - a^2 - b^2 = 0$, which is equivalent to $a^2 + b^2 = c^2$. With the value of c as the first entry, the sequence can be used to check any 3 numbers. If the resulting display is 0, the numbers satisfy the Pythagorean Relation; if the display is not zero, they do not.

3. **a.** By the Pythagorean Theorem, $a^2 + b^2 = c^2$

$$(3x)^2 + (4x)^2 = 361$$
$$9x^2 + 16x^2 = 361$$
$$25x^2 = 361$$
$$x^2 = \tfrac{361}{25}$$

Area $(3x)(4x) = 12x^2 = 12\left(\tfrac{361}{25}\right)$

$$= 173.28 \text{ sq. in.}$$

b. $(3x)(5x) > 173.28$
$$15x^2 > 173.28$$
$$x^2 > 11.552$$

By the Pythagorean Theorem
$$(3x)^2 + (5x)^2 = d^2$$
$$9x^2 + 25x^2 = d^2$$
$$34x^2 = d^2$$
$$34(11.552) = d^2$$
$$392.768 = d^2$$
$$19.82 = d$$

The screen must be 20 inches or more.

---

1. (B) 3. (B) 5. (A) 7. (A) 9. (E) 11. (D)
13. (A) 15. (B) 17. (A) 19. (E) 21. (D)

---

1. $-4x + 10$ 3. 60 cm 5. $\tfrac{3}{2}$ 7. 1
9. $y = 0, y = 5, y = -3$ 11. (1)
13. $x = 3$ $y = -1$ 15. 2:46 17. 10

1. -5 2. 5 3. $(r + 8)(r - 8)$ 4. 30 ft.
5. 36 6. $\dfrac{7x}{15}$ 7. 3 8. 6.7 9. 1 10. $n - 2$
11. $20°$ 12. 200 13. $x^2 + 2x + 11$ 14. 100
15. B 16. $6y - 1$ 17. 0 18. (2) 19. (1)
20. (2) 21. (3) 22. (4) 23. (4) 24. (2)
25. (3) 26. (4) 27. (1) 28. (3) 29. (4)
30. (1)
31.

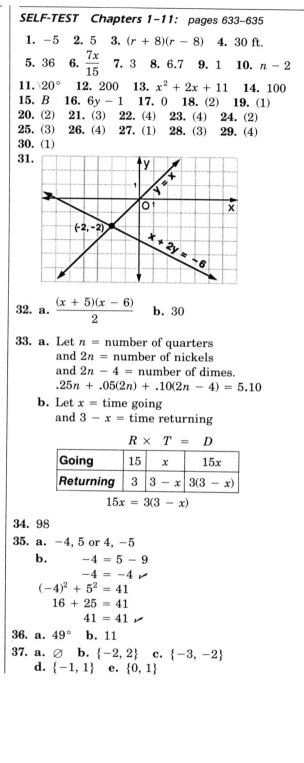

32. **a.** $\dfrac{(x + 5)(x - 6)}{2}$ **b.** 30

33. **a.** Let n = number of quarters
and $2n$ = number of nickels
and $2n - 4$ = number of dimes.
$.25n + .05(2n) + .10(2n - 4) = 5.10$

b. Let x = time going
and $3 - x$ = time returning

	R	× T	= D
Going	15	x	$15x$
Returning	3	$3 - x$	$3(3 - x)$

$$15x = 3(3 - x)$$

34. 98
35. **a.** $-4, 5$ or $4, -5$

b. $-4 = 5 - 9$
$-4 = -4$ ✓
$(-4)^2 + 5^2 = 41$
$16 + 25 = 41$
$41 = 41$ ✓

36. **a.** $49°$ **b.** 11
37. **a.** $\varnothing$ **b.** $\{-2, 2\}$ **c.** $\{-3, -2\}$
d. $\{-1, 1\}$ **e.** $\{0, 1\}$

INDEX

Abscissa, 474
Absolute inequality, 179
Absolute value(s), 63
  graph of an equation
    involving, 480
  of numbers, 63
  solving equations involving,
    182–183
  of zero, 63
Acute angle, 81, 217
Acute triangle, 222
Addition
  associative property of, 101–
    102
  closure property of, 95, 101
  commutative property of, 101–
    102
  of matrices, 128
  of rational expressions with
    different denominators, 334–
    336
  of rational expressions with
    same denominators, 334
  of opposites, 63
  of polynomials, 259–261
  of radicals, 579–580
  of real numbers, 99–100
  of real numbers on a number
    line, 98–99
Addition property
  of equality, 149
  of inequality, 173
  of opposites, 101–102
  of zero, 101
Additive
  identity, 101–102
  inverse, 101–102
Age problems, 245, 538
Algebraic
  fraction, 323
  logic, 46
  models, 66

  solution of system of linear
    equations, 530–533
  strategies, 156, 247
  translating—language into
    verbal phrases, 66, 67
Algebraic expression(s), 43
  evaluating, 45, 120
  simplifying, 48–49
Alphanumeric, 20–21
Altitude, 228
Angle(s), 45
  acute, 81, 217
  base—of isosceles triangle, 222
  complementary, 217
  corresponding—of similar
    triangles, 414
  cosine of, 420
  of depression, 430
  of elevation, 430
  measure of, 45
  obtuse, 176, 217
  problems, 217, 219
  right, 217
  sine of, 420
  straight, 217
  sum of—of triangle, 222
  supplementary, 217
  tangent of, 420
  vertex of, 217
  vertex—of isosceles triangle,
    222
  vertical, 217
Annual income, 354
Apollo, 8, 68
Approximating square root, 574
Archimedes, 198
Area(s)
  formulas, 232
  problems, 232, 599, 613
  surface, 236
  writing formulas for, 232–233
  using coordinates, 475–476

Ascending powers, 259
Associative property
  of addition, 101–102
  of multiplication, 106–107
Average problems, 346
Axes, coordinate, 473
Axis of symmetry, 599

Base, 48, 228, 233
  angle of isosceles triangle, 222
  of isosceles triangle, 222
  of trapezoid, 233
Between, 74
Bhaskara, 57
Binomial, 259
Binomials, multiplication of
  two, 277–279
Braces, 54
Brackets, 44
Business problems, 543

Calculator connection, 21, 46,
  51, 112, 121, 273, 343, 357,
  422, 575
Cancellation, 323, 325
Chance occurrence, 439
Circle, 158, 229, 233
Circumference, 158, 229
Closure, 95, 110, 111
  property of addition, 101
  property of multiplication,
    106–107
Coefficient, 48
Coin problems, 4, 205, 539
Collinear points, 496
Combined variation, 412
Combining like terms, 119, 160–
  161
Common
  greatest—factor, 291–292
  lowest—denominator, 335

Common *(cont.)*
  monomial factor, 292–293
  ratio factor, 387
  solution, 527
Commutative property
  of addition, 101–102
  of multiplication, 106
Comparing numbers, 74
Complementary angle problems, 219
Complementary angles, 217
Complementary events, 455–457
Completing square, solution of quadratic equation by, 607–608
Completing trinomial square, 605–606
Composite numbers, 291
Compound events, 448–457
Compound inequalities, 175–176
Cone, 236
Congruent, 10, 82
Consecutive, 17
  even integers, 199
  integer problems, 200
  integers, 199
  odd integers, 199
Consistent equations, 534
Consistent equations, system of, 534
Constant, 48, 259
Constant of variation, 441
Continued ratio, 318
Convention, 43
Conversion fraction, 387
Coordinate
  axes, 473
  of point, 60
Coordinates, 474
Correspondence, one-to-one, 60
Corresponding
  angles of similar triangles, 414
  sides of similar triangles, 414
Cosine of angle, 420
Counting numbers, 55
Counting principle, 447
Cryptarithm, 20–21
Cube of number, 573–574
Cube root of number, 573–574

Customary units, 70
Cylinder, 236

Dead Sea, 65
Decimal(s)
  expressing—as rational number, 56
  expressing rational number as, 56
  repeating, 56
  terminating, 56
Defining an operation, 24–25
Degree
  in angle measure, 45
  of polynomial, 259
  of polynomial equation, 304
Denominator
  lowest common, 335
  rationalizing irrational radical, 586
Dependent equations, system of, 534
Depression, angle of, 430
Descartes, René, 473
Descending powers, 259
Diameter, 229
Difference, 66
  factoring—of two squares, 292
  multiplication of sum and—of two numbers, 43
Digit, 12
  problems, 12–13, 547
Direct variation, 400–401
Distance, 207, 475–476
Distance formula, 616–617
Distributive property of multiplication, 116–117
Dividend, 111
Divider, plane, 507–508, 554–555
Dividing zero by nonzero number, 111
Division
  of fractions, 111
  of monomial by monomial, 285
  of polynomial by monomial, 285
  of polynomial by polynomial, 285–287

  of powers of same base, 267
  of radicals, 584–585
  of real numbers, 111
Division property
  of equality, 149
  of exponents, 267–268
  of inequality, 173
Divisor, 111
Domain
  of relation, 479–480, 483
  of variable, 68
Double root, 304, 598
Drawing a diagram, 1, 5–8

Edges, 238
Einstein, Albert, 66, 274
Element of set, 54
Elevation, angle of, 430
Empty set, 54
Equal sets, 54
Equality, 149
  addition property of, 149
  division property of, 149
  multiplication property of, 149
  reflexive property of, 150
  subtraction property of, 149
  symmetric property of, 150
  transitive property of, 150
Equation(s), 68
  algebraic solution of system of linear, 530–533
  consistent, 534
  degree of polynomial, 304
  dependent, 534
  equivalent, 153, 528
  fractional, 342–343
  graph of linear—by intercepts method, 488–489
  graph of linear—by slope-intercept method, 503
  graph of linear—in two variables, 490
  graph of quadratic, 596–599
  graphic solution of system of linear, 527–528
  identity, 163
  inconsistent, 534
  involving rational expressions, 358–361, 364

Equation(s) *(cont.)*
  involving several variables,
    166–168
  linear, see Linear functions
  open sentences, 68
  quadratic, 304, 596
  slope of linear, 499
  solution of, 153–154
  solution of quadratic—by
    completing square, 607–608
  solution of quadratic—by
    factoring, 304–306, 603
  solution of quadratic—by
    formula, 609–611
  solution of radical, 591–593
  solution set of, 68
  solving, 68, 153–154
  system of linear, 527, 530–
    534
  transforming, 167–168, 360–
    361
  writing—for line, 499–501
  writing verbal sentences as,
    66–67, 166
  y-intercept of linear, 499
Equilateral triangle, 222
Equivalent
  equations, 153, 528
  inequalities, 183
  sets, 54
Estimation, 2, 23, 87, 574
Euler, Leonhard, 31
Evaluating
  algebraic expressions, 45, 120
  formula by solving equation,
    167
  subject of formula, 166
Even integer, 199
Even integers, consecutive, 199
Events, 440
  certain, 441
  comlementary, 455–457
  compound, 448–457
  impossible, 441
  independent, 448
Exponent, 48, 266–268
  on calculator, 51
  negative, 267–268
  zero, 267–268
Expressing
  decimal as rational number, 56

rational number as decimal, 56
  sentences using mathematical
    symbols, 66–67
Expression(s), 43
  algebraic, 45
  evaluating algebraic, 45
  mixed, 334
  numerical, 43
  rational, 323
Extraneous value, 360
Extremes, 394

Faces, 236
Factor(s), 48, 291
  common monomial, 292–293
  greatest common, 292–293
  prime, 291
  of product, 48
Factoring, 292–309
  common monomial factor,
    292–293
  completely, 300–302
  difference of two squares, 292
  grouping, 292–293
  number, 291
  polynomial, 297–302
  solution of quadratic equation
    by, 304–306
  trinomials, 297–300
Fermat, Pierre, 437
Fibonacci, 59
Fibonacci sequence, 59
Finding a pattern, 15–17
Finite set, 54
Flowchart, 50–51, 78, 100, 161,
  287, 301, 337, 427
  symbols, 50
Formula(s), 66
  area, 232–233
  distance, 616–617
  evaluating—by solving
    equation, 166–167
  evaluating subject of, 166
  perimeter, 228
  quadratic, 609–611
  subject of, 166
  for surfaces of solids, 238–239
  transforming, 167–168
  using—to study related
    changes, 239, 241–242

for volumes of solids, 240–241
  working with, 166–168
  writing—for areas, 232–233
Fraction(s), algebraic, 323
Fractional equation, 358–360
Fulcrum, 214
Function, 483
  graph of linear, 488–489
  graph of quadratic, 596–600
  line test for, 484
  linear, 488
  notation, 484–485

Galileo, 409
General framework, 1
Geometric problems, 219, 222,
  224, 227, 236, 238, 240, 306
Golden ratio, 396
Gram, 70
Graph
  of inequalities, 174, 182–183,
    507–509, 554–555, 600
  of linear equation by
    intercepts method, 488–489
  of linear equation by slope-
    intercept method, 503
  of linear equation in two
    variables, 490
  of linear function, 488–489
  of lines parallel to x-axis, 488–
    489
  of lines parallel to y-axis, 488–
    489
  of number, 60
  of ordered number pairs, 473
  of a point, 60
  of quadratic equation,
    599–600
  of quadratic function, 596–600
  of relations, 479–480
  of set of numbers, 77, 79
  of solution set of system of
    inequalities, 554–555
Graphic solution
  of quadratic equation, 596
  of system of linear equations,
    527–528
Graphing
  on a number line, 60
  in a plane, 473

Graphing a line
 using intercepts, 489–490
 using slope and intercept,
  502–503
 using a table of values, 480
Greatest common factor, 292–
 293
Grouping symbols, 44
 bar, 44
 brackets, 44
 parentheses, 44, 46
Guess and check, 1–3

Half-plane, 507–509, 554, 555
Height, 228
Hexagon, 230
Hypotenuse of right triangle,
 222

Identity
 additive, 101
 equation, 163
 multiplicative, 106
Image, 474
Income, annual, 354
Inconsistent equations, system
 of, 534
Independent events, 448
Index of radical, 573
Indirect measurement
 using similar triangles, 415–
  416
 using trigonometry, 420
Inequality(ies), 74, 173
 absolute, 179
 addition property of, 173
 compound, 175–176
 equivalent, 183
 graphing, 182–183, 507–509,
  600
 graph of solution set of system
  of, 554–555
 multiplication property of, 173
 problems, 173–174
 properties of, 173
 solving, 174–176
 symbols of, 74
 transitive property of, 174
Infinite set, 54

Integer(s)
 consecutive, 199
 even, 199
 negative, 55
 odd, 199
 positive, 55
 problems, 199
Intercepts method, graph of
 linear equation by, 488–489
Interest, annual rate of, 354
Intersection of sets, 126
Inverse, 19
 additive, 101–102
 multiplicative, 106
 operation, 153
 variation, 404–405
Investment problems, 354, 545
Irrational
 rationalizing—radical
  denominators, 586
 set of—numbers, 57
Isosceles triangle, 163, 222
 base of, 222
 base angle of, 222
 leg of, 222
 vertex angle of, 222

Joint variation, 410–411

Kepler, Johannes, 409
Kilogram, 70

Law of lever, 214
Leg
 of isosceles triangle, 222
 of right triangle, 222
 of a trapezoid, 223
Lever, 214
 law of, 214
 problems, 215
Like
 combining—terms, 119
 radicals, 579
 terms, 119, 160–161
Lilavati, 57
Line(s),
 graph of—parallel to x-axis,
  488–489

graph of—parallel to y-axis,
  488–489
 number, 60
 of reflection, 597–599
 of sight, 430
 slope of, 492–493
 writing equation for,
  499–501
 y-intercept of, 499–501
Linear equation(s), see Linear
 functions
 algebraic solution of system of,
  527–528
 graph of—by intercepts
  method, 488–489
 graph of—by slope-intercept
  method, 503
 graph of—in two variables,
  490
 graphic solution of system of,
  527–528
 matrix solution of system of,
  561–564
 slope intercept form,
  499–501
 system of, 527, 530–533
 system of simultaneous, 530
 system of three, 557–559
 y-intercept of, 499
Linear inequalities
 graphing, 507–509, 554–555
Linear function, 488
Linear function, graph of, 488–
 489
List of set, 54
Literal equations
 see Equations involving
  several variables
 see also Formula(s), working
  with
Lowest common denominator,
 335
 of rational expressions, 335–
  336
Lowest terms, reducing rational
 expressions to, 324–325

Making a list, 1, 11–13
Making a table, 1, 11–13
Mapping diagram, 483

Matrices, 128
  operations with, 129–131
  used in solving a system of
    equations, 561–564
Mathematical symbols,
  expressing verbal phrases
  using, 66–67
Max, 138
Maximum point, 597
Means, 394
Measure of angle, 45
Member of set, 54
Metric units, 70
Milligram, 70
Minimum point, 597
Miscellaneous problems, 156–
  157, 164–166, 245–246, 552–
  553
Mixed expression, 335
Mixture problems, 203–204,
  540–541
Model, algebraic, 66
Money-value problems, 203–205
Monomial(s), 259
  division of—by monomial, 285
  division of polynomial by, 285
  multiplication of—by
    monomial, 276
  multiplication of polynomial
    by, 277
  square root of, 579
Monomial factor, common, 291–
  292
Money-value problems, 203–205
Motion problems, 207, 209–210,
  366–367, 549–550, 613
Multiplication
  associative property of, 106–
    107
  closure property of, 106
  commutative property of, 106–
    107
  distributive property of, 149
  of matrices, 129–130
  of monomial by monomial, 276
  of polynomial by monomial,
    276
  of polynomial by polynomial,
    277–279
  of powers of the same base,
    266

of radicals, 583–584
of real numbers, 105–107
of sum and difference of two
  numbers, 281
of two binomials, 277–279
Multiplication property
  of equality, 149
  of exponents, 266
  of inequalities, 173
  of one, 106
  of reciprocals, 106
  of zero, 106
Multiplicative
  identity, 106
  inverse, 106

Natural numbers, 55
Negative
  exponents, 267–268
  integers, 55
  real—numbers, 57
  slope, 494
Network, 31
Newton, Isaac, 66
Notation
  function, 484–485
  scientific, 271–272
Null set, 54
Number(s)
  absolute value of, 63
  comparing, 74
  composite, 291
  consecutive, 17
  cube of, 573–574
  cube root of, 573–574
  expressing decimal as
    rational, 56
  expressing rational—as
    decimal, 56
  factoring, 291
  graph of, 60
  irrational, 573
  integers, 55
  natural, 55
  opposite of directed, 62–63
  ordered—pair, 387, 473
  ordering—on number line, 74–
    75
  ordering real, 74–75
  prime, 291

problems, 160–161, 164, 364,
  612–613
properties of real, 95, 101,
  106, 116
rational, 55
real negative, 55, 57
real positive, 55, 57
root of, 573
set of irrational, 57
set of rational, 55
set of real, 57
signed, 57
square of, 573
square root of, 573–574
successor of, 55
symbols for, 43
whole, 55
Number line, 60
  addition of real numbers on,
    98–100
  ordering real numbers on,
    74–75
  real, 60
Numeral(s), 43
Numerical
  coefficient, 48
  expression, 43
Numerical expressions,
  simplifying—containing
  powers, 48–49

Obtuse angle, 176, 217
Obtuse triangle, 222
Odd integer, 199
Odd integers, consecutive, 199
Odds, 456
One, multiplication property of,
  106
One-to-one correspondence, 60
Open sentence, 68
Operation(s), 43
  defining, 124–125
  order of, 44
  symbols for, 43
Opposite of real number,
  62–63
Order of operations, 44
Ordered
  graph of—number pair, 473
  number pair, 387, 473

Ordering
    numbers on number line, 74–75
    real numbers, 74–75
Ordinate, 474
Origin, 474
Outcome, 440

Palindrome, 59
Parabola, 596
Parallel lines, 495
    slope of, 495
Pascal, Blaise, 437
Pascal's triangle, 59
Parentheses, 44
Pentagon, 164
Pentagonal right prism, 236
Percent mixture problems, 351, 364
Percentage problems, 346
Perfect square, 574
    square root of, 574
Perimeter
    formulas, 228
    problems, 9, 227, 552
Plane
    coordinate, 511
    divider, 507–508, 554–555
Point(s)
    collinear, 496
    coordinate of, 60
Polygon, 227
Polyhedron, 236
Polynomial(s)
    addition of, 259–261
    degree of, 259
    division of—by monomial, 285
    division of—by polynomial, 285–287
    factoring, 297–302
    multiplication of—by monomial, 276
    multiplication of—by polynomial, 277–279
    subtraction of, 261
Polynomial equation
    degree of, 259
    standard form of, 259
Positive
    integers, 55

numbers, real, 57
slope, 494
Power(s), 48
    ascending, 259
    descending, 259
    division of—of the same base, 267–268
    finding—of power, 266–267
    multiplication of—of same base, 266, 268
    raising to a, 573
    simplifying numerical expressions containing, 48–49
Prime
    factor, 291
    number, 291
Principal square root, 573
Prism, 236
Probability, 439–447
    chance occurrence, 439
    complementary events, 455–457
    compound events, 448–457
    counting principle, 449
    event, 440
    experimental, 439
    independent events, 448
    odds, 456
    predicting outcomes, 440
    problems, 443–457
    sample space, 440
    theoretical, 440
Problem solving, 1–41, 156
Problems
    age, 245, 538
    angle, 217, 219
    area, 232, 599, 613
    average, 346
    business, 543
    coin, 4, 205, 539
    complementary angles, 219
    consecutive integer, 199–200
    digit, 12–13, 547
    geometric, 219, 222, 224, 227, 236, 238, 240, 306
    inequality, 173–174
    investment, 354, 545
    involving rational expressions, 346
    lever, 215

miscellaneous, 156–157, 164–166, 245–246, 552–553
mixture, 203–204, 540–541
money-value, 203–205
motion, 207, 209–210, 366–367, 549–550, 613
number, 160–161, 164, 364, 612–613
per cent mixture, 351, 364
perimeter, 9, 227, 552
probability, 443–457
proportion, 395–396, 556
pulley, 215
ratio, 388–389, 556
similar triangles, 414–416
solving, 1–41, 156, 537–541
sum of angles of triangle, 222, 224
supplementary angles, 219
surface area, 238
trigonometry, 420–438
variation, 401, 404–405, 411
vertical angles, 219
volume, 240
work, 369–371, 613
Product, 43, 66
    factors of, 48
Property
    addition—of equality, 149
    addition—of inequalities, 173
    associative—of addition, 101–102
    associative—of multiplication, 106
    closure—of addition, 95, 101–102
    closure—of multiplication, 106
    commutative—of addition, 101–102
    commutative—of multiplication, 106
    distributive—of, 112
    division—of equality, 149
    exponent, 266–268
    of inequalities, 173–174
    multiplication—of equality, 149
    multiplication—of inequalities, 173
    multiplication—of one, 106
    multiplication—of zero, 106

Property *(cont.)*
  of real numbers, 95, 101–102
  reflexive—of equality, 149
  substitution, 95
  subtraction—of equality, 149
  symmetric—of equality, 149
  transitive—of equality, 149
  transitive—of inequalities, 174
Proportion(s), 394
  problems, 395–396, 556
  solving, 395
  similar triangles, 414
Protractor, 219
Pulley, 215
Pyramid, 236
Pythagorean Relation, 614–617

Quadrants, 474
Quadratic equation, 304
  graph of, 596–600
  solution of—by completing
    square, 607–608
  solution of—by factoring, 304–
    306, 603
  solution of—by formula, 609–
    611
  standard form of, 304, 596
Quadratic formula, 609–611
Quadratic function, graph of,
  596–600
Quadrilaterals, 323
Quotient, 43, 66, 111

Radical(s), 573–595
  addition of, 579–580
  division of, 584–585
  index of, 573
  multiplication of, 583–584
  rationalizing irrational—
    denominator, 586
  simplifying, 578–579
  subtraction of, 579–580
Radical equations, solution of,
  591–593
Radical sign, 573
Radicand, 573
Radius, 229
Range of relation, 479–480

Rate, 208
  annual—of interest, 354
  of work, 369
Ratio, 387–393
  common—factor, 387
  continued, 318
  problems, 388–389, 556
  terms of, 387
Rational expression(s)
  addition of—with different
    denominators, 335–336
  addition of—with same
    denominator, 334
  decimal, 56
  division of, 329
  multiplication of, 328
  problems, 346
  reducing to lowest terms, 323–
    325
  simplifying, 323–325
  subtraction of—with different
    denominators, 335–336
  subtraction of—with same
    denominator, 334
Rational number(s), 55
  expressing decimal as, 56
  expressing—as decimal, 56
  set of, 55, 57
Rationalizing irrational radical
  denominator, 586
Real number line, 60
Real numbers
  addition of, 99–100
  addition of—on a number line,
    98–100
  division of, 111
  multiplication of, 105–107
  negative, 57
  ordering, 74–75
  ordering—on a number line,
    74–75
  positive, 57
  properties of, 95, 101, 106
  set of, 57
  subtraction of, 110
Reciprocal, 106, 111
Rectangular right prism,
  236
Reflection, 62, 474
Reflexive property of equality,
  150

Relations, 479
  domain of, 479–480
  graphs of, 479–480
  range of, 479–480
Repeating decimals, 56
Replacement set, 68
Richter, Charles, 52
Richter scale, 26, 52
Right angle, 217
Right prism, 236
Right triangle, 222
  hypotenuse of, 222
  leg of, 222
Root
  cube—of number, 573–574
  double, 304, 598
  extraneous, 360
  of equation, 304
  of number, 573
  square—of monomial, 579
  square—of number, 573
Row-by-column multiplication,
  129–130

Sample space, 440
Scalene triangle, 222
Scientific notation, 271–272
Seismograph, 52
Sentence(s)
  expressing—using
    mathematical symbols, 68–
    69
  open, 68
Set(s)
  element of, 54–55
  empty, 54
  equal, 54
  equivalent, 54
  finite, 54
  graph of—of numbers, 77, 79
  infinite, 54
  of irrational numbers, 57
  list of, 54
  meaning of, 54
  member of, 54
  notation, 54
  null, 54
  of rational numbers, 55
  of real numbers, 57
  replacement, 68

Set(s) *(cont.)*
  set-builder notation, 54
  solution, 68
  subset, 54
  union of, 126
Side(s)
  corresponding—of similar
    triangles, 414
Similar triangles, 414–416
  corresponding angles of, 413
  corresponding sides of, 413
  problems, 414–415
  using—in indirect
    measurement, 415–416
Simplifying, 43
  algebraic expressions, 44–49
  numerical expressions
    containing powers, 48–49
  radical, 578–579
  rational expressions,
    323–325
Simultaneous linear equations,
  system of, 530
Sine of angle, 400
Slope, 492–506
  of line, 492–493
  of linear equation, 499–501
  negative, 494
  no, 494
  of parellel lines, 495
  positive, 494
  zero, 494
Slope-intercept
  form of equation, 499–501
  method, graph of linear
    equation by, 503
Solid, 236
Solution
  algebraic—of system of linear
    equations, 530–533
  common, 527
  of equation, 68
  graphic—of system of linear
    equations, 527–528
  of quadratic equation by
    completing square, 607–608
  of quadratic equation by
    factoring, 304–306, 603
  of quadratic equation by
    formula, 609–611
  of radical equation, 591–593

Solution set, 68
  of equation, 68
  graph of—of system of
    inequalities, 554–555
Solving
  equations, 68, 150, 153–154,
    160–161
  inequalities, 173–175
  problems, 1–41, 156
Spheres, 236
Square(s)
  completing trinomial, 605–606
  of number, 573
  perfect, 574
Square root(s), 573
  approximating—by estimation,
    574
  of monomial, 579
  of number, 573–574
  of perfect square, 574
  principal, 573
Squaring a binomial, 279
Standard form
  of a linear equation, 488
  of a quadratic equation, 304,
    596
Straight angle, 217
Strategy, 1
  algebraic, 156
  draw a diagram, 1, 5–8
  find a pattern, 1, 15–17
  guess and check, 2–3
  make a list, 11–13
  make a table, 11–13
  work backward, 1, 19–21
Subject of formula, 166
Subset, 54
Substitution property, 95, 101,
  149
Subtraction
  of polynomials, 261
  of radicals, 579–580
  of real numbers, 110
Subtraction property
  of equality, 149
  of inequality, 173
Successor of number, 15, 55, 199
Sum, 43, 66
  of angles of triangle, 222
  of angles of triangle problems,
    222, 224

multiplication of—and
    difference of two numbers,
    281
Supplementary angles, 217
Supplementary angles problems,
  219
Surface area problems, 238
Surfaces of solids, formulas for,
  236, 238, 239
Symbol(s)
  calculator, 21
  of equality, 68
  expressing mathematical
    sentences using
    mathematical, 66–67
  flowchart, 50–51
  grouping, 44–46
  of inequality, 74
  for numbers, 43
  for operations, 43
Symmetric property of equality,
  149
System
  algebraic solution of—of linear
    equations, 530–533
  of consistent equations, 534
  of dependent equations, 534
  graph of solution set of—of
    inequalities, 554–555
  graphic solution of—of linear
    equations, 527–528
  matrix solution of—system of
    linear equations, 561–564
  of inconsistent equations, 534
  of linear equations, 527
  of three linear equations, 557–
    559

Table
  trigonometric, 423
Taj Mahal, 62
Tangent of angle, 420
Term(s), 48
  like, 119
  unlike, 119
Terminating decimals, 56
Theory of Relativity, 66
Time, 209
Tower of Babel, 53
Transformation, 474

Transforming
  formulas, 167–168
  mixed expressions, 335
Transitive property
  of equality, 150
  of inequalities, 174
Translating verbal phrases into
  algebraic language, 66–67,
  166
Translation, 98, 475, 506
Trapezoid, 233
Tree diagram, 6, 8, 446
Triangle(s)
  acute, 222
  equilateral, 222
  hypotenuse of right, 222
  isosceles, 163, 222
  properties of, 222
  right, 222
  scalene, 222
  similar, 414–415
  sum of angles of, 222
Triangular right prism, 236
Trigonometry, 420–438
  problems, 427–438
  table, 423
Trinomial, 259
  completing—square, 605–606
  factoring, 297–300
  square, 605–606
Turning point, 597

Unit
  cubes, 240
  squares, 232
Union of sets, 126
Unlike terms, 119

Value, 43
Variable, 45
  domain of, 68
  values of, 45
Variation, 400–413
  combined, 412
  constant of, 400
  direct, 401–402
  inverse, 404–405
  joint, 410–411
  problems, 401, 404–405, 411
Venn diagram, 6–7, 87, 175–176
Verbal sentences, writing—as
  equations, 66–67, 166
Vertex
  of angle, 217
  angle of isosceles triangle, 222
Vertical
  angle problems, 219
  angles, 217
  line test for function, 484
Vertices, 238
Volumes of solids, formulas for,
  240–241

Whole numbers, 55
Work backward, 1, 19–20
Work problems, 369–371, 613
Work, rate of, 369

x-axis, 473
x-coordinate, 474
x-intercept, 488–490

y-axis, 473
y-coordinate, 474
y-intercept, 488–490
y-intercept of line, 499–501

Zero
  absolute value of, 63
  addition property of, 101–102
  dividing—by non-zero number,
  111
  exponents, 267–268
  multiplication property of, 106
  slope, 494

PHOTO CREDITS